SYSTEMS ANALYSIS AND DESIGN METHODS

THIRD EDITION

SYSTEMS ANALYSIS AND DESIGN METHODS

Jeffrey L. Whitten, MS, CDP

Professor

Lonnie D. Bentley, MS, CDP

Professor

Victor M. Barlow, MBA, CDP

Assistant Professor

All at Purdue University
West Lafayette

IRWIN

Burr Ridge, Illinois
Boston, Massachusetts
Sydney, Australia

Cover illustration: Chris Sheban
Richard D. Irwin, Inc., recognizes that certain terms
in the book are trademarks, and we have made every
effort to reprint these throughout the text with the
capitalization and punctuation used by the holders of
the trademark.

Senior sponsoring editor: Rick Williamson
Associate editor: Rebecca J. Johnson
Marketing manager: Jim Lewis
Project editor: Susan Trentacosti
Production manager: Bob Lange
Interior designers: Tara L. Bazata/Jeanne M. Rivera
Cover designer: Jeanne M. Rivera
Art manager: Kim Meriwether
Art studio: ElectraGraphics, Inc.
Compositor: Progressive Typographers, Inc.
Typeface: 10/12 ITC Garamond Light
Printer: Von Hoffmann Press

Library of Congress Cataloging-in-Publication Data

Whitten, Jeffrey L.
    Systems analysis and design methods/Jeffrey L. Whitten, Lonnie
D. Bentley, Victor M. Barlow.—3rd ed.
        p.   cm.
    Rev. ed. of: Systems analysis & design methods. 2nd ed. 1989.
    Includes bibliographical references and index.
    ISBN 0-256-09360-1.—ISBN 0-256-10132-9 (instructors ed.)   0-256-10831-5 (international ed.)
II. Barlow, Victor M.   III. Whitten, Jeffrey L. Systems analysis &
design methods.   IV. Title.
QA76.9.S88W48   1994
005.1—dc20
                                                                93–34091

Printed in the United States of America
        4   5   6   7   8   9   0   VH   0   9   8   7   6   5

To my lovely wife, Deb. Your love and encouragement have given new significance to my every accomplishment. Also, to my mother and father. You have always been a source of inspiration. — *Jeff*

To my wife and best friend, Cheryl. And to my children, Coty, Robert, and Heath. God blessed me with a wonderful family. — *Lonnie*

To my wife, Linda, the love of my life. To my sons, Kevin and Eric, the joys of my life. And to my parents, the source and mentors of my life. — *Vic*

To the students of the Computer Technology Department at Purdue University, West Lafayette, and Statewide Technology sites. May all your experiences be successful. — *Jeff, Lonnie, and Vic*

THE INTENDED AUDIENCE FOR THIS BOOK

Systems Analysis and Design Methods, third edition, is intended to support a practical course in information systems development. This course is normally taught at the sophomore, junior, senior, or graduate level. It is taught in vocational trade schools, junior colleges, colleges, and universities. The course is taught to both information systems and business majors. Previous editions have been used successfully in one-, two-, and three-term courses, in both semester- and quarter-based systems.

We recommend that students should have taken either the computer- or information-literacy course. One or more programming courses enhances the experience for information systems majors.

WHY WE WROTE THIS BOOK

Today's students are "consumer-oriented," due in part to the changing world economy that promotes competition and professional currency. They expect to walk away from a course with more than a grade and the promise that they'll someday appreciate what they've learned. They prefer to "practice" the application of concepts, not just study applications of the concepts. We wrote the first and second editions of the book to deal with the following perceived problems:

- Some books were (and are still) *too* conceptual. While most reinforce concepts with examples, those examples are too few or too late to generate student interest.
- Some books were (and still are) *too* mechanical. They totally ignored concepts in favor of mechanics. We have sought a balance.
- To our surprise, many books still perpetuate the myth that classical, structured, and modern techniques are mutually exclusive. All techniques have their relative advantages and disadvantages, but few books attempt to integrate both coverage and techniques.
- Most books still lack sufficient examples to adequately demonstrate concepts and techniques.

Our goal was to write a textbook that overcomes these problems. Additionally, we wanted to write a book that serves information systems majors as a valuable post-course reference guide.

Consistent with the first two editions, we have written the book using a lively, conversational tone. Today's students often lose interest in textbooks because they are written in the more traditional, academic, factual tone. Our conversational approach results in a somewhat longer text; however, the "talk with you, not at you" style seems to work well with a wider student audience. We hope that our style does not offend or patronize any specific audience. We apologize if it does.

CHANGES FOR THE THIRD EDITION

We believe that we preserved all of the features you liked in the first and second editions. Additionally, we have made numerous changes for the third edition. They include:

- Concepts, applications, and practice have been even better blended to provide a solid learning path from concepts to tools to techniques.
- The importance of people skills and human interaction during systems development has been further improved through categorization of systems development audiences: owners, users, designers, and builders—each with its own perspectives on the development process and deliverables.
- The Next Generation boxes have been updated appropriately to deal with emerging techniques and technologies such as object-oriented methods and AD/Cycle.
- The conceptual framework or pyramid model has been significantly improved. The model is derived from John Zachman's internationally acclaimed *Framework for Information Systems Architecture.* The model, as presented in the pyramid, is much more elegant and applicable to the comprehensive subject of systems development. Consistent with Zachman's framework, geographic or network models are introduced as a complement to data and process modeling.
- The text is full color. More important, the text makes pedagogical use of color. Using John Zachman's *Framework for Information Systems Architecture,* we associate color with various elements as follows:

  indicates something to do with "people."

  indicates something to do with "data."

  indicates something to do with business or system "processes" and "activities."

  indicates something to do with "geography" or "networks."

  indicates something to do with "technology."

  indicates something to do with systems "development activities or deliverables."

- More emphasis is placed on client, end-user, and business perspectives—due, in part, to the consistent application of Zachman's framework.
- We offer improved integration of systems development methodologies—especially, information engineering, structured analysis and design, and prototyping (and RAD). Information engineering-compatible techniques are subtly emphasized since they have replaced structured analysis and design as the preferred techniques in progressive shops. In reality, information engineering encompasses structured techniques, applying them with a data-centered focus to the enterprise as a whole.
- We include an entire chapter on Computer-Aided Systems Engineering (CASE). More important, the CASE coverage includes atypical CASE tool categories such as code generators, central repositories, code/test/debug tools, and reverse engineering, to name a few. CASE application remains integrated throughout the book.
- The life cycle coverage has been improved in several ways. First, most of the life cycle diagrams are now repository-based as opposed to data flow-based. Not only is this a more realistic approach, it also eliminates the dated "waterfall" look of the life cycle. Second, we added new chapters on strategic systems planning (for aligning information systems projects with corporate goals) and existing systems support (maintenance, enhancement, reengineering, and reverse engineering).
- We've changed our approach to logical versus physical design. First, we renamed them to essential (= logical) and implementation (= physical), consistent with progressive literature on the subject. Second, we've eliminated the chapter on physical or implementation modeling of the existing system since such modeling is discouraged, even in modern structural analysis thinking. Subsequently, we shift implementation or physical DFDs to the design unit where they are used to model the proposed implementation of processes in the

target system (in lieu of systems flowcharts, which are still briefly covered).

- Since data analysis (normalization) is viewed by many as a design technique (that prepares data models for implementation), we shifted that subject to the design unit. In the analysis unit, coverage is limited to data modeling (without analysis). In retrospect, we agreed with many experts and critics that normalized data models are not necessarily user-friendly, and thus, not very analysis-friendly. We recognize that this point is arguable and suggest that users should have no difficulty in covering the chapters back-to-back during the systems analysis unit of their course.

- Driven by trends toward distributed and client/server applications development, a new chapter on modeling system geography was introduced to the systems analysis unit. Subsequently, the process design and distribution chapter deals with implementation modeling for distributed and client/server applications.

- The data dictionary and procedure description chapters were merged.

- An analysis-to-design transition chapter for process models (DFDs) is introduced. The omission of a formal approach to transitioning process models from analysis to design is often cited as a weakness in classical process modeling approaches.

- The file and database design chapters were merged. Greater emphasis is placed on relational database design than on conventional files or other database structures. Distributed database concepts and issues are introduced; however, it was determined that a more thorough discussion of data distribution is best left to a database textbook.

- The input and output design chapters were merged.

- State transition diagrams, a widely used behavior modeling technique, replace our proprietary terminal dialogue charts as the tool for designing terminal dialogue. Also, the implications of graphical user interfaces are now explored.

- Each chapter and module includes end-of-chapter key terms, cross-referenced to page numbers in the chapter or module.

- A special *Instructor's Edition* provides additional insights into changes for the third edition. The Instructor's Edition also includes key term definitions and common synonyms, and instructor references.

HOW TO USE THIS BOOK

Systems Analysis and Design Methods, third edition, is divided into five parts. Past experience suggests that instructors can easily omit chapters they feel are less appropriate to their own students and resequence chapters (especially within a part) to meet their own course goals.

Part One, Fundamentals of Systems Development, presents the information systems development situation and environment. The chapters introduce the analyst and client, information systems building blocks, the systems development life cycle, systems development techniques and methodologies, and computer-aided systems engineering. A visual model, the information systems pyramid, is used to organize concepts and applications. Part One can be covered quickly or in depth, depending on the students' backgrounds and instructor preferences and schedule constraints.

Part Two, Systems Planning and Analysis Methods, covers the process of planning and analysis, as well as the requisite tools and techniques. The systems planning chapter is optional but highly recommended, given recent trends toward businesses asking for alignment of information system goals with overall business goals and priorities.

Part Three, Systems Design Methods, covers the process of general and detailed design, as well as the requisite tools and techniques.

Part Four, Systems Implementation and Support, covers the process of implementation and support. Tools and techniques are generally covered in programming courses. The chapters are provided here mostly to place programming into the perspective of the entire life cycle and to introduce important activities that usually do not find their way into the programming courses.

Part Five, Cross Life Cycle Activities, covers processes and techniques that span multiple phases of the life cycle. Included are modules on project management, fact finding, feasibility analysis, and interpersonal communications. As cross life cycle activities, these modules can be interwoven into the course at various points in time. Minimum module prerequisites are clearly listed at the beginning of each module.

SUPPLEMENTS

It is our purpose to provide our users with a complete course, not just a textbook. Therefore, we are committed to providing a comprehensive package of supporting materials. The supplements for the third edition of *Systems Analysis and Design Methods* include the following components.

For the Instructor

Instructor's Edition   One of the major improvements for the third edition is the Instructor's Edition of the textbook. For each chapter, it provides an instructor's overview, summary of changes for the third edition, instructor references, and key terms and common synonyms.

Instructor's Guide   For those users who are more comfortable using a separate instructor's supplement or who desire more detailed supplemental material, an Instructor's Guide is also available. For each chapter and module in the textbook, it includes the following sections:

- *Instructor Overview.* This section provides a brief, high-level overview of the chapter and may be useful for determining the inclusion or exclusion of a chapter and the degree of coverage desired.
- *What's Different Here and Why.* This section was intended to provide a transition from the second edition to the third. It highlights the changes made in the third edition.
- *Instructor References.* This section provides additional references the instructor may use as background reading or to obtain additional instructional material.
- *Key Terms and Common Synonyms.* This section lists the key terms used in the chapter and indicates common synonyms or aliases that may be used in other textbooks.
- *Instructional Objectives and Lesson Planning Ideas.* This section offers suggestions for using classroom time. Lessons suggestion ideas are designed around student learning objectives.

Alternative classroom approaches and evaluation mechanisms are offered.

- *Discussion Guidelines for Chapter Opening Minicase.* This section contains answers for the discussion questions at the end of each minicase. We have found this to be a useful point of departure for stimulating class discussion.
- *Module and Running Case Assignment Suggestions.* This section offers suggested assignments for the appropriate modules and running case episodes to accompany that particular chapter. Since the material in the modules and the running case spans several chapters, this section may help you decide the most appropriate time to introduce them into the course.
- *Problems and Exercises Answers.* This section contains the answers to the problems and exercises at the end of the chapter.
- *Projects and Minicases Answers.* This section contains the answers to the projects and minicases at the end of the chapter. Projects and minicases differ from problems and exercises in that they are intended to be more involved, complex, and difficult.
- *Projects and Cases Supplement Guidelines.* This section suggests appropriate Systems Analysis and Design Milestones in the Projects and Cases supplement to accompany that particular chapter. This should help keep the students' systems development project synchronized with the material covered in the textbook.

Full-Color Transparency Acetates   More than 50 full-color transparencies are available to textbook adopters. Each highlights important principles or concepts taught in the textbook and has been reproduced with the pedagogical use of color as it appears in the textbook.

Transparency Masters   All of the more than 300 figures from the textbook have been reproduced as transparency masters that you may use to make overhead transparencies.

Test Bank   A written test bank covering all the chapters and modules includes approximately 3,000 questions. Each chapter and module has questions in the following formats: true/false, multiple choice, sentence completion, and matching. The test bank includes the answer and page references for each question and an explanation or rationale for each true/false question that is false.

For the Student

Projects and Cases (Student Workbook)   More than the average case book! The third edition has retained all of the best features of the second edition and has added new value.

We still believe the Build Your Own Case option to be the most creative and valuable feature of the casebook because it allows the students and instructor to overcome many of the limitations of the traditional case study method. Because of the difficulty of attempting to distill the complexities (including real political issues) into the English language, many "canned" case studies contain discrepancies and omissions that confuse both students and instructors. This approach allows students to build their own systems development project from their own work or life experience (not necessarily in computer or information systems). This ensures that each project will be different, interesting, and individually tailored to each student. The project will also reinforce the textbook concepts while preventing the scope of the project from growing unreasonable. We have used this approach in our classes for a number of years and student response has been very favorable.

However, we realize that some instructors don't have enough time to create their own cases or help students to create their own. (Admittedly, the Build Your Own Case option does require considerable amount of time and effort for both the student and instructor). Therefore, despite the previously mentioned limitations of the case study approach and with some reservations, we offer a second option with the third edition. We have included two *complete* canned cases for use with the textbook. The canned case option was designed to emphasize all of the important concepts in the textbook. Unlike the cases included with the second edition which were never intended to be complete cases but merely guidelines to use as a point of departure, the two cases in the third edition are complete and comprehensive.

Regardless of which approach you choose to use, the casebook includes a Systems Analysis and Design Milestones section that breaks the entire systems development project down into a number of smaller pieces (i.e., data modeling, output design, etc.) which the students can complete in a week or two. Some milestones are optional and, depending on the emphasis of your course, you may want to omit some and emphasize others.

CASE Tools   The third edition can be packaged with affordable student versions of Visible Analyst Workbench for DOS or Visible Analyst Workbench for Windows, both developed by Visible Systems, Inc. The student versions include fully functional software, as well as a tutorial to get the student started.

In addition, a tutorial for INTERSOLV's Excelerator for Windows has been developed by Anthony Connor of Clemson University and Margaret Batchelor of Furman University. This tutorial is comprised of short step-by-step lessons that teach mastery of most of Excelerator's features.

ACKNOWLEDGMENTS

We are indebted to many individuals who have contributed to the development of three editions of this textbook. First, we wish to thank those many individuals who reviewed the prior editions of the textbook — we have tried to retain all of the content and features that you endorsed with your adoptions.

Second, we wish to thank the reviewers and critics of this edition. Your patience, constructive criticism, and suggestions were essential and much appreciated. As you might guess, opinions from different individuals were varied. We have incorporated as many as possible into this edition. And we have already started a file of suggestions for the fourth edition. We eagerly look forward to additional guidance as we begin the fourth edition!

Special thanks is given to Dorothy Jane Miller, our secretary, who provided encouragement, overtime, patience, and "freedom from interruption" at appropriate times during the project.

We also thank our students. You are truly special and make teaching a rewarding experience. This book is for you!

Finally we acknowledge the contributions, encouragment, and patience of the staff of Irwin. We would offer special thanks to the following Irwin editors: Larry Alexander, Rebecca Johnson, Susan Trentacosti, and Rick Williamson—all of them *professionals* with whom any authors should be privileged to work!

We hope we haven't forgotten anyone. And we assume full responsibility for any inadequacies or errors in this text. Any comments, suggestions, corrections, or ideas are welcome. Write to us in care of Irwin.

To those who used our prior editions, thank you for your continued support. For new users, we hope you'll truly see a difference in this text. We hope you enjoy this edition as much as we enjoyed writing it. And until we unveil *fourth edition,* enjoy!

Jeff Whitten
Lonnie Bentley
Vic Barlow

Brief Contents

Contents

PART TWO

Systems Planning and Systems Analysis

PART THREE

Systems Design

PART FOUR

Systems Implementation and Systems Support

P A R T   F I V E

Cross Life Cycle Activities

SYSTEMS ANALYSIS
AND DESIGN METHODS

This is a practical book about information systems development methods. This first part focuses on "the big picture." Before you learn about specific activities, tools, techniques, methods, and technology, you need to understand this big picture.

Systems development isn't a mechanical activity. There are no magic secrets for success; no perfect tools, techniques, or methods. To be sure, there are skills that can be mastered. But the complete and consistent application of those skills is still somewhat of an art.

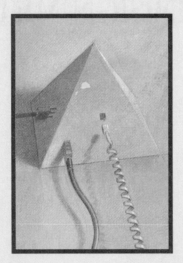

We start in Part One with fundamental concepts, philosophies, and trends . . . the basics! If you understand these basics, you will be better able to apply, with confidence, the practical tools and techniques you will learn in Parts Two through Five. Furthermore, you will find yourself able to adapt to new situations and methods.

Five chapters make up this part. Chapter 1 introduces you to the modern systems analyst, the professional most likely to facilitate systems development and practice the techniques presented in this book. You'll also learn about the relationships among systems analysts, end-users, managers, and other information systems professionals. Finally, you'll learn to prepare yourself for a career as an analyst (if that is your goal). Regardless, you will understand how you will interact with this important professional.

Chapter 2 introduces the product we will teach you how to build—information systems. Specifically, you will learn to examine information systems in terms of common information systems building blocks: PEOPLE, DATA, ACTIVITIES, NETWORKS, and

TECHNOLOGY. A visual pyramid model will help you organize these building blocks so that you can see them applied in the subsequent chapters.

Chapter 3 introduces a high-level (meaning general) process for developing information systems. This is called a systems development life cycle. The life cycle consists of five logical stages: planning, analysis, design, implementation, and support. Each of these stages consists of phases. The life cycle presented in this chapter will provide the framework for the remaining parts and chapters in the book.

Chapter 4 introduces systems development techniques and methodologies used to develop systems. These techniques and methodologies are gradually transforming the "art" of systems development into an "engineering-like" discipline. The chapter will introduce a diverse set of techniques and methodologies, including the popular structured techniques, information engineering and prototyping.

Chapter 5 introduces computer-aided systems engineering, or CASE. CASE technology assists systems analysts, programmers, and other developers with the proper application systems development techniques and methodologies. CASE technology promises to improve tomorrow's systems development productivity and the quality of the resulting information systems.

As we stated, these five chapters teach the fundamentals. We think you'll find the presentation practical, up-to-date, and down-to-earth. Part One will give you the solid foundation on which to build your study of the tools and techniques presented in Parts Two through Five.

SYSTEMS DEVELOPMENT FUNDAMENTALS

1

The Modern Systems Analyst

Chapter Preview and Objectives

No matter what your occupation or position in any business, you will likely encounter systems analysts. Many of you will become systems analysts. This chapter introduces you to the systems analyst, the person who usually performs the activities described in this book. You will know that you understand the modern systems analyst when you can:

Define
the systems analyst's role and responsibilities in a typical organization or business.

Define
the computer end-user's and manager's roles in a business as they relate to the systems analyst.

Differentiate
between a *systems analyst,* a *programmer/analyst,* and other common synonyms.

Define
systems planning, systems analysis, systems design, systems implementation, and *systems support*—the five principle activities performed by the systems analyst.

Describe
where the modern systems analyst fits into a progressive organization or business.

Develop
a plan of study for education/training that will prepare professionals for a career as a systems analyst.

Understand
the career prospects for future systems analysts.

M I N I C A S E

J. B. HOFFMAN AND COMPANY

J. B. Hoffman and Company is a manufacturer of medical and pharmaceutical products. It is headquartered in Minneapolis, Minnesota.

Scene: *Lobby of the hotel where Joe, a college recruit, spent the night. Cathy, an employee of Hoffman, approaches Joe, who appears to be waiting for someone.*

Cathy: Joe Elston?

Joe: That's me. Are you Cathy?

Cathy: Cathy Mennon. Hi, Joe. How was your trip?

Joe: Not too bad. I assume you are my host for the recruiting visit?

Cathy: That's right. Shall we just have breakfast here at the hotel?

[while walking toward the cafe]

Let me explain what today is going to be like. As your host, I will not be evaluating you. I will not participate in any way in the employment decision. My job is to tell you what it's like to work at J. B. Hoffman and to answer your questions.

[noting Joe understands, she continues . . .]

I guess I should begin by telling you a little about myself. I'm a department head in manufacturing information systems. That means I don't work at the central complex—I work at the Meridian Avenue plant where you'll have your afternoon interviews. Seven systems analysts work in my department. We build and support the material requirements planning systems for our biochemical plants that are distributed throughout the world.

Joe: Do you have any programmers working for you?

Cathy: Yes and no. All of our information systems specialists are called systems analysts. You start as an analyst and work your way up to senior analyst. Some analysts, like my staff, work on applications. Others work on corporate and regional databases. Still others work in networking and telecommunications. There are many different technical specialities. But all of them are called systems analysts and most all of them do at least some programming.

[A hostess escorts Cathy and Joe to their table.]

Joe: I thought systems analysts wrote specifications *for* programmers.

Cathy: Perhaps in some companies, but I doubt that there are many. There is a big difference in the classical, academic definition of a systems analyst and the real world responsibilities. The academic definition is usually restricted to a person who does systems analysis and design. In reality, most systems analysts do at least some programming.

Joe: I was taught that systems analysts do the analysis and design for new information systems. Does a Hoffman analyst get to analyze and design new systems?

Cathy: Absolutely! We expect all of our analysts to work closely with their customers to analyze and design their systems. But they also do some programming. Most of the programming you will do is in support for existing programs, but . . .

Joe: Excuse me for interrupting. You mean maintenance programming?

Cathy: Yes, but that term is so negative. Most of what the industry calls maintenance programming is *not* correcting mistakes and bugs. Most of it involves fine tuning and enhancing the software— such as writing a program to generate a new report.

Joe: Who writes the software for an entirely new system?

Cathy: It varies. We use contract programmers for a lot of the new software.

Joe: Contract programmers?

Cathy: Yes, contract programmers are non-Hoffman employees. They freelance their programming expertise to many companies. We often find them in colleges, professional societies, and through references. We sign them to a contract to write specific software based on specifications developed by our analysts. The use of contract programmers gives our systems analysts more time to do true systems analysis and design.

Joe: You said "it varies." I assume then that you don't always use contract programmers for new systems development?

Cathy: That's right. Sometimes it may be too expensive. Or perhaps the system is too secret to involve third parties. In those cases, our own systems analysts write the programs. Still, I'd guess that the average Hoffman analyst doesn't write too much code. We spend most of our time working with users and designing systems and making improvements. That's why we call all of our people systems analysts, not programmer/analysts.

Joe: This is interesting. So after I go through the training program, I'd jump right into doing systems analysis, systems design, or support programming?

Cathy: Actually, there is no full-time training. In fact, most new hires are doing productive work their first week. Based on your interviews, you would be assigned to a team of systems analysts. Someone in your team would be your mentor. That person would be responsible for showing you the ropes and helping you adjust to the corporate culture. Your department head would plan your first three months to include on-the-job experiences and formal course work.

Your team would be responsible for specific systems. You would likely inherit one or more of those systems as part of your on-the-job training. Your mentor would help you learn your assigned systems and introduce you to your customers. Most of your early work would involve supporting your assigned systems—that may,

of course, involve some simple analysis and design. But you won't develop new systems until you go through our methodology and standards courses.

Joe: I love it. I thought that I'd be in classes for six solid weeks. What else can I expect?

Cathy: You can expect your days to consist of varied experiences. Look at my Day Timer *[an appointment book].* As you see, this week I am conducting user interviews, attending project meetings, attending classes to upgrade my own skills, and working in various committees.

Joe: You know, there's a lot more to this systems analysis job than I ever imagined. I'm looking forward to meeting some of your people.

         [Waiter arrives at their table.]

Waiter: Are you ready to order?

Discussion Questions

1. Why would a company eliminate programmers and give all their information systems staff a title of systems analyst? Do you see any advantages or disadvantages not stated in the case? Why do you suppose Hoffman wouldn't eliminate their systems analysts instead, or as well?

2. Some people enjoy programming; others despise it. Fortunately, there are many computing careers that involve little or no programming. But why do you suppose that programming is an essential skill to most nonprogramming jobs?

3. Students often proclaim, "I can't wait to graduate. The biggest difference between school and the real world is that, when you are working on a project in the real world, you don't have to worry about a test in this other course, homework in that other course, and another project in still another course." Comment on this statement.

4. Another student proclamation: "I can't wait to graduate. Just think, no more school. No more studying." Comment on this statement.

WHO SHOULD READ THIS BOOK?

Today, it is hard to imagine any industry or business that has not been affected by computers. One way or another, you will either help to develop computer applications, use computer applications, or be affected by computer applications—probably all three!

Clearly, you expect computer specialists to develop computer applications. But in this age of personal computers and friendlier computer software, both end-users and managers frequently develop (or enhance) their own computer applications.

This book, as suggested by its title, is concerned with the development of systems and applications. (Throughout this book, we will use the terms *application* and *system* interchangeably.) Whether you are preparing for a career as a computer professional or as an end-user of computers, application and system development is an essential subject to know and understand. It's hard to imagine any career or job that would not benefit from this understanding.

This first chapter is about the professional most typically associated with developing computer applications, the systems analyst. It doesn't matter whether or not you want to become a systems analyst. Your personal motivation to learn about the systems analyst should be driven from one of the following realities (read the column that best describes your career or goals):

As a Future Computer or Information Systems Specialist

You may have to learn, from end-users and managers, the business terminology, procedures, and problems associated with a system.

You may have to solicit, from end-users and managers, the objectives and requirements for a new and improved system. You will also be asked to verify your understanding of business problems, objectives, and requirements.

You may have to propose and defend technical solutions for business problems and requirements.

You may have to design various inputs, outputs, files, databases, programs, and manual procedures for a new system. You may also have to develop working prototypes of a system or its components.

You may have to test, in cooperation with its end-users, a new system that has been developed.

You may have to train end-users and managers to use a new or unfamiliar computer-based system.

You may have to respond to normal day-to-day problems and mistakes that arise from using a computer-based system.

You may help end-users develop their own systems using microcomputers or fourth-generation programming languages.

You may be called upon to write computer programs that fulfill the requirements and specifications that were defined by a systems analyst.

All of the above activities are typically performed by systems analysts.

As a Future End-User of Computers and Information Systems

You will likely have to explain to systems analysts the current business terminology, procedures, problems, and ways a system might be improved.

You will likely have to explain to systems analysts the objectives, requirements, and priorities for an improved system. You will also be asked to verify the analyst's understanding of your business problems, objectives, and requirements.

You will likely have to evaluate, and possibly approve, a systems analyst's proposed technical solution to your business requirements.

You may have to evaluate a systems analyst's design of a new system. You may also have to test and evaluate a systems analyst's working prototype of a new system or component.

You may have to help a systems analyst test a new system before it is placed into day-to-day operation.

You may frequently be trained by a systems analyst on how to use a new or improved system.

As you use any computer-based system, you will probably refer problems, mistakes, and concerns to the systems analyst responsible for supporting that system.

You may develop or enhance your own systems (using microcomputers and software tools).

You may have to write your own programs that generate new reports from data stored in existing systems.

All of the above activities are typically performed by users of computers. For the latter two possibilities, you would be performing many of the same tasks performed by systems analysts.

It is equally important to understand the motivations of the other audience; therefore, go back now and read the other column. The similarities between the two columns implies a high degree of interaction between the specialists and users. *Systems development is about people working with people!* That will be a continuous theme in this book.

SYSTEMS ANALYSTS—MODERN BUSINESS PROBLEM SOLVERS

Just by reading the above lists, you could probably define the term *systems analyst.* Nevertheless, we will take a closer look at the job and career. First, we will examine the origins of the title. Then we will look at a more formal definition of the systems analyst and classify the activities performed by the systems analyst. Finally, we will describe where systems analysts work.

Origins of the Systems Analyst

The first systems analysts were born out of the industrial revolution. They were not concerned with computers or computer-based systems. Instead, they were industrial engineers whose responsibilities centered around the design of efficient and effective manufacturing systems. Information systems analysts evolved from the need to improve the use of computer resources for the information-processing needs of business applications. In other words, they designed computer-based systems that manufacture information.

Despite all of its current and future technological capabilities, the computer still owes its power and usefulness to *people.* Businesspeople define the applications and problems to be solved by the computer. Computer programmers and technicians apply the technology to well-defined applications and problems.

Computers are simply tools that offer the opportunity to collect and store enormous volumes of data, process business transactions with great speed and accuracy, and provide timely and relevant information for management. Unfortunately, this potential has not been fully or even adequately realized in most businesses. Business users may not fully understand the capabilities and limitations of modern computer technology. Likewise, computer programmers and technicians frequently do not understand the business applications they are trying to computerize. Worse still, some computer professionals become overly preoccupied with computer technology. A communications gap has always existed between those who need the computer and those who understand the technology. The systems analyst bridges that gap. You can (and probably will) play a role as either a systems analyst or one who works with systems analysts.

Job Descriptions for the Modern Systems Analyst

In simple terms, systems analysts are people who understand both business and computing. Systems analysts transform the business and information requirements of computer users into the computer-based technical solutions (systems) that are implemented by computer programmers and other computer specialists. A more formal definition follows.

A **systems analyst** studies the problems and needs of a business to determine how people, processes, data, communications, and technology can best accomplish improvements for the business.

When computer technology is used, the analyst is responsible for the efficient capture of data from its business source, the flow of that data to the computer, the processing and storage of that data by the computer, and the flow of useful and timely information back to the business and its people.

To make things simpler, a systems analyst is a business problems solver. How so? Computers are only of value to a business if they help solve problems. (We use the term *problem* here to collectively include both correcting bad situations as well as exploiting opportunities to improve the business in any way.) Thus, a systems analyst helps the business by solving its problems. Furthermore, the process of creating or improving a system is itself a problem to be solved. Consequently, the very act of systems development is a problem-solving activity.

Modern systems analysts develop both an organization's **business systems** and its **information systems.** Business systems support, and sometimes automate, day-to-day business operations. Information systems generate information to help managers solve problems and make intelligent decisions. In reality, business and information systems are rapidly becoming one and the same. Throughout this book, we use the terms rather interchangeably.

Systems analysts sell users and managers the services of computing. More importantly, they sell change (which doesn't always make systems analysts popular). Every new system changes the business. Increasingly, the very best systems analysts literally change their organizations—providing information that can be used for competitive advantage, finding new markets and services, and even dramatically changing and improving the way the organization does business.

Unfortunately, the title *systems analyst* is not used consistently in industry. Different organizations assign this title to anyone from a computer programmer to a sophisticated designer of computer applications. What is even more confusing is that, in most organizations, systems analysts actually perform some or all of the duties of a computer programmer, and vice versa. So how can you identify a true systems analyst? One way would be to compare their responsibilities against a classical job description such as the one illustrated in Figure 1.1.

There are several legitimate, but often confusing, variations on the title *systems analyst*. The most common of these is the **programmer/analyst** (or analyst/programmer), whose job includes the responsibilities of both the programmer and the analyst. In reality, most systems analysts do some programming and most programmers do some systems analysis and design, regardless of titles. Other common variations on the title are listed in the margin. Some of these titles represent systems analysts whose job descriptions are restricted to certain technical specialties. For example, a **database analyst** is a systems analyst who plans, analyzes, designs, implements, and supports databases, but not necessarily the applications that use those databases.

The Systems Analyst as a Problem Solver

Essentially, the systems analyst uses a problem-solving approach to develop systems.

Problem solving is the act of studying a problem environment in order to implement corrective solutions. *Problem* is a collective term that in-

———— ✓ ————

Synonyms for Systems Analyst

Analyst
Business analyst
Consultant
Data analyst
Database analyst
Information analyst
Management consultant
Operations analyst
Systems consultant
Systems engineer
Systems integrator

JOB TITLE:	Systems analyst (multiple-job levels)
REPORTS TO:	Systems development team manager or assistant director of systems development
DESCRIPTION:	A systems analyst shall be responsible for studying the problems and needs set forth by this organization to determine how computer hardware, applications software, files and databases, networks, people, and procedures can best solve these problems and improve business and information systems.
RESPONSIBILITIES:	1. Evaluates projects for feasibility. 2. Estimates personnel requirements, budgets, and schedules for systems development and maintenance projects. 3. Performs interviews and other fact gathering. 4. Documents and analyzes current system operations. 5. Defines user requirements for improving or replacing systems. 6. Identifies potential applications of computer technology that may fulfill requirements. 7. Evaluates applications of computer technology for feasibility. 8. Recommends new systems and technical solutions to end users and management. 9. Identifies potential hardware and software vendors, when appropriate. 10. Recommends and selects hardware and software purchases (subject to approval). 11. Designs system inputs, outputs, on-line dialogue, flow, and procedures. 12. Designs files and databases (subject to approval by Data Administration). 13. Writes, tests, and/or supervises applications software development. 14. Trains users to work with new systems and versions. 15. Converts operations to new systems or versions. 16. Supports operational applications.
EXTERNAL CONTACTS:	1. Assigned end users of mainframe computers and applications. 2. Assigned owners (end user management) of mainframe computers and applications. 3. Data Administration Center personnel. 4. Network Administration Center personnel. 5. Information Center personnel. 6. Operations Center personnel. 7. Methodology/CASE expert and staff. 8. Computer hardware and software vendors. 9. Other systems analysis and development managers.
MINIMUM QUALIFICATIONS:	Bachelor or Masters Degree in Computer Information Systems or related field. Programming experience preferred. Prior experience with business applications considered helpful. Prior training or experience in systems analysis and design, preferably structured methods, preferred. Good communications skills—oral and written—are mandatory.
TRAINING REQUIREMENTS:	Analysts must complete or demonstrate equivalent backgrounds in the following in-hours training courses: STRADIS Methodology and Standards, Joint Application Design (JAD) Techniques, Systems Application Architecture (SAA) Standards, Fundamentals, DB2 Database Design Techniques, CSP Prototyping Techniques, Excelerator/IS Computer Aided Design Techniques, Project Management Techniques, Microcomputer Software Tools, and Interpersonal and Communications Skills for Systems Analysis.

JOB LEVELS:	Initial assignments are based on programming experience and training results. The following job levels are defined:
	    Programmer/analyst: 30% analysis/design—70% programming
	    Analyst/programmer: 50% analysis/design—50% programming
	    Analyst: 70% analysis/design—30% programming
	    Senior analyst: 30% management—60% analysis/design and 10% programming
	    Lead analyst: 100% analysis/design or consulting

FIGURE 1.1   Job Description for a Systems Analyst   The job description for a systems analyst will vary from firm to firm. This description is representative of a systems analyst.

cludes (1) situations, real or anticipated, that require corrective action; (2) opportunities to improve a situation despite the absence of complaints; and (3) directives to change a situation regardless of whether anyone has complained about the current situation.

Do not assume that all problems are bad. A systems analyst might implement improvements to a system about which nobody has complained. How can we call that a problem? Simple—anytime you see an opportunity to improve a system and do nothing about it, it is a problem.

Planning is the ongoing study of a problem environment to identify problem-solving possibilities. Ideally, the projects that are selected will provide the greatest long-term benefit to the business. Thus, planning of information systems cannot be separated from the planning of the business itself. It should be noted that many businesses have not yet placed a premium on this planning activity.

Analysis is the study of the problem environment and the subsequent definition and prioritization of the requirements for solving the problem. Throughout analysis, the emphasis is on the business, not the computer.

Design is the evaluation of alternative problem solutions, and the detailed specification of the final solution. Throughout design, the emphasis usually shifts from the business to the computer solution. Design specifications are typically sent to programmers for systems implementation.

Implementation is the construction or assembly of the problem solution, culminating in a new environment based on the solution. Once implemented, the new system is said to be "in operation" or "in production."

Support is the ongoing maintenance and enhancement of the solution during its lifetime (which may be weeks, months, or years).

When performed by the systems analysts, the above problem-solving activities are collectively called a systems development life cycle.

A **systems development life cycle** is a systematic and orderly approach to solving business problems, and developing and supporting resulting information systems. It is sometimes called an **applications development life cycle.**

The term *cycle* refers to the natural tendency for systems to cycle through these phases (Figure 1.2). The systems analyst is the one computer or information systems specialists who is involved in all of these activities.

Do all systems analysts perform all of these activities? It varies from one organization to another. Systems planning is usually performed only by experienced analysts and only in mature organizations that have realized the importance of such planning. Although most analysts do both systems analysis and design, some are restricted to only analysis or only design. As noted earlier, most analysts do some programming, at least as part of ongoing systems support. Programmer/analysts also do some programming as part of systems implementation.

Many (if not most) of you have written computer programs. How do the activities of the programmer compare with those of the systems analyst? Programmers tend to be most concerned with technology—the hardware and

FIGURE 1.2 A Systems Development Life Cycle This diagram depicts the typical phases or activities that a systems analyst uses to solve problems. Collectively, the phases are known as a systems development life cycle. Different information systems shops adopt different variations on this life cycle concept.

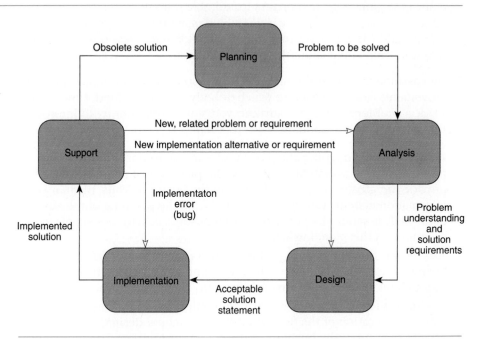

software. Systems analysts, on the other hand, are responsible for the system as a whole. The whole system includes:

- PEOPLE (users, management, and information specialists).
- DATA (how it is captured, edited, used, and stored).
- ACTIVITIES (both automated and manual, that combine to form useful business and information applications).
- NETWORKS (where data is stored and processed and how data is exchanged bewteen different locations).
- TECHNOLOGY (the hardware used, the software used, and the software written).

Systems analysts, unlike programmers, also become much more involved with business people and organizational behavior and politics.

Consequently, the systems analyst's job presents a fascinating and exciting challenge to many individuals. It offers high management visibility and opportunities for important decision making and creativity that may affect an entire organization. Furthermore, this job can offer these benefits relatively early in your career (compared to other jobs and careers at the same level).

Where Do Systems Analysts Work?

Systems analysts can be found in most businesses. Most systems analysts work for the information systems (IS) unit in an organization. Others work for management consulting firms that build systems for other organizations. Still others work for software houses that create software packages for resale to other organizations. Let's examine each of these in more detail.

The Information Systems Unit in an Organization

The IS unit in an organization may have any of the names listed in the margin. Depending on the organization, IS may report directly to the president, to an executive vice president, or to the vice president or director of a key business unit (e.g., Finance). Many organizations have learned that operating the IS shop under any one business unit results in conflicts with other business units that need computer support. Thus, in many organizations, the IS manager has achieved vice president status or, more recently, the designation of chief information officer (CIO). This places the IS function at the same level as each of the business units (e.g., Finance, Human Resources, Sales and Marketing) and ensures that no business unit monopolizes the services provided by IS.

The internal organization of information systems can be depicted by organization charts. The structure of the IS function varies from company to company. Figure 1.3 is an organization chart for a relatively progressive IS unit. It reflects changes that are currently being driven by trends in database, microcomputers, and data communication usage.

The figure illustrates four key centers of IS activity: systems development and support, data management, telecommunications, and computer operations. The systems development unit includes systems analysts and programmers who develop and support key systems for the organization's users and management. The figure reflects the common practice of dividing development and support into teams that are responsible for specific groups of applications. Each analyst would be normally be assigned to support specific applications (and provide backup support for others). Additionally, analysts and programmers will normally be assigned to temporary work units called *project teams.* These teams will tackle larger development projects (for instance, new systems or major system redesigns).

A development and support unit may also include the relatively modern concept of a development center.

> A **development center,** staffed by experienced systems analysts, provides tool and methodology consultation and expertise to systems analysts who work on systems development projects.

These consultants might include, for example, tool experts, methodology experts, and technology experts. That group may also include auditors and quality control specialists who enforce the standards of systems development.

It should be noted that some organizations are starting to decentralize the development of information systems to independent business units in the organization. This trend is discussed in greater detail in The Next Generation box.

The development and support unit also includes an information center.

> An **information center** supports end-user computing — helping the organization's users and management develop and maintain their own systems, usually smaller and less strategic than the systems developed by analysts and programmers.

Most of these end-user systems involve personal computers and networks. Others involve helping the users find and access mainframe data by generating their own reports or downloading the data to personal computers. Information centers employ systems analyst-like consultants to help the end-users.

✓

Synonyms for Information Systems Department

Computer Information Systems
Data Processing
Data Services
Data Systems
Electronic Data Processing
Information Resources
Information Services
Management Information Systems
Management Systems

FIGURE 1.3 Organization of the Information Systems Function Every information systems shop develops its own unique structure; however, this structure is fairly typical of a progressive shop. Programmers and systems analysts are organized into project teams. Various other information specialists may be added to those teams as necessary. The teams are frequently created and disbanded as projects are started, completed, or cancelled.

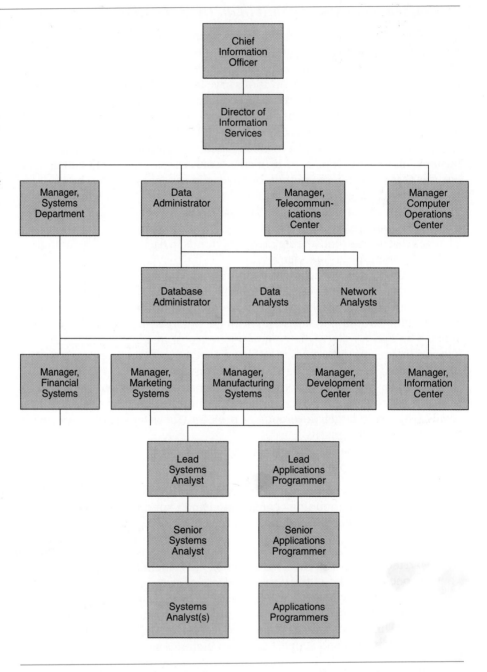

Both the development and information centers might be involved in systems development for personal computers. The difference is that the development unit would build a system for the user, whereas the information center would help the user build his or her own system.

The *data management center* tracks and manages the enormous volume of data that is stored in the average organization. It is frequently divided into two

---

The Next Generation

INDEPENDENT BUSINESS UNITS:
A NEW HOME FOR TOMORROW'S ANALYSTS?

Today's business world is increasingly characterized by reorganization and decentralization. How will this affect IS and systems analysts? The impact of this trend is most evident in what has come to be known as independent business units.

The concept behind **independent business units** (IBUs) is to decentralize the information systems development group directly into the user community. In other words, systems development managers, their team leaders, systems analysts, and programmers no longer report to the IS function. Instead, they report directly to the managers and users (called a *business unit*—e.g., manufacturing, marketing, sales, engineering, accounting, human resources, finane, etc.) that they serve. Each business unit operates independently from the others—hence the name *independent business units*.

The central IS function still exists. It provides mainframe computer access and support, data and database administration, and technical support for data communications and networking.

Most IBUs will still use the central IS computing facilities; however, many IBUs downsize appropriate applications to run on local minicomputers, personal computers, and personal computer networks.

Why decentralize information systems development and support? The benefits are numerous. They include:

- Improved relationships with end-users and business managers—the loyalty is to the business unit and its users.
- Improved communication between the business units and the central IS function.
- Getting technology closer to the users who benefit from it.
- Developing in systems managers, systems analysts, and programmers a better understanding of the business unit and their needs.
- Transferring to business unit managers the control of IS budgets and priorities for their own business units. They lobby for their own systems budgets and staff and they decide how to distribute effort between support of existing systems and development of new systems.
- Making each IS business unit pay for itself as a part of that overall business unit.
- Changing systems developers from a project orientation to a full-time service orientation.

A reorganization into IBUs is not without its problems. The coordination and integration of technology and information systems becomes more difficult. The coordination and administration of corporatewide data is also complicated by IBUs. Finally, personnel transfers between units is complicated by the reality that each unit is autonomous.

Some organizations solve these problems by assigning each IBU an account executive. Account executives have a dual reporting relationship. The business unit manager (e.g., vice president or director of the business function) provides business direction and sets all applicatons and systems development budgets and priorities. The IS director or manager provides technology direction and sets standards for communication and consistency.

What impact will these changes have on systems analysts? It is entirely possible that systems development in the next generation will more commonly be done in IBUs. That means that more analysts will work directly for their users, not for a central computer or IS group. In this environment, systems analysts will be expected to show greater business savvy and interpersonal skills.

Additionally, IBUs will likely achieve a greater balance in the skill sets of their systems analysts. Today, most analysts come from computer and IS college programs. Tomorrow's analysts will likely include greater numbers of business and noncomputer graduates (probably with IS minors that include courses about systems development).

Does all this sound improbable? It's happening now! For more information, try reading "The New IS Force," *Datamation,* August 1, 1989, pp. 18–23.

functions: data administration and database administration. Data administration is concerned with what data is stored and where. Database administration is concerned with the technology used to store data and performance of that technology. These units employ analyst-like specialists called data analysts or database analysts.

The **telecommunications center,** another relatively new center of activity, is concerned with how computers and databases are distributed and interconnected. We expect to see growth in demand for systems analysts who specialize in design of distributed applications, integration of distributed applications, and distributed databases. For the time being, we will call them **network analysts**.

The **computer operations center** provides central computing resources and services such as data entry, operations, systems programming, and other technical support.

Management Consulting Firms

Management consulting firms build and sometimes manage systems for other organizations. Why wouldn't an organization build all systems through its own IS unit? Perhaps the IS unit is understaffed. Perhaps the IS unit's management is looking for technical expertise that their own staff doesn't possess. Perhaps the IS unit's management is looking for an unbiased opinion and fresh ideas. The list of reasons is endless. Examples of well-known management consulting firms are listed in the margin. You may recognize that many of these firms also provide accounting and auditing services.

The systems analysts employed by management consulting firms are usually called **management consultants** or *systems consultants*. They are loaned (for a fee) to the client for *engagements* (a consulting term that means the same as projects) that result in a new system for the client. Management consulting firms represent one of the more attractive employment options for the aspiring systems analyst who doesn't mind travel and occasionally long hours. The engagements tend to be very challenging and provide a wide variety of exposure and experiences. Also, management consulting firms keep their consultants on the cutting edge of technology and techniques in order to compete for business.

One of the fastest-growing markets for management consulting firms is **systems integration.** This involves helping organizations integrate applications that were not well designed for integration or that run on very different technical platforms from different computer manufacturers. Systems analysts that specialize in systems integration are frequently called **systems engineers** or **systems integrators**.

Software Houses

Software houses develop packages for resale to other organizations. These packages are rarely customized to the specific need of any one customer (although they might offer limited customization features at the time of installation). Software packages may be geared either to specific business functions (e.g., accounting, payroll, purchasing) or a specific industry/business (health, retail, education, government etc.). Examples of software houses are listed in

---

✓

Representative Management Consulting Firms

American Management
  Systems
Anderson Consulting
Electronic Data Services
  (EDS)
Ernst & Young
James Martin & Associates
Peat Marwick & Company
Touche Ross Company

✓

Representative Software Houses

Computer Associates
Lotus Development
  Corporation
IBM
McCormack & Dodge
Microsoft Corporation
MSA

the margin. Notice that some software houses are also hardware manufacturers (e.g., IBM).

The development of software packages follows a path similar to that of developing information systems within an organization. The only difference is the customer. Software packages must be developed to meet the needs of as many customers as possible, whereas information systems are developed to meet the requirements of a specific organization and its users. Software houses usually refer to their systems analysts as **software engineers.**

USERS AND OTHER CONTACTS

Systems analysts would not exist without the needs of their clients. By far, the most commonly used term to describe the client is *user.*

> A **user** is a person, or group of persons, for whom the systems analyst builds and maintains business information systems.

Users don't always refer to themselves as users — some actually take offense at the term. Synonyms are listed in the margin. As Ed Yourdon points out, "the user is the 'customer' in two important respects: (1) as in many other professions, 'the customer is always right,' regardless of how demanding, unpleasant, or irrational he or she may seem; and (2) the customer is ultimately the person paying for the system and usually has the right or ability to refuse to pay if he or she is unhappy with the product received."[1]

Based on Yourdon's definition, we can identify at least two specific user/customer groups: end-users and owners.

> **End-users** are those who either have direct contact with computer-based system (e.g., they use a terminal or PC to enter, store, or retrieve data) or they use information (reports) generated by a system.
>
> **Owners** provide management sponsorship of a computer-based system. In other words, they pay to have the system developed, they approve the technology used in the system, and they pay to keep the system running. Most importantly, they expect the system to return lifetime benefits that exceed the lifetime costs.

Clearly, an owner can also be one of the end-users of a system. A third type of user, the **indirect user,** is equally important. An indirect user does not directly use a system, but is affected by it.

Beginning in the next chapter, and continuing throughout the book, we will address the values and concerns of today's users. Young systems analysts are frequently surprised to learn that users don't always have a positive attitude about the IS function in their organization. Tomorrow's systems analysts must work to improve the image of IS. That can only occur when the user is viewed as a true customer and participant in the development process.

Clearly, tomorrow's users need an understanding of the tools and methods employed by systems analysts. This book will provide the necessary survey for that audience.

───────── ✓ ─────────

Synonyms for Users

Clients
Customers
End-users
Owners
Participants

─────────────────────

---

[1] Edward Yourdon, *Modern Structured Analysis* (Englewood Cliffs, N.J.: Prentice Hall, Yourdon Press Computing Series, 1989), p. 41.

**FIGURE 1.4 People with
Whom the Analyst Must
Work** As facilitators of
systems development, the
analyst must work with
many types of people,
both technical and non-
technical. The joint efforts
of these professionals,
as coordinated by the
analysts, will result in
successful information
systems applications.

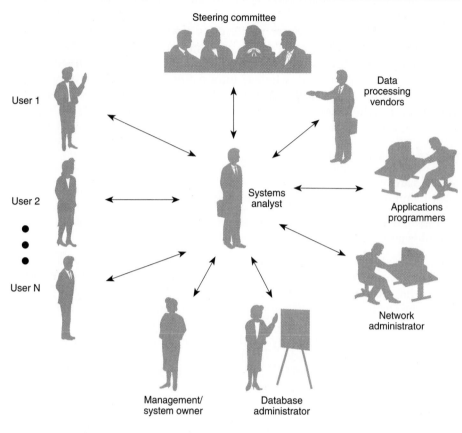

It takes more than systems analysts and users to build effective systems. In
addition to clients, the systems analyst works with a number of peers. These
include programmers, database specialists, networking specialists, computer
center specialists, and hardware and software sales representatives. As illustrated in Figure 1.4, the systems analyst's role in the typical project is to act as a
facilitator. The analyst may well be the only individual who sees the system as a
whole. As such, the systems analysts must possess a unique set of skills and
abilities to accomplish the complex task of facilitating systems.

PREPARING FOR A CAREER AS A SYSTEMS ANALYST

This book will not make you a competent systems analyst any more than a first
programming book or course would make you a competent computer programmer. You will, however, be able to immediately apply the skills and concepts you learn in this book to small systems projects, although you may need
some supervision. Furthermore, you will have a solid foundation on which to
base additional systems training.

That brings us to the obvious question: What does it take to become a successful systems analyst? One writer suggests the following:

I submit that systems analysts are people who communicate with management and users at the management/user level; document their experience; understand problems before proposing solutions; think before they speak; facilitate systems development, not originate it; are supportive of the organization in question and understand its goals and objectives; use good tools and approaches to help solve systems problems; and enjoy working with people.[2]

This seems like quite a tall order, doesn't it? It is often difficult to pinpoint those skills and attributes necessary to succeed. However, the following subsections describe those skills most frequently cited by practicing systems analysts.

Working Knowledge of Information Systems and Technology

The systems analyst is "an agent of change." He or she is responsible for showing end-users and management how new technologies can benefit their business and its operations. To that end, the analyst must be aware of both existing technologies and techniques. Such knowledge can be acquired in college courses, professional development seminars/courses, and in-house corporate training programs. Some of the technologies and topics that you should be studying today are listed in the margin. Colleges offer courses that include many of these topics.

One of the best ways to keep up on what's happening is to develop a disciplined and organized habit of skimming and reading various trade periodicals on computers and information systems. Examples of helpful trade publications are found in the margin (we purposefully restricted this list to periodicals that cover the entire computing and IS spectrum, not just one facet such as personal computers). Most of these magazines should be available in your college or corporate library.

It is often difficult to keep up with all the periodicals, even by skimming them. For this reason, we call special attention to *Information Week* as an outstanding periodical that summarizes the week's news, includes numerous "trend" and "applications" articles, and surveys other periodicals and articles of interest to the busy IS professional. We also note the monthly *IS/Analyzer,* which uses insightful case studies and observations to cover, in depth, a single trend or issue.

Another way to keep current is through professional association. Consider joining a professional association such as the Data Processing Management Association (DPMA), the Society for Information Managers (SIM), or the Association for Systems Management (ASM).

Computer Programming Experience and Expertise

Whether or not a systems analyst writes programs, it is clear that they must know how to program. Why? Because, the systems analyst is the principle link between business users and computer programmers. Consequently, many, but by no means all, organizations consider experience in computer programming to be a prerequisite to systems analysis and design.

——— ✓ ———

Current Information Systems and Technology Topics for Future Analysts

Computer graphics
Computer-integrated
 manufacturing
Distributed databases
End-user computing
Executive information
 systems
Expert systems
Keyless data entry
 technology
Microcomputers and
 software
Networking
Object-oriented
 programming
Fourth-generation languages

——— ✓ ———

Useful Trade Publications

ComputerWorld
Computer Decisions
Datamation
Information Week
IS/Analyzer

---

[2] Michael Wood, "Systems Analyst Title Most Abused in Industry: Redefinition Imperative," *ComputerWorld,* April 30, 1979, pp. 24, 26.

You should not, however, assume that a good programmer will become a good analyst or that a bad programmer could not become a good analyst. There is no such correlation. Unfortunately, many organizations insist on promoting good programmers who become poor or mediocre systems analysts. Worse still, mediocre programmers are often passed over in the belief that they cannot become good analysts.

Regardless of opinions concerning the need for programming experience, it is difficult to imagine how systems analysts would prepare specifications for a programmer if they didn't have some programming background. This background can be obtained at virtually any college or vocational school. Most systems analysts need to be proficient in at least one high-level programming language. For business applications, this language should probably be COBOL. For engineering or scientific applications, FORTRAN, C, or Pascal may be the better choice. Soon, all companies doing business with the Department of Defense may need Ada programming expertise. Other applications such as computer-assisted manufacturing may also switch to Ada. Software houses are increasingly using the language C, especially for microcomputer software packages. Object-oriented programming with languages like C++ are becoming popular.

Today's analyst should also become familiar with at least one **fourth-generation language** or **applications generator.** Analysts use these to prototype or quickly generate working models of systems in order to more quickly establish system requirements. For example, IBM's application generator, Cross System Product (CSP), can not only generate prototypes, but also convert approved prototypes into more efficient COBOL programs as the system moves closer to implementation. Examples of fourth-generation langauges and application generators are listed in the margin.

General Business Knowledge

In most instances, the systems analyst need not be an expert in a specific business application or function (such as accounting or production). However, analysts should be able to communicate with business experts to gain knowledge of problems and needs. Much of the knowledge about any functional business area is gained while on the job. Still, we strongly suggest you include courses in as many of the subjects as possible that are listed in the margin. Specializations such as accounting or manufacturing can be very valuable in some instances. These subjects are taught in many colleges.

It should be noted that by working with business experts, systems analysts gradually acquire business expertise. It is not uncommon for analysts to develop so much expertise over time they move out of IS and into the user community.

Problem-Solving Skills

The systems analyst must have the ability to take a large business problem, break that problem down into its component parts, analyze the various aspects of the problem, and then assemble an improved system to solve the problem. Engineers call this problem-solving process *analysis and synthesis.* The analyst must also learn to analyze problems in terms of causes and effects rather than in

---

✓

Representative Fourth-Generation Languages and Applications Generators

Cross System Product (CSP)
Focus
Ideal
Mantis
Natural
ObjectView
Powerbuilder
RAMIS
SAS

✓

Important Business Subjects

Business law and ethics
Business statistics
Economics
Financial accounting
Human resources
   management
International business
Inventory control
Managerial or cost
   accounting
Manufacturing management
Marketing
Operations
Organizational behavior
Principles of business
Principles of management
Production scheduling
Quantitative methods

terms of simple remedies. Being well organized is also part of developing good problem-solving skills.

Analysts must also be able to creatively define alternative solutions to problems and needs. Creativity and insight are more likely to be gifts than skills, although they can certainly be developed to some degree. Perhaps the best inspiration for students and young analysts comes from the late USN Rear Admiral Grace Hopper, mother of the COBOL language. She suggests that "the most damaging phrase in the English language is 'We've always done it that way.' " Always be willing to look beyond your first idea for other solutions.

Interpersonal Communications Skills

An analyst must be able to communicate effectively, both orally and in writing. The analyst should actively seek help or training in business writing, technical writing, interviewing, presentations, and listening. A good command of the English language is considered essential. These skills are learnable, but most of us must force ourselves to seek help and work hard to improve them. College recruiters and business managers will emphatically tell you that communications skills are the single most important ingredient for success. This is because of the number and complexity of the interpersonal relationships previously discussed. Some of the topics in this book survey the importance of communications skills for the systems analyst.

Interpersonal Relations Skills

It has been suggested that analysts "need to exercise the boldness of Lady Godiva, the introspection of Sherlock Holmes, the methodology of Andrew Carnegie, and the down-home common sense of Will Rogers."[3] In other words, systems work is people-oriented and systems analysts must be extroverted or people-oriented. Interpersonal skills help us work effectively with people. Although these skills can be developed, some people simply do not possess an extroverted personality. The interpersonal nature of systems work is demonstrated in the running SoundStage case study that appears at intervals throughout this book.

Interpersonal skills are also important because of the political nature of the systems analyst's job. The analyst's first responsibility is to the business, its management, and its workers. Individuals frequently have conflicting goals and needs. They have personality clashes. They fight turf battles over who should be responsible for what and who should have decision authority over what. The analyst must mediate such problems and achieve benefits for the business as a whole.

Another aspect of interpersonal relations is recognition of the analyst's role as an agent of change. The systems analyst is frequently as welcome as an IRS auditor! Many individuals feel comfortable with the status quo and therefore resent the change that the systems analyst brings. An analyst should study the theory and techniques of effecting change. Persuasion is an art that can be

---

[3] Kenniston W. Lord, Jr., and James B. Steiner, *CDP Review Manual: A Data Processing Handbook,* 2nd ed. (New York: Van Nostrand Reinhold, 1978), p. 349.

learned. Begin by studying sales techniques—after all, systems analysts sell change.

Finally, systems analysts work in teams composed of IS professionals, end-users, and management. Being able to cooperate, to comprise, and to function as part of a team, is critical for success in most projects. Because development teams include people with dramatically different levels of education and experience, group dynamics is an important skill to develop.

Flexibility and Adaptability

No two systems development projects encountered by a systems analyst are identical. Each project offers its own unique challenges. Thus, there is no single, magical approach or solution applicable to systems development. Successful systems analysts recognize this fact and learn to be flexible and adapt to special challenges or situations presented by specific systems development projects.

Many organizations have standards that dictate specific approaches, tools, and techniques that must be adhered to when developing a system. Although these standards should be followed as closely as possible, the systems analyst must be able to recognize when variations upon (or single-instance exceptions to) those standards are necessary and beneficial to a particular project. At the same time, the analyst must be aware of the implications of not following the standards. It's a balancing act that usually improves with experience.

Character and Ethics

The nature of the systems analyst's job requires a strong character and sense of ethics. Because systems analysts require information about the organization in order to develop systems that properly support the organization, those analysts are often privy to sensitive plans and secrets. Consequently, the analyst must be very careful not to share that information with others, either within or outside the organization.

Systems analysts also frequently uncover dissent in the ranks of employees and gain access to sensitive and private data and information (through sampling of memos, files, and forms) about customers, suppliers, employees, and the like. The analyst must be very careful not to share such feelings or information with the wrong people. Trust is sacred! Confidence is earned!

Systems analysts also design systems and write programs. But who owns such intellectual property? In most cases, the design and programs are the property of the organization since they paid for the services of the analyst and programmers. It would be unethical to take (or sell) such designs and programs to another company.

Finally, systems analysts are a key interface between the computing industry and end-users and management. They have a moral obligation to set a good example for end-users, especially in the area of software copyrights. Systems analysts should help end-users and management appreciate the importance of honoring the terms of software licensing agreements.

Systems Analysis and Design Skills

All analysts need *thorough ongoing training* in systems analysis and design. Systems analysis and design skills can be conveniently factored into three subsets—concepts and principles, tools, and techniques. This book begins or extends that training.

When all else fails, the systems analyst who remembers the basic concepts and principles of systems work will still succeed. No tool, technique, process, or methodology is perfect in all situations! Concepts and principles will help you adapt to new and different situations and methods as they become available. We have purposefully emphasized applied concepts and principles in this book. This is not a mechanical, "monkey see, monkey do" book! We believe that if you carefully study the concepts presented in Part One, you will be better able to communicate with potential employers, business users, and computer programmers alike. Also note the references at the end of this chapter for books that emphasize problem solving.

Not too long ago, it was thought that the systems analyst's only tools were paper, pencil, and flowchart template. Over the years, several tools and techniques have been developed to help the analyst. Today, a new generation of computer-based tools are emerging. Tools and their associated techniques help the analyst build systems faster and with greater reliability. This book comprehensively covers the modern tools and techniques of the trade.

Techniques are specific approaches for applying tools in a disciplined manner to successfully develop systems. There are numerous popular techniques —each has its own supporters. We will present the most popular techniques throughout this book. You will learn to use and integrate these techniques and avoid the pitfall of blind devotion to any one technique.

Systems analysis and design techniques are constantly evolving. Sensible systems analysts avail themselves of any opportunity to improve their skills. Other books provide the easiest source of self-improvement. On the other hand, forward-thinking organizations are willing to invest in courses and seminars to keep their analysts current.

WHAT ARE THE JOB AND CAREER PROSPECTS?

The life of a systems analyst is both challenging and rewarding. But what are the prospects for the future? Do organizations need systems analysts? Will they need them into the foreseeable future? Is the job changing for the future, and if so, how? These questions are addressed in this concluding section.

Career Prospects

Is it worth preparing yourself for a career as a systems analyst? Absolutely! The job outlook is bright. According to the Bureau of Labor Statistics, opportunities for systems analysts are expected to increase much faster than the average for all professions; even more than for programmers. Depending on the economy, businesses will need between 173,000 and 264,000 new systems analysts by the year 2000—an increase of 24 to 37 percent since 1988. During this same period of time, the overall job force is expected to increase only 8 to 22 percent.

Systems analyst is ranked as the 12th fastest-growing occupation between now and the year 2000. In terms of total demand, it is ranked 4th. The demand is increasing because industry needs systems analysts to meet the seemingly endless demand for more computer-based systems.[4]

Interestingly, as we write this section, the *supply* of systems analysts may be decreasing. Why? The pool of properly educated applicants is declining (as evidenced by the declining numbers of computer and information systems majors in the country). Perhaps too many young people think all a computer graduate does is program computers. You'll learn otherwise in this book. Regardless, your basic economics principles should tell you that if the demand for analysts is increasing, and the supply for analysts is decreasing, the opportunities and salaries for analysts should increase much faster than the average!

Suppose the supply problems continue. Where will businesses find the much-needed talent? The economy has a tendency to correct itself. As salaries increase, more students will be attracted to the profession. In the interim, many companies will attempt to "grow their own" analysts—train and educate other professionals to fill the need. More likely, however, industry will turn to the growing ranks of computer and information systems minors—graduates whose background includes some information systems training. Those with background from a prior systems development course offer the greatest potential.

Another solution lies in technology. In the early decades of telephone service, the operator had to connect every call. It eventually became apparent that if demand for telephone service continued to expand as rapidly as it had, then everybody would eventually have to become a telephone operator just to make the system work. And that is exactly what happened! Through the development of automatic-switching technology, the caller became the operator, placing his or her own call by dialing the desired numbers. The same phenomenon is occurring in systems analysis. As systems analysts struggle to keep up with demand for applications, the computer industry is responding with technology that helps analysts design systems faster (and of higher quality). That same technology promises to ultimately generate the programs automatically. That could free up many programmers to join the ranks of systems analysts. That technology is here today; we'll discuss the current state of the art throughout this book.

What happens to the successful systems analyst? Does a position as a systems analyst lead to any other careers? Indeed, there are many career paths. Some analysts leave the IS field and actually join the user community. Their experience with developing business applications, combined with their total systems perspective, can make an experienced analyst a unique business specialist. Alternatively, analysts could become IS managers or move into technical specialities (such as database, telecommunications, microcomputers, and so forth). The opportunities are virtually limitless.

How Will Tomorrow's Analysts Differ from Today's?

As with any profession, systems analysts can expect change. What kind of changes can be predicted? It is always dangerous to play the futurist, but we'll take a shot at it.

---

[4] *Monthly Labor Review* 112, no. 11 (November 1989), pp. 42–47.

We believe that a greater percentage of tomorrow's systems analysts will not work in the IS department. Instead, they will work directly for their end-users. This will enable them to better serve their users. It will also give users more power over what systems are built and supported. Evidence of this trend was presented earlier in The Next Generation box about independent business units.

We also believe that a greater percentage of systems analysts will come from noncomputing backgrounds. At one time most analysts were computer specialists. Today's computer graduates are becoming more business literate. Similarly, today's business and noncomputing graduates are becoming more computer literate. Their full-time help and insight will be needed to meet demand and to provide the business background necessary for tomorrow's more complex applications.

Summary

Systems analysts are people who understand both business and computing. A systems analyst studies the problems and needs of an business to determine how people, processes, data, communications, and technology can best accomplish improvements for the business. When computer technology is used, the analyst is responsible for the efficient capture of data from its business source, the flow of that data to the computer, the processing and storage of that data by the computer, and the flow of useful and timely information back to the business and its people.

Modern systems analysts develop both an organization's business systems and its information systems. Business systems support, and sometimes automate, day-to-day business operations. Information systems generate information to help managers solve problems and make intelligent decisions.

Essentially, the systems analyst performs one or all of the following activities: systems planning, systems analysis, systems design, systems implementation, and systems support. Collectively, these activities are called the systems development life cycle or the applications development life cycle. Most analysts do some programming, at least as part of ongoing systems support. Programmer/analysts also do some programming as part of systems implementation.

Most systems analysts work for the development center of an information systems or data processing unit of a business. Other analysts (with different titles) can be found in management consulting firms and software houses.

Regardless of whom systems analysts formally report to, they are ultimately responsible to users. Users define problems to be solved and requirements to be fulfilled by information systems. Users also include managers who sponsor and pay for systems development projects. Users are an important partner in modern systems development. Knowledge of systems development tools and methods is becoming increasingly important.

Systems analysts must have a wide range of knowledge and skills, including (1) working knowledge of information systems and information technology; (2) computer programming experience and expertise; (3) general business knowledge; (4) problem-solving skills; (5) interpersonal communications; (6) interpersonal skills; (7) flexibility and adaptability; (8) good character and ethics; and (9) formal systems analysis and design principles, tools, and techniques.

For the remainder of this century, the demand for systems analysts is expected to grow much faster than average. The supply, however, may not grow so quickly. Businesses will seek creative responses to fill the gap. Increased end-user systems development computing and increased use of systems development technology offer some hope.

In the next chapter we'll prepare you to learn about the process of systems development by teaching more about the product, information systems.

Key Terms

analysis, p. 11

applications development life cycle, p. 11

application generator, p. 20

business system, p. 9

computer operations center, p. 16

data (database) analyst, p. 9

design, p. 11

development center, p. 13

end-user, p. 17

end-user computing, p. 13

fourth-generation language, p. 20

implementation, p. 11

independent business unit, p. 15

indirect user, p. 17

information center, p. 13

information system, p. 9

management consultant, p. 16

network analyst, p. 16

owner, p. 17

planning, p. 11

problem solving, p. 9

programmer/analyst, p. 9

software engineer, p. 17

software house, p. 16

support, p. 11

systems analyst, p. 9

systems development life cycle, p. 11

systems engineer (integrator), p. 16

systems integration, p. 16

telecommunications center, p. 16

user, p. 17

Questions and Problems

1. Explain why a noncomputer professional (for instance, engineer, business manager, accountant, and the like) needs to understand systems development.

2. What is the role of the systems analyst when developing a computer application? To whom is the systems analyst responsible?

3. Make an appointment to visit with a systems analyst or programmer/analyst in a local business. Try to obtain a job description from the analyst. Compare and contrast that job description with the job description provided in this chapter.

4. Using the definitions of *planning, analysis, design, implementation,* and *support* provided in this chapter, write a letter to your instructor that proposes the development of an improved personal financial management system (to plan and control your own finances). Tell your instructor what has to be done. Assume that your instructor knows nothing about computers or systems analysts. In other words, be careful with your use of new terms.

5. Differentiate between the five centers of activity in modern data processing: end-user computing, data, development, telecommunications, and computer operations.

6. Visit a local data processing department in your business community. Compare and contrast its organization with the generic organization described in this chaper. Are the five centers of activity present? What are they called? How is their organization different? How is it better? Do you see any disadvantages? Where do the systems analysts fit in?

7. Visit a local data processing department in your business community. How are its project teams formed? How are those teams organized? How does this compare and contrast with the generic structure described in this chapter?

8. Kathy Thomas has been asked to reclassify her systems analysts. Virtually all of her analysts perform planning, analysis, design, implementation, and support.

However, depending on their experience, the percentage of time in the five phases varies. Younger analysts do 80 percent programming, whereas the most experienced analysts do 80 percent systems planning and analysis—largely because of their greater understanding of the business and its users and management. How should Kathy reclassify her personnel?

9. Diversified Plastics, Inc., has adopted an unusual data processing organization. The Management Systems group consists of systems analysts who perform only planning, analysis, and very general design. The Technical Systems group consists of programmers who perform only detailed design, implementation, and support. All analysts must come from a business, engineering, or management background, with no computer experience requirement. Programmers must come from a computing background. Transfers between the groups are discouraged and in most cases not allowed. What are the advantages and disadvantages of such an organization and its policies?

10. Based on the systems analyst's job characteristics and requirements described in this chaper, evaluate your own skills and personality traits. In what areas would you need to improve?

11. Federated Mortgages' corporate information officer (CIO) is facing a budget dilemma. End-users have been buying microcomputers at an alarming rate. In a sense, the business users of these microcomputers, who have little or no DP background, are developing their own application systems. Because of this, the CIO has been asked to justify the continued growth of his budget, especially the growth of his programming and systems analysis staff. There is even some feeling that the number of programmers and analysts should be reduced. How can the DP manager justify his staff? Will the roles of his programmers and analysts change? If so, how? Can the users completely replace the programmers and analysts?

12. The students in an introductory programming course would like to know how systems analysis and systems design differs from computer programming. Specifically, they want to know how to choose between the two careers. Help them out by explaining the differences between the two and pointing out factors that might influence their decision.

13. Your library probably subscribes to at least one big-city newspaper. Additionally, your library, academic department, or instructor may subscribe to a data processing newspaper such as *ComputerWorld* or *MIS Week.* Study the job advertisements for systems analysts and programmer/analysts. What skills are being sought? What experience is being required? How are those skills and experiences important to the role of the analyst as described in this chapter?

14. You need to hire two systems analysts. Explain to a Personnel department recruiter the characteristics and background you seek in an experienced systems analyst.

15. Prepare a curriculum plan for your education as a systems analyst. If you are already working, prepare a statement that expresses your personal need for continuing education to become a systems analyst.

Projects and Minicases

1. A systems analyst applies new technologies to business and industrial problems. As a prerequisite to this "technology transfer," the analyst must keep abreast of the latest trends and techniques. The best way to accomplish this is to develop a

disciplined reading program. This extended project will help you develop this program.

   a. Visit your local school, community, or business library. Make a list of all computing, data processing, and systems-oriented publications.

   b. Skim two or three issues of each publication to get a feel for their contents and orientation. Select the five periodicals that you find most interesting and helpful. We recommend that you select five publications that address the areas of microcomputers, mainframes, data communications, applications, and management issues.

   c. Set up a browsing schedule. This should consist of one or two hours a week that you will spend browsing the list of journals. You should try to maintain this schedule for 10 to 15 weeks. If you miss a day, make it up within one week.

   d. Set up a journal to track your progress. Record the date, the journals browsed, the title or subject of the cover story or headlines, and the title of one other article that caught your eye.

   e. Learn to browse. You won't have time to read. If you try to read everything, you will get discouraged and quit the program. Instead, study the table of contents. Read only the first paragraph or two of each article along with any highlighted text in the article. Read the conclusion or last paragraph of the article. Then move on to the next article no matter how interesting the present one. Note any article which you want to fully read after browsing your reading list.

   f. After browsing each of your selected publications, select at least two articles to read thoroughly. The number you read is limited only by your interest and available time. Record these articles in your journal.

This project will show you how to keep up with a rapidly changing technological world without consuming excessive time and effort.

2. For whom should the systems analyst work? Rolland Industries is facing a data processing reorganization dilemma. Non-DP management is pressing for a new structure whereby most systems analysts would directly report to their application user group (such as Accounting, Finance, Manufacturing, Personnel) as opposed to reporting to Data Processing management. Non-DP management feels that, in the existing structure, systems analysts are too influenced to "change everything" for the sake of computing because they report to Data Processing. To ensure that systems meet Data Processing standards, a small contingent of analysts would remain in Data Processing as a quality assurance group that has final signoff on all systems projects.

   Data Processing managers are resisting this change. They feel that systems analysts will become technologically "out-of-tune" if removed from DP. They also feel that separating the systems analysts from one another will result in less sharing of ideas, and, subsequently, reduce innovation. They also feel that data files and programs will be unnecessarily duplicated. Conflicts between analysts and programmers, who will remain in DP, will likely increase. DP also feels that users will hire new systems analysts without regard to programming and technical experience or familiarity with DP's technical environment.

   Systems analysts, themselves, are split on the issues. They see the benefits of users being more directly in control of their own systems destinies; however, they are concerned that users and user management will be less forgiving when faced with budget overruns and schedule delays that have historically plagued DP. Analysts are also concerned that they will become more prone to technological obsolescence if they are physically relocated outside of Data Processing and its more technically oriented staff.

   The decision will likely be made at a higher level than DP. What do you think should be done?

——————————— *Suggested Readings* ———————————

MacDonald, Robert D. *Intuition to Implementation: Communicating about Systems Toward a Language of Structure in Data Processing System Development.* Englewood Cliffs, N.J.: Prentice Hall, Yourdon Press Computing Series, 1987. This is a good problem-solving textbook.

Martin, James. *Application Development without Programmers.* Englewood Cliffs, N.J.: Prentice Hall, 1982. This book describes the trend toward the use of nonprocedural languages and the impact that they will have on systems analysts.

Martin, James. *An Information Systems Manifest.* Englewood Cliffs, N.J.: Prentice Hall, 1982. This book describes several important trends in information systems, technology, management, and systems development. It also describes how educators, students, and various computer professionals, including analysts, should prepare for this coming age.

Yourdon, Edward. *Modern Structured Analysis.* Englewood Cliffs, N.J.: Prentice Hall, Yourdon Press Computing Series, 1989. This book is the source of our classification scheme for end-users. Yourdon's coverage is in Chapter 3, "Players in the Systems Game."

E P I S O D E   1

SoundStage Entertainment Club
A Preview of Your Demonstration Case

This is the story of Sandra Shepherd and Robert Martinez, systems analysts for SoundStage Entertainment Club.

Systems analysis and design is more than concepts, tools, techniques, and methods. It is people working with people. Although experience is the best teacher, you can learn a great deal by observing other systems analysts in action. Ms. Nancy Piccard, Director of Information Systems Services (ISS) for SoundStage Entertainment Club, has kindly consented to let you watch two of her analysts on a typical project.

Sandra Shepherd, a senior systems analyst and project manager, has volunteered for this demonstration. She has successfully implemented several information systems for SoundStage and should be able to provide you with a valuable learning experience. Bob Martinez, Sandra's partner, is a new programmer/analyst at SoundStage. In fact, today is his first day! Bob has to go through orientation today, and Nancy has invited you to observe the orientation. It'll be a good way for you and Bob to get acquainted with Sound Stage.

Welcome to SoundStage Entertainment Club

SoundStage Entertainment Club is one the fastest-growing music and video clubs in America. The company headquarters, central region warehouse, and sales office are located in Indianapolis, Indiana. Other sales offices are located in Baltimore and Seattle. We begin the preview by joining Bob in Nancy's office.

"Hi, Bob!" Nancy extended her hand. "It's great to have you aboard. My name's Nancy Piccard, and I'm the Director of Information Systems Services. I didn't get a chance to meet you when you interviewed last month. Why don't you tell me a little about yourself?"

"Well, I just graduated from college," replied Bob, somewhat nervously. "I received my bachelor's degree in information systems from State University. I was president of my local student chapter of the Data Processing Management Association and hope to get involved in the Indianapolis professional chapter as well. My career goals are oriented toward applying the systems analysis and design skills I learned in college. That's the main reason I accepted this job. It looked like I'd get a chance to do some analysis and design here — not just programming."

"That's why we hired you, Bob," Nancy replied. "You may be interested to know that we were especially impressed by your classroom experience with computer tools for systems analysis and design. We just bought a package called Excelerator from Intersolv Corporation. We want you to learn and use that package on your first project and report your experiences back to the management staff. You'll learn more about this technology from your partner, Sandra Shepherd, a senior systems analyst who is very familiar with the tools and techniques you learned in college."

"I learned Excelerator in my systems analysis courses. I'm going to enjoy using it on a real project. And I met Sandra during the interview."

Nancy continued, "Sandra will show you the ropes and help you learn about SoundStage and our way of doing things. You'll meet with Sandra soon."

Nancy stood up and motioned Bob to the door. "Okay, let's take a tour of the building."

As they walked up to the second floor, Nancy continued her orientation. "The second floor houses several departments including Personnel, Building Services, Accounting, Marketing, and Budget. These outside wall offices belong to executive managers, staff, and assistants."

"I don't know how much you know or remember about SoundStage, so I'll give you a quick overview. We used to be called SoundStage Record Club. It used to be a record subscription service. Today, records have been replaced with cassette tapes and compact discs. We've also expanded into videotapes, audio/video magazines, and entertainment paraphernalia. That's why we changed our name to SoundStage Entertainment Club."

"Customers join the club through advertisements and member referrals. The advertisements typically dangle a carrot such as, 'Choose any 10 cassettes for a penny and agree to buy 10 more within two years at regular club prices.' I'm sure you've seen such offers."

"Yes," replied Bob.

Nancy continued, "Club members receive monthly promotions and catalogs that offer a selection of the month. They must respond to the offer within a few weeks or that selection will automatically be shipped and billed to their account. Customers can also order alternative selections and special merchandise from the catalogs. After members fulfill their original subscription agreement, they are eligible for bonus coupons that may be redeemed for free merchandise from our catalogs."

"How many members are there?" asked Bob.

"The last number I heard was 340,750," replied Nancy. "Of that total, about 180,000 accounts are active, having purchased merchandise in the last 12 months. As I said, we recently reorganized and

are diversifying our product mix. We expect that compact disc sales will eventually overtake cassettes. We are also expecting videotapes, and perhaps videodiscs, to become a sizable percentage of sales. There is even talk of selling home computer software, mostly the entertainment stuff. The range of entertainment media for the modern household has forced us to diversify."

They walked along the hallway by the executive offices. Nancy continued, "The office on your left belongs to our president and chief executive officer, Steven Short."

As they moved past Steve's office, Nancy handed Bob a piece of paper (see Figure E1.1) and continued. "This is our organization chart, Bob. As you can see, SoundStage is divided into three main divisions including the Information Systems Services division. The offices you walked through on your way to this office belong to the Administrative Services division. They include offices for the managers and staff of Accounting, Budgeting, Marketing, Personnel, and Building Services."

"Down on the first floor we have the Operations division, which handles day-to-day operations including customer services, purchasing, inventory control, warehousing, and shipping and receiving. I will soon take you through those facilities."

"In the basement, you'll find the Information Systems Services division, where you'll have your office. Our mainframe computers are located down there, along with my staff, which totals about 35 people."

They walked down to the first floor. Once again, Bob was confronted with a maze of offices. Nancy explained, "These are the offices for the Operations division, including Purchasing and Inventory Control. They buy the merchandise we will resell to the customer."

"And this office area we are approaching is the Customer Services division. Those clerks are processing orders, backorders, follow-ups, and other customer transactions. That's where your first project is going to be. Sandra will be taking you there to meet your end-users."

The mention of his first project got Bob excited. "Where to next?"

Nancy guided Bob through a pair of double doors. "This is the warehouse. This is where the action is!"

The warehouse was quite large. Bob was somewhat surprised by the size of the operation. Clerks were loading and unloading trucks, stocking shelves, and filling orders. As Bob and Nancy walked through the warehouse, Bob commented, "This must be a difficult operation to coordinate."

Nancy answered "I wouldn't want to do it. We haven't done much in the way of information systems support for the warehouse. We are just starting to investigate that possibility. We do support Purchasing and Inventory Control, but that is as close as we come to supporting the warehouse."

Nancy continued, "Well, let's go downstairs and tour our own Information Systems Services division, and then I'll take you to Sandra."

The Information Systems Services Division at SoundStage

Bob and Nancy walked into the Computer Room, where Nancy removed another sheet of paper from her folder (see Figure E1.2) and handed it to Bob.

"This is an organization chart for Information Systems Services, Bob. You might want to refer to it as we complete this part of the tour. I now report directly to Steve Short. That reflects the recent reorganization. I used to report to the vice president of the Administrative Services division, but it caused some problems with prioritizing requests from the other divisions."

"We are entering the Computer Operations Center right now. You are looking at our IBM AS/400 computer. We have an audiovisual self-study course that you will take to learn more about the IBM computer and its OS/400 operating system. For now, it should suffice to say that this machine supports most of our computer-processing needs. It has several disk drives capable of storing several million bytes of data. We also have a full complement of tape drives, optical character readers, and printers. If you have any questions, just ask."

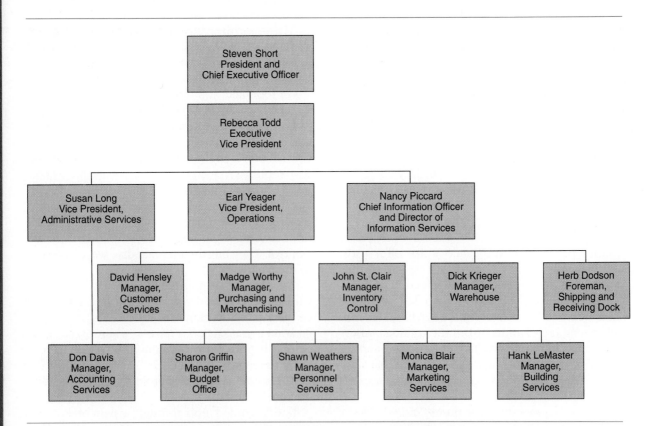

FIGURE E1.1   SoundStage Organization Chart   This organization chart identifies the management structure for SoundStage Entertainment Club. Note the relatively high stature of the Information Systems Services. This reflects a modern, forward-thinking attitude toward the information services function in the company.

Bob replied, "I've never used this particular computer before. Is that going to be a problem?"

"No," answered Nancy. "You have a good, solid education in computers. No matter what machines you learn in school, the technology is constantly changing. You should be able to learn any new system fairly fast. Most people do. I don't know what your expectations were when you graduated, but I think you'll find that your education is only starting now that you are out of school. You'll catch on quickly. Trust me."

They moved into an adjacent office area. Nancy said, "And this is where you'll be working. It's called the Development Center. Two groups of systems analysts and programmers are located in here. The East group builds and supports administrative applications. The West group supports operations applications. You are assigned to the Customer Services team in the Operations Support group."

Bob interrupted, "I see a couple of groups on the organization chart that I'm not familiar with:

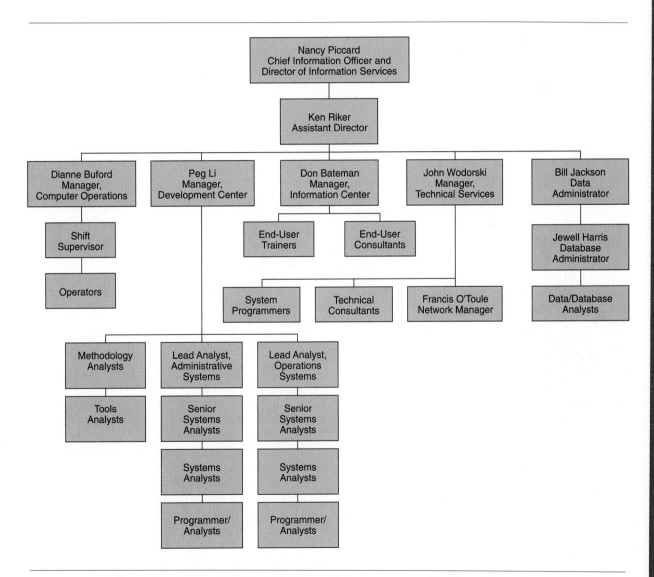

FIGURE E1.2 **Information System Services Division** This organization chart depicts SoundStage's Information Services Division.

Technical Services and the Information Center. What are they?"

"Good question," Nancy said. "This middle section of the facility is staffed by various support groups including Technical Services, the Information Center, and Data Administration."

"The Technical Services group doesn't really serve users—it serves us. When you have a technical question or problem concerning the IBM computer system or our computer networks, that group will help you out. They're also the people who will be teaching you about that IBM AS/400 computer system."

"The Information Center group's primary function is to help all SoundStage end-users learn how to apply microcomputers to do some of their own work. They teach the users how to do word processing, spreadsheets, and things like that. To be honest, many of us old-timers are new to the PC game; therefore, they help us as well."

"And this area?" asked Bob.

"These offices are assigned to our Data and Database administrators. They manage our corporate data model and our database management system. You'll work with his staff on any information systems projects that use the database. We are mandated by policy to implement all new systems using that database; therefore, I'm sure you'll soon be visiting his staff."

Once again, Bob turned his attention to the organization chart and asked, "What are methodology and tool analysts?"

"More good questions," answered Nancy. "I'm surprised that you weren't told about our FAST methodology when you interviewed."

Bob replied, "I was told you used a structured methodology, similar in concept to the structured analysis and design methods I learned in school."

"That's entirely correct," answered Nancy. FAST stands for *Framework for the Application of Structured Techniques*. It is a comprehensive, step-by-step methodology for building systems using the structured techniques. You'll be going to our next FAST overview class. And your partner, Sandra, is an experienced FAST user."

"Our methodology analyst, Susan Clark, coordinates FAST training and helps our analysts with their FAST questions. She has two tools analysts working for her. They help our analysts apply tools like Excelerator within the FAST methodology.

Our goal is to provide analysts with computer-based tools for all phases of FAST."

Bob seemed excited, "That would be great! I hope I'll get a chance to use tools in addition to Excelerator!"

Nancy nodded, "Of course you will!"

"I can't help but notice that you don't have programmers on the organization chart," said Bob.

"That's right. We subcontract or outsource most of our programming these days. It costs less than doing the work in-house. It keeps our own staff focused on the business and technology instead of program details. Programmer/analysts, like yourself, do some of the maintenance programming, but we have even started outsourcing that. In the long run, I hope to eliminate the need for outsourcing through automatic code generation technology—as part of that idea of using tools for all phases of our methodology!"

Bob responded, "Obviously, I've got a lot to learn."

Bob and Nancy returned to the Development Center and walked into an office. Nancy continued, "And this is your office—I wonder where . . . Here she is!"

A woman entered the office. Nancy continued, "Sandra, I want you to meet Bob Martinez, your new partner. Bob, . . . "

Sandra interrupted, "Bob and I met during his interview, Nancy. Welcome, Bob. I'm glad you accepted our offer. I really wanted to work with you —so much so that I asked to be your partner."

Nancy replied, "Terrific! I didn't know you had already met. I don't know what you've told Bob about yourself, Sandra, but I'd like to do a little biographical sketch."

"Sandra has been with us for seven years. She was recently promoted to senior systems analyst because she has proven herself to be one of our most competent, progressive, and personable analysts. Sandra has a bachelor's degree in business. She had little formal computing education but she has done well because she always seeks out opportunities to learn more through reading, seminars, and company training courses. Sandra's credentials also include recognition as a Certified Systems Professional or CSP."

Bob interrupted, "Is that anything like the Certificate in Data Processing or CDP?"

Sandra answered, "Yes, but the CSP is a relatively new certification program that recognizes systems professionals. I was required to pass examinations on various aspects of systems work and provide evidence of my systems experience. I am certified for three years, after which I must be re-certified through evidence of continuing education or reexamination."

Nancy responded, "So you see, Bob, you'll be learning from one of our best people!" With that, Nancy started to leave the office. "Once again, Bob, welcome! I'll leave you with Sandra now. She'll help you get organized and start teaching you all the things you'll need to start learning. We'll sit down in a week or so to set some goals for your first six months. Bye!"

HOW TO USE THE DEMONSTRATION CASE

You've just been introduced to a case study that will be continued throughout this book. It's important that you understand the purpose of the case study:

> The purpose of the continuing case is to show you that tools and techniques alone do not make a systems analyst. Systems analysis and design involves a commitment to work for and with a number of people.

When we started writing this book, we wanted to make sure that the chapters would teach you the important concepts, tools, and techniques. But we were also afraid that you might begin to believe that, if you knew those tools and techniques, you'd have all the knowledge necessary to be a systems analyst. Chapter 1 emphasized the importance of communications and interpersonal skills; however, this continuing case demonstrates the *people* side of the job.

The case study is divided into episodes that represent various stages of a typical project. Although each episode will introduce new tools and techniques, the episodes are not intended to teach the tools and techniques. Instead, they introduce a new situation in which you will need to use a new tool or technique. The episodes show that need in terms of what Sandra and Bob must do to develop a new computer information system for SoundStage. In almost all cases, you will see Sandra and Bob working closely with their business users.

The chapters that follow an episode will teach you how to use the tools and techniques that were introduced in that episode. And those chapters will apply the tools and techniques to a Sound-Stage project.

A brief transition called Where Do You Go From Here? concludes each subsequent episode and introduces the chapters that follow it. We hope you'll find these demonstrations interesting and informative. We think they'll help you place the subject of systems analysis and design into its most practical setting—people working with people.

2

Information Systems
Building Blocks

Chapter Preview and Objectives

Systems analysts develop information systems for organizations. Before learning the process of building systems, you need a clear understanding of the product you are trying to build. Most of you were introduced to information systems in an earlier course (for instance, Introduction to Information Systems or Introduction to Computers and Applications). However, since this is a "development" course, you need to understand information systems in terms of five essential "building blocks" — PEOPLE, DATA, ACTIVITIES, NETWORKS, and TECHNOLOGY. You will be prepared to study information systems "development" after you can:

Define
the product called an *information system.*

Define
information worker and describe the relationship between
information workers and information systems.

Describe
four groups of PEOPLE who are "players in the information systems game."

Describe
five categories of information systems users and their unique
data and information needs.

Describe
the difference between *data* and *information.*

Describe
four views of the DATA building block.

Describe
four views of the ACTIVITIES building block.

Describe
six information functions commonly provided by information systems.

Explain
why networks have become an important building block in information systems.

Describe
four views of the NETWORKS building block.

Describe
four views of the TECHNOLOGY building block.

ARCHER INDUSTRIES, INC.

Archer Industries, Inc. is a manufacturer of sporting goods. It was founded as Archer Tents and Awnings, but expanded through the acquisition of various other sporting goods manufacturers. It is headquartered in Cary, North Carolina.

Scene: *A conference room where Becky, a senior systems analyst in the Information Systems department at the Cary plant, has called a meeting. Sitting around the conference table are: Todd, a plant manager; Walter, a production scheduler; Harold, a production manager; and Jane, a database analyst who works with Becky.*

Becky: As you know, I have been assigned as project manager for the new production scheduling and control system project. I called this meeting to establish a "vision" of that project as a preface to any formal systems analysis and design activities.

Todd: Before what activities?

Becky: I'm sorry. I promised myself that I wouldn't get carried away with computer terminology. Systems analysis is the activity during which my team will learn about your business problems and requirements. Systems design is the activity during which my team will design your new system. But for right now, I just want to get some sense of what you think that system will look like.

Todd: Good enough! The way I see it, the system should provide several essential functions. First, the system should help us plan production lots [authors' note: production lots are orders to produce some quantity of a given product]. Second, the system should schedule production lots and materials handling for each day's manufacturing. Third, the system should monitor production activities as they occur. Fourth, the system should help production managers adjust the schedule when slippages occur due to machine malfunctions and other problems. Finally, the system

should generate production information for various management uses.

Walter: That's certainly one way to look at this system. From my perspective, the current manufacturing environment's biggest flaw is that data is out of control. My top priority would be to get control of the data.

Becky: Please explain.

Walter: Production Scheduling and Control requires us to bring together data from many sources. Product Engineering provides the bill of materials and product specifications. Industrial Engineering provides the manufacturing process specifications. Cost Accounting provides the manufacturing cost standards. Stores provides the inventory data. Marketing provides the market forecasts. Sales provides actual customer orders. Our suppliers provide constantly changing materials pricing. The big problem is harnessing all of this data in an organized fashion. If we get control of that data, all of Todd's functions can be built around that data.

Jane: We can do all of that. Becky and I will implement a normalized DB2 database that consolidates all that data into a highly organized repository. We'll write SQL programs to properly maintain that data and provide users with QMF to . . .

Todd: Hold on! You're speaking a foreign language to me!

Harold: Can I interrupt? I think we might be missing the big picture here. We are only one of seven manufacturing sites in the Southeast. I'm sure that we each do things somewhat differently, but we all do a lot of the same things. Shouldn't our vision include a plan to duplicate this system at all the manufacturing sites? Why "reinvent the wheel" at each site?

Becky: Good point. For all we know, they may already have a system that we can use or adapt.

Jane: No, I checked it out. They are using the same dated package system we use here. They were very interested in our project. We might also want to consider providing them access to our new system once it is built. We could easily connect them via a phone link into our CICS environment . . .

Todd: More terminology! Ugh! But I like the idea of sharing or duplicating the system. We do have to remember, however, that our site does consolidate manufacturing data for the entire company. Whether we like it or not, we cannot exclude their need to interface to any new system we might develop.

Becky: It seems we all have different visions for the project and system. Believe it or not, I think that you have all helped me establish a high-level vision. You want a system that does certain things, that harnesses data for different information needs, and that can be shared or duplicated at other manufacturing plants. I think we have a project vision. Let's take a five-minute break and then plan this project.

Discussion Questions

1. Why did the different participants in this meeting have entirely different views of the same system?

2. Each of the participants were concerned with different aspects of the system. Briefly characterize each participant's perspective of the ideal system.

3. Why did Jane's view of the system cause communications problems with Todd?

4. How do these different views affect Becky's job? How should she deal with such diverse views?

WHAT ARE INFORMATION SYSTEMS?

There is a legend that suggests that the first systems analyst appeared on the scene some 6,000 years ago, during the construction of the Egyptian pyramids. This self-made analyst, concerned about the inefficient methods used to construct the great monuments, offered the following suggestion to Khufu, builder of the Great Pyramid at Cheops:

> "O Noble Khufu, it's time we got organized. We've been pushing this rock through the desert in the wrong direction for seven years. What we need is this Pyramid Erection and Routing Technique." Rumor also has that he was flogged on the spot and never heard from again—at least not until the mid-20th century.[1]

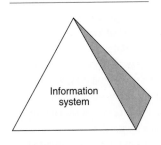

Information Systems Pyramid Model Note: If you are familiar with our pyramid model as presented in our other texts, you will notice both similarities and differences. The faces will prove to be the same; however, in this book, the details about the faces are presented from the perspective of systems development.

That first systems analyst was looking for a better way to build a product—a pyramid. Today's systems analyst is also looking for a better way to build a product—in this case, information systems. As a tribute to that first systems analyst, we will use a pyramid to illustrate information system concepts and building blocks. The pyramid consists of four sides and a square base (see margin). Each face plus the base represents a different building block of information systems and raises concepts and issues that you must consider when developing an information system.

What are information systems? Some people will tell you that an information system can't be defined, but that you'll know one when you see it. If you can't define an information system, how can you justify its need or costs to management? So that we can all speak the same language, let's define *information system.*

---

[1] Kenniston W. Lord, Jr., and James B. Steiner, *CDP Review Manual: A Data Processing Handbook,* 2nd ed. (New York: Van Nostrand Reinhold, 1978), p. 349.

An **information system** is an arrangement of components that are integrated to accomplish the purpose of fulfilling the information needs of an organization.

The primary purpose of an information system is to collect, process, and exchange information among business workers. The information system is designed to support the business system's operations. In most cases it is difficult to distinguish between information and business systems. For purposes of this book, the components of an information system are grouped into five building blocks — PEOPLE, ACTIVITIES, DATA, NETWORKS, and TECHNOLOGY. Thus, we can refine our definition as follows:

An **information system** is an arrangement of people, activities, data, networks, and technology that are integrated for the purpose of supporting and improving the day-to-day operations in a business, as well as fulfilling the problem-solving and decision-making information needs of business managers.

Whenever people get together in an organization, they work out some sort of system to collect, process, and exchange information. Those systems do not require computers to make them work. However, the power of computer technology makes modern information systems feasible. The computer's power amplifies the potential of the other building blocks! Consequently, we introduce a technology base to your pyramid (see margin). Furthermore, when we use the term *information system,* we generally mean **computer-based information system** or **computer information system.**

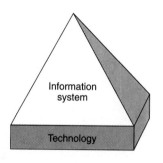

Computer Information Systems Pyramid Model

Today, there are two types of information systems, personal and multiuser.

Personal information systems are those designed to meet the personal information requirements of a single user. They are designed to boost an individual's productivity.

Personal information systems are most frequently implemented on personal computers although they can also be developed on larger computers. Most personal information systems are developed by end-users, not by information systems professionals. Preferably, these users employ the same basic analysis and design techniques that are used by systems analysts (albeit on a smaller scale).

Multiuser information systems are those designed to meet the information requirements of work groups (e.g., departments, offices, sections, and divisions) or entire organizations.

Examples include order entry, accounting, payroll, production scheduling, inventory control, and so forth. The primary difference between the personal and multiuser systems is that multiuser systems must provide for sharing of data, information, and other resources. Most such systems are implemented on minicomputers, mainframe computers, or networks of personal computers (or possibly a combination of all three). Multiuser information systems are usually designed and built by systems analysts and other information systems professionals.

In order to build either personal or multiuser business and information systems, systems analysts and users must effectively combine the building blocks of those systems. These include:

- PEOPLE — the users, managers, and developers of information systems.
- DATA — the raw material used to create useful information.
- ACTIVITIES — business activities (including management) and the data processing and information generation activities that support the business activities.
- NETWORKS — the decentralization of the business, the distribution of the other building blocks to useful locations, and the communication and co-ordination between those locations.
- TECHNOLOGY — the hardware and software that supports the other building blocks.

Each of these building blocks is further examined in this chapter.

Note   Throughout this book, we have used a consistent color scheme for both the pyramid model and the various tools that relate to, or document, the building blocks. The color scheme is based on the building blocks as follows:

Lavender represents PEOPLE.   Tan represents NETWORKS.

Gold represents DATA.   Aqua represents TECHNOLOGY.

Rose represents ACTIVITIES.

BUILDING BLOCK — PEOPLE

The first and most vital building block of information systems is PEOPLE. The overriding philosophy of systems development should be that the system is for the people. In this section, we begin to reveal the pyramid model by identifying the players in the information systems game. As we identify these players, we will examine their unique perspectives on information systems and the implications for systems analysts.

All of the players in the information systems game share one thing in common — they are what the U.S. Department of Labor now calls information workers.

> The term **information worker** (also called *knowledge worker*) was coined to describe those people whose jobs involve the creation, collection, processing, distribution, and use of information.

The livelihoods of information workers depend on information and the decisions made from information. Some information workers (such as systems analysts and programmers) create systems that process and distribute information. Others (such as clerks, secretaries, and managers) primarily capture, distribute, and use data and information. Today, more than 60 percent of the U.S. labor force is involved in the production, distribution, and usage of information. Not surprisingly, an information services industry (which includes the computer, software, networks, and information-consulting industries) has developed to support the growing information needs of businesses.

Systems analysts view people in terms of their roles in developing and supporting information systems. These roles are illustrated in the context of the pyramid model in Figure 2.1. Let's examine their roles.

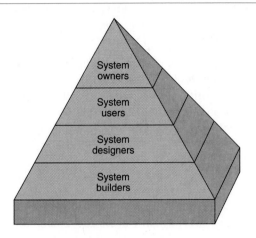

FIGURE 2.1 **The PEOPLE Building Block of Information Systems** This face of your pyramid model identifies the players in the information systems game. They are classified according to their development roles. It should be noted that any one person's responsibilities may overlap more than one of these categories.

System Owners

For any system, large or small, there will be one or more system owners. Owners usually come from the ranks of management. For medium-to-large multiuser information systems, the owners are usually middle or executive managers. For smaller multiuser systems, the owners may be middle managers or supervisors. For personal information systems, the owner may actually double as the system's user—anyone who has a personal computer or terminal and takes the initiative to create his or her own system.

> **System owners** are an information system's sponsors and chief advocates. They are usually responsible for budgeting the money and time to develop and support the information system, and for the ultimate acceptance of the information system.

System owners tend to think in very general terms, not in details. You'll see this when we examine the system owners' view of the other building blocks.

In the pyramid model (Figure 2.1), notice that system owners are the furthest removed from the technology base. They tend to be the least interested (or impressed) with the technical solution for any system—if the system works properly, they may not even care which technology is used. On the other hand, they are aware of the costs of technology. Increasingly, they expect the costs of any technology to be offset by equivalent (or greater) benefits to the business.

Systems analysts usually find it necessary to secure commitment of a system owner or sponsor for every systems development project. That commitment is crucial to subsequently getting the same commitment from their subordinates. Management commitment is also crucial to the acceptance of an information system's costs. This book will teach you several tools and techniques for dealing with system owners on their levels of interest.

System Users

System users make up the vast majority of the information workers in any information system.

System users are the people who use (and directly benefit from) the information system on a regular basis — capturing, validating, entering, responding to, and storing data and information.

System users are the people for whom systems analysts develop information systems. System users define (1) the problems to be solved, (2) the opportunities to be exploited, (3) the requirements to be fulfilled, and (4) the business constraints to be imposed by or for the information systems. They also tend to be concerned with the "look and feel" of the user interface to any computer-based system (e.g., how the display screens and reports look, how easy or hard it is to learn and use the system).

System users are usually less concerned with costs and benefits than system owners. Their primary concern is to "get the job done." They are usually concerned with the business requirements of the information system. In the pyramid model (Figure 2.1), system users are closer to, but still relatively removed from, the technology base. Although users have become more computer literate over the years, their primary concern is still the business job to be done — it is on that basis that they are evaluated. Consequently, the systems analyst should strive to keep discussions with users at the business detail level as opposed to the technical detail level. Much of this book is dedicated to teaching analysts and system users how to effectively communicate with one another.

If at all possible, systems analysts should have direct contact with system users. Managers sometimes try to isolate users from a systems analyst because they are "too busy." This approach rarely produces acceptable results. There is one case, however, where the systems analyst cannot work directly with users. This is the case of an analyst who is working for a software house — there are no real users. Instead, the analyst is attempting to design software packages that will appeal to representative users in a typical business or industry,

It should be noted that system users rarely see themselves in terms of information systems roles. Instead, they usually see themselves in terms of more traditional structures such as organization charts or job categories. Organization charts have the same general shape as your pyramid. The users illustrated in Figure 2.2 can be classified according to their level of responsibility. Each group presents unique data processing and information needs to the systems analyst. Let's briefly examine each class of system users.

Clerical Workers as System Users

---------- ✓ ----------
Clerical Workers

Bookkeeper
Clerk
Office worker
Salesperson
Secretary

Clerical workers perform the day-to-day information activities in the business. Such activities include screening and filling orders, typing correspondence, and responding to customer inquiries. Examples of clerical workers are listed in the margin. Data is captured or created by these workers, many of whom perform manual labors in addition to their information roles. For example, the warehouse clerk who fills an order may be packing the products for shipment as well as recording data about the transaction on a packing slip.

Clerical workers initiate or handle the bulk of any organization's data. They often make routine decisions based on data and generate information for managers. The volume of data in the average organization is staggering. Clerical workers are constantly in need of systems that help them process more data with greater speed and fewer mistakes.

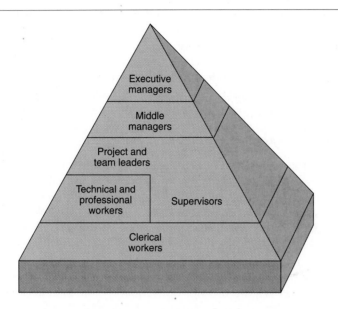

FIGURE 2.2 **Different Types of System Users** This alternate view of the PEOPLE face of your pyramid represents how system users view themselves. Most system users fall into one of the five groups shown on the pyramid.

Clerical workers tend to have a very limited, personal view of the business. They understand their own jobs very well; however, they frequently do not see the "big picture." When dealing with clerical workers, the analyst must find ways to verify their local views and show them the big picture (with as few details as possible). Clerical workers also frequently see the world in terms of "how" things are currently done (with or without technology). They tend to be the users who feel most threatened and intimidated by computers. They are not always receptive to new approaches—they often proclaim, "but we've always done it *this* way." The challenge for the systems analyst is to get clerical workers to see the world in terms of "what" those workers do (or want to do), and together explore alternative ways to do those things.

Technical and Professional Staff as System Users

Technical and professional staff consists largely of business and industrial specialists who perform highly skilled and specialized works. Because their work is based on well-defined bodies of knowledge, they are sometimes called knowledge workers.

> **Knowledge workers** are a subset of information workers whose responsibilities are based upon a highly specialized body of knowledge.

Examples of technical and professional staff are listed in the margin. They are usually college educated. Their jobs involve generating ideas and information. And of course, they also use information. Many of their information needs are only in response to their *current* projects or assignments. The next project or assignment may require an entirely different set of information. It should be noted that technical and professional staff include systems analysts, programmers, and other information systems professionals.

Technical and professional workers present special challenges to the systems analyst. Why? Because, in a very real sense, they are equals. They are

✓

Technical and Professional Staff

Accountant
Engineer
Financial analyst
Human resource specialist
Information systems
  professional
Lawyer
Marketing specialist
Scientist
Statistician

educated and creative. They are also more computer literate than clerical workers (or think they are) and much less threatened by technology. They frequently present solutions before fully analyzing problems and requirements. This is because they, like clerical workers, tend to have a very personal view of the business. They understand their own jobs very well; however, they frequently do not always see the "big picture." Again, the analyst must find ways to verify those personal views *and* show them the bigger picture.

Supervisors as System Users

Supervisors, the lowest level of management, control the day-to-day operations in the organization. Examples of supervisors are cited in the margin. Supervisors directly oversee the work of clerical workers, technical and professional staff, or both groups. Most supervisors are concerned with no more than the current day's or the next week's operations. Most supervisors came to that position from the ranks of clerical workers or technical and professional staff. Still, they may not be familiar with the details of the work they supervise—indeed, they are often "out of touch" with details because details may have changed since they last did the job. Consequently, the systems analyst should be wary of depending on supervisors for such details. Notice that the supervisors for technical and professional staff usually go by titles such as *project leader* or *team leader*.

Recall that supervisors can be system owners or sponsors for small multiuser information systems projects. But they can also be system users of an information system. As users, supervisors tend to be interested in "budgets for" and "efficiency of" the operations they supervise. Consequently, they require day-to-day, detailed, and historical information about those activities performed by their subordinates. For example, an order-entry supervisor might need an order register that indicates the status of all orders recently processed. A production foreman might need the detailed production schedule for any given day. Supervisors may also be interested in information that summarizes operational efficiency—they are frequently evaluated on such efficiency. Finally, supervisors may need information that identifies specific operational problems.

Middle Managers as System Users

Middle managers are concerned with relatively short-term (sometimes called *tactical*) planning, organizing, controlling, and decision making. Examples of middle managers are listed in the margin. They are less interested in the detailed, day-to-day operations that involve supervisory staff. Instead, they are concerned with a longer time frame, perhaps a month or quarter. Some of their functions include gathering operating information for higher-level management and developing tactical strategies and plans that implement executive management's wishes.

Many middle managers were never clerks or supervisors in the areas they manage. They tend not to be interested in details about operations. Nor are they always concerned with operational efficiencies. They see a "bigger picture." Their goals and priorities frequently conflict with those of their clerks and supervisors. Why? Because middle managers look at longer-term problems and solutions. For example, they might dramatically change a system (against the wishes of supervisors and clerks) because they are trying to reverse an unacceptable business trend (such as rising costs or declining business volume).

Finally, middle managers are frequently forced to compete with other managers for scarce resources (e.g., money, people, space). As such, they frequently disagree with one another. Systems analysts should anticipate and recognize this conflict—even the possibility that some managers may strongly support a new system, whereas others strongly resist it.

Again, middle managers sponsor many information systems projects as system owners. But they can also be users of information systems. As system users, middle managers tend to be interested in summarized information. Often, they need comparisons between the same information over different time periods. They may also seek information that identifies exceptions to some standard or rule. Such information identifies problem-solving and decision-making opportunities. Finally, middle managers need information that helps them test solutions and play "what if . . . ?" games.

Executive Managers as System Users

Executive managers are responsible for the long-term (sometimes called *strategic*) planning and control for the organization. Executive managers frequently look a year or more into the past and future. They examine trends, establish long-range plans and policies for the organization, and then evaluate how well the organization carries them out. They allocate the scarce resources of the organization, including land, materials, machinery, labor, and capital (money). Examples of executive management are listed in the margin.

Like middle managers, executive managers sponsor many information systems as system owners. They frequently determine or influence the strategic direction of most or all information systems. But like lower levels of management, they can also be system users. Because they are concerned primarily with the overall condition of the organization, executive managers usually want highly summarized information to support planning, analysis, and strategic decisions.

———————— ✓ ————————

Executive Managers

Chief executive officer
Chief information officer
Chief operating officer
College dean
Comptroller
Member of a board of
  directors
Partner
President
Principal
Vice president

—————————————————————

System Designers

Let's return to the role-oriented pyramid model (Figure 2.1). The next category to be introduced is system designers.

> **System designers** translate users' business requirements and constraints into technical solutions. They design the computer files, databases, inputs, outputs, screens, networks, and programs that will meet the system users' requirements. They also integrate the technical solution back into the day-to-day business environment.

In your pyramid model, systems designers are clearly located closer to the technology base. Frequently, they must make technology choices and design systems within the constraints of those choices. The systems analyst is the principle system designer of most multiuser information systems. (The end-user, possibly working with information center consultants, is the system designer for most personal information systems.) But systems analysts and end-users are not the only system designers. Designers also include various other hardware and software specialists who provide additional expertise and technology to the systems development process. Examples include database analysts, network analysts, microcomputer specialists, and hardware and software vendors. Most of this text is intended to prepare systems analysts to fulfill their role as system designers.

System Builders

System builders represent our final category of system development roles.

> **System builders** construct the multiuser information system based upon the design specifications from the system designers. (End-users, with help from information center consultants, usually construct their own personal information systems).

The applications programmer is the classic example of a system builder. However, other technical specialists may also be involved, such as systems programmers, database programmers, network administrators, and microcomputer software specialists.

In your pyramid model (refer back to Figure 2.1), system builders are located directly above the technology base — they directly use the technology to build the technical solution. It should be noted that the systems analyst may also be a system builder. At the very least, the systems analyst must work with the system builders and be aware of the detailed requirements of their jobs. In other words, the systems analyst must ensure that the design is clear, consistent, and sufficiently detailed such that the system builder can implement the design. Although this book is not directly intended to educate or train the system builder, it is intended to teach the systems analyst how to better communicate with the systems builders.

BUILDING BLOCK — DATA

When engineers design a new product, they must create a *bill of materials* for that product. A bill of materials says nothing about what the product is intended to do. It states only that certain raw materials and subassemblies make up the finished product. The same analogy can be used for information systems. Data can and should be thought of as the raw material used to produce information. Consequently, we consider data to be one of the fundamental building blocks of an information system.

Most people use the terms *data* and *information* interchangeably. However, they are not really the same thing. The distinction is important when defining and developing information systems. Throughout the book, we will make the following distinction:

> **Data** is a collection of raw facts in isolation. Data describes the organization. These isolated facts convey meaning but generally are not useful by themselves.
>
> **Information** is data that has been manipulated so it is useful to someone. In other words, information must have value, or it is still just data. Information tells people something they don't already know or confirms something that they suspect.

One person's information may be another person's data. For example, consider a report that lists all customer accounts. If customers call to find out their current balance due, a clerk can answer their questions by looking at the report. To the clerk, this report is information. But suppose a manager wants to know

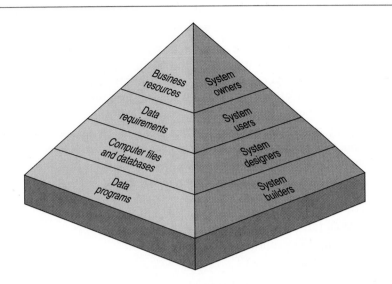

FIGURE 2.3 The DATA Building Block of Information Systems This face of your pyramid model illustrates how different players in the information systems game view data. Remember, the data face of your pyramid represents data independent of how it is used. Think of it as an "inventory" of data that is available for use to make information.

the total dollar amount of delinquent accounts. The manager would have to identify delinquent accounts on the report and then sum their balances. Thus, the report (as it is) represents data to that manager. On the other hand, a report that listed those delinquent accounts and their summed balances would represent information to the manager.

In this chapter, we'll deal exclusively with data, independent of how it is used. In other words, we're not concerned with how data gets transformed into useful information. Think of data as an "inventory of data" that can be used in creative ways to produce many different information-oriented products.

The DATA face of your pyramid model is illustrated in Figure 2.3. Notice that different people have different views of data. Let's examine those views and discuss their relevance to the systems analyst. The systems analyst who truly understands these views will be better equipped to deal with the tools and techniques that are taught in this book.

The System Owner's View of Data

Strictly speaking, the average system owner is not interested in raw data. They are interested in the things that data describes. These things are business resources.

> **Business resources** are (1) things that are essential to the system's purpose or mission; or (2) things that must be managed or controlled in order to achieve business goals and objectives.

To system owners, data is itself a resource that helps to better manage these other business resources. What business resources are important to system owners? For any given business or information system, most fall into one of the following categories:

- Tangible things (such as materials, supplies, machines, vehicles, and products).
- Roles (such as customers, suppliers, employees, and account holders).

- Events (such as orders, requisitions, contracts, trips, accidents, or sales).
- Places (such as sales offices and warehouse locations).

What does this mean to the systems analyst? When dealing with system owners, systems analysts try to identify the resources that are important to the system. Analysts, however, frequently use the term **entity** to describe a resource. They also seek to learn about problems, opportunities, and constraints that apply to these entities. They might ask the owner,

> Information is descriptive of "things." What things in this system are of interest to you? About what things do you seek information? For what things do you make decisions or have expectations, goals, or objectives?

An order entry manager might respond with the following list:

CUSTOMERS (a role).

FORECASTS (an event).

ORDERS (another event).

SALES REGIONS (a place).

COMPLAINTS (still another event).

PRODUCTS (a tangible thing).

To the analyst, each of these things is an entity that can be described by data. On the other hand, the system owner may not be knowledgeable about, or even interested in, all of the data that describes these entities. Thus, when dealing with system owners, analysts should generally avoid tools and techniques that focus on such data details as fields, record layouts, file organizations, and database structures. Instead, the analyst might focus on simple lists of the entities of interest and any problems, concerns, or issues that relate to these entities. It might also be useful to examine the natural business relationships between these entities (see Figure 2.4). Complex diagrams may actually hinder communication with owners.

The System User's View of Data

The users of an information system are usually the data experts. They understand the business's data better than anyone else. As information workers, they capture, store, process, edit, and use the data on a daily basis. Unfortunately, they frequently see the data only in terms of how data is currently implemented or how they think it should be implemented. To them, data is recorded on forms, stored in file cabinets, recorded in books and binders, organized into spreadsheets, or stored in computer files and databases. The challenge for the systems analyst is to identify and verify users' data requirements exclusively in business terms (so that alternative implementations might be considered).

> **Data requirements** are a representation of users' data in terms of entities, attributes, and rules. Data requirements are expressed in a form that is independent of considering how that data will be physically stored or implemented (e.g., forms, files, spreadsheets).

Modern systems analysts try to get their system users to see their data in terms of these entities, attributes, and rules. To the systems analyst, an *entity* is something about which data is important. The system users know them as *forms, files,* and *records* (although not necessarily the computer context of those terms). These entities are much the same as those of interest to system owners—in

Entity	CUSTOMER	ORDER	PRODUCT	SALES REGION	FORECAST
CUSTOMER		Places		Located in	
ORDER	Placed by		Requests		
PRODUCT		Requested on			Has
SALES REGION	Serves				
FORECAST			Applies to		
COMPLAINT	Registered by	Registered for	Registered about		

FIGURE 2.4 **A View of Data Suitable for System Owners** System owners are generally "put off" by complex pictures or diagrams. This simple view of their data focuses on business entities and how they are related.

other words, *tangible things, roles, events,* and *places.* But because system users are generally more knowledgeable about the data, they can frequently identify the need for more of these entities (files and records) than the typical system owner. For instance, whereas a system owner may only refer to an entity called CUSTOMER, system users might differentiate between PROSPECTIVE CUSTOMERS, CURRENT CUSTOMERS, and FORMER CUSTOMERS. Why? Because they know that slightly different types of data describe each type of customer.

System users, especially clerical workers, are also knowledgeable about the data attributes that describe the entities. **Attributes** are facts that describe all or most instances of an entity. For example, system users can tell you that an ORDER (an entity) is uniquely identified by an ORDER NUMBER. They might also know that ORDER NUMBER is derived from the date it is received (e.g., 012492 for January 24, 1992) and a sequential counter (e.g., 001, 002, 003, etc.) for that date. And they might tell you that an ORDER is also described by an ORDER DATE, a CUSTOMER NUMBER and NAME, a SHIPPING and BILLING ADDRESS, and that it sometimes comes prepaid by an enclosed check (which is called the PREPAID AMOUNT).

System users are also knowledgeable about the **rules** (also called *relationships*) that govern data and entities. For example, every ORDER must have one and only one CUSTOMER. Every ORDER must include at least one PRODUCT. A PRODUCT doesn't necessarily have to be on any current ORDER. Other examples might include (1) the CUSTOMER CREDIT RATING (an attribute of the entity CUSTOMER) must be A, B, C, or D; (2) the CUSTOMER CREDIT RATING for all new customers is B; and (3) never delete a CUSTOMER record if there is an outstanding INVOICE for that customer. These are all data rules that would be known to the system users. Any information system's integrity will be based on how well it manages data within the limits of these rules.

The systems analyst works closely with users to learn about their data requirements. These requirements are expressed in terms of the aforementioned entities, attributes, and rules. Figure 2.5 illustrates a tool used by some systems

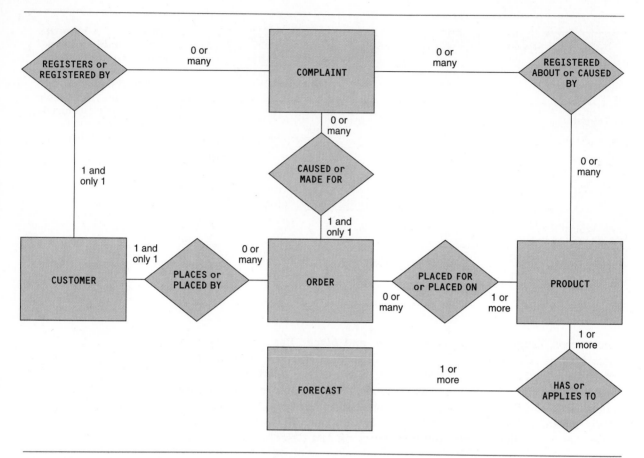

FIGURE 2.5   A View of Data Suitable for System Users   System users can generally understand relatively simple diagrams of data if they are properly presented and explained. This diagram illustrates entities, relationships, and a few rules.

analysts to document system users' view of data (entities, relationships, and some rules). Some of this book focuses on such tools and techniques. Don't worry about understanding the diagram—just notice that the picture does not in any way represent or suggest that the data is stored on forms, in files, or in a computer system.

The System Designer's View of Data

System users define the data requirements for an information system. System designers translate those requirements into **computer files and databases** that will support the ACTIVITIES building block in the information system. The system designers' view of data is constrained by the limitations of specific technology. Sometimes, the analyst is able to choose that technology. Often, the choice has already been made and the analyst must use the technology that is available. For instance, a system might be constrained to use an available file organization (such as VSAM) or a specific database management system (such as DB2).

In any case, the system designer's view of data consists of record layouts, data structures, database schemas, file organizations, fields, indexes, and other

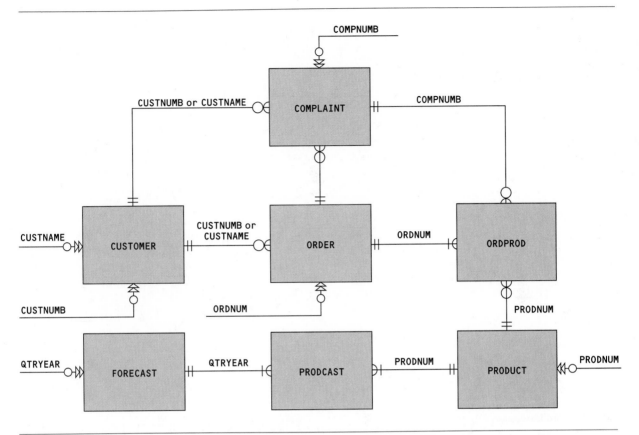

FIGURE 2.6   A View of Data Suitable for System Designers   System designers usually illustrate data in somewhat greater technical detail than system owners or users. Notice that the language and abbreviations begin to take on a technical jargon.

technical items. Some of these, if presented properly, can be interpreted by system users—most cannot!

The systems analyst and other information systems specialists design and document these technical views of the data. For instance, Figure 2.6 illustrates a system designer's view of a database to be implemented using a database management system. The designers' intent is to represent the design such that: (1) it fulfills the data requirements of the users; and (2) it provides sufficient detail and consistency for communicating the data design to the system builders. Ideally, the system's builders will never require clarification from the system's designers. (Unfortunately, this rarely happens!) Once again, this book will cover tools and techniques for transforming user data requirements into a technical design.

The System Builder's View of Data

The final view of data is relevant to the system builders. In your pyramid model, system builders are closest to the technology foundation. Not surprisingly, they are forced to represent data in very precise and unforgiving languages. They write data programs to implement the computer files and databases.

FIGURE 2.7 **A View of Data Suitable for System Builders** System builders must use precise languages to describe data to a computer system.

```
DATA DIVISION.
FILE SECTION.
FD    PRODUCT-FILE
      RECORD CONTAINS 64 CHARACTERS.
01    PRODUCT-RECORD.
      05 PRODUCT-NUMBER  PIC 9(6).
      05 PRODUCT-DESCRIPTION  PIC X(30).
      05 PRODUCT-PRICE  PIC 9(3)V99.
      05 QUANTITY-ON-HAND  PIC 9(6).
      05 QTY-ON-ORDER  PIC 9(6).
      05 REORDER-POINT  PIC 9(6).
      05 DISCOUNT-VOLUME PIC 999.
      05 DISCOUNT-RATE  PIC V99.
```

```
CREATE TABLE CUSTOMER
    (CUSTNUMB  CHAR (10) NOT NULL,
     CUSTNAME  CHAR (32) NOT NULL,
     CUSTADDR  CHAR (20) NOT NULL,
     CUSTCITY  CHAR (10) NOT NULL,
     CUSTST  CHAR (2) NOT NULL,
     CUSTZIP  CHAR (9) NOT NULL,
     CUSTAREA CHAR (3),
     CUSTPHN  CHAR (7),
     CUSTBAL  DECIMAL (7,2)
     CUSTRATG  CHAR (1)
     CUSTLIMT  SMALLINT)
CREATE INDEX CNUMBIDX ON CUSTOMER (CUSTNUMB)

CREATE INDEX CNAMEIDX ON CUSTOMER (CUSTNAME)
```

—————— ✓ ——————

Database Management Systems (and languages)

DB2
SQL/DS
IMS
CA-IDMS/DB
CA-DATACOMB-DB
dBASE
Rbase
Paradox

Data programs (or subprograms) are technical, machine-readable descriptions of how data is to be stored in files or databases.

Perhaps you are familiar with the FILE CONTROL paragraphs in a COBOL ENVIRONMENT DIVISION. If so, you are also familiar with record descriptions for those files — expressed as PICTURE clauses in a COBOL DATA DIVISION (see top-left example in Figure 2.7). A more sophisticated representation of data is provided by a data definition language (DDL) such as those provided in database management systems (see margin for languages and bottom-right of Figure 2.7 for an example). It is not the intent of this book to teach these data languages — only to place them into the context of the DATA building block.

BUILDING BLOCK — ACTIVITIES

Let's start the same way we did with the DATA building block. When engineers design a new product, that product should serve some function. It must do something useful. Prospective customers define the desired functionality of the product and the engineer creates a design to provide that functionality. ACTIVITIES define the functionality of an information system.

Business activities are day-to-day processes that support their purpose, mission, goals, and objectives. Most businesses organize themselves around business activities such as marketing, sales, warehousing, shipping and receiving, personnel, accounting, and production.

Information system activities are processes that support business activities by (1) providing data and information processing; and (2) improving upon and streamlining the business activities. Some activities may be implemented as software. Others may be performed by people.

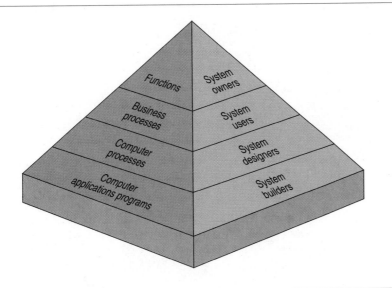

FIGURE 2.8 **The ACTIVI-
TIES Building Block of
Information Systems**
This face of your pyramid
model illustrates how different players in the information systems game
view activities. The activities face of your pyramid
illustrates how data gets
transformed into useful
information. As such, it
builds on the data inven-
tory presented in the data
face of your pyramid.

Essentially, activities are the work performed by people and machines for the business. Some activities are repetitive. Other activities occur less frequently or rarely.

Systems analysts are interested in both business and information system activities. But as we just learned with data, different representations of activities are suited to different people. Figure 2.8 illustrates different people's views of information system activities. Once again, the systems analyst who understands these views will be better equipped to deal with the tools and techniques taught in this book.

The System Owner's View of Activities

System owners are usually interested in the big picture — in this case, groups of activities called functions.

> **Functions** are ongoing activities that support the business. They usually include many distinct activities that share something in common.

Business system functions include sales, service, manufacturing, shipping, receiving, accounting, and so forth. **Information system functions** support these ongoing business functions. Examples include data processing, decision support, and office automation. System owners determine which potential information systems functions offer the best payoff and benefits. How do they do this?

Essentially, system owners see their business functions in terms of various planning parameters such as goals and objectives.

> **Goals** are general statements of intent — something that the business wants to achieve.

For example, manufacturing (a business function) may set a goal to minimize the cost of goods produced.

> **Objectives** are more specific targets that help to achieve goals.

FIGURE 2.9 **A System Owner's View of Activities** System owners usually see business and information systems in terms of functions—void of details such as inputs, outputs, and procedures.

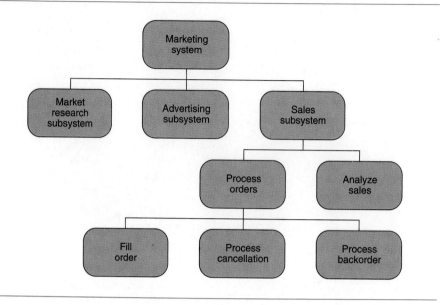

For instance, manufacturing can achieve its goal by (1) decreasing scrap (wasted material) by 12 percent by September 15; (2) increasing product output by 40 tons per quarter by consolidating similar orders into the production schedule; and (3) decreasing total machine downtime by 50 hours per month through preventative maintenance.

System owners (working with information systems managers) try to determine what types of information system functions would best support the goals and objectives for the business functions. The best information systems provide the biggest payoff toward achieving business goals and objectives.

When working with system owners, systems analysts can use various tools to model functions. For example, an analyst might draw a simple top-down picture of information systems in terms of subsystems and their related functions (see Figure 2.9). Notice that this diagram is very abstract. There is no evidence of inputs to or outputs from the functions. There is no evidence of procedures or methods used by the functions. This makes the diagram very readable to managers who want to see the big picture. You'll learn how to properly develop and use similar diagrams later in the book.

Information systems provide different levels of support for different business functions and different users. Figure 2.10 identifies typical information systems applications. The functions are aligned with the information workers (users) who typically use those functions. Let's briefly examine each of these functions. Keep in mind that these functions are most typically developed using the tools and techniques that you will learn in this book.

Transaction (or Data) Processing

Transaction or **data processing** was the earliest business computing application.

Transactions are business deals or arrangements. They are significant in that they bring new data into a business (and information) system.

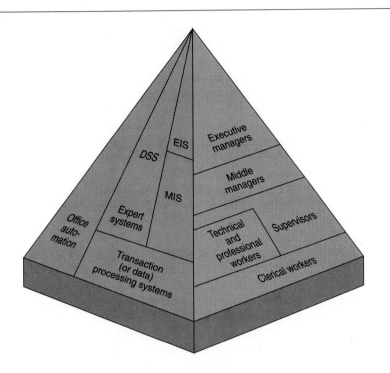

FIGURE 2.10 **The Typical Information System Applications** Information systems provide different levels of support for different business functions and users. This version of your pyramid demonstrate which typical applications serve which users.

Examples of transactions include *orders, reservations,* and the like. Transactions can be initiated from outside the organization (such as a customer placing an order) or within the organization (a secretary requisitioning office supplies from the storeroom). Think of transactions as events that bring new data into a business. That data must be processed.

Transaction processing systems are information system applications that capture and process data about (or for) transactions. They are also called *data processing systems.*

Some experts consider transactions processing systems as being separate from information systems. On the other hand, to produce any information, we must first capture some data. Transactions represent 90 percent or more of all data that must be captured. Consequently, we consider transaction processing as a major function of information systems.

There are three types of transaction processing. The most common is the processing of **input transactions.** Examples of input transactions are listed in the margin. Each of these transactions inputs new data into the information system. A second type of transaction processing focuses on **output transactions.**

Output transactions trigger responses from or confirm actions to those who eventually receive those transactions. Examples of output transactions are also listed in the margin. In each case, the output transaction confirms an action or triggers a response (such as confirming a flight reservation or acknowledging the receipt of payment of a bill).

A third type of transaction processing, **data maintenance,** provides for routine updates to stored data (see margin on p. 56 for examples). For example,

---

✓

Input Transactions

Accounting vouchers
Airline reservations
Bank deposit slips
Charge card vouchers
Course registrations
Customer orders
Grocery checkouts
Hotel checkouts
Merchandise returns
Payments
Purchase orders
Time cards

✓

Output Transactions

Airline reservation
  confirmation, tickets, and
  boarding passes
Customer invoices
Customer account
  statements
Course schedules
IRS 1040s
Merchandise picking and
  packing orders
Paychecks
Sales receipts

—————— ✓ ——————
**Data Maintenance
Transactions**

Transactions
Applications (for anything)
Catalog price changes
Change-of-address cards
Discontinued products
New products
——————————————

most businesses store data about their customers. Customer data maintenance provides for adding new customers, deleting customers with whom the company no longer does business, and modifying attributes of existing customers (such as name, address, phone number, credit rating, and so forth).

As shown in Figure 2.10, transaction processing is usually performed by and for clerical workers. To some extent, the clerical tasks performed by these workers can be replaced by automation, which is faster and more accurate. These workers can, in turn, assume more stimulating and productive assignments. The reduction of clerical tasks is frequently offset by an increase in data entry for the system. One trend in transaction processing is on-line support. On-line systems displace fewer workers because clerks must enter data on computer terminals that are on-line with the computer. Data is captured and processed immediately. In this manner, discretionary decisions can be handled by people rather than machines.

Despite the fact that transaction processing was the first generation of information system support, it still provides exciting new applications. For instance, there are still older batch systems that need to be converted to on-line and/or database versions. Also, new technologies for keyless data capture will force the redesign of many transaction processing systems, such as image processing and electronic data interchange (EDI). Finally, many companies are directly linking their customers and suppliers into their own transaction processing systems to reduce transaction response time, increase transaction throughput, and gain a competitive advantage in their market.

Management Reporting and Executive Support

Management reporting is required to plan, monitor, and control business operations. Because the information is intended for the three levels of management (supervisors, middle managers, and executives), the term *management information system* was coined to describe applications that produce this type of information.

A **management information system** (MIS) is an information system application that provides for management-oriented reporting, usually in a predetermined, fixed format. MIS provides well-defined information responses based on (1) predefined uses of captured and processed transactions and data; or (2) predefined management or statistical models (such as those encountered in "operations research").

The MIS concept includes the production of information based upon accepted management or mathematical models. For instance, *materials requirements planning* (MRP) is a formal model for determining production schedules and raw material purchasing schedules. An MIS for MRP would generate these schedules based on sales projections, product composition, inventory, and the MRP model. Management information systems typically produce three types of information: detailed, summary, and exception reports.

Detailed reports present information with little or no filtering or restrictions. An example would be a detailed listing of all customer accounts or products in inventory. Some detailed reports are historical in nature. They confirm and document the successful processing of transactions and serve as an audit trail for subsequent management inquiry. These reports assist management planning and controlling by generating schedules and analysis. Other detailed reports are regulatory, that is, required by government.

Summary reports categorize information for managers who do not want to wade through details. The data is categorized and summarized to indicate trends and potential problems. The use of graphics (charts and graphs) is rapidly gaining acceptance because it more clearly summarizes trends at a glance.

Exception reports filter data before it is presented to the manager as information. Only exceptions to some condition or standard are reported. Classic examples of exception reports include a report that identifies items that are low in stock (soon to run out), and a report that identifies customers whose accounts are overdue or delinquent.

Notice in Figure 2.10 that management reporting primarily supports middle management and supervisory staff. Management reporting is always fertile ground for new and improved systems. Management-oriented users seem to have limitless imagination for new or better reports. Similarly, government regulation of many industries may continue to exert new reporting requirements on businesses.

Executive information systems are an extension of the MIS concept.

> An **executive information system** (EIS) is an information system application that provides high-level executives with sophisticated tools for consolidating and summarizing data at a very high level. It is sometimes called an *executive support system.*

EIS provides flexible access to summary-level data and information from multiple files and databases that were created through transaction processing and MIS. Additionally, an EIS may tap commercial databases (those not owned by the company) for access to stock market, economic, and industry data. An EIS will frequently manipulate the data using statistical and management analysis techniques to identify trends and issues. The final information is often presented graphically.

Executive information systems are usually purchased, not built. What, then, are the implications for the systems analyst? First, the analyst may be involved in the selection of the EIS software packages. Second, the analyst must specify and implement the "links" to the relevant files and databases that will feed the EIS.

Decision Support and Expertise

Sometimes managers don't know what information will help them until the need to make a decision arises.

> A **decision support system** (DSS) is an information system application that provides its users with decision-oriented information when a decision-making situation arises.

A DSS does not make decisions or solve problems—people do! DSS is simply concerned with providing useful information to support the decision-making process.

Most decisions fall between two extremes, structured and unstructured. **Structured decisions** are those that can be predicted. Analysts and users can't always predict *when* they will happen, but we can predict *that* they will happen. Users can usually define the information requirements to support a structured decision. For example, we can provide an on-line credit check to help a manager or clerk decide whether or not to let someone charge a sale at a local department store. We know those charge sales will occur. We can also define what information is needed to check a person's credit.

Unstructured decisions cannot be predicted. We don't know when the decision-making need will occur, and we also don't know the nature of such a decision. Can you think of an unstructured decision? If you could, it wouldn't be unstructured. Because the unstructured decision can't be predefined or predicted, you can't predefine what information will be necessary to assist the decision-making process.

DSS is primarily intended to support the relatively unstructured decisions. Does this seem impossible? The concept behind DSS is that the data/information needed for many unstructured decisions has already been captured by the transaction-processing and management information systems. Additional data may be available in national and international databanks around the world. As seen in Figure 2.10, DSS can support all user levels, clerks through executives. They are all decision makers, although their decisions vary in type and importance. DSS provides one or more of the following types of support to the decision maker:

- Identification of problems or decision-making opportunities (similar to exception reports).
- Identification of possible solutions to problems.
- Generation of information needed to make the decision (or pass the decision authority to a higher level of management).
- Processing of "what if" analyses of different variables in a decision or possible solution (provided in most spreadsheets).
- Simulation of possible solutions and their results.

Expert systems are an extension of the decision support concept.

An **expert system** is an information system application that captures the knowledge or expertise of a specialist and then simulates the "thinking" of that expert for those with less or no expertise. Expert systems are also called *knowledge-based systems* or *articial intelligence-based systems.*

Expert systems address the critical need to duplicate the unique expertise of experienced managers, professionals, technicians, and clerical workers. These individuals often possess, through experience, knowledge and expertise that cannot be replaced by merely hiring new people. It is equally difficult to rely on these individuals as consultants since they can only serve in one location at any given time and because they deserve to be promoted like other capable workers.

Expert systems capture the knowledge and expertise of individuals, and then make that knowledge and expertise available to others. They imitate the reasoning of the experts in their respective fields. A more formal definition of an expert system is a computer-based information system that has been encoded with human knowledge and experience to achieve expert levels of problem solving. The following examples are real:

- A food manufacturer uses an expert system to preserve the production expertise of experienced engineers who are nearing retirement.
- A major credit card broker uses an expert system to accelerate credit screening that requires data from multiple sites and databases.

- A plastics manufacturer uses an expert system to determine the cause of quality control problems associated with a sophisticated machine press.

Virtually any organization can benefit from automation of internal or external expertise. Most expert systems are designed to support technical and professional workers, supervisors, and some middle managers (see Figure 2.10).

Office Automation

To many, the term *office automation* simply means word processing. It is, in fact, much more than that.

> **Office automation** (OA) is an information systems application that provides for improved communications between all levels of information workers (whether or not they work in "an office").

OA is concerned with the complete communication of information to and from various workers. The key idea is *complete communication.* Whereas the other functions have been mostly concerned with data processing and information generation, OA is concerned with getting relevant information to all those who need it. Clearly, the term *office* is used rather loosely in the term *office automation.* Every information worker (as seen in Figure 2.10) is a potential OA user regardless of whether they are physically located in an office.

Office automation functions include word processing, electronic mail, voice mail, electronic calendaring, audio/video conferencing, facsimile, and image processing. These applications are normally covered more fully in the introduction to computers or information systems course.

What are the implications of OA for the systems analyst? There is a definite trend toward the integrated office system, which integrates the aforementioned OA functions with those of transaction processing, MIS, DSS, and so forth. For example, an MIS or DSS function might trigger an electronic mail message to a relevant decision maker. Or perhaps a transaction processing function might produce a highly customized and personalized word processing response to customers.

Integration of Information Systems Applications

We conclude our discussion of applications and functions by noting a trend in business toward systems integration. Historically, most applications and functions have been developed in relative isolation from one another. This has resulted in what some experts call *islands of automation.* Systems integration attempts to better link these islands such that they can share common data and information.

> **Systems integration** is the problem-solving process associated with making various business and information systems applications work cooperatively—sharing available data, information, and other resources.

Nowhere is this trend more evident than in the manufacturing industry. Manufacturers have raced to automate complex production processes using robots and controllers. Manufacturing information systems have, for the most part, been relatively isolated from such factory floor automation. However, in progressive firms, information systems are being integrated with factory floor systems using a concept called *computer-integrated manufacturing* (CIM). To learn more about CIM, see The Next Generation box for this chapter.

The Next Generation

COMPUTER-INTEGRATED MANUFACTURING SYSTEMS

The trend toward information systems integration is most evident in the manufacturing industry. Worldwide competition among manufacturers is forcing them to move toward a "factory of the future" concept characterized by (1) flexible and technically sophisticated production machines; (2) instant quality recognition and zero-defect goals; and (3) highly coordinated production scheduling combined with optimal raw materials inventory management.

The factory of the future will require that machines on the factory floor be highly integrated with information systems, especially those systems that directly affect production activities. Information systems include order entry, materials purchasing, inventory control, production planning, and production control. Shop floor manufacturing systems include robots and machines under the direct control of environmental sensors, numerical controllers (NCs), and programmable logic controllers (PLCs).

When the world of manufacturing-oriented information systems is "married" to the world of shop floor automation, the result is **computer-integrated manufacturing.** The goal is simple—to fully automate every aspect of the factory of the future! It combines various computer technologies with the ability to manage and control the entire business. Some of the major areas of a CIM system include:

- *Computer-aided design* (CAD)—the engineer's use of computer tools to automate the drafting and design of new products.
- *Computer-aided engineering* (CAE)—an extension of CAD that helps engineers test and analyze new products (e.g., test for strength and durability).
- *Computer-aided process planning* (CAPP)—an extension of CAD that allows engineers to simulate the manufacturing process for a new product before that product's design has been approved (such that design changes might be made before the costly process of setting up the manufacturing process).
- *Computer-aided manufacturing* (CAM)—the programming of NCs, PLCs, and robots to automate machines and tools used to manufacture the product. Ideally, CAM programs could be generated directly from CAD/CAE/CAPP tools (giving us our first level of integration—between the engineers and the shop floor).
- *Automated material handling*—the use of technology to automatically move materials and product subassemblies from one location on the shop floor to another. Examples include automated guided vehicles (AGVs) and automatic storage and retrieval systems (AS/RS).
- *Production control systems*—includes sensors and devices that automatically monitor shop floor processes and measure quality control. Examples include automatic testing equipment (ATEs) and geometric coordinate measurement machines (CMMs).
- *Manufacturing planning and control systems*—the use of mathematical models to coordinate sales and orders with production and inventory control. Examples include materials resource planning (MRP), just-in-time (JIT) inventory, and statistical quality control (SQC) techniques. Clearly, these techniques are information oriented and supported by information systems. These applications derive much of their basic data from other information systems applications (such as order entry and purchasing).

The above areas are usually implemented on a wide variety of technical platforms. For example, the CAD/CAM/CAE tools may use UNIX-based Sun workstations while the shop floor controllers may come from ALLEN-BRADLEY. The manufacturing processes may be controlled from DEC minicomputers. And most information applications are implemented on IBM computers.

Do you see the problem? Much of the CIM problem requires integration across sophisticated information networks. Engineers are trained to deal with the factory floor automation details. But systems analysts deal with the manufacturing planning and control information systems. The two groups must work together to implement the underlying information networks needed by CIM.

When students seek information systems employment, they are often given many applications areas as options. They may wonder, "Where is the next generation of 'hot' applications?—That's where I want to be." We can think of few opportunities more golden than manufacturing information systems in support of CIM.

The System User's View of Activities

Returning to Figure 2.8, we are ready to examine the system user's view of activities. Users see activities in terms of distinct business processes.

> **Business processes** are distinct activities that have inputs and outputs. Specific methods and step-by-step procedures often underlie these processes. Business processes may be implemented by people, machines, or computers.

Clearly, processing data and generating information are business processes that are performed by information systems.

Unfortunately, users tend to see their processes only in terms of how they are currently implemented or how they think they should be implemented. There is a great need in business to totally rethink processes to eliminate redundancy, increase efficiency, and streamline the entire business. Once again, the challenge for the systems analyst is to identify and express user process requirements exclusively in business terms (such that alternative methods and implementations might then be considered).

Modern systems analysts have been successful at getting their users to view their process requirements in terms of simple input-process-output models (**process models,** for short). Figure 2.11 is representative of such models. It includes **processes** (the rounded rectangles), information and **data flows** (the arrows), and **data stores** or files (the open-ended boxes). Notice how the picture suggests *what* the system is doing, but not *how* the system is implemented (e.g., no reference to computers). The picture also hides many details such as the contents of the data flows and stores, and the step-by-step procedures of the processes. Analysts use other tools to document those user requirements.

What does this have to do with the systems analyst? It is the analyst's responsibility to identify and prioritize the user's process requirements. This includes determining the detailed information needed, the detailed data needed to produce that information, and the correct business procedures needed to transform the data into the information. One of the biggest hurdles facing the systems analyst is trying to express an understanding of the system user's processing requirements in a form that can be easily understood and verified. Any tool used must be relatively intuitive and void of technical implications. Figure 2.11 has proven to be a suitable model for expressing user process requirements. You'll learn to use that model in this book. Other tools will be needed to represent details not shown in such figures. They too will be covered in the book.

The System Designer's View of Activities

As was the case with the DATA building block, the system designer's view of activities is constrained by the limitations of specific technology. Sometimes, the analyst is able to choose that technology. But often, the choice has already been made and the analyst must use the technology that is available. In either case, the designer's view of activities is more technical.

The designer must first determine which activities to computerize and how to automate them. For example, will the information system process data in batches or process data on-line? From that point forward, the designer tends to think in terms of computer processes.

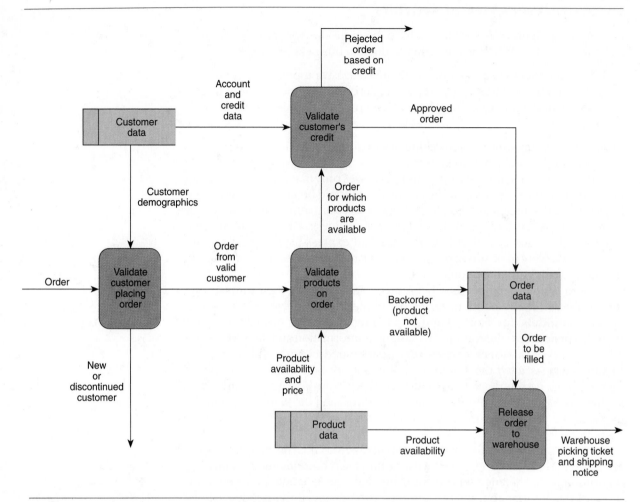

FIGURE 2.11   **A View of Activities Suitable for System Users**   System users respond positively to simple input-process-output models of activities.

Computer processes are business processes that are, or will be, implemented on the computer. Computer processes may automate the business process or simply support the business process.

The designer's view of the computer processes is usually described in terms of computer program specifications such as program *structure charts, logic flow-charts,* and *record layout charts.* As was the case with data, some of these technical views of processes can be understood by users, but most cannot.

The systems analyst and other information systems specialists address and document these technical views of system activities. For instance, Figure 2.12 illustrates a designer's view of a computer program to be written. You may have drawn similar diagrams in a programming course. The designers' intent is to prepare specifications that (1) fulfill the process requirements of the users; and (2) provide sufficient detail and consistency for communicating the process design to the system builders. Once again, this book covers tools and techniques for transforming user business requirements into a technical design.

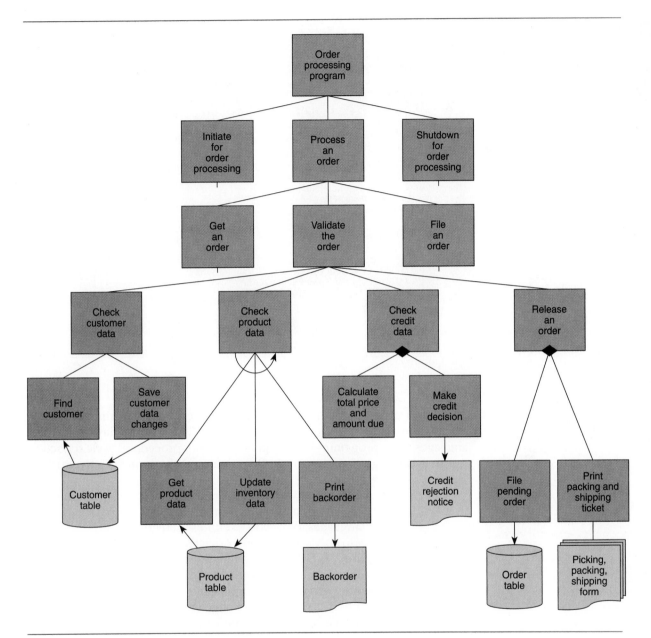

FIGURE 2.12   A View of Activities Suitable for System Designers   System designers tend to focus only on those activities that will become software. As such, they tend to represent these software components in terms of top-down structures suitable for programmers.

The System Builder's View of Activities

System builders represent activities using precise computer programming languages that describe inputs, outputs, and logic. These languages are used to write applications programs.

   Applications programs are language-based, machine-readable representations of what a computer process is supposed to do, or how a computer process is supposed to accomplish its task.

FIGURE 2.13 A View of Activities Suitable for System Builders System builders use precise computer languages to describe processes and activities to a computer system.

```
PROCEDURE DIVISION.
A000-CREATE-SALES-REPORT.
    OPEN INPUT SALES-INPUT-FILE
       OUTPUT SALES-REPORT-FILE.
       WRITE SALES-REPORT-LINE FROM FIRST-HEADING-LINE AFTER ADVANCING TO-
          TOP-OF-PAGE.
       WRITE SALES-REPORT-LINE FROM SECOND-HEADING-LINE AFTER ADVANCING 1
          LINES.
    MOVE SPACE-TWO-LINES TO PROPER-SPACING.
    READ SALES-INPUT-FILE
       AT END
          MOVE 'NO ' TO ARE-THERE-MORE-RECORDS.
    PERFORM B010-PROCESS-SALES-RECORD
       UNTIL THERE-ARE-NO-MORE-RECORDS.
    CLOSE SALES-INPUT-FILE
       SALES-REPORT-FILE.
    STOP RUN.
       .
       .
       .
```

```
TABLE FILE PRODUCT
PRINT PRODNUMBER AND PRODDESCRIP AND
PRODUCTPRICE AND DISCOUNTVOL AND
DISCOUNTRATE
HEADING
"PRODUCT PRICE LIST REPORT"
END
```

Most of you are at least somewhat familiar with the syntax of a third-generation programming language such as COBOL, BASIC, C, or Pascal. Third-generation languages force programmers to express detailed, procedural logic for activities. Such languages represent the builder's view (as shown in the left portion of Figure 2.13) of information systems.

Or perhaps you are familiar with fourth-generation languages such as those provided in most database management systems (such as dBASE or Focus). Fourth-generation languages allow you to express activities in terms of "what you want the process to do," independent of details about logic (see the bottom right portion of Figure 2.13). These are sometimes called *nonprocedural languages*. Regardless of your familiarity with computer languages, they are used to represent activities (including their inputs and outputs) to computers. It is not the intent of this book to teach these languages—only to place them into the context of the ACTIVITIES building block.

BUILDING BLOCK—NETWORKS

Once again, let's start the same way we did with the DATA and ACTIVITIES building blocks. When engineers design a new product (like an automobile engine), that product has spatial properties (or *geometry*) that describe the product in three dimensions. In a similar vein, so do information systems— except that we use the term *networks*.

Networks are (1) the distribution structure of people, data, activities, and technology (the other building blocks) to suitable business locations, and (2) the movement of data between those locations.

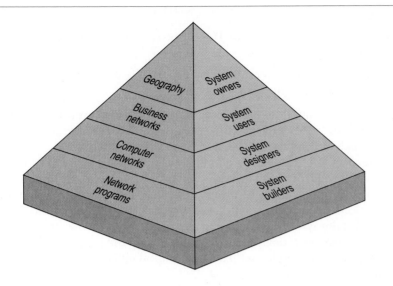

FIGURE 2.14 **The NET-WORKS Building Block of Information Systems** This face of your pyramid model illustrates how different players in the information systems game view networks.

The intent of networking is to provide for cooperative processing between systems, computers, and people. Networking continues to grow in importance because of the following factors:

- The pace of business is increasing. Organizations that can gain faster access to critical information have a competitive advantage. This is sometimes accomplished by duplicating systems in multiple locations or distributing portions of systems to the locations that have the greatest need for those portions.

- The reach of business is expanding. Organizations that can extend their information systems to include their customers and suppliers have a competitive advantage. For example, if a retailer can directly tie its purchasing information systems into its supplier's order information systems, it can bypass the delays caused by mailing, phoning, and keying orders.

- The complexity of business is increasing. Organizations operate in more locations, national and international, then ever before. Different laws, different markets, and different languages make it difficult for organizations to coordinate their efforts and consolidate important corporatewide data.

- Data is being carelessly duplicated at an alarming rate, driven by the availability and installation of large numbers of relatively inexpensive minicomputers and personal computers. The data is, in many cases, out of control, inconsistent, and lacking in security.

The term network has a technical meaning to most computer professionals; however, as we learned with data and activities, different representations are suited to different people. There are business networks, and there are computer networks. Figure 2.14 summarizes different people's views of information systems networks. Once again, the systems analyst who understands these views will be better equipped to deal with the tools and techniques taught in this book.

The System Owner's View of Networks

To the system owner, networking is not a technical issue. In fact, it is merely geographical.

> The **geography** of a system is those geographic locations in which the business chooses or needs to operate.

Try not to restrict your thinking to geography in the classic sense. True, geography includes such locations as cities, states, and countries. But it also includes locations such as campuses, sites, buildings, floors, and rooms.

With respect to locations, systems analysts must learn which business functions are performed at which locations. Are some functions duplicated at some locations? Are some functions provided from a single (or few) location(s)? What locations are to be served or affected by the information system? These are concerns that the systems analysts must negotiate with the owners. The system owners will ultimately decide the degree to which the system will be centralized, distributed, or duplicated. There are important cost and political considerations in these decisions.

Geography has taken on a new dimension in many businesses that strategically seek to integrate their information systems into the businesses of their suppliers and customers. For example, Sears, the retailing giant, has directly linked its purchase order information systems with the order-processing information systems of many of its primary suppliers. Thus, Sears can avoid the delays associated with mailing or phoning orders to replenish stock. This system also gives the retailer a competitive advantage — keeping products on the shelves with minimum lead time — which, in turn, reduces inventory carrying costs and increases profits.

Another geographic location being tested by some executive managers is the cottage office. In this method, employees (permanent and temporary) are allowed to work out of a home office with complete access to company databases and communications tools. Some experts predict that cottage offices will become a permanent and important component of the work force by the end of this decade. After all, many individuals (both students and workers) honestly feel that they are more productive in the quiet of their own home. So long as employees remain accountable for their productivity and quality, cottage offices may yield benefits over traditional business office architecture.

When dealing with system owners, an analyst might communicate an information system's geography with simple maps or floor plans (Figure 2.15). They might also use matrices to determine what business functions are performed at each location. As with the other building blocks, this book will provide you with effective techniques for defining and documenting system geography for systems owners.

The System User's View of Networks

Consistent with the user views of data and processing, system users are the experts about the requirements for any given location. Like system owners, system users are interested in working locations. But because they are closer to the day-to-day locations, they might identify locations that are unknown or forgotten by system owners. Users also tend to have a more microscopic view of

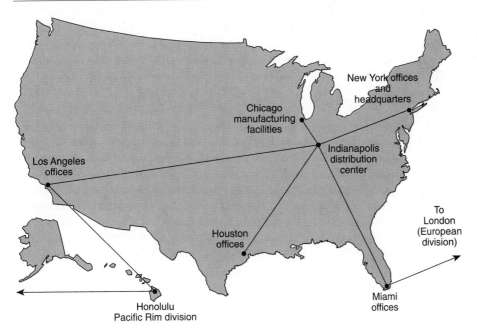

FIGURE 2.15 A View of Networks Suitable for System Owners System owners tend to view networks in terms of business operating locations on a simple map. Other representations include state maps, city maps, campus maps (for multibuilding sites), building floor plans, etc. Note that any connections only suggest loosely defined interactions between locations.

locations. For instance, owners might think of buildings, whereas users become concerned with individual offices or areas within buildings.

Users might also identify some truly unique definitions of location. For example, consider a traveling salesman as part of an order entry information system. That salesman's location varies. He might be in his car, in his office, in the office of a client, or working from a home office. How do you define such a location? One possibility would be to define a single, generic location that represents "wherever the salesman is at any given time." This forces the system users, designers, and builders to consider creative possibilities such as the use of portable computers and computers with cellular modems.

More importantly, system users think in terms of business networks.

A **business network** defines detailed working locations, specific resources available at each location, and the business communications requirements between those locations. A business network does not necessarily require a computer network. A business network is sometimes called a *logistic network*.

The location specific resources are expressed in terms of the building blocks you've already learned— PEOPLE, DATA, and ACTIVITIES that are needed at the location (we purposely left out TECHNOLOGY). The business communications requirements are expressed in terms of location-to-location data flows. For example, the flow of data between offices might be important.

How does this affect the modern systems analyst? Most analysts are at least somewhat familiar with computer networks. They must, however, initially divorce themselves from that technology to consider the business network. First, the analyst must identify all locations pertinent to the system. Whereas system owners tend to have a macro view of the system's locations (for instance,

Building B, 3rd floor), system users frequently have a more detailed, micro view of those same locations. For instance:

> Most purchasing agents are in Building B, 3rd floor, Offices 302 through 315. The purchasing managers are in Offices 320 and 325. Don't forget to include our special agents who share space with Engineering in Building A, 1st floor, Room 101. Oh yes, and what about the 10 procurement agents who are usually on the road—working with our many suppliers? When they are not on the road, they share space in Room 201 of Building B.

It can get far more complicated for systems that must support different cities, states, or countries. It can also be complicated by systems that directly interface to other businesses (for example, suppliers and customers).

Even as systems analysts define system users' data and activity requirements (or building blocks), they should also begin defining how those requirements should be distributed, duplicated, or shared with the different locations. In other words, the user's view of networks depends on the PEOPLE, DATA, and ACTIVITY building blocks. A systems analyst might represent the user's **logistic network** for an information system using a flow-like diagram as shown in Figure 2.16. Notice that the communications requirements between locations are expressed as data flows. Although more detailed than a map, this picture is still not technical. You'll learn to use similar tools and techniques for user-oriented network representation in this book.

The System Designer's View of Networks

As was the case with the DATA and ACTIVITIES building blocks, the system designer's view of networks is constrained by the limitations of specific technology. The emphasis shifts to a computer network that can support the business network.

> A **computer network** is a technical arrangement that interconnects computers and peripherals such that they can exchange data and share technical resources. It is sometimes called a **distributed systems architecture.**

Once again, sometimes the analyst is able to choose this technology. Often, the choice has already been made and the analyst must use the networking technology that is available. In either case, the designer's view of networks is technical.

Based on the system user's view of the business network, the system designer must create a new (or use an existing) computer network that implements the desired distributed information system.

> **Distributed information systems** are those whose data and processes have been divided into subsets that are technically distributed to, or duplicated in, multiple locations. The opposite term is *centralized information systems*.

It may surprise you that even the classical, centralized information system is really distributed. Think about it. The computer is in one location (the IS shop), and the users are in other locations (anywhere they do business). Thus, all information systems are, in the classical sense of the word, distributed. Distribution decisions require the system designers to determine:

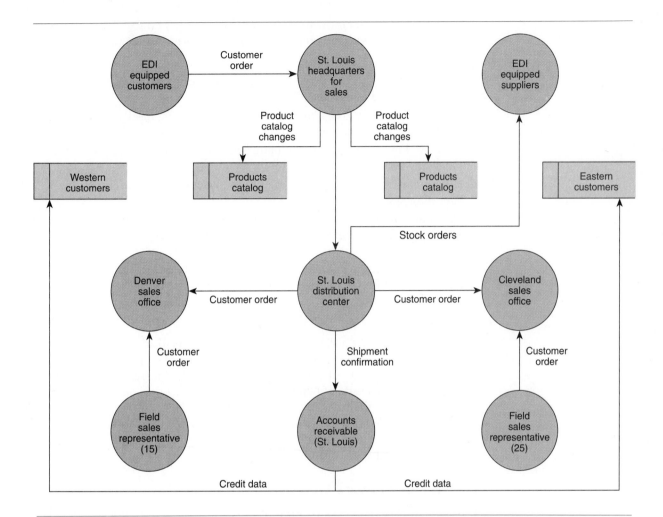

FIGURE 2.16 A View of Networks Suitable for System Users This user-oriented flow diagram documents business data flow requirements between multiple locations. It also suggests that regional sales offices will maintain their own regionally based customer files and duplicated product catalog files.

- Which of the system user's location requirements must be provided through common access to a central or shared location.
- Which of the user's location requirements should be distributed to a unique location.
- Which location requirements should be duplicated at several locations.

As with the system user's view of networks, the system designers' view is also expressed in terms of technology locations and the people, data, and activities and technology needed at those locations. The designer's view includes computer technology needed at each location, technical people needed at each location, the design of data files and databases for each location, and the selection or design of software (activities) for each location. Again, this is called a distributed systems architecture. Clearly, these decisions have cost/benefit implications that ultimately need to be resolved by system owners.

The system designer must also determine (or accept) how each location will physically connect with the other locations. Today's **connectivity** options are numerous. Examples include:

- Terminals directly connected to a mainframe computer using a teleprocessing monitor (such as CICS).
- Terminals connected to a mainframe computer using modems, phone lines, and a teleprocessing monitor.
- Minicomputers connected to a mainframe via a network operating system (such as SNA).
- Minicomputers connected to a mainframe via a satellite and network operating system.
- Personal computers connected to a mini- or mainframe computer via an ASCII port and a terminal emulation software package (such as Crosstalk or Kermit).
- Personal computers connected to other personal computers, minicomputers, or mainframe computers using a local or wide area network and network operating system.
- Laptop computers connected via internal modem to a mainframe computer.

The systems analyst and telecommunications specialists address and document these technical views of networks. For instance, Figure 2.17 illustrates a system designer's view of a network. As always, the designer's intent is to prepare specifications that (1) fulfill the business network requirements of the users; and (2) provide sufficient detail and consistency for communicating the network design to the system builders. And once again, this book covers tools and techniques for transforming user business requirements into a technical design for a distributed information system.

The System Builder's View of Networks

The final view of networks is relevant only to the system builders. They use telecommunications languages and standards to write network programs.

> **Network programs** are machine-readable specifications of computer communications parameters such as node addresses, protocols, line speeds, flow controls, and other complex, technical parameters.

Examples include CICS, SNA, TCP/IP, and Novell Netware/386. Regardless of your familiarity with network languages and terminology, they are used to implement computer-based networks for information systems. It is not the intent of this book to teach these languages—only to place them into the context of the information system building block called NETWORKS. Most information systems programs offer courses that can expand your technical understanding of networking.

BUILDING BLOCK — TECHNOLOGY

We'll conclude our study of the information systems building blocks with TECHNOLOGY, the base of the information systems pyramid (see margin). Each side of the foundation represents the technology used to implement or support its corresponding face of the pyramid. Collectively, these technologies are

Information system

Technology

Information Technology

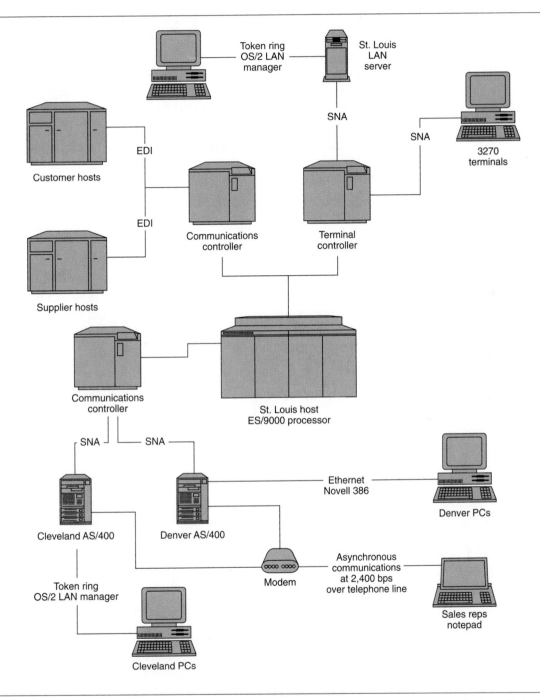

FIGURE 2.17 A View of Networks Suitable for System Designers Designers view networks in technical terms —that is, what technology is used at each location, what data and activities are distributed to which locations, and what technology is used to implement data flows between locations.

often called *computer* or *information technology*. The latter has become the term of choice.

> **Information technology** refers to the modern merger of computer technology with telecommunications technology. Information technology includes computers, peripherals, networks, fax machines, telephony, intelligent printers, and any other technology that supports either information processing or business communications.

Let's briefly examine the information technology base for information systems.

Data Technology

The DATA building blocks of information systems is implemented using data technology.

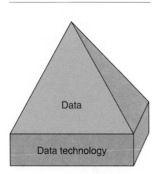

> **Data technology** (see margin) includes all hardware and software required to capture, store and manage the data resource.

The hardware includes keyboards, printers, tape drives, diskette drives, fixed magnetic disk drives, optical disk drives, and the like. Data technology also includes the software for data storage, manipulation, and management. This includes file management systems such as ISAM and VSAM, database management systems such as DB2 and dBASE, and spreadsheets such as Lotus 1-2-3 and Excel. We refer you to almost any introduction to computers or information systems textbook for more about data technology. This technology is also covered extensively in programming and database courses and textbooks. In this text, we will only address the analysis and design decisions that lead to the selection and use of specific data technology.

Processing Technology

The ACTIVITIES building blocks of information systems are implemented using processing technology.

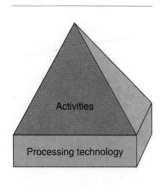

> **Processing technology** (see margin) includes all hardware and software required to support business and information system activities. Clearly, this includes computers and their input/output peripherals (e.g., printers, displays, mice, scanners, and optical mark readers). It also includes applications programs.

Applications programs, as you are probably aware, are written in a variety of languages such as COBOL, BASIC, C, CSP, Focus, and dBASE, to name a few. Most of you have written applications programs. Process technology also includes operating systems and other system software.

Finally, processing technology also includes any machine or device that contains a computer processor. The use of microprocessor technology extends to robots, manufacturing equipment, environmental control systems, and so forth. The integration of information systems with machines and devices has become a high priority in many businesses, especially manufacturing organizations.

This is not a programming course or text. Furthermore, most of you have been exposed to simple computer technology and architecture in an introductory computer or information systems course. We refer you to texts on those subjects for more information. We will, however, address the analysis and design decisions and methods that result in the selection of computer technology and the design of applications software.

Communications Technology

The NETWORKS building blocks of information systems are implemented using communications technology.

> **Communications technology** (see margin), also called *networking* or *telecommunications technology,* includes the hardware and software used to interconnect data and process technology at different locations.

This is a broad category of technology. In its simplest form, it includes the technology to connect office terminals at a single site to a host computer at that same site. Local and wide area networks of personal computers in a single site represent a more complex form of network technology. Still more complex are telecommunications networks that link large computers and/or networks at one site to large computers and/or networks at another site (perhaps separated by thousands of miles).

Most information systems and computer science schools include at least one complete course on this technology. We'll refer you there for more about the technology. However, once again, this textbook will address the analysis and design decisions that determine the locations of data and activities, and the interconnections between those locations.

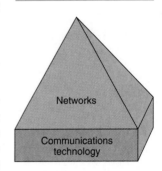

Networks

Communications technology

Technical Specialists

System owners, users, designers, and builders must be supported by various technical specialists who understand the technology.

> **Technical specialists** (see margin) sell, configure, repair, and maintain information technology for systems owners and users.

Some technical specialists work in the information technology industry. They sell and subsequently support the technology. Some technical specialists work for the business. The configure, purchase, install, and support the technology. Still other technical experts sell their technical services to the business as third parties. Perhaps they have a special expertise or can offer less expensive services.

Clearly, systems developers, programmers, and consultants are also technical specialists in their own right. We have already covered their perspectives on information systems development throughout the chapter.

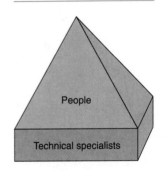

People

Technical specialists

Summary

An information system is an arrangement of people, activities, data, networks, and technology that are integrated to accomplish the purpose of supporting and improving the day-to-day operations in a business, as well as fulfilling the problem-solving and decision-making information needs of business managers.

Today, there are two types of information systems, personal and multiuser. Personal information systems are those designed to meet the personal information requirements of a single user. Multiuser information systems are those designed to meet the information requirements of work groups (e.g., departments, offices, sections, divisions) or entire organizations. In order to build either personal or multiuser business and information systems, systems analysts and users must effectively combine the building blocks of those systems.

The first and most vital building block of information systems is PEOPLE. The term information worker was coined to describe those people whose jobs involve the creation, collection, processing, distribution, and use of information. Systems analysts view people in terms of their roles in developing and supporting information systems. For every information system, there will be one or more system owners. System owners are an information system's sponsors and chief advocates. They are usually responsible for spending the money to develop and support the information system, and for the ultimate acceptance of the information system. System users make up the vast majority of the information workers in any information system. These are the people who use (and directly benefit from) the information system on a regular basis—capturing, validating, entering, responding to, and storing data and information. There are five classes of users: clerical workers, technical and professional workers, supervisors, middle managers, and executive managers. Each of these user classes presents unique perspectives and information needs to the systems analyst. System designers translate users' business requirements and constraints into technical solutions consisting of computer files, databases, inputs, outputs, screens, computer networks, and programs that will meet the requirements. System builders then construct the information system based upon the design specifications from the system designers. The other information system building blocks are viewed differently by the system owners, users, designers, and builders.

The second building block of information systems is DATA. Data is a collection of raw facts in isolation. Data describes the organization. These isolated facts convey meaning but generally are not useful by themselves. Data is used to build information. Information is data that has been manipulated so it is useful to someone. The DATA building block is concerned primarily with the data resource, not how the data is transformed into information.

The average system owner is not interested in raw data. He or she is interested in the things that data describes. These things are called business resources (or entities). When dealing with system owners, systems analysts try to identify the business entities that are important to the system. Conversely, system users view data in more detailed terms. Modern systems analysts try to get their system users to see their data requirements in terms of entities, relationships, attributes, and rules. An entity is something about which data is important. Attributes describe facts of interest about an entity. Rules are conditions that govern entities and attributes. System designers view data within the limitations of specific technology. That view of data is usually specified in terms of computer files and databases. System builders specify data programs using very precise, and unforgiving, programming and database languages.

The third building block of information systems is ACTIVITIES. Business and information system activities put the DATA building block to use, capturing and transforming data into useful information. System owners view activities in terms of high-level functions provided to the business. Examples include transaction processing, management reporting, decision support, simulation of expertise, generation of executive information, and office automation. System owners view these functions in general terms of how well they fulfill business goals and objectives. On the other hand, system users view activities in terms of distinct processes with specific inputs and outputs, data files, and business policies and procedures that must be executed by the processes. System designers view activities in terms of computer processes (program specifications). And, of course, system builders view these software components in terms of computer application programs.

The fourth building block of information systems is NETWORKS. Networks make possible (1) the distribution of people, data, activities, and technology (the other building blocks) to suitable locations; and (2) the communication of data between those locations. Most system owners are exclusively concerned with the geography of systems — geographic locations in which the business chooses to operate. System users are interested in business or logistic networks — working locations, resources at those locations, and business communications needed between locations. System designers view networks in technical terms of distributed systems architecture and computer networks that implement the business network. This architecture describes the technical connectivity between locations (hardware and software). Accordingly, designers talk about connectivity technology such as local and wide area networks and teleprocessing software. Finally, system builders specify network programs in terms of technical jargon such as addresses, protocols, line speeds, flow controls, and other often complex parameters. They may also use telecommunications standards and teleprocessing/networking languages to describe these computer-compatible networks.

The fifth and final building block of information systems is TECHNOLOGY. Data technology includes all hardware required to capture, store, and manage the data resource. Processing technology includes all hardware and software required to transform data (inputs) into useful information (outputs). Communications technology includes the hardware and software used to interconnect data and process technology at different locations. Finally, technical specialists select, install, and support all of this technology.

Given this fundamental understanding of information systems building blocks, you are now ready to examine the process of building information systems.

--- *Key Terms* ---

application program, p. 63

attribute, p. 49

business activity, p. 52

business network, p. 67

business process, p. 61

business resource, p. 47

business system function, p. 53

clerical worker, p. 42

communications technology, p. 73

computer-based information system, p. 39

computer files or database, p. 50

computer information system, p. 39

computer-integrated manufacturing, p. 60

computer network, p. 68

computer process, p. 62

connectivity, p. 70

data, p. 46

data flow, p. 61

data maintenance, p. 55

data processing, p. 54

data program, p. 52

data requirement, p. 48

data store, p. 61

data technology, p. 72

decision support system, p. 57

detailed report, p. 56

distributed information system, p. 68

distributed systems architecture, p. 68

entity, p. 48

exception report, p. 57

executive information system, p. 57

executive manager, p. 45

expert system, p. 58

function, p. 53

geography, p. 66

goal, p. 53

information, p. 46

information system, p. 39

information system activity, p. 52

information system function, p. 53

information technology, p. 72

information worker, p. 40

input transaction, p. 55

knowledge worker, p. 43

logistic network, p. 68

management information system, p. 56

management reporting, p. 56

middle manager, p. 44

multiuser information system, p. 39

network program, p. 70

network, p. 64

objective, p. 53

office automation, p. 59

output transaction, p. 55

personal information system, p. 39

process, p. 61

process model, p. 61

processing technology, p. 72

rule, p. 49

structured decisions, p. 57

summary report, p. 57

supervisor, p. 44

system builder, p. 46   technical and professional   transaction processing
system designer, p. 45   staff, p. 43   system, p. 55
system owner, p. 41   technical specialist, p. 73   unstructured decision,
systems integration, p. 59   transaction, p. 54   p. 58
system user, p. 42

Questions, Problems, and Exercises

1. Define *information system*. Information systems are all around you. From your last job, data processing or otherwise, give an example of a completely manual information system. Describe how a computer might improve that information system.

2. Identify and briefly describe two general types of information systems. Give examples of each.

3. Define *information worker*. Why are information workers considered the most vital building block of information systems?

4. Briefly describe the five building blocks of every information system.

5. Systems analysts must identify system users for an information systems development project. Who are system users? What is their role in systems development projects? How do system users differ from system owners?

6. Give three examples of systems users from each of the following classifications: clerical workers, technical and professional staff, supervisory staff, middle management, and executive management. Explain the job responsibilities of each example you provided and state why your example represents that particular classification of system user. For each system user, describe a situation that would make that worker a client of an analyst.

7. For each of the following classifications of system users, identify the individual as clerical staff, technical and professional staff, supervisory staff, middle management, or executive management. Defend your answer.
   a. Receptionist.
   b. Shop floor foreman.
   c. Financial manager.
   d. Assistant store manager.
   e. Chief operating officer.
   f. Manufacturing control manager.
   g. Terminal operator/data entry.
   h. Applications programmer.
   i. Programmer/analyst.
   j. Warehouse clerk who fills orders.
   k. Stockholder.
   l. Product engineer.
   m. Consultant.
   n. Broker.
   o. Accountant.

8. Obtain an organization chart from a local company or at your library. Classify each person and/or job position appearing on the chart according to the type of system user (such as clerical staff). (Alternative: Substitute the organization charts appearing in the SoundStage episode that preceded this chapter.)

9. Consider an organization by which you were, or are, employed in any capacity. Identify the information workers at each level in the organization according to their classification of system user. (Alternative: Substitute your school for the organization. Students are part of the clerical staff classification. Do you see why?)

10. Describe four views of the PEOPLE building block. How do the roles and responsibilities of each information worker differ?

11. What is the relationship between data, information, input, processing, and output? Give an example of this relationship.

12. Differentiate between data and information. Identify each of the following as data or information. Explain why you made the classification you did.
    a. A report that identifies, for the purchasing manager, parts that are low in stock.
    b. A customer's record in the customer master file.
    c. A report your boss must modify to be able to present statistics to his boss.
    d. Your monthly credit card invoice.
    e. A report that identifies, for the inventory manager, parts low in stock.

13. The systems owner's view of data is primarily concerned with entities. If the owner of a course scheduling information system is the registrar, brainstorm the entities that this registrar might identify.

14. Differentiate between *entities, relationships, attributes,* and *rules.* Using the course registration example in the last exercise, give examples of each. What information workers (class of users) are most concerned with relationships, attributes, and business rules of data entities?

15. Explain how the systems designer's and systems builder's views of data differ.

16. Define the terms *business function* and *information systems function.* Give examples of each, citing both the business and information function provided by each system.

17. Make an appointment to visit a systems analyst at a local data processing installation. Discuss one of the information systems projects the analyst has worked on. What were some of the functions (e.g., transaction processing, decision support, office automation) being supported?

18. Define the term *transaction.* Identify three types of transaction processing. Give examples of each.

19. Identify the type of information support provided for each of the following applications.
    a. A customer presents a deposit slip and cash to a bank teller.
    b. A teller gets a report from the cash register that summarizes the total cash and checks that should be in the drawer.
    c. Before cashing a customer's check, the teller checks the customer's account balance.
    d. The bank manager gets an end-of-day report that shows all tellers whose cash drawers don't balance with the cash register summary report.
    e. The system prints a report of all deposits and withdrawals for a given day.
    f. A chief executive officer is using his computer to obtain access to the latest stock market trends.
    g. A doctor keys in data describing symptoms of a patient; he receives a report that suggests what illness the patient is likely suffering from and a detailed explanation concerning the rationale as to why the symptoms suggest that particular illness.
    h. An employee is electronically sending a memorandum concerning a scheduled committee meeting to all committee members that are to attend.

20. Transaction processing is usually performed by and for which class of system users?

21. Define the term *management information system* applications. Identify three types of information provided by management information systems. Give three examples of each.

22. What are decision support system applications? Differentiate between structured and unstructured decisions.

23. What are executive information system applications?

24. What are expert system applications? Give an example of an expert system application. How do expert system applications differ from decision support system applications?

25. Briefly define the term *office automation systems.* Give several examples of office automation applications. What trend is developing in office automation?

26. What is meant by the term *systems integration?* What is its importance to the classical system functions (e.g., transaction processing, management reporting, etc.)?

27. Briefly explain the different activity views of systems owners, users, designers, and builders.

28. Briefly define *networking.* Give several reasons why networking is becoming increasingly important.

29. Briefly explain the differing views of networking according to owners, users, designers, and builders.

30. Why are systems owners primarily concerned with the geography of systems?

31. Define the term *logistic network.* What information worker is primarily concerned with logistic networks?

32. Explain the relationship between TECHNOLOGY and the other four information systems building blocks: PEOPLE, ACTIVITIES, DATA, and NETWORKS.

Projects and Minicases

---

1. Knowledgeable University plans to support course registration and scheduling on a computer. The following client community has been designated:
   a. Curriculum deputy—one per department, responsible for estimating demand by that department's own students for each course offered by the university. This person may revise demand estimates from time to time.
   b. Schedule deputy—one per department, responsible for deciding which courses from that department will be offered, at what times, by what teachers, and with what enrollment limits. These parameters may change during the registration period. This is the only person who can increase or decrease enrollment limits for a course (including adding or deleting a course to or from the schedule).
   c. Schedule director—in charge of allocating classroom and lecture hall space and time to departments. Also prints the schedule of classes to show students what will be offered and when.
   d. Students—submit course requests and revisions and receive schedules and fee statements.
   e. Counselors—advise students and approve all course requests and revisions. Also help students resolve time conflicts (where student has registered for two courses that meet at exactly the same time).

   This is the cast of characters (which may be revised or supplemented by your instructor to more closely match your school). For each client, brainstorm and describe different types of functional support that might be provided.

2. James Oliver, president of Oliver Pest Control, was under the impression that computerization of his clerical functions would effectively reduce the size of that staff. Since installing computer support eight years ago, he has seen his clerical staff increase 10 percent, whereas organization staffing has remained steady. His friends in the local chamber of commerce have experienced the same phenomenon. Mr. Oliver is particularly puzzled by another curious trend. Although the use of computers did not decrease his need for clerical staff, it has reduced his dependence on middle management and professional staff. This seems exactly the opposite of his expectations; however, it is typical.

   Do background research into two local data processing shops and some of their implemented computer-based systems. Were their clients expectations similar to Mr. Oliver's prior to completion of the system? How did the analysts deal with the expectations, if they did so at all? How would you have dealt with Mr. Oliver's expectations prior to beginning the systems project? Were the outcomes of the implemented system similar? What factors might explain these outcomes?

3. Last year, Hologram, Inc.'s comptroller purchased an IBM personal computer and the spreadsheet, Lotus 1-2-3. He learned how to use the product to support his own budgeting and cost control decision-making needs. He likes Lotus, but he's getting tired of re-entering the same data into different spreadsheets. He wants to know why he can't store the data in a database on his microcomputer. Better still, he knows that much of the data currently resides on the Unisys mainframe computer in data processing. Why not tap that database to minimize the amount of data he has to input personally? In terms of the ACTIVITIES building block, what application is the comptroller requesting? Ignoring technical implications, are his expectations reasonable?

4. Liz, an account collections manager for the bank card office of a large bank, has a problem. Each week, she receives a listing of accounts that are past due. This report has grown from a listing of 250 accounts (two years ago) to 1,250 accounts (today). Liz has to go through the report to identify those accounts that are seriously delinquent. A seriously delinquent account is identified by several different rules, each requiring Liz to examine one or more data fields for that customer. What used to be a half-day job has become a three-days-per-week job. Even after identifying seriously delinquent accounts, Liz cannot make a final credit decision (such as a stern phone call, cutting off credit, or turning the account over to a collections agency) without accessing a three-year history on the account. Additionally, Liz needs to report what percentage of all accounts are past due, delinquent, seriously delinquent, and uncollectible. The current report doesn't give her that information. What kind of report does Liz have—detail, summary, or exception? What kind of reports does Liz need? What kind of decision-support aids would be useful?

--------------------- *Suggested Readings* ---------------------

Davis, Gordon B. "Knowing the Knowledge Workers: A Look at the People Who Work with Knowledge and the Technology That Will Make Them Better." *ICP Software Review,* Spring 1982, pp. 70–75. This article provided our first exposure to the concept of information workers as the new majority.

Davis, Gordon B, and Margrethe H. Olson. *Management Information Systems: Conceptual Foundations, Structure, and Development.* 2nd ed. New York: McGraw-Hill, 1985. This is our all-time favorite book on information systems concepts and principles.

Lord, Kenniston W., Jr., and James B. Steiner. *CDP Review Manual: A Data Processing Handbook.* 2nd ed. New York: Van Nostrand Reinhold, 1978.

Zachman, John A. "A Framework for Information System Architecture." *IBM Systems Journal* 26, no. 3 (1987). We adapted the pyramid model for information systems building blocks from Mr. Zachman's conceptual framework. We first encountered John Zachman on the lecture circuit where he delivers a remarkably informative and entertaining talk on the same subject as this article. Mr. Zachman's framework has drawn professional acclaim and inspired at least one conference on his model. His framework is based upon the concept that architecture means different things to different people. His model suggests that information systems consist of: (1) three distinct "product-oriented" views—data, processes (which we call activities), and networks; and (2) six different audience-specific views for each of those product views—the ballpark and owner's views (which we renamed as owner's and user's views, respectively), the designer's and builder's views (which we combined into our designer's view), and an out-of-context view (which we called the builder's view). Any combination of a product-oriented view and a audience-oriented view is someone's definition of information system architecture. Our adaptations make the framework an ideal conceptual model for this entire textbook.

EPISODE 2

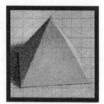

A New Information Systems Project Is Started

Sandra and Bob get an assignment: "How can we improve order processing and Member Services to support new strategic products and services through improved information system support?"

The Project Directive

Sandra entered her office with the Monday morning "blahs." Things had been pretty slow since the new Accounts Receivable system had been placed into operation. Very few errors had been made and the end-users were happy. But she eagerly wished for a new project. She wouldn't have to wait long.

Debbie Lopez phoned at 8:45 A.M. Debbie was administrative assistant to David Hensley, Customer Services manager for SoundStage.

"Good morning, Sandra! How was your weekend?"

"Just fine, Debbie. What can I do for you?"

"Can I come and see you for a moment?" asked Debbie.

"Sure," Sandra replied.

A few minutes later, Debbie entered Sandra's office. She gave Sandra a piece of paper (Figure E2.1). It was an administrative memorandum from Rebecca Todd, executive vice president of Sound-Stage.

Debbie said, "This is a directive from Ms. Todd in her capacity as chairperson of the Strategic Planning group. Are you familiar with the activities of that group?"

"Yes," answered Sandra." That group hired the IBM Business System Planning [BSP] consultants to come up with a strategic plan for the business and its information services. I believe they documented management's business plan and then developed an overall architecture for our future databases, networks, and applications."

"That's right," replied Debbie. "And they developed a prioritized list of information systems development projects based on perceived value to the business plan. Well, the Order Entry and Member Services system is first on the list, and I wouldn't trust it to anyone but you. That Accounts Receivable system you created is a godsend. Bill is so pleased with the system! Meanwhile, this one is even more important. The business plan suggested a major expansion of marketing and customer-oriented services. The Order Entry and Member Services system must be completely overhauled to enable the business plan."

Ken Riker, assistant director of Information Services, knocked at the open door. "Good morning Debbie. Sorry I'm late. And good morning to you too, Sandra. I trust Debbie has told you about the new project."

Noting both women's affirmative nods, he continued, "We've hand-picked you for this one, Sandra. It's mission critical. The information systems portion of our strategic plan calls for us to eventually redesign all mission critical systems as an integrated whole. We want a well-controlled FAST

FORM IS-100-A.RFSS

Request for System Services

SUBMITTED BY: **Strategic Planning Group** Date: **March 15, 1993**

DEPARTMENT: **Member Services**

TYPE OF REQUEST:
(check one)

☒ NEW SYSTEMS DEVELOPMENT

☐ EXISTING SYSTEM ENHANCEMENT

☐ EXISTING SYSTEM MODIFICATION

☐ NOT SURE

BRIEF STATEMENT OF PROBLEM OR OPPORTUNITY (Attach additional documentation as necessary):

The strategic planning group has targeted the member services business processes and information systems applications for reengineering. The new system must provide greater club, membership, and contract flexibility while addressing a major capacity problem.

BRIEF STATEMENT OF EXPECTED SOLUTION:

We envision completely new business processes that would be supported by a superset of the business area data model developed for member services as part of the strategic plan.

We envision a client/server application that can interface with existing host-based applications.

ACTION (To be completed by Steering Committee or Strategic Planning Committee)

☒ REQUEST APPROVED: ASSIGNED TO: Sandra Shepherd
     START DATE: ASAP
     BUDGET: $30,000

☐ REQUEST DELAYED: BACKLOGGED UNTIL: _____

☐ REQUEST REJECTED

Ken Riker	*Strategic Planning Group*	*3-16-93*
Signature	Representing	Date

LAST REVISED: October 1992

FIGURE E2.1

project on this one — it'll be a great learning experience for your rookie partner. The strategic plan documentation includes several high-level system models that should provide you a running start."

"Get me a list of your current support responsibilities. I'll have Peg reassign those tasks — this is your top priority. I want you to stress quality first, then productivity. This system will be the role model for several more strategic applications to come. Like I said, we'll be reengineering all mission critical systems over the next five years or so. I have another meeting in three minutes, so I have to go. See me if you have any questions . . . Oh yes, and congratulations on a choice assignment!"

Sandra wasn't sure whether she was thrilled or scared. Debbie picked up the conversation. "Told you this one was big! In a nutshell, we want to see what you can do to improve our Order Entry and Follow-Up system. As you know, our product mix is rapidly changing, especially with the addition of compact digital discs. We want to disband the current club membership structure that ties members to a particular medium such as compact discs, audiocassettes, or videotapes. In its place, we want a flexible membership club that is not dependent on type of merchandise. Marketing has the details."

Sandra interrupted, "Our group doesn't support Marketing."

Debbie replied, "The business plan suggested the development of cross-functional information systems. We're ignoring organizational boundaries and territories. The goal is to design systems across multiple organization boundaries according to common data needs and functional efficiency. Clearly, Marketing and Order Services functions need to be integrated, regardless of where we place them in the organization chart."

Sandra interrupted again, "An exciting concept. Has everybody bought in on that concept?"

Debbie answered, "Almost everybody! So long as Information Services doesn't personally change the organization chart, we are OK. But make no mistake about it — there may be some turf battles in this project. We'll have to count on management to address those conflicts. Actually, we believe the biggest conflict may come from within your own Information Services division."

"The Administrative Systems group?" queried Sandra.

"You guessed it," replied Debbie. "Clearly, you will be reengineering some of their information systems and integrating them into your own. Nancy and Ken will have to defuse that bomb and smooth any ruffled feathers. I suspect they may actually assign one of their analysts, part-time, to your team. Maybe it's time to break down some of the territorialism in IS, too."

Debbie continued, "I've only touched the tip of the iceberg in describing the Marketing/Order Services business plan to you. We're looking to new markets, new marketing strategies, new membership and sales goals, even new order technologies. They want you to prototype a new phone-based member response technology as part of this project. And we want to solve some of the existing system's problems while we're at it."

"Whew! I guess I better get started. First, I want to talk this over with David," said Sandra. "I think we have a priority project here, but I need to do some preliminary analysis. I take it this project doesn't go through the steering committee?"

"Correct" answered Debbie. "The steering committee only evaluates user-initiated system requests to assess feasibility and priority. This project comes from the planning committee, which has already assessed importance and feasibility — they assigned it top priority for new systems development."

Debbie stood up. "Well, I just wanted to deliver the request personally. Thanks! I'll see you later, Sandra."

Sandra carefully studied the memorandum and her notes. Then she phoned David, the manager of Customer Services.

"David? Hi. Listen, I got the new Order Services system directive a few minutes ago. Can we meet over lunch to discuss the current order entry system? Say, about 11:30?"

"Great!" said David. "I've got an earful for you! See you at 11:30!"

An Early Meeting

Sandra invited Bob to join her for the luncheon meeting. She knew Bob was anxious to start a real project. Besides, she wanted to give Bob a tour of the Member Services offices after the meeting. The meeting started on schedule.

"David, I'd like you to meet Bob Martinez, my new partner. Bob, this is David Hensley, the manager of Member Services."

"I'm glad to meet you, David," said Bob. "How are you today?"

"Just fine, Bob. So, Sandra, you've been assigned to the new Order Services project? It's got to feel good to get picked for the first real IS project to come out of the business strategic plan." David was openly enthusiastic.

They ate lunch and then directed the conversation to the proposed project. Sandra opened the questioning. "Tell me something about how you do order entry and approval today. I'm just trying to get a general feel for your situation. We'll study the system in detail later."

"Well," David responded, "I'll try to give you an overview. First, I took the liberty of working up a standard Request for System Services form (see Figure E2.1) even though the project came out of the strategic planning committee. I thought it might be useful to see the project initiative in the usual form."

Sandra replied, "Thank you," as she passed a copy to Bob.

David continued, "We process three kinds of orders: dated orders, priority orders, and merchandise orders for the record/tape club and forthcoming compact disc club. Dated orders are those that are automatically filled if a member doesn't return a dated order card to cancel or change the monthly title of the month's offer. Priority orders are for those cards that were returned. Merchandise orders are for any other type of merchandise we sell, including shirts, posters, computer software, videotapes, and so on. The orders are manually screened for accuracy and completeness. Some orders have to be transcribed to . . ."

"Transcribed?" asked Bob.

"Yes, copied to our standard order forms. Most orders are then sent to ISS to be processed. ISS performs credit checks but cannot reject orders officially. We do the follow-up letters to customers if they must pay on their account before we can release their order. Most orders pass the credit check and are split into multiple orders corresponding to merchandise ordered."

Sandra reentered the conversation. "What kind of problems exist in the current system, David?"

"First off, I purposefully simplified my description of order entry and approval. You need to spend time in Member Services to really appreciate the magnitude of my operation. But, to answer your question, I've got a big problem coming! I had a meeting with Rebecca Todd yesterday. I was told to expect an aggressive new marketing program over the next three years . . . TV, radio, newspapers, and magazines. Also, a single integrated club will replace all existing clubs. Instead of tying members to an agreement based on a certain number of purchases over some time period, they will be tied to agreements based on a certain number of *credits* over a period of time. This will allow members to purchase different types of merchandise with various levels of credit toward fulfilling the purchase agreement."

Bob interrupted, "I'm not sure I see the difference."

"All right," David answered. "Let's say you join the compact disc club today. You must buy six discs in the next two years. You can also purchase cassettes and videotapes, but those don't count against your membership agreement. Under the new approach, you will simply join the club—not the compact disc club—just the club. Each compact disc you buy will establish a certain number of credits toward your membership agreement. But, so will cassettes, videotapes, videodiscs, computer games, and any other merchandise. And when you fulfill your membership agreement, you'll receive SoundStage dollars with subsequent purchases. Those dollars can be credited towards the purchase of any of our merchandise."

Bob nodded that he now understood. "Wow! Where do I sign up?"

"That's the problem," answered David. "The current system cannot handle any of this. In order to give customers this new level of service, and to give Marketing the go-ahead to start advertising the service, we have to totally redesign the supporting information systems, in both Marketing and Order Services."

David continued, "Management expects sales to increase at least 25 percent over the first three years of the new services. I can't handle that load with my existing staff, and I have no place to put new staff. I'm convinced we need better computer information support. Meanwhile, my response time to orders is getting worse each month. Orders are getting more complicated. New A/R policies allow customers to carry balances forward on account. We didn't use to do that. You see, Accounts Receivable's new system has complicated my operation. By the way, I forgot to mention that we are taking over the subscription function that enrolls new members through advertisements and special orders."

"I'm beginning to see the magnitude of your problem, and you're right; we're going to have to spend some time in Member Services to fully understand. I'm a bit lost already. Can you explain what you think you need?" Sandra asked.

"Yes. I would like more significant support of order processing. I want to be able to better follow up on outstanding orders. I want some way to prioritize backorders. I'd also like to get some decent information that shows me where I can improve my operation. I don't really know what information—I'm kind of hoping that you can help me figure that out. At least help me harness my data so I can get at it when I want to. And—of course—I need to support these new business initiatives we've discussed." David's tone indicated some frustration with his current operation.

"Whew!" said Sandra. "Bob and I will need to gather some more facts. We should talk with you and your staff very soon. David, we'd like to utilize a new technique in gathering information from you and your staff. It's called joint applications development. Essentially, it requires group participation of your staff in rather lengthy, intensive meetings. It replaces many time-consuming interviews that can result in conflicting information. What do you think?"

David thought about it a second and then answered. "Well, the participation idea is ideal. I do have some concerns with taking too many people off their regular jobs at the same time, but I can imagine some of the benefits. I do like the idea. Let's see if we can work out some ground rules for balancing your need to gather information against my need to get our regular work done. I'm sure we can work something out."

Sandra nodded and then looked at Bob. "We'd better get going, Bob. We have a final review conference on the Accounts Receivable system in 15 minutes. And I do want to walk you through Member Services first. Thanks for lunch, David. We'll get back to you soon!"

Project Assessment

Phase 1 of their FAST methodology required a preliminary study of the existing system and a feasi-bility assessment. Since the project's feasibility and priority had already been established by the planning committee, they did not seek approval from the IS steering committee.

A FAST preliminary study is based primarily on the input of one or two managers in the user community. Sandra and Bob elected to conduct one more in-depth interview with David. They presented their findings in a memorandum (see Figure E2.2).

Where Do We Go from Here?

Sandra and Bob now have a project. Where should they start? If you were assigned to write a proposal that outlines a step-by-step process to develop the new system, what would your steps be? Take a moment to jot down your ideas.

In Chapter 3 you will study a general process for developing an information system—Sandra and Bob are being asked to develop an Order Services Information system. The process you'll be studying is called a systems development life cycle. The steps are called phases. Each phase may consist of several tasks. Those tasks will be studied in later chapters. The systems development life cycle is the basis for all of the tools and techniques you will learn in the remainder of this book.

In Chapter 4 you will be introduced to the structured techniques. The structured techniques are used during the systems development life cycle to precisely analyze and design information systems. Sandra and Bob will use these techniques extensively in future chapters and episodes. You will learn the techniques by studying their work products.

Then, in Chapter 5, you will learn about computer-aided systems engineering—systems development technology used by systems analysts to properly apply the structured techniques. They improve the quality of systems. In some cases they also improve the productivity of systems analysts. You'll see the results as Sandra and Bob progress through this project.

These next three chapters complete your introduction to systems development fundamentals. Notice we used the term *introduction*. The subsequent parts and chapters build on the introduction to teach practical skills and techniques. We'll return at opportune times to see how Sandra and Bob are coming along with their new project.

InterOffice Memo

To:	Information Systems Services Steering Committee
From:	Sandra Shepherd, Systems Analyst
	Robert Martinez, Programmer Analyst
Date:	April 15, 1993
Subject:	Assessment for New Member Services Information System

We have just completed the initial investigation for a proposed member services information system in accordance with our FAST methodology. This Initial Study Report is a deliverable of FAST. Much of the material was gleaned from the Business Strategic Plan as developed by our consultants. Since this project received high priority in that plan, the issue of "feasibility" analysis has been predetermined. Consequently, this report merely reaffirms the project and expands somewhat on its focus.

We reaffirm the BSP finding that stated that "A re-developed member services information system will substantially improve support for the Member Services and peripheral business units, and our customers."

Problem Statement

Member Services handles membership subscriptions and member orders. Subscription and order processing is, for the most part, a manual process. Users have developed rudimentary databases and spreadsheet applications on non-integrated microcomputer systems to help with some aspects of the system; however, there are many more clerks than there are micros. Furthermore, the BSP study found that the limited computerization was merely automating what appeared to be outdated business processes. The following specific problems were identified or verified in this preliminary study:

1. A constantly changing product mix (from records to cassettes to compact discs and beyond) has led to jury-rigged systems and procedures.

2. The changing product mix has led to a strategic decision to factor memberships according to diversified media choices. The current system will not support such a change.

3. Directives to increase membership and orders through aggressive advertising will shortly overload the systems ability to process transactions on a timely basis. Customer shipments and payments could be delayed.

4. Response times to orders have already doubled during peak periods. This reduces member satisfaction and cash flow.

5. Management has suggested a "Preferred Member Program" that cannot be implemented with current data.

6. Unpaid orders have increased from 2%, only two years ago, to 4%. This has been attributed to a poor credit check interface.

7. Member defaults on contracts have increased 7% in recent years. It is believed that the current system does a poor job of reminding members of their obligations and notifying them of their status.

8. Members have started to complain about the automatic cancellation of memberships after contract fulfillment plus account inactivity. Members claim bonus credits are lost in the system. An independent consultant's audit verified this problem and traced it to a data integrity cause.

FIGURE E2.2

Assessment for New Member Services Information System Page 2

9. Backorders are not receiving proper priority. Some backorders go three months before they are filled (or canceled). New orders frequently deplete new inventory before the backorders are considered.

Scope of the Project

The strategic IS plan (and the involved users and management) are requesting a system that will:

1. Expedite the processing of subscriptions and orders through improved data capture technology, methods, and decision support.
2. Reduce unpaid orders to 2% by 1994.
3. Reduce defaulted contracts by 5% by 1995.
4. Support constantly changing club structures and member contracts.
5. At least triple the capacity of transaction processing by 1993.
6. Reduce response time by 50% by 1993.
7. Rethink underlying business processes that have a high impact on customer satisfaction and complaints.
8. Provide improved marketing analysis of promotional programs.
9. Provide improved follow-up mechanisms on orders and backorders.

The primary users of the new system would be the management and staff of the Member Services Division. The improved system will either affect or interface to the following:

1. Purchasing
2. Warehouse
3. Distribution Centers
4. Accounts Receivable
5. Marketing.

Constraints

The system must be operational in 9 months.

The system must be developed in accordance with the strategic plan's development infrastructure, methodology, standards, and approved CASE technology. Requests for variance must be approved through the Development Center.

The system cannot alter any existing file or database structures in the accounts receivable system.

The system must conform with the technology infrastructure of the business strategic plan. In particular, the system must accommodate the planned change to bar coding technology in the warehouse, optical mark sensing for orders, and telephone response technology as a terminal option.

The system has been categorized as "department" level and should therefore be implemented with client/server technology as defined in the strategic plan.

Recommendations

We highly recommend that this project be expedited for systems analysis. The basis of our recommendation is that this system is essential to the strategic mission of SoundStage. Unless the member services system is overhauled and dramatically improved, it is unlikely that the system will be

FIGURE E2.2 *Continued*

Assessment for New Member Services Information System Page 3

able to support the strategic business initiatives that SoundStage has already set in motion. Indeed, it is also likely that the current system will soon lead to a declining membership base and reduced orders.

Since the strategic plan has already performed a detailed study of this business area, we request a waiver on the detailed Study Report as called for in the FAST methodology. We suggest the project proceed immediately to requirements analysis using a Joint Application Development and Rapid Application Development approach. The requested variance is in accordance with FAST 2.2.3 guidelines for methodology variance.

Resource Requirements for Systems Analysis

The definition of user requirements will require approximately $14,950, budgeted as follows:

2.5 person months for JAD sessions and verification	$9,500
1.0 person months of user release time	3,550
0.5 person months of secretarial support	1,000
overhead	900

TOTAL	$14,950

The User Requirements Report will include estimates for design and implementation.

If you have questions, our phone extensions are 355 and 356. You can also contact us through electronic mail using either SSHEPHERD or RMARTINEZ as an address.

cc: Ken Riker, Assistant Director for Systems Development
Peg Li, Manager, Development Center

FIGURE E2.2 *Concluded*

3

A Systems Development Life Cycle

Chapter Preview and Objectives

This chapter introduces a systems development life cycle (SDLC) as the framework for information systems development. Systems development is not a hit-or-miss process! As with any product, information systems must be carefully developed. Successful systems development is governed by some fundamental, underlying principles that we will introduce in this chapter. We also introduce a systems development life cycle as a disciplined approach to developing information systems. Although such an approach will not guarantee success, it will improve the chances of success. Most experts agree that there is a life cycle, but beyond that, there's little agreement. There are as many versions of the SDLC as there are authors and companies. Although their terminology differs, they are more often alike than different. You will know that you understand the SDLC when you can:

Describe
eight basic principles of systems development.

Define
problems, opportunities, and *directives*—the triggers for systems development projects.

Describe
a framework that can be used to categorize problems, opportunities, and directives.

Describe
a phased approach to systems development. For each phase or activity, describe its purpose, trigger, people roles, inputs, deliverables, and decision-making opportunities.

Describe
the cross life cycle activities that overlap the entire life cycle.

Describe
variations on the life cycle for end-users who elect to develop their own personal information systems.

CENTURY TOOL AND DIE, INC.

Century Tool and Die, Inc., is a major manufacturer of industrial tools and machines. It is located in Newark, New Jersey.

Scene: *Conference room of Century Tool and Die, Inc., where Valerie and Larry have just sat down to discuss their current project — improving the recently implemented Accounts Receivable (A/R) information system. Larry is the assistant A/R manager. Valerie is a systems analyst for the Information Systems department. As they start to discuss the project, they are interrupted by Robert Washington, the executive vice president of Finance, and Gene Burnett, the A/R manager.*

Larry suddenly looks very nervous. And for good reason — he had suggested the new system, and it has not turned out as promised. Gene's support had been lukewarm, at best. Mr. Washington initiates the conversation.

Robert: We've got big problems. This new A/R system is a disaster. It has cost the company more than $625,000, not to mention lost customer goodwill and pending legal costs. I can't afford this when the board of directors is complaining about declining return on investment. I want some answers. What happened?

Gene: I was never really in favor of this project. Why did we need this new computer system?

Larry: Look, we were experiencing cash flow problems on our accounts. The existing system was too slow to identify delinquencies and incapable of efficiently following up on those accounts. I was told to solve the problem. A manual system would be inefficient and error-prone. Therefore, I suggested an improved computer-based system.

Gene: I'm not against the computer. I approved the original computer-based system. And I realized that a new system might be needed. It's just that you and Valerie decided to redesign the system without considering alternatives — just

like that! In my opinion, you should always analyze options. And let's suppose that a new computerized system was our best option. Why did we have to build the system from scratch? There are good A/R software packages available for purchase.

Robert: Gene has a point, Larry. Still, the system you proposed was defended as feasible. And yet it failed! Valerie, as lead analyst, you proposed the new system, correct?

Valerie: Yes, with Larry's help.

Robert: And you wrote this feasibility report early in the project. Let's see. You proposed replacing the current batch A/R system with an on-line system using a database management package.

Valerie: Strictly speaking, the database management package wasn't needed. We could have used the existing VSAM files.

Robert: The report says you needed it. I paid $15,000 to get you that package!

Valerie: Bill, our database administrator, made that recommendation. The A/R system was to be the pilot database project.

Robert: You also proposed using a network of microcomputers as a front end to the mainframe computer?

Valerie: Yes, Larry felt that a mainframe computer-based system would take too long to design and implement. With microcomputers, we could just start writing the necessary transaction programs and then transfer the data to the database on the mainframe computer.

Robert: It seems to me that some sort of design work should have been done no matter what size computer you used . . . *[brief pause]* In any case, the bottom line in this report is that your projected benefits outweighed the lifetime costs. You projected a 22 percent annual return on investment. Where did you get that number?

Valerie: I met with Larry four times—about six hours total, I'd say—and Larry explained the problems, described the requirements, made suggestions, and then projected the costs, benefits, and rate of return.

Robert: But that return hasn't been realized, has it? Why not? *[long pause]* Valerie, what happened after this proposal was approved?

Valerie: We spent the next nine months building the system.

Robert: How much did you have to do with that, Larry?

Larry: Not a lot, sir. Valerie occasionally popped into my office to clarify requirements. She showed me sample reports, files, and screens. Obviously, she was making progress, and I had no reason to believe that the project was off schedule.

Robert: Were you on schedule, Valerie?

Valerie: I don't believe so, Mr. Washington. My team and I were having some problems with certain business aspects of the system. Larry was unfamiliar with those aspects and he had to go to the account clerks and the accountants for answers.

Robert: I don't get it! Why didn't *you* go to the clerks and accountants?

Gene: I can answer that. I designated Larry as Valerie's contact. I didn't want her team wasting my people's time—they have jobs to do!

Robert: Something about that bothers me, Gene. In any case, when this project got seriously behind, Larry, why didn't you consider canceling it, or at least reassessing the feasibility?

Larry: We did. Gene expressed concern about progress about seven months into the project. We called a meeting with Valerie. At that meeting, we learned that the new database system wasn't working properly. We also found that we needed more memory and storage on the microcomputers. And to top it off, Valerie and her staff seemed to have little understanding of the business nature of our problems and needs.

Valerie: As I already pointed out, I wasn't permitted contact with the users during the first seven months. Besides, we were asked by Larry to start programming as quickly as possible so that we would be able to show evidence of progress.

Robert: I'm no computer professional; however, my engineering background suggests that some design or prototyping should have been done first.

Valerie: Yes, but that would have required end-user participation, which Larry and Gene would not permit.

Larry: As I was saying, we considered canceling the project. But I pointed out that $150,000 had already been spent. It would be stupid to cancel a project at that point. I did reassess the feasibility, and concluded that the project could be completed in four more months for another $50,000.

Robert: Nobody asked me if I wanted to spend that extra money.

Larry: We realize that the system hasn't worked out as well as we had hoped. We are trying to redesign . . .

Robert: As well as you hoped? That's an understatement! Let me read you some excerpts from Gene's last monthly report. Customer accounts have mysteriously disappeared, deleted without explanation. Later, we discovered that data-entry clerks didn't know that the F2 key deletes a record. Also, customers have been legally credited for payments that were never made! Customers have been double billed in some cases! Reports generated by the system are late, inaccurate, and inadequate. Cash flow has been decreased by 35 percent! My sales manager claims that some customers are taking their business elsewhere. And the Legal Department says we may be sued by two customers and that it will be impossible to collect on those accounts where customers received credit for nonpayments. You tell me, what would you do if you were in my shoes?

Discussion Questions

1. What would you do if you were in Mr. Washington's shoes? How would you react to Larry's performance? Valerie's? Gene's?

2. What did Valerie do wrong? Was she in control of her own destiny on this project? Why or why not?

3. What did Larry or Gene do wrong? Can either be held responsible for the failure of a computer project when they have limited computer literacy or experience?

4. What was wrong with the feasibility report? Did Valerie and Larry meet often enough? Was the input to that report

sufficient? Did the team commit to a solution too early? Did programming begin too soon? Why or why not?

5. Why were Valerie and her staff uncomfortable with the business problem and needs?

6. Should the project have been canceled? What about the $150,000 investment that had already been made?

7. If you were Valerie or Larry, what would you have done differently?

ESSENTIAL PRINCIPLES FOR SUCCESSFUL SYSTEMS DEVELOPMENT

This chapter introduces a process called a systems development life cycle.

> A **systems development life cycle (SDLC)** is a process by which systems analysts, software engineers, programmers, and end-users build information systems and computer applications. IBM and others refer to it as an *applications development cycle.*

Either way, it is a project management tool used to plan, execute, and control systems development projects.

Before we study the life cycle, let's introduce some general principles that should underlie all systems development.[1] Because the characters in the minicase violated most of those principles, we'll frequently refer back to that minicase as an example of how *not* to execute a systems development project.

Principle 1: Get the User Involved

Analysts and programmers frequently refer to "my system." This attitude has, in part, created an "us-versus-them" attitude between analysts/programmers and their users. Although programmers and analysts work hard to create technologically impressive solutions, those solutions often backfire because they don't address the real organization problems or they introduce new organization or technical problems. For this reason, user involvement is an absolute necessity for successful systems development. The individuals responsible for systems development must make time for users, insist on user participation, and seek agreement from users on all decisions that may affect them.

In the minicase, Valerie was not permitted to work directly with the users of the system because Gene was concerned about the amount of time that users would be taken away from their jobs. Larry didn't understand enough about their jobs to be helpful. Consequently, Valerie was not able to develop a system that truly met the needs of the users.

Misunderstandings continue to be a significant problem in systems development. However, user involvement and education minimizes such misunderstandings and helps to win user acceptance of new ideas and change. Because people tend to resist change, the computer is often viewed as a threat. Through education, information systems and computers can be properly viewed by users as tools that will make their jobs less mundane and more enjoyable.

---

[1] Adapted from R. I. Benjamin, *Control of the Information System Development Cycle* (New York: Wiley-Interscience, 1971).

Principle 2: Use a Problem-Solving Approach

The systems development life cycle is, first and foremost, a problem-solving approach to building systems. The term *problem* is used here to include real problems, opportunities for improvement, and directives from management. The classical problem-solving approach is as follows:

1. Identify the problem (or opportunity, or directive).
2. Understand the problem's environment and the problem's causes and effects.
3. Define the requirements of a suitable solution.
4. Identify alternative solutions.
5. Select the "best" solution.
6. Design and implement the solution.
7. Observe and evaluate the solution's impact. Refine the solution accordingly.

The notion here is that systems analysts should approach all projects using some sort of problem-solving approach.

In the minicase, Valerie did not appear to use a problem-solving approach. She did not identify the problem(s); understand the environment, causes, and effects; define requirements; or evaluate alternatives. She seemed to assume that only one alternative (computerization) was available and that systems development should start with the design of that computer-based solution.

Principle 3: Establish Phases and Activities

Most SDLCs consist of *phases.* In its simplest, classical form, the SDLC consists of four phases (Figure 3.1, left): systems analysis, systems design, systems implementation, and systems support. Modern variations have added another phase, systems planning (Figure 3.1, right).

Notice that your pyramid model has been layered to illustrate these phases. Each layer has greater volume just as each phase introduces greater detail. What does this added detail describe? The information systems building blocks, of course — PEOPLE, DATA, ACTIVITIES, NETWORKS, and TECHNOLOGY. Also notice that as you progress top-to-bottom through the phases, you are moving closer to the technology base of the pyramid — suggesting that your early concerns are with the business and your later concerns become more technical.

Note The color blue is introduced here. It will be used consistently throughout the text to identify systems development phases, activities, and tasks.

Because projects may be quite large and each phase usually represents considerable work and time, the phases are usually broken down into *tasks* that can be more easily managed and accomplished. This chapter will focus on high-level phases. The underlying tasks will be covered in later chapters.

In the chapter minicase, there was no indication that the project had been broken down into phases or activities. Consequently, there were no natural checkpoints in the process where the entire project was reevaluated.

Generally speaking, the phases of a project should be completed top-to-bottom, in sequence. This may leave you with the impression that once you finish a phase, you are done with that phase for good. That is, however, a false impression. At any given time, you may be performing tasks in more than one phase

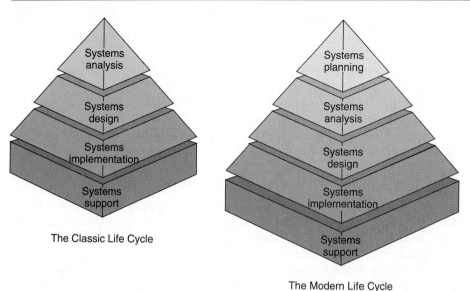

FIGURE 3.1 **The Classic and Modern Systems Development Life Cycle** Your pyramid model can be sliced to reveal the classic and modern phases of the systems development life cycle. The primary difference is that the modern life cycle is driven by systems planning. As you move top-to-bottom on the pyramid, your interests gradually shift from business-only issues to technical issues.

simultaneously. Furthermore, you may have to backtrack to previous phases and activities to make corrections or to respond to new requirements. Obviously, you can't get carried away with this type of backtracking or you might never implement the new system.

Principle 4: Establish Standards for Consistent Development and Documentation

An organization has many information systems that may include thousands of programs and software packages. If each analyst/programmers were to adopt their own preferred SDLC and use their own tools and techniques to develop and document systems, a state of chaos would quickly result. Why? In medium to large information systems shops, systems analysts and programmers (and users) come and go. They may be promoted or transferred; some may quit the organization. In order to promote good communication between this constantly changing base of users and information systems professionals you must develop standards to ensure consistent systems development.

Systems development standards usually describe (1) activities, (2) responsibilities, (3) documentation guidelines or requirements, and (4) quality checks. These four standards should be established for every phase in the life cycle.

The need for documentation standards underscores a common failure of many analysts—the failure to document as an ongoing activity during the life cycle. Most students and practitioners talk about the importance of documentation, but talk is cheap! When do you really place comments in your computer programs? Honestly? After you finish, of course! Therein lies the problem: Most of us tend to *post-document* software. Unfortunately, we often carry this bad habit over to systems development.

Documentation should be a working by-product of the entire systems development effort. Documentation reveals strengths and weaknesses of the system to others — *before* the system is built. It stimulates user involvement and reassures management about progress. Be wary of any SDLC that has a *documentation phase* — a planned post-documentation approach frequently causes communication breakdowns. This book teaches technique, but it also teaches documentation. Learn to use the tools and techniques to communicate with users *during* the life cycle, not after!

In the minicase, Valerie did not have any standards for development and no documentation was produced.

Principle 5: Justify Systems as Capital Investments

Information systems are capital investments, just as are a fleet of trucks or a new building. Even if management fails to recognize the system as an investment, you should not. When considering a capital investment, two issues must be addressed.

First, for any problem, there are likely to be several possible solutions. The analyst should not accept the first solution that comes to mind. The analyst who fails to look at several alternatives is an amateur. Second, after identifying alternative solutions, the systems analyst should evaluate each possible solution for feasibility, especially for cost-effectiveness.

> **Cost-effectiveness** is defined as the result obtained by striking a balance between the cost of developing and operating a system, and the benefits derived from that system.

Cost-benefit analysis is an important skill to be mastered.

If you look back at the project at Century Tool and Die, Inc., you will see that alternatives were not evaluated. In fact, it seems that the technology was selected early and that the choice was not motivated by the problem. With respect to cost-effectiveness, the costs were grossly underestimated. Also, developers of the system decided to continue the project on the basis of the money that they had already spent, not on whether or not the project was still feasible. This dilemma leads us to our next principle.

Principle 6: Don't Be Afraid to Cancel or Revise Scope

A significant advantage of the phased approach to systems development is that it provides several opportunities to reevaluate feasibility. There is often a temptation to continue with a project only because of the investment already made. Century Tool and Die, Inc., made this mistake and threw more good money into a system that still failed. In the long run, canceled projects are less costly than implemented disasters! This is extremely important for young analysts to remember.

Similarly, many analysts allow project scope to increase during a project. Sometimes this is inevitable because the analyst learns more about the system as the project progresses. Unfortunately, most analysts fail to adjust estimated costs and schedules as scope increases. As a result, the analyst frequently and needlessly accepts responsibility for cost and schedule overruns.

The authors of this text advocate a creeping commitment approach to systems development.[2]

> Using the **creeping commitment approach,** multiple feasibility checkpoints are built into the systems development life cycle. At any feasibility checkpoint, all costs are considered *sunk* (meaning *irrecoverable*). They are, therefore, irrelevant to the decision. Thus, the project should be reevaluated at each checkpoint to determine if it is *still* feasible.

At each checkpoint, the analyst should consider (1) cancellation of the project if it is no longer feasible, (2) reevaluation of costs and schedule if project scope is to be increased, or (3) reduction of scope if the project budget and schedule are frozen, but not sufficient to cover all project objectives.

The concept of sunk costs is more or less familiar to some financial analysts and managers, but it is frequently forgotten or not used by the majority of practicing analysts and users.

Principle 7: Divide and Conquer

All systems are part of larger systems (called *supersystems*). Similarly, virtually all systems contain smaller systems (called *subsystems*). These facts are significant for two reasons. First, systems analysts must be mindful that any system they are working on interacts with its supersystem. If the supersystem is constantly changing, so might the scope of any given project change as the analyst learns more about the supersystem. Most systems analysts can still be faulted for underestimating the size of projects. Most of the fault lies with not properly studying the implications of a given system relative to its larger whole — its supersystem.

If the supersystem were better understood, the analyst would realize that the project is larger and be able to make better estimates of the costs and time required to build the new system. The addition of a planning phase to the modern SDLC addresses this concern.

To understand the second division, consider the old saying that "if you want to learn anything, you must not try to learn everything — at least not all at once." For this reason, we divide a system into its subsystems in order to more easily conquer the problem and build the larger system. By dividing a larger problem (system) into more easily managed pieces (subsystems), the analyst can simplify the problem-solving process. We'll be applying this principle throughout this book.

Principle 8: Design Systems for Growth and Change

There is a critical shortage of information systems professionals needed to develop systems. Combined with the ever increasing demand for systems development, many systems analysts have fallen into the trap of developing systems to meet only *today's* user requirements. Although this may seem to be a necessary approach at first glance, it actually backfires in almost all cases.

---

[2] Thomas Gildersleeve, *Successful Data Processing Systems Analysis,* 2nd ed. (Englewood Cliffs, N.J.: Prentice Hall, 1985), pp. 5–7.

FIGURE 3.2 **Systems Entropy Occurs Sometime During the Systems Support Phase** This flow diagram of the life cycle demonstrates what happens when a system is placed into operation—it enters the support phase of the life cycle. From that point on, it is subject to modification and enhancement, until such a time as the system entropy sets in and the system has become obsolete or too costly to maintain.

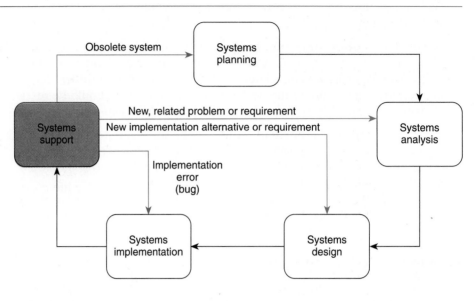

Entropy is the term systems experts use to describe the natural and inevitable decay of all systems. Entropy is illustrated in Figure 3.2. Notice that, after a system is implemented, it enters the support phase of the life cycle. During the support phase, the analyst encounters the need for changes that range from correcting simple mistakes, to redesigning the system to accommodate changing technology, to making modifications to support changing user requirements. As indicated by the blue arrows, many of these changes direct the analyst and programmer to rework former phases of the life cycle. Eventually, the cost of maintenance exceeds the costs of starting over—the system has become obsolete. This is indicated by the red arrow in the figure.

Unfortunately, systems that are designed to meet only current requirements are usually difficult to modify in response to new requirements. The systems analyst is frequently forced to duplicate files and "patch" programs in ways that make the system very costly to support over the long run. As a result, many systems analysts become frustrated with how much time must be dedicated to supporting existing, patchworked systems and how little time is left to work on important, new systems development.

Even if you design the system to easily adapt to change (our last principle), at some point in time, it will become too costly to simply support the existing system. Why? Perhaps the organization itself has changed too dramatically to be supported by the system. Or perhaps the requirements have become too complex to be patched into the existing system. In either case it is time to start over! This situation puts the term *cycle* into the term *systems development life cycle.* No system lasts forever (although many do last for a decade or longer).

It doesn't have to be that way! Today's tools and techniques make it possible to design systems that can grow and change as requirements grow and change. This book will teach you many of those tools and techniques. For now, it's more important to simply recognize that flexibility and adaptability do not happen by accident—they must be *built* into a system.

- Get the user involved.
- Use a problem-solving approach.
- Establish phases and activities.
- Establish standards for consistent development and documentation.
- Justify systems as capital investments.
- Don't be afraid to cancel or revise scope.
- Divide and conquer.
- Design systems for growth and change.

FIGURE 3.3 **Principles of Systems Development** These eight principles should underlie any version of a systems development life cycle.

We have presented eight principles that should underlie any SDLC. These principles, summarized in Figure 3.3, can be used to evaluate any life cycle, including ours.

Reaction versus Planning

Traditionally, users and management have initiated most projects, because they are closer to the organization's activities that need improvement. Analysts, on the other hand, are frequently expected to survey the organization for possible improvements. When users or analysts initiate projects, they are said to be *reacting* to situations.

Regardless, the impetus for most projects is some combination of problems, opportunities, or directives.

Problems are undesirable situations that prevent the organization from fully achieving its purpose, goals, and objectives.

For example, an unacceptable increase in the time required to fill an order can trigger a project to reduce that delay. Problems may either be current, suspected, or anticipated.

An **opportunity** is a chance to improve the organization even in the absence of specific problems. (Note: You could argue that any unexploited opportunity is, in fact, a problem.)

For instance, management is always receptive to cost-cutting ideas, even when costs are not currently considered a problem. Opportunistic improvement is expected to be the source of today's most important systems development projects.

A **directive** is a new requirement that's imposed by management, government, or some external influence. (Note: You could argue that until a directive is fully complied with, it is, in fact, a problem.)

For example, the Equal Employment Opportunity Commission, a government agency, may mandate that a new set of reports be produced each quarter. Similarly, company management may dictate support for a new product line or policy. Some directives may be technical. For instance, systems may be strategically directed to convert from batch to on-line processing or from conventional files to database. Such measures are appropriate if the current technology is obsolete, difficult to maintain, slow, or cumbersome to use.

Finally, and increasingly more common, many projects are triggered by an intensive information systems planning directive. Systems planning is a special directive that applies formal long-range planning methods to chart an information systems plan that mirrors and supports corporate business plans. Such a directive requires joint participation of business executives, information systems management, and experienced systems analysts (often called *planning analysts*). The output of this planning activity is a schedule of projects (in the form of directives) to develop specific applications, databases, and networks.

The PIECES Framework

There are far too many potential problems, opportunities, and directives to list them all in this book. However, James Wetherbe has developed a useful framework for classifying problems, opportunities, and directives.[3] He calls it **PIECES** because the letters of each of the six categories, when put together, spell the word *pieces.* These categories are:

P   The need to improve *performance.*

I   The need to improve *information* (or data).

E   The need to improve *economics* or control costs.

C   The need to improve *control* and security.

E   The need to improve *efficiency* of people and machines.

S   The need to improve *service* to customers, partners, employees, and so on.

Figure 3.4 expands on each of these categories.

The categories of the PIECES framework are related. Any given project can be characterized by one or more categories. Furthermore, any given problem, opportunity, or directive may have implications in more than one PIECES category. PIECES is a practical framework—not just an academic exercise.

The PIECES framework is significant because it teaches you to always examine project triggers in terms of their bottom-line impact on the organization. When you begin a systems project, consider using Figure 3.4 to list the problems, or better still, identify problems that the user has yet to see.

THE LIFE CYCLE—A PROBLEM-SOLVING APPROACH

In this section we'll examine a modern systems development life cycle. We'll begin with a high-level picture that omits many details. Then we'll progress to more complete figures of various phases.

An Overview of the Life Cycle

Figure 3.5 illustrates a typical, but modern systems development life cycle. The symbology, used throughout this chapter, is defined in the legend and described as follows:

• The rounded rectangles represent key activities or phases in systems development. While many people play roles in each phase, each phase has a

---

[3] James Wetherbe, *Systems Analysis and Design: Traditional, Structured, and Advanced Concepts and Techniques,* 3rd ed. (St. Paul, Minn.: West, 1988), p. 114.

The PIECES Problem-Solving Framework and Checklist

The following checklist for problem, opportunity, and directive identification uses Wetherbe's PIECES framework. Note that the categories of PIECES are not mutually exclusive; some possible problems show up in multiple lists. Also, the list of possible problems is not exhaustive. The PIECES framework is equally suited to analyzing both manual and computerized systems and applications.

PERFORMANCE Problems, Opportunities, and Directives

A. Throughput—the amount of work performed over some period of time.
B. Response time—the average delay between a transaction or request and a response to that transaction or request.

INFORMATION (and Data) Problems, Opportunities, and Directives

A. Outputs
 1. Lack of any information
 2. Lack of necessary information
 3. Lack of relevant information
 4. Too much information—"information overload"
 5. Information that is not in a useful format
 6. Information that is not accurate
 7. Information that is difficult to produce
 8. Information is not timely to its subsequent use
B. Inputs
 1. Data is not captured
 2. Data is not captured in time to be useful
 3. Data is not accurately captured—contains errors
 4. Data is difficult to capture
 5. Data is captured redundantly—same data captured more than once
 6. Too much data is captured
 7. Illegal data is captured
C. Stored data
 1. Data is stored redundantly in multiple files and/or databases
 2. Stored data is not accurate (may be related to #1)
 3. Data is not secure to accident or vandalism
 4. Data is not well organized
 5. Data is not flexible—not easy to meet new information needs from stored data
 6. Data is not accessible

ECONOMICS Problems, Opportunities, and Directives

A. Costs are unknown
B. Costs are untraceable to source
C. Costs are too high

CONTROL (and Security) Problems, Opportunities, and Directives

A. Too little security or control
 1. Input data is not adequately edited
 2. Crimes are (or can be) committed against data
   a. Fraud
   b. Embezzlement
 3. Ethics are breached on data or information—refers to data or information getting to unauthorized people
 4. Redundantly stored data is inconsistent in different files or databases
 5. Data privacy regulations or guidelines are being (or can be) violated
 6. Processing errors are occurring (either by people, machines, or software)
 7. Decision-making errors are occurring
B. Too much control or security
 1. Bureaucratic red tape slows the system
 2. Controls inconvenience customers or employees
 3. Excessive controls cause processing delays

EFFICIENCY Problems, Opportunities, and Directives

A. People, machines, or computers waste time
 1. Data is redundantly input or copied
 2. Data is redundantly processed
 3. Information is redundantly generated
B. People, machines, or computers waste materials and supplies
C. Effort required for tasks is excessive
D. Materials required for tasks is excessive

SERVICE Problems, Opportunities, and Directives

A. The system produces inaccurate results
B. The system produces inconsistent results
C. The system produces unreliable results
D. The system is not easy to learn
E. The system is not easy to use
F. The system is awkward to use
G. The system is inflexible to new or exceptional situations
H. The system is inflexible to change
I. The system is incompatible with other systems
J. The system is not coordinated with other systems

FIGURE 3.4 **The PIECES Framework** The PIECES framework can be used to group problems, opportunities, and directives.

key facilitator. The key facilitator's typical job title is indicated in parentheses.

- The thick blue arrows represent major information flows that *trigger* projects or major phases. References to these triggers in the text are underlined.

- The thick black arrows represent the **major deliverables** (or *outputs*) of the systems development phases. Each deliverable contains important documentation and/or specifications. Notice that the deliverable of one

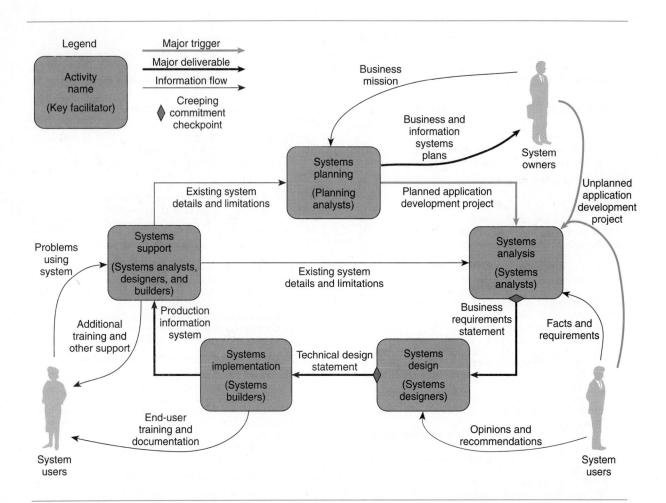

FIGURE 3.5   A Systems Development Life Cycle   This flow diagram of the life cycle demonstrates the "big picture" view of systems development. In subsequent figures and explanations, will expand on this diagram by looking "inside" each of the major activities to identify specific phases.

phase may serve as input to another phase. References to these deliverables in the text are underlined.

- The thin black arrows represent other input and/or output information flows. These flows can take the form of conversations, meetings, letters, memos, reports, and the like. In the interest of readability, we have omitted several information flows on this diagram. Many more will be included in subsequent diagrams in the chapter and textbook. References to these inputs and outputs in the text are underlined.

- The people silhouettes indicate people or organizations with whom the analyst may interact. Notice that the roles introduced in Chapter 2 have been carried over to this figure. For instance, we see system owners and system users participating in the life cycle. Once again, we did not show all people interactions on this overview diagram of the life cycle.

- Finally, consistent with our creeping commitment principle, the red diamonds indicate checkpoints at which the project participants should re-evaluate feasibility and/or project **scope.**

As you can see, the modern life cycle consists of five high-level functions. We'll get into details later, but each of these functions deserves a brief explanation:

1. *Systems planning.* The scope of systems planning is the entire business, a division, or some other significant organization unit. The purpose is to identify and prioritize those information systems applications whose development would most benefit the business as a whole. This phase is indicative of a relatively mature information systems operation.

    The inputs are the business mission and any existing system details and limitations. The key outputs or deliverables are business and information systems plans and planned application development projects.

    Planned projects subsequently "cycle" through the remaining phases.

2. *Systems analysis.* The scope of systems analysis is a single information system application. The purpose is to analyze the business problem or situation, and then to define the business requirements for a new or improved information system. The business requirements do not specify a computer-based solution.

    The input trigger is either the planned application development project (from the systems planning phase) or an unplanned application development project (in response to an unanticipated problem, opportunity, or directive). Other inputs include existing system details and limitations and business-related facts and requirements. The key deliverable is a business requirements statement that explains "what" the users need, but not "how" we plan to design or implement those requirements.

3. *Systems design.* The scope of systems design remains the single information system application from systems analysis. The purpose is to design a computer-based, technical solution that meets the business requirements as specified in systems analysis.

    The input trigger is the business requirements statement. Other inputs include design-related opinions and recommendations from system users. The key deliverable of systems design is a technical design statement. This deliverable states (or demonstrates) "how" the information system will technically fulfill the user's business requirements.

4. *Systems implementation.* The scope of systems implementation is defined by the technology-related components of the information systems application that was designed in the previous phase. The purpose is to construct and/or assemble the technical components and deliver the new or improved information system into operation.

    The input trigger is the technical design statement from systems design. The key deliverable is a production information system. The term *production* is used to describe a system that has been delivered into day-to-day operation. Another output is the end-user training and documentation necessary to use the production system.

5. *Systems support.* The scope of systems support is the production information system delivered from systems implementation. The purpose of sys-

tems support is to sustain and maintain the system for the remainder of its useful life.

The input to this phase is the <u>production information system</u>. Various support activities are then triggered by <u>problems using the system</u>.

At some point in time, the production system will become too expensive to maintain, or it will cease to provide adequate business support. At that time, the life cycle will start over, returning to systems planning or systems analysis.

That completes our brief overview of the life cycle. In the next five sections, we more closely examine each of these high-level functions. You'll learn that each function actually consists of phases and activities. As we examine those phases and activities, we'll also describe participants, additional inputs and outputs, and the feasibility/scope checkpoints.

Systems Planning

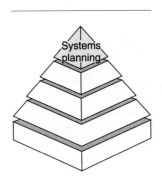

As suggested earlier, not all organizations include systems planning in their SDLC. Even today, many or most development projects are initiated only in response to management and user requests. Those managers and users who "scream the loudest" (or have the most money to spend) dictate which projects are begun, regardless of the overall impact on the business. But systems planning is becoming increasingly common as businesses learn that information systems should not randomly evolve—they should be planned.

> The **systems planning** function of the life cycle seeks to identify and prioritize those technologies and applications that will return the most value to the business. Synonyms include *strategic systems planning* and *information resource management.*

Systems planning is driven by the cooperation of systems owners. Hence, in the pyramid model, it addresses PEOPLE, DATA, ACTIVITIES, and NETWORKS from the system owners' perspectives (which were introduced in Chapter 2). Information systems participants do, however, usually exploit the opportunity to educate system owners about TECHNOLOGY directions and potential (the base of the pyramid).

Before we begin, we should point out that systems planning is an ongoing process. It must be repeated regularly in order to ensure (1) that information systems are developing according to the plan, and (2) that management decisions or external factors have not changed the plan.

Figure 3.6 shows the phases of systems planning. This is the first of several diagrams that expand on Figure 3.5, the overview diagram for the SDLC. Notice that the interconnections (depicted by arrows) between systems planning, systems support, and systems analysis were carried down from Figure 3.5 to this new diagram.

Also notice that there are no information-oriented triggers (blue arrows) on the diagram. What, then, triggers systems planning phases? Generally, systems planning is triggered by an agreement between information systems management and top business executives that information systems and the use of technology should be carefully planned.

There are many different systems planning methodologies; however, this relatively simple approach is representative. Study Figure 3.6 as we discuss the three phases in greater detail.

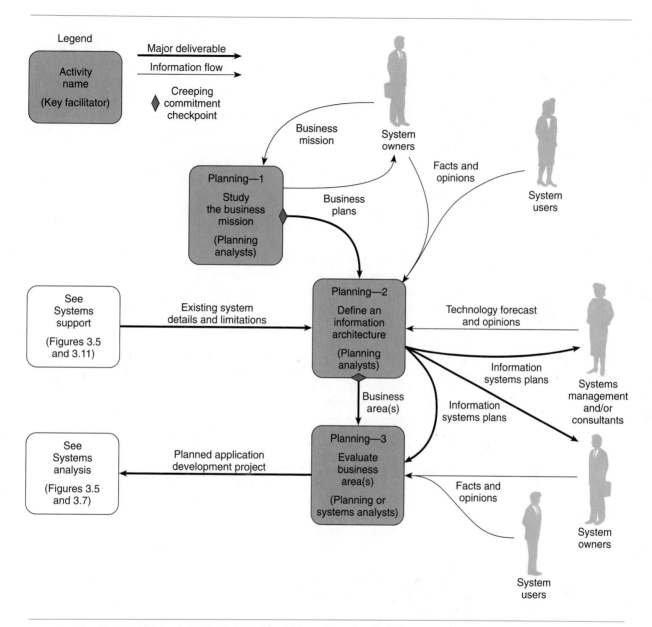

FIGURE 3.6 **Systems Planning Phases in the Life Cycle** This flow diagram expands on the systems planning phases of systems development (which were introduced in Figure 3.5). Systems planning focuses almost exclusively on business and management concerns, as opposed to technical concerns.

Planning Phase 1: Study the Business Mission

Although many businesses haven't formally documented their mission, they all have one. If information systems are to truly return value to the business, they need to directly address that mission. Thus, the first phase of systems planning is to study the business mission.

Ideally, the scope of the phase should be the entire business. For some companies, that is much too large. Consequently, the scope might be reduced to a more manageable level—a division, a plant, or some other significant operating unit. For other companies, the scope of the phase is limited by the level of top management support received. Top executives of the organization must be willing to participate in the development of any strategic plan. Otherwise, the phase is useless.

Members of the planning team include information systems managers and the key business executives (system owners in the figure). Notice that the key facilitator of this phase (in parentheses) is a **planning analyst.**

Planning analysts are specially trained information systems planning professionals. Their job is similar to that of systems analysts; however, they must be even more *business*-oriented than the average systems analyst. Planning analysts must be familiar with the planning methodology to be used and the deliverables to be produced. They require a unique blend of skills and experiences, including business management, systems analysis and design, data management, and networking.

Many IS shops have difficulty finding the correct mix of these skills. Particularly, IS professionals tend to be either too applications-oriented, too database-oriented, or too network-oriented. In this case, the business usually hires management consultants to serve as the planning analysts. These consultants are widely available through IS consulting firms (e.g., Ernst & Young, James Martin & Associates, or IBM)

The input to this phase is the business mission, as "discovered" through interviews and group sessions with system owners. The business mission is usually defined in terms of customers, products and services, material resources, human resources, geographic operating locations, management structures and philosophy, corporate goals and objectives, unavoidable business constraints, critical business success factors, and other management-oriented criteria.

The key deliverable is business plans. Hopefully, those plans already exist; this phase merely translates them into terms or formats that are useful to the system owners and planning analysts in subsequent planning phases. (All too often, that plan does not exist!)

The red diamond on this phase suggests a feasibility/scope checkpoint. Based on the findings of this phase, the planning effort could be canceled due to a lack of management commitment or funding. It is more likely, however, that the project will continue to the next phase—possibly with a reduced business scope.

Planning Phase 2: Define an Information Architecture

Given an understanding of the business mission, you can now develop a plan for information systems that truly mirrors and supports that business mission. The next phase of systems planning is to define an information architecture for information systems.

> An **information architecture** is a plan for selecting information technology and developing information systems needed to support the business mission. Synonyms include *information systems plan* and *master computing plan.*

This architecture phase can take six months or longer to complete.

Once again, this phase is facilitated by planning analysts. The team also includes the same IS managers and business executives included in the previous planning phase. Additionally, the team should normally include at least one database, networks, and applications management representative—the reason will become apparent in a moment.

The key input to this phase is the business plans from the last phase. Other inputs include existing system details and limitations (from the documentation maintained during the Systems Support phase), facts and opinions (from appropriate system owners and system users), and technology forecasts and opinions (from information systems management and/or consultants).

These inputs are used to build the information architecture. Your information systems pyramid model defines the key components in an information architecture:

- A DATA architecture that identifies and prioritizes the databases that need to be developed. These databases should be highly flexible so that they can support several areas of the business. (Note: A data or database manager usually helps define the data architecture.)

- A NETWORK architecture that identifies and prioritizes computer networks that need to be developed. These networks must optimize information systems support at all appropriate business operating locations. (Note: A networks manager usually helps define the network architecture.)

- An ACTIVITIES architecture (more appropriately called an *applications architecture*) that identifies and prioritizes business areas for which business processes and/or information systems applications must be redesigned. (Note: One or more applications development managers usually help define the applications architecture.)

- A PEOPLE architecture (actually, an *IS organization structure*) necessary to develop and support the databases, networks, and applications.

- A TECHNOLOGY architecture that identifies the information technologies that should be used for applications, and possibly for applications development. (Note: Information systems management, along with the applications, database, and network managers, usually provides the technology forecast and opinions needed to develop the technology architecture.)

The information architecture is packaged in the deliverable, information systems plans. These plans will ultimately influence the development and support for all future information systems. Thus, they must be made available to information systems management, any contracted consultants, and all current and future information systems owners.

Some planning methodologies call for another deliverable, distinct business areas (to be passed on to the next planning phase).

> **Business areas** are groups of logically related business functions and activities, independent of organization structure. The term was invented as part a planning methodology called *information engineering*.

This definition requires some clarification. Business areas define major functions of the business, for example, PROCESS AND FILL ORDERS. Several organization units may play a role in any given business area. For example, the processing and filling of orders normally requires activity in at least the following organization units: Sales, Order Processing, Accounts Receivable, Warehousing, and Shipping. Ideally, information systems should be built around the integration of these units' activities as they relate to the common business function—in this case, processing and filling orders.

Business areas are usually prioritized according to their perceived importance and value to the business. The next planning phase deals with one business area at a time (in order of priority).

Once again, the red diamond for this phase suggests a feasibility/scope checkpoint. The planning project could be terminated due to a lack of either funds or management commitment. Even if this happens, you still have the information architecture to guide future systems development. Alternatively, the project can continue to the next phase, business area analysis.

Planning Phase 3: Evaluate Business Area

The information architecture is a good high-level IS plan. But some IS shops seek to further refine that plan to define specific systems development projects. Thus, the next phase of systems planning is to evaluate business area(s) to identify and prioritize specific development projects. A project may trigger development of any of the following:

1. A network—subsequent projects would build databases and/or applications around that network.
2. A database—subsequent projects would build applications around that database.
3. An information systems application—which may include building an applications-oriented database or network if a network or database is not completed first.

This textbook focuses primarily on projects to develop information systems applications (3 above).

Business area analysis (BAA) can be a time-consuming phase that takes six months or longer (per business area) to complete. But many companies are willing to spend that time to ultimately develop highly integrated information systems around their business areas. Because BAA takes so long, most businesses analyze only one or two areas at a time, preferably those identified as most crucial in the strategic information architecture.

Note   One challenge facing those organizations that analyze business areas is simply keeping up with application demand. For any given business area, the planing analysts may identify several applications development projects. While systems analysts and applications programmers tackle those projects, the planning analysts generally move on to another busines area, defining still more projects, and so forth. The growth in planned projects can easily exceed resources for those developing the applications. Fortunately, modern systems development technology (Chapter 5) offers some hope for keeping up with this demand.

Once again, this phase is facilitated by the same planning analysts who facilitated development of the information architecture. Systems analysts who will ultimately develop the applications are also frequently added to the team. Although most executive managers are excused from the phase, all managers in the specific business area must be involved in order to identify the applications needed and prioritize the projects to develop those applications. Some system users may also become involved.

The key inputs are the <u>information systems plans</u> and <u>business area(s)</u> from the previous phase. Additionally, the analyst collects <u>facts and opinions</u> from appropriate system owners and system users.

The key deliverable of the phase is <u>planned applications development proj-ects</u> that will eventually be passed on to systems analysis. This deliverable will normally include documentation that can serve as a useful first draft for many systems analysis deliverables. The plan often calls for rather dramatic changes to how the business area will conduct business (fewer people, less bureaucracy, etc.).

There is no feasibility/scope checkpoint at the end of the phase. Why? The planning process has defined and prioritized projects. Only one question re-mains: When do we (can we) commit resources to the development of those planned applications?

This completes our survey of the systems planning phases. As a closing note, detailed coverage of systems planning is not generally included in the first systems analysis and design course. At some point it will probably become mandatory. For the time being, we direct you to Chapter 6 for more about systems planning tools, techniques, and methodologies.

Systems Analysis

Systems analysis is the classic first step toward building an information systems application. The scope of systems analysis is the single application.

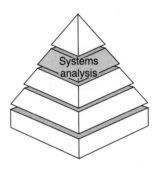

> **Systems analysis** is the study of a current business and information sys-tem, and the definition of user requirements and priorities for a new in-formation system. Synonyms include *business problem analysis, require-ments analysis,* and *logical design.*

Systems analysis is driven by the cooperation of system users. Hence, in the pyramid model, it addresses PEOPLE, DATA, ACTIVITIES, and NETWORKS from the system users' perspectives (which were introduced in Chapter 2). Systems analysis builds on any knowledge derived from systems planning.

Figure 3.7 shows the phases of a typical systems analysis. Notice the blue arrows—systems analysis may be triggered either by the <u>planned ap-plication development project</u> (from systems planning) or an <u>unplanned ap-plication development project</u> (in response to an unanticipated problem, opportunity, or directive). It is important to recognize that regardless of how well you plan information systems, there will almost always be unplanned projects.

Let's examine the analysis phases in somewhat greater detail. Study Figure 3.7 as we discuss these three phases.

Analysis Phase 1: Survey Project Feasibility

Systems development can be very expensive. Thus, it pays to answer the impor-tant question, "Is this project worth looking at?" The first phase of systems analysis, survey project feasibility, answers this question. The **survey phase** is also sometimes called a *preliminary investigation* or *feasibility study.* It

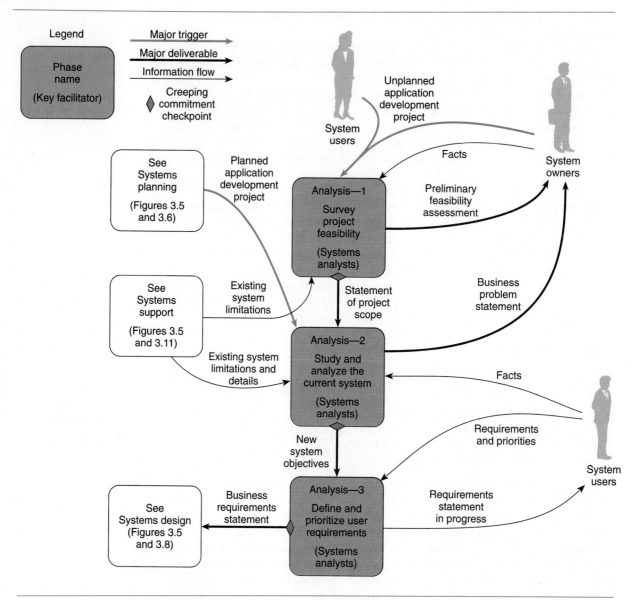

FIGURE 3.7 Systems Analysis Phases in the Life Cycle   This flow diagram expands on the systems analysis phases of systems development (which were introduced in Figure 3.5). Like systems planning, systems analysis focuses mostly on business concerns, albeit at the system user's level of detail, not the system owner's level of detail.

amounts to a "quick-and-dirty" (usually two to three days) preliminary investigation of the plans, problems, opportunities, and/or directives that triggered the project.

One very important task in this phase is to define the scope or size of the project. Projects that fail do so most often because of poor scope management. If uncontrolled, scope tends to grow (along with costs and missed deadlines). You can't manage scope if you don't define it. Think of scope as a statement of the users' expectations of the project. Once defined, you can renegotiate for additional resources if users increase that scope.

Project scope includes the identification of users, managers, and sponsors (at all levels of responsibility); identification of perceived problems, opportunities, and directives; identification of any business and technical constraints; and possible or perceived solutions and expectations. Given this information, the analyst will then assess the *initial*—we stress the word *initial*—feasibility of the project scope.

Note   Do you always assess project worthiness? NO. For instance, in the case of "planned" projects, someone has already determined that the project is worth doing. All that remains is to officially "scope" the project. For another example, if a vice president insists you look at an application, it might not be wise to question worthiness at this point in the project. Wait until after the detailed study phase to "suggest" unworthiness. And for yet another example, if the problem, opportunity, or directive to be studied is very apparent or beyond question, worthiness is not a valid question—you should define scope and get on with the project.

The systems analyst plays the key role in this phase. Indeed, the entire phase usually involves one systems analyst and one or few system owners. System users are rarely involved until the next phase.

The key input to the phase is the unplanned application development project. Some businesses have standard information forms for this input. Additional inputs to this phase are facts collected from system owners, and any existing system limitations from the systems support phase documentation.

The key deliverable phase is a preliminary feasibility assessment. This might be a report or a verbal presentation. The report version is sometimes called an *initial study report*. The analyst's recommendation may prescribe (1) a "quick fix," (2) an enhancement of the existing system and software, or (3) a completely new information system. For the latter possibility, a statement of project scope should be prepared as a deliverable to the next phase.

The findings must be usually reviewed by system owners (or a **steering committee** that includes system owners). This is especially common when many proposed projects are competing for the same resources. The red feasibility/scope diamond results in one of four possible decisions: (1) approve the project to continue to the next phase, (2) change the scope and continue on to the next phase, (3) reject the project outright, or (4) delay the project in favor of some other project.

In the latter case, the project is added to a file of projects commonly called the **backlog.** Unfortunately, many businesses have a backlog of projects that would require years to complete—even if all other development work and systems support were halted! Backlogged projects are frequently ideal candidates for end-user application development. End-user applications development variations are described at the end of the chapter.

Analysis Phase 2: Study and Analyze the Current System

There's an old saying that suggests, "Don't try to fix it unless you understand it." With those words of wisdom, the next phase of a systems analysis is to study and analyze the current system. There is always a current business system, regardless of whether it currently uses a computer. The **study phase** provides the systems analyst with a more thorough understanding of the problems, opportunities, and/or directives that triggered the project. Indeed, the analyst frequently uncovers new problems and opportunities. The study phase may answer the questions, "Are the problems worth solving?" and "Is a new system worth building?"

Do the problems and opportunities really exist? If so, how serious are they? Many times, the initial problems were mere symptoms, frequently of more serious or subtle problems. During the study, you need to address the causes and effects of problems, opportunities, and directives. The PIECES framework presented earlier in this chapter can be used as an effective tool for evaluating various aspects of the current system.

Note   Can you ever skip the study phase? RARELY! You almost always need some understanding of the existing system. But there may be reasons to complete the phase as quickly as possible. First, if the project was triggered by systems planning (business area analysis), the worthiness of both the project and system was determined in that phase; the study phase requirements are reduced to "understanding the current system." As another example, if the project was triggered by a management directive (e.g., "We must have this system by February 1 in order to comply with new federal regulations."), then worthiness is definitely not in question. In fact, we may want to get through the study phase rather quickly to increase the likelihood of meeting the deadline.

Once again, the systems analyst plays the key role in this phase. But system users are actively involved in the study. System owners must visibly support the study to ensure that all system users actively participate. We'll study specific approaches to the study phase in Chapter 6.

As shown in the figure, the key input is either the statement of project scope from the survey phase or a planned application development project from systems planning. The analysts study the system by collecting facts from system users. Existing system limitations and details (from systems support) can also provide valuable information for the study phase. From this information, the analysts seek to understand the system's problems and limitations.

The findings of the study phase are documented as a deliverable called a business problem statement (or *detailed study report*) passed along to system owners. This problem statement may take the form of a formal business report, an updated feasibility assessment, or a formal presentation to management.

The system owners will either agree or disagree with the findings of the study. The red feasibility checkpoint (diamond) for this phase suggests that the project can be (1) canceled if the problems prove not worth solving, or a new system is not worth building, (2) approved to continue to the next systems analysis phase, or (3) reduced in scope or increased in budget and schedule, and then approved to continue to the next systems analysis phase.

If the project continues to the next phase, the systems analysts should define new system objectives to be passed on to the next phase. These objectives do not define inputs, outputs, or processes. Instead, they define the business criteria on which any new system will be evaluated. For instance, we might define an objective that states that the new system must "reduce the time between order processing and shipping by three days," or the system must "reduce bad credit losses by 45 percent." Think of objectives as the "grading curve" for evaluating any new system that you might eventually design and implement.

Analysis Phase 3: Define and Prioritize Users' Requirements

Given approval of the business problem statement, now you can design a new system, right? No, not yet! What capabilities should the new system provide for its users? What data must be captured and stored? What performance level is expected? Careful! This requires decisions about *what* the system must do, not *how* it should do those things.

The next phase of systems analysis is to define and prioritize users' requirements. It is sometimes called a *requirements analysis phase* or **definition phase.** Simply stated, the analyst approaches the users to find out what they need or want out of the new system. This is perhaps the most important phase of the life cycle. Errors and omissions in the definition phase result in user dissatisfaction with the final system, and costly modifications to that system.

Essentially, the purpose of requirements analysis is to identify the data, process, and network requirements for the users of a new system. (Did you notice the correspondence to the faces of your pyramid model? You already know the PEOPLE who will use the system.) Most importantly, the purpose is to specify these requirements without expressing computer alternatives and TECHNOLOGY details; at this point, keep analysis at the business level!

Note   Can you ever skip the definition phase? NO! One of the most common complaints about new systems and applications is that they don't really satisfy the users' needs. This is usually because the analysts found it difficult to separate "what" the user needed from "how" the new system would work. In other words, the analysts became so preoccupied with the technical solution, that they failed to consider the users' essential requirements. Requirements are a statement of "what" the system must do "no matter how" you design and implement it. The definition phase formally separates "what" from "how" to properly define and prioritize those requirements.

Once again, the systems analyst facilitates the definition and prioritization of users' requirements. However, system users play the absolutely essential role in the phase.

As shown in the figure, the analysts collect requirements and priorities from system users. This information is collected by way of interviews, questionnaires, and group meetings. The challenge to the analysts is to validate those requirements.

Clearly, the new system objectives from the study phase provides some validation. But detailed validation requires the analyst to translate the user's words into a more precise representation of the requirements. The system users can then validate this requirements statement in-progress and offer corrected requirements and priorities. The most popular approach to translating and validating users' requirements is modeling.

Modeling is the act of drawing one or more graphical representations of a system. The resulting picture represents the user's data, processing, or network requirements from a business point-of-view.

This is the "picture is worth a thousand words" approach. Different diagrams will describe data requirements, process requirements, and geographic requirements. You are probably familiar with program models such as structure charts, flowcharts, and pseudocode. They model the modular structure and logic of a program. In this book you will learn several tools and techniques for modeling business requirements.

Another approach to translating and validating requirements is prototyping.

Prototyping is the act of building a small-scale, representative or working model of the users' requirements for purposes of discovering or verifying those requirements.

This is the "they'll know what they need when they see it" approach. The analyst uses powerful prototyping tools to quickly build computer-based **prototypes**. The users can then react to the prototype to help the analyst refine or add to the requirements. Prototypes can also be used to develop or refine the aforementioned system models.

The final models and prototypes are usually organized into a deliverable called a business requirements statement. Some approaches call for great detail in this requirements statement, whereas others emphasize "the big picture." We'll examine some of these approaches in Chapters 4 and 6. The requirements statement becomes the trigger for systems design.

Although it is rare, the project could still be canceled at the end of this phase. More realistically, the project scope (or schedule and budget) could be adjusted if it becomes apparent that the new system's requirements are much more substantive than originally anticipated—hence, the red diamond feasibility checkpoint for this phase.

Systems Design

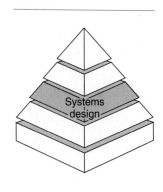

Given a reasonable understanding of the users' requirements, the systems analysts can turn their attention to systems design. It is during systems design that systems analysts finally begin addressing technology issues and details—in other words, *how* the new system will be implemented.

Systems design is the evaluation of alternative solutions and the specification of a detailed computer-based solution. It is also called *physical design.*

Systems design is driven by various system designers, including the systems analyst. Hence, in the pyramid model, it addresses PEOPLE, DATA, ACTIVITIES, and NETWORKS from the system designers' perspectives (which were introduced in Chapter 2). What about the TECHNOLOGY building block? Often, that technology is in place, or specified by a predefined technology architecture (from Systems Planning). In other cases, the analyst must select or supplement the technology. In all cases, systems design builds on the knowledge derived from systems planning and systems analysis (thus, the volume of our pyramid increases).

Figure 3.8 shows the phases of a typical systems design. Notice the blue arrows—systems design is triggered by the business requirements statement from systems analysis. Now study Figure 3.8 as we discuss these three phases in greater detail.

Design Phase 1: Select a Design Target (from Candidate Solutions)

Design is a detailed, technical, and potentially time-consuming process. There are usually numerous alternative ways to design any given system. Some of the pertinent questions include the following:

- How much of the system should be computerized?

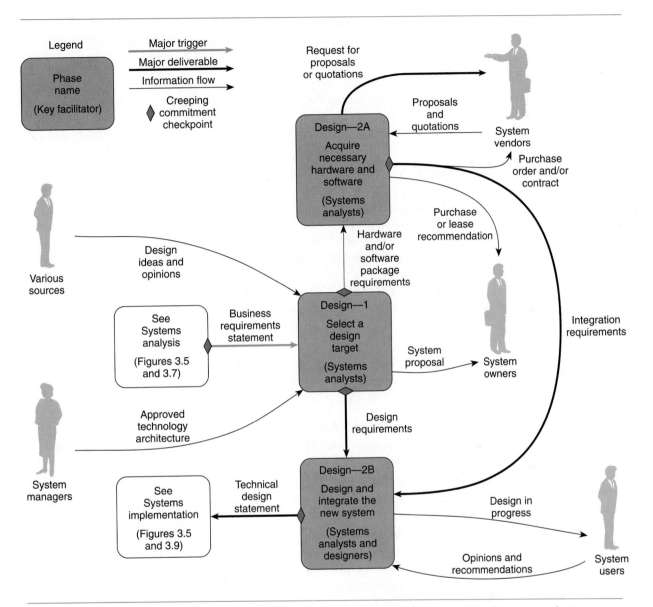

FIGURE 3.8 **Systems Design Phases in the Life Cycle** This flow diagram expands on the systems design phases of systems development (which were introduced in Figure 3.5). Systems design changes the focus from purely business issues (which were the domain of planning and analysis) to more technical issues.

- Should we purchase software or build it ourselves (called the **make-versus-buy decision**)?
- If we decide to make the system, should we design an on-line or batch system?
- Should we design the system for a mainframe computer, minicomputer, microcomputers, or some combination?
- What information technologies might be useful for this application?

The first phase of systems design is to select a feasible design target. Implicit in this phase is the need to first identify *candidate design solutions.*

As shown in the figure, the **selection phase** is triggered by the *business requirements statement* from a systems analysis. Some candidate solutions will be proposed as design ideas and opinions by various sources (e.g., systems analysts and system designers, other information systems managers and staff, technical consultants, system users, or system vendors). Some technical choices may be limited by a predefined approved technology architecture provided by systems managers.

After defining candidates, each candidate is evaluated by the following criteria:

- **Technical feasibility.** Is the solution technically practical? Does our staff have the technical expertise to design and build this solution?
- **Operational feasibility.** Will the solution fulfill the user's requirements? To what degree? How will the solution change the user's work environment? How do users feel about such a solution?
- **Economic feasibility.** Is the solution cost-effective (as defined earlier in the chapter)?
- **Schedule feasibility.** Can the solution be designed and implemented within an acceptable time period?

Infeasible candidates are eliminated from further consideration; however, several alternatives usually prove to be feasible. The analyst is usually looking for the *most* feasible solution—the solution that offers the best *combination* of technical, operational, economic, and schedule feasibility.

The key deliverable of the selection phase is a formal systems proposal to systems owners, who will usually make the final decision. This proposal may be written or verbal and is subject to negotiation (hence, the doubled-end arrow). Clearly, several outcomes are possible (notice the red feasibility/scope diamonds for this phase). Specifically, system owners might choose any one of the following options:

- Approve and fund the systems proposal (possibly including an increased budget and timetable if scope has significantly expanded).
- Approve or fund one of the alternative system proposals.
- Reject all of the proposals and either cancel the project, or send it back for new recommendations.
- Approve a reduced-scope version of the proposed system.

Additional deliverables depend on the final decisions. Assuming approval of at least one design solution, the decision will require the analyst to do any of the following:

- Acquire necessary hardware and/or software (the "buy" decision).
- Design a system and its software (the "make" decision).
- Some combination of both of the options.

If you decide to "buy" components of the new system, the appropriate hardware and/or software package requirements (inclusive of business requirements) must be delivered to design phase 2A: acquire necessary hardware and software.

If you decide to "make" components of the new system, the appropriate design requirements (inclusive of business requirements) must be delivered to design phase 2B: design and integrate the new system. It should be noted that even in those cases where you purchase all of the software, that software must be integrated into other information systems as well as the business. In other words, there is almost always some design and integration to be performed.

Design Phase 2A: Acquire Necessary Hardware and Software

Here is a phase that is missing from many SDLCs and methodologies. College graduates are often shocked to discover the high percentage of computer software that is purchased (or leased) rather than built. Also, any new system may present the need to acquire additional hardware, such as personal computers or printers. Recall that the make-versus-buy decision was made in the last phase. If the decision includes a *buy* component, then the next phase of systems design is to acquire necessary hardware and software. It is sometimes called the **acquisition** (or *procurement*) **phase**.

Why include this phase in the life cycle? The selection of hardware and software takes time. Much of that time occurs between order and delivery. This time lag must be figured into the life cycle in order to schedule the subsequent life cycle phases!

The key facilitator in the acquisition phase is still the systems analyst; however, several other parties get involved. Clearly, the system vendors (who sell hardware and/or software) get involved. Also, because these purchases exceed the authorized spending limits of the average systems analyst, the purchases must be approved by some level of management.

The key input to the phase is the hardware and software package requirements from the last phase of systems design. The systems analyst normally communicates these requirements to possible system vendors as a written request for proposals or quotations. Then, system vendors respond with proposals and quotations.

The analyst's job is to evaluate proposals and quotes to determine (1) which ones meet requirements and specifications, and (2) which one is the most cost effective. The analysts make a purchase or lease recommendation to the system owners (and/or information system managers). This recommendation may also be negotiated. Finally, the analyst intiates a purchase order and/or contract for the approved hardware and software packages.

Design Phase 2B: Design and Integrate the New System

Given the approved, feasible solution from the selection phase, you can finally design and integrate the new system. You understand *what* the requirements are from the definition phase, and *how* you plan to fulfill those rquirements from the selection phase. Thus, you can now justify the time and cost to design the new system.

Ideally, any new system should work in harmony with other current information systems. Similarly, if we have purchased software packages, those packages must work in harmony with any components of the systems which are to be built in-house. For these reasons, we not only design the new system — we integrate the new system as well.

The key facilitator of the **design and integration phase** is still the systems analyst. But various other design specialists play important roles. For instance, database specialists might design or approve the design of any new or modified databases. Network specialists might design or modify the structure of any computer networks. Microcomputer specialists may assist in the design of workstation-based software components. And as always, the system users must be involved—they evaluate the new system's ease-of-learning, ease-of-use, and compatibility with the stated business requirements.

The design phase uses the design requirements from the selection phase as its primary input. If the system will include purchased software packages, the acquistion phase may provide integration requirements that specify how those packages should be interfaced to other systems and packages.

As the analysts complete different aspects of the design, they offer them to system users as a design in progress. The system users respond with opinions and recommendations. The final deliverable is a technical design statement.

The technical design statement is frequently divided into two parts: general and detailed design.

The **general design** serves as an outline of the overall design.

The **detailed design** focuses on the detailed specifications for components in the outline.

General design can take several forms, but the most common approach is modeling. (Modeling was defined in analysis phase 3: define and prioritize user's requirements). Normally, general design models will depict:

- The structure of files and database (a data structure-oriented diagram is common).
- Processing methods and procedures (a flow-oriented diagram is common).
- The structure of the computer network (if applicable—again, a flow-oriented diagram is common).

The detailed design is also frequently broken into two parts: external and internal design. External and internal design frequently occur in parallel.

The **external design** is the specification of the system's interface to its users. The external design includes inputs, outputs, screens, and screen-to-screen transitions.

The **internal design** is the specification of the system's software (structure and logic), files, and database—those features that are less visible to the users.

Most detailed designs employ some combination of system modeling and prototyping.

Prototyping was introduced in analysis phase 3: define and prioritize users' requirements. Design-by-prototyping allows the analyst to quickly create scaled-down, but working versions of a system or subsystem. In the interest of speed, certain features and capabilities, such as input editing, may be left out of prototypes. Prototypes will typically go through a series of iterations and user reviews until they evolve into an acceptable design.

A project is rarely canceled in the design phase, unless it is hopelessly over budget or behind schedule. However, the red diamond suggests a feasibility/ scope checkpoint. If the design is too expensive to implement in its entirety, the scope can be reduced prior to implementation. Indeed, implementation frequently occurs in versions, each version adding new functionality to the system.

Systems Implementation

Systems implementation is the next step in the life cycle.

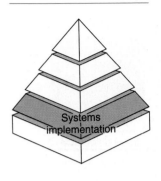

> **Systems implementation** is the construction of the new system and the delivery of that system into "production" (meaning "day-to-day operation"). Unfortunately, *systems development* is a common synonym. (Note: We dislike that synonym since it is more frequently used to describe the *entire* life cycle.)

In your pyramid model, the implementation phase implements the DATA, ACTIVITIES, and NETWORKS building blocks of the system. The PEOPLE focus changes from the system designer to the system builder. The perspectives of system builders were introduced in Chapter 2.

Figure 3.9 illustrates the phases of a typical systems implementation. Notice that the trigger for system implementation is the technical design statement from systems design. Once again, we ask that you study this figure as we walk through these phases in somewhat greater detail.

Implementation Phase 1: Build and Test Networks and Databases (if Necessary)

In many cases, new or enhanced applications are built around existing networks and databases. If so, skip this phase. However, if the new application calls for new or modified networks or databases, they must normally be implemented prior to writing or installing computer programs. Why? Because applications programs will use those networks and databases. Thus, the first phase of some implementations is to build and test networks and databases.

The key facilitators are various systems designers, not programmers. For instance, computer networks are usually implemented by the same networking specialists who designed them. Similarly, corporate databases are usually implemented by the database specialists who designed them. (Alternatively, systems analysts may be called upon to design noncorporate, applications-oriented databases, with possible design approval from database specialists.)

The key input to the phase is the subset(s) of the technical design statement that reflects the network or database designs. The deliverables are networks and unpopulated databases. The term *unpopulated* means that the database structure is implemented, but data has not been loaded into that database structure. The programmers will eventually write programs to populate and maintain the database.

Implementation Phase 2: Build and Test the Program

Now we come to the SDLC phase with which you are probably most familiar — build and test the programs. This is also sometimes called the *construction phase*. Actually, you are probably most familiar with the principal effort of this phase, applications programming.

Obviously, the **build/test phase** is only applicable to software components that you've decided to write, not purchase. On the other hand, you may also have to write enhancement programs to augment a purchased software package.

Building and testing programs is frequently the most time-consuming and tedious phase of the life cycle. However, the time required for construction is often longer than it should be because the preceding phases and activities were

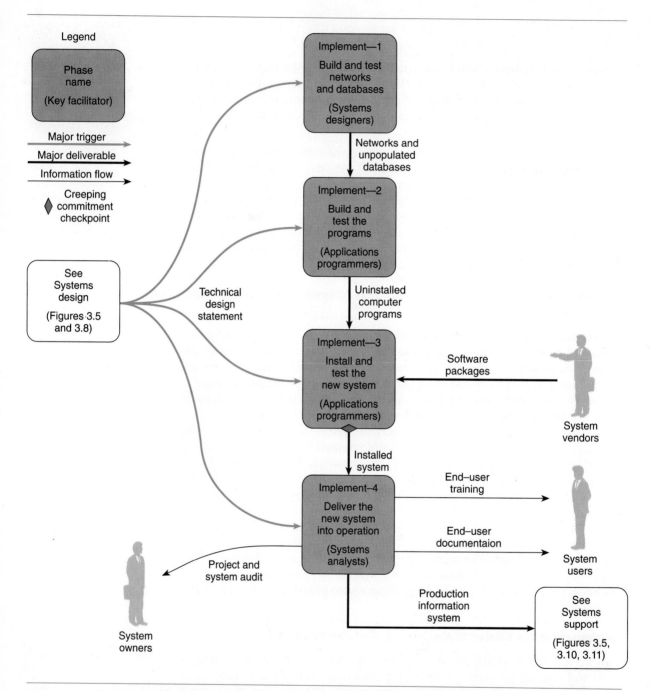

FIGURE 3.9   Systems Implementation Phases in the Life Cycle   This flow diagram expands on the systems implementation phases of systems development (which were introduced in Figure 3.5). The implementation phases construct the technical building blocks of the system and then place the system into daily operation.

completed hastily or not at all. Programmers should work from specifications that have been developed and refined through the earlier phases and activities. If the specifications are unclear, incomplete, inaccurate, or otherwise faulty, the construction phase will be much more complicated and time consuming.

The key facilitator in the build/test phase is the applications programmer, not the analyst. The analyst is still involved to the extent that specifications may need to be clarified; the perfect specification probably doesn't exist.

The key input to this phase is the subset of the <u>technical design statement</u> that contains the program specifications. If new or modified networks and databases are to be used, the implemented <u>networks and unpopulated databases</u> are input from the previous implementation phase.

The deliverable for this phase is <u>uninstalled computer programs</u>. The term *uninstalled* means that the programs have been thoroughly debugged and unit tested, but not installed for production use.

Unit tests ensure that the applications programs work properly when tested in isolation from other applciations programs.

There are three ways to accelerate the build/test phase.

- The programs can be written in a fourth-generation language (4GL). Many businesses have realized significant productivity gains when switching from third-generation languages (like COBOL, BASIC, and C) to 4GLs (like Focus, SAS, and CSP).
- If prototypes were developed in the design phases, they might be expanded into the final working systems. The programmer would simply add the data editing, security, features, and capabilities that were left out of the prototype. (Note: Most prototypes are written in the aforementioned 4GLs.)
- Design specifications could be input to an automatic code generator that can create programs in minutes and hours instead of days and weeks. Examples of code generators include INTERSOLV's APS/PW and Knowledge Ware's Construction Workbench.

One product, IBM's CSP *(Cross System Product),* can do both prototypes and code generation; it can create and refine prototypes in its native CSP fourth-generation language, and then generate COBOL programs from those final CSP programs.

The expanded use of code generators could eventually change the key facilitator of this phase from the applications programmers to the systems analysts. Why? Because generated code is debugged at the model or specification level, not the language. Clearly, the systems analysts are more familiar with the models and specifications because they developed them. Could this eventually make applications programmers obsolete? Such a development is not impossible—it should certainly be considered by anyone contemplating a career as an applications programmer.

Implementation Phase 3: Install and Test the New System

The next phase of systems implementation is to install and test the new system. Obviously, the programs completed in the last phase must be installed. But why is another test necessary?

System tests ensure that applications programs written in isolation work properly when they are integrated into the total system.

It is not at all uncommon for programs that work perfectly by themselves to fail to work at all when combined with other, related programs. If this happens, the programmer must often return to the build/test phase.

The **install/test phase** is also used to install purchased or leased software packages. Even those packages must be system tested to ensure that they properly interact with other programs and packages. Furthermore, the programmer may have built special programs to connect the package with other programs, files, and databases. These integrated solutions must also be system tested.

Once again, the applications programmers are the key facilitators for this phase. And once again, systems analysts must be available to clarify requirements and design specifications.

A key input to this phase is the subset of the technical design statement that specifies how the various purchased software packages, built and tested programs, and files and databases are to be assembled into an integrated system. Other key inputs include any purchased software packages as delivered from system vendors, and uninstalled computer programs from the build/test phase.

The deliverable is an installed system ready to be delivered into production.

The red feasibility checkpoint diamond represents our last chance to not implement the system. Why would we build a system and then not deliver it into production? This is a rare occurrence, but it does happen. One possibility might be that the production system would be too expensive to run or maintain.

Implementation Phase 4: Deliver the New System into Operation

What's left to do? New systems usually represent a departure from the way business is currently done; therefore, the analyst must provide for a smooth transition from the old system to the new system and help the users cope with normal start-up problems. Analysts must also train users, write various manuals, and load files and databases. Thus, the last phase of implementation is to deliver the new system into operation. Much of the work in the **delivery phase** can overlap the build/test and install/test phases.

The systems analyst usually facilitates this phase, involving other systems professionals, system users, and system owners as required.

The diagram shows that the key input to this phase is the installed system from the install/test phase. The analysts deliver any necessary end-user training and end-user documentation that will assist system users as they convert to the new system.

The key deliverable is, of course, a production information system that can now support the business on a day-to-day basis. Different strategies are used to cross over to a new system. They'll be discussed later in the text.

Some time after the new system is placed into production, a project and system post-evaluation (called a *post-audit*) might be conducted. The purpose is to evaluate and improve both the systems development process, and the new information system. This project and system audit is usually conducted with the system owners.

Why isn't there a feasibility/scope checkpoint after this phase? There would be no justification for placing a new system into production, and then choosing not to support it.

Systems Support

Once the system is placed into operation, the analyst's role changes to systems support. In fact, a significant portion of most systems analysts' time and effort is spent providing ongoing support for existing systems.

> **Systems support** is the ongoing maintenance of a system after it has been placed into operation. This includes program maintenance and system improvements.

In your pyramid model, systems support maintains all of the building blocks (and their documentation) for a production information system. Systems analysts usually coordinate systems support, calling on the services of maintenance programmers and system designers as necessary. Systems support doesn't consist of phases so much as it does ongoing activities. But before we study these activities, we must revise our life cycle overview diagram. Why? Because systems support is frequently accomplished by revisiting other phases in the life cycle. Figure 3.10 clarifies this claim. The blue arrows demonstrate that during systems support, various situations can re-trigger systems analysis, systems design, and systems implementation.

Now let's take a look at the ongoing activities that make up systems support. Figure 3.11 illustrates these activities. Notice that the trigger for systems support is a production information system from systems implementation. The disk and hard copy symbols represent the programs, files and databases, and supporting documentation for the new system. The ongoing support activities maintain these components. Eventually, the system becomes obsolete and the life cycle starts over again. At that time, existing system details and limitations are sent to either systems planning or systems analysis (possibly both).

The support activities are not numbered since they are not really sequential. Study the figure as we introduce these activities in somewhat greater detail.

Support Activity: Correct Errors

Once a system is in operation it is inevitable that it will fail, from time to time, due to software errors and misuse. Consequently, one ongoing activity in systems support is to correct errors.

System users report errors encountered when using the system. Systems analysts normally act as liaisons to the users to identify the cause and effects of the system errors. The analyst may implement a quick fix and documentation. More commonly the analyst specifies procedures or specifications necessary to correct these software "bugs" and sends them back to the systems implementation phase. Why go back to systems implementation? Maintenance programmers must not only fix the bugs. They must also repeat unit and system tests to ensure that the "fixes" don't create new bugs. Only then can the system be returned to production status.

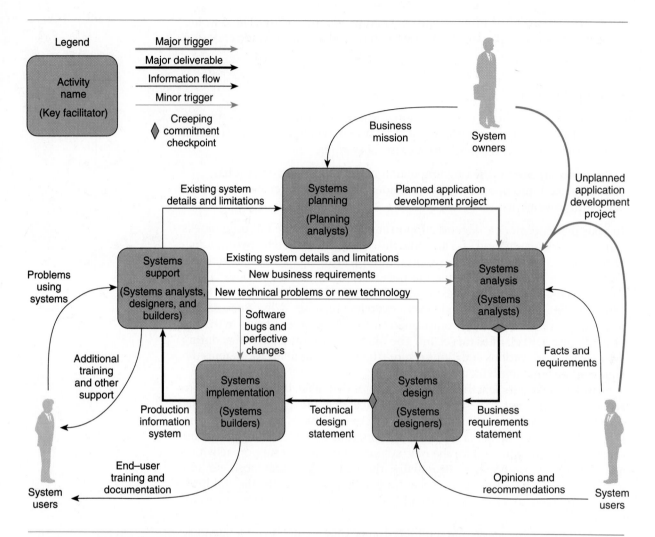

FIGURE 3.10 Revised SDLC Overview   This flow diagram represents a revision to our overview SDLC diagram presented in Figure 3.5. Notice that the systems support phase may trigger the need to revisit the systems analysis, design, and implementation phases.

Support Activity: Recover the System

From time to time, a system failure will result in an aborted program or loss of data. This may have been caused by human error or a hardware or software failure. The system users notify their analyst of the system "crash." The systems analyst may then be called upon to recover the system—that is, to restore a system's files and databases, and to restart the system.

Support Activity: Assist the Users of the System

Just as a successful company must stand behind its products, successful systems analysts must stand behind their information systems. Regardless of how well the users have been trained, and how good the end-user documentation is,

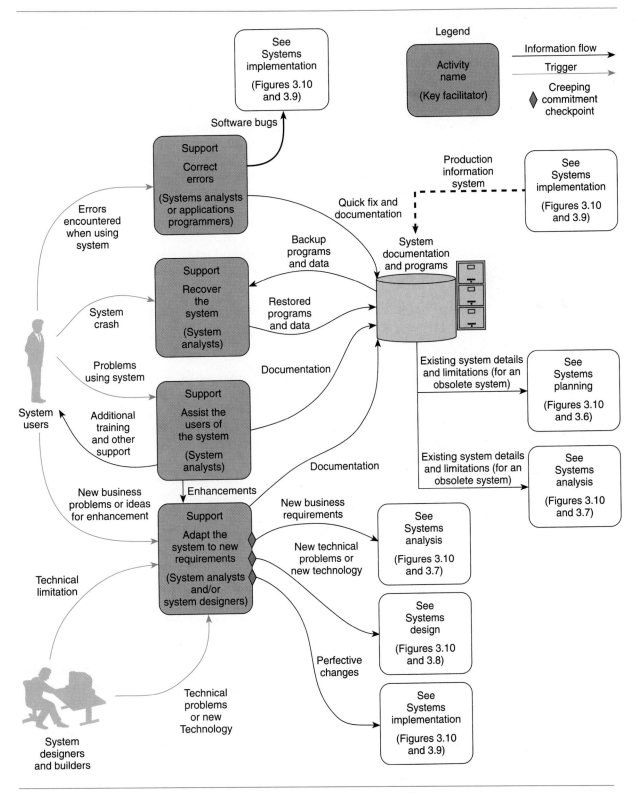

FIGURE 3.11 Systems Implementation Phases in the Life Cycle This flow diagram expands on the systems support activities of systems development (which were introduced in Figure 3.10).

users will eventually require additional assistance—unanticipated problems arise, new users are added, and so forth. Throughout this chapter and book, we have stressed the importance of user participation in systems development. Similarly, we now stress the importance of systems analysts' participation in systems support.

A good systems analyst maintains open communications with users and management, constantly assessing the user's perceptions about the system. How do they "feel" about the system? Additionally, a good systems analyst periodically "experiences" systems. From time to time, analysts stand over the user's shoulder. Really good analysts will even try their hand at using their own systems. This type of participation with users during support usually proves beneficial to credibility and confidence, as well as proving to be a real eye-opener to the analyst.

In this activity, the system users inform the analyst about problems using the system. The analyst responds with (1) changes to operating procedures and documentation, (2) additional training and other support, and (3) suggested enhancements. The latter is handled in the next support activity.

Support Activity: Adapt the System to New Requirements

The most time-consuming support activities may be triggered by the following factors:

- New business problems as identified by system users (for example, an un-explained increase in credit defaults—have customers found a way to beat our system?)
- Ideas for enhancement as identified by system users (for example, the need for a new business report or the need to process a new type of business transaction).
- New technical problems as identified by system designers and system builders (for example, date fields that cannot accommodate arithmetic operations after the year 2000).
- New technology as identified by system designers and system builders (for example, a decision to discontinue support for the database management system used by the information system).

Adaptive maintenance forces an analyst to analyze the situation and return to the appropriate analysis, design, and implementation phases. If the enhancement is driven by new business problems or requirements, the analyst returns to systems analysis phases. If the enhancement is driven by new technology or technical problems, the analyst returns to the appropriate systems design phases. If the enhancement merely perfects the system (a "quick fix"), the analysts sends perfective changes directly to the systems implementation phases. Progression through the subsequent life cycle phases is dependent on the scope and complexity of the changes.

Often overlooked is the expanded role of the end-user in maintenance. The end-users can often respond to their own requirements by using query languages, report generators, and other user-friendly tools with existing system files and databases. Additionally, they can frequently download data from those

Systems Development Life Cycle (Representative project)							
Name	March	April	May	June	July	August	September
Systems planning (BAA)	██	█					
Systems analysis		██	█				
Systems design			██	██	█		
Systems implementation				██	██	█	
Systems support						██	██
Cross life cycle activities	██	██	██	██	██	██	██
Fact finding	██	██	██	██	█		
Documentation	██	██	██	██	██	██	██
Presentation	██	██	██	██	██	██	
Estimation			██	█			
Measurement		██	██	██	██	██	
Feasibility analysis		██	██	██	█		
Project management	██	██	██	██	██	██	██
Process management	██	██	██	██	██	██	██

FIGURE 3.12 **Cross Life Cycle Activities** This bar diagram depicts the partial schedule for a typical project. Specifically, the bar chart identifies numerous cross life cycle activities and the life cycle phases they may overlap.

files and databases into personal computer spreadsheets and databases to fulfill new requirements.

CROSS LIFE CYCLE ACTIVITIES

Systems development involves a number of activities called cross life cycle activities.

> **Cross life cycle activities** are activities that overlap many or all phases of the life cycle — in fact, they are normally performed in conjunction with several phases of the life cycle.

Cross life cycle activities include fact finding, documentation and presentation, estimation and measurement, feasibility analysis, project management, and process management. Let's briefly examine each of these activities. The "cross life cycle" nature of these activities is illustrated in the time bar chart in Figure 3.12.

Note Notice how the systems analysis, systems design, and systems implementation phases overlap. They are completed sequentially, but the overlap is common.

Cross Life Cycle Activity: Fact Finding

There are many occasions for fact finding during the life cycle.

Fact finding—also called *information gathering* or *data collection*—is the formal process of using research, interviews, meetings, questionnaires, sampling, and other techniques to collect information about systems, requirements, and preferences.

Fact finding is most crucial to the systems planning and systems analysis phases. Why? It is during these phases that the analyst learns about a business's and system's vocabulary, problems, opportunities, constraints, requirements, and priorities.

Fact finding is also used during the systems design and support phases, but to a lesser extent. For instance, in systems design, fact finding becomes technical as the analyst attempts to learn more about the technology selected for the new system. Ultimately, it is during systems support that, through fact finding, an analyst determines that a system has "decayed" to a point that the system needs to be redeveloped from scratch.

Cross Life Cycle Activity: Documentation and Presentations

Communication skills are essential to the successful completion of a project. Two forms of communication that are common to systems development projects are documentation and presentation.

Documentation is the activity of recording facts and specifications for a system.

Presentation is the related activity of formally packaging documentation for review by interested users and managers. Presentations may be either written or verbal.

Clearly, documentation and presentation opportunities span the entire life cycle and support all of the classical phases: planning, analysis, design, implementation, and support.

As described throughout this chapter, documentation for a project is built during various phases and activities. Over time, a sizable base of documentation for various systems and applications is compiled. Many businesses try to get control over such documentation so that it is kept and maintained for future use. **Version control** over documentation has become a critical success factor; it involves keeping and tracking multiple versions of a system's documentation. At a minimum and at any given time, most information systems shops want to keep documentation for all of the following versions:

1. One or more previous versions of the system.
2. The current production version of the system.
3. Any version of the system going through the build and test activity.
4. Any version going through the life cycle to create a new version.

Many companies use library managers and version control software (see Chapter 5) to manage documentation.

In the figures in this chapter, the arrows on the flow-oriented models have been used to represent the typical use and updating of project documentation, and the presentation of documentation to various audiences by systems analysts and programmers.

Cross Life Cycle Activity: Estimation and Measurement

Information systems are significant capital investments. For this reason, estimation and measurement activities are commonly performed to address the quality and productivity of systems.

> **Estimation** is the activity of approximating the time, effort, costs, and benefits of developing systems. The term *guesstimation* (as in "make a guess") is used to describe the same activity in the absence of reliable data.
>
> **Measurement** is the activity of measuring and analyzing developer productivity and quality (and sometimes costs).

Estimating is an extremely important activity because information systems, as noted earlier in the chapter, are capital investments. You don't want to spend $25,000 of time and effort to solve a problem that is costing your organization $2,000 per year. The payback would take more than a decade.

Estimating can be a difficult and frustrating activity, because it is difficult to translate an abstract problem statement into a precise estimate of the time, effort, and costs needed to solve that problem. Multiple factors influence the estimate (see margin for examples). There are two common approaches to estimation. First, some analysts avoid estimation out of fear, uncertainty, or lack of confidence. In this case, the analyst may resort to what are jokingly called "guesstimates." Alternatively, better analysts draw on experience and data (both their own and the collective experience of others) from previous projects to continually improve their estimates.

Measurement has become important because of the productivity and quality problems that plague systems development. In response to those problems, the industry has developed methods (Chapter 4) and tools (Chapter 5) to improve both quality and productivity. These methods and tools can be costly. Formal measurement of development productivity may be the only way to justify the cost of these expenses.

The field of software and systems metrics offers hope for the future.

> **Software and systems metrics** provides an encyclopedia of techniques and tools that can both simplify the estimation process and provide a statistical database of estimates versus performance.

─────── ✓ ───────

Factors that Affect Estimates

Size of project team
Experience of project team
Number of users and
  managers
Attitudes of users
Management commitment
Availability of users and
  managers
Other projects in progress

─────────────────

Cross Life Cycle Activity: Feasibility Analysis

A system development life cycle that supports our creeping commitment approach to systems development recognizes feasibility analysis as a cross life cycle activity.

> **Feasibility** is a measure of how beneficial the development of an information system would be to an organization.
>
> **Feasibility analysis** is the activity by which feasibility is measured.

Too many projects call for premature solutions and estimates. This approach often results in an over-commitment to the project. If analysts are so accurate in feasibility estimates, why then are so many information systems projects late and over budget? Systems analysts tend to be overly optimistic in the early stages of a project. They underestimate the size and scope of a project because they haven't yet completed a detailed study.

A project that is feasible at any given stage of systems development may become less feasible or infeasible at a later stage. For this reason, we use the creeping commitment approach to reevaluate feasibility at appropriate checkpoints (indicated by small diamonds) in this chapter's flow-oriented models of phases and activities.

Various measures of feasibility were introduced in design phase 1: select a design target. These measures included technical feasibility, operational feasibility, economic feasibility, and schedule feasibility.

Cross Life Cycle Activity: Project and Process Management

Systems development projects may involve a team of analysts, programmers, users, and other IS professionals who work together.

> **Project management** is the ongoing activity by which an analyst plans, delegates, directs, and controls progress to develop an acceptable system within the allotted time and budget.

Failures and limited successes of systems development projects far outnumber very successful information systems. Why is that? One reason is that many systems analysts are unfamiliar with or undisciplined in the tools and techniques of systems development. But most failures are attributed to poor leadership and management. This mismanagement results in unfulfilled or unidentified requirements, cost overruns, and late delivery.

The systems development life cycle provides the basic framework for the management of systems projects. Because projects may be quite large and complex, the life cycle's phased approach to the project results in smaller, more measurable milestones that are more easily managed.

> **Process management** is an ongoing activity that establishes standards for activities, methods, tools, and deliverables of the life cycle.

Process management is a relatively new concept in systems development. The intent is to standardize both the way we approach projects, and the deliverables we produce during projects. For more on process management, see The Next Generation box. Many shops now set up quality assurance teams to evaluate deliverables from projects. This measure is not intended to constrain creativity, but rather, to ensure that documentation will be transferable to the next generation of programmers and analysts who work on any given system.

SYSTEMS DEVELOPMENT BY END-USERS

Not all systems and applications are developed by the team of analysts, designers, and builders described in this chapter. End-users now develop many of their own systems. Because many of our readers will become part of the user community, we will briefly describe variations on the life cycle for end-user systems development. These variations will be described by the major phases involved.

The Next Generation

PROCESS MANAGERS: AN ON-LINE SDLC

The systems development life cycle is a "process" that must be managed. A typical company's chosen life cycle may consist of hundreds of activities, thousands of steps, dozens of deliverables, and numerous variations to adapt the life cycle to various situations and scenarios. The entire process can prove most complex to document, implement, and manage. The technology of process managers may become the only practical way to manage and implement the life cycle.

A *process manager* is an on-line technology for defining and managing a life cycle. A process manager documents the life cycle in great detail. In addition to phases, activities, and tasks, the documentation includes:

- Deliverables and their structure.
- Standards for quality control.
- Recommendations for tools and techniques.

A minimal process manager is an on-line computer application that guides various developers through the accepted SDLC. Eventually, this capability should develop "intelligence" to provide sophisticated decision support and expertise to the developers and help them customize the life cycle to specific situations or scenarios. It this sense, it acts as an expert system. The expert system can intelligently customize the SDLC in ways no human consultants might imagine.

A good process manager will also help you plan management systems development. It will log planned activities, timetables, actual time spent, and other variables as developers work their way through the life cycle. Eventually, process managers should be able to measure and analyze developers and development effort.

A superior process manager will actually "invoke" other technical tools to perform tasks. For instance, at an appropriate point in the life cycle the process manager may invoke another computer tool that helps developers draw various system models. In a sense, it could eventually serve as a "technical umbrella" over the entire systems development process.

This technology is not yet prevalent. Indeed, the base technology may not be usable until systems development management commits sufficient resources to document the life cycle and all its variations. Consultants can help. And methodology vendors will sell prepackaged life cycles based on their biases. Still, these prepackaged solutions rarely offer the flexibility and customization options that are truly needed to deal with a given organization's culture, politics, organization, and technology choices.

On-line process managers will eventually become part of most companies' systems development infrastructure. This technology will improve with advances in expert system technology and tool interpretability technology. Meanwhile, those companies experimenting with on-line process managers today, may achieve some form of competitive systems development advantage in the next generation.

End-User Systems Planning

End-user systems development is almost never driven by formal systems planning. Where, then, do end-user systems development projects originate?

Most end-user applications are **personal information systems** designed to meet the needs of one (or few) user(s). Such systems can rarely be cost justified for traditional systems development by analysts and programmers. But with the advance of the personal computer and 4GLs such as personal database managers (such as dBASE) and spreadsheets (such as Lotus 1-2-3), these systems can be developed by end-users.

Some end-user development projects come from backlogged projects that were originally submitted for traditional systems development, but were abandoned for more important projects.

End-User Systems Analysis

Systems analysis is just as important in end-user systems development as it is in traditional systems development. There are, however, some notable differences.

The end-user plays the important roles of systems analyst, system owner, and system user. Clearly, this simplifies the problem statement and requirements definition.

Phase 1: Assess project feasibility is normally bypassed, because you wouldn't start the project if you didn't already think the problem was worth looking into.

Phase 2: Study the current system is still important. True, as the system user, you already understand your own system better than the average systems analyst. Still, it is important to (1) understand how your application interacts with others; and (2) analyze the problems you want to solve to identify their true causes and effects (and subproblems).

Phase 3: Define and prioritize the users' requirements is also very important. True, you are the user and should have a pretty good idea of your requirements without a formal definition. Still, a formal definition of requirements helps you identify subtle requirements and important relationships between requirements.

Many of the systems analysis tools and techniques taught later in this book are equally useful for end-user development projects.

End-User Systems Design

Some end-users have a natural tendency to skip design and jump directly to programming in their favorite 4GL, database manager, or spreadsheet. Don't make this mistake. End-user applications should be designed according to the same criteria used by good systems analysts and designers — perhaps more so because end-users have to support their own systems.

Once again, the end-user plays the important roles of systems analyst, system designer, system owner, and system user. But for systems design, the end-user can frequently call on the advice of special end-user consultants in the business. Many information systems shops organize an end-user consulting service under the name information center.

> An **information center** is a training and consulting service center for end-users who develop their own systems. It is usually part of the information systems department. Synonyms include *end-user computing center* and *end-user development center*.

These centers provide equipment and software standards, end-user tool and development training, and end-user consulting services.

Phase 1: Select a design target. This phase is often simplified by standards that specify which personal computers, peripherals, and PC software will be supported by the technical specialists. For example, the information center may inform you that they will only provide training and consulting services for (1) Compaq and IBM PCs; (2) dBASE III and IV databases and application generators; (3) Lotus 1-2-3 Version 2 and 3 spreadsheets and graphics; and (4) Word-Perfect word processors. These constraints limit your candidate systems. You still need to evaluate the technical, operational, economic, and schedule feasibility of your design target.

End-users can sometimes find prewritten applications that suit their needs. The decision to buy an application package would be considered in this phase. Representative packages would be identified for analysis in the next phase.

Phase 2A: Acquire necessary hardware and software. This phase is sometimes simplified for the reasons described in the last paragraph. After selecting the appropriate hardware, software tools (e.g., dBASE or Lotus 1-2-3), or prewritten package (e.g., an accounts receivable package), the phase is generally reduced to requesting that your management buy the necessary hardware and/or software. If management agrees, it places the order. Otherwise, the project comes to a rather abrupt conclusion.

Note Users should never recommend prewritten applications packages for purchase without first using the package, evaluating the documentation, considering the training issues, and evaluating costs (acquisition and maintenance).

Phase 2B: Design and integrate the new system. This phase should consume the majority of the end-user's time. The prototyping approach, using a database manager (such as dBASE) and/or a spreadsheet (such as Lotus 1-2-3), is normally used. The final production system can evolve directly from this prototype.

The systems design tools and techniques taught later in this textbook, adapted for use with PC software, can help end-users perform their own systems design. More importantly, the system design chapters provide insight into features and controls that should be designed into any system, regardless of size.

End-User Systems Implementation

Obviously, end-user applications are implemented directly by their users. Once again, the end-user may find appropriate training and consultation from an information center or similar end-user consulting unit.

Phase 1: Build and test networks and databases. End-users rarely design or build networks — they use those networks provided by networking specialists. Networks might be used to download data from multiuser information systems. End-user consultants can usually help accomplish this capability.

End-users do frequently implement personal files and databases using tools like dBASE. Simple databases can be built with little training and effort; however, even moderate or complex databases can be built with some additional training and consultation. End-user consultants can provide feedback about the quality of database design and implementation.

Phase 2: Build and test the programs. This phase is normally completed by end-users. Actually, the programs evolve out of the prototypes from system design. For programs generated by or written in a database language (again, we cite dBASE as an example), this phase may replace prototype input screens with forms-oriented screens. Input data might be more carefully checked for errors. Outputs and queries might similarly be reformatted for readability. New functions might be added, and menus might be developed.

In a similar vein, for programs written using a spreadsheet (again, we cite Lotus 1-2-3 as an example), this phase may simplify data input through macros. Outputs might also be formatted by macros. Some spreadsheets allow the construction or modification of menus.

The importance of unit testing cannot be overstressed. End-users have a bad habit of implementing programs that have not been fully tested. Sometimes, those programs generate incorrect data for months or years before those errors are detected (with considerable embarrassment or worse for the end-user).

Phase 3: Install and test the new system. End-users rarely face the difficult integration issues of traditional systems development. Why? End-user systems are smaller, with fewer interfaces. If the application does receive or send data to other systems or applications, it is still important to test those interfaces.

Phase 4: Deliver the new system into operation. Just as systems analysts deliver multiuser systems into operation, end-users deliver personal systems into operation. The phase is simplified since the end-user is already intimately familiar with the system (having developed it). Still, end-user documentation is encouraged for substitute users and future users. People do, after all, frequently transfer to new jobs ot leave the business.

End-User Systems Support

End-users generally support their own systems. Repeat! End users generally support their own systems! Why do we stress this point? Management should not encourage end-user systems development unless end-users understand and commit to the support of those systems. When support breaks down, management is frequently forced to pay substantial monies to hire external support for what has become an essential information system.

Responsibility for all of the support activities rests with the end-users. End-user consulting is generally available only to advise — not to support production systems. Thus, the end-user must

- Correct errors (be their own maintenance programmers).
- Recover the system(do their own backup and recovery of programs and data).
- Assist other users of the system (provide training and revised documentation as new versions evolve).
- Adapt the system to new requirements (which becomes more difficult as a system approaches entropy).

Managing these activities is not an impossible task. But problems occur when end-users fail to consider the time and costs needed to maintain personal information systems. This fact may explain why the majority of today's end-user PCs are only used for word processing, simple spreadsheets, and simple file management.

The bottom line on end-user systems development is simple: End-user systems development should be more similar to, than different from, traditional systems development. Only the project scope and roles change significantly.

Summary

Eight basic principles apply to all systems development projects:

1. Actively involve the users in systems development.
2. Use a problem-solving approach to systems development.

3. Establish phases and activities so that progress can be determined and feasibility reevaluated.
4. Establish standards to ensure consistent development and documentation.
5. Realize that systems are capital investments and should be justified as such.
6. Establish checkpoints to allow the option of canceling the project if it has become infeasible.
7. Break large systems down into smaller pieces and be aware that all systems interact with other systems.
8. Design systems for growth and flexibility, because they will change over time.

The two general origins for systems projects are reaction to situations and planning. Systems development projects are usually triggered by problems, opportunities, and directives. Problems exist when the current system is not fulfilling the organization's purpose, goals, and objectives. Opportunities are chances to improve a system despite the absence of specific problems. Directives are decisions imposed on a system by management or government. All problems, opportunities, and directives can be evaluated in terms of their bottom-line impact relative to performance, information and data, economy, control and security, efficiency, and service.

A systems development life cycle (SDLC) is a disciplined approach to developing information systems. It is a project management tool that breaks a large project down into manageable pieces (usually called phases) and provides opportunities for feasibility reevaluation. A modern life cycle typically consists of five high-level phases, including systems planning, systems analysis, systems design, systems implementation, and systems support.

Systems planning seeks to identify and prioritize information systems applications whose development would return the most value to the business. Once an information systems project is identified in systems planning, the project goes through a systems analysis phase. Systems analysis involves studying the current business and information system, and defining user requirements for a new information system. Once requirements have been defined, the project may proceed to the systems design phase. Systems design involves evaluation of alternative solutions to the user requirements and the specification of detailed computer-based components. The technical design is then used to complete the systems implementation phase. Systems implementation includes the construction and delivery of the new system into operation. Finally, the implemented system proceeds to the systems support phase. Systems support involves ongoing maintenance of the production system.

There are numerous activities that overlap many or all phases of a life cycle. Such activities are referred to as cross life cycle activities. Cross life cycle activities include fact finding, documentation and presentations, estimation, feasibility analysis, project management, and process management.

Today, end-users develop some of their own information systems. They will likely develop many more in the future. End-users should perform most of the same activities performed in the classical systems development life cycle. The projects tend to be smaller, with fewer technical options. But the activities, for the most part, are the same —except that the end-user becomes the facilitator of most phases. Many IS shops have business units dedicated to helping end-users develop their own systems.

Key Terms
---

acquisition phase, p. 115
backlog, p. 109
build/test phases, p. 117
business area, p. 105

business area analysis,
   p. 106
cost-effectiveness, p. 94

creeping commitment
   approach, p. 95
cross life cycle activity,
   p. 125

definition phase, p. 111

delivery phase, p. 120

design and integration
   phase, p. 115

detailed design, p. 116

directive, p. 97

documentation, p. 126

economic feasibility,
   p. 114

entropy, p. 96

estimation, p. 127

external design, p. 116

fact finding, p. 126

feasibility, p. 127

feasibility analysis, p. 127

general design, p. 116

information architecture,
   p. 104

information center, p. 130

install/test phase, p. 120

internal design, p. 116

key facilitator, p. 99

major deliverable, p. 99

make-versus-buy decision,
   p. 113

measurement, p. 127

modeling, p. 111

operational feasibility,
   p. 114

opportunity, p. 97

personal information
   systems, p. 129

PIECES, p. 98

planning analyst, p. 104

presentation, p. 126

problem, p. 97

process management,
   p. 128

project management,
   p. 128

prototype, p. 112

prototyping, p. 112

schedule feasibility,
   p. 114

scope, p. 101

selection phase, p. 114

software and systems
   metrics, p. 127

steering committee,
   p. 109

study phase, p. 109

survey phase, p. 107

systems analysis, p. 107

systems design, p. 112

systems development life
   cycle, p. 91

systems development
   standards, p. 93

systems implementation,
   p. 117

systems planning, p. 102

systems support, p. 121

system test, p. 120

technical feasibility,
   p. 114

unit test, p. 119

version control, p. 126

Questions, Problems, and Exercises

1. What are the eight fundamental principles of systems development? Explain what you would do to incorporate those principles into a systems development process.

2. Make an appointment to visit a systems analyst at a local information systems installation. Discuss the analyst's current project. What problems, opportunities, and directives triggered the project? How do they relate to the PIECES framework?

3. Using the PIECES framework, evaluate your local course registration system. Do you see any problems or opportunities? (*Alternative:* Substitute any system with which you are familiar.)

4. Assume you are given a programming assignment that requires you to make some modifications to a computer program. Explain the problem-solving approach you would use. How is this approach similar to the phased approach of the systems development life cycle presented in this chapter?

5. Which phases of the SDLC presented in this chapter do the following tasks characterize?
   a. The analyst demonstrates a prototype of a new sequence of work order terminal screens.
   b. The analyst observes the order-entry clerks to determine how a work order is currently processed.
   c. The analyst develops the internal structure for a database to support work order processing.
   d. An analyst is teaching the plant supervisor how to inquire about work orders using the new microcomputer.

  e. A plant supervisor is describing the content of a new work order progress report that would simplify tracking.

  f. The analyst is reading an inquiry concerning whether or not a computer system might solve the current problems in work order tracking.

  g. The analyst is installing the microcomputer and database management system needed to run work order–processing programs.

  h. The analyst is reviewing the company's organizational chart to identify who becomes involved in work order processing and fulfillment.

  i. The analyst is comparing the pros and cons of a software package versus writing the programs for a new work order system.

  j. An analyst is testing a computer program for entering work orders to the system.

  k. The analyst is correcting a program to more accurately summarize weekly progress.

  l. The analyst is buying a microcomputer and some available software.

  m. Information systems management and top business executives are identifying and prioritizing business area applications that should be developed.

6. Visit a local information systems installation. Compare their SDLC with the one in this chapter. Evaluate the company's SDLC with respect to the eight systems development principles. (*Alternative:* Substitute the SDLC used in another systems analysis and design book, possibly assigned by your instructor.)

7. Management has approached you to develop a new system. The project will last seven months. It wants a budget next week. You will not be allowed to deviate from that budget. Explain why you shouldn't over-commit to early estimates. Defend the creeping commitment approach as it applies to cost estimating. What would you do if management insisted on the up-front estimate with no adjustments?

8. You are an independent consultant. Write a proposal letter that offers your services as a consultant to solve the problems in the minicase. The user, who is skeptical of all computer people, is turned off by computer buzzwords. Be sure your proposal explains in step-by-step fashion how you will build a system that meets this user's needs.

9. You have a user who has a history of impatience—encouraging shortcuts through the systems development life cycle and then blaming the analyst for systems that fail to fulfill expectations. By now, you should understand the phased approach to systems development. For each activity, compile a list of possible consequences to use when the user suggests a shortcut through or around that activity.

10. When describing the phased approach that you plan to follow in developing a new information system, your client asks why your approach is missing a feasibility analysis and project management phase. How would you respond?

11. Identify and briefly describe the five high-level phases that are common to most modern systems development life cycles. Which of the five phases were omitted from the traditional life cycle?

12. There are still many organizations whose life cycle does not include a systems planning phase. Explain the overall importance of doing systems planning.

13. For each of the high-level phases of the modern systems development life cycle, identify and briefly describe their lower-level phases.

14. Explain the difference between external and internal design.

15. During the Select a Design Target (from Candidate Solutions) phase of systems design, candidate solutions are evaluated according to several feasibility criteria. Identify and briefly define four feasibility criteria that are used. Assuming that

several alternative candidate solutions prove feasible, how would you attempt to identify the best candidate?

16. What is the difference between a *system test* and a *unit test*?
17. What is the difference between *estimation* and *measurement*?
18. What is the difference between *feasibility* and *feasibility analysis*?
19. Redraw the chapter's flow-oriented pictures for end-user system development.

Projects and Minicases

1. Jeannine Strothers, investments manager, has submitted numerous requests for a new investment tracking system. She needs to make quick decisions regarding possible investments and divestments. One hour can cost her thousands of dollars in profits for her company.

   She has finally given up on Information Systems for not giving her requests high enough priority to get service. Therefore, she goes to a computer store and buys a microcomputer along with spreadsheet, database, and word processing software. The computer store salesperson suggests that she build a database of her investments and options, subscribe to a computer investment databank (accessed via a modem in the microcomputer), feed data from her database and the bulletin board into the spreadsheet, play "what if" investment games on the spreadsheet, and then update the database to reflect her final decisions. The word processor will draw data from the database for form letters and mailing lists.

   After discussing her plans with Jeff, a systems analyst at another company, he suggests she take a systems analysis and design course before beginning to use the spreadsheet and database. The local computer store, on the other hand, says she doesn't need any systems analysis and design training to be able to develop systems using the spreadsheet and database programs. Their reasoning is that spreadsheets and database tools are not programming languages; therefore, she shouldn't need analysis and design to build systems with them. Is the computer store correct? Why or why not? Can you convince Jeannine to take the systems analysis and design course? What would your arguments be?

2. Jeannine Strothers, the impatient manager in the earlier minicase, did not take Jeff's advice. She built the new system, but she can't get top management to allow her to use it. And she's run into a number of other problems.

   First, the financial comptroller has been reevaluating company investment strategies and policies. Jeannine wasn't aware of that. The new system does not account for many of the policies that are being considered.

   Her own staff has rejected the investment and divestment orders generated by the system. She used Information Systems' existing file structure to design those orders, only to find out that her clerks had abandoned those files two years ago because they didn't include the data necessary to execute order transactions. Her staff is also critical of the design, saying that minor mistakes send them off into the "twilight zone" with no easy way to recover.

   The computer link to the investment databank has been useless. The data received and its format are not compatible with systems requirements. Although other databanks are available, the current databank has been prepaid for two years. Additionally, Jeannine is now skeptical of such services.

   Some of her subordinate managers are insisting on graphic reports. Unfortunately, neither her database management nor spreadsheet package supports graphics. She's not sure how to convert the data of either package to a graphic format (assuming that it is even possible).

   To top off her problems, she isn't sure that her existing database structure can be modified to meet new requirements without having to rewrite all the programs, even those that appear to be working. And her boss is not sure that he wants to invest the money in a consultant to fix the problems.

Jeannine's analyst friend, Jeff, is not very sympathetic to her problems: "Jeannine, I don't have any quick answers for you. You've taken too many shortcuts through the project life cycle. When we do a system, we go through a carefully thought-out procedure. We thoroughly study the problem, define needs, evaluate options, design the system and its interfaces, and only then do we begin programming."

Jeannine replies, "Wait a minute. I only bought a microcomputer. It's not the same as your mainframe computer. I didn't see the need for going through the ritual you guys use for mainframe applications. Besides, I didn't have the time to do all those steps."

Jeff's parting words are philosophical: "You didn't have the time to do it right. Where will you find the time to do it over?"

What principles did Jeannine violate? Why do so many people today fall prey to the belief that the life cycle for an application is somehow different when using microcomputers? What conclusions can you draw from Chapter 2 (the information systems chapter) that might help Jeannine learn from her mistakes? What would you recommend to Jeannine if she were to decide to approach her manager with a plan to salvage the system?

3. Evaluate the following scenarios using the PIECES framework. Do not be concerned that you are unfamiliar with the application. That situation isn't unusual for a systems analyst. Use the PIECES framework to brainstorm potential problems or opportunities you would ask the user about.

   a. The staff benefits and payroll counselor is having some problems. Her job is to counsel employees on their benefit options. The company has just negotiated a new medical insurance package that requires employees to choose from among several health maintenance options (HMOs). The HMOs vary according to employee classifications, contributions, deductibles, beneficiaries, services covered, and service providers permitted. The intent was to provide the most flexible benefits possible for employees, to minimize costs to the company, and to control costs to the insurance agency (which would affect subsequent premiums charged back to the company).

   The counselor will be called on to help employees select the best plan for themselves. She currently responds manually to such requests. But the current options are more straightforward than those under the new plan. She can explain the options, what they do and do not cover, what they cost and may cost, and the pros and cons. However, current employee distrust of the new plan suggests that she will need to provide more specific suggestions and responses to employees.

   She may have to work up scenarios—possibly worst-case scenarios—for many employees. The scenarios will have to be personalized for each employee's income, marital and family status, current health risks, and so on. In working up a few sample scenarios, she discovered first that it takes one full day to get salary and personnel data from the Information Systems Department. Second, employee data is stored in many files that are not always properly updated. When conflicting data becomes apparent, she can't continue her projections until that conflict has been resolved. Third, the computations are complex. It often takes one full day or more to create investment and/or retirement scenarios for a single employee. Fourth, there are some concerns that projections are being provided to unauthorized individuals, such as former spouses or nonimmediate relatives. Finally, the complexity of the variations in the calculations (there are a lot of "If this, do that" calculations) results in frequent errors, many of which probably go undetected.

b. The manager of a tool and die shop needs help with job processing and control. Jobs are currently processed by hand. First, a job number is established. Next, the job supervisor estimates time and materials for the job. This is a time-consuming process, and delays are common. Then, the job is scheduled for a specific day and estimated time.

On the day the job is to be worked on, materials orders have to be issued to Stores. If materials aren't available, the order has to be rescheduled.

Time cards are completed in the shop when workers fulfill the work order. These time cards are used to charge back time to the customer. Time cards are processed by hand, and the final calculations are entered on the work order. The work orders are checked for accuracy and sent to CIS, where accounting records are updated and the customer is billed.

The problem is that the customer frequently calls to inquire on costs already incurred on a work order; but it's not possible to respond because CIS sends a report of all work orders only once a month. Also, management has no idea of how good initial estimates are or how much work is being done on any tool or machine, or by any worker.

c. State University's Development Office raises funds for improving instructional facilities and laboratories at the university. It has uncovered a sensitive problem: The data is out of control.

The Development Office keeps considerable redundant data on past gifts and givers, as well as prospective benefactors. This results in multiple contacts for the same donor—and people don't like to be asked to give to a single university over and over!

To further complicate matters, the faculty and administrators in most departments conduct their own fund-raising and development campaigns, again resulting in duplication of contact lists.

Contacts with possible benefactors are not well coordinated. Whereas some prospective givers are contacted too often, others are overlooked entirely. It is currently impossible to generate lists of prospective givers based on specific criteria (e.g., prior history, socioeconomic level), despite the fact that data on hundreds of criteria have been collected and stored. Gift histories are nonexistent, which makes it impossible to establish contribution patterns that would help various fund-raising campaigns.

Suggested Readings

Benjamin, R. I. *Control of the Information System Development Cycle.* New York: Wiley-Interscience, 1971. Benjamin's 16 axioms for managing the systems development process inspired our adapted principles to guide successful systems development.

Gildersleeve, Thomas. *Successful Data Processing Systems Analysis.* 2nd ed. Englewood Cliffs, N.J.: Prentice Hall, 1985. We are indebted to Gildersleeve for the creeping commitment approach.

Hammer, Mike. "Reengineering Work: Don't Automate, Obliterate." *Harvard Business Review,* July–August 1990, pp. 104–11. Mike Hammer is a noted consultant and author (and entertaining speaker!) who talks of a new development paradigm that suggests systems analysts take a more "active" (as opposed to "passive") role in reshaping underlying business processes instead of merely automating existing and dated business approaches. This article documents several classic examples where the business process was successfully redefined (during systems analysis) resulting in a more competitive business and information system.

Wetherbe, James. *Systems Analysis and Design: Traditional, Structured, and Advanced Concepts and Techniques.* 3rd ed. St. Paul, Minn.: West, 1988. We are indebted to Wetherbe for the PIECES framework.

Zachman, John A. "A Framework for Information System Architecture," *IBM Systems Journal* 26, no. 3, 1987. This article presents a popular conceptual framework for information systems planning and the development of an information architecture.

4

Systems Development Techniques and Methodologies

Chapter Preview and Objectives

This chapter introduces you to **systems development techniques and methodologies,** a subject separate from, but related to, the systems development life cycle (SDLC). The chapter first differentiates between techniques, methodologies, and the life cycle. Most of the chapter then focuses on those techniques known as the "structured techniques." The increasingly popular prototyping and joint application design techniques are also introduced. Finally, the chapter introduces commercial methodologies as an attractive alternative for businesses to acquire a package of techniques and a step-by-step approach for using them. You will understand the development techniques and methodologies when you can:

Differentiate
between the life cycle, techniques, and methodologies.

Compare and contrast
several popular systems development techniques
and their principal supporters, strategy, tools, and approach.

Justify
the integrated use of several techniques as opposed to adopting a
single technique, and describe how multiple techniques are packaged
into methodologies.

Describe
the common characteristics of a commercial methodology.

Make
a lifelong commitment to avoid blindly following any one technique
or methodology.

MINICASE

THE THREE OSTRICHES

Most of the chapter minicases in this book represent realistic people and situations written in a playscript style to help you follow the people as the story unfolds. This minicase is different. We believe there are numerous ways to teach and learn. The following fable, which appears in a book by Jerry Weinberg called Rethinking Systems Analysis and Design, *is an entertaining yet effective minicase. Read and enjoy the fable, but pay attention to its moral. There are many systems analysts who have discovered the moral of this fable the hard way!*

Three ostriches had a running argument over the best way for an ostrich to defend himself. Although they were brothers, their mother always said that she couldn't understand how three eggs from the same nest could be so different. The youngest brother practiced biting and kicking incessantly, and held the black belt. He asserted that ''the best defense is a good offense.'' The middle brother, however, lived by the maxim that ''he who fights and runs away, lives to fight another day.'' Through arduous practice, he had become the fastest ostrich in the desert — which you must admit is rather fast. The eldest brother, being wiser and more worldly, adopted the typical attitude of mature ostriches: ''What you don't know can't hurt you.'' He was far and away the best head-burier that any ostrich could recall.

One day a feather hunter came to the desert and started robbing ostriches of their precious tail feathers. Now, an ostrich without his tail feather is an ostrich without pride, so most ostriches came to the three brothers for advice on how best to defend their family honor. ''You three have practiced self-defense for years,'' said their spokesman. ''You have the know-how to save us, if you will teach it to us.'' And so each of the three brothers took on a group of followers for instruction in the proper method of self-defense — according to each one's separate gospel.

Eventually, the feather hunter turned up outside the camp of the youngest brother, where he heard the grunts and snorts of all the disciples who were busily practicing kicking and biting. The hunter was on foot, but armed with an enormous club, which he brandished menacingly as the youngest

brother went out undaunted to engage him in combat. Yet fearless as he was, the ostrich was no match for the hunter, because the club was much longer than an ostrich's leg or neck. After taking many lumps and bumps, and not getting in a single kick or bite, the ostrich fell exhausted to the ground. The hunter casually plucked his precious tail feather, after which all his disciples gave up without a fight.

When the youngest ostrich told his brothers how his feather had been lost, they both scoffed at him. ''Why didn't you run?'' demanded the middle one. ''A man cannot catch an ostrich.''

''If you had put your head in the sand and ruffled your feathers properly,'' chimed the eldest, ''he would have thought you were a yucca and passed you by.''

The next day the hunter left his club at home and went out hunting on a motorcycle. When he discovered the middle brother's training camp, all the ostriches began to run — the brother in the lead. But the motorcycle was much faster, and the hunter simply sped up alongside each ostrich and plucked his tail feather on the run.

That night the other two brothers had the last word. ''Why didn't you turn on him and give him a good kick?'' asked the youngest. ''One solid kick and he would have fallen off that bike and broken his neck.''

''No need to be so violent,'' added the eldest. ''With your head buried and your body held low, he would have gone past you so fast he would have thought you were a sand dune.''

A few days later, the hunter was out walking without his club when he came upon the eldest brother's camp. ''Eyes under!'' the leader ordered and was instantly obeyed. The hunter was unable to believe his luck, for all he had to do was walk slowly among the ostriches and pluck an enormous supply of tail feathers.

When the younger brothers heard this story, they felt impelled to remind their supposedly more mature sibling of their advice. ''He was unarmed,'' said the youngest. ''One good bite on the neck and you'd never have seen him again.''

''And he didn't even have that infernal motorcycle,'' added the middle brother. ''Why, you could have outdistanced him at half a trot.''

But the brothers' arguments had no more effect on the eldest than his had on them, so they all kept practicing their own methods while they patiently grew new tail feathers.

Source: From Gerald M. Weinberg, *Rethinking Systems Analysis and Design*, pp. 23–24. Copyright © 1982 by Scott, Foresman and Company. Reprinted by permission.

Discussion Questions

1. What is the moral of the fable?

2. Methodologies are specific strategies and techniques for developing systems. Each uses specific tools and steps and each claims to address the needs of systems development better than the other methodologies. Today's analyst is faced with an almost overwhelming choice of methodologies. Before we study and evaluate those choices, how might the fable and its moral relate to the study of methodologies?

TECHNIQUES AND METHODOLOGIES COMPLEMENT THE LIFE CYCLE

Systems development techniques and methodologies are frequently confused with the life cycle. Is there a difference? Some experts claim that they have replaced the life cycle. In this section, we hope to dispel that myth and demonstrate that systems development techniques and methodologies are intended to complement the life cycle, not replace it.

Long Live the Life Cycle!

Has the life cycle become obsolete? NO! The life cycle is not dead! Recall the intent of the life cycle—to plan, execute, and control a systems development project. The life cycle defines the phases and tasks that are *essential* to systems development, no matter what type or size system you may try to build. Note the emphasis on the word essential. For instance, we should *always* study and analyze the current system (at some level of detail!) before defining and prioritizing user requirements. It's common sense!

Where, then, did all the myths about the life cycle's death get started? The answer lies in the popularity of techniques and methodologies. Let's formally define those terms:

A **technique** is an approach that applies specific tools and rules to complete one or more phases of the system development life cycle. A common synonym is *paradigm* (pronounced "pear-a-dime").

Most techniques are only applicable to a part of the total life cycle. Therefore, they cannot, by themselves, replace the life cycle! Some of the more popular systems development techniques are intended to introduce engineering-like precision and rigor into the life cycle.

One example of a development technique, with which you may be familiar, is structured programming. Structured programming is applicable to the systems implementation and support phases of the life cycle. It provides no support for systems planning or systems analysis, and minimal support for systems design. Thus, you must combine structured programming with other development techniques to fully address the life cycle.

Development techniques are frequently invented by researchers, consultants, and academicians. Eventually, they become well documented in the form of courses, seminars, textbooks, industry books, and the like. Because each expert tries to improve on a technique, many variations of the same technique become common. Over time, the more successful development techniques become practiced by a noticeable percentage of systems developers. This chapter will examine some of the most popular techniques.

What about methodologies?

> A **methodology** is a comprehensive and detailed version of an *entire* systems development life cycle that incorporates (1) step-by-step tasks for each phase, (2) individual and group roles to be played in each task, (3) deliverables and quality standards for each task, and (4) development techniques (as defined above) to be used for each task. (It may also incorporate specific technology for developers — this technology is covered in the next chapter.)

Two points are important. First, a true methodology should encompass the entire systems development life cycle, including systems support (which is frequently omitted). Second, most modern methodologies incorporate the use of several development techniques (and their associated tools).

Many businesses purchase their development methodology. A methodology vendor (e.g., Structured Solutions) sells a methodology (e.g., STRADIS) in the form of manuals, software, training, and consulting. This chapter will describe and list some of the more popular commercial methodologies that you may encounter.

> ***Note*** Given these two definitions, we must now admit that the terms *technique* and *methodology* are used rather loosely and interchangeably in computing literature and industry. For example, *structured analysis* is frequently called a methodology, when in fact it is a technique. That technique is part of several true methodologies, such as STRADIS.

Thus, the life cycle isn't dead — it is merely disguised in the form of popular commercial methodologies, and those methodologies incorporate popular techniques. This chapter is about those techniques and methodologies.

THE STRUCTURED TECHNIQUES

In this section we introduce the popular structured techniques.

> The **structured techniques** are formal approaches to breaking up a business problem into manageable pieces and relationships, and subsequently reassembling those pieces and relationships (possibly with additions and deletions) into a useful business (and computer) solution to the problem. A synonym is *structured methods*. (Regrettably, such synonyms contribute to the confusion between the terms *technique* and *methodology*.)

In a sense, the structured techniques are a "divide and conquer" approach to problem solving as it relates to systems and software development. We'll survey the following structured techniques:

- Structured programming.
- Structured design (various approaches).
- Modern structured analysis.

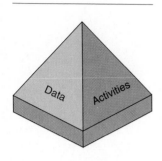

- Data modeling.
- Information engineering.

In this chapter we'll only introduce the concepts that underlie these approaches. Don't be concerned if you don't fully understand the examples presented; other chapters in the book are intended to teach you how to perform these techniques.

A Framework for the Structured Techniques

Your pyramid model can be used to classify the structured techniques. Most of today's structured techniques focus on two system perspectives. ACTIVITIES and DATA—which are also two faces of the pyramid (see Figure 4.1).

Recall that the ACTIVITIES building blocks were based on the input-process-output (IPO) concept. Because the process is at the center of the input-to-output transformation, these techniques are commonly called **process-oriented techniques.** Process techniques build models of systems based on studying processes and/or their inputs and outputs. Structured design and modern structured analysis are examples of such techniques

Data-oriented techniques build models of systems based on the ideal organization and access of system's data, independent of how that data will (or might) be used to fulfill information (output) requirements. Data modeling and information engineering are examples of data-oriented techniques.

It should be noted that most process-oriented techniques recognize the importance of data. Similarly, data-oriented techniques generally recognize the importance of processes. Why? In Chapter 2 you learned that all systems contain both data and process components. The two approaches differ only in the building block that "drives" (or is emphasized by) the technique. For this reason, these two general approaches are sometimes called *process-driven* or *data-driven.*

What about the other building blocks in the pyramid model? Shouldn't development techniques deal with those building blocks also? Absolutely! Unfortunately, few modern techniques have directly dealt with the NETWORKS building block as presented in Chapter 2. However, that should soon change. Today, many system developers talk about downsizing appropriate applications to smaller, distributed computers and networks. The development of these distributed, cooperative, and client/server applications will force development techniques to evolve to more adequately address the NETWORKS building blocks. (Note: This book will offer some suggestions of its own.)

What about the PEOPLE building block? Later in this chapter, you'll learn that some development techniques, like joint application development (JAD), do address the PEOPLE issues of systems development (e.g., fact finding, requirements determination, politics, human engineering).

Now let's examine some specific structured techniques in greater detail.

Structured Programming

Structured programming has evolved into a de facto (meaning unwritten) standard in much of the computing industry.

FIGURE 4.1   IS Dimensions Emphasized by Today's Structured Techniques   One basis for classifying the structured techniques is the systems dimension(s) that they "most" emphasize. Most popular techniques seek to focus attention upon either systems activities or systems data, or upon both.

Structured programming is a process-oriented technique for designing and writing programs with greater clarity and consistency. Essentially, structured programming suggests that the logic of any program can and should be written with a limited set of control structures.

Some of structured programming's early principle advocates were Corrado Böhm, Guiseppe Jacopini, J. Edgar Dykstra, and Harland Mills.

After all these years, the technique is still often misunderstood. Structured programming deals *only* with program logic and code. It suggests that programs should be designed such that they can be read from top-to-bottom, with minimum branching. Specifically, well-structured programs are written exclusively with various combinations of three **restricted control structures.** The three basic structures, as you are probably aware, are

- A *sequence* of instructions or group of instructions.
- A *selection* of instructions or group of instructions based on some decision criteria (this construct is often referred to as the if-then-else or case construct).
- An *iteration* of instructions or group of instructions that are repeated based on some criteria (this construct comes in two basic forms: repeat-until and do-while).

These constructs can be repeated (or nested) within one another. An important characteristic of the structures is that each construct *must* exhibit a *single-entry, single-exit* property. This means that there can only be one entry point and one exit point from any given structure. Structured code reads like this page — top to bottom, with no backward references. This makes the code easier to read, test, debug, and maintain. The structured programming constructs and properties are illustrated in the flowchart in Figure 4.2.

The structured programmer designs logic using modeling tools such as flowcharts, box charts (Nassi-Schneiderman diagrams), pseudocode, or action diagrams — carefully restricting the structures to those constructs described above. Then, the programmer will code that structure, carefully preserving the restricted control structures and single-entry, single-exit properties.

Depending on the language used, the programmer may still use GOTO statements or their equivalents. Contrary to some popular belief, structured programming is *not* GOTO-less programming. It seeks only to control the frequently undisciplined use of the GOTO statement, especially the GOTO that sends control backward (bottom to top) through the program (thus creating more than one entry point into a sequence of instructions).

As a structured technique, structured programming supports only portions of the design, implementation, and support phases of the life cycle. Structured programs are easier to write, test, and debug, and much easier to maintain (due to their start-to-finish, book-like readability).

Structured programming has been popular for many years; however, you may be surprised to learn that most IS shops' software libraries are still suffering from years of programs written before structured programming became widely practiced. These "legacy" systems are a maintenance nightmare. Many shops are searching for techniques and technology to rapidly reengineer these unstructured programs into structured programs.

FIGURE 4.2   Structured Programming Concepts   This flowchart demonstrates the basic concepts behind structured programming, restricted control structures, and single-entry, single-exit flow.

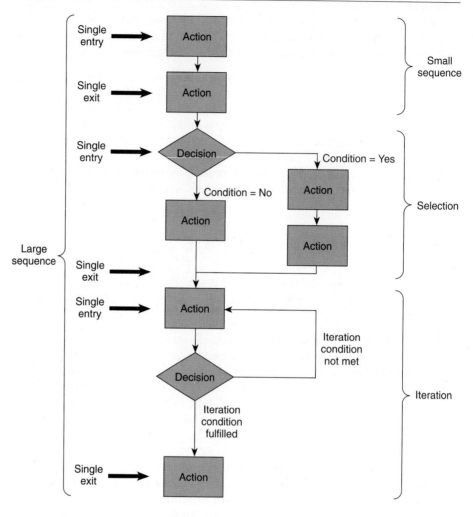

Structured Design

Structured design techniques help developers deal with the size and complexity of programs.

Structured design is another process-oriented technique for breaking up a large program into a hierarchy of modules that result in a computer program that is easier to implement and maintain (change). Synonyms (although technically inaccurate) are *top-down program design* and *structured programming.*

The principal advocates of structured design include Larry Constantine, Ed Yourdon, Meiler Page-Jones, Jean Dominique Warnier, Ken Orr, and Michael Jackson.

The concept is simple. Design a program as a top-down hierarchy of modules. A module is a group of instructions—a paragraph, block, subprogram, or subroutine. The top-down structure of these modules is developed according to various design rules and guidelines. (Thus, merely drawing a hierarchy or structure chart for a program is *not* structured design.)

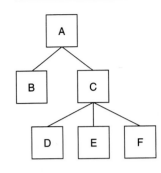

Ideally, each module's internal logic would be written using the structured-programming technique. Thus, we see that techniques can be used in combination to improve problem solving.

Separate schools of thought have developed on the proper technique for accomplishing a well-structured design. These include

- *Yourdon-Constantine.* This technique derives the ideal software structure by studying the flow of data through necessary program functions. (Note: This technique depicts the top-down structure of modules as an inverted tree—see margin.)
- *Warnier-Orr.* This technique derives the ideal software structure by studying the contents of the outputs and inputs. (Note: This technique depicts the top-down hierarchy as a series of left-to-right braces—see margin.)
- *Jackson.* This technique also derives the ideal software structure by studying the contents of the outputs and inputs. (This technique, popular in Europe, depicts the top-down hierarchy of modules as an inverted tree—similar to Yourdon-Constantine.)

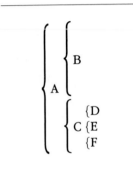

All of these techniques are considered process techniques because their purpose is to design processes—specifically, software processes. Because the Yourdon technique is more widely known and practiced, the remainder of this discussion will focus on that approach.

Yourdon structured design seeks to factor a program into the top-down hierarchy of modules that have the following properties:

- Modules should be highly **cohesive;** that is, each module should accomplish one and only one function. This makes the modules reusable in future programs.
- Modules should be loosely **coupled;** in other words, modules should be minimally dependent on one another. This minimizes the effect that future changes in one module will have on other modules.

The software model derived from Yourdon structured design is called a **structure chart** (Figure 4.3). The structure chart is derived by studying the flow of data through the program. Structured design is performed during the design phase of your life cycle. It does not address all aspects of design—for instance, structured design will not help you design inputs, databases, or files.

The benefits of a structured design are numerous. First, programs that are factored according to structured design can be more easily written and tested by multiple programmer teams. Why? Because the interfaces between modules are both well defined and limited by rules, the modules that test correctly by themselves should test correctly when brought together as a system. Top-down program structures also simplify programming effort because they lend themselves to top-down coding and stub testing (a technique familiar to experienced programmers). Second, systems and programs developed with struc-

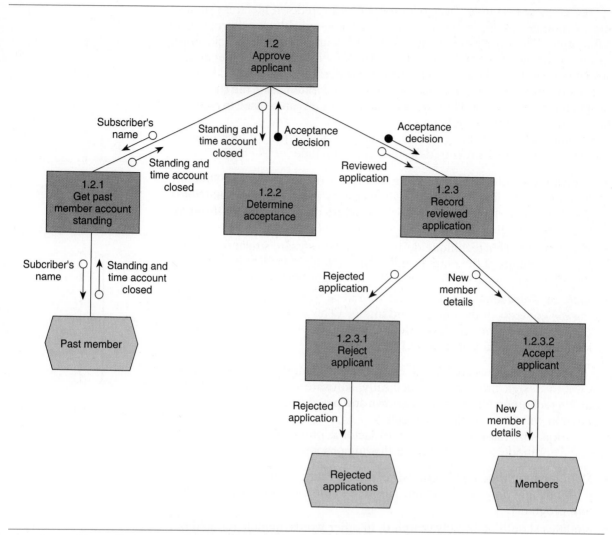

FIGURE 4.3 The End Product of Structured Design   This structure chart is typical of the process models that characterize techniques like Yourdon/Constantine structured design.

tured design are more easily maintained. A major consequence of most program maintenance is the *ripple effect,* which occurs when a change in one module necessitates changes in numerous other modules. (The ripple effect can occur recursively.) Structured design intentionally seeks to reduce the ripple effect by minimizing inter-module connections and dependence.

A third, often overlooked benefit of structured design is that program modules developed according to the technique tend to be reusable. This is because they are built to be cohesive. Code reusability has become a major issue in the industry because companies cannot afford to repeatedly "reinvent the wheel." What some fail to recognize is that modules will not be reusable unless they are intentionally designed for reusability.

Modern Structured Analysis

Structured analysis is the most popular and most widely practiced process-centered technique. (At one time, it was the most popular technique of any type.)

> **Structured analysis** is a process-centered technique that is used to model user requirements for a system. Structured analysis breaks up a system into processes, inputs, outputs, and files. It builds an input-process-output, flow-oriented model of a business problem or solution.

The principal advocates of structured analysis are Tom DeMarco, Chris Gane, Trish Sarson, and Ed Yourdon.

The structured analysis technique is simple in concept. A new system model evolves from a series of flow-oriented diagrams called **data flow diagrams** or DFDs (Figure 4.4). These are not program flowcharts! They only show flow of data, storage of data, and the processes that respond to and change data. Because data flow diagrams depict data flows between processes, structured analysis is sometimes referred to as a *data flow approach.* On the other hand, most experts consider DFDs to be a *process model* and place emphasis on those processes — hence, structured analysis is truly a process-oriented approach.

The analyst might produce several DFDs in structured analysis. These DFDs may differ with respect to

- Whether they model the current system or the proposed system.
- Whether they model the implementation details of the system (sometimes called the *physical* system) or the essence of the system (sometimes called the *logical* system).

Thus, any given DFD models one of the following: (1) the current system's implementation, (2) the current system's essence, (3) the proposed system's essence, or (4) the proposed system's implementation.

The concept of a **logical system,** sometimes called an **essential system,** was created by and is crucial to structured analysis. It addresses the following problem. We tend to limit our own creativity by prematurely thinking about a new system in terms of *how* it should be designed and implemented (the **physical system**). Structured analysis encourages (or requires) the analyst to define *what* the system should do (the logical system) before deciding *how* the system should be designed or implemented. Thus, the technique forces the analyst to first consider "the business solution" — and only later consider the "technical" solution. Advocates insist that by separating logical and physical concerns, the following benefits are realized:

- The analyst more accurately identifies business and end-user requirements by not prematurely worrying about technology.
- The analyst is more inclined to conceive more creative alternative solutions instead of solutions that are based on the existing system (avoiding the "we've always done it that way" syndrome). Creative solutions can establish competitive advantage in today's complex economy.

Structured analysis was introduced to the information systems community in 1978. It enjoyed a rapid and widespread growth in popularity through the mid-1980s. The popularity of data flow diagrams is attributed to the emergence

FIGURE 4.4 **An End Product of Structured Analysis** This data flow diagram is typical of the process models that characterize methodologies like structured analysis. These models are particularly useful for documenting either the current business system, or the requirements for a new and improved business and/or computer system.

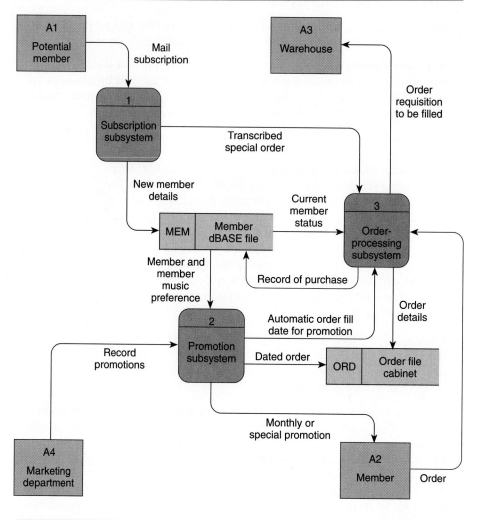

FIGURE 4.4 **An End Product of Structured Analysis** This data flow diagram is typical of the process models that characterize methodologies like structured analysis. These models are particularly useful for documenting either the current business system, or the requirements for a new and improved business and/or computer system.

of the technique. In Gane and Sarson's approach and in DeMarco's approach, the analyst draws four complete sets of data flow diagrams:

1. Physical data flow diagrams of the current system (that show *how* the current system works).
2. Logical data flow diagrams of the current system (derived from 1 above — but showing "what" the system currently does).
3. Logical data flow diagrams for the new system (additions, deletions, and modifications to 2 above — showing "what" the new system must do).
4. Physical data flow diagrams of the target system (that show "how" the new system implements its requirements).

You may wonder how well structured analysis and data flow diagrams have worked in the industry. For most analysts, the results have been mixed. On the one hand, data flow diagrams have proved easy to draw and even easier to read. The methodology, although rarely performed with the intended rigor, has improved the results of systems analysis. On the other hand, analysts have learned that data flow diagrams and structured analysis cannot, by themselves, ensure completeness, consistency, and accuracy in "business and user requirements definition." Furthermore, the very act of going through all the various stages of data flow diagraming slows systems analysis—in an era when productivity has become a major issue.

What was wrong with the original technique? First, analysts frequently spent too much time modeling the current system (steps 1 and 2). This led to a condition known as *analysis paralysis* and considerable user and management impatience. Second, many analysts had difficulty making the transition from physical to logical modeling (step 1 to step 2). Third, structured analysis was an incomplete technique: DFDs model only one dimension of a system—its processing. Data and network modeling were, for the most part, ignored.

Recently, Ed Yourdon introduced an improved version of structured analysis called **modern structured analysis.** This approach eliminates detailed modeling of the current system (both physical and logical). Instead, it calls for

1. A very simple physical data flow diagram of the proposed system (that shows the context or boundaries of the system—called an *environmental model*).
2. A logical data model (discussed in greater detail later).
3. Bottom-up, logical data flow diagrams for the new system (that show "what" the system must do in response to specific business events and desired responses to those events).
4. Top-down logical data flow diagrams for the new system (that show "what" the system must do—actually a rearrangement of the bottom-up DFDs from 3 above).
5. Physical data flow diagrams of the new system (that show "how" the new system will implement the requirements from 4 above).

Notice that the emphasis in all steps has clearly shifted to the new system—little attention is given to the current system. Different variations on structured analysis fall between the original approach and the modern approach. Most experts now admit that, at the very least, less emphasis should be spent on detailed modeling of the current system. This is because (1) the amount of time is often not cost justified, (2) the current system tends to bias or limit creativity when changing context from the current system to the new system, and (3) users and managers have little patience with modeling a system that they fully expect to change or replace.

Structured analysis was the first structured technique to specifically address the systems analysis phases of the life cycle. But it only addresses those phases from the specification or modeling viewpoints. The analyst must rely on other techniques to collect facts, analyze the system for problems and opportunities, identify user requirements, identify candidate solutions, and evaluate those candidates for feasibility. This fact is not well understood by many who have tried and failed with the technique.

It should be noted that structured analysis and structured design are integrated techniques. The Yourdon structured design approach provides formal strategies for deriving program structure charts from properly constructed data flow diagrams of structured analysis. Collectively, they are often called a *software engineering* technique.

Data Modeling

Today, process-oriented techniques have been complemented by data-oriented techniques.

> **Data modeling** is a data-oriented technique that represents a system in terms of the system's data, independent of how that data will be processed or used to produce information.

Once again, the concept is simple. If the data is captured and stored in flexible file and database structures, all current and future information needs can be fulfilled using that data. Data modeling has evolved out of the techniques used by database designers. Information modeling, as created by Matt Flavin, and object-oriented analysis (in part), as created by Stephen Mellor and Sally Shlaer, can be classified as data modeling techniques.

The data modeling techniques are briefly described as follows. First, identify those business "things" (called *entities*) about which the business or application collects data. Entities might be any of the following:

- Tangible things (e.g., materials, supplies, machines, vehicles, and products).
- Roles (e.g., customers, suppliers, employees, and account holders).
- Events (e.g., orders, requisitions, contracts, trips, accidents, or payments).
- Places (e.g., sales offices and warehouse locations).

Note   Does the above list look familiar? It should. The same list was introduced as the system owner's view of the DATA building block in Chapter 2. Data modeling techniques represent the DATA building block.

Next, identify those *attributes* that describe an instance of each entity. For example, consider the entity, STUDENT. Some of the attributes that describe an instance of STUDENT include: STUDENT ID NUMBER, STUDENT NAME, CAMPUS ADDRESS, CAMPUS PHONE, HOME ADDRESS, HOME PHONE, SEMESTER CLASSIFICATION, CREDIT HOURS COMPLETED, GRADE POINT AVERAGE, and so forth. These are the attributes, or fields, that we eventually want to store in a student file or database. Inputs will eventually be designed to capture these attributes. And a wide variety of outputs can be designed to report on these and other attributes.

Next, identify the business activities that occur between entities—for instance, STUDENTs "ENROLL IN" COURSEs. STUDENT and COURSE are entities. "ENROLL IN" is a business activity that associates instances of STUDENT and COURSE. Why is this important? Because we don't only want to store data about STUDENTs and COURSEs—we will also likely want to know which STUDENTs are in which COURSEs. The analyst will normally draw a picture or **data model** of these entities, relationships, and attributes (Figure 4.5).

In some data modeling approaches, the analyst also uses formal techniques to ensure that the data model will be flexible enough to adapt to current and

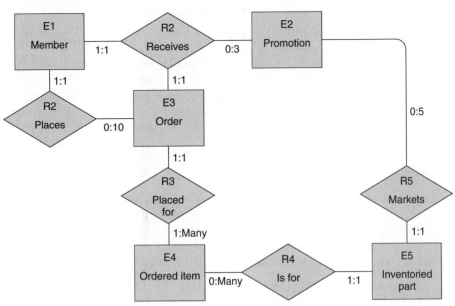

FIGURE 4.5   An End-Product of Data Modeling   This model, called an *entity-relationship-attribute (ERA) diagram,* is typical of the data models that result from data-modeling techniques.

future requirements based on the same data. This technique, called *normalization,* is covered later in the textbook.

Some data modeling approaches also require that the analyst identify the business events that cause instances of data entities to be created, deleted, or modified. For example, data about a STUDENT might be "created" when (1) a new student is admitted, (2) a former student is readmitted, or (3) a student transfers from another campus. These are three distinct business events in a school. Such an event analysis is useful for eventually identifying process models that can properly capture and store the data represented on the data model.

The benefits of data modeling are important. If the files and databases for the new system are built according to the data model, they will do all of the following:

- Contain accurate and up-to-date data.
- Meet all of today's output requirements.
- Meet future output requirements without drastic changes to the system, because the data is already there or can be easily added to the appropriate entities.

In other words, processes and applications can evolve around the data model, with little change to the data model itself.

Data modeling techniques have become very popular. Unfortunately, there are problems to overcome. Many books represent data modeling only as a *database* technique. Actually, it works equally well with conventional files and distributed data. Most books advocate building very large corporate databases and systems. Actually, the technique works equally well for incrementally building smaller application databases that can be easily integrated at a later date.

Although data modeling is useful, sooner or later the processes of a system must also be designed. Process-oriented tools and techniques, such as modern structured analysis, can be used to address these process requirements and thereby complement the data modeling techniques. This blending of data and process-modeling techniques has evolved into another technique, information engineering.

Information Engineering

In addition to blending data and process modeling, information engineering places new emphasis on the importance of systems planning. Arguably, information engineering has replaced structured analysis and design as the most popular technique in current practice. (Note: Information engineering actually includes most of the concepts, tools, and techniques of structured analysis.)

> **Information engineering** is a data-driven, but process-sensitive technique that is applied to the organization as a whole (or across some major sector of the organization), rather than on an ad hoc, project-by-project basis. Although the technique suggests a balance between data- and process-oriented methods, it is clearly data-driven; data models are built first and process models are built later.

Information engineering's principle advocates are James Martin and Clive Finkelstein.

This is the first structured technique we've covered that addresses most of the life cycle. In fact, the only phase not included in information engineering is systems support. Of particular and unique interest is information engineering's emphasis on systems planning. Information engineering defines **systems planning** as the improvement of the organization through information technology. The approach seeks to identify mission critical data and functions that should be supported and integrated through technology.

The following steps of information engineering are illustrated in Figure 4.6.

1. The analysts initially engage in strategic systems planning for the organization (or, more typically, for a major sector — such as a division, campus, site, etc.).
2. Based on the resulting strategic plan, the analysts "carve" out subsystems that information engineering call *business areas.* Structured systems analysis is performed on that entire business area. (Information engineers call it *business area analysis.*)
3. The analysts then "carve" out another subsystem that represents a high priority application (for the business area) that will go through further application analysis and design (using structured techniques).
4. The analyst implements the designed application.

Subsequent projects will address other applications in the same business area — until such a time that the business area is completely supported by a set of integrated applications. These applications are integrated on the business area's database.

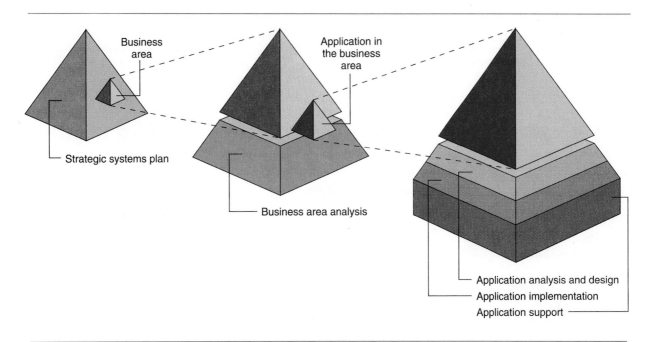

FIGURE 4.6 **Information Engineering** Your pyramid model can be adapted to illustrate the information engineering technique. A strategic plan is completed for the entire organization (or some major subset). The plan identifies multiple business areas. Each of those business areas is subjected to a thorough business area analysis. This analysis identifies multiple applications to be designed, implemented, and supported.

Other business areas would go through a similar analysis, design, and implementation process.

In information engineering, the center of the universe is stored data (see Figure 4.7). Analysts, programmers, and other computer professionals are responsible for designing all data stores and ensuring that data is captured, stored, and properly maintained. They also design and implement the major information outputs of the system. End-users fulfill many additional information requirements by learning and applying user-friendly report writers and query languages.

Information engineering arguably represents the most radical revision of the classical structured techniques. Still, information engineering techniques appear to be gaining momentum. This may be because the technique truly balances and integrates the data and process perspectives.

Although we choose not to blindly follow any one technique, this book embraces much of the concept and flavor of information engineering. Fortunately, the information engineering technique also serves to teach the fundamental tools and approaches of structured analysis, structured design, data modeling, and other equally important approaches.

THE JOINT APPLICATION DEVELOPMENT (JAD) TECHNIQUE

Whereas the previously discussed techniques place significant emphasis upon processes and data, an increasingly popular technique called joint application development places significant emphasis upon *people*.

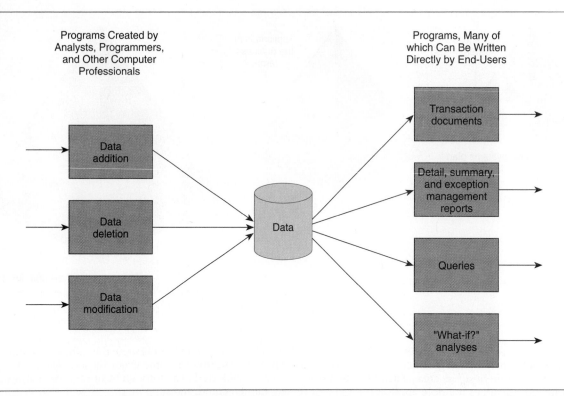

Programs Created by
Analysts, Programmers,
and Other Computer
Professionals

Programs, Many of
which Can Be Written
Directly by End-Users

FIGURE 4.7 The Information Engineering Philosophy In the information engineering philosophy, the center of the universe is data. The systems developers build a powerful, integrated, and flexible database for each identified business area. Applications are built around the database. In most cases, the applications depicted to the left of the database would be built by the systems developers. In contrast, many of the applications depicted to the right of the database can be built by the end-users themselves.

Joint application development (JAD) is a highly structured workshop that brings together users, managers, and information systems specialists to jointly define and specify user requirements, technical options, and external designs (inputs, outputs, and screens).

Note The original acronym stood for joint application *design;* however, even in its original form, it was much more oriented to supporting the systems analysis phases, especially requirements definition. Because we now place so great an importance on user involvement throughout the life cycle, we prefer to use the term *development.*

JAD attempts to solicit greater user and management participation in the systems development life cycle. There are numerous benefits to this increased participation:

- JAD tends to improve the relationship between users, management, and information systems professionals.

- JAD tends to improve the computer literacy of users and managers as well as the business and application literacy of information systems specialists.

- JAD places the responsibility for conflict resolution where it belongs—on the shoulders of both users and management.

- JAD usually decreases elapsed systems development time by consolidating multiple interviews into the structured workshop.

- JAD usually lowers the cost of systems development by getting the requirements correctly defined and prioritized the first time.
- JAD usually results in greater system value and user/management satisfaction. It increases user and management confidence and support of the project.
- JAD usually results in systems that cost less to maintain because the first version already fulfills requirements.

To what degree does JAD accelerate the life cycle? Most JAD sessions are scheduled as three- to five-day workshops of four- to eight-hour duration. Led by an experienced JAD systems analyst, the workshops can more quickly identify vocabulary, problems, requirements, priorities, and alternative solutions, and select a final technical solution. Experienced JAD analysts can even use structured techniques with users, carefully avoiding technical buzzwords and complicated rules. A three- to five-day JAD workshop can replace a one- to six-month calendar of interviews and less-structured meetings. Fewer conflicts arise as the system moves into design and implementation because everybody (users, managers, and information systems specialists) has agreed on all of the important system building blocks.

To succeed, JAD calls for several prerequisites. First, management must be willing to release workers from their day-to-day jobs (or pay overtime) to participate in sessions. Second, management must be willing to participate in the sessions themselves. Management must also foster an environment of cooperation and listening when working with subordinates during the sessions. Finally, recorders and session leaders must be well trained to focus and redirect discussion, and mediate conflicts and disputes.

JAD opportunities and techniques will be more fully described at appropriate points in this book.

THE PROTOTYPING AND RAPID DEVELOPMENT TECHNIQUES

The final techniques to be introduced in this chapter are prototyping and rapid application development (RAD). Let's begin with prototyping.

> **Prototyping** is a popular engineering technique used to develop a small-scale working (or simulated) model of a product or its components. When applied to information systems development, prototyping involves building an iterative, working model(s) of the system or a subsystem.

Prototyping can be used in various phases of the life cycle. There are four types of information systems prototyping:

- **Feasibility prototyping** is used to test the feasibility of a specific technology that might be applied for an information system. Consider the following example. A college placement center currently types student information into their database from applications submitted by the students. The director of the placement center is curious as to whether or not students can directly enter the application data into personal computers (for later editing and processing on their minicomputer). The analyst creates a technology prototype (not fully functional) to test the students' reactions to entering their own data into the personal computer system. Based on the results of the prototype, a decision can be made as to whether or not this technical solution is worth further development.

- **Requirements prototyping** (sometimes called *discovery prototyping*) is used to "discover" the users' business requirements. It is intended to stimulate the users' thinking. The concept is simple — the users will recognize their requirements when they see them. During requirements prototyping, the analyst may "paint" sample screens or reports, and solicit user feedback regarding their content (but not their format).

  It is important to emphasize that discovery prototypes are rarely intended to simulate the "look" of the final system. In fact, the danger of requirements prototyping is that users may (1) become overly concerned with format of screens and reports, and/or (2) that users will consider the form of the prototype to be the form of the final system.

- **Design prototyping** (sometimes called *behavioral prototyping*) is used to simulate the design of the final information system. Whereas requirements prototyping focuses on content only, design prototyping focuses on form and operation of the desired system. When an analyst creates a design prototype, the users are expected to evaluate that prototype as if it were part of the final system. Thus, users should evaluate how easy the system is to learn and use, as well as the "look and feel" of the screens and reports, and the procedures required to use the system. These prototypes may serve either partial design specifications, or they may evolve into implementation prototypes.

——————— ✓ ———————

Database Languages for Prototyping

ADS/O
dBASE
IDEAL
Focus
MANTIS
Natural
Paradox
Progress
RAMIS
R:BASE
SAS

————————————————

- **Implementation prototyping** (sometimes called *production prototyping*) is an extension of design prototyping where the prototype evolves directly into a production system. Initially, implementation prototypes usually omit details like data editing, security, and help messages. These details may be added later if the prototype is intended to evolve into a production system.

  Implementation prototyping became popular with the increased availability of fourth-generation languages (4GLs) and applications generators. These database languages and generators provide powerful tools for quickly generating prototypes of files, databases, screens, reports, and the like. Fourth-generation languages tend to be less procedural than traditional languages like BASIC, COBOL, FORTRAN, and C. By less procedural we mean that the code is more English-like and, in many cases, allows you to specify "what" the system should do without specifying "how" to do it. This makes it possible to develop the prototypes more quickly. Examples of database languages suitable for prototyping are listed in the margin.

The benefits of prototyping are numerous:

- End-users become more active participants in systems development. They tend to be more excited by working prototypes than paper design specifications.

- Requirements definition is simplified through realization that many end-users will not understand or be able to state their detailed requirements until they see a prototype.

- The likelihood that end-users will approve a design or paper only to reject the implementation of that design is greatly reduced.

- Design by prototyping is said to reduce development time; however, some experts dispute these savings.

But prototyping can also be dangerous. What are the disadvantages?

- Prototypes tend to skip through analysis and design phases too quickly. This encourages the analyst to jump too quickly into code without understanding problems and requirements.
- Prototyping can discourage the consideration of alternate technical solutions. The analyst tends to go with the first prototype alternative that receives a reasonably favorable reaction from users.
- Systems implemented from prototypes frequently suffer from lack of flexibility to adapt to changing requirements—they were developed "quick and dirty."
- Prototypes are not always easy to change. Several experts have noted the growing libraries of poorly designed, unstructured, unreadable, and inadequately documented 4GL code.
- Prototypes are rarely polished. The technology used can actually inhibit user comprehension and, subsequently, discourage participation.

A fundamental question is "Can you prototype without a specification?" In other words, "Is prototyping a substitute for structured techniques?" Based on how 4GLs are being used by many analysts, you might get the impression that paper specifications—for example, the structured techniques—are unnecessary. The answer, however, is an emphatic *NO!* No competent engineer would prototype without a specification. Prototyping should complement appropriate specification techniques. The rapid application development technique addresses this problem.

> **Rapid application development (RAD)** is the merger of various structured techniques (especially the data-driven information engineering) with prototyping techniques and joint application development techniques to accelerate systems development.

RAD calls for the interactive use of structured techniques and prototyping to define the users' requirements and design the final system. Using structured techniques, the developer first builds preliminary data and process models of the requirements. Prototypes then help the analysts and users to verify those requirements, and to formally refine the data and process models. The cycle of models, then prototypes, then models, then prototypes, and so forth ultimately results in a combined business requirements and technical design statement.

COMMERCIAL METHODOLOGIES

Let's now take a closer look at methodologies.

> **Commercial methodologies** are "off-the-shelf," step-by-step procedures, individual and group roles, deliverables, quality standards, preferred techniques, and tools for completing the entire systems development life cycle. Examples of commercial methodologies are listed in the margin.

———— ✓ ————
Popular Commercial Methodologies

CARA
The GUIDE
Method/1
Navigator
PRIDE
SDM
SPECTRUM
STRADIS

Admittedly, many surveys show that the vast majority of IS shops have not yet invested in a standard commercial methodology. That doesn't mean that the programmers and analysts do not use the popular techniques such as structured analysis or information engineering (and some of them mistakenly call those techniques their "methodology"). On the other hand, progressive IS shops are increasingly investing in commercial methodologies in order to fully exploit today's popular techniques and the technology that supports those techniques. (Note: Systems development technology is introduced in the next chapter.)

Most modern, commercial methodologies are based on some combination of techniques. For example, Structured Solutions' STRADIS methodology is based largely on structured analysis and design techniques. Ernst & Young's Navigator methodology is based on the information engineering technique. Many commercial methodologies also include, require, or recommend the use of CASE technology (covered in the next chapter).

Commercial methodologies originated in the 1970s when information systems management realized that most systems were being built with haphazard methods and variable documentation. Most projects were behind schedule and over budget. If and when the projects were completed, users were frequently less than satisfied with the results, and maintenance costs were excessive.

Although they tend to be very expensive, commercial methodologies offer numerous advantages. First, they standardize the phases and deliverables of systems development. This makes it possible to react to turnover on project teams (for instance, promotions, transfers, resignations, and terminations). Second, methodologies promote consistent documentation of systems. This reduces the lifetime maintenance costs for systems. Third, methodologies shift the burden for maintaining and improving the methodology to the methodology vendor. These vendors can afford to (and, to be competitive, must) invest greater resources to keep up with the latest techniques and trends for systems development.

Recall that the commercial methodology vendor sells its methodology in the form of manuals, software, consultants, and training courses. When not properly managed, methodologies can quickly become "shelfware"—a collection of manuals collecting dust on an out-of-the-way shelf. How does one manage a methodology? Most IS shops that commit to a methodology appoint a methodology manager (or coordinator or expert). This person coordinates training, methodology implementation, project selection, standards, problem resolution, and other methodology-related problems.

THE BOTTOM LINE

What was the moral of the fable of the three ostriches? As its author, Gerald Weinberg stated it, "It's not know-how that counts; It's know-when." In other words, you should look for the strengths in all techniques and methodologies to determine how they can be applied to each individual project at appropriate times. If you look hard enough, you'll find that most techniques have strengths that improve almost any project.

We suggested that prototyping without a specification is dangerous. But which specification-oriented, structured technique best complements prototyping? Both process-oriented and data-oriented methods have their strong points. We believe the question is moot. You should always use both modeling approaches. Why? Because, as we have seen, all systems have a data and a process dimension. One is not necessarily more important than the other.

A useful analogue to this suggestion can be found in the accounting profession. No accountant would suggest that management choose between a balance sheet (position statement), income statement (profit-loss statement), and a cash flow statement (statement of change in financial position). Why? Each of these documents describes the business from a different but equally important viewpoint.

The structured techniques, prototyping, and JAD clearly complement one another. Similarly, RAD combines information engineering, JAD, and prototyping techniques. For this reason, we will not teach any specific, single technique. Instead, we will teach a wide variety of systems analysis and design tools and techniques, demonstrating how they can and should be used together to develop superior systems. The book's running case study has integrated these techniques into a methodology that the organization calls Framework for the Application of the Structured Techniques (FAST).

However, we are not trying to start a new camp of FAST ostriches. You should make every effort to keep up to date on new techniques and approaches. For instance, The Next Generation box for this chapter introduces what many experts think will become the next "preferred" technique, *object-oriented development*. Almost certainly, commercial methodologies will rush to incorporate this new technique when it becomes more popular.

However, don't blindly align yourself with any one camp, as the ostriches did when they steadfastly followed the favorite brother. Look for ways to integrate and complement new techniques into your preferred or standard methodology. And always remember, the basic life cycle is always there to guide your use of all techniques.

Summary

Systems development techniques and methodologies are frequently confused with the systems development life cycle. Methodologies are complete systems of techniques including step-by-step procedures, deliverables, roles, tools, and quality standards for completing the entire systems development life cycle. Techniques are methods that apply specific tools and rules to the orderly completion of one or more phases of the systems development life cycle. Thus, systems development techniques and methodologies are intended to complement (or implement) the life cycle.

The most popular systems development techniques in use today are the so-called structured techniques, which are characterized as either process- or data-oriented. Process-oriented techniques build models of systems based on studying the flow of inputs to processes to outputs. Data-oriented techniques build models of systems based on the ideal organization and access of data, independent of how that data will be used to fulfill information requirements.

The most popular process-oriented structured techniques include structured programming, structured design, and structured analysis. Structured programming is a technique for designing and writing programs with greater clarity and consistency. Structured design is a technique and set of guidelines for designing a hierarchy of logical modules that represents a computer program that is easier to implement and maintain. Structured analysis is a technique that translates users' requirements for a system into a picture that identifies system functions, activities, inputs, outputs, and data stores. The structured analysis technique has recently been revised. The revised version of structured analysis, referred to as modern structured analysis, places more emphasis upon data modeling and less emphasis upon studying the current system.

Data modeling and information engineering are two of the more popular data-oriented structured techniques. The data modeling technique represents users' require-

The Next Generation

OBJECT-ORIENTED TECHNIQUES

At any given time, a new technique or collection of techniques is usually poised to change the next generation of methodologies. Today, that new direction is being set by the *object-oriented techniques.* Like many new techniques, they are being touted as methodologies. But according to terms established in this chapter, they are truly techniques that will slowly be incorporated in tomorrow's methodologies.

Object-oriented techniques can be thought of as the marriage of data-oriented and process-oriented techniques. Data and processes are encapsulated into objects. An object contains data and the processes that can use or update that data. Only the processes (sometimes called *services*) defined for the data in an object can use or update that object. Different instances and types of objects interact with one another by sending "messages" that instruct them to execute specific processes in an object.

Objects are defined from the abstract to concrete. For example, we may define an abstract information systems object called REPORT which defines data attributes common to all instances of any type of report (e.g., REPORT NAME, REPORT DATE, and PAGE NUMBER). We might then define another object for a specific report. It will automatically *inherit* the data attributes and processes of REPORT and add to it the unique data attributes and processes needed for the specific report. That report may become an object in yet another report. (*Note:* The above

explanation greatly oversimplifies the object approach. The literature on this approach is quite extensive and growing.)

What is the advantage of this approach? A well-defined object library, once implemented, will contain reusable objects and code. Objects will be self-contained and, thus, easily maintained. Object-oriented techniques promise to achieve the benefits that structured design techniques sought to attain.

The object-oriented revolution has already begun. Its historical roots are in object-oriented programming languages like Smalltalk and C++. We've already learned that object-oriented programming is dependent on good analysis and design—but a different type of analysis and design than is associated with current techniques.

We are already seeing several first-cut object-oriented analysis and object-oriented design techniques from experts such as Yourdon, Coad, and Rumbaugh. The transition to these new methods is complicated by the "revolutionary" change as opposed to "evolutionary" change. In other words, object-oriented techniques requires a radical change in thinking that is slowed by our knowledge of current methods.

There is some hope on the horizon. James Martin, pioneer in information engineering techniques, has recently teamed with James O'Dell to develop an evolutionary object-oriented approach that builds on information engineering concepts. Perhaps it will ultimately

be called *object engineering* or *information object engineering.*

Object engineering will dramatically change the way we develop systems. Our approach will become much more of an information factory approach where each new requirement is thought of as a new object to be designed and built.

However, instead of building the new object from scratch (today's typical approach), we will examine our existing objects inventory for subassemblies that might prove useful in building the new object. We'll only design and implement new subassemblies to fulfill aspects of the requirement that cannot be satisfied by the existing objects (subassemblies).

And, of course, the new object and any unique subobjects used to build it are now available for use in future new objects!

This approach is truly an information factory approach. It emulates the way a manufacturer builds many new products from existing designs and subassemblies.

It is not a question as to if object-oriented techniques will provide the basis for tomorrow's best methodologies; it's only a question of when. Future editions of this textbook will almost certainly present object-oriented methods as the mainstream of systems analysis and design.

In the meantime, do not underestimate the significance of these techniques. Start reading the literature today. Otherwise there may eventually come a day when your existing techniques are obsolete! Really!

ments for a system in terms of the system's data, independent of how that data will be processed or used to produce information. The information engineering technique applies the structured techniques (both data- and process-oriented) to the organization as a whole, rather than on an ad hoc, project-by-project basis.

Two techniques that have also enjoyed an increase in popularity are joint application design (JAD) and prototyping. JAD is a highly structured workshop that brings together users, managers, and information systems specialists to jointly define and specify user requirements, technical options, and external designs. Prototyping is a technique that involves developing a small-scale working (or simulated) model of a system or a sub-system.

Most organizations acquire commercial methodologies from outside vendors rather than custom build their own. Commercial methodologies are "off-the-shelf," step-by-step procedures, individual roles, tools, and quality standards for completing the entire systems development life cycle.

In general, the modern analyst should be familiar with all popular techniques and methodologies, especially the structured techniques, JAD, and prototyping. But instead of choosing a methodology, the analyst should integrate new and old approaches, taking advantage of the strengths of each.

Systems development techniques and methodologies impose rigor, detail, and precision on the systems development life cycle — so much so that it is easy for developers to get lost in the detail and rigor. To help developers with these techniques and methodologies, we now have systems development technology — software tools that help us deal with rigor and details, and ultimately build systems better and faster. That is the subject of the next chapter.

Key Terms

cohesive, p. 147

commercial methodologies, p. 159

coupled, p. 147

data flow diagram (DFD), p. 149

data model, p. 152

data modeling, p. 152

data-oriented techniques, p. 144

design prototyping, p. 158

essential system, p. 149

feasibility prototyping, p. 157

implementation prototyping, p. 158

information engineering (IE), p. 154

joint application development (JAD), p. 156

logical system, p. 149

methodology, p. 143

modern structured analysis, p. 151

physical system, p. 149

process-oriented techniques, p. 144

prototyping, p. 157

rapid application development (RAD), p. 159

requirements prototyping, p. 158

restricted control structures, p. 145

structure chart, p. 147

structured analysis, p. 149

structured design, p. 146

structured programming, p. 145

structured technique, p. 143

systems planning, p. 154

technique, p. 142

Questions, Problems, and Exercises

1. Define the term *methodology*. Why don't most methodologies address the make-versus-buy issue relative to software?

2. Define the terms *techniques* and *methodologies*. What is the relationship between the two?

3. Why don't most organizations develop their own in-house methodology?

4. Explain why some experts argue that the systems development life cycle has become obsolete. Why are they wrong?

5. Differentiate between a process-oriented versus a data-oriented systems development techniques. Give an example of each.

6. Assume you are being interviewed by a college recruiter. The college recruiter is annoyed by the liberal use of buzzwords in the computing field. The recruiter is particularly annoyed at hearing people use the word "structured" all too loosely. How would you define *structured programming* to the recruiter?

7. What are the three restricted control structures of structured programming? What is one important characteristic common to the three structures?

8. What phases of the systems development life cycle does structured programming support?

9. Briefly describe the concept of structured design. Identify three separate schools of thought on structured design. Which is the more popular of the three?

10. What benefits does structured design offer to systems development?

11. The Yourdon structured design technique seeks to factor a program into a top-down hierarchy of modules that is highly cohesive and loosely coupled. What is meant by the term *highly cohesive?* What is meant by the term *loosely coupled?*

12. What improvements are suggested by Ed Yourdon's new version of structured analysis, called modern structured analysis?

13. Briefly describe the concept of structured analysis. What benefits does structured analysis offer to systems development? What systems development life cycle phases are addressed by structured analysis? What are some of the disadvantages of the structured analysis technique?

14. Explain why the concept of a logical system is crucial to structured analysis. What are the benefits of separating logical and physical concerns for a system?

15. Briefly define the data modeling technique for systems development. What are the benefits of data modeling? What are some of the problems with this technique?

16. Briefly describe the concept of information engineering. What phases of the systems development life cycle are addressed by information engineering?

17. Briefly describe the joint application design (JAD) technique for systems development. What benefits does JAD offer to systems development?

18. Briefly describe the concept of prototyping. What benefits does prototyping offer to systems development? What are some of the dangers or disadvantages of prototyping? Identify four types of information systems prototypes.

19. What is RAD? What techniques does it combine? How would you characterize the benefits of RAD?

20. What are commercial methodologies? What are some of the advantages offered by commercial methodologies?

Projects and Minicases

1. In Chapter 3 you learned that systems projects originate because of the need to address one or more problems, opportunities, and directives. In this chapter you learned about a number of systems development techniques and their advantages and disadvantages. For each systems development technique, characterize a systems project that would be ideally suited for applying the technique. Characterize a systems development project that might benefit from an integrated application of one or more of the techniques. Finally, assume your boss has assigned you the responsibility of studying existing systems development techniques and making a recommendation about which systems development technique the organization should adopt. What recommendation would you make?

2. You are teaching a course on the systems development techniques presented in this chapter. The students are somewhat confused over the objectives of each

technique. Each uses different tools and seemingly focuses on achieving different results. Prepare a study guide that briefly describes each technique in terms of its concepts, advantages, and disadvantages. Using the systems development life cycle (SDLC) as a framework, describe the scope of emphasis that each technique places on the SDLC phases.

3. Find an information systems shop in the local community that uses or has used a commercial methodology. How was it selected? Were any advantages realized? Were any disadvantages realized? If they could start over, what would they look for today?

--------------------------- *Suggested Readings* ---------------------------

All of the following books go into much greater detail than we did in this chapter. They are provided in the event that you want to or need to delve further into any given technique.

Connell, John L., and Linda Brice Shafer. *Structured Rapid Prototyping: An Evolutionary Approach to Software Development.* Englewood Cliffs, N.J.: Yourdon Press, 1989. This book presents an appropriate perspective on prototyping techniques.

Connor, Denis. *Information System Specification and Design Roadmap.* Englewood Cliffs, N.J.: Prentice Hall, 1985. A dated, but nonetheless valuable comparative analysis of various "techniques."

DeMarco, Tom *Structured Analysis and System Specification.* Englewood Cliffs, N.J.: Prentice Hall, 1978. The classic and still popular book on structured analysis, but new readers should be aware that many of its philosophies and approaches have since been refuted or updated in Yourdon's book on modern structured analysis.

Gane, Chris. *Rapid Systems Development.* Englewood Cliffs, N.J.: Prentice Hall, 1989. This book presents a RAD-like technique. It suggests a nice blend of data modeling, process modeling, and prototyping.

Gane, Chris, and Trish Sarson. *Structured Systems Analysis: Tools and Techniques.* Englewood Cliffs, N.J.: Prentice Hall, 1978. This is the other classic book on structured analysis.

Flavin, Matt. *Fundamental Concepts of Information Modeling.* Englewood Cliffs, N.J.: Yourdon Press, 1981. This was one of the earliest books on data-modeling approaches.

Martin, James. *Information Engineering: Volumes 1–3.* Englewood Cliffs, N.J.: Prentice Hall, 1989 (Vol. 1), 1990 (Vol. 2, 3). Volume 1 is an introduction. Volume 2 covers planning and analysis. Volume 3 covers design and construction (implementation).

Martin, James, and Carma McClure. *Structured Techniques: The Basis for CASE.* Englewood Cliffs, N.J.: Prentice-Hall, 1988. This is the most comprehensive book about the structured techniques that we've seen. We don't agree with all of the authors' criticisms of certain tools and techniques, but the coverage is comprehensive, balanced, and includes some tools that are not covered in most books.

Page-Jones, Meiler. *A Practical Guide to Structured Systems Design.* 2nd ed. Englewood Cliffs, N.J.: Yourdon Press, 1988. This is our favorite reference on the structured design technique.

Rumbaugh, James; Michael Blaha; William Premerlani; Frederick Eddy; and William Lorensen. *Object-Oriented Modeling and Design.* Englewood Cliffs, N.J.: Prentice Hall, 1991. This book presents what we think may be one of the first comprehensive, nonacademic approaches using true object-oriented techniques. Object-oriented analysis and design is still searching for its champion (in the spirit of Yourdon, DeMarco, Gane, or Martin).

Shlaer, Sally, and Stephen J. Mellor. *Object-Oriented Analysis.* Englewood Cliffs, N.J.: Yourdon Press, 1988; and *Object Lifecycles: Modeling the World in States.* Englewood Cliffs, N.J.: Yourdon Press, 1992. The first title actually represents the data-modeling technique more than it does modern "object" thinking. The second title extends some true object-oriented techniques to derive process models from data models. This interesting technique deserves mention as a candidate to incorporate object thinking into the framework of existing structured techniques.

Yourdon, Edward, and Larry Constantine. *Structured Design: Fundamentals of a Discipline of Computer Program and System Design.* Englewood Cliffs, N.J.: Yourdon Press, 1986. This classic text on structured design methodology is still must reading for instructors who want to teach the subject. The book is difficult reading for all but the best students. See the Meiler Page-Jones book for a more student-oriented text.

5

Computer-Aided Systems Engineering

Chapter Preview and Objectives

In this chapter, you will complete the overview of systems development fundamentals by learning about systems development technology—software tools that help systems analysts, applications programmers, and other information systems professionals build systems. This technology is relatively new, but most information systems shops are at least experimenting with the technology. Some companies have made major commitments to deploying and using the technology. Collectively, this technology and its use is called computer-aided systems engineering (CASE). You will be prepared for the CASE revolution when you can:

Define
CASE in its broadest context of automating or assisting the entire systems development life cycle.

Differentiate
between upper-CASE, lower-CASE, and cross life cycle CASE products.

Describe
representative CASE tools for systems planning, systems analysis and design, systems design and implementation, systems support, and cross life cycle activities.

Differentiate
between application programmers' workbenches, component generators, and code generators.

Define and differentiate
between reengineering and reverse engineering.

Describe
the architecture of a CASE tool in terms of repositories and facilities.

Differentiate
between local and centralized repositories, and describe how each works in the CASE tool environment.

Describe
three ways to integrate CASE tools.

Explain
the benefits and costs of CASE technology, and justify the expenditure on CASE tools.

Explain
guidelines for successfully creating or improving a CASE environment for systems developers.

INTERNATIONAL FLAVORS, INC.

International Flavors, Inc., is a manufacturer of flavors for food products and medicines. They sell their product to the end-producers of food products and medicines. The company is headquartered in Little Rock, Arkansas.

Scene: *A conference room, where several information systems managers and team leaders have assembled for a biweekly staff meeting. The participants include Mike Crup, director of Information Services; Todd Hamblin, manager of Systems Development; Sharon Sarandon, Team leader for Production and Inventory Information Systems; Peg Greenkorn, team leader for Financial Information Systems; Maggie Wu, team leader for Marketing Information Systems; Herb Capone, team leader for Administrative Information Systems; and JoBeth White, methodology and CASE coordinator.*

Mike: Let me begin with a directive. Everything we are about to discuss stays in this room until further notice.

[*Mike pauses to confirm that everyone understands his directive.*]

Well, you've all known this was coming. Our consultants have recommended a reorganization, including Information Services. Mr. Bell [*the chief operating officer*] has approved that plan.

Peg: Are we going to like this?

Mike: Maybe. Maybe not! As you know, the company has posted three poor financial quarters in a row. Management is convinced that we must become "leaner and meaner" in order to compete in this industry. That means downsizing wherever possible; and increasing productivity and increasing quality, but with fewer people.

Herb: You've got to be kidding! I understand the desire to increase productivity, and we've been expecting a "total quality management" initia-

tive, but fewer people? I assume this applies to Information Systems?

Mike: Yes, and they're not kidding. It's not just IS. The whole company is downsizing. Details are still being worked out, but here's what you can expect over the next year. Each of you, except for JoBeth, will be expected to identify one or two staff for probable layoff. I know that's tough. And you should each expect to lose one additional person through normal attrition over the next year. To make it simple, we probably won't replace the next person who retires, quits, or is terminated from your staff.

JoBeth: Not that I'm complaining, but how did I escape this downsizing?

Mike: There is a logical explanation. Despite the downsizing initiative, we all realize that demand for information services will not end. If anything, the consultants are mapping out a strategic plan for information systems development that will increase development efforts.

Herb: I repeat: You've got to be kidding!

Mike: This is no joke, Herb. But back to JoBeth's question. Over the next two years, we will be expanding JoBeth's staff into a full-fledged development center. Her job will be to reevaluate our systems development methodology — either replacing it or improving it to increase developer productivity without sacrificing information systems quality. I also expect her to find and deploy whatever development technology that is necessary to improve productivity, quality, and reliability in systems development — thereby eventually making up for the staff we expect to lose.

JoBeth: Will I be taking on some of the displaced programmers and analysts?

Mike: Not likely. I am assuming that the most likely candidates for the development center are the

Herb:
: same people that your colleagues in this room can least afford to lose, especially now! On the other hand, the individuals who get laid off would probably be the least adaptable to the new way of doing things. Besides, the consultants believe we need to search for unique skills in your new staff.

Herb:
: Is this reasonable? Do you really think we can assimilate technology that fast?

JoBeth:
: According to my research, some companies are doing just that! There are a growing number of CASE success stories out there. . .

[Author's note: CASE is a systems development technology.]

Herb:
: And a lot of horror stories that CASE hasn't lived up to promises.

JoBeth:
: Like any tool or technology, CASE can be misapplied. A lot of companies try to use CASE without a methodology or as a substitute for a methodology. Most experts point to that practice as a "formula for failure."

Mike:
: That concurs with what the consultants are telling us. More importantly, corporate management seems to feel the same. They expect the chemists and engineers to roll out new products in less time, and suggest that if Information Services can't keep up, we should be replaced.

Todd:
: Are you suggesting that IS might be outsourced to another company?

Mike:
: It is a possibility. But I believe we have enough good people that we can reduce that risk. In fact, I personally believe that through CASE technology and improved methods, we can double or triple our development productivity over the next five years, and move towards a "zero defect" goal in quality!

Sharon:
: But we also have some people who have resisted our current methodology and CASE tools.

Mike:
: Yes, I realize that. That fact should weigh prominently into your decision on who to keep and who to recommend for layoff.

Herb:
: I, for one, am going to have trouble with that decision. I have some crackerjack programmers who will absolutely resist this technology, but they are my best people.

Mike:
: Herb, I'd like to talk to you about that after this meeting is adjourned.

Discussion Questions

1. Why is Herb concerned with the prospects of trying to increase both productivity and quality at the same time?

2. How do you suppose that technology can be used to address productivity and quality? What kinds of technology might be useful in a systems development life cycle?

3. What "unique skills" do you think JoBeth must look for in her new staff. How should she organize her development center?

4. What did Mike mean when he said, "That fact should weigh prominently in your decision . . ."?

5. This is a frightening scenario, but one which can happen in anyone's career. Most of the people who will get laid off in the minicase are probably college educated. How can you protect yourself against the type of layoffs described in this minicase?

6. What do you think is going to happen to Herb?

COMPUTER-AIDED SYSTEMS ENGINEERING (CASE)

You may be familiar with the old story of the cobbler (shoemaker) whose own children had no shoes. That situation is not unlike the one faced by systems developers. For years we've been applying computer technology to solve our users' business problems; however, we've been slow to apply that same technology to our own problem—that is—the problem of *developing* information systems. In the not-too-distant past, the principle tools of the systems analyst were paper, pencil, and flowchart template.

Today, an entire technology has been developed, marketed, and installed to assist systems developers. Chances are that your future employer is using or will be using this technology to develop systems. That technology is called *computer-aided systems engineering.*

Computer-aided systems engineering (CASE) is the application of computer technology to systems development activities, techniques, and methodologies. CASE "tools" are *programs* (software) that automate or support one or more phases of a systems development life cycle. The technology is intended to accelerate the process of developing systems and to improve the quality of the resulting systems.

Some people refer to this as computer-aided *software* engineering, but software is only one component of information systems. (Recall your information systems building blocks.) Thus, we prefer the broader context of the term *systems.*

CASE is not an alternative to the techniques or methodologies described in the last chapter. Instead, it is an enabling technology that supports those techniques and methodologies.

The term **systems engineering** is based on a goal of CASE technology: that systems development can and should be performed with engineering-like precision and rigor. The vision is to change the way we build and maintain information systems. In its broadest context, the CASE approach calls for automation across the entire systems development life cycle. Let's take a look at the CASE concept and survey some representative product categories and tools.

The History and Evolution of CASE Technology

The concept of using computers to automate systems development is not new. In a sense, common language compilers and interpreters can be thought of as CASE tools. Let's briefly examine the history and evolution of this technology, a technology that will almost certainly affect your future.

The true history of CASE dates back to the early- to mid-1970s. The ISDOS project, under the direction of Dr. Daniel Teichrowe at the University of Michigan, developed a language called *Problem Statement Language* (PSL) for describing user problems and solution requirements for an information system into a computerized dictionary. A companion product called *Problem Statement Analyzer* (PSA) was created to analyze the problem and requirements statement for completeness and consistency. PSL/PSA ran on large mainframe computers that consumed precious and expensive machine resources. Few companies could afford to dedicate computer resources to PSL/PSA.

The real breakthrough came with the advent of the IBM Personal Computer. Not long thereafter, in 1984, an upstart company called Index Technology (now known as INTERSOLV) created a PC software tool called Excelerator. Its success established the CASE acronym and industry. Today, hundreds of CASE products are available to various systems developers.

Today, most CASE products run on personal computers or intelligent workstations. In some environments, these workstations are networked to shared CASE tools, information, and peripherals. CASE workstations may also be connected to a host computer (minicomputer or mainframe). A typical CASE workbench is shown in Figure 5.1.

CASE technology bears a remarkable similarity to another engineering technology: *computer-aided design/computer-aided manufacturing* (CAD/CAM). Modern engineers use CAD tools to design and analyze new products. CAM tools then automatically generate the computer programs that will run the shop floor machinery needed to produce the design. CASE seeks to do for

information systems developers what CAD/CAM does for engineers: help them design better products (systems) and automatically generate the computer programs.

It is important to realize that modern CASE technology is still very young. But the technology is improving at a staggering rate. New products (called *tools*) are emerging monthly. The best existing products are improving annually. And, of course, some products fail in the marketplace. The following survey is intended to provide a framework for classifying CASE tools. Examples are provided for cross-reference only. Check the popular trade literature to compile the latest CASE tools and trends. (Note: The suggested readings at the end of the chapter include some CASE-specific publications.)

A Survey of CASE Tools and Representative Products

An entire industry has been created to develop and manufacture CASE tools. There are hundreds of tools available from dozens of tool vendors. It is not surprising, then, that many information systems professionals and managers feel overwhelmed by the number of choices! This section tries to place CASE tools into context. We begin with a simple framework. Then we'll look at representative CASE tools within the framework's categories.

A CASE Tool Framework

Let's base our framework on a concept that is already familiar to you, the systems development life cycle. In other words, we will classify tools according to which phases of the life cycle they support. We'll use the simplified life cycle model in Figure 5.2. Our CASE framework is based on the following popular terminology:

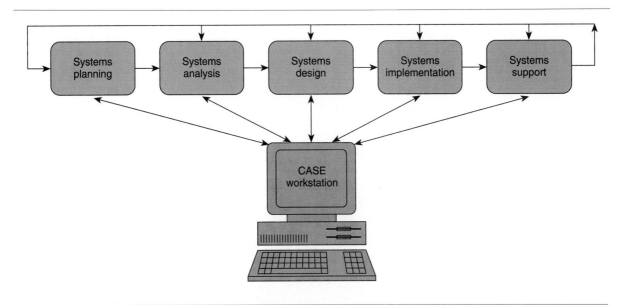

FIGURE 5.2   A Life Cycle–Based Framework for CASE Tools   CASE tools can be classified according to the phases in which they automate or assist developers. We'll plug CASE tools into this simplified life cycle model. All tools will be accessed via a typical CASE workstation.

The term **upper-CASE** describes tools that automate or support the "upper" or front-end phases of the systems development life cycle; namely, systems planning, systems analysis, and general systems design.

The term **lower-CASE** describes tools that automate or support the "lower" or back-end phases of the life cycle; namely, detailed systems design, systems implementation, and systems support.

The term **cross life cycle CASE** refers to tools that support activities across the entire life cycle. This includes activities such as project management and estimation.

Notice that there is some overlap between upper- and lower-CASE tools. This is because our profession has never reached agreement on when systems analysis ends and when systems design begins.

A typical business's complete CASE tool kit should include one or more products from each category. Although some CASE vendors offer an integrated CASE product family that rather comprehensively covers all categories, it is highly unlikely that any firm will find a single source for every CASE tool they need, or might want to use. With that in mind, let's look at some of the categories, subcategories, and representative CASE tools.

CASE Tools for Systems Planning (Upper-CASE)

Recall that the systems planning phases of the life cycle are used to chart long- and/or short-term direction for information systems, which complements the long- and/or short-term directions of the business itself. Upper-CASE tools for systems planning are intended to help analysts and consultants capture, store,

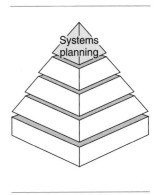

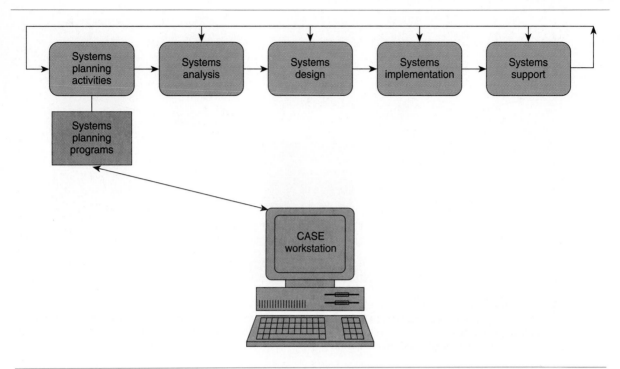

FIGURE 5.3   CASE Tools for Systems Planning   The first tools we'll add to our CASE tool kit are for systems planning.

organize, and analyze models of the business. These models and their evaluation help the information systems planners define and prioritize

- Business strategies that are being (or will be) implemented.
- Complementary information systems and information technology strategies to be implemented.
- Databases that need to be developed.
- Networks that need to be developed.
- Applications that need to be developed around the databases and networks.

Planning tools (programs) are added to our CASE workbench model in Figure 5.3.

For example, KnowledgeWare's ADW Planning Workbench provides workstation tools for describing a business in terms of its goals, objectives, critical success factors, problems, organization structure, geographic locations, information needs, functional needs, and so forth. As shown in Figure 5.4, this information can be entered in the form of models (pictures), descriptions, and matrices. The matrices are of special importance since they are used to associate information about different categories of planning information. For example, a matrix can be used to show which organization units (such as departments) are responsible for which organization goals.

Most planning tools can perform an analysis on different matrices to identify logical "groups" of data, functions, locations, and other planning information. This analysis reveals logical groupings that can be prioritized into development projects. Ideally, the planning tools develop a global **enterprise model** that can be passed on to any systems analysis and design CASE tools that might be

---
✓

Representative Planning Tools

ADW Planning Workbench by KnowledgeWare
DevelopMate™ by IBM
PC/Prism by INTERSOLV

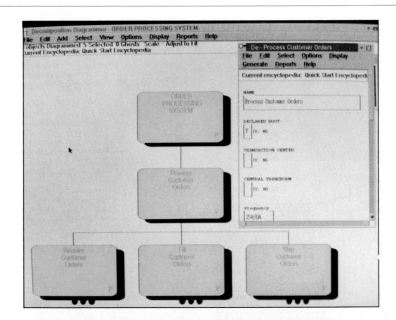

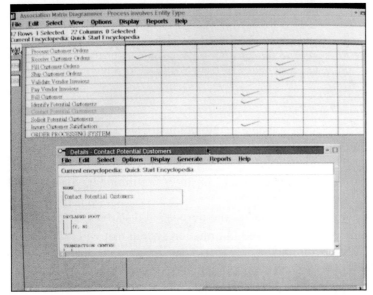

FIGURE 5.4 **A Typical CASE Tool for Systems Planning** These screens from KnowledgeWare's ADW Planning Workbench demonstrate a representative CASE tool for systems planning. The key to many planning tools is the association matrices that help planners group data, activities, and locations into logical databases, applications, and networks. **(Both photos courtesy of Purdue University)**

used to help develop the planned databases, networks, and information systems. Representative systems planning tools (and their vendors) are listed in the margin on page 172.

CASE Tools for Systems Analysis and Design (Upper-CASE)

This category "invented" the term CASE. Upper-CASE tools for systems analysis and design are intended to help systems analysts better express users' requirements, propose design solutions, and analyze the information for consistency, completeness, and integrity. This information helps analysts

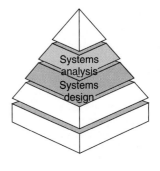

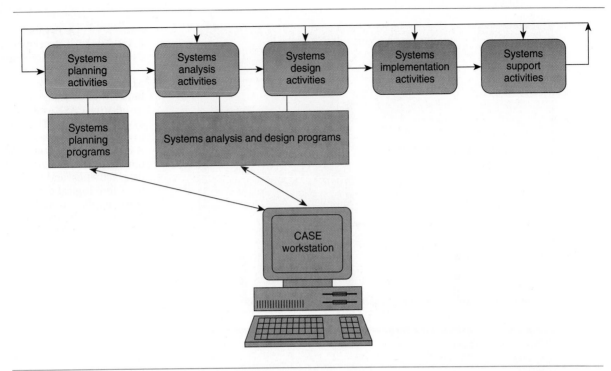

FIGURE 5.5 CASE Tools for Systems Analysis and Design We now add systems analysis and design tools to our CASE tool kit. These are the first CASE tools encountered by many young information systems professionals.

- Define project scope and system boundaries.
- Model and describe the current information system (if required in the methodology).
- Model and describe the users' business requirements for a new information system.
- Prototype requirements for the purpose of discovery or verification.
- Design a computer-based information system that will fulfill the user's business requirements.
- Prototype specific design components (such as screens and reports) for the purpose of verification and ease-of-use.

Representative tools are listed in the margin. For the most part, these tools were developed to support the structured techniques such as structured analysis and design, data modeling, and information engineering (all introduced in Chapter 4). Analysis and design CASE tools are added to our model in Figure 5.5.

One of the first analysis and design CASE tools, INTERSOLV's Excelerator/IS, is used throughout this book to demonstrate the use of CASE technology in the running case study of SoundStage Entertainment Club. This choice is not a product endorsement. Rather, it is a recognition that it is widely available to academic institutions via an academic grant/computer laboratory program.

For another example, KnowledgeWare's ADW Analysis, Design, and Rapid Application Development (RAD) Workbenches (three separate products) pro-

---

✓

Representative Analysis and Design CASE Tools

ADW Analysis, Design, and RAD Workbenches by KnowledgeWare

Bachman Analyst and Bachman Designer by Bachman Information Systems

Excelerator/IS and Excelerator II by INTERSOLV

Systems Architect by Popkin Software

Information Engineering Facility by Texas Instruments

The Visible Analyst Workbench by Visible Systems

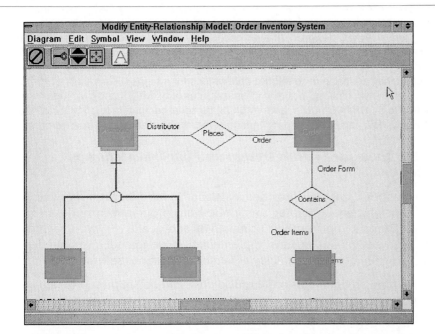

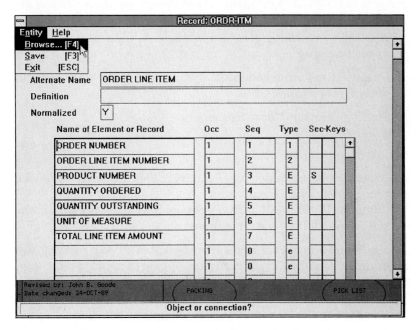

FIGURE 5.6 **Typical CASE Tools for Systems Analysis and Design** These screens from IN-TERSOLV's Excelerator for Windows workbench demonstrate a representative CASE tool for systems analysis and design. The screens show a typical system model and details. (Both photos courtesy of INTER-SOLV, Inc.)

vide PC-based tools for analyzing and designing information systems. As shown in Figure 5.6, these tools help analysts express requirements and designs as graphical models (pictures) and associated descriptions. Like most tools in this class, they provide analytical capabilities to evaluate the quality and completeness of the models and descriptions. These analytical tools are based on accepted rules of information engineering and structured analysis and design.

Important note   You may have already noticed that many CASE tools are PC-based. Before we continue, it is important to realize that applications do *not* have to be planned, analyzed, designed, or even necessarily implemented or supported on the same computer that will "run" the final applications software. Indeed, the largest number of CASE tools run on PC- or MS-DOS, OS/2® or UNIX™ workstations—even though they are being used to build applications for a much wider variety of computer sizes and environments, including (1) mainframes (e.g., MVS™ and VM™), (2) minicomputers (e.g., OS/400®, UNIX, and VMS), (3) personal computers (e.g., DOS, Windows, Windows/NT, and OS/2), and (4) various network and client/server arrangements.

CASE Tools for Systems Design and Implementation (Lower-CASE)

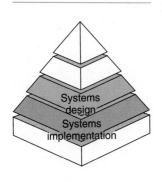

Early upper-CASE tools provided assistance to the systems analyst and designers. But what about helping the applications programmers and other system implementers improve their productivity and quality? Lower-CASE tools for detailed design and systems implementation are intended to help designers and programmers more quickly generate applications software. This includes

- Helping programmers more quickly test and debug their program code.
- Helping programmers or analysts to automatically generate program code from analysis and design specifications.
- Helping designers and programmers to design and automatically generate special or detailed system design components like screens and databases.
- Automatically generating complete application code from analysis and design specifications.

Representative tools are listed in the margin. The lower-CASE tools fall into three general subcategories:

- Applications programmer workbenches.
- Component generators.
- Code generators.

These subcategories are added to our model in Figure 5.7. Let's discuss each in somewhat greater detail.

The simplest lower-CASE tools address only systems *implementation* and the needs of the applications programmer.

> An **application programmer's workbench** is a CASE tool that supports the coding, compiling, testing, and debugging of applications programs on an intelligent workstation (e.g., PC). Ideally, the resulting code should be transportable to a variety of personal computer, minicomputer, mainframe computer, and computer networking environments.

For example, the Micro Focus COBOL/2 Workbench provides a complete, PC-based coding, compiling, testing, and debugging environment for COBOL programmers that far exceeds the capabilities of many (if not most) mainframe COBOL compilers. The workstation supports the off-loading of mainframe COBOL program development to the intelligent programmer workstation. Although the code is developed on the workstation, it is fully compatible with mainframe COBOL standards. Additionally, available options can emulate mainframe services such as CICS and DB2.

✓

Representative Design and Implementation CASE Tools

ADW and IEW Design and Construction Workbenches by KnowledgeWare
APS by INTERSOLV (formerly by Sage Software)
Bachman Database Administrator for DB2, IDMS, and IMS by Bachman Information Systems
CSP by IBM
EASEL by Easel Corporation
ObjectView by KnowledgeWare
Powerbuilder by Powersoft
Synon by Synon Corporation
Telon by Computer Associates (formally by Pansophic)

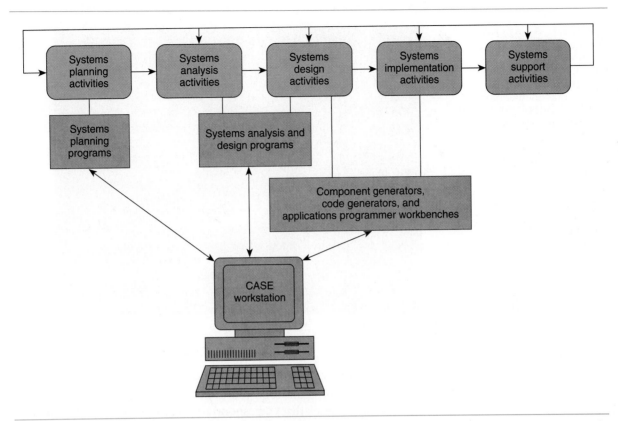

FIGURE 5.7   CASE Tools for Systems Design and Implementation   The next tools we'll add to our CASE tool kit are for detailed systems design and implementation.

The power of the COBOL/2 Workbench lies in its testing and debugging tools. Its ANIMATOR feature (see Figure 5.8) allows you to "watch" your program execute, line-by-line; either forward or backward! Color and graphics help you follow the logic as it executes. Also, data variables can be monitored in windows as the program executes. The ANIMATOR can also create a graphical structure chart for any program. As a program executes, the modules change color to demonstrate the flow of control between modules. Finally, the ANA-LYZER feature helps programmers evaluate how thoroughly a program is tested by identifying program instructions not executed by the test data file.

Let's move on to a different subcategory of lower-CASE tool for detailed systems design and implementation.

A **component generator** assists with the design of specialized information system components, and then generates the special code required to implement those components.

Most component generators are intended to deal with a specific design problem. For example, the Bachman Database Administrator (DBA) Workbench from Bachman Information Systems is a PC-based CASE tool used to design, redesign, and fine-tune a mainframe database for information systems. Its use would normally be restricted to data management professionals responsible for

FIGURE 5.8 **An Application Programmer's Workbench** This screen, from Micro Focus's COBOL Programmer's Workbench demonstrates its ANIMATOR feature that simplifies and enhances testing and debugging of COBOL programs. The ANIMATOR allows the programmer to "watch" the program execute and trace changing values of variables during execution. (Courtesy of Micro Focus)

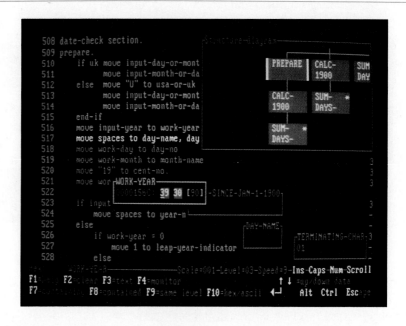

designing shared databases. The DBA workbench can also generate the "data definition language" program required to create or modify a database. It is offered in several versions that correspond to popular mainframe-based database management systems like DB2, IMS, and IDMS.

Similarly, Easel Corporation's EASEL Workbench is a PC-based tool that helps a system designer create a **graphical user interface** (GUI) for any application. (Note: A GUI consists of screen designs that conform to a consistent standard such as Microsoft's Windows, IBM's Presentation Manager™, or Apple's MultiFinder. GUIs are believed to make applications easier to learn and use.) After helping the designer create the GUI screens, the workbench can also generate the program code to implement those screens. After that, the programmer need only link the application programs to that code.

CASE tools to design GUIs are becoming an increasingly important product niche in response to the growing acceptance and popularity of the GUI interface. Also fueling this popularity is the movement toward **client/server application** design in which applications programs are distributed across multiple computers (transparently to the end-user). In a client/server application, the GUI programs (for input and output) usually run on the user's own PC or intelligent workstation, called a *client*. Other parts of the same application program (such as database reads and writes) run on another computer(s), called the *server*. The server might be a PC file server, a minicomputer, or a mainframe. Regardless, to the end-user it appears as if the entire application is running on his or her own PC.

Let's move on to yet another subcategory of lower-CASE tools for systems design and implementation. Programming is a "logical" process. Did it ever occur to you that someone might be able to write a program that writes programs?

A **code generator** automatically generates application programs from analysis and design specifications. It is sometimes called a *program generator.*

Because code generators quickly reveal errors and omissions in the design specifications, most either include (a) their own detailed design capabilities, or (b) a direct interface to an upper-CASE design tool. Lower-CASE tools for detailed systems design and systems implementation are intended to help analysts and/or programmers more quickly generate applications programs than the traditional hand-coding process — even faster than using a 4GL!

There are three types of code generators on today's market. Knowledge-Ware's ADW Construction Workbench is representative of an integrated code generator that is integrated into an upper- and lower-CASE product family. It directly feeds off the other CASE tools in the family to generate COBOL code. It even generates the necessary calls to some database management systems (such as DB2) and some teleprocessing monitors (such as CICS). If the workbench reveals design errors or omissions, you must return to the ADW Design Workbench to make the corrections.

INTERSOLV's APS/PW code generator is representative of a stand-alone code generator. It can stand alone from any upper-CASE tool. How so? APS/PW works on a prototyping-like approach called *painting.* The analyst or programmer "paints" the application. The "painting" is then regenerated as a COBOL program, complete with database and teleprocessing function calls. Like the KowledgeWare code generator, APS can invoke the testing and debugging tools of the Micro Focus COBOL Programmer's Workbench to test and debug programs at the "painting" level. (Note: PSW/PW can also accept input from INTERSOLV's and KnowledgeWare's upper-CASE tools to further accelerate the painting of the final application.)

IBM's Cross System Product (CSP) is usually thought of as a mainframe fourth-generation language; however, it is also representative of a third type of code generator. After writing and testing a program in CSP, you can regenerate it as a COBOL program for a variety of IBM computers and operating systems. Why not just place the CSP programs into production? Perhaps the COBOL programs provide superior performance given the expected volume of transaction data. Or perhaps the business seeks to maximize its greater expertise in COBOL to maintain the programs. (Most firms do have more COBOL programmers than CSP programmers.) Or perhaps the firm just wants to take advantage of CSP as a prototyping language prior to generating COBOL programs.

CASE Tools for Systems Support (more Lower-CASE)

Finally, we come to lower-CASE tools that support the maintenance activities for production information systems. This is fertile new ground for the CASE tool industry. It is estimated that there are between 60 and 80 billion lines of COBOL code in business systems alone. Lower-CASE tools for systems support are intended to help analysts, designers, and programmers react to inevitable, ever-changing business and technical environments. This includes

- Helping programmers restructure existing or old program code to be more maintainable.
- Helping programmers and analysts react to changing user requirements.

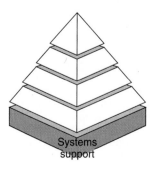

Systems support

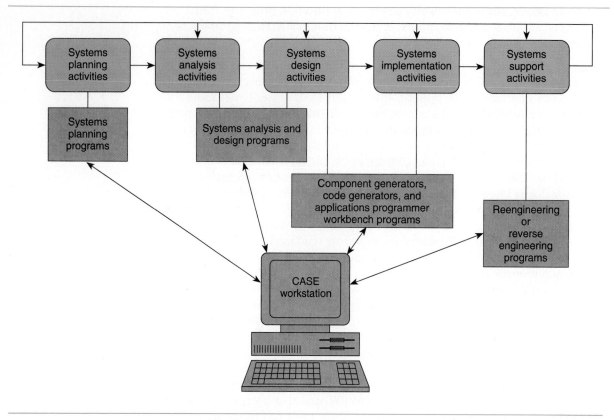

FIGURE 5.9 CASE Tools for Systems Support   The next tools we'll add to our CASE tool kit are for maintenance and reengineering of current systems.

- Helping analysts and programmers reengineer programs to accommodate newer technology (such as changing the "preferred" database management system).

- Helping analysts and programmers determine when the costs of maintaining a system exceed the benefits of maintaining the system. (In other words, "Is it time to start over?")

- Helping analysts recover any reusable information from obsolete programs as a preface to taking that information back to upper-CASE tools and redeveloping a major, new information system.

Representative tools are listed in the margin. These tools are added to our workbench model in Figure 5.9.

The focus in systems support technology is *reengineering*. This term has come to mean different things to different people; therefore, we should provide our definition as a point of departure for further discussion.

Reengineering is the process of altering or restructuring an existing system. Reengineering should be distinguished from redevelopment (starting over). Popular synonyms are *renovation* and *reclamation*.

This is the broadest possible definition of the term and consistent with CASE authority Charles Bachman's opinion that all systems development is, in a sense, reengineering. Reengineering includes the following scenarios:

- Improving programs without changing either user requirements or the technical design. Changing a unstructured or poorly structured program into a well-structured program is an example of this type of reengineering. This is called **restructuring**.

- Improving programs by changing technical design but not changing user requirements. In this case, the program does the same thing but the technology has changed. Examples include (1) changing from VSAM file processing to database processing, (2) changing from one database management system to another (e.g., IMS to DB2), or (3) changing from one language standard to another (e.g., VS/COBOL to COBOL/2). This is also called *restructuring*.

- Improving programs by changing user requirements, but not necessarily technical design. This is classic maintenance programming in response to new business needs. It is assumed that the analyst must perform a mini–systems development life cycle to meet the requirements; however, changes will be restricted to the new requirements and their impact on existing requirements.

- Recognizing that the system is obsolete and that it would be beneficial to start over with a complete systems analysis. In this situation, it is highly desirable to recapture as much of any reusable analysis and design information that is buried in the obsolete programs as is possible. That would reduce the amount of information that must be captured in the upper-CASE tool.

Let's discuss a few types of systems support and reengineering tools.

A program maintenance tool seeks to (1) simplify existing program maintenance, (2) extend the useful lifetime of existing systems, and (3) identify programs or program modules that should be replaced instead of maintained.

IBM's COBOL Structuring Facility (COBOL/SF) is a mainframe-based tool that reads unstructured or poorly structured COBOL programs and automatically reengineers them into structured code. The intent is to reduce maintenance costs for, and extend the life of, those programs. COBOL/SF can find "dead code" (code that is never executed—probably a remnant of testing and debugging) and code that reflects poor programming practice (e.g., use of the ALTER verb). COBOL/SF also attempts to measure the maintainability of a program giving some insight into when it might be better to start over.

Another type of maintenance tool tests the impact of changes on a program's performance. For instance, IBM's Workstation Interactive Test Tool (WITT) can perform **regression analysis** on any COBOL program. It measures the performance of the program before and after maintenance changes to determine if performance degrades (or improves) and by how much. This can effectively extend a program's useful life despite numerous maintenance changes.

Recently, reverse engineering tools have emerged to build a bridge "backwards" to upper-CASE tools.

> **Reverse engineering** is the process of analyzing existing application programs and database code to create higher-level representations of the code. It is also sometimes called *design recovery*.

Think of reverse engineering as *reverse programming*—that is—finding the programmers' original ideas in the code. What kinds of ideas? Reverse engineering can produce structure charts, flowcharts, decision trees, and other

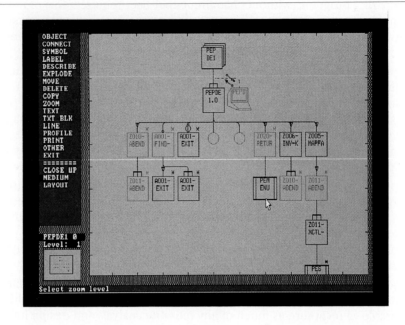

information from the code. With these specifications, the programmer might be able to make changes and enhancements that would have been impossible to do at the code level.

Reverse engineering is not just for maintenance programming. Program code doesn't just contain the programmer's ideas, but also the ideas of the analysts who designed the system. Furthermore, the code is a "snapshot" of the business and user requirements (at some point in time). Some of those requirements may be reusable in a new system. Thus, the ultimate purpose of reverse engineering a system is to capture reusable information for subsequent systems analysis and design.

For example, INTERSOLV's Excelerator for Design Recovery reads existing COBOL and DB2 code and recovers reusable models and information from that code. The product can recapture structure charts, database structure diagrams, and field names and properties. The screen in Figure 5.10 was "captured" from COBOL code using Excelerator for Design Recovery. It also analyzes the code to determine maintainability and identify serious errors. Once captured, the design can be either modified, or exported to Excelerator for Planning, Analysis, and Design, where the analyst can then redevelop the system.

We want to emphasize two important points about the emerging technology of reverse engineering:

- There is little consensus agreement as to how much useful information is buried in yesterday's application programs. In many cases, those programs are based on poor or obsolete design approaches. Many experts fear that recapturing those mistakes may encourage the same sins to be repeated. On the other hand, it is safe to assume that some business information is buried in any program. For instance, most data structures, fields, and arrays represent business information. It would be nice to automatically capture and study that information.

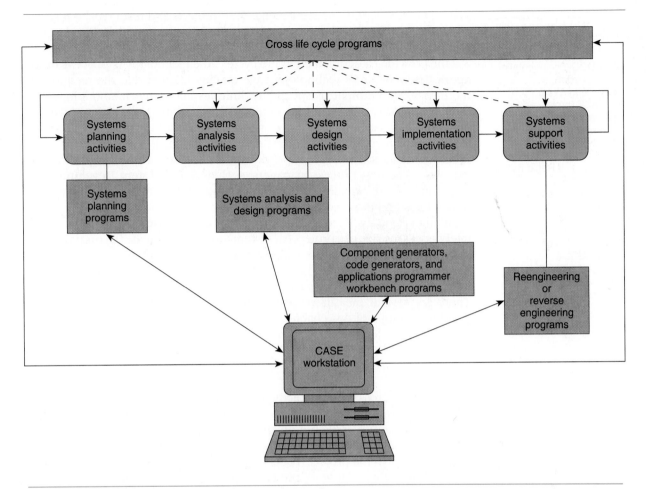

FIGURE 5.11   CASE Tools for Cross Life Cycle Activities   The last tools we'll add to our CASE tool kit are for activities that span the entire life cycle. These programs are depicted above the life cycle phases.

- Any information captured will be, by its very nature, in physical or implementation terms. For instance, a reverse engineering tool might capture a data field called ORD_DATE_2. In the upper-CASE tool, the analyst should rename the field to a more logical, business-oriented name, such as Customer Order Promise Date. Notice that, in this hypothetical example, the business name is apparently very different from the implementation name given in the program.

CASE Tools that Support Cross Life Cycle Activities

A wide variety of CASE tools support activities across the entire systems development life cycle. Cross life cycle tools are added to your workbench model in Figure 5.11. Representative tools are listed in the margin on page 184. The list of activities supported is representative, but not comprehensive. Let's discuss some of the more common tools.

Project management is one cross life cycle activity common to most projects. A wide variety of project management software packages exist because numerous professions use them. Project management tools help managers

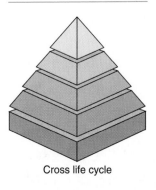

Cross life cycle

——————— ✓ ———————
**Representative Cross Life
Cycle CASE Tools**

ADW/DOC (documentation
  facility) from
  KnowledgeWare
firstCASE (project and
  process management,
  estimation) by AGS
  Management Systems
HyperAnalyst (process
  manager) by Rapid System
  Development
Project Bridge (estimator) by
  ABT
Project Workbench (project
  manager) by ABT
PVCS (version control) from
  INTERSOLV
SPQR/20 (estimator) by
  Software Productivity
  Research

plan, schedule, report on, and manage their projects and resources. But some project management tools have crossed into CASE by virtue of their interfaces to other CASE tools.

One growing category of cross life cycle CASE technology is **process managers.** Users of CASE technology have quickly learned that the technology requires a commitment to specific techniques and some methodology. Although people can be educated in techniques and methodologies, even the best require occasional assistance with the methodology, and most require assistance with applying the right CASE tools and facilities at the right time. Process management software provides the necessary on-line guidance and expertise. The best process managers, such as Rapid System Development's HyperAnalyst, are actually capable of invoking (starting) CASE tools at appropriate times in the methodology.

Another category of cross life cycle CASE is **estimation.** Attempting to accurately assess the size of a project (or system) and then estimate the time and cost to complete the project is very difficult. The creeping commitment approach that you learned in Chapter 3 is only a technique that calls for gradual commitment to multiple estimates in a project. But how do you estimate size and cost? There are several measurement techniques, like **function points** that are beyond the scope of this chapter. Needless to say, if there is a formal, mathematically-based technique, there are probably software tools available to assist developers with the estimates. Examples include ABT's Project Bridge and Software Productivity Research's SPQR/20. Each of these tools estimate function points.

Yet another cross life cycle activity is **documentation.** The deployment of CASE technology creates a wealth of documentation. At various points in any project, you want to assemble documentation that may have been created in different phases of the life cycle. This documentation becomes work products or deliverables in the project. Tools like KnowledgeWare's ADW/DOC allows you to design a custom work product or deliverable, and then automatically retrieves the appropriate diagrams, specifications, or information. It can even incorporate word processing and spreadsheet files. It provides headers, footers, page numbers, table of contents, and other important features for document production. You can even build reusable templates that standardize certain deliverables or work products.

CASE ARCHITECTURE

Now that you have a better idea of what CASE tools are available, let's take a look at the architecture of CASE tools. At the center of that architecture is a database called a *repository.* Around that repository are facilities. In this section we'll briefly examine this architecture and then evaluate the importance of integrating CASE tools.

Repositories

Our experience suggests that most people are first attracted to CASE tools by their graphics capabilities (e.g., the ability to draw flowcharts, data models, or data flow diagrams). But the real power of a CASE tool is derived from its repository.

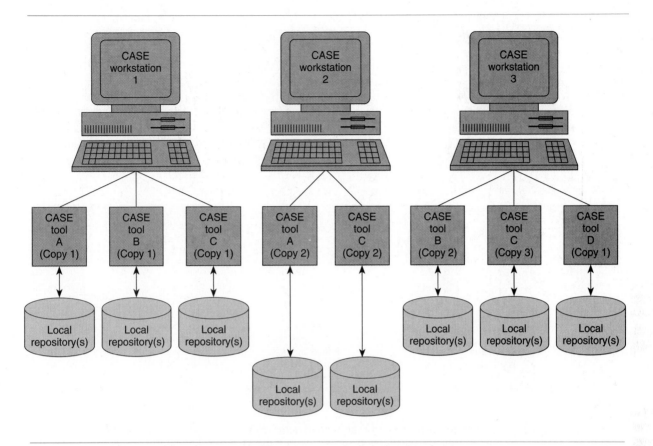

FIGURE 5.12 **Stand-Alone CASE Repositories** In first generation CASE tools, each tool had its own repository. If a single vendor had two CASE products, each product had its own repository.

A **repository** is a developers' database. It is a place where the developers can store diagrams, descriptions, specifications, applications programs, and any other working by-products of systems development. Synonyms include *design database,* **dictionary,** and **encyclopedia**.

Most first generation CASE tools had stand-alone, proprietary repositories. A CASE tool could only read and write from its own repository (see Figure 5.12).

Second generation CASE tools were frequently built around a shared (but still proprietary) repository. For example, KnowledgeWare has built a family of six CASE workbenches (Planning, Analysis, RAD, Design, Construction, and DOC) around a single, integrated repository (which they called an encyclopedia). These tools not only share a repository; they make the workbenches highly dependent on one another. (That's not necessarily bad!).

For example, the KnowledgeWare Construction Workbench requires repository data from the Design Workbench in order to properly generate code. Such product families have become known as **i-CASE** or **integrated-CASE** (see Figure 5.13).

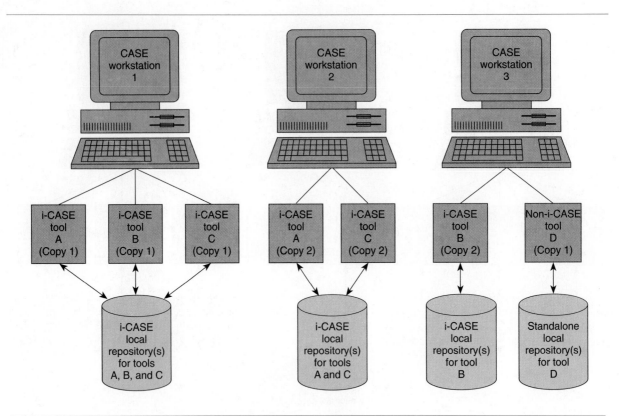

FIGURE 5.13   **Integrated CASE Repositories**   In the second generation, many tool vendors integrated their tools (products) over a common repository. That allowed all tools within the same product family to share information. Notice that some products still use stand-alone repositories.

Some second-generation CASE tools take a different approach to sharing a common repository. Although data could be shared across the common repository, the tools can also function independently. For example, INTERSOLV's code generator, APS, can accept repository data from its own Excelerator upper-CASE product. But APS can also generate code through prototyping without using an Excelerator or upper-CASE information. This approach might be thought of as **m-CASE** or **modular-CASE**. Pick and use only the modules you really want or need.

In both Figures 5.12 and 5.13, we show one repository per CASE tool or i-CASE tool family. Most current tools support this "local" repository concept.

A **local repository** contains the developers' data for a single systems development project. Synonyms include *project repository, project dictionary,* and *project encyclopedia.*

In any given IS shop, several projects are in progress at any given time; therefore, there are at least as many local repositories as there are projects. More than one project can be stored on a single workstation's hard disk.

Most of today's local repositories are accessible to only a single developer at a time. That means that only one person at a time can be using *any* CASE tool against the same local repository. This is terribly inconvenient: As we've sug-

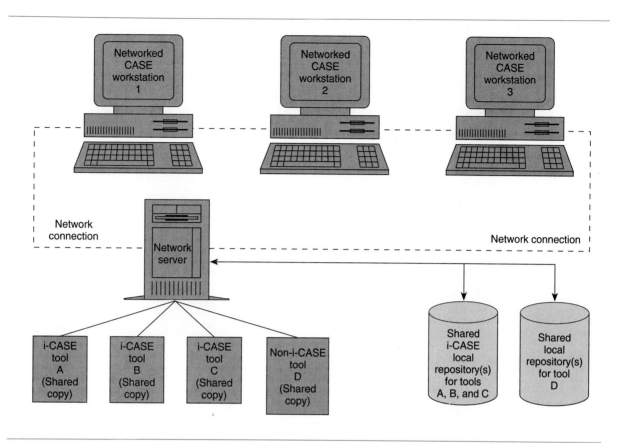

FIGURE 5.14 Networked Local Repositories   Soon, local repositories will routinely be stored on network servers and shared by work groups (teams) of systems developers. In many instances, the CASE tools themselves will be installed on the network server and shared by multiple CASE workstations.

gested on several occasions, analysts and developers work in groups and teams on projects.

Soon, local repositories will be enabled for **work groups.** This will allow a single local repository to be simultaneously shared by several team members and workstations via a network file server (see Figure 5.14). Database technology will prevent two individuals from trying to change the same piece of information at the exact same time. This will be a true, shared database implementation of the repository concept.

Even in work group configurations, local repositories will present problems. There will certainly be redundant data stored in all these local repositories since, at any given time, many projects may require access to certain global information and specifications. What we really need is a single "master" repository for all data.

A **central repository** is a database that includes information about all past, present, and future information systems, databases, networks, and technology for an entire organization (or a major subset of an organization). It is inclusive of all current projects-in-progress as well as current production systems (if that information has been entered).

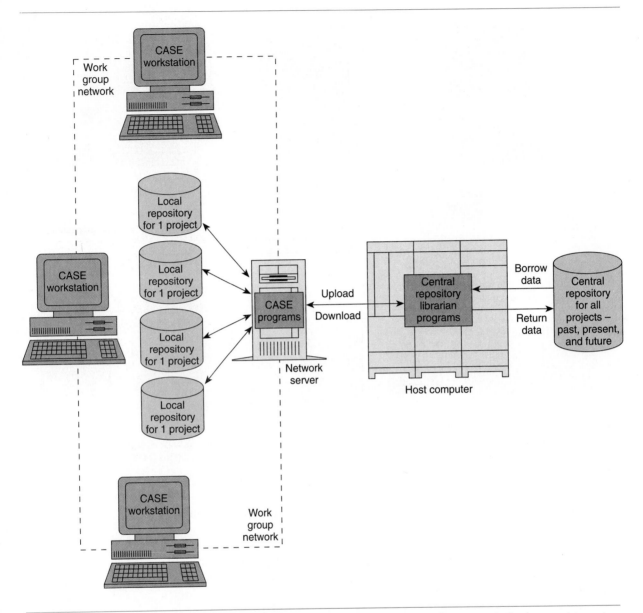

FIGURE 5.15 **A Central Repository** Ideally, a company would have a single (or relatively few) central repository(s) for all projects. Local repositories exchange data with the central repository on an "as needed" basis.

Ideally, every project would share the central repository. That would eliminate the need for local repositories. Unfortunately, database technology can't provide reasonable access times for such a large repository. Consequently, we expect local repositories to coexist with a central repository. Local repositories would then be required to borrow (check out) and return (check in) information to and from that central repository (see Figure 5.15). Notice that a "librarian program" (our term) coordinates the use of the central repository. IBM's Repository manager/MVS is one example of a central repository.

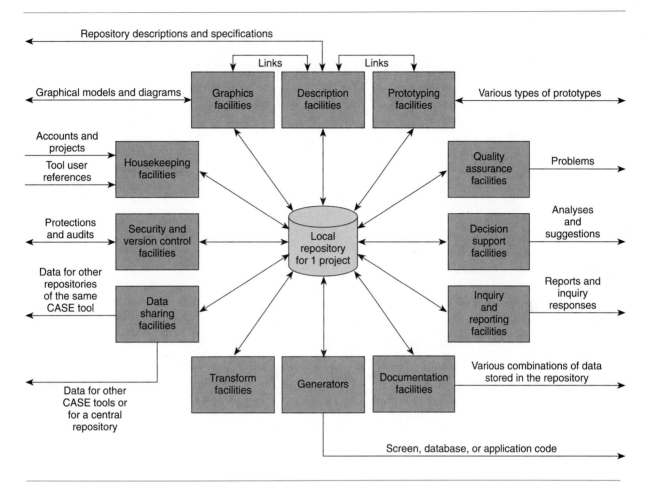

FIGURE 5.16 **CASE Tool Facilities** CASE tool facilities are built into tools to interact with the tool's local repository.

Facilities and Functions

To use a repository, we obviously need input and output facilities. Representative CASE tools include some subset of the following facilities, illustrated in Figure 5.16:

- **Graphics facilities** are used to diagram or model information systems using various techniques. Examples include tools for drawing process and data models. Usually, objects on one graph can be linked to new graphs to show greater detail. This supports top-down development techniques (such as structured analysis).

- **Description facilities** are used to record, delete, edit, and output non-graphical information and specifications. Examples include business definitions; contents of files, inputs, and outputs; properties of data fields; and procedures and logic for processes. As shown in the figure, these facilities can be started by themselves, or invoked directly by the graphics tools. For instance, after drawing a process on a diagram, you could immediately describe the logic for that process to the repository.

- **Prototyping facilities** are used to analyze or design components such as inputs, outputs, screens, or forms. The prototyping facilities of a given CASE tool may be non-functional or functional.
  - *Nonfunctional* prototyping tools *simulate* the implementation of design components. There is no real application program code. Such prototypes are used primarily to "discover" user requirements.
  - *Functional* prototyping tools are real applications programs that implement the design component. Such prototypes are used to "test" a possible system implementation. The prototype's program code may or may not become part of the final production system.

  As shown in the figure, some prototyping tools can directly use, maintain, or create descriptions in the repository.
- **Inquiry and reporting facilities** are used to extract information and specifications out of the repository. They can support simple inquiries such as, "tell me about the input called ORDER"; or, more complex inquiries such as, "provide me a listing of every input or file that contains any field that includes a two-character date field."
- **Quality assurance facilities** analyze graphs, descriptions, and/or prototypes for consistency, completeness, and conformance to generally accepted "rules" of systems development. For the most part, quality assurance facilities only identify errors. Error correction is still the responsibility of the analyst, designer, or programmer.
  - Most CASE tools provide *on-demand* analyses that generate predefined quality reports. The better tools provide exception and summary reports that only identify quality problems. Some tools allow you to design your own custom reports using the repository's contents.
  - Some CASE tools provide *real-time* quality analysis; that is, they provide immediate feedback as you create and edit graphs, descriptions, and prototypes. This feature is akin to having an expert systems developer looking over your shoulder.
- **Decision support facilities** analyze information in the repository to provide support for decisions. For example, an estimation tool might count function points to help an analyst reevaluate scope. Likewise, a planning tool might analyze data and process models to identify data entities that might be implemented as a single database.
- **Documentation facilities** are used to assemble graphs, repository descriptions, prototypes, and quality assurance reports into formal documents or deliverables that can be reviewed by project participants. Good documentation tools give you the flexibility to perform the following functions while customizing your own documents: (1) develop your own outline, (2) retrieve the relevant graphs, descriptions, and the like from the repository, (3) organize those items according to the outline, and (4) format the output for your intended audience (e.g., users or managers or technical staff).
- **Transform facilities** automate or assist the transformation of something into another form. For example, a CASE tool might automatically transform a business-oriented data model into a technically-oriented data model that could be implemented. Another example might be a facility that transforms an unstructured program into a structured program. Any

given transform tool is usually based on widely accepted rules and procedures for a common systems development transformation.

A special type of transform works in reverse. It analyzes existing application programs and transforms them into higher-level graphical models and descriptions. Such a transform is used in reverse engineering tools.

- **Generators** automatically translate user requirements and/or technical designs into working applications and programs. An example would be a lower-CASE tool that generates COBOL, database, or GUI screen code graphical representations and/or repository descriptions.

- **Data-sharing facilities** provide export and import repository information between different local repositories of the same CASE tool. Other data-sharing facilities are geared towards exporting and importing data to and from different CASE and non-CASE tools. (Note: Word processors and spreadsheets are examples of non-CASE tools into which you might want to transfer data from a CASE tool.)

- **Security and version control facilities** maintain integrity of repository information. For instance, many CASE tools allow you to "lock" information that you don't want anyone to change without your permission. On the other hand, central repositories provide tools for identifying changes made to any specifications that might affect existing production information systems. This is called *version control.*

- **Housekeeping facilities** establish user accounts, project directories, user privileges, tool defaults and preferences, backup and recovery, and so forth.

CASE Integration

Developers' early experiences with CASE tools have revealed that "one size rarely fits all" information shops or projects. Many information systems shops have already selected major tools for (1) planning, (2) analysis and design, (3) prototyping and/or code generation, (4) testing and debugging, and (5) project management. These major tools may or may not be provided by a single vendor.

In some cases multiple major tools for competing vendors might have been selected by different, autonomous groups in the business. And in many cases, the major tools might be supplemented by minor, add-on tools from entirely different vendors.

Note It is important that a CASE tool kit be integrated such that the tools share data and cooperate with one another. This has proven to be a formidable task for both systems developers and tool vendors.

Consider the problem of *non*integrated tools. Information from a planning workbench would have to be re-input into an analysis and design workbench. Then, the information produced with the analysis and design workbench would have to be re-input to the code generator.

One solution is to buy an i-CASE product family. But you may not like some of the tools in a given i-CASE product line. Furthermore, you may still need other tools to fill identified gaps in the i-CASE product line.

Let's examine the requirements for CASE integration, starting with the following goal:

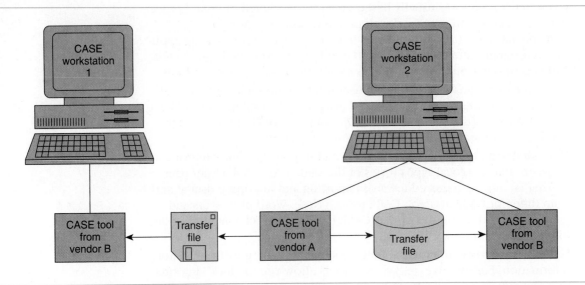

FIGURE 5.17   Data Integration by File Transfer   The most common way for two different vendors to exchange data between their tools is to build in a file transfer facility. The transfer file's structure may be proprietary or it may conform to some accepted standard.

A truly integrated CASE tool kit would allow systems developers to "mix and match" CASE tools according to their development requirements. Developers could plug in a new tool from a new tool vendor and be assured that it would accept any required information from their existing tools. They could change tools or tool vendors and be assured that information from their old tools will work with their new tools. The tools would have a common "look and feel" to facilitate learning and use. Finally, the tools would talk to and cooperate with one another.

This is an ambitious goal, but one necessary to achieve the ultimate payoff in CASE technology. To this end, we will describe a general framework for CASE tool integration. This framework calls for three types of integration: data, presentation, and tool.

Data Integration

The ability to share the results of work on one CASE tool with another CASE tool is perhaps the most important type of CASE integration. Obviously, i-CASE and m-CASE tool vendors provide this type of integration within their own products, but we are talking about providing that same level of integration regardless of a tool's vendor. There are several approaches to achieving **data integration:**

- *File transfer or data interchange.* In this common approach, information is exported directly from one tool to another tool (see Figure 5.17). The receiving tool will put the data in its own local repository. The export file's structure must be well defined using one of the following formats:
  - · *Tool-specific,* as required by the vendor of one or both tools. It is most important that the receiving tool(s) accept the data into its own local repository. This is how the Bachman Analyst upper-CASE workbench accepts information from KnowledgeWare's ADW/Planning Workbench (also an upper-CASE tool).

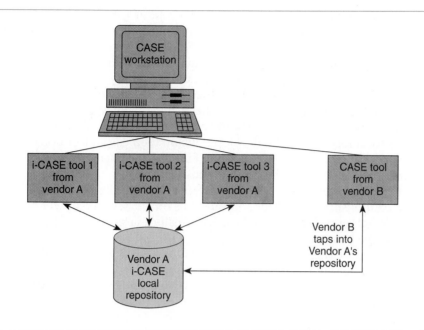

FIGURE 5.18 Data Integration through a Data Link A data link occurs when one vendor "taps" into another vendor's proprietary repository to use or update the data in that repository.

The above approach can work fine; however, if either tool vendor changes the structure of their local repository, the interface may no longer function until one or both vendors revise the file structure.

· *Proprietary standard,* as defined by a single vendor. CASE tool vendors can then enable their tools to import data to, and export information from, their own local repositories using this standard. An example is IBM's **External Source Format (ESF).** A variety of upper-CASE products can generate ESF files for IBM's fourth-generation language and code generator, Cross System Product (CSP).

· *Industry standard,* such as the **CASE Data Interchange Format (CDIF),** as developed by the Electronic Industries Association. Industry standards are formed through the cooperation of CASE tool vendors and experts. INTERSOLV is an example of a company that intends to enable its products to read data and write information in the CDIF format.

• *Repository linkage.* If a tool vendor publishes the structure of its own local repository, other tool vendors can design their tools to use the information stored in that repository. This approach has been taken by many minor tool vendors who build add-on tools to be used with the products of major tool vendors. The approach is illustrated in Figure 5.18.

Unfortunately, if you use a tool that reads data from some other tool vendor's local repository, you accept the risk of change. If the repository's tool vendor changes that repository (which happens frequently), the minor tool may no longer work (until updated by its own tool vendor).

• *Central repository interface.* The central repository concept was introduced earlier in the chapter, and most experts consider it the ultimate in data/information integration. A central repository should be defined by a consortium of CASE tool vendors who seek to make that repository's data structure "open" to any and all CASE tool vendors. Tool vendors must then "enable" their tools to store data into the central repository, and retrieve data from the central repository. Finally, developers must commit to storing information about all production information systems and systems development projects into the central repository.

As was shown in Figure 5.15, if developers wanted to do systems development work, they would borrow (check out) the necessary data from the central repository. That data would be moved to the appropriate CASE tool's local repository. The developer would then use and/or update the local data via the CASE tool(s). When finished, the data would be returned (checked in) to the central repository.

The central repository requires software that acts as a librarian and version control supervisor. This software runs on whatever computer where the central repository is stored (today, usually a large minicomputer or mainframe computer).

IBM's AD/Cycle Repository Manager is an example of a central repository and librarian. IBM's AC/Cycle is highlighted in The Next Generation box as one possible future for integrated CASE using the central repository concept.

Presentation Integration

Because a complete CASE tool kit includes many different products, possibly from several different tool vendors, a consistent user (screen) interface is highly desirable. A consistent "look and feel," accomplished by **presentation integration,** reduces the learning curve and makes the tools easier to use.

Most CASE tools have adopted or are moving towards standardized graphical user interfaces (GUIs). Most UNIX-based CASE tools have standardized on the Motif interface. Most PC-based CASE tools have moved to OS/2's Presentation Manager interface because the OS/2 operating system offers power features needed by the CASE tools. We expect Windows/NT to become a popular alternative platform, because its operating system proposes to offer many of the same power features as OS/2 and UNIX.

Tool Integration

In a perfect CASE world, not only can tools exchange data with one another, but they can also interact with one another. **Tool integration** makes it possible for tools to

- Send messages to one another (e.g., signal the end of one tool's activity so that another tool can start a related activity).
- Invoke (meaning "start") another tool.
- Invoke or use another tool's facilities.

This capacity is sometimes called *control integration.*

For example, consider a tool interface between a documentation tool and an analysis and design tool. Suppose the documentation tool contains a template that you built for a specific deliverable in your systems development methodology. A tool interface might automatically import any changes made to a diagram from the analysis and design tool—if that diagram is defined as part of the template. In other words, there would be no need to update the information with the document tool.

Current research in this area should soon result in greater tool integration and intelligence.

IMPLEMENTING CASE

CASE is still a relatively young technology that is not widely used in all organizations. How can you become a CASE activist and make CASE a career booster? How do you get management to invest in CASE? First, you need to understand the benefits and costs of CASE. More importantly, you need to cost justify the technology. Finally, you need to ensure that your organization has a sound CASE implementation plan. This section addresses all of these implementation issues.

Benefits of CASE

Like any other tool, CASE can be misapplied or not used to its fullest potential. However, when properly used, CASE can result in the following benefits.

Increased Productivity

CASE automates many of the most tedious clerical activities of developers. It reduces the time needed to complete many tasks, especially those involving diagraming and associated specifications. Estimates of improved productivity through application of CASE technology range from 35 to more than 200 percent.

Unfortunately, CASE productivity gains never immediately manifest themselves. In other words, there is a learning curve. True productivity gains usually come after using the technology on several projects. If management is looking for early tangible benefits, it should focus its attention on *quality*.

Improved Quality

Quality is measured in many ways. Does an information system fulfill user requirements? Can it easily adapt to ever-changing requirements? How many bugs are in the programs? Can programs be easily modified or reused?

CASE can eliminate or substantially reduce omissions and defects that would prove very costly to correct during systems implementation or support. Assuming analysts, designers, and programmers are applying sound development techniques — such as those discussed in the last chapter — CASE can provide almost immediate quality improvement benefits. (In contrast, if poor techniques are used, CASE merely helps you to develop poor quality systems much faster!)

Better Documentation

An early benefit of CASE is higher-quality documentation. CASE tools also make it easier to maintain that documentation. Indeed, we've also noticed an increased willingness of developers to maintain documentation if they are provided with CASE tools.

The Next Generation

AD/CYCLE: IBM'S BLUEPRINT FOR APPLICATION DEVELOPMENT AND INTEGRATED CASE

Few announcements were greeted with more enthusiasm and curiosity than IBM's September 1989 announcement of **AD/Cycle.** Different experts called it "IBM's endorsement of CASE," "the shot heard around the IS world," and "the most significant event in systems development since the invention of COBOL."

AD/Cycle is a comprehensive blueprint for implementing and integrating methodologies and CASE. It is a relatively open framework into which one can "plug" a large number of tools from IBM and several CASE tool vendors, including most of the major players in the CASE industry.

To better integrate CASE tools and methodologies, AD/Cycle's master plan calls for a new development environment that

- Covers the entire systems development life cycle.
- Can incorporate any modern methodology and techniques.
- Uses automated systems development technology (CASE) across the entire life cycle.

- Exploits the relative strengths of a developer's workstation and the shared mainframe in a client/server–like arrangement.
- Achieves data integration via a central repository.
- Achieves presentation integration through OS/2 operating system's Presentation Manager GUI for all AD/Cycle-compliant CASE tools.

In this feature, we'll briefly review some of these basic capabilities.

In establishing AD/Cycle, IBM realized that it could not provide all CASE tools for all people. To that end, it partnered with leading CASE tool vendors to provide a relatively complete (and growing) suite of tools. But IBM also opened up its blueprint to all CASE tool vendors, and many have committed to enabling their products within the IBM blueprint.

IBM's framework is based upon the existence of a comprehensive systems development life cycle that includes planning and support. AD/Cycle marketing literature uses IBM's version of the life cycle;

however, *any* version could be overlaid.

IBM's AD/Cycle is absolutely dependent on the use of rigor and precision in the life cycle. Most firms can only achieve such rigor and precision by building or buying a methodology to support the life cycle. Contrary to some popular belief, AD/Cycle can be used within the context of virtually any methodology.

For the most part, IBM's current AD/Cycle framework is targeted towards developing applications for IBM environments: MVS and VM (mainframe), OS/400 (minicomputer), and OS/2 (personal computers). Given the current trends toward "open systems," we can speculate that AD/Cycle will eventually target applications development for non-IBM platforms (such as Windows/NT and UNIX) as well.

AD/Cycle seeks to automate the entire life cycle through CASE. Currently, some AD/Cycle CASE tools are installed on workstations or network servers, whereas others are installed on host computers

continued

Reduced Lifetime Maintenance

The net benefit of higher systems quality and better documentation should be reduced costs and effort required for maintaining systems. This, in turn, creates more time and resources for new systems development. How so?

Most information systems shops are unable to find enough time to do all the new systems development (or major redevelopment) projects that are proposed. They simply spend too much time, effort, and resources to maintain existing systems. By reducing the amount of time spent on maintenance, a shop creates equivalent time to reengineer older applications, and focus efforts on

---

The Next Generation

concluded

(minis and mainframes). Regardless, all tools are executed from an intelligent CASE workstation. Looking ahead, and consistent with current downsizing trends, we expect that most host-based tools will eventually be downsized to the workstation or network server platforms.

AD/Cycle achieves data integration through a central repository called the Repository. Repository is currently implemented as a DB2 database to be executed on an MVS mainframe computer. As the downsizing trend continues in industry, IBM will downside Repository to run on AIX and OS/2 (or equivalent) file servers. Conceivably, Repository will be distributed across several computers. It is also conceivable that market demand for "open systems" will ultimately result in versions of Repository for non-IBM computers, operating systems, and networks.

The Repository database is based on an "open" *information model*. To make the information model flexible enough to accommodate information from as many CASE tools as possible, IBM has created a consortium of CASE tool vendors (partners) to jointly design the information model. Tool

vendors can extend the model to store any unique information captured or created in their own tools. Repository also contains "links" to application program libraries and database dictionaries.

IBM's Repository Manager provides the software tools and services to (1) help tool vendors "enable" their tools to exchange data with Repository, and (2) enable systems developers to exchange data between Repository and the local repositories of their installed CASE tools.

AD/Cycle achieves presentation integration through the OS/2 operating system's Presentation Manager GUI. All tools are encouraged to conform to this standard. Additionally, all OS/2 CASE tools are expected to "register" to an OS/2 program called Workstation Platform. Workstation Platform provides a menu system for all AD/Cycle workstation tools and links them, as appropriate, to AD/Cycle host tools.

AD/Cycle's blueprint calls for cooperative processing between many PC-based and host-based CASE tools (including Repository Manager). This creates a client/server-like arrangement where the systems developers see everything running on the PC even though

certain functions might be "subcontracted" to a mainframe or minicomputer.

The degree to which CASE tools from different vendors might exchange signals or invoke one another within AD/Cycle is unclear at this time. On the other hand, cooperative processing between tools indicates that some degree of tool integration is possible.

AD/Cycle is an ambitious project, even for a company as large and influential as IBM. AD/Cycle is a long-term strategy. The framework and implementation is coming along slower than industry would like (causing normal criticisms and concerns in both trade literature and customer circles).

But the task is formidable. To create an open architecture into which *all* tool vendors can plug their tools requires an unprecedented degree of cooperation between IBM and the CASE tool vendors to agree on suitable standards. And some would argue that IBM is the only computer company large enough to "drive" development of the independent standards.

If successful, AD/Cycle CASE could be right around the corner. (See the Suggested Readings at the end of the chapter for additional material about AD/Cycle.)

---

new applications that can return greater value and competitive advantage to the business!

Methodologies that Really Work

CASE has generated renewed interest in the importance of methodologies. Commit the following truths to memory:

- True CASE success is dependent on properly using a methodology!
- True methodology success may also be dependent on properly using CASE!

Too many businesses have realized little value from CASE because their developers weren't skilled with structured techniques and/or did not use a methodology. Similarly, many methodologies originally failed (or achieved limited popularity) because developers were unable to (1) implement the precision required to use the methodology, or (2) manage the volume of information and specifications that are required in the methodology. CASE has made it possible to realize the benefits of various methodologies' insistence on higher precision and detailed specifications.

Costs of CASE

CASE doesn't come cheap! In fact, it is usually very expensive. The cost of outfitting every systems developer with a preferred CASE tool kit is still prohibitive to many, if not most, information systems shops. Let's briefly examine the costs of CASE.

Hardware and Systems

Most CASE tools require a PC or workstation. The PC or workstation must be configured to acommodate the requirements of the most demanding CASE tool in the tool kit. Here are some considerations:

- Today's better CASE tools require a high performance graphics workstation (80386 workstation or better). Most require a lot of random access memory (4 to 16 MB). Most current tools require their programs and repositories to be stored on the local hard disk; however, the trend is to move that storage to a network file server.
- Most tools require a high-resolution graphics display standard such as VGA, SVGA, XGA, or 8514.
- Networked CASE workstations require appropriate network adapter cards and networking software.
- If you are developing applications for a mainframe or minicomputer, you may need a communications adapter and terminal emulation software. This gives you access to host-based development tools (e.g., database management systems, 4GLs, and compilers) and application software libraries (which may contain reusable code). It may also provide access to a central repository.
- Better CASE tools also require a sophisticated operating system such as UNIX, OS/2, or DOS with Windows.

Software

Most tool vendors offer quantity discounts and special purchase programs. Many offer introductory discounts as well. But there is also an important hidden cost—extended maintenance contracts. This maintenance contract is *almost* a mandatory extra-cost item. It buys you free upgrades to new versions (which can be quite expensive if purchased separately). More importantly, it gives you access to "hotline" phone support that can be critical when you encounter tool problems. Extended maintenance must normally be renewed annually and is usually priced on the basis of the number of tools you have purchased (less applicable quantity discounts).

Training and Consulting

The proper application of CASE tools generally requires some degree of training. Most vendors offer their own courses in larger cities. Some will bring courses to your site if you are wiling to pay travel and lodging costs for the instructors. Third-party training may be available for some of the more widely used CASE tools.

Most organizations establish a goal of internalizing CASE training. In other words, they utilize vendor and third-party training only to train their own instructors. Then, they use their own instructors to train most of their own staff. Periodically, the internal instructors must have their skills updated for new releases of tools.

Finally, a CASE consulting industry has evolved to support users of CASE tools. The consultants offer training and on-site services that can be crucial to accelerating the learning curve and to deployment and use of the tools. Representative services include

- Orientation of management and staff to CASE technology.
- Selection of the standard CASE tool kit and/or methodology.
- Integration of CASE tools into your methodology.
- Customization of your methodology to support selected tools (e.g., changing and documenting your methodology's data model diagram conventions to match the capabilities of the CASE tool).
- Adaptation of your CASE tools to support the requirements of your methodology (to get around inadequacies of the CASE tools).
- Pilot project supervision or guidance (but NEVER let the consultant do the project for you. Insist on participation of your own staff to move toward independent skill development).

Justification for CASE

Can the costs of CASE be justified? Do the benefits exceed the costs? What level of benefit is required to offset the costs? Many managers' first reaction to CASE is, "I can't afford to do that. It will never pay for itself!" Too many information professionals cannot argue their case since they are not skilled in the techniques of financial analysis. Here is a simple financial analysis that might help.

Figure 5.19 is a spreadsheet we set up for a cost/benefit analysis. You could easily set up a similar (or more complex and comprehensive) spreadsheet on your own. A spreadsheet allows you and your management to challenge the assumptions and estimates and still quickly perform the necessary decision analysis: "Do we, or don't we, invest in this technology?"

Our analysis is based on some worst-case assumptions that are designed to prove CASE is *not* cost justifiable. That's right: *not* cost justifiable! Our approach says, "Let's overestimate the costs and underestimate the tangible benefits. If CASE can be defended under these conditions, it will really return greater value with more 'reasonable' estimates and assumptions."

As you see in the figure, even with overestimated costs and underestimated tangible benefits, CASE yields a positive return on investment. When you make the numbers reflect quantity discounts, realistic salaries and overhead, and

CASE Costs and Benefits

Costs of CASE

80486 Personal Computer 2MB RAM & 200MB Disk	($3,500)
Representative Operating System	(200)
Memory Upgrade from 2MB to 8MB	(300)
High Resolution Graphics Adapter and Display	(1,500)
Mouse	(100)
CASE Software (budget for two tools)	(12,000)
Training on the Two CASE Tools (with travel & lodging)	(3,000)
Four Years Extended Maintenance on CASE Software	(4,000)
Total Costs	**($20,600)**

Derivation of Tangible CASE Benefits

Annual cost of a Systems Analyst (underestimated)		30,000
Overhead for Analyst (e.g., pension, insurance, vacation, etc.)		15,000
Total Annual Employee Cost		$45,000
Estimated Productivity Enhancement (underestimated)	15%	
Annual Productivity		6,750
Total Productivity Benefits (over 5 years)		**$33,750**

Cost/Benefit Analysis

Payback Period Analysis (lifetime costs/annual benefits)	3.1 years
Net Present Value Analysis	
Time Adjusted Benefits (discounted at 10%)	$25,583
Time Adjusted Costs (again, discounted at 10%)	($13,431)
Net Present Value (positive number indicates GOOD investment)	$12,152

better productivity gains, the net present value grows significantly. Also, we haven't figured in the most important intangible benefits: higher quality systems, better documentation, and reduced maintenance.

Here's an interesting situation. Suppose you plug in costs for your own company's tools and the net present value comes back negative. Is the investment bad? Not necessarily! You simply need to ask yourself one more question: "Will the intangible benefits that were not quantified at least equal or offset the negative net present value?" If so, it is still a good investment. In most cases, we think that the intangible benefits will far exceed the tangible benefits and that, in the final analysis, you cannot afford *not* to take advantage of this technology.

Implementation Recommendations

Given a management commitment to try CASE, how do you ensure that CASE technology is successfully implemented. We offer the following suggestions:

- *Get your methodology in order.* Remember, without commitment to a good methodology, CASE offers little improvement over traditional, manual techniques. Curiously, many firms invest in CASE before a methodology. Most experts suggest you reverse the order — first, methodology; second, CASE.

- *Select the right people.* Select analysts, designers, and programmers who really want CASE to succeed. Enthusiasm is often more important than experience. Also, only provide CASE tools to those who commit to using them within the framework of your methodology. Otherwise, benefits will come very slowly. Keep tools in the hands of its real users and take the technology away from those who don't use the technology.

  Also, get management commitment to CASE. It is especially important, however, to "manage" management's expectations (see next item).

- *Develop reasonable expectations.* Using CASE technology and techniques is like learning to drive a car with a manual transmission. While you're learning, you are constantly thinking about a seemingly impossible combination of details: releasing the clutch, watching the RPMs, listening to the engine, shifting the gears, not stalling the engine, and so forth. It seems like you'll never learn. But eventually, you stop thinking about it. You just do it!

  Early on, it won't look like you are achieving significant benefits with any new CASE tool. That's because you are still learning. Concentrate on fully using the facilities to improve the quality of your work. Productivity will come in later projects. With each successive project, you'll see some productivity improvement.

- *Choose an appropriate pilot project.* When using CASE technology (or a methodology) for the first time, *never* select a mission-critical, multianalyst, multiyear project. Select a small-scale pilot project that isn't critical and can afford the delays associated with learning the new technology (or methodology).

- *Create or use support mechanisms.* Many CASE vendors have established user groups. Join your local user group to "network" (meaning "communicate") with those who use the same tools you're using. User groups also provide input to improving CASE tools. The major CASE vendors each sponsor annual conferences.

  Consider forming a local users group within your company or community.

- *Learn from others.* CASE is no longer the new kid on the block. There are several well-documented implementations. Start by reading Barbara Bouldin's *Agents of Change: Managing the Introduction of Automated Tools,* in which Barbara details her experiences of introducing CASE to AT&T.

  There are now several CASE industry newsletters. Some of these are listed in the suggested readings at the end of your chapter. Your business, school, or library may subscribe to some of these publications.

  The CASE consulting industry now sponsors a number of general interest CASE conferences to present latest trends and experiences. Conferences such as CASE World and CASE War Games provide interested professionals with opportunities to see many different CASE vendors and product lines, and hear noted academic and industry CASE experts.

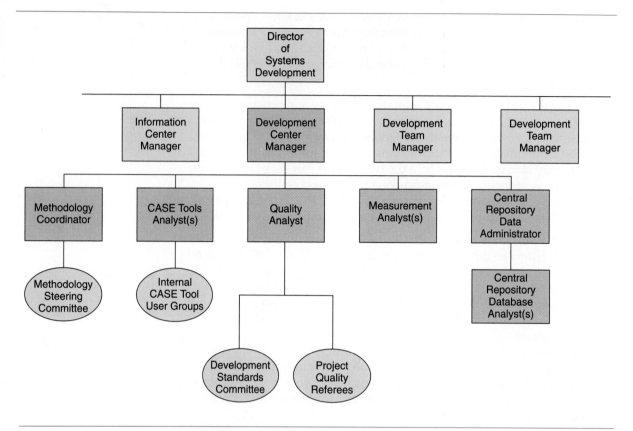

FIGURE 5.20   A Development Center's Organization   A development center provides support services for information systems developers.

- *Implement a continuous process improvement plan.* Continuous process improvement is a "total quality management" concept. It states that you can only improve overall quality in continuous, incremental steps. A CASE continuous process improvement plan would evaluate every project to subsequently fine-tune the methodology and CASE standards.
- *Implement a development center.* The ultimate implementation vehicle for both CASE and methodologies is a development center.

    A **development center** is a central group of information systems professionals who plan, implement, and support a systems development environment for other information systems developers. Just as an information center provides consulting services to end-users who develop their own personal information systems, a development center provides consulting services to IS professionals developing shared information systems.

    Development center consultants do not develop systems. Rather, they provide consulting services to those who do develop systems (including systems analysts and programmers).

    As shown in Figure 5.20, a development center reports to the manager in charge of all systems development. It provides consulting services to all systems development teams. A fully staffed and mature development center might include the following specialists:

- **Methodology coordinator** who is responsible for methodology acquisition, installation, training, customization, and support. A steering committee of managers and developers may help this individual.
- **CASE tools analysts** who are specialists in one or more CASE tools, and who provide CASE tool installation, training, and support services.
- **Quality analysts** who establish and document methodology, technique, and tool standards, and assist with quality checks for all projects. Supporting committees help this individual create standards and audit projects for conformance to these standards.
- **Measurement analysts** who establish standards for measuring productivity, quality, and value returned on methodologies and CASE.
- **Central repository administrator** who manages the data stored in a corporate central repository. (Note: Some organizations may prefer to locate this individual in the data management group since most central repositories double as the database dictionary.) Database experts may also be needed to support a central repository.

Summary

Computer-aided systems engineering (CASE) is the application of computer technology to systems development activities, techniques, and methodologies. CASE "tools" are programs (software) that automate or support one or more phases of a systems development life cycle.

CASE tools can be classified according to their support of the systems development life cycle. Upper-CASE tools automate or support the "upper" or "front-end" phases of the systems development life cycle; namely, systems planning, systems analysis, and general-to-detailed systems design.

Lower-CASE tools automate or support the "lower" or "back-end" phases of the life cycle; namely detailed systems design, systems implementation, and systems support. There are several types of lower-CASE tools. An application programmer's workbench supports the coding, compiling, testing, and debugging of applications programs on an intelligent workstation. A component generator assists with the design of specialized information system components, and then generates the code required to implement those components. A code generator automatically generates application programs from analysis and design specifications.

Some lower-CASE tools support systems support and maintenance through reengineering. Reengineering is the process of altering or restructuring an existing system. It is the opposite of redevelopment (starting over). Eventually, when a system becomes obsolete, lower-CASE tools support reverse engineering. Reverse engineering is the process of analyzing existing application programs and database code to create higher-level representations of the code. In many cases, these representations are graphs and diagrams. Reverse engineering is sometimes called *design recovery.*

The term *cross life cycle CASE* refers to tools that support activities across the entire life cycle. This includes diverse activities such as project management and estimation.

CASE tools share a common general architecture. They are built around a predefined repository. A repository is a developer's database. It is a place where the developers can store diagrams, descriptions, specifications, applications programs, and any other working by-products of systems development. Synonyms include *design database, data dictionary,* and *encyclopedia.* Most CASE tools require one instance of a repository for each unique systems development project—these instances are called *local repositories.* Some firms are beginning to consolidate these local repositories into a central repository.

CASE tools build facilities to store data in, analyze data in, and retrieve data from their repository. Typical facilities include graphics, description, reporting and inquiry, prototyping, quality assurance, decision support, documentation, transformation, generation, data sharing, and housekeeping.

Integration of CASE tools has become a critical factor for the future. CASE tools are evolving toward three types of integration. Data integration allows CASE tools with different local repository structures to exchange data. Presentation integration give all CASE tools the same "look and feel." Tool integration allows CASE tools to "invoke" one another automatically.

To better manage a CASE implementation, many firms implement a development center. A *development center* is a central group of information systems professionals who plan, implement, and support a standard systems development environment, including methodology and CASE, for other information systems developers. Development center staff provide consulting services to those who apply the methodology and CASE tools.

Key Terms

AD/Cycle, p. 196

application programmer's workbench, p. 176

CASE data interchange format (CDIF), p. 193

CASE tools analyst, p. 203

central repository, p. 187

central repository administrator, p. 203

client/server application, p. 178

code generator, p. 179

component generator, p. 177

computer-aided systems engineering (CASE), p. 169

cross life cycle CASE, p. 171

data integration, p. 192

data-sharing facilities, p. 191

decision support facilities, p. 190

description facilities, p. 189

development center, p. 202

dictionary, p. 185

documentation, p. 184

documentation facilities, p. 190

encyclopedia, p. 185

enterprise model, p. 172

estimation, p. 184

External source format (ESF), p. 193

function points, p. 184

generator, p. 191

graphics facilities, p. 189

graphical user interface (GUI), p. 178

Housekeeping facilities, p. 191

inquiry and reporting facilities, p. 190

integrated-CASE (i-CASE), p. 185

local repository, p. 186

lower-CASE, p. 171

measurement analyst, p. 203

methodology coordinator, p. 203

modular-CASE (m-CASE), p. 186

presentation integration, p. 194

process managers, p. 184

project management, p. 183

prototyping facilities, p. 190

quality analyst, p. 203

quality assurance facilities, p. 190

reengineering, p. 180

repository, p. 185

regression analysis, p. 181

restructuring, p. 181

reverse engineering, p. 181

security and version control facilities, p. 191

systems engineering, p. 169

tool integration, p. 194

transform facilities, p. 190

upper-CASE, p. 171

work groups, p. 187

-------------------- *Questions, Problems, and Exercises* --------------------

1. Define computer-aided systems engineering in its broadest context of automating or assisting the entire systems development life cycle.

2. In a staff meeting, a senior manager proposes, "I say we dump our methodology and invest in CASE." Respond to this manager.

3. Briefly describe the history and evolution of CASE technology. What is the similarity between the development of CASE technology and CAD/CAM technology.

4. Management wants to build a CASE "tool kit." Identify and briefly differentiate between the five general categories of CASE tools that should be included in the tool kit.

5. How do upper-CASE tools support systems planning? How do upper-CASE tools also support systems analysis and design? Which upper-CASE tools, those for systems planning or for systems analysis and design, would you expect to encounter as a young information systems professional? Explain your answer.

6. How do lower-CASE tools support detailed design and systems implementation?

7. What are three types of lower-CASE tools? Briefly distinguish between the three categories.

8. Some people believe that code generators will obsolete the role of *programmer* as we know it. What do you think? Why? Now suppose that such a change does occur. What do you think will happen to applications programmers? How about systems analysts?

9. How do lower-CASE tools support the maintenance activities (systems support) for production information systems?

10. Bedico & Company invested in upper-CASE tools four years ago. Based on their experiences, they like what they see; however, they are frustrated because the tool isn't used enough. They feel that because maintenance and system support activities consume 80 per cent or more of most development time and resources, they are prevented from convincing management to further invest in an expanded CASE tool kit. Can you help them out with some ideas?

11. Briefly describe the differences between reengineering and reverse engineering. How would you respond to a manager's belief that all systems development is reengineering?

12. You have been assigned to analyze and design a completely new order processing system for Tides Manufacturing, Inc. What types of information might you recover from the existing order-processing applications programs by using a reverse-engineering tool? What limitations do you expect to find in the captured information?

13. Identify four cross life cycle activities that may be supported using CASE tools.

14. What role does a repository serve in CASE architecture?

15. Explain the difference between a local and central repository. Which is better? Why?

16. List several common CASE tools facilities that might be implemented around a repository.

17. Visit an information systems shop that is using CASE tools. Pick a CASE tool and compare its facilities against the typical facilities described in this chapter. Did you discover any new capabilities? (Don't be surprised if you do. The industry is constantly improving the tools.)

18. Why is it important that CASE tools be integrated (work cooperatively)? What should be the goal of CASE integration?

19. Identify and explain three ways in which CASE tools can be integrated.

20. List three different approaches to data integration. What are the advantages and/ or disadvantages of each.

21. What is presentation integration? Why is it important for applications? Why is it important for CASE tools? What major trend is driving this type of integration?

22. Identify and explain several of the benefits of implementing CASE technology.

23. What is the relationship between CASE technology success and methodology success?

24. Describe several of the costs of CASE.

25. Visit an information systems shop that has invested in CASE. Make a list of the costs that they have incurred so far. Unless such information is confidential, try to get a sense of the total dollar commitment made so far. How are these costs being justified to non–information systems managers?

26. List several recommendations to successfully implement CASE.

27. Identify and differentiate between the types of specialists that might work for a development center.

Projects and Minicases

1. Assume that you are the newly hired CASE coordinator for a major company. You are to make a presentation to the senior management proposing that large sums of money be spent to buy a methodology and CASE tools for your company. Prepare a spreadsheet that details the projected costs and benefits. Use Figure 5.19 as a starting point; however, try to make the analysis more realistic by expanding assumptions, cost categories, and benefit categories. Consider, for example, that a developer is unlikely to spend all his or her time developing systems; tools can only improve productivity during times that they are being applied to their intended tasks.

2. For upper-, lower-, and cross life cycle CASE tools, perform a market survey and analysis. Go to your library and research the topic in trade journals and periodicals. What are the names of the most popular CASE tools for each category and by whom are they sold? How much do they cost? For each of those tools, what operating systems and hardware are required? Are they integrated to a common repository? Are there any vendor-supplied interfaces to other CASE tools?

3. Review some of the trade literature from current users of CASE tools. What is the general level of satisfaction among these users? What are their major complaints? Make a list of the features that current users would like to see incorporated into the next generation of CASE tools to make them even more powerful. What recommendations can you offer to solve some of the major complaints or concerns about CASE?

Suggested Readings

Bouldin, Barbara. *Agents of Change: Managing the Introduction of Automated Tools.* Englewood Cliffs, N.J.: Prentice Hall, 1989. This book documents the experiences of a "CASE champion" as she introduced CASE tools in a large company.

C/A/S/E Outlook. CASE Consulting Group Incorporated, 11830 Kerr Parkway, Suite 315, Lake Oswego, Oregon 97035. This is a bimonthly journal on CASE.

CASE Strategies. Cutter Information Corporation, 37 Broadway, Arlington, MA 02174-5539. This is a monthly newsletter on computer-aided systems engineering. Each issue is approximately 20 pages long. Most issues begin with a management-oriented current issue or topic. Each issue reviews at least one CASE product and summarizes CASE industry news. This newsletter is our favorite, quick method of staying on top of the latest trends and news in the CASE industry.

CASE Trends. Software Productivity Group, Inc., P.O. Box 294-MO, Shrewsbury, MA 01545-0294. This is a CASE periodical published nine times yearly.

Gane, Chris. *Computer-Aided Software Engineering: the Methodologies, the Products, and the Future.* Englewood Cliffs, N.J.: Prentice Hall, 1990. This book provides a conceptual overview of CASE and structured techniques, as well as a market survey.

Martin, James, and Carma McClure. *Structured Techniques: The Basis for CASE.* Englewood Cliffs, N.J.: Prentice Hall, 1989. This book stresses the importance of understanding the structured techniques in order to maximize the return on investment for upper- and lower-CASE tool.

Montgomery, Stephen L. *AD/Cycle: IBM's Framework for Application Development and CASE.* New York: Van Nostrand Reinhold, 1991. This book thoroughly explains IBM's integrated-CASE blueprint.

The six chapters in Part Two introduce you to the systems planning and systems analysis activities of systems development. Chapter 6 introduces **systems planning,** a development process that seeks to identify and prioritize the technology and applications that will return the most value to the business. Not all organizations include systems planning in their systems development life cycle; however, systems planning is becoming increasingly common as businesses learn that information systems should not randomly evolve—they should be planned.

Systems analysis is the most critical process of information systems development. It is during systems analysis that we learn about the existing business system, come to understand problems, define objectives and priorities for improvement, and define business requirements. Clearly, the quality of any subsequent systems design, implementation, and support is dependent on good systems analysis. In fact, those three phases are useless if they don't solve the correct problems, fulfill objectives, meet requirements, and provide flexibility in a cost-effective fashion! Systems analysis is often short-changed in the development process because most analysts and users are not well schooled in systems analysis methods. Chapter 7 provides an overview of systems analysis phases, activities, roles, deliverables, tools, and techniques.

The remaining chapters introduce specific tools and techniques for completing systems planning and analysis. In Chapter 8 we introduce **data modeling,** a technique for organizing and documenting a system's data. The chapter introduces entity relationship diagrams as a tool for depicting the associations among different categories of data within a business or within information systems. It does not imply ''how'' that data is implemented, created, modified, used, or deleted. These data models eventually serve as the starting point for designing files and databases.

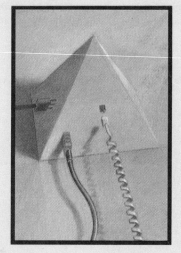

Chapter 9 introduces **process modeling.** It explains how data flow diagrams can be used to depict information system problems and solutions in terms of the flow of data through the system and the activities performed on that data. These process models serve as a starting point for designing computer-based methods and application programs.

Chapter 10 explains the concept of **network modeling.** You'll learn about location connectivity diagrams and how they map people, data, and activities to geographic locations. These network models serve as a starting point for designing the communications systems for distributing systems data, activities, and people to various geographic locations.

Finally, Chapter 11 introduces the concept of a **project repository** as a place where documents, documentation, and specifications that are associated with an application and project are kept. This chapter also introduces tools and techniques for specifying data, information, and process requirements in the project repository.

SYSTEMS PLANNING AND SYSTEMS ANALYSIS

6

Systems Planning

Chapter Preview and Objectives

In this chapter you will learn more about the systems planning phases in the systems development life cycle: *study, definition,* and *analysis*. It is not very likely that a young systems analyst or user will become directly involved in a systems planning project. On the other hand, it is increasingly likely that a young systems analyst or user will become involved in application development projects that were triggered by systems planning. Therefore, it is useful to understand the planning process and its deliverables. You will know that you understand the process of systems planning when you can

Define
systems planning and relate the term to its phases: *study, definition,* and *analysis*.

Describe
the role of a repository in systems planning.

Describe
the study, definition, and analysis phases in terms of your information system building blocks.

Differentiate
between "business area" analysis and "systems" analysis.

Explain
how business areas and business area analysis can lead to more highly integrated applications and application value.

Describe
the study, definition, and analysis phases in terms of objectives, activities, roles, inputs, and outputs, and tools and techniques.

Recognize
a number of system modeling techniques and describe their roles in systems planning.

Recognize
a number of matrix analysis techniques and describe their roles in systems planning.

Identify
those chapters and modules in this book that can help you interpret documentation that results from systems planning.

ARKANSAS TECHNICAL INSTITUTES

The Arkansas Technical Institutes is a technical college with campuses located throughout its parent state. The main campus is in Little Rock, Arkansas.

Scene: *Jennifer Clinton, director of Information Systems, is meeting Jesse Brown, chancellor of the Arkansas Technical Institutes. Mr. Brown called the meeting to address some concerns he has about the performance of Information Systems on recent projects.*

Jenny: Good morning, chancellor.

Jesse: Good morning. I prefer "Jesse." I understand you prefer "Jenny." I don't think we've run into each other for at least a couple years, no?

Jenny: Right. Mr. Buchanan said you have a number of questions about Information Systems that he couldn't answer?

Jesse: That's right. I was expecting Robert [Buchanan] at this meeting, but it looks like he may not show. Why don't we just get started.

[Jenny nods in agreement.]

I have some questions and concerns about some recent projects that are late. What's happened to the new fund-raising decision-support system?

Jenny: I know its late. We had to pull people off that project to make major changes to the budget system.

Jesse: Who made that decision?

Jenny: Paul Bush! Both systems report to his area and are supported by my group's Financial Systems team.

Jesse: Well, I can understand his decision. The budget cycle starts soon and he has to be ready for state legislature. But we also start a major fund-raising campaign next month, but your support systems are not even close to ready. Am I right?

Jenny: I'm not happy to admit it, but yes, you are correct.

[Jesse does not respond.]

We've had a couple other problems with the development system. It has to interface with the alumni system, which is 10 years old and is maintained by a different development team and vice president. That team needs to make some changes to make the interface work, but that project has been given a lower priority by Bill Quayle *[the vice president who decides that team's budget].* He's got that entire team working on the new student record system.

Jesse: I know that project. We don't need that student record system just yet!

Jenny: Yes, but it is important to him. Until it is done, the new on-line registration system project cannot be started.

Jesse: So what! I'll admit that the current system is dated, but it does work. Let me understand this. We've mailed out campaign literature to prospective donors. We've paid for radio and television advertising. We've contracted with a professional fund-raising firm. But our campaign is going to stall because Bill Quayle won't pay for a small but vital alumni information system change that will enable our support systems?

Jenny: That's it!

Jesse: Can't you reallocate people to the Financial Systems team? An expanded team might be able to do the alumni system modifications themselves.

Jenny: That won't help much. The rest of the team would still be working on the budget system. That system is just about as important, right?

Jesse: You're right, of course. So why isn't the Financial Systems team larger given that it has two very high priority projects at this time?

Jenny: I can't answer that. I inherited the team structure from my predecessor. I don't think the current team sizes have changed much over the past several years. No vice president wants his or her support reduced.

Jesse: That doesn't make much sense to me. How do your development teams decide what to do and not do?

Jenny: Each team supports a well-defined group of managers and users that report to a single vice president. Each team has its own budget, and the vice president determines what can and cannot be done within that budget. There's always more to do than resources to do it.

Jesse: Yes, but some things are more important than others. How often do the budgets get reallocated?

Jenny: I don't think any team's budget has ever been decreased. Some increase a little faster than others.

Jesse: But don't you think changing business priorities should decide where development budgets and information systems staff should go?

Jenny: That makes some sense, yes. But if team composition changed drastically from year to year, how would that affect our ability to support pro-duction systems on a daily basis? Also, if teams and budgets change every year, who will make those decisions? I'm not knowledgeable enough about tactical and strategic institute directions to make those decisions.

Jesse: I'd like to respond to that comment with a different question: Shouldn't you be?

Discussion Questions

1. How would you answer Jesse's final question? Is it reasonable to expect Jenny to understand the business that well?

2. What kinds of problems might arise if information systems budgets and teams changed drastically from year to year? How might systems support be affected?

3. How do you think you might improve this (very common) situation without making dramatic budget and team changes every year?

WHAT IS SYSTEM PLANNING?

Systems planning was first defined in Chapter 3:

> The **systems planning** function of the life cycle seeks to identify and prioritize those technologies and applications that will return the most value to the business. Synonyms include *strategic systems planning, information strategy planning,* and *information resource management.*

Not all organizations include systems planning in their SDLC. But systems planning is becoming increasingly common as businesses learn that information systems should not randomly evolve—they should be planned. Why systems planning? Consider the following argument offered by information systems planning expert Michael Porter:

> The importance of the information revolution is not in dispute. The question is not whether information technology will have a significant impact on a company's competitive position; rather the question is when and how this impact will strike. Companies that do not anticipate the power of information technology will be forced to accept changes that others initiate and will find themselves at a competitive disadvantage.[1]

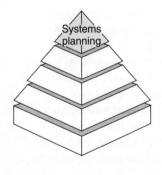

Systems planning is driven by the cooperation of *system owners.* Hence, in your pyramid model, it addresses PEOPLE, DATA, ACTIVITIES, and NETWORKS from the system owners' point-of-view (which was introduced in Chapter 2). System owners come from the ranks of executive management and higher-level middle management.

---

[1] Michael E. Porter, "How Information Gives You Competitive Advantage," *Harvard Business Review* 63, no. 4 (July–August 1985), pp. 149–160.

Historically, information systems have focused on back-office functions like accounting, payroll, order processing, and so forth. Through strategic planning, businesses try to redirect that focus on using information technology and systems to truly streamline the business, increase business value, and gain competitive advantage.

In Chapter 3 (and Figure 3.6), you learned that systems planning consists of three phases:

1. Study the business mission (also called the study phase).
2. Define an information architecture (also called the definition phase).
3. Analyze a business area (also called the **business area analysis** or BAA phase).

Systems planning is similar to systems analysis. The primary difference is in "scope." Systems planning deals with a larger portion of the enterprise, but in lesser detail than systems analysis. But the phase names are similar. So are many of the strategies, tools, and techniques. For this reason, systems planning responsibilities are frequently assigned to senior systems analysts. They are sometimes called **planning analysts.**

Figure 6.1 is a life cycle diagram that illustrates the three planning phases. We've changed the diagram somewhat since Chapter 3. The work products and deliverables no longer flow from phase to phase. Instead, they flow between the phase and a planning repository.

> A **planning repository** is a central location where information systems professionals store all documents and documentation associated with a systems planning project. Synonyms include *central repository.*

The introduction of the planning repository promotes reuse of documentation and allows phases to overlap. The repository also promotes rework that may have to be done if errors and omissions are discovered later.

Planning repositories are usually implemented using CASE technology. As shown in the figure, a planning repository will eventually be used to create and populate several *project repositories;* one for each application development project initiated through the plan.

This chapter examines each of the planning phases in greater detail. There are numerous systems planning techniques and methodologies (see margin). Our presentation is a hybrid of several of these techniques and methodologies, but is most consistent with the popular *information engineering* methods. To give you "the feel" of a true systems development methodology, each phase is described in terms of the: (1) *purpose* of the phase, (2) *activities* that should be performed, (3) *roles* played by various people in each activity, (4) *inputs* and *outputs* for each activity, and (5) *techniques* or *skills* that can be used to complete each activity.

THE STUDY PHASE OF SYSTEMS PLANNING

The first phase of systems planning is to *study the business mission.* It is surprising that many businesses haven't formally documented their mission, but they all have one. Good information systems planning is dependent on good busi-

—————— ✓ ——————

Systems Planning Techniques

Critical success factor (CSF) analysis
Enterprise modeling
Information engineering (several variations)
Value chain analysis

—————— ✓ ——————

Systems Planning Methodologies

Business Systems Planning/ Strategic Alignment (BSP/ SA) by IBM
Foundation by Arthur Andersen (also supports other life cycle phases)
Navigator by Ernst & Young (also supports other life cycle phases)

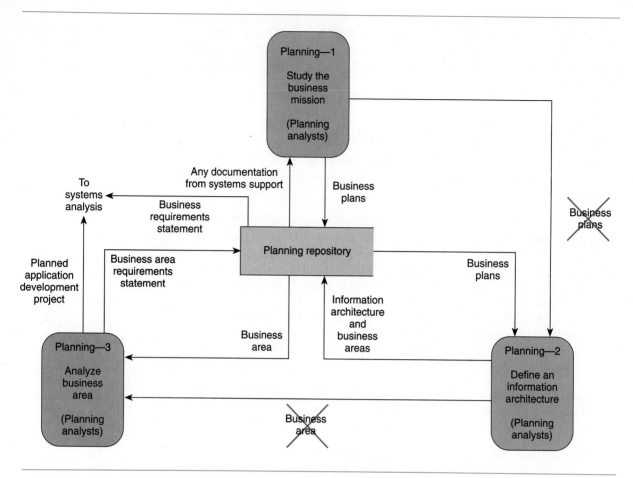

FIGURE 6.1 **Systems Planning**   The three phases of systems planning were introduced in Chapter 3. Planning information and results are recorded in a repository for later use in application development projects.

ness planning. Accordingly, information systems management frequently plays a key role in suggesting and driving this phase. To provide an unbiased analysis, consultants are frequently engaged to direct this phase.

If information systems are to truly return value to the business, they need to be aligned with the business mission. Furthermore, information systems should address those business activities that will return the greatest value to the business. This phase establishes the business foundation for IS planning.

Building Blocks for the Study Phase

The fundamental objectives of the study phase are

- To establish the mandate for strategic systems planning. (This may involve convincing top management of the importance of strategic systems planning.)
- To build a working partnership between information systems management and top business management.
- To analyze enterprise strategies that may affect information systems.

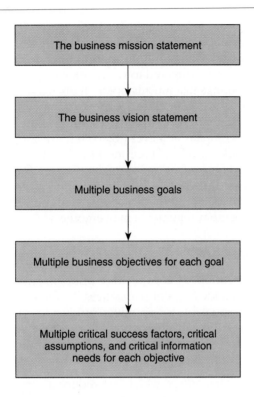

FIGURE 6.2 **Critical Success Factor Hierarchy** Information engineering advocates frequently employ the technique of critical success factor (CSF) analysis to measure business performance. CSF analysis defines this hierarchy of measures.

To achieve these objectives, you need an executive's understanding of the business, not the information systems. Consequently, this is one of the few life cycle phases in which your information systems building blocks are *not* very useful.

Fortunately, there are alternative frameworks that can help. For example, information engineering methods provide frameworks for identifying business performance measures. Also, Michael Porter's *Competitive Forces, Competitive Strategies* and *Value Chain* provide useful frameworks for studying and analyzing a business's ability to compete in today's market.

Critical Success Factor Analysis

Most businesses evaluate their success or failure by some set of performance measures. The information engineering methodology advocates a popular measurement technique sometimes called *critical success factor analysis.*

Critical success factor analysis is the identification of a hierarchy of performance measures that lead to identification of critical factors and issues that will determine a business's success. The hierarchy is depicted in Figure 6.2.

By way of explanation, a **business mission statement** is a one- or two-paragraph statement of a business's reason for existing. A **business vision statement** is a brief statement of the business's definition of success as envisioned by its top executives. **Business goals** are general measures of success such as "maximize market share" or "minimize defective products." These goals are directed toward achieving the business vision. **Business objectives** are measurable milestones with specific time frames such as "reduce defective manufactured parts to 3.4 per million opportunities for defects by January 1994." These objectives are directed toward achieving goals. **Critical success factors** (CSFs) are "things that must happen" if objectives are to be achieved. **Critical assumptions** are "things you expect to happen or not change" if objectives are to be achieved. **Critical information needs** are needed to measure how well a company achieves objectives.

CSF analysis provides a useful baseline not only for evaluating business performance, but also for evaluating information systems' contribution to the value of the business. Clearly, the best information systems are those that directly contribute to achievement of business goals and objectives. Indeed, specific information system statements may emerge as CSFs during analysis.

Competitive Analysis

Analysis of the competition, or **competitive analysis,** may be appropriate if the planning scope includes the entire business, or if it includes any operating unit that is "solely" responsible for a product, product line, or service. In either situation, analyzing the threat of competition is a useful exercise. According to Michael Porter, five forces determine an organization's ability to compete in the economy. Those forces include

- The *threat of new competitors,* which can reduce a company's share of the market. How easy would it be for a new competitor to join the industry?
- The *threat of substitute products or services,* which could reduce a company's share of the market, or even obsolete the company's products and services. What is the potential that some business might invent or develop such a product or service?
- The *bargaining power of customers,* which could reduce a company's profit margin or market share. How dependent are customers on the company's products and services?
- The *bargaining power of suppliers,* which could also reduce a company's profit margin or market share. How dependent is the business on its suppliers?
- The *rivalry among existing competitors.* What is the useful marketing life of the company's products and services?

These competitive forces on any business can be studied and analyzed as part of the study phase.

Value Chain Analysis

The value chain is another concept from Michael Porter that can help management and analysts gain an understanding of the business.

> The **value chain** is a diagrammatic model of business activities and how they add value to a business. By analyzing the value chain, an analyst gains insight into how an organization works. In particular, the value chain helps the analyst appreciate those activities that, if improved

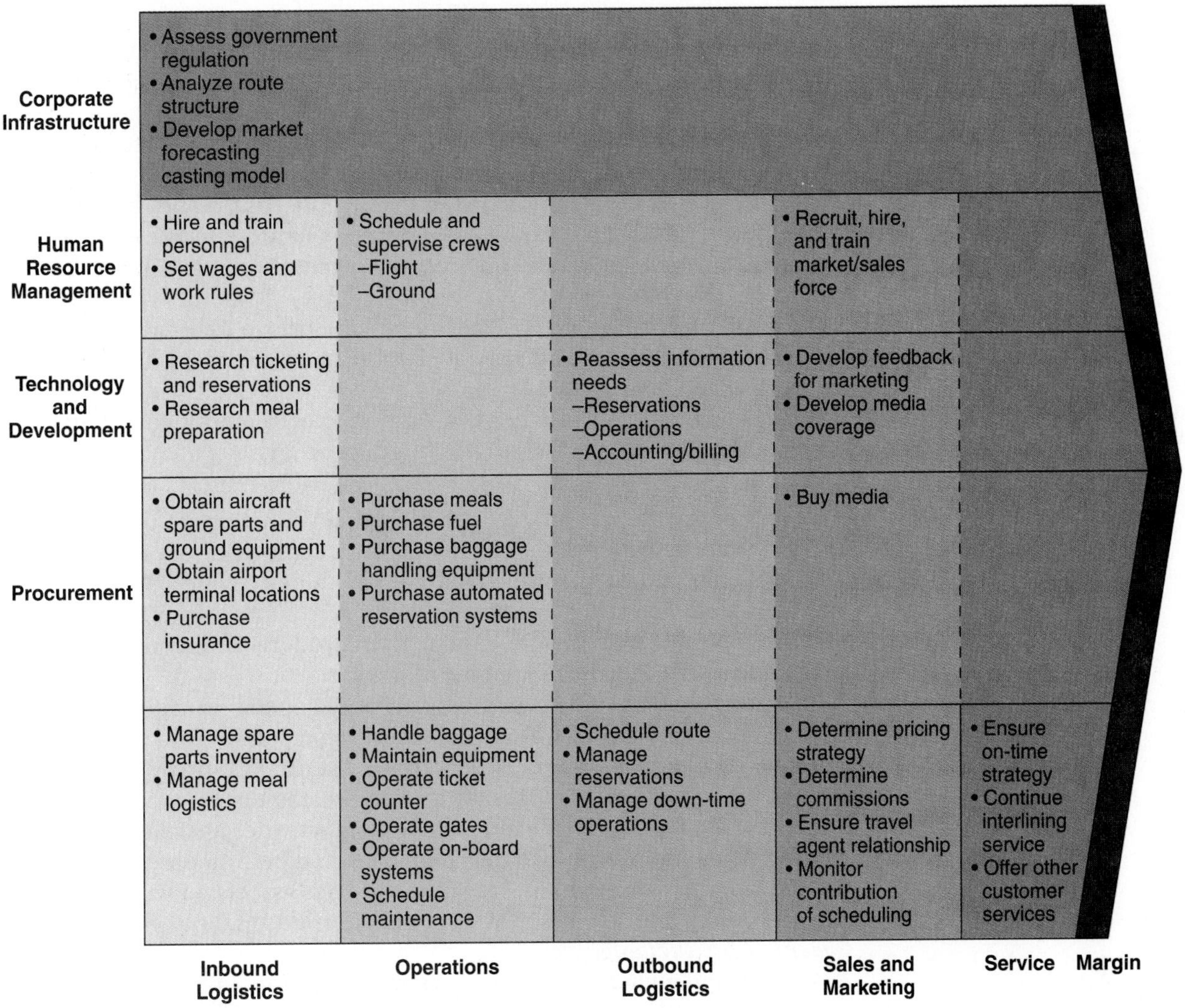

	Inbound Logistics	Operations	Outbound Logistics	Sales and Marketing	Service
Corporate Infrastructure	• Assess government regulation • Analyze route structure • Develop market forecasting casting model				
Human Resource Management	• Hire and train personnel • Set wages and work rules	• Schedule and supervise crews –Flight –Ground		• Recruit, hire, and train market/sales force	
Technology and Development	• Research ticketing and reservations • Research meal preparation		• Reassess information needs –Reservations –Operations –Accounting/billing	• Develop feedback for marketing • Develop media coverage	
Procurement	• Obtain aircraft spare parts and ground equipment • Obtain airport terminal locations • Purchase insurance	• Purchase meals • Purchase fuel • Purchase baggage handling equipment • Purchase automated reservation systems		• Buy media	
	• Manage spare parts inventory • Manage meal logistics	• Handle baggage • Maintain equipment • Operate ticket counter • Operate gates • Operate on-board systems • Schedule maintenance	• Schedule route • Manage reservations • Manage down-time operations	• Determine pricing strategy • Determine commissions • Ensure travel agent relationship • Monitor contribution of scheduling	• Ensure on-time strategy • Continue interlining service • Offer other customer services

Margin

Figure from FOUNDATIONS OF BUSINESS SYSTEMS by Arthur Andersen & Co., Per O. Flatten, Donald J. McCubbrey, P. Declan O'Riordan, and Keith Burgess, copyright © 1989 by the Dryden Press, reprinted by permission of the publisher.

FIGURE 6.3   Value Chain Analysis   Porter's value chain analysis can be useful in understanding which activities return the most value to a business. This is a value chain diagram for an airline company.

through the use of information technology, can return the most value to the business as a whole.

A sample value chain is shown in Figure 6.3. Notice that the value chain for any company consists of primary and support activities:

• *Primary value activities* support the creation of products and services, their sale to the customer, and any support provided after the sale. This may include the following:

- *Inbound logistics* are any activities that collect the inputs needed to produce products and services. Inputs may include raw materials, supplies, technology, or data. Representative activities include receiving, warehousing, and inventory control.
- *Operations* are any activities that transform inputs into products or services. Representative activities include all manufacturing processes, product assembly, and product packaging.
- *Outbound logistics* are any activities that handle finished products or transport products and services to customers. Representative activities include finished goods warehousing and inventory control, order processing, and shipping.
- *Sales and marketing* are any activities that persuade customers to buy products and services and make it possible to place orders. This includes advertising and retailing.
- *Service* includes any activities that support after-sale or ongoing service enhancing the value of the original product or service purchased by the customer.

- *Support value activities* provide complementary functions to one or more primary value activities. Although necessary in most cases, these support activities often return less value than primary value activities. Many organizations are downsizing these support activities because they believe that, in some cases, they subtract value from products and services. Support value activities may include the following:
  - *Corporate infrastructure* activities are simply general management and professional services (such as accounting, finance, legal, and information systems services — exclusive of those that produce hardware and software products and services for resale to customers.)
  - *Human resource management* includes any activities associated with hiring, firing, training, evaluating, and managing personnel.
  - *Technology development* includes any activities associated with creating new products and services. This is typically called *research and development* or *product engineering.*
  - *Procurement* includes any activities that acquire the inputs necessary to produce products and services. Procurement triggers the primary value activities described earlier.

- *Margin* is the difference between the revenue generated by all activities and the costs of those activities. Margin reminds us that the purpose of all business activities is to either maximize profit (in profit-making businesses) or minimize costs of quality products and services (in nonprofit organizations).

What's the point? After documenting the value chain(s) for a business or significant portion of a business, the planning analysts and managers can evaluate the value chain to identify areas for further study or business improvements. (Further study is accomplished in the evaluation phase of systems planning, and/or in the phases of systems analysis.)

Specifically, analysts and managers should try to identify (1) primary activities that could add significant value to the business's products and services, and (2) primary or support activities that have costs growing significantly faster than other activities. These indicate possible areas for improvement. Ultimately, the

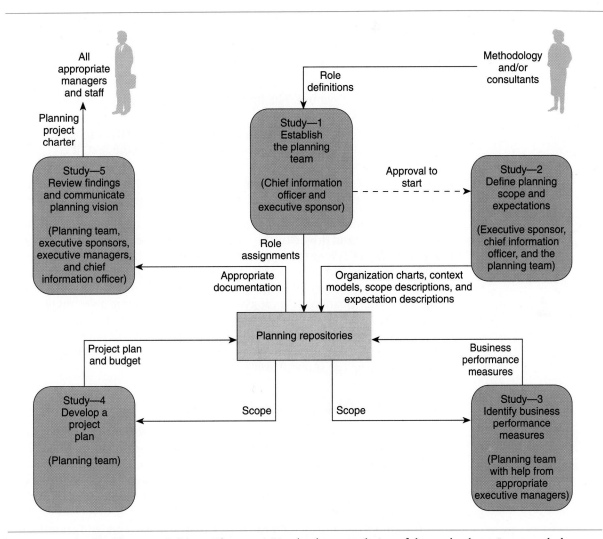

FIGURE 6.4 **Study Phase Activities** These activities lead to completion of the study phase. In general, the activities are completed in a clockwise sequence from the top. On the other hand, the project repository allows you to overlap activities or return to any previous activity to do rework.

information systems plan will look for ways that IS can contribute to that improvement! Today, IS contributions come in two forms:

- Helping the business redesign its business activities and processes (independent of information technology).
- Delivering new or improved information systems that support business activities and processes.

Study Phase Activities, Roles, and Techniques

Figure 6.4 illustrates the typical activities of the study phase. Since this is the first activity-level diagram of the chapter, let's establish the following guidelines for reading all such diagrams:

- Silhouettes represent people or departments that initiate projects, phases, and activities.
- The rounded rectangles represent *activities*. Each activity is numbered solely for identification purposes. The activity name is printed in the upper half of the symbol. The participants in the activity are printed in the lower half of the symbol. The first participant is always the person who "guides" the activity.
- The solid arrows reflect inputs and outputs for an activity. Each is named. When one of these inputs or outputs is referred to in the text, it is underlined.
- Dashed arrows represent events that trigger an activity. (There are no triggers on Figure 6.4; however, they will appear in other diagrams.)

Systems planning is rarely triggered by a formal request. Increasingly, it is first suggested by and driven by the chief information officer (or equivalent) in the firm. Today's CIOs have become more sensitive to the need to strategically align information systems with the business. In other situations, top management may suggest the need for planning in response to concerns about the value of their information technology and information system investments. And in some cases, systems planning is suggested by a management consulting firm that helps the business reorganize itself and improve business processes.

Regardless of the motivation, let's now examine the activities that must be completed during the study phase.

Activity 1: Establish the Planning Team

The first activity in the study phase is to *establish the planning team*. This may seem obvious; however, it is crucial to the success of the phase. Appointments are usually made by the project's executive sponsor and the CIO (or equivalent). The executive sponsor is the highest-level executive manager in the business or major portion of the business to be studied. Ideally, this should be the CEO or president.

The required participants and roles to be played in most planning methodologies are standardized by the methodology or management consulting firm used. Ideally, the first appointments should be the CEO and CIO themselves. Why? Because it visibly demonstrates top management's commitment to the project! Subsequently, the CEO would likely appoint all vice presidents and above to represent executive management's interests. The CIO should then appoint data analysts (two or more) to collect data to be recorded in the planning repository. Other CIO appointments may include a data administrator and a networks administrator.

Consultants are frequently employed to facilitate group sessions using JAD-like techniques and to provide an unbiased analysis of findings. If so, the CEO and CIO should jointly engage and contract the management consultants. At least one of those consultants would facilitate most team efforts. Otherwise, a planning analyst skilled in JAD-like techniques would be designated to facilitate team efforts.

The inputs to this phase are role definitions as defined by the planning methodology or consultants. The output of this phase is the role assignments.

(Note: Throughout this discussion, outputs are typically stored in the repository to support their reuse at any subsequent and appropriate time.)

There are no special techniques or skills required to complete this activity.

Activity 2: Define Planning Scope and Expectations

The first planning "team" activity in the study phase is to *define planning scope and expectations.* Clearly, the ideal scope is the entire business. Unfortunately, this is sometimes difficult because of geography, size and complexity, politics, or the lack of top management commitment. It is particularly important to define scope if it is less than the entire system. Purdue University serves as an example: its strategic-planning initiative limited scope to business services and physical facilities. For various reasons, the academic schools were omitted from the scope.

This activity will be facilitated by the consultant(s) or planning analyst(s), but it should clearly be driven or supported by the executive sponsor and chief information officer. Thus, executive commitment to the planning initiative is reinforced.

Based on interviews and facilitated group discussions within the planning team, the scope of the project is defined. Various outputs such as organization charts, context models (scope-oriented pictures), and scope descriptions will be recorded in the repository.

One of the easiest ways to define scope is to draw context model(s), simple pictures that reflect the boundaries and scope of the system. Context models should be constructed as a group task. For example, Figure 6.5 is a **context diagram** that depicts the net inputs and outputs of a business with the outside world. This particular context model uses named arrows to represent business inputs and outputs. This diagram serves no design purpose. It merely gets people thinking and talking about "the big picture" — the business as a whole (or the defined subset and its connections to the rest of the business).

Yet another option is to draw a context diagram that depicts competitive forces using Porter's model. Such a diagram would use incoming arrows to name specific competitive forces and outgoing arrows to name barriers that the business could erect to prevent or stall competition.

Finally, a value chain diagram could be constructed as an alternative or additional context diagram. See the readings list at the end of the chapter for more information on actually constructing or applying value chain analysis.

In addition to scope or context, the expectation descriptions or "visions" of the planning initiative are defined by the team and recorded in the repository. This usually requires that those familiar with the importance of systems planning educate those who are not. Similarly, those who are familiar with the planning methodology must orient those who are not.

The following techniques and skills are applicable to this activity:

- Interviewing and other fact finding (see Part Five, Module B, Fact-Finding Techniques).
- Joint application development or JAD (see Part Five, Module B, Fact-Finding Techniques).
- Context or scope modeling (see Chapter 9, Process Modeling).

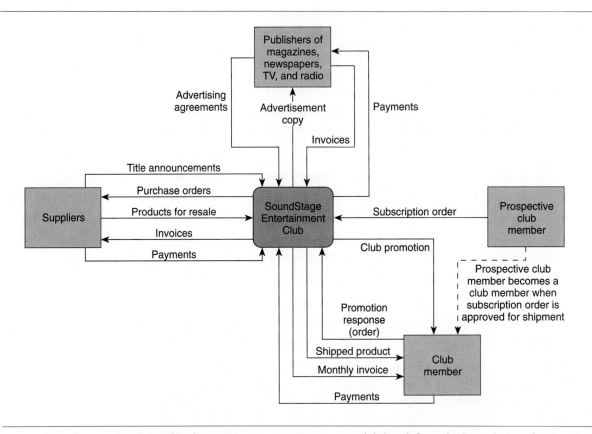

FIGURE 6.5 **Context Model** This diagram is a context or scope model that defines the boundaries of a planning project. In this case, the model depicts the business as the center of the universe. The squares are external agents with which the business interacts. The arrows depict important inputs, outputs, and influences between the system and its outside world.

Activity 3: Identify Business Performance Measures

The planning team should next *identify business performance measures.* How do the executives evaluate the business's performance? How do they evaluate their own performance? It is *not* the intent of this activity to actually evaluate the business or its management. Rather, it is merely to identify those measures that are used, or should be used.

The activity is best accomplished through JAD-like discussions among the planning team. Appropriate executive managers who are not on the team may be interviewed prior to the JAD sessions or included in the JAD discussions. The input is the scope as defined in the previous activity. The outputs are the business performance measures that are recorded in the repository.

Different methodologies suggest different performance measures. One of the most common approaches is critical success factor analysis, described earlier in the chapter. Some CASE tools, such as KnowledgeWare's ADW Planning Workbench, provide the ability to directly record goals, objectives, critical assumptions, critical success factors, and other performance measures into its repository.

The following techniques and skills are applicable to this activity:

- Interviewing and other fact finding (see Part Five, Module B, Fact-Finding Techniques).
- Joint application development or JAD (see Part Five, Module B, Fact-Finding Techniques).

Activity 4: Develop a Project Plan

Even a planning project must be planned. Based on the scope described in Activity 2, the planning team develops a project plan and budget for completing the next phase of the project, *define an information architecture*. (Note: It would be difficult or impossible to plan further ahead to include the analysis phase, since that phase is repeated for each of several business areas which have yet to be defined.)

The project plan and budget is developed by the planning team. The input is scope. Often, the information systems professionals on the team play the lead role since they have more recent experience in project planning and scheduling techniques. The principle technique and skill required to complete this activity is *project management*. Project management, another cross life cycle activity, is covered in Part Five, Module A, Project Management.

Activity 5: Review Findings and Communicate Planning Vision

The findings and recommendations of the planning team should be reviewed in a group forum that includes executive managers not originally on the planning team. The planning team should then make a decision as to whether or not to continue the project to the next phase, and whether or not to accept the proposed project plan and budget. If the project is "a go," the planning vision should be communicated to a wider business audience.

The planning team performs this activity. The inputs include any appropriate documentation from the repository. The final output is a planning project charter. An outline for the charter is shown in Figure 6.6. The written report is normally shared with all executive and middle managers in the business, as well as all information systems managers and appropriate staff. A cover letter from the executive sponsor and chief information officer reaffirms management commitment and authority to continue to the next phase. That letter encourages an equal commitment on behalf of all personnel who will become involved in subsequent planning phases.

One of the following decisions results from the review activity:

- Accept the charter and then continue to the definition phase.
- Amend the charter to adjust scope, schedule, or budget. The amended charter would receive authorization to continue to the definition phase.
- Rework the work products of any activity to reflect changes or issues raised during the review. A follow-up review and revised charter may or may not be required.
- Terminate the project due to lack of resources or commitment to continue the project.

The techniques and skills needed to complete this activity are all cross life cycle skills:

FIGURE 6.6 An Outline for a Planning Project Charter This outline is typical of a planning project charter. Different organizations and/or methodologies may impose different standards.

Cover letter from executive sponsor serves as letter of authority.

I. Executive summary (maximum: 1–2 pages).
II. Background information (maximum: 1–2 pages).
    A. Purpose of the planning project.
    B. Scope of the planning project.
    C. Expectations of the planning project.
    D. Brief explanation of report contents.
III. Findings of the study phase (2–3 pages).
    A. Business mission and vision.
    B. Competitive analysis (if completed).
    C. Value chain analysis (if completed).
    D. Critical success factors analysis.
        1. Business goals.
        2. Business objectives.
        3. Critical success factors.
        4. Critical assumptions.
        5. Critical information needs.
    E. Problems and opportunities.
IV. Definition phase plan.
    A. A framework for information architecture.
    B. The plan.
        1. Phase objectives.
        2. The planning team.
        3. Activities and deliverables.
        4. Detailed schedule.
        5. Budget.
V. Appendixes (if necessary).

- Feasibility assessment (see Part Five, Module C, Feasibility Analysis).
- Report writing (see Part Five, Module D, Interpersonal Skills).
- Verbal presentations (once again, see Part Five, Module D).

THE DEFINITION PHASE OF SYSTEMS PLANNING

The second phase of systems planning is to *define an information architecture*—the definition phase.

> An **information architecture** is a vision and plan for using information technology and developing information systems needed to support a business's mission. Synonyms include *information systems plan, information strategy plan,* and *master computing plan.*

An information architecture can take six months or longer to complete. In this section, we will examine the purpose of the definition phase. Then we will, once again, walk through its activities, participants, inputs, outputs, and techniques. Techniques will be cross-referenced to the chapters and modules that teach those techniques.

Building Blocks for the Definition Phase

The fundamental objectives of the definition phase are

- To define a data, application, network, information services, and technology infrastructure for future information systems.

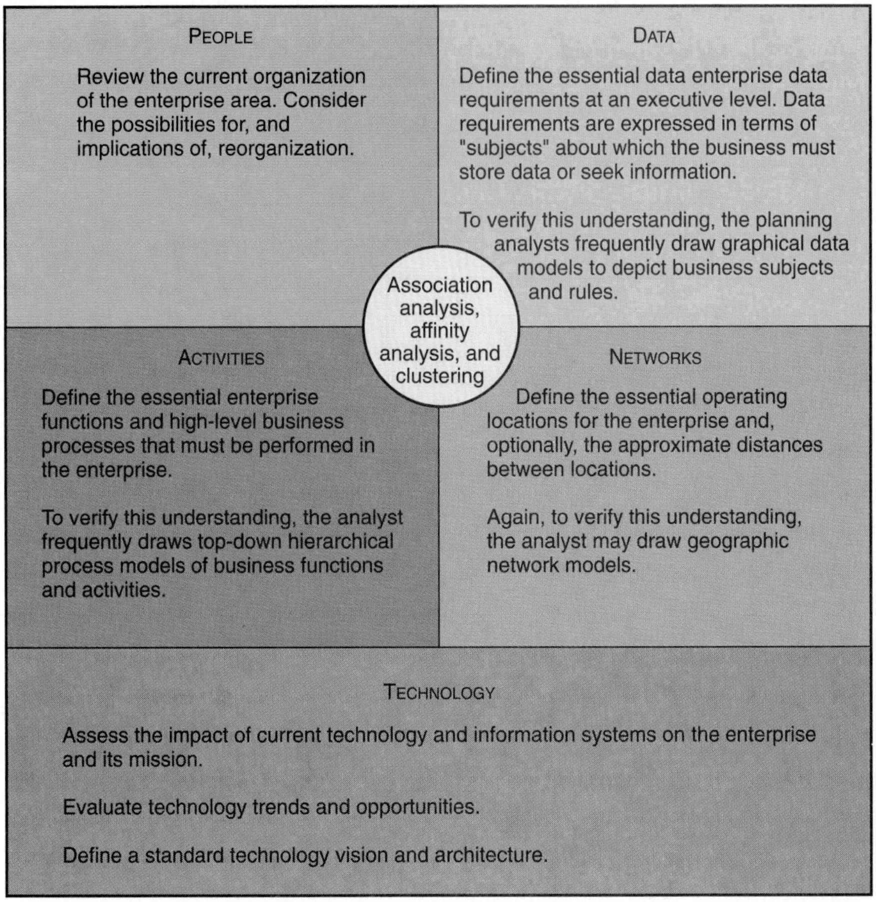

PEOPLE	DATA
Review the current organization of the enterprise area. Consider the possibilities for, and implications of, reorganization.	Define the essential data enterprise data requirements at an executive level. Data requirements are expressed in terms of "subjects" about which the business must store data or seek information.

To verify this understanding, the planning analysts frequently draw graphical data models to depict business subjects and rules. |
| ACTIVITIES | NETWORKS |
| Define the essential enterprise functions and high-level business processes that must be performed in the enterprise.

To verify this understanding, the analyst frequently draws top-down hierarchical process models of business functions and activities. | Define the essential operating locations for the enterprise and, optionally, the approximate distances between locations.

Again, to verify this understanding, the analyst may draw geographic network models. |

Association analysis, affinity analysis, and clustering

TECHNOLOGY

Assess the impact of current technology and information systems on the enterprise and its mission.

Evaluate technology trends and opportunities.

Define a standard technology vision and architecture.

FIGURE 6.7 Definition Phase Building Blocks The definition phase identifies the short- and long-term strategies for each of the information systems building blocks: PEOPLE, DATA, ACTIVITIES, NETWORKS, and TECHNOLOGY. These strategies are identified based on the analyses listed in the circle.

- To identify and prioritize logical business areas for further planning, or to identify and prioritize application development projects.

The required level of "definition" can be derived from your information system building blocks (Figure 6.7). Notice that, for most of the building blocks, it is possible to draw graphical models to "define" the information architecture. The ideal architecture can be derived using techniques listed in the circle.

Association Analysis

Association analysis is a technique that examines the natural associations or relationships between any two objects or ideas. The technique uses simple and highly effective tools called *association matrices* to examine these relationships.

An **association matrix** documents the relationships between any two performance measures or enterprise model objects. A matrix is a table of rows and columns, like a spreadsheet. The rows and columns correspond to instances of two different "things" to be compared. The cells document the relevance of the instance of one "thing" (row) to another "thing."

A. Function-to-Organization Unit Association Matrix

Organization Unit / Business Function	Research and Development	Warehouse	Production	Sales	Accounting
Marketing	S			P	S
Product Engineering	P		S	S	S
Process Engineering	P	S	S		S
Inventory Control		P	S		S
Cost Accounting	S	S	S		P

P = Organization unit has primary responsibility for this business function.
S = Organization unit plays some role in this business function.

B. Data-to-Function Association Matrix

Data Subject or "Data Entity" / Business Function	Customer	Order	Product	Production Lot	Production Process
Marketing	CRUD	CRUD	CRUD		
Product Engineering			RU		R
Process Engineering			R		CRUD
Inventory Control			RU		R
Cost Accounting			RU	R	RU

C = The business function can create instances of this data subject in a database.
R = The business function can read instances of this data subject in a database.
U = The business function can update (change) instances of this data subject in a database.
D = The business function can delete (archive) instances of this data subject in a database.

FIGURE 6.8   Association Matrices   A matrix is the perfect tool for analyzing or comparing any two objects or ideas to learn their relationships. Two examples are presented to demonstrate the diversity of the tool.

The matrix actually sounds more difficult than it is. A couple of simple examples should help clarify the concept. Figure 6.8A demonstrates a function–to–organization unit association matrix. It tells us which ongoing business functions are performed by which organization units. Figure 6.8B demonstrates a data-to-function association matrix. It identifies which business functions create, delete, modify, and use what data.

Depending on the methodology used, matrices may be constructed and studied to answer the following questions:

- Which organization units use what data?
- Which organization units perform which business functions?
- Which organization units are at which locations?
- What data is used or maintained by what organizational units?
- What data is used or maintained by what business functions?
- What data is needed at what locations?

- Which business objectives are pertinent to which business functions?
- Which organizational units are responsible for which business objectives?
- Which critical success factors apply to which business functions?
- Which critical information needs require what data?

The list of building blocks and measures that could be constructed is quite large. The methodology and planning analysts would identify appropriate matrices and construct them using CASE tools that support association matrices (or their equivalent).

Affinity Analysis and Clustering

"Divide and conquer!" That is the age-old approach for dealing with size and complexity. You do it every time you write an outline for a paper or report. But how do we divide and conquer databases, networks, technology, and applications? Historically, information systems architecture has been divided along organization-structure lines. Applications were built for single (or a few) organization units in the organization chart, according to the units' ability to pay for applications.

The result of this divide-and-conquer approach is common to most businesses. The applications are not integrated. They do not interface or interact "efficiently" or "effectively" with applications developed for other organization units. The result is sometimes called "islands of automation." There is a better approach, which uses various systems planning techniques, such as affinity analysis and clustering.

Affinity is a measure of what different "things" have in common. Affinity values are always between 0 and 1. If two "things" are never used or happen together, their affinity is 0. If two "things" are always used or happen together, their affinity is 1. Accordingly, an affinity of 0.65 means that the compared "things" are used or happen together 65 percent of the time.

Affinity analysis is a formal technique for identifying properties that different "things" have in common such that those "things" might be combined into logical groups.

Affinity analysis begins with an association matrix that compares two "things." In our case, those things would correspond to the information systems building blocks:

- PEOPLE (organization units).
- DATA (specifically, things about which management may seek information).
- ACTIVITIES (specifically, business functions or processes).
- NETWORKS (specifically, geographic locations where activities occur).

For example, we may want to construct matrices to compare variables such as the following:

- What data is created, maintained, or used by what processes.
- What data is used at what geographic locations.
- What activities are performed at what locations.
- What people are at what locations.

Affinity is calculated for each cell in the matrix. The formula is as follows:

Affinity (entity 1 to entity 2)

$$= \frac{\text{Number of times entity 1 and entity 2 occur together}}{\text{Number of times entity 1 occurs all together}}$$

Fortunately, you don't have to do this by hand. Most planning CASE tools include affinity analysis routines that perform the calculations to produce an *affinity matrix.* (Note: This process can take hours, even on a good CASE tool!) This affinity matrix looks the same as the input matrix except that each cell includes an affinity number (between 0 and 1).

By itself, an affinity matrix is of little value. *Clustering* adds value to the matrix.

> **Clustering** is a technique that forms logical groups of "things" based on affinity measures. A threshold between 0 and 1 is established. A threshold of 0 would result in one big group containing all "things." A threshold of 1 would result in a group for each "thing." In practice, an analyst experiments with different thresholds (e.g., 0.50, 0.75, 0.65) until he or she achieves acceptable groupings.

Planning CASE tools can also do this clustering for you. You simply input a threshold, and the CASE tool rearranges the rows and columns of the matrix to clearly identify the logical groups. There are no steadfast rules to determine how many groups make sense or whether any specific groupings make sense —the team must make that determination.

So, how is this applied to planning? Clustering can be used to group data subjects into logical databases, or processes into logical business areas. As you'll soon learn, this technique becomes important to defining an information architecture.

Definition Phase Activities, Roles, and Techniques

Figure 6.9 illustrates the typical activities of the definition phase. The definition phase is triggered by the authority and funding to continue the planning project from the study phase. The definition phase activities are much more intense than those in the study phase. Before we examine definition phase activities, let's discuss some possible adjustments in the planning team.

The planning team for the study phase consisted mostly of very high-level executives. Because the definition phase requires more familiarity with business processes, these top executives generally appoint subordinate executive managers to represent their interests in the definition phase. The top executives may be reassigned to a steering subcommittee that oversees the definition phase and reviews its results. (Note: Some executives may elect to remain on the main working team.) The information systems representatives generally carry over from the study phase; however, their numbers may be expanded to accommodate the increased time demands.

Now, let's examine each of the definition phase activities.

Activity 1: Model the Enterprise

The first activity in the definition phase is to *model the enterprise.* What is a model?

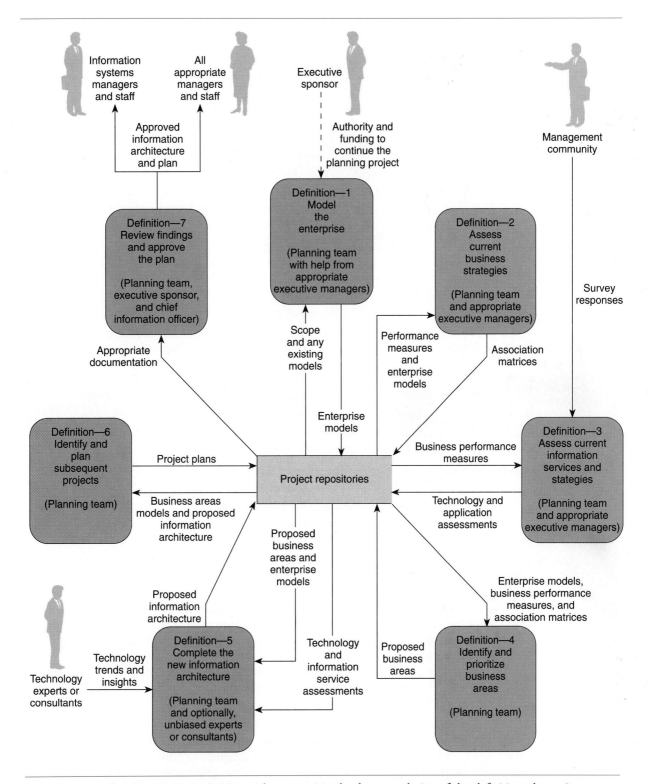

FIGURE 6.9 **Definition Phase Activities** These activities lead to completion of the definition phase. Once again, the activities are generally completed in a clockwise sequence from the top. On the other hand, the project repository allows you to overlap activities or return to any previous activity to do rework.

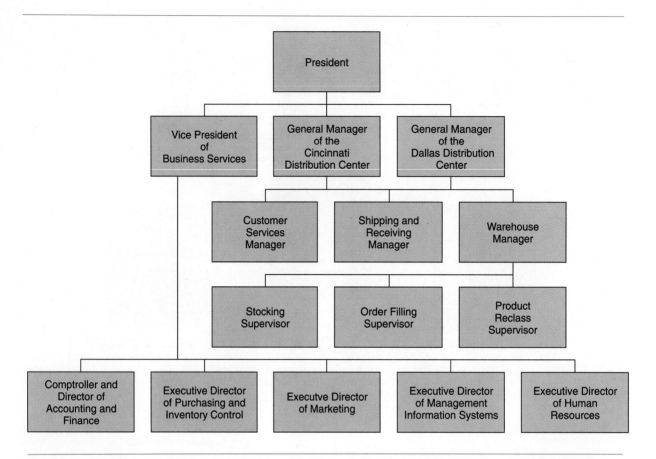

Organization Charts   Organization charts model the reporting relationships in the business.

A **model** is a representation of reality. In the classic tradition that "a picture is worth a thousand words," most models are graphical representations of reality.

Models developed in the definition phase depict the business system at a general level of detail suited to the executive managers who helped develop the models. These models are sometimes called **enterprise models.**

Enterprise models become an integral part of the information architecture or information strategy plan. Later, during systems analysis and design of specific information systems applications, the enterprise models provide a point of departure for more detailed systems and application models.

The pyramid model provides a framework for developing the enterprise models. Appropriate business system models to be added to the repository may include the following:

- PEOPLE — Organization charts may be used to model the responsibility and management structure within the business. A simple example is shown in Figure 6.10. Executive managers may take this opportunity to propose a restructuring of the business. The current trend is toward fewer levels of management at the top of the business and self-managed work groups or teams at the bottom of business. This is called *downsizing* of the organization.

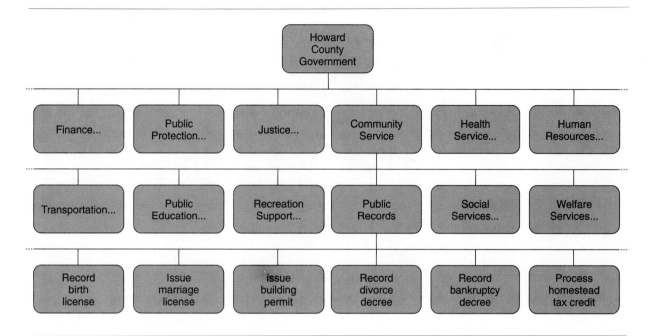

FIGURE 6.11   Decomposition Diagram   A decomposition diagram is a useful tool that breaks factors of an enterprise into functions and processes. It does not show inputs and outputs. This keeps the diagram simple enough to verify by executive managers.

- ACTIVITIES—Most methodologies (including information engineering) suggest a top-down hierarchy chart or *decomposition diagram* that identifies business functions and business processes, independent of the organization chart. A typical decomposition diagram is shown in Figure 6.11.

    Business functions are ongoing activities (or subsystems) in the business. They are named with nouns such as "marketing" and "inventory control." Business functions may consist of other business functions; however, they eventually decompose into business processes.

    Business processes are distinct activities that have a beginning (input) and an end (output). They are not ongoing—they just "happen"! Business processes are named with verb phrases such as "process order," "purchase raw materials," and "process time card."

- DATA—For purposes of this phase, most planning methodologies require a high-level (management-oriented) *subject data model* that identifies those "things" or *subjects* about which management believes the company should store data (or generate information).

    The **subject data model** is the most important component in the enterprise model and information architecture. It is the key to integrating future applications through shared databases. A small but representative data model is shown in Figure 6.12. A complete subject data model for an enterprise is typically much larger—it could fill a small bulletin board!

- NETWORKS—Some methodologies require a simple map, list, or outline of geographic operating locations of the business. Alternatively, a more formal network model (picture) may communicate the same information.

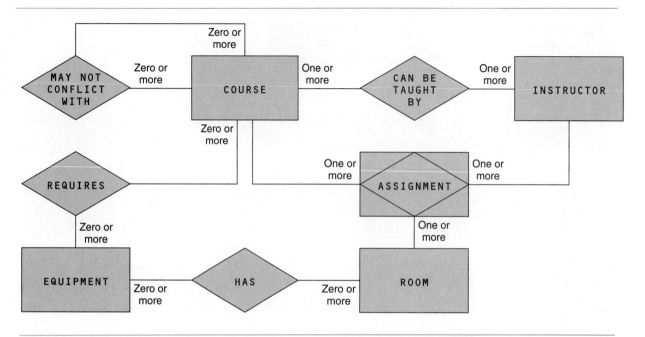

FIGURE 6.12 **A Subject Data Model** A subject data model is an important tool for defining information needs and shared data requirements.

All the above enterprise models are developed jointly by the entire planning team. The information systems professionals on the team generally construct and maintain the models, since they are usually more skilled in the modeling techniques and CASE tools used to record the models and details into the repository. In particular, a data administrator plays a key role in starting the corporate data model, and a network administrator plays a key role in starting the corporate network model.

Inputs to this modeling activity include scope (from the study phase) and any existing models. The output, stored in the repository and shared with any interested staff, are the resulting enterprise models.

The following techniques and skills are applicable to this activity:

- Fact finding (see Part V, Module B, Fact-Finding Techniques)
- Joint application development (see Part V, Module B, Fact-Finding Techniques).
- Data modeling (see Chapter 8, Data Modeling).
- Process modeling (see Chapter 9, Process Modeling).
- Network modeling (see Chapter 10, Network Modeling).

Activity 2: Assess Current Business Strategies

We cannot complete a new information architecture until we *assess current business strategies.* Which critical success factors (CSFs) affect which goals and objectives? Do any CSFs affect multiple goals and objectives? What organization units are responsible for which goals, objectives, and CSFs? What information needs support which business functions and processes? These and other similar questions are addressed in this activity.

Some business strategies can be evaluated before enterprise modeling. Thus, this activity can be started at the same time as Activity 1. Other assessments require the enterprise models. Different methodologies call for different assessments. The popular information engineering methods analyze the relationships between various combinations of the following:

- Organization units (PEOPLE).
- Information needs (DATA).
- Business functions and processes (ACTIVITIES).
- Geographic locations (NETWORKS).
- Performance measures (e.g., goals, objectives, CSFs).

This analysis is typically accomplished by constructing and analyzing various association matrices. Association analysis and association matrices were described earlier. Their use is common to many techniques, including information engineering and business systems planning. By constructing and analyzing association matrices, the planning team learns how information system building blocks should support the business mission.

The association matrices are constructed by the planning team. Planning analysts usually enter the data into the repository using a CASE tool that supports matrices. The inputs are the performance measures (from the study phase) and the enterprise models (from Activity 1 of this definition phase). The outputs are the association matrices that are added to the repository.

The principle techniques and skills required to complete this activity are interviewing, observations, sampling, studying, and joint application design. These techniques are covered in Part Five, Module B, Fact-Finding Techniques. Matrix techniques are relatively straightforward; therefore, they are not further covered in this text.

Activity 3: Assess Current Information Services and Strategies

At the same time that we assess current business strategies, it is also useful to assess current information services and strategies. For the information systems professionals on the planning team, this activity can be "humbling." It is common to learn that information services is *not* providing the value expected by top management. It is also common to learn that information services is *not* committing resources to those business functions that will truly improve the performance measures of the business. That is precisely why this activity is important!

The planning team usually conducts surveys of the management community to collect necessary opinions and data. The survey instrument (usually a questionnaire) must be carefully designed not to lead or bias the findings. The tabulated results are compiled and reviewed by the planning team.

Association matrices might be constructed in this activity also. Such matrices suggest relationships between information systems, information technology, and business performance measures.

The inputs include business performance measures (from the study phase) and survey responses. The outputs are technology and application assessments in various formats such as questionnaire response summaries, association matrices, and narrative analyses.

Once again, the principle techniques and skills required to complete this activity are interviewing, observations, sampling, studying, and joint application design. These techniques are covered in Part Five, Module B, Fact-Finding Techniques.

Activity 4: Identify and Prioritize Business Areas

This activity is unique to the popular information engineering methods. Although information systems applications are built one at a time, analysts prefer that the resulting applications be highly integrated. By "integrated" we mean that those applications evolve around a common database, network, and technology architecture. This almost never happens by accident! To accomplish this goal, the business models must be partitioned into integrated business areas.

> A **business area** is a logical grouping of business processes, data, and locations that will be supported by highly integrated information systems applications. A business area usually includes business processes from multiple organization units (e.g. departments). A business area almost never includes all business processes from a single organization unit.

In other words, business areas are "cross-functional." They break down the barriers of applications built for or around single organization units. For example, an order-filling business area might include people, data, processes, locations, and technology from several departments (such as Sales, Warehousing, Manufacturing, Shipping, and Accounts Receivable). Instead of building separate systems for those departments, a streamlined and integrated system is built for all those activities and departments involved in that business area.

How are business areas identified? A business area should be large enough to share a common database. That database should have a high degree of data sharing among the business area's processes, and a much lower degree of data sharing with other business areas. All or parts of a business area may be duplicated at multiple locations. For example, the aforementioned order-processing business area may include business processes that must be duplicated at each sales office. The most common approach to identifying business areas involves affinity analysis and clustering, described earlier.

Affinity analysis and clustering are used to identify subject data models around which a business area can be built. They can also be used to distribute data and processes to specific locations. Affinity analysis and clustering are almost always performed by a CASE tool. The inputs are appropriate objects from the enterprise models (specifically, data subjects, business processes, locations, and organization units), business performance measures, and all association matrices. The resulting affinity analysis "suggests" business areas. These business areas should be discussed by the planning team and refined as necessary according to "soft" factors such as politics.

After the business areas are identified, they should be ranked using the business and information systems assessments from Activities 2 and 3. Rankings should reflect business areas' relative importance to the business as a whole, ideally using the performance measures identified in the study phase. (Notice that we are gradually moving toward redirecting application development to those areas that will most improve the business—that was the goal for this entire strategic planning initiative!) The outputs of this activity are proposed business areas.

The principle techniques required to complete this activity are affinity analysis and clustering. This text primarily teaches techniques common to systems planning, analysis, and design. Because affinity analysis and clustering are exclusively planning techniques, further discussion is deferred to the James Martin readings listed at the end of this chapter.

Activity 5: Complete the New Information Architecture

Much of the information architecture is complete. The enterprise models are a major component. These models define a future vision of the business's DATA, ACTIVITIES, and NETWORKS. But what about the TECHNOLOGY and PEOPLE building blocks of your pyramid model? Also, what impact will the defined business areas have on the building blocks?

Based on the proposed business areas, the enterprise models should be divided into submodels that reflect those business areas. In other words, each business area should have its own **business area models** that document the DATA, ACTIVITIES, and NETWORKS for that business area. These business area models become part of the proposed information architecture. This will make it easier to export business area documentation for subsequent systems planning, analysis, and design.

But we also need to complete the information architecture for each business area. Specifically, what TECHNOLOGY will be used to build and implement information systems for each business area? Also, how should information systems specialists and services (PEOPLE) be organized to best develop and support applications for each business area?

The TECHNOLOGY building block is defined after a thorough survey of technology trends. Consultants and technology information services (such as the Gartner Group) can provide technology trends and insights. They can also help the planning team identify applications that might benefit by emerging technologies and help to weigh the risks associated with specific technologies. Combined with the technology and information services assessments (from Activity 3), a TECHNOLOGY architecture is added to the proposed information architecture. The **technology architecture** may specify some or all of the following:

- Database management systems to be used for all applications in the business and business areas.
- Network topology and protocols to be used for all applications in the business and business areas.
- Emerging technology (e.g., imaging, electronic data interchange, bar coding) to be used by appropriate applications.
- Processor technology to be used by specific types of applications (e.g., mainframe, minicomputer, personal computer, or combinations thereof).
- Operating systems and system software to be used on specific computers (e.g., MVS, VM, UNIX, OS/400, DOS, OS/2, Windows/NT).
- Software development methods to be used for application development (e.g., information engineering, rapid application development, STRADIS).
- Software development technology to be used for application development (e.g., COBOL, C, SAS, Focus, Excelerator, APS).

Defining the information services architecture (PEOPLE building block) requires the planning team to determine the best manner in which to organize information services. Should all application development services be centralized in an IS shop? Or, should developers report directly to the organization units for the business area? If so, should certain development support services such as data management, network management, and the development center be centralized?

These questions might best be answered by consultants, tempered by the politics and realities of the planning team. The information services assessments (from Activity 3) should also be factored into the decisions.

The combination of the enterprise models (factored into business areas), the technology architecture, and information services organization completes our proposed information architecture.

There are no special techniques to be mastered for this activity.

Activity 6: Identify and Plan Subsequent Projects

Depending on the planning methodology used, the next activity proceeds in one of two directions:

- *Identify and plan subsequent application development projects.* Those projects would then be sent directly to systems development for systems analysis and design. (Thus, systems planning is finished—there will be no business area analysis phase!)
- *Identify and plan subsequent business area analysis projects.* This approach is required in the popular information engineering–based methods. The business areas, in order of priority, are sent to the analysis phase of systems planning. This analysis results in a plan for highly integrated applications and development in that business area. That plan would guide future systems development for that area. (Note: The remainder of this chapter is based on this option.)

The analysis of a business area can take several months. Business areas should be evaluated in order of their priority, which, if you recall, is based on their importance in achieving business goals, objectives, and CSFs. More than one business area can be evaluated at a given time. The number is limited by the number of information systems staff the company can afford to pull from normal application development projects to perform the business area analysis. The planning team prepares the project plans, but the chief information officer plays a key role since he or she must commit the IS staff.

The inputs to this activity include the business areas, models, and proposed information architecture. The outputs are project plans. These plans should include

- A general schedule for analyzing all business areas. This schedule could span several years, since an enterprise may include a dozen or more business areas.
- A detailed schedule for completing each business area analysis that will be started immediately after this phase.

The principle technique required to complete this activity is *project planning and scheduling,* which is covered in Part Five, Module A, Project Management.

Activity 7: Review Findings and Approve the Plan

As with the study phase, the definition phase findings and recommendations should be reviewed. The review should be performed by the planning team and

I. Executive summary (maximum: 1–2 pages).
II. Background information (maximum: 1–2 pages).
   A. Purpose of the planning project.
   B. Scope of the planning project.
   C. Planning methodology.
   D. Brief explanation of report contents.
III. The information architecture.
   A. Business strategy.
     1. Business mission and vision.
     2. Business goals and objectives.
     3. Critical success factors, assumptions, and information.
   B. Enterprise models.
     1. Data model and descriptions.
     2. Process model and descriptions.
     3. Network model and descriptions.
   C. Business areas and descriptions.
   D. Technology architecture.
     1. For enterprise-wide computing.
     2. For business area computing.
   E. Information services organization.
     1. Centralized services.
     2. Distributed services.
IV. Business area analysis planning.
   A. General plan and schedule for all areas.
   B. Detailed plan(s) for immediate business area analysis.
     1. Schedule.
     2. Budget.
V. Appendixes (as appropriate).

FIGURE 6.13 An Outline for an Information Architecture and Plan This outline is typical of a project information architecture and plan, the key deliverable of the definition phase of systems planning. Different organizations or methodologies may impose different standards.

include the top executives who were on the study phase planning team. Two levels of review are possible.

- A **quality review** ensures that the phases were completed in accordance to the methodology and that resulting documentation (in the repository) is complete, consistent, and in compliance with standards.
- A **feasibility review** reassesses the feasibility of continuing the project.

In both cases the planning team assembles the appropriate documentation from the repository. The desired output is an approved information architecture and plan. An outline for a representative written report is shown in Figure 6.13. One of the following decisions must be made:

- Authorize the project to continue, as is, to the analysis phase.
- Adjust the scope, cost, or schedule for the project and then continue to the analysis phase.
- Cancel the project due to either (1) lack of resources to further plan the business areas, or (2) lack of management commitment to further planning.

The techniques and skills needed to complete this activity are all cross life cycle skills:

- Feasibility assessment (see Part Five, Module C, Feasibility Analysis).
- Report writing (see Part Five, Module D, Interpersonal Skills).
- Walkthroughs (again, see Part Five, Module D).
- Verbal presentations (again, see Part Five, Module D).

THE BUSINESS AREA ANALYSIS (BAA) PHASE OF SYSTEMS PLANNING

What?! An "analysis" phase within systems "planning"! No, it is not a mistake. The information engineering methods do indeed call for an analysis-like phase within systems planning. By contrast to classical systems analysis, a business area analysis requires: (1) a broader scope — an entire business area as opposed to one application, and (2) less detail in the analysis. The tools and techniques of business area analysis (BAA) and systems analysis are more alike than different.

The purpose of BAA is to devise a plan that will lead to highly integrated information systems applications for a business area. The analysis phase triggers projects to

1. Develop a single, integrated database for the entire business area.
2. Develop a common network infrastructure for the business area (including connections to other networks and the outside world).
3. Develop planned application development projects, priorities, and schedules for the business area. These information system applications will evolve around, and be highly integrated through, the shared database and the network infrastructure.

In this section we will examine a typical BAA. Then we will, once again, walk through activities, participants, inputs, outputs, and techniques. Techniques will again be cross-referenced to the chapters and modules that teach those techniques.

Building Blocks for the BAA

The fundamental objectives of a BAA are

*   To identify business-level requirements for a shared database for the business area.
*   To identify business-level requirements for a shared network for the business area.
*   To refine technical requirements for the business area database and networks.
*   To identify high-level business requirements for integrated applications in the business area.

Here again, the information system building blocks (Figure 6.14) serve as a useful framework for identifying business area requirements. Notice that, for most of the building blocks, we are refining the enterprise models from the definition phase. The resulting business area models represent management's ideal "vision" of a highly streamlined and integrated system.

A modern trend in both business area analysis and systems analysis is business process reengineering.

Business process reengineering is a requirement to study fundamental business processes, independent of organization units and information systems support, to determine if the underlying business processes can be significantly streamlined and improved. It is sometimes called *business process redesign.*

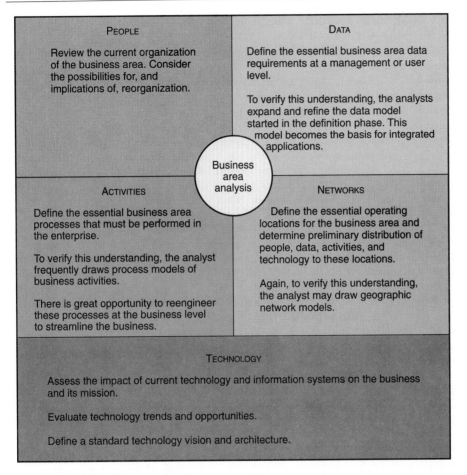

PEOPLE

Review the current organization of the business area. Consider the possibilities for, and implications of, reorganization.

DATA

Define the essential business area data requirements at a management or user level.

To verify this understanding, the analysts expand and refine the data model started in the definition phase. This model becomes the basis for integrated applications.

Business area analysis

ACTIVITIES

Define the essential business area processes that must be performed in the enterprise.

To verify this understanding, the analyst frequently draws process models of business activities.

There is great opportunity to reengineer these processes at the business level to streamline the business.

NETWORKS

Define the essential operating locations for the business area and determine preliminary distribution of people, data, activities, and technology to these locations.

Again, to verify this understanding, the analyst may draw geographic network models.

TECHNOLOGY

Assess the impact of current technology and information systems on the business and its mission.

Evaluate technology trends and opportunities.

Define a standard technology vision and architecture.

FIGURE 6.14 Analysis Phase Building Blocks The analysis phase examines proposed information system building blocks to gain a high-level understanding of the business's requirements.

This concept goes far beyond integration of business processes across organizational boundaries. It is based on the discovery that many of an average business's processes are based on decades-old, sequential, assembly-line thinking. Many companies are discovering that these processes can be significantly restructured to improve efficiency, regardless of computer automation. Business process reengineering has become an important new objective in business area analysis. For more on this trend, see The Next Generation box in this chapter.

BAA Activities, Roles, and Techniques

Figure 6.15 illustrates the typical activities of a BAA. Let's examine the activities.

Activity 1: Establish the Analysis Team

The old planning team has been disbanded. Their executive managers cannot provide the level of detail needed in this phase. For the analysis phase we establish a new team, which we will call the *analysis team*. The team must be cross-functional; therefore, it will consist of representative managers and/or supervisors from each organization unit that participates in the business area.

BUSINESS PROCESS REENGINEERING

Business process reengineering. The term is confronting business professionals and information systems professionals everywhere they turn. Is it hype? Or is it truly the focus of the next generation of systems development. This chapter introduced the term as follows:

Business process reengineering is the study and redesign of fundamental business processes, independent of organization units and information systems support, to determine if the underlying business processes can be significantly streamlined and improved.

Why has the term suddenly entered the vocabulary of business and information systems professionals alike?

The problem lies fundamentally in history and our human tendency is not only to resist change, but to close our minds to thinking about change. Most of our business processes are based on the sequential processing approaches that date back to early assembly lines and precomputer paper flows. Has the computer helped to redesign those processes? For the most part, no.

What we have done with computers is to automate the antiquated, sequential, flow-oriented processes of the past. In fact, in many cases we complicated the sequential flow by introducing new sequential data processing processes in between many existing sequential business processes.

The problem was first brought to our attention by Michael Hammer and Tom Davenport in the summer of 1990. Their groundbreaking articles (see the suggested readings at the end of the chapter) called everyone's attention to the need to redesign business processes before you automate them. The articles describe real world case studies where business achieved orders of magnitude improvement in business process efficiency through radical rethinking of business processes.

Today, everybody seems poised to jump on the business process reengineering bandwagon. What does this have to do with information systems and strategic systems planning?

Business process reengineering begins with a reality check—information systems has *not* become the cure-all for business that it professes to be! True, computers and information systems can contribute to radical improvement of business processes, but not by merely automating or slightly improving the status quo. What information systems development offers is the unique opportunity to engage in the process of rethinking business processes—especially in a systems planning project. Let's examine that claim.

Strategic information systems planning requires top management commitment. So does business process reengineering. Systems planning requires an uncommon cooperation between business management and information systems management and staff. So does business process reengineering. Systems planning requires cross-functional thinking that breaks down organization structure barriers. So does business process reengineering. Systems planning often calls for radical change in information systems architecture, organization, and delivery. Business process reengineering does the same for organization structures, job definition, value systems, and the like. The fit is natural!

It is so natural, in fact, that techniques and methodologies for integrating business process reengineering into systems planning and systems analysis are starting to emerge. For example, you will soon be hearing about a new class of enterprise engineering methods from James Martin, one of the founders of information engineering. Enterprise engineering appears to be evolutionary. It combines

- Information engineering, a preferred set of techniques for planning, analyzing, designing, and implementing the data resource available to the enterprise.

- Business process reengineering, the subject of this feature, which will provide complementary techniques for planning, analyzing, redesigning, and reimplementing efficient and effective business processes in the enterprise.

- Total quality management, a formal approach to ensuring and continuously improving both the business processes and their supporting information resources.

Thus, business process reengineering is hardly a fad. It will undoubtedly affect and change future systems development approaches, tools, and techniques. And it will probably change future editions of this book. For more about the impact of business process reengineering, read the May 1992 issue of the *American Programmer.* The entire issue is dedicated to the subject. It offers several insights into the next generation of techniques and methodologies as they relate to the subject.

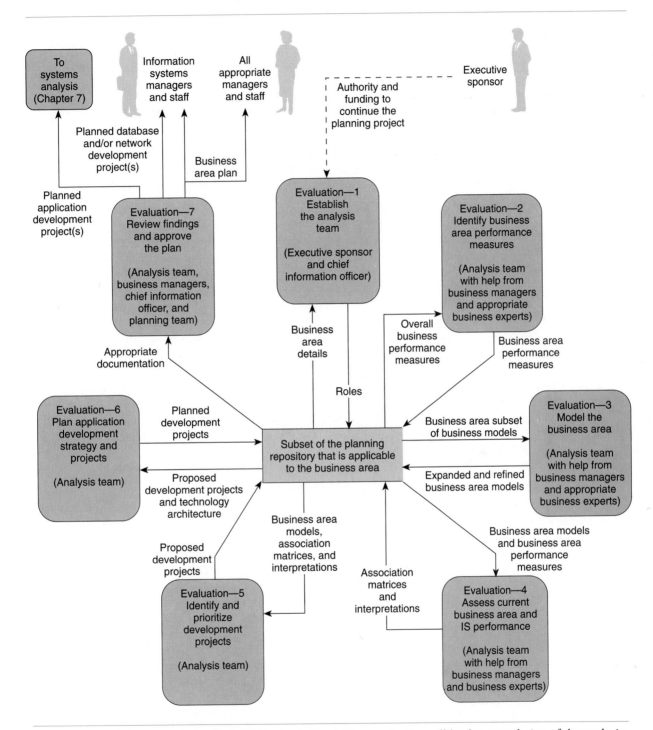

FIGURE 6.15 **Business Area Analysis Phase Activities** These activities will lead to completion of the analysis phase. Once again, the activities are generally completed in a clockwise sequence from the top. On the other hand, the project repository allows you to overlap activities or return to any previous activity to do rework.

Cross-functional representation promotes our goals of streamlining and integration.

Information systems should be represented by its best systems analysts. Some IS shops form SWAT teams for business area analyses.

> **SWAT** stands for **specialists with automated tools.** It is a highly cohesive team that includes skilled facilitators for JAD sessions, at least one analyst or consultant knowledgeable in the methodology, experienced CASE tool users, experienced system modelers, data analysts, network analysts, and the like. SWAT teams are rarely aligned with traditional information systems development teams that support only their assigned users. Instead, SWAT teams are assigned to go wherever they are needed.

For a BAA, the SWAT team is assigned to that business area. After the phase is completed, the SWAT team will either (1) be reassigned to analyze a totally different business area, or (2) be retained to develop all information systems applications in the business area (and then move on to another business area).

Team assignments are usually made jointly by an executive sponsor and the chief information officer. The executive sponsor is an executive manager to whom the team officially reports. That role should be filled by the highest level executive to whom all participating managers commonly report.

The inputs to this activity are business area details including organization charts and business area models (both from the definition phase). The outputs are assigned roles. Ideally, the team should be managed by one of the non-IS managers. On the other hand, the team is usually led by an IS professional who understands the methodology to be used.

There are no special skills or techniques to be learned for this activity other than knowledge of the role definitions from the methodology to be used.

Activity 2: Identify Business Area Performance Measures

Just like the company has performance measures (identified in the study phase), each business area also has (or should have) its own performance measures. In order to ensure that future information systems applications contribute maximum value to the business area, we must *identify business area performance measures.* Appropriate measures include mission, vision goals, objectives, critical success factors, critical assumptions, and critical information needs—all of which were introduced earlier in the chapter.

This task is performed by the analysis team with additional help as appropriate from other managers in the business area. Clearly, the team should first review the overall business performance measures that were recorded in the repository during the study phase. Why? Because a business area's performance measures should build on and support applicable measures for the business as a whole. Some of these measures may not apply. Others may serve as a point of departure for defining the business area measures.

The result of this activity is business area performance measures. The business area derives its goals from the goals and objectives of the business as a whole, tempered by the business area's own mission. Once the goals are established, supporting objectives, critical success factors, assumptions, and information needs can be established, just as they were for the business as a whole.

The following techniques and skills are applicable to this activity:

- Interviewing and other fact finding (see Part Five, Module B, Fact-Finding Techniques).

- Joint application development or JAD (see Part Five, Module B, Fact-Finding Techniques).

Activity 3: Model the Business Area

The next activity in the analysis phase is to *model the business area.* Recall that enterprise models were created and then partitioned into business areas during the definition phase. In this activity we retrieve those models to expand and refine them. The analysis team's lower-level managers are closer to and more familiar with the business area's building blocks than were the higher-level managers who drafted the earlier models.

Models developed in this phase still depict the business system at a fairly general level of detail. The models are generally suited to the middle managers who help develop them. Obviously, supervisors and clerks could develop more detailed and refined models, but that level of detail would not serve the planning interests. Business area models will eventually serve as a point of departure for applications development (systems analysis and design).

Your pyramid model provides a framework for the business area models to be developed. The most typical models include

- ACTIVITIES — **Process models** (Figure 6.16) model the flow of data through business processes in the system. This is the only business area model developed from a blank sheet of paper. Details about the processes and data flows are not usually described until the "systems" analysis phases.
- DATA — The business area's **data model** was started in the definition phase. Now it is modified or expanded based on input from the analysis team. A typical model was shown previously in Figure 6.12. In this activity we also begin defining the types of data to be stored for each subject on the diagram.
- NETWORKS — The business area's network description was also started in the definition phase. Now it is expanded to identify missing locations, relevant sublocations (e.g., rooms in a building), distances between locations, and, possibly, essential data flows between locations.

All the above business area models are developed jointly by the entire analysis team. As was the case with enterprise models, IS professionals tend to take the lead in constructing and maintaining the models, since they are usually more skilled in the modeling techniques and CASE tools used to record the models and details into the repository. In particular, the data administrator and network administrator continue to play key roles in the evolution of the data and network models, respectively. Systems analysts play the key roles for process models.

The input to this activity is the business area's subset of the enterprise model (from the definition phase). The output is the expanded and refined business area models.

The following techniques and skills are applicable to this activity:

- Fact finding (see Part Five, Module B, Fact-Finding Techniques).
- Joint application development (see Part Five, Module B, Fact-Finding Techniques).
- Data modeling (see Chapter 8, Data Modeling).
- Process modeling (see Chapter 9, Process Modeling).
- Network modeling (see Chapter 10, Network Modeling).

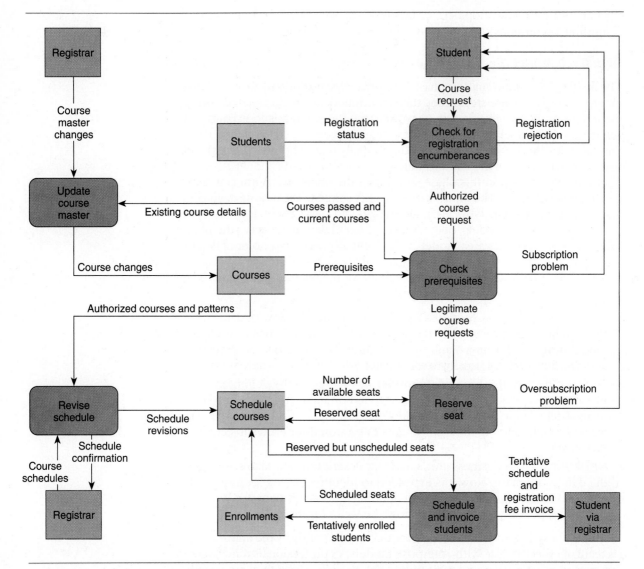

FIGURE 6.16 **Data Flow Diagrams** A data flow diagram is a useful tool that tracks the flow of data through business processes.

Activity 4: Assess Current Business Area and IS Performance

Our ultimate goal is to identify and plan applications for development. But we should first *assess current business area and IS performance.* This assessment is similar in procedure to the one we did for the business as a whole during the definition phase.

This analysis is typically accomplished by constructing and analyzing various association matrices. Association analysis and association matrices were described earlier in the chapter. By constructing and analyzing association matrices, the analysis team learns how well existing building blocks support the business area's performance measures.

It is also appropriate to assess current information services and strategies. The analysis team usually conducts new surveys of the management and user community for the business area to collect data. The results are compiled and reviewed by the team.

The inputs to this activity are the business area models (from Activity 3) and the business area performance measures (from Activity 2). The outputs are the association matrices and interpretations.

Once again, the principle techniques and skills required to complete this activity are interviewing, observations, sampling, studying, and joint application design. These cross life cycle skills and techniques are covered in Part Five, Module B, Fact-Finding Techniques.

Activity 5: Identify and Prioritize Development Projects

The business area models represent a vision of an integrated business system. The association matrices and interpretations tell us how well current systems support the business area mission. Given these two inputs, we can now *identify and prioritize development projects* that will help us achieve the vision. Several types of projects may be defined in order to accomplish any of the following aims:

• Develop a subject database.
• Develop a network.
• Modify existing applications ("quick fixes" only).
• Develop new information systems applications.

This activity is relatively straightforward. The business area models already illustrate the integration of business processes, data, and networks. The analysis team simply identifies or partitions appropriate development projects on the models and prioritizes them into some sequence that makes sense. Frequently, the data model becomes the target for a database development project and the network model becomes the target for a network development project. Application development projects are discovered on the process model(s).

The output of this activity is proposed development projects. There are no special skills or techniques required to complete this activity.

Activity 6: Plan Application Development Strategy and Projects

Systems planning is rapidly coming to a conclusion. Now we can plan real development projects. The inputs to this activity include the proposed development projects and the technology architecture. Recall that the technology architecture was approved in the definition phase. The outputs are planned development projects.

The plan is developed by the analysis team, with appropriate help from information systems managers. The plan should include a general schedule or sequence for all development projects in the business area. This schedule could span several years, since a business area may generate many projects. Information engineering and CASE experts expect that this time frame could be substantially reduced through rapid application development (RAD) techniques and CASE technology.

The principle technique required to complete this activity is *project planning and scheduling,* which is covered in Part Five, Module A, Project Management.

Activity 7: Review Findings and Approve the Plan

As with the study and definition phases, the analysis phase findings and recommendations should be reviewed. The review should be performed by the analy-

sis team and include all management-level staff that will be affected by the plan.
As always, two levels of review are possible.

- A **quality review** ensures that the analysis phase was completed in accordance to the methodology and that resulting documentation (in the repository) is complete, consistent, and in compliance with standards. This
  is especially important since this documentation will be passed on to
  those who will develop databases, networks, and applications according
  to the plan.
- A **feasibility review** reassesses the feasibility of starting the highest-pri-
  ority development project(s).

In both cases the systems analyst or system owner assembles the appropriate
documentation from the repository. The ultimate outputs include (1) the business area plan itself, (2) planned database and/or network development proj-
ect(s), and (3) planned application development project(s). Recall that the
latter is one of the major triggers for systems analysis. The process of systems
analysis is explained in the next chapter.

The techniques and skills needed to complete this activity are all cross life
cycle skills:

- Feasibility assessment (see Part Five, Module C, Feasibility Analysis).
- Report writing (see Part Five, Module D, Interpersonal Skills).
- Walkthroughs (again, see Part Five, Module D).
- Verbal presentations (again, see Part Five, Module D).

--------------------------------- *Summary* ---------------------------------

Strategic systems planning seeks to identify and prioritize those technologies and applications that will return the most value to the business. Systems planning is driven by
systems owners' concerns and strategic business issues. It consists of three distinct
phases that address the PEOPLE, DATA, ACTIVITIES, and NETWORKS building blocks. The
three phases are: study the business mission (or the study phase), define an information
architecture (or the definition phase), and analyze business areas (or the analysis
phase). The results of each phase are stored in a planning repository.

The purpose of the study phase is to understand the business scope and mission. The
business mission can be analyzed in the context of several frameworks, including critical
success factor analysis, competitive analysis, and value chain analysis. A complete study
phase includes these activities: (1) establish the planning team, (2) define planning
scope and expectations, (3) identify business performance measures, (4) develop a
project plan, and (5) review findings and communicate planning vision. The deliverable
for the study phase is a project-planning charter that communicates the planning vision
and mandates the forthcoming definition phase.

The definition phase is triggered when the planning project's executive sponsor
gives authority to continue. The purpose of the definition phase is to define a vision and
plan for using information technology and developing information systems needed to
support a business's mission. To build this information architecture, a planning team
completes the following activities: (1) model the enterprise, (2) assess current business
strategies, (3) assess current information services and strategies, (4) identify and prioritize business areas, (5) complete the new information architecture, (6) identify and plan
subsequent projects, and (7) review findings and approve the plan. Some activities make
use of special tools and techniques such as association matrices, affinity analysis, and
cluster analysis. The end-product of the definition phase is called the information architecture and plan. It sets forth a high-level strategic plan for information systems. This

architecture includes a number of pictorial enterprise models. The key enterprise model is called a subject data model, around which business functions and business processes can be integrated.

The analysis phase, or business area analysis (BAA), adds detail to the high-level plan. The purpose of a BAA is to expand on the plan for a single business area. A business area is a logical grouping of business processes, data, and locations, independent of organizational boundaries or structure. The objective of BAA is to build a plan for highly integrated applications for the business area.

The BAA activities include (1) establish the analysis team, which often includes specialists with automated tools and is nicknamed a SWAT team, (2) identify business area performance measures, (3) model the business area, (4) assess current business area and IS performance, (5) identify and prioritize development projects, (6) plan development projects, and (7) review findings and approve the plan. Increasingly, a BAA is likely to attempt business process reengineering to redesign existing business processes for increased efficiency and effectiveness ''prior to'' applying information technology.

The end-products of the BAA are a business area plan, planned database and network development projects, and planned application development projects. The latter end-product triggers the systems analysis phase, which is covered in the next chapter.

Key Terms

affinity, p. 227

affinity analysis, p. 227

association matrix, p. 225

business area, p. 234

business area analysis (BAA), p. 213

business area models, p. 235

business function, p. 231

business goal, p. 215

business mission statement, p. 215

business objective, p. 216

business process, p. 231

business process reengineering, p. 238

business vision statement, p. 215

clustering, p. 228

competitive analysis, p. 216

context diagram, p. 221

critical assumption, p. 216

critical information need, p. 216

critical success factor (CSF), p. 215

critical success factor analysis, p. 216

data model, p. 243

enterprise model, p. 230

feasibility review, p. 237

information architecture, p. 224

model, p. 230

planning analyst, p. 213

planning repository, p. 213

process model, p. 243

quality review, p. 237

specialists with automated tools (SWAT), p. 242

subject data model, p. 231

systems planning, p. 212

technology architecture, p. 235

value chain, p. 216

Questions, Problems, and Exercises

1. Identify and briefly describe the purpose of the three systems planning phases.

2. What is a planning repository? How does it differ from a project repository? How is a planning repository typically implemented? How might it be implemented otherwise?

3. What are the fundamental objectives of the study phase? Why aren't the information systems building blocks very useful in this phase?

4. Name and briefly describe three frameworks for studying a business mission.

5. Identify three possible triggers of the study phase.

6. Your company has never formally done a business systems plan. You want to do an information systems plan. Convince your CEO to allow you to start a project that does both.

7. Which of the two groups of value chain activities should have the greatest impact on the business? (Note: Don't look for the definitive answer in the text.)

8. You have a goal—to get an "A" in this course. Establish objectives for your goal as well as critical success factors, assumptions, and information needs for each objective.

9. Do a competitive analysis for a typical video rental store. (Alternative: Do a competitive analysis for your current or previous employer's business, or for your school.)

10. Draw a value chain diagram for your current employer, past employer, or school.

11. Differentiate between the study and definition phases of systems planning. Then do the same for the definition and analysis phases.

12. What are the fundamental objectives of the definition phase? Explain how the information systems building blocks aid in identifying the required level of understanding to be obtained in the study phase.

13. What are the components of the enterprise model? Which one is generally considered most important? Why?

14. Explain the concept of a business area and how it differs from an organization unit. Why is the concept important to systems planning?

15. You have used affinity analysis and clustering to divide your business into eight business areas, each consisting of business processes built around a single database. Each database consists of data subjects that have a minimum .70 affinity relative to the processes that use or update those data subjects. Your executive sponsor doesn't understand and won't approve the proposed business areas until you can explain how they were defined in a clearer fashion. Explain the business areas accordingly.

16. Name at least five typical components of a technology architecture. To what does a technology architecture apply—the business or business area?

17. Why might a business area have a different technology architecture? (Note: Don't look for the answer to this question in the book.)

18. Differentiate between business area analysis and systems analysis. Why is business area analysis part of the systems planning function?

19. What are the fundamental objectives of the analysis phase? How might the information systems building blocks be used to identify what requirements must be specified to fulfill these objectives?

20. What is the value of modeling during the business area analysis phase? How do business area models differ from enterprise models? Name three types of business area models.

21. How might business process reengineering help the business regardless of information systems support?

22. What is a SWAT team? How does it differ from the typical application development team described in Chapter 1?

23. Name four types of development projects that may emerge from a business area analysis.

24. What is the end product of each systems planning phase? Explain the purpose and content of each of these products.

Projects and Minicases

1. You are the chief information officer of a medium-sized insurance broker. Your president suggests that the business does not need a strategic systems plan because: (1) you are required to attend regular staff meetings with top executives so you should know what is and is not important, (2) you chair a middle-manage-

ment steering committee that screens worthiness of all requests independent of budget and politics, (3) top middle managers are fully capable of making their own project decisions based on the above worthiness measure and how much they're willing to pay to have Information Systems develop the applications, and (4) the company does not exploit the latest technology; therefore, it doesn't need a technology plan. Prepare a formal recommendation to change the CEO's mind. He is not very computer literate. He thinks you are just trying to substantially increase your budget to build your own empire. He thinks no computer person could possible appreciate his business.

2. There are a number of CASE tools specifically oriented to strategic systems planning. Through research, do a market survey to identify the characteristics and capabilities of these tools. Briefly compare several products. Select one product and write the vendor. Request product literature, company information, success stories, and the like. Complete your report with an in-depth discussion of the CASE tool.

3. There are a number of commercial methodologies that include systems planning within their approach. Through research, do a market survey to identify the characteristics and capabilities of the planning capabilities of these methodologies. Select one methodology and write the vendor. Request product literature, company information, success stories, and the like. Compare and contrast the methodology with the generic methodology presented in this chapter. Force yourself to identify at least three major features or capabilities that you like better, and three that you have some concerns about.

4. Make an appointment with a top information systems executive in a local business. Are information systems developing according to a master plan? Why or why not? If not, how are information systems priorities determined? Is it working? How could the priority-setting approach be improved?

    If the business has a master information systems plan, how was it developed? How long did it take? Was the master plan influenced by business executives? Was there a master plan for the business? Did it exist prior to the information systems planning initiative?

    Present your findings and your personal analysis of those findings.

Suggested Readings

Flaatten, Per O., Donald J. McCubbrey, P. Declan O'Roirdan, and Keith Burgess. *Foundations of Business Systems.* Hinsdale, Ill.:1 Dryden Press, 1989. Endorsed by Andersen Consulting of Arthur Andersen & Company, this text offers several chapters that introduce and expand on systems planning. Andersen Consulting's Foundation methodology supports systems planning, presumably through the methods described in this book. Chapters 1 and 2 lay a foundation for systems planning. Chapters 4 through 6 discuss the Andersen methodology for planning. Finally, Appendix 1 details the technique of value chain analysis.

Martin, James. *Information Engineering: Book II. Planning and Analysis.* Englewood Cliffs, N.J.: Prentice Hall, 1990. Information-engineering techniques underlie today's most popular methodologies. Chapters 1 through 14 represent the most comprehensive treatment of information engineering for systems planning and business area analysis.

Porter, Michael E. "How Information Gives You Competitive Advantage." *Harvard Business Review* 63, no. 46 (July–August 1985, pp. 149–60. Porter is considered a genius of systems planning. His books and papers are must reading for students who want to learn more about strategic alignment of information systems with business needs.

Zachman, John. "A Framework for Information System Architecture." *IBM Systems Journal* 26, no. 3, 1987. The Zachman framework inspired our pyramid model, and also IBM's AD/Cycle information model. But before that, it was viewed as a strategic-planning model. John Zachman is viewed by many as IBM's strategic-planning guru and chief advocate of their Business Systems Planning (BSP) methodology.

7

Systems Analysis

Chapter Preview and Objectives

In this chapter you will learn more about the analysis phases in the systems development life cycle: survey, study, and definition. These three phases are collectively referred to as *systems analysis.* You will know that you understand the process of systems analysis when you can

Define
systems analysis and relate the term to the survey, study, and definition phases of the life cycle.

Describe
the survey, study, and definition phases in terms of your information system building blocks.

Describe
the survey, study, and definition phases in terms of objectives, activities, roles, inputs and outputs, and tools and techniques.

Recognize
a number of system modeling techniques and describe their roles in systems analysis.

Describe
and use the PIECES framework for problem solving in systems analysis.

Identify
those chapters and modules in this textbook that can help you actually perform the activities of systems analysis.

MINICASE

COLLEENS FINANCIAL SERVICES

Colleens Financial Services is a nationwide financial services company headquartered in Omaha, Nebraska.

Scene: *Senior systems analyst Fred McNamara is meeting with Ken Borelli, the MIS manager at Colleens Financial Services. Fred has just completed an evaluation of a new software package, a fourth-generation programming language (4GL). In addition to evaluating the 4GL, Fred has been asked to learn about a popular systems development approach—called prototyping—that makes use of 4GLs to build working models of systems. Fred is meeting with Ken to give him his assessment of the 4GL product and prototyping as an alternative approach to systems development.*

Ken: So what do you think about that new software product—is it worthy of being called a 4GL?

Fred: Without a doubt! Third-generation programming languages like COBOL can't compare to it! I think this product can do wonders for the systems staff. It is very user friendly, and it provides a number of facilities that assist in developing a complete system.

Ken: What kind of facilities does it provide?

Fred: To give you some idea, I used a facility to develop a database containing actual data, another facility to produce a relatively complex printed report against that database, and other facilities to develop menus and other input and output screens—all in a fraction of the time that would have been required with COBOL.

Ken: You said it was very user friendly. Does that mean you didn't have to spend a lot of time referencing manuals?

Fred: I hardly used the manuals. The facilities simply led me through a series of questions or prompts. All I had to do was answer the questions. The 4GL generated the program code, which I subse-

quently executed. The productivity implications are tremendous!

Ken: Sounds like the product is a good investment. And what about prototyping? Do you think prototyping is something we should consider doing as an alternative to our current approach to developing systems?

Fred: Well, prototyping certainly takes advantage of tools such as 4GLs. The strategy is very simple. It emphasizes the development of a working model of the target system, instead of traditional paper specifications. You begin building the model by first defining the database requirements for the new or desired system. Identifying the database requirements is relatively simple, since the data for most systems already exists, either in computer files or manual forms. Afterward, you then build and load a database using the 4GL. Once you have the database built, the rest is easy. You can use various facilities to quickly generate the menus, reports, and input and output screens. The analyst does not have to worry about whether the working model is totally complete or accurate. The end-user is encouraged to review the model and provide the analyst with feedback. If the model, say, a report, is not acceptable, the analyst simply makes requested changes and reviews it again with the end-user at some later time. This repetitive process and active end-user participation are considered essential and to be encouraged.

Ken: Now that's a new one! I certainly agree with the idea of encouraging end-user participation. But this attitude of encouraging or expecting to keep redoing work would be difficult to adjust to.

Fred: I'm sure some of us old-timers will have some difficulty adjusting to this type of thinking.

Ken: Well, it sounds like this new 4GL and prototyping should be pursued further. Both 4GL and prototyping seem to offer some productivity gains. I've been looking for a way to get the end-users more involved in the systems development process, and I think this prototyping approach is the answer. There is one other thing: I suspect my staff's morale would be improved. I believe my staff would be motivated by this new 4GL product and by prototyping's emphasis on building a model as a basis for performing systems development.

Fred: I agree! I can't wait to start my next project. I've got just the project picked out. I received a request for a new employee benefits system from the Personnel Department. I thought I'd use the 4GL in conjunction with the prototyping approach to complete this project. I won't be needing any programmers since I'll be developing the system myself while I'm working with the end-users. I've already drafted a memo asking Personnel to provide me with some sample records, forms, reports, and other materials that will help me identify their data storage requirements. From those samples, I'll be able to build a database for the new system. Then I'll start meeting with the end-users to define and implement screens and reports.

Ken: Hold your horses! I do have some concerns about this approach to systems development. Neither the tool nor the approach justifies a departure from the systems development life cycle concept we follow here at Colleens. And that's exactly what you're proposing. You're proposing to select a project, define some basic requirements, and jump right into the design and construction of a new system. That I won't have. I want you to reconsider things.

[Ken reaches for a systems development standards manual on his bookshelf. He turns to a figure that depicts the systems development life cycle phases.]

Notice that our systems development life cycle includes several systems analysis phases, including a survey or preliminary investigation phase. Do you fully understand why we require this first phase?

Fred: Sure, that's where we perform a very quick study of the proposed project request. We try to gain a quick understanding of the size, scope, and complexity of the project.

Ken: You're half right, but why must we complete the phase? I'll tell you why! Because we receive numerous project requests from our end-users! We have a limited number of resources. We can't take on all the requests. It is the purpose of this phase to address the seriousness of the problems and to prioritize the project request against other requests.

Fred: I see what you're saying. I sort of jumped the gun by picking this Personnel project without considering other project requests that might be given higher priority.

Ken: Good. Now let's consider the second phase, the detailed study phase. If the project is to be pursued further, we then conduct a detailed study or investigation of the current system. We want to gain an understanding of the causes and effects of all problems, and to appreciate the benefits that might be derived from any existing opportunities. We don't want to bypass this phase. This phase ensures that any new system we propose solves the problems encountered in the current system.

Fred: That brings us to the definition phase, right?

Ken: Right. For all practical purposes, this is where you were proposing to begin your project. You were going to define data storage requirements for a new employee benefits system. I assume you would also attempt to identify other requirements. What particularly bothered me was that I didn't get an impression that you intended to study the database requirements to ensure data reliability and flexibility of form. And what about completeness checks? What were you going to do to ensure that processes were sufficient to ensure data will be properly maintained?

Fred: I'm beginning to lose my confidence in this prototyping approach. I was about to make some big mistakes by trying to bypass several important problem-solving phases and tasks.

Ken: Listen, there's no reason to write off prototyping. So long as we base our prototypes on some sound design principles, I think we can still achieve all the benefits that you described earlier. I have another concern, though. We will eventually want to look at alternative solutions such as manual versus computer-based systems, on-line versus batch systems, and the like. These options should be evaluated for technical, operational, and economical feasibility. It seems to me that this activity should precede extensive prototyping. We don't want to prematurely commit to less feasible or infeasible solutions.

Fred: I understand, it looks like prototyping has potential, especially in the definition phase of analysis. And I suspect that prototyping will greatly accelerate our systems design and implementation phases.

Ken: I think we both understand that we need to give this prototyping approach some more thought. I want to establish a task force to study prototyping and its implications for our internal systems development standards. And I want you to be in charge of that committee. In the meantime, I see no reason not to try the new 4GL in conjunction with our usual development standards.

Discussion Questions

1. Do you think Fred did a thorough evaluation of the fourth-generation software product? What benefits do you believe can be derived from using such a tool?

2. Did Fred view prototyping as an alternative to the traditional systems development life cycle? If so, how should he have viewed it?

3. What systems analysis phases would have been skipped by the prototyping approach Fred proposed to follow? What do you think would have been the results of the employee benefits project if Fred had approached the project in the manner he originally envisioned?

WHAT IS SYSTEMS ANALYSIS?

Systems analysis was first defined in Chapter 3:

Systems analysis is the study of a current business and information system application and the definition of user requirements and priorities for a new or improved application.

Systems analysis should be thought of as "business problem solving, independent of technology."

Systems analysis is driven by system users' concerns; hence, it addresses the PEOPLE, DATA, ACTIVITIES, and NETWORKS building blocks from a user perspective. Emphasis is placed on business issues, not technical or implementation concerns. Hence, the TECHNOLOGY building block in the pyramid is, for the most part, not relevant.

In Chapter 3 (and Figure 3.7) you learned that systems analysis consists of three phases:

1. *Survey project feasibility* (or the survey phase).
2. *Study and analyze the current system* (or the study phase).
3. *Define and prioritize users' requirements* (or the definition phase).

Figure 7.1 is a life cycle diagram that illustrates three analysis phases. Notice the big red Xs. In Chapter 3 we depicted work products and deliverables as "flowing" from one phase to the next. That was, however, an oversimplification. Actually, work products and deliverables are stored in project repositories.

A **project repository** is a collection of those places where we keep all documents, documentation, and programs that are associated with the application and project.

Although we show only one project repository in the figure, it is normally implemented as some combination of the following:

- Notebooks and/or file folders of various documents and documentation.
- A disk or directory of word processing, spreadsheet, and other

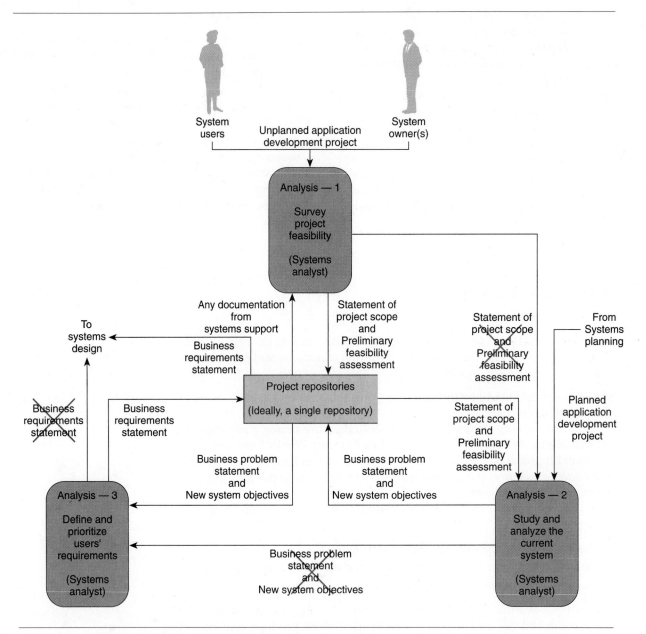

FIGURE 7.1   Systems Analysis   The three phases of systems analysis were introduced in Chapter 3. We've made some revisions on this reproduction. The deliverables don't really pass from phase to phase. Instead, they are recorded into various project repositories for use in later phases.

computer-generated files that contain project correspondence, reports, and data.

- One or more CASE local repositories (as discussed in Chapter 5).
- A library of computer programs and documentation.

Hereafter, we will refer to these as making up a singular project repository. (Note: A single repository is actually the ideal implementation!)

If you study Figure 7.1 carefully, you'll notice that the "X'd-out" deliverables are still flowing between the phases — only now they flow by way of the repository. This is a better and more realistic view of systems development. Why? Because systems development phases and activities are not really sequential! Work in one phase can and should overlap work in another phase, so long as the necessary information is already in the repository. Furthermore, this model permits the developer to backtrack when an error or omission is discovered.

This chapter examines each of the above phases in greater detail. To give you "the feel" of a true systems development methodology, each phase is described in terms of the (1) *purpose* of the phase, (2) *activities* that should be performed, (3) *roles* played by various people in each activity, (4) *inputs* and *outputs* for each activity, and (5) *techniques* or *skills* that can be used to complete each activity.

THE SURVEY PHASE OF SYSTEMS ANALYSIS

Systems development can be very expensive. Thus, it pays to answer the important question, "Is this project worth looking at?" The survey phase of systems analysis answers this question. In different methodologies it is sometimes called the *preliminary investigation, initial study,* or *feasibility study* phase.

Some projects do not require this evaluation. For example, projects initiated by formal systems planning have already been justified by the planning team and management. Similarly, the worthiness of some projects may be predetermined by a high-ranking requester, such as a vice president or senior director. Thus, some projects skip directly to the study phase.

But most unplanned projects do go through the "quick-and-dirty" survey phase. In this section we will examine the purpose of the survey phase. Then we will walk through its activities, participants, inputs, outputs, and techniques. Techniques will be cross-referenced to the chapters and modules that teach those techniques.

Building Blocks for the Survey Phase

The fundamental objectives of the survey phase are:

- To identify problems, opportunities, and/or directives that initiated this project request.
- To determine whether solving the problems, exploiting the opportunities, and/or satisfying the directives will benefit the business.

To achieve these objectives, you need a very general understanding of the current system. Your information system building blocks (Figure 7.2) provide a framework for understanding the current system during the survey phase. The desired information represents the system owner's view of the application, which includes very few details. More detailed information will be collected if the project is approved to continue into the study phase.

Survey Phase Activities, Roles, and Techniques

Figure 7.3 illustrates the typical activities of the survey phase. Because this is the first activity-level diagram of the chapter, let's establish the following guidelines for reading all such diagrams:

FIGURE 7.2 Survey Phase Building Blocks
The survey phase examines an information system at a very general level of detail.

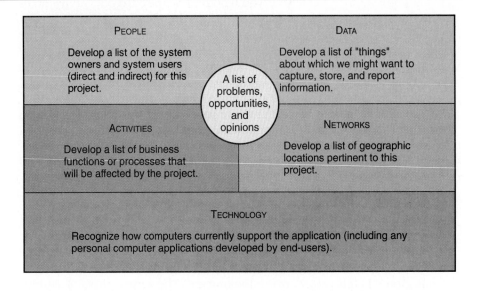

PEOPLE

Develop a list of the system owners and system users (direct and indirect) for this project.

DATA

Develop a list of "things" about which we might want to capture, store, and report information.

A list of problems, opportunities, and opinions

ACTIVITIES

Develop a list of business functions or processes that will be affected by the project.

NETWORKS

Develop a list of geographic locations pertinent to this project.

TECHNOLOGY

Recognize how computers currently support the application (including any personal computer applications developed by end-users).

- Silhouettes represent people or departments that initiate projects, phases, and activities.
- The rounded rectangles represent *activities.* Each activity is numbered solely for identification purposes. The activity name is printed in the upper half of the symbol. The participants in the activity are printed in the lower half of the symbol. The first participant is always the person who "guides" the activity.
- The solid arrows reflect inputs and outputs for an activity. Each is named. When one of these inputs or outputs is referred to in the text, it is underlined.
- Dashed arrows represent events that trigger an activity. (There are no triggers on Figure 7.3; however, they will appear in other diagrams.)

In Chapter 3 you learned that the survey phase is usually initiated by an unplanned application development project from a system owner or system user. This input can be formal or informal. In some organizations, formal project requests are documented with a standard form such as the one shown in Figure 7.4.

In response to the request, a systems analyst is usually assigned to complete the survey phase. In most cases the entire survey phase should require no more than two to four days. This is because substantial time and resources should not be committed until the project is deemed feasible and appropriate.

Activity 1: Conduct Initial Interview(s)

The survey phase begins with one or two interviews with the project requester. In most cases the requester is either the system owner or a system user. In preparation for the interview, the analyst may seek existing system documents and documentation. Representative interview questions are listed in Figure 7.5. The actual interview typically lasts 45 to 60 minutes.

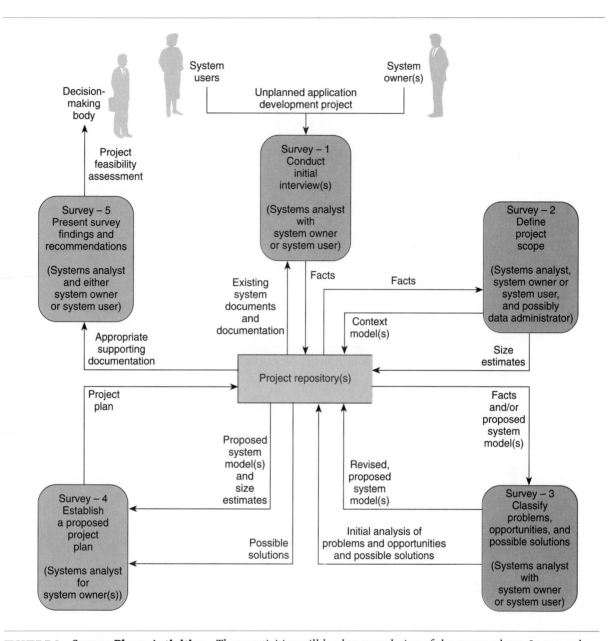

FIGURE 7.3 **Survey Phase Activities** These activities will lead to completion of the survey phase. In general, the activities are completed in a clockwise sequence from the top. On the other hand, the project repository allows you to overlap activities or return to any previous activity to do rework.

As shown in Figure 7.3, facts obtained from the interview are recorded in the appropriate project repository. These facts include

- Lists of PEOPLE, DATA, ACTIVITIES, locations and NETWORKS, and existing TECHNOLOGY that pertain to this project.
- Lists of problems and opportunities that led to the project request.
- Any relevant constraints, ideas, and opinions.

FORM IS-100-A.RFSS

Request for System Services

SUBMITTED BY: | Barbara Rushin | Date: | July 25, 1993

DEPARTMENT: | Transportation Fleet Services

TYPE OF REQUEST:
(check one)

☒ NEW SYSTEMS DEVELOPMENT

☐ EXISTING SYSTEM ENHANCEMENT

☐ EXISTING SYSTEM MODIFICATION

☐ NOT SURE

BRIEF STATEMENT OF PROBLEM OR OPPORTUNITY (Attach additional documentation as necessary):

Currently, we have no way of attributing all direct and indirect departmental costs to a company vehicle. Hence, we are unable to assess the return on investment for any vehicle, the cost efficiency of our service group, or the best date to retire a vehicle.

BRIEF STATEMENT OF EXPECTED SOLUTION:

We envision a computer-based cost-accounting system that allows all direct and indirect costs of my department to be assigned to a fleet vehicle. The system would generate various daily, monthly, and quarterly costing reports, both for vehicles and cost categories.

Service is not excited about this project. Expect lukewarm reception, but know that I have authorized the study and will insist on their cooperation.

ACTION (To be completed by Steering Committee or Strategic Planning Committee)

☒ REQUEST APPROVED: | ASSIGNED TO: Wayne Tatlock
| START DATE: ASAP
| BUDGET: $30,000

☐ REQUEST DELAYED: | BACKLOGGED UNTIL: _____

☐ REQUEST REJECTED

Susan S. Bodkin   *ISS Steering Committee*   *August 15, 1993*
Signature   Representing   Date

LAST REVISED: October 1992

FIGURE 7.4 **Request for System Services** In some organizations a project request must be submitted on a standard request form. Notice that this form allows for both planned and unplanned projects.

About PEOPLE

- Who would be the end-users (*direct users*) of any system that we might build?
- Would anyone be indirectly affected by the system (*indirect users*)?
- Who developed the existing system?
- What people or political problems, opportunities, or directives triggered this project request?
- How will management and users feel about this project if it is approved for application development?
- How do the managers and users feel about working with computers and computer professionals?

About DATA

- What are the key inputs to this system?
- What are the key outputs from this system?
- Is any data currently being captured and stored in computer files and/or databases?

About ACTIVITIES

- What is the purpose or mission of this business area?
- What are the goals and objectives of this business area?
- Has any of this system been computerized (possibly by end-users with personal computers)?

About NETWORKS

- Will this project provide support for multiple locations? If so, where are they?
- How do the locations currently communicate?
- Are any computer networks currently in use?

FIGURE 7.5 Sample Questions for Survey Phase Interview These questions, organized around information system building blocks, are representative of those asked in a first interview for a project.

The principle technique and skill required to complete this activity is *interviewing*. Interviewing, a cross life cycle activity, is covered in Part Five, Module B, Fact-Finding Techniques.

Activity 2: Define Project Scope

Based on the facts from the initial interview(s), the systems analyst should attempt to define the initial scope of the proposed project. Absolute accuracy is neither essential nor possible! You'll soon see that we reevaluate scope several times during a project.

One of the easiest ways to define scope is to draw context model(s), simple pictures that reflect the boundaries and scope of the system. For example, Figure 7.6 is a **context diagram** that depicts the net inputs and outputs of a proposed system and their relationships with other business subsystems, computer systems, departments, people, and the outside world. Similar context pictures could be drawn for the PEOPLE, DATA, ACTIVITIES, and NETWORKS building blocks. Models are typically drawn by systems analysts, ideally with some help and verification by the system owner or user (who was interviewed).

Note   Figure 7.3 implies that a data administrator may also be involved in this activity. How so? In many businesses, the data administration group has defined (or is defining) a "corporatewide" data model. During the scoping activity for any project, the data administrator can help the systems analyst and system owner recognize and "carve out" that portion of the corporatewide data model that applies to a specific project. This data submodel serves as a point of departure for later data modeling activities.

FIGURE 7.6   Context Model   This diagram is a context or scoping model that defines the boundaries of a system project. In this case the model depicts the activity view of the system. The system is the rounded rectangle in the center of the diagram. The squares are other business subsystems, computer systems, departments, people, and organizations that are outside the boundary and scope of the project. The arrows depict important inputs and outputs between the system and its boundary.

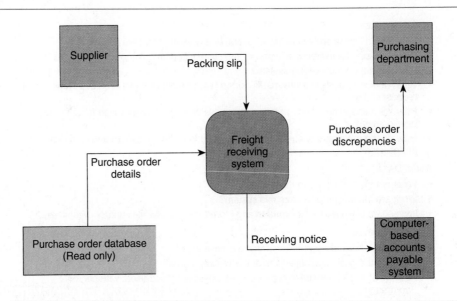

Some systems analysts also generate size estimates for a project. One popular technique requires the analyst to count **function points** based on projected numbers of inputs, outputs, files, and processes. These function point estimates can be compared to a history of previous projects to more accurately assess the project's size and complexity.

The following techniques and skills are applicable to this activity:

- Data scope modeling (see Chapter 8, Data Modeling).
- Process scope modeling (see Chapter 9, Process Modeling).
- Network scope modeling (see Chapter 10, Network Modeling).
- Function point analysis (see Part Five, Module A, Project Management).

Activity 3: Classify Problems, Opportunities, and Possible Solutions

Based on the initial interview facts, and possibly the proposed system models, the systems analyst and system owner/user quickly classify the problems, opportunities, and possible solutions. Figure 7.7 is a sample tool that documents an initial analysis of problems and opportunities, and possible solutions. Problems and opportunities could be classified in terms of:

- *Urgency.* In what time frame must/should the problem be solved or the opportunity exploited? A rating scale could be developed to answer this question.
- *Visibility.* To what degree would a solution to the problem or opportunity be visible to top management? Again, a rating scale could be developed to answer this question.
- *Annual benefits.* Approximately how much would a solution to this problem or opportunity increase revenues or reduce costs? If this question cannot be easily answered, then try this alternate question: How much would the system owner be willing to pay to solve this problem or exploit this opportunity?

Problem/Opportunity	Urgency?	Visibility?	Annual Benefits?	Priority?	Solution
1. Response time to bid on sporting events is excessive. We lose a lot of possible contracts.	Fix within six months	High	$250,000	2	New development
2. Number of potential events is growing faster than our ability to bid on those events. The opportunity to bid on additional, profitable events exists.	Fix within one year	Medium	$125,000	6	New development
3. Difficult to calculate estimated costs for a bid. If you underestimate a bid, you cannot charge the customer for excess costs.	Fix within three months	High	$50,000	4	Enhancement, then new development
4. There is no historical database on which to base future estimates.	Fix within six months	Low	$20,000	5	New development
5. We have recently purchased a competitor; however, we have since discovered fundamental incompatibilities between our respective event-scheduling data and systems.	Need immediate fix, if possible	High	$75,000	1	Quick fix, then new development
6. We have overbooked vehicles and equipment for events and subsequently incurred costly rental expenses to legally cover obligations.	Need within six months	High	$2,500	3	Leave well enough alone
7. We have occasionally booked events only to discover that we didn't have the "properly skilled" staff matched to the obligations.	Need within two years	Medium	$10,000	7	New development

FIGURE 7.7 **Problem/Opportunity Survey Matrix** This simple template allows analysts to document the classification of problems and opportunities. The completed form(s) would then be added to the project repository for future reference.

- *Priority.* Based on the answers to the above questions, rank the problems and opportunities (1 = Most important).
- *Solution.* For each problem or opportunity, the systems analyst will classify its solution as one or more of the following:
  - Leave well enough alone: Either (1) there is no problem, (2) the problem is not as serious as first perceived, (3) the problem would cost too much to fix, or (4) management is not sufficiently committed to a solution.
  - A quick fix: A solution that can be quickly implemented without going through the entire systems development life cycle. This may or may not require changing any application programs!
  - An enhancement: A solution that will build on existing application programs and procedures. This might involve adding new inputs or outputs to the existing computer programs.
  - New development: A solution that will replace the existing business and information system with new application programs and procedures.

Notice, in the example, that a solution may have two components—a temporary fix, followed by a more permanent solution (e.g., see problem #5).

There are no tangible skills required for this activity except for experience. For the young systems analyst, lack of experience can be compensated for by working with more experienced analysts.

Activity 4: Establish a Proposed Project Plan

If recommending either an "enhancement" or "new development," the systems analyst (or project manager) needs to develop an initial project plan (and, if necessary, revise the original size estimates). The word "initial" cannot be overstressed! At this stage in the life cycle, we know too little to accurately predict costs or schedules. An initial project plan should consist of the following:

• A first-draft master plan and schedule for completing the entire project. This schedule will be modified at the end of each phase of the project.

• A detailed plan and schedule for completing the next phase of the project (the study phase). In most cases this schedule will be more accurate, but still subject to a lack of detailed knowledge about the current system and user requirements.

The project plan is derived from experience plus the proposed system models and size estimates and possible solutions.

The principle technique and skill required to complete this activity is *project management*. Project management, another cross life cycle activity, is covered in Part Five, Module A, Project Management.

Activity 5: Present Survey Findings and Recommendations

The findings and recommendations of the systems analyst and system owner or user must usually be presented to a decision-making body for approval to continue the project into the study phase. In many businesses a steering committee makes this decision.

> A **steering committee** is a decision-making body that prioritizes potential information systems projects. Ideally, such a committee is dominated by business managers, but includes a few information systems managers.

The systems analyst and system owner or user assemble the appropriate supporting documentation from the repository. The resulting project feasibility assessment presentation may be written, verbal, or both. An outline for a typical written report is shown in Figure 7.8. The decision maker(s) will make one of the following decisions:

• Accept the recommendation of the systems analyst and system owner. The project would receive authorization to continue to the study phase.

• Amend the recommendation of the systems analyst and system owner. The amendment(s) may adjust scope or proposed solution. The amended project would receive authorization to continue to the study phase.

• Send the recommendation back to the systems analyst and system owner

I. Executive summary (maximum: 1 page).
   A. Summary of recommendation.
   B. Brief statement of anticipated benefits.
   C. Brief explanation of report contents.
II. Background information (maximum: 1–2 pages).
   A. Brief description of project request.
   B. Brief explanation of summary phase activities completed.
III. Findings (2–3 pages).
   A. Problems and analysis.
   B. Opportunities and analysis.
   C. Directives and implications.
IV. Detailed recommendation.
   A. Narrative recommendation (1 page).
      1. Quick fixes (as necessary).
      2. Enhancements (as necessary).
      3. New systems development (as necessary).
   B. Project scope (0–3 pages).
      1. Process context model (optional).
      2. Data context model (optional).
      3. Network context model (optional).
   C. Project plan (3–4 pages).
      1. Initial project objectives.
      2. Initial master plan and assumptions.
      3. Detailed plan for study phase.
   D. Appendices (if necessary).

FIGURE 7.8 An Outline for a Project Feasibility Assessment This outline is typical of a project feasibility assessment or initial study report. Different organizations and/or methodologies may impose different standards.

for revision. For example, the decision maker may suggest that the scope or problem solution be reconsidered or clarified.

- Reject the project due to either (1) lack of resources to develop the system, or (2) low priority compared to other recommended projects.

The techniques and skills needed to complete this activity are all cross life cycle skills:

- Feasibility assessment (see Part Five, Module C, Feasibility Analysis).
- Report writing (see Part Five, Module D, Interpersonal Skills).
- Verbal presentations (once again, see Part Five, Module D).

THE STUDY PHASE OF SYSTEMS ANALYSIS

You may recall the old saying, "Don't try to fix it unless you understand it." The next phase of a systems analysis is to study and analyze the current system. There is always a current system, regardless of whether or not it currently uses computers. The study phase provides the analyst with a more thorough understanding of problems, opportunities, and/or directives. The study phase answers the questions, "Are the problems really worth solving?" and "Is a new system really worth building?" In different methodologies it may be called the *detailed study* or *problem statement* phase.

Can you ever skip the study phase? Rarely! You almost always need some understanding of the current system. But there may be some reasons to accelerate the study phase. First, if the project was triggered by systems planning, the worthiness of the project is not in doubt—the study phase is reduced to understanding the current system, not analyzing it. Second, if the project was initiated by a directive (such as, "process financial aid applications according to new regulations that go into effect on July 1"), then worthiness is, once again, not in doubt.

Note   What if there is no current system? True, in most situations, there is at least a current non-computer-based system. But in one special case, there really is no current system. If the project goal is to create an applications package for resale (such as an inventory control package), then there may be no system to study. On the other hand, this special case has its own substitute study phase: Study the Current Market.

In this section we will examine the purpose of the study phase. Then we will, once again, walk through its activities, participants, inputs, outputs, and techniques. Techniques will be cross referenced to the chapters and modules that teach those techniques.

Building Blocks for the Study Phase

The fundamental objectives of the study phase are

- To understand the business environment of the system.
- To understand the underlying causes and effects of problems.
- To understand the benefits of exploiting opportunities.
- To understand the implications of noncompliance with directives.

The required level of understanding can be derived from your information system building blocks (Figure 7.9). Notice that, for most of the building blocks, it is possible to draw graphical models to verify the analyst's understanding of the users' business environment. This information represents the system users' view of the application, which is much more detailed than the system owner's view captured in the survey phase.

Several frameworks exist for analyzing the problems, opportunities, and directives. The circle inside the matrix illustrates the authors' favorite, James Wetherbe's **PIECES** framework. Each letter in the PIECES acronym represents a class of problems, opportunities, and directives:

> **P**erformance.
> **I**nformation.
> **E**conomics.
> **C**ontrol and security.
> **E**fficiency.
> **S**ervice.

Let's examine each of the PIECES categories in greater detail.

Performance Analysis

Performance problems occur when business activities are performed too slowly to achieve objectives. Performance opportunities occur when **performance analysis** uncovers a way to speed up a business task that is otherwise achieving objectives. A performance directive may occur if, for example, management

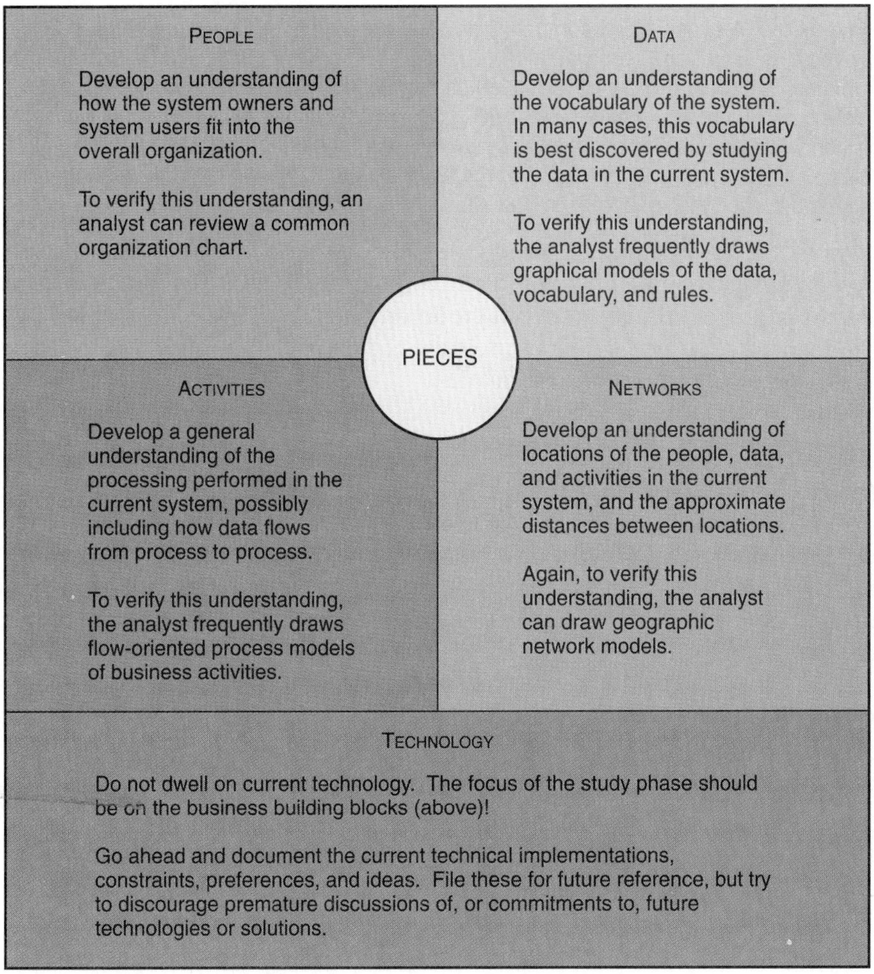

PEOPLE	DATA
Develop an understanding of how the system owners and system users fit into the overall organization. To verify this understanding, an analyst can review a common organization chart.	Develop an understanding of the vocabulary of the system. In many cases, this vocabulary is best discovered by studying the data in the current system. To verify this understanding, the analyst frequently draws graphical models of the data, vocabulary, and rules.

PIECES

ACTIVITIES	NETWORKS
Develop a general understanding of the processing performed in the current system, possibly including how data flows from process to process. To verify this understanding, the analyst frequently draws flow-oriented process models of business activities.	Develop an understanding of locations of the people, data, and activities in the current system, and the approximate distances between locations. Again, to verify this understanding, the analyst can draw geographic network models.

TECHNOLOGY

Do not dwell on current technology. The focus of the study phase should be on the business building blocks (above)!

Go ahead and document the current technical implementations, constraints, preferences, and ideas. File these for future reference, but try to discourage premature discussions of, or commitments to, future technologies or solutions.

FIGURE 7.9 **Study Phase Building Blocks** The study phase examines current information system building blocks to gain a general level understanding of the system and its problems and opportunities. The PIECES framework can be used to examine problems and opportunities.

decides that all transactions are to be done on-line to the computer. Performance is measured by throughput and response time.

Throughput is the amount of work performed over some period of time. Most throughput projects are concerned with transaction processing throughput.

Consider the following scenario.

Example A local credit union has been studying data about consumer loan applications. Over the past year, loan applications have increased 124 percent. The manager realizes that, if this growth rate continues, the current loan officers will not be able to keep up with the demand. The throughput of the current system must be increased.

Response time is the average delay between a transaction and a response to that transaction.

The following scenario illustrates the point.

Example   A construction company has been contracted to perform repairs and improvements for a large corporate site consisting of many buildings. The corporate site submits work orders to the construction company. The work orders go through a processing cycle that may include Information Systems, Purchasing, Accounting, Personnel, and Operations. Currently, an average delay of 62 days occurs between the submission of the order and the arrival of the work crew to fulfill the order. Management wants to reduce this response time as much as possible.

Although throughput and response time are considered separately in the preceding discussion, they should also be considered jointly. For instance, one way to improve throughput in our credit union example would be to improve the loan officers' average response time for each loan application by giving those officers timely credit information.

Information and Data Analysis

Information is a crucial commodity for management and end-users. The information system's ability to produce useful information should be evaluated for problems and opportunities. Improving information is not a matter of generating large volumes of information. On the contrary, *information overload* is a major problem in many businesses — one that is easily recognized by its piles of computer outputs!

Using **information** and **data analysis,** situations that call for information improvements might be detected, including:

- *Lack of any information concerning the decision or current situations.* For example, the Accounting Department suspects that air travel reimbursements do not reflect minimum costs and bargains that could be obtained. However, it has no information to support its suspicions; therefore, it cannot justify possible changes to its procedures.

- *Lack of relevant information concerning the decisions or current situations.* For example, a personnel manager must allocate scarce overtime dollars to the supervisors of three manufacturing departments. The report that predicts the amount of work to be done does not break the information down to the department level.

- *Information that is not in a form useful to management.* For example, an inventory control clerk for a large printing business must reorder paper and supplies each Monday. The clerk is given an inventory report. However, the report includes all 3,000 inventory items. The clerk has to compare quantity in stock and projected usage for each item on the report — just to identify items that need to be reordered. An exception report that has already made the comparison between supply and demand and that reports only those items needing to be reordered would be more convenient.

- *Lack of timely information.* Consider this example: A hotel chain allows customers to make reservations for any hotel in the chain from any other hotel in the chain. However, when a reservation is made or canceled, it takes three days to get that information to the hotel that is affected. Meanwhile, that hotel may overbook or underbook rooms because the information is not timely.

- *Too much information.* This was previously referred to as information overload. End-users frequently complain about receiving numerous, large reports that contain unnecessary data.
- *Inaccurate information.* Information contains errors that lead to bad decisions or other problems.

Information may also be the focus of a directive; the classic example of this is a new reporting requirement imposed by a local, state, or federal government agency.

Whereas information analysis examines the outputs of a system, data analysis examines the data stored within a system. Problems frequently encountered include the following:

- *Data redundancy.* The same data is captured and/or stored in multiple places. Data redundancy consumes valuable storage space and creates problems with data integrity. Consider, for instance, a manager who receives notification that he has eight registered copies of a word processor. These copies are eligible for a free upgrade to the new version. He sends in the order and is promptly notified that the software vendor cannot upgrade the copies because he is not legally registered. Obviously, the notification and order were processed across data files with inconsistent data.
- *Data inflexibility.* The data is captured and stored; however, it is organized such that certain reports and inquiries are impossible or difficult to support. Consider a large collection of data describing researchers who have had contact with radioactive isotopes. The files — held in numerous file cabinets — are organized by researcher. Now the government requests data on every researcher who has used a particular isotope. The request can only be fulfilled by going through all of the researchers' files one by one. (Did you also notice the performance problem in this scenario?)

Economic Analysis

Economics is perhaps the most common motivation for projects; **economic analysis** is therefore an important part of the study phase. The bottom line for most managers is dollars and cents. Economic problems and opportunities pertain to costs. Consider the following examples.

Example   The Marketing Department needs to establish the new prices for products in its catalog. In order to establish a price that will recover manufacturing costs and overhead and provide an acceptable profit margin, it needs a cost breakdown by product — including materials, direct labor, and overhead (for instance, utilities and plant maintenance). Although the department has access to budgeted cost standards, it needs historical data on actual costs because those costs may be exceeding the budgets.

Example   A purchasing manager has been ordered to reduce the costs of raw materials. There are two ways to reduce costs: first, by carefully comparing the alternative pricing structures offered by different suppliers; second, by taking advantage of bulk quantity discounts. However, it is necessary to strike a balance between reducing purchasing costs and increasing inventory costs. (Yes, it costs money to store and handle inventory!). An information system can greatly assist the decision-making process.

Control and Security Analysis

Business activities need to be monitored and corrected when substandard performance is detected. This is made possible by using **control** and **security analysis.** Controls are installed to improve systems performance, prevent or detect systems abuses or crime, and guarantee the security of data, information, and equipment. Two types of control situations trigger projects: systems with too few controls and systems with too many controls.

A system that has too few controls may result in discrepancies between the information system and the business system. The following example is typical.

Example A distribution warehouse for farm machinery parts is experiencing a stock problem. The computer information system is releasing orders after checking the inventory file to ensure that the products ordered are in stock. However, when the warehouse clerk tries to fill the order, the parts are not always in stock. An analysis reveals that, when the stock clerks place new inventory on the shelves, they do not count that inventory. They simply accept the supplier's word that the quantity shipped is accurate. This system has too few controls to ensure the accuracy of inventory counts.

On the other hand, it is possible to go overboard. A system with too many checks and balances slows the throughput of the system. The red tape associated with decision making in some firms causes long delays and other problems.

Society's greatest concern about the Information Revolution regards the privacy and security of data, information, and programs. Privacy problems center around who has read-access to data and information. It has become increasingly subject to government regulation. Many companies have become sensitive to the possibility of electronic espionage — incidents where one company electronically "steals" the ideas of another.

Security problems center around who has write-access to data. An increasingly common problem is the ability of unauthorized individuals to execute transactions. Stories of "hackers" who make free phone calls and execute transactions against other peoples' bank accounts are typical examples.

Efficiency Analysis

Efficiency is sometimes confused with economy. There is, however, a distinction between the two: whereas economy is concerned with the amount of resources used, efficiency is concerned with how those resources are used with minimum waste. Efficiency is defined as output divided by input; therefore, **efficiency analysis** detects opportunities to increase output, decrease input, or both. The commodities to be increased or decreased can be people, money, materials, or any other resource. The idea is to get more from less or at least to get more out of what you have.

Example A manufacturing facility consists of 125 workstations of various types. Different products go through different types of workstations during production. Management is concerned with the need to expand production, but there is no money to expand facilities. (Did you recognize the throughput nature of this problem?) Management has observed two major limitations in current operations. First, separate orders for the same product are not consolidated. This causes workstations to be set up and broken down for the same product several times each day. Second, management has noticed that some workstations seem to be idle during some parts of the day and overworked during other parts of the day. Obviously, the production scheduling and control system is not making efficient use of production workstations.

Service Analysis

Service improvements represent a diverse category. Projects triggered by service improvements seek to provide better service to the business, to its customers, or to both. Improved services are intended to increase the satisfaction of customers, employees, or management. Like our other categories, service improvements may be intended to solve specific service problems, exploit opportunities to improve service, or fulfill a management directive. **Service analysis** is used to improve one or more of the following:

- *Accuracy.* Accuracy is concerned with correctness of processing results. For instance, a company may want to reduce the number of billing errors on customer invoices. We've all heard stories about customers who get a $100,000 phone bill by mistake. Another example is order processing. Instead of sending the customer 10,000 pencils, a system sends 10,000 boxes of pencils, each box containing 100 pencils. Or instead of sending part A-4666-L-G (man's tweed sport coat, size 46 long, color gray), a customer receives D-4666-L-G (woman's dress, size 6, long sleeve, color gray).

- *Reliability.* No, reliability is not the same as accuracy. Whereas accuracy is concerned with *correctness,* reliability is concerned with *consistency* of processing and results. For example, an order-processing system may be denying credit to some customers but allowing credit to others with equivalent credit ratings and payment histories.

- *Ease of use.* Today, much concern is expressed about the *user friendliness* (or, as is more often the case, lack of user friendliness) of computer-based systems. A system, whether manual or computerized, should be as easy to use as possible. Many projects are initiated to improve the ease of use of computer systems. Similarly, many projects are initiated to overcome manual systems that have become too complex and awkward.

- *Flexibility.* Flexibility is concerned with a system's ability to handle exceptions to normal processing conditions. For instance, management may encounter a situation in which installment payments are promptly posted to customer accounts. However, prepayments or overpayments against accounts are sent through a lengthy process that delays customer orders even though those customers are not delinquent in their payments. The system isn't flexible enough to respond to special payment alternatives.

- *Coordination.* A business system consists of many functions that must coordinate their activities to achieve goals and objectives. The desired result is *synergy,* meaning that the whole organization receives a benefit greater than the sum of the parts. A classic example of coordination occurs between the production and inventory functions of a business. When production is scheduled, it is important that raw materials be delivered to the workstation at the scheduled time. However, the workstations may not have enough floor space to hold an entire day's requisition for raw materials. Clearly, an information system is needed to coordinate the flow of raw materials to workstations.

It should be obvious from the preceding discussion that the categories of the PIECES framework are related. Any given project can be characterized by one or more categories. Furthermore, the cause of any specific problem may be another problem itself. PIECES is a practical framework. As we study the systems analysis phases, you'll learn how to apply it.

Study Phase Activities, Roles, and Techniques

Note The project repository may be populated with various documentation and programs that have accumulated over time. This includes documentation and programs from the last time that the system went through the systems development life cycle.

Figure 7.10 illustrates the typical activities of the study phase. The study phase may be triggered by an authority to continue from survey phase, or by a planned application development project originating out of the systems planning phase. Both the survey and study phases focus on the users' "current system"; however the study phase is a much more in-depth investigation that may last several weeks.

Activity 1: Assign Project Roles

You've learned that systems analysts frequently work with project teams composed of a variety of information systems professionals. Given the planned application development project or authority to continue from the survey phase, the systems analyst, IS management, and system owner must work together to formulate the initial project team and its assigned project roles.

Most methodologies define roles. We defined relatively few roles in our diagrams; however, some methodologies define dozens of roles. Roles are not job titles. A single person may actually be assigned to serve many roles. Similarly, several individuals may serve the same role. (Users are classic examples of the latter.)

If you look closely at Figure 7.10, you'll see that system owners and users are "active" participants in activities. This is modern thinking! In the not-too-distant past, system owners and users played "passive" roles of mere information providers. The trend towards active participation of system owners and users is intended to increase their sense of responsibility and ownership in the final information system.

The IS management and system owners should jointly assign roles. IS management knows which individuals have the necessary systems development and technical skills. System owners know who the users and decision makers are. It is important to involve as many users as possible. Also, don't forget to invite indirect users — those representatives of departments who, although not direct users of the system, might be affected by the system.

Together, IS management and system owners should determine who the project manager will be. Some methodologies suggest project management responsibilities be shared by the lead systems analyst and the system owner or designee. There are no special skills or techniques for this activity.

Activity 2: Learn about Current System

Once individuals and role assignments for the project have been established, we can begin to learn more about the current system. Learning about the current system requires cooperation between the analysts, system owners, and system

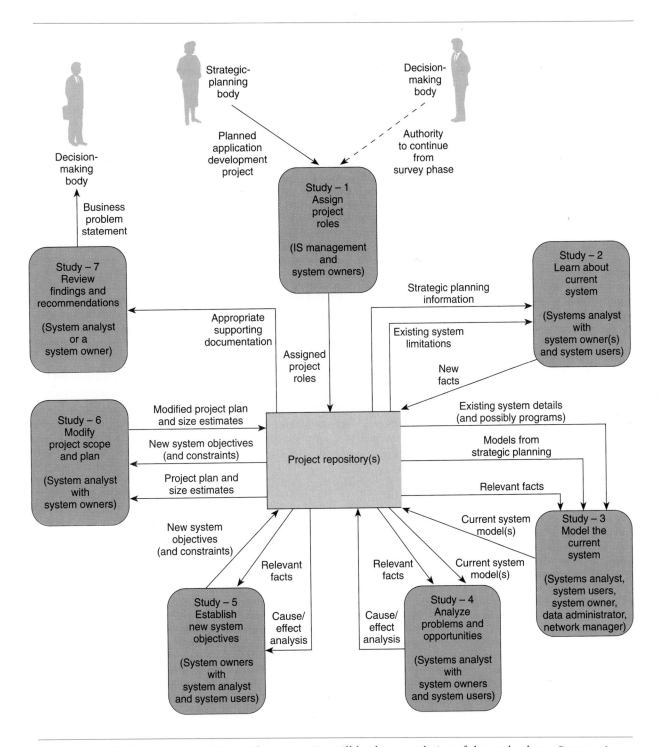

FIGURE 7.10 **Study Phase Activities** These activities will lead to completion of the study phase. Once again, the activities are generally completed in a clockwise sequence from the top. On the other hand, the project repository allows you to overlap activities or return to any previous activity to do rework.

users. Success is highly dependent upon system owners recognizing that their subordinates (system users) must be provided with the release time necessary to share with the analyst their knowledge about the current system. On the other hand, the analyst must recognize that the system users have their jobs to do and that their time is valuable.

The analyst will need to draw upon a number of alternative fact-finding techniques that can be used to minimize the necessary user-release time, while maximizing the amount of information obtained. To obtain facts, the analyst may use one or more fact-finding techniques, including:

- Interviewing.
- Meetings and group discussions.
- Research (for example, the repository).
- Observation.
- Sampling of files and forms.
- Surveys and questionnaires.

As shown in Figure 7.10, the analyst should always look for existing information in the repository. If the project was triggered by systems planning, the existing system was likely studied during the planning phases. Regardless of strategic planning, you are likely to find sources and information about the existing system limitations in some file.

Any new facts learned during interviews, discussions, and the like should be added to the repository. Almost all information, regardless of source, can be classified into your pyramid building block categories: PEOPLE, DATA, ACTIVITIES, NETWORKS, and TECHNOLOGY. Our goal will be to obtain a basic understanding of the system as a whole, while focusing most of our attention upon the problem and opportunity areas of the current system.

The principle techniques and skills required to complete this activity are interviewing, observations, sampling, and surveying. These techniques are covered in Part Five, Module B, Fact-Finding Techniques.

Activity 3: Model the Current System

How do you verify your understanding of the current system? One way is to draw models.

> A **model** is a representation of reality. In the classic tradition that "a picture is worth a thousand words," most models are graphical representations of reality.

The value of modeling in the study phase has come under some debate in recent years. Many early techniques, such as DeMarco's structured analysis, called for very detailed models of the current system. This consumed considerable time. Users and managers became frustrated with endless reviews of a system that they wanted to replace anyway! This problem has sometimes been referred to as *analysis paralysis.*

A second controversy centers around which models best aid the analyst in understanding the current system. Some experts advocate **process models,** such as *data flow diagrams.* Other experts insist that **data models,** such as

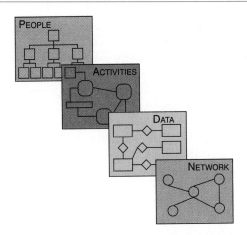

FIGURE 7.11 **Current System Models** Your information system building blocks can serve as a basis for drawing models that represent various aspects of the current system: PEOPLE, DATA, ACTIVITIES, and NETWORKS.

entity-relationship diagrams, are superior for learning business vocabulary and rules.

Best current practice and the pyramid model suggest that both data and process models, as well as people and network models, provide unique and equally valuable insight during the learning process. At the same time, current system models should be restricted to a general level of detail to avoid the aforementioned *analysis paralysis.*

Appropriate current system models (see Figure 7.11), may include:

- PEOPLE — Organization charts, although notoriously out-of-date, provide insight into the organization and management style in a business.
- ACTIVITIES — Today, it is widely accepted that a one- or two-page data flow diagram should prove sufficient to get a feel for the current system processing.
- DATA — A one-page data model diagram is useful for learning business vocabulary and rules.
- NETWORKS — A one- or two-page network model identifies the current locations of activities and data, and the approximate distances between those locations.

All the above models are developed by cooperation between the systems analyst, system users, and system owner. The data administrator and network manager may assist in the modeling of data and networks.

Inputs to this modeling activity may include relevant facts, previously drawn models from strategic planning, and existing system details. In the future, *reverse engineering* (described fully in Chapter 5) may offer the hope of automatically capturing existing system models directly from existing applications programs.

The following techniques and skills are applicable to this activity:

- Data modeling (see Chapter 8, Data Modeling).
- Process modeling (see Chapter 9, Process Modeling).
- Network modeling (see Chapter 10, Network Modeling).

Activity 4: Analyze Problems and Opportunities

Now that they know more about the current system, the analysts must work with system owners and system users to analyze relevant facts and analyze problems and opportunities. The analyst may use the current system model(s) to serve as a blueprint that identifies those aspects of the current system to be analyzed. (Note: Some problems and opportunities may be apparent on the models, but our experience suggests that many will not.)

You might be asking, "Weren't problems and opportunities identified earlier in the survey phase?" Yes, they were — in a sense. But as is frequently the case, initial problems may be only symptoms of other problems, perhaps not well known or understood by the users. Besides, we haven't really "analyzed" those problems in the classical sense.

Problem analysis is a difficult skill for beginning analysts. Experience suggests that most new analysts (and many more experienced analysts) try to solve problems without really analyzing them. They might state a problem like this: "We need to . . ." or "We want to . . ."

Do you see what's wrong with such an approach? They are stating the problem in terms of a solution. Successful problem solvers have learned to state the problem, not the solution. Then they analyze that problem for causes and effects.

The PIECES framework becomes most useful for **cause/effect analysis.** As you collect facts, note problems and limitations according to the PIECES categories. Remember, a single problem may be recorded into more than one category of PIECES. Also, don't restrict yourself to only those problems and limitations noted by end-users. As the analyst, *you* may also identify potential problems! Next, for each problem, limitation, or opportunity, ask yourself the following questions and record answers to them.

1. What is causing the problem? What situation has led to this problem? Understanding *why* is not as important. Many current systems were never "designed." They simply "evolved." It is usually pointless to dwell on history. In fact, you should be careful not to insult system owners and users who may have played a role in how things evolved.
2. What are the negative effects of the problem or failure to exploit the opportunity? Learn to be specific. Don't just say, "excessive costs." How excessive? You don't want to spend $20,000 to solve a $1,000 problem.
3. The effect sometimes identifies another problem. If so, repeat steps one and two.

Opportunities do not have causes. But they do have effects (benefits) that will eventually have to be weighed against the cost of implementing the opportunity.

This analysis technique results in a cause/effect analysis of this study phase activity and will be used to establish new and revised objectives for a new and improved system.

The principle techniques and skills required to complete this activity are interviewing, observations, sampling, surveying, and joint application design. These techniques are covered in Part Five, Module B, Fact-Finding Techniques.

Activity 5: Establish New System's Objectives

How can we determine the success of a systems development project? Success should be measured in terms of the degree to which objectives are met for the new system.

An **objective** is a measure of success. It is something that you expect to achieve, if given sufficient resources.

Objectives represent the first attempt to establish expectations for any new system.

Objectives should be precise, measurable statements of "business" performance that define the expectations for the new system. Some examples might include the following:

- Reduce the number of uncollectible customer accounts by 50 percent within the next year.
- Increase by 25 percent the number of loan applications that can be processed during an eight-hour shift.
- Decrease by 50 percent the time required to reschedule a production lot when a workstation malfunctions.

The following is an example of a poor objective: "Create a delinquent accounts report." This is a poor objective because it only states a requirement, not an actual objective. Now, let's reword that objective: "Reduce credit losses by 20 percent through earlier identification of delinquent accounts." This gives us more flexibility. Yes, the delinquent accounts report would work. But a customer delinquency inquiry might provide an even better way to achieve the same objective.

Objectives must be tempered by known constraints.

A **constraint** is something that will limit your flexibility in defining a solution to your objectives. Essentially, constraints cannot be changed.

Examples of constraints include:

- *Schedule:* The new system must be operational by April 15.
- *Cost:* The new system cannot cost more than $350,000.
- *Technology:* The new system must be on-line, or all new systems must use the DB2 database management system.
- *Policy:* The new system must use double-declining balance inventory techniques.

The output of this activity is a statement of new system objectives (and constraints). The relevant facts about the current system and the issues identified by the cause/effect analysis serve as inputs for establishing those objectives. Figure 7.12 suggests a template that might prove useful to record problem and opportunity analysis from Activity 4, and system objectives from Activity 5.

Recall that the PIECES framework was used earlier to identify problems and opportunities. Presumably, our new system objectives should address those problems and opportunities with questions such as, "What are the specific performance objectives?" and "What are the specific information and data objectives?" and so forth.

Problem/Opportunity	Causes and/or Effects	System Objectives	System Constraint
1. Problem: Response time to bid on sporting events is excessive. 2. Opportunity: Number of potential events is expected to grow faster than the company's ability to bid on those events. 3. Problem: Difficult to calculate costs for a bid. 4. Problem/opportunity: No historical database on which to base future estimates.	**Causes:** There is no historical database of actual costs incurred for any type of event. There is currently no way to estimate revenue from advertisers. Resource requirements and availability data are inconsistent and unreliable. All estimates are made manually, using a variety of individual methods; therefore, if one estimator gets behind schedule, he can't pass off the estimate to another estimator. **Effects:** The company loses $150,000 per year due to inability to bid within specified deadlines. The company also loses the opportunity to bid on approximately $125,000 worth of events per year. The company loses $60,000 per year in contracts because of its tendency to overbid on most events to avoid absorbing excess costs.	1. Standardize the method and rules for estimation by December 1, 1993. 2. Create appropriate databases for estimation factors (e.g., costs, revenues, and resources) by March 1, 1994. 3. Reduce time required for any estimate from two weeks to two days by May 1, 1994.	1. There is a maximum budget for the entire project (meaning all system objectives for all problems and opportunities) of $150,000.
5. Problem: Fundamental incompatibilities between the company's and recently acquired competitor's current event-scheduling data and systems.	**Causes:** Traditional business line has been "sporting events." Acquisition of Eventron, Inc., adds new types of events such as "the performing arts" and "political" events to the agenda. **Effects:** It is feared that the company's bidding and scheduling problems may infect the successful business as acquired. The loss potential exceeds $900,000 per year of standing or repeat contracts.	1. Standardize the method and rules for estimation by December 1, 1993. 2. Preserve existing databases for appropriate estimation factors (e.g., costs, revenues, and resources) by December 1, 1993. 3. Retain the current system's ability to bid on its unique events within five working days.	1. Many of Eventron's contracts cover the same event over several years. The system must retain the ability to work within defacto contract standards for certain types of events. 2. The new, merged system must run on the same Apple Macintosh computers and AppleTalk LAN.

FIGURE 7.12   A Problem/Opportunity/Objective/Constraint Matrix   This template provides a vehicle for recording and analyzing problems and opportunities, and recording new system objectives. The results would be added to the project repository.

There are no tangible skills required for this activity except for experience. For the young systems analyst, lack of experience can be compensated for by working with more experienced analysts.

Activity 6: Modify Project Scope and Plan

Recall that project scope is a moving target. Based upon our initial understanding and estimates from the survey phase, it may have grown or diminished in size and complexity. (Note: Growth is much more common!) Now that we're approaching the completion of the study phase, we should reevaluate project scope and revise the project plan accordingly. The systems analyst and system owner are the key individuals in this activity.

The analyst and system owner will consider the possibility that not all objectives may be met by the new system. Why? The new system may be larger than expected, and they may have to reduce the scope to meet a deadline. In this case the system owner will rank the objectives in order of importance. Then, if the scope must be reduced, the higher-priority objectives will tell the analyst what's most important.

To accomplish this activity, the systems analyst and system owners will refer to the project plan and size estimates established during the survey phase and the new system objectives (and constraints) identified in the previous study phase activity. The result of this activity is a modified project plan and size estimate for the project. The project plan will include revision to the overall project schedule and a detailed plan and schedule for completing the next phase of the project (the definition phase).

The following techniques and skills are applicable to this activity:

- Managing growing user expectations (see Part Five, Module A, Project Management).
- Project management (again, see Module A).

Activity 7: Review Findings and Recommendations

As with the survey phase, the study phase findings and recommendations should be reviewed. Two levels of review are possible.

- A **quality review** ensures that the phase was completed in accordance with the methodology and that resulting documentation (in the repository) is complete, consistent, and in compliance with information systems standards.
- A **feasibility review** reassesses the feasibility of continuing the project. The steering committee is rarely involved; however, the system owner or some higher level of management must be made aware of any unusual increase in possible costs or delayed schedules.

In both cases the systems analyst or system owner assembles the appropriate supporting documentation from the repository. The resulting business problem statement may be written as a report, presented verbally to management, or inspected by an auditor or peer group (called a **walkthrough**). An outline for a representative written report is shown in Figure 7.13.

FIGURE 7.13 **An Outline for a Business Problem Statement** This outline is typical of a project feasibility assessment or detailed study report. Different organizations and/or methodologies may impose different standards.

I. Executive summary (maximum: 2 pages).
   A. Summary of recommendation.
   B. Summary of problems, opportunities, and constraints.
   C. Brief statement of objectives for a new system.
   D. Brief explanation of report contents.
II. Description of study phase activities and methods (2–3 pages).
   A. Interviews/JAD sessions conducted.
   B. Samples collected.
   C. Surveys conducted.
   D. Observations performed.
III. Overview of current system (5–7 pages).
   A. Strategic planning implications (if appropriate).
   B. Models of current system.
      1. Organization chart.
      2. Data model(s).
      3. Process model(s).
      4. Network model(s).
IV. Cause/effect analysis of current system (5–10 pages).
   A. Performance analysis.
   B. Information and data analysis.
   C. Efficiency analysis.
   D. Control and security analysis.
   E. Economic analysis.
   F. Service analysis.
V. Detailed recommendations.
   A. Proposed objectives for new system (2–3 pages).
      1. Prioritized list of objectives.
      2. List of constraints.
   B. Proposed project scope (1–2 pages).
   C. Proposed project plan (2–4 pages).
      1. Overall project plan and schedule.
      2. Detailed plan and schedule for definition phase.
VI. Appendices (if necessary).

One of the following decisions must be made:

- Authorize the project to continue, as is, to the definition phase.
- Adjust the scope, cost, and/or schedule for the project and then continue to the definition phase.
- Cancel the project due to either (1) lack of resources to further develop the system, (2) realization that the problems and opportunities are simply not as important as anticipated, or (3) realization that the benefits of the new system are not likely to exceed the costs.

The techniques and skills needed to complete this activity are all cross life cycle skills:

- Feasibility assessment (see Part Five, Module C, Feasibility Analysis).
- Report writing (see Part Five, Module D, Interpersonal Skills).
- Walkthroughs (again, see Part Five, Module D).
- Verbal presentations (again, see Part Five, Module D).

Using JAD Techniques in the Study Phase

The study phase activities could be completed in the general sequence just described. The project repository approach does allow for some overlapping of the activities. Depending on the size of the system and number of users, the study phase might take anywhere from a couple of weeks to a couple of months. But there is a faster approach.

In Chapter 4, joint application development (JAD) was introduced as a technique for accelerating various phases of systems development.

> **Joint application development** is a highly structured workshop that brings together users, managers, and information systems specialists to jointly analyze, design, and/or prototype information systems.

JAD can reduce the time requirement for the study phase to approximately one week of preparation and a one- to three-day workshop.

The systems analyst would facilitate the JAD workshop, keeping it on schedule and mediating conflicts. The workshop attendees would be identified in Activity 1. The workshop would combine Activities 2 through 7. The systems analyst and systems owner may still follow up on the JAD workshop by completing Activity 7.

JAD techniques and skills are covered in Part Five, Module B, Fact-Finding Techniques.

THE DEFINITION PHASE OF SYSTEMS ANALYSIS

Many analysts make a critical mistake after completing the study of the current information system. The temptation at that point is to begin looking at alternative solutions, particularly computer solutions. The most frequently cited error in new information systems is illustrated in the statement, "Sure the system works, and it is technically impressive, but it just doesn't do what we wanted (or needed) it to do." Did you catch the key word? The key word is *what!* Analysts are frequently so preoccupied with the computer solution that they forget to define the business requirements for that solution. The definition phase answers the question, "What does the user need and want from a new system?" The definition phase is critical to the success of any new information system!

In different methodologies the definition phase might be called the *requirements analysis* or *logical design* phase. Some methodologies split the requirements definition into two phases. For example, Structured Solutions's STRADIS calls for a *draft requirements* phase and a *total requirements* phase, separated by a high-level design phase. That approach is based on the notion that some requirements, such as auditing and internal controls, only become apparent after some of the design decisions have been made.

Can you ever skip the definition phase? Absolutely not! New systems will always be evaluated, first and foremost, on whether or not they fulfilled business objectives and requirements—regardless of how impressive or complex the technological solution might be!

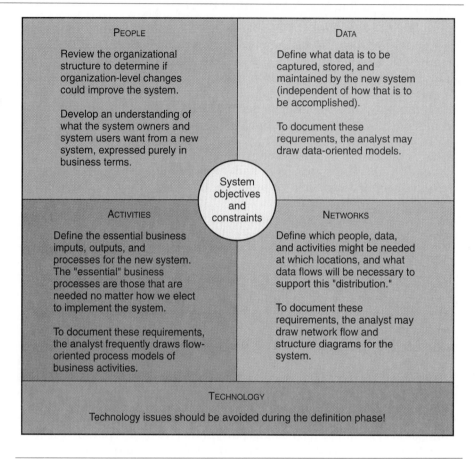

In this section we will examine the purpose of the definition phase. Then we will, once again, walk through its activities, participants, inputs, outputs, and techniques. Techniques will again be cross referenced to the chapters and modules that teach those techniques.

Building Blocks for the Definition Phase

The fundamental objectives of the definition phase are:

- To define business (nontechnical) requirements that address problems identified with the current system.
- To define business requirements that exploit opportunities identified with the current system.
- To define business requirements that fulfill directives.
- To offer system designers absolute flexibility with regard to upcoming design choices (such as centralized versus distributed, on-line versus batch, and so forth).

Here again, your information system building blocks (Figure 7.14) can serve as a useful framework for identifying what information systems requirements need to be defined. Notice that for most of the building blocks, it is possible to draw graphical models to document the requirements for a new and improved system. This information represents the system users' view of the new application. Notice that the TECHNOLOGY building block is de-emphasized during the definition phase.

Definition Phase Activities, Roles, and Techniques

Figure 7.15 illustrates the typical activities of the definition phase. Once again, let's examine the typical activities conducted during this phase, the individuals' roles in each activity, the inputs and outputs, and techniques that are commonly used to complete each activity.

Activity 1: Identify Requirements

This activity sounds simple, right? Unfortunately, few activities have frustrated analysts more! Requirements identification is triggered by an approval to continue from the study phase. The purpose of this activity is to solicit the requirements for a new application. This activity is performed jointly by the system users, system owners, and systems analyst. Requirements should be driven by the new system objectives (and constraints) identified during the Study phase and recorded in the repository. The resulting business and user requirements may also be recorded in the project repository, usually in a narrative format.

There are two commonly used strategies for getting the users to identify their business requirements for a new and improved system. The first approach is based on the classic *input-process-output* model. Users are asked to identify their inputs and outputs. Every input is traced forward to its resulting outputs. Similarly, every output is traced backwards to its original inputs. The tracing process reveals the necessary processes and files required in or by the new system. This is called the *process-driven* approach for requirements definition.

The second approach is based on the belief that inputs, outputs, and process requirements are too volatile. In other words, these requirements change too frequently to be useful. Stored data, on the other hand, is generally considered to be more stable over a longer period of time. This suggests an approach based on defining stored data requirements. Theoretically, if you capture and flexibly store the data that describes an application, you should be able to fulfill any current or future information need that would be derived from that data. This is called the *data-driven* approach for requirements definition.

Today most systems analysts use some combination of these two approaches to identify overall requirements. Various fact-finding techniques are used to solicit requirements using either approach. These fact-finding techniques, discussed in Part Five, Module B, Fact-Finding Techniques, include:

- Interviewing.
- Group meetings and discussions.
- Surveys (using questionnaires).
- Research (especially of software packages that address similar requirements).
- Brainwriting and brainstorming.

Identifying requirements is only half the challenge. We must also find a way to express the requirements such that they can be verified and communicated to both technical and nontechnical audiences. Technicians must understand re-

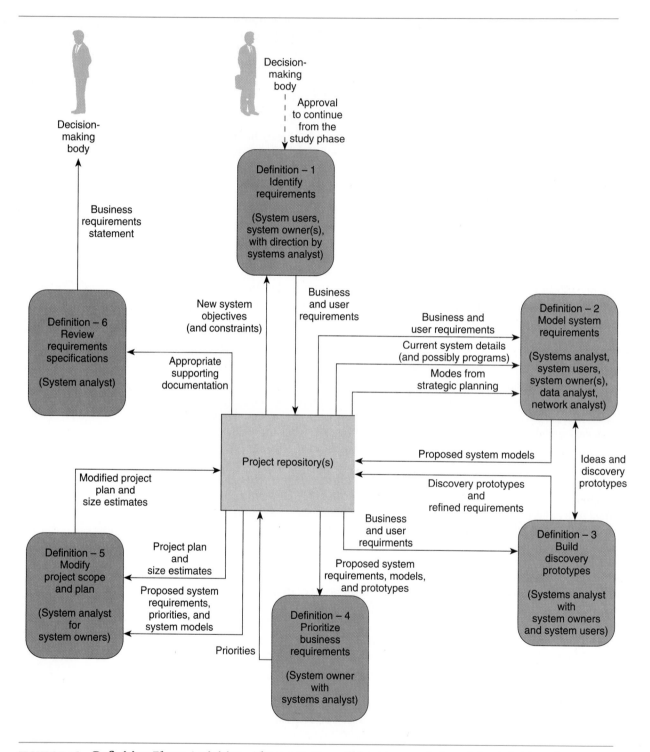

FIGURE 7.15 Definition Phase Activities   These activities will lead to completion of the definition phase. Once again, the activities are generally completed in a clockwise sequence from the top. On the other hand, the project repository allows you to overlap activities or return to any previous activity to do rework.

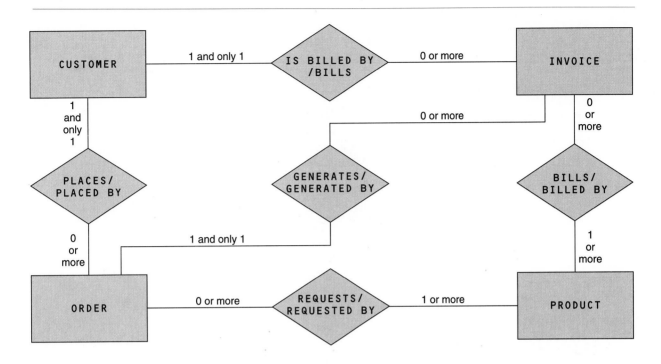

FIGURE 7.16 **Data Model Diagrams** Data model diagrams (also called entity-relationship diagrams) are a favorite tool for modeling the data requirements for a new system. The rectangles are "things" about which users need to store data. The diamonds and connections are associations or links between instances of different entities.

quirements so they can transform them into appropriate technical solutions. Nontechnicians must understand requirements so they can prioritize the needs and justify the expenditures for *any* technical solution.

Activity 2: Model System Requirements

Best current practice suggests that requirements be expressed in terms of system models. System models can be verified by users and analyzed by CASE tools and systems analysts. Once again, your pyramid model identifies the need for four requirements models:

- PEOPLE — Organization charts, carried forward and updated from the study phase.
- DATA — **Data model diagrams** (Figure 7.16) are used to model the data requirements for many new systems. These data models eventually serve as the starting point for designing files and databases.
- ACTIVITIES — **Data flow diagrams** (Figure 7.17) are frequently used to model the processing requirements for most new systems. These process models serve as a starting point for designing computer-based methods and application programs.
- NETWORKS — **Connectivity diagrams** (Figure 7.18) map the above people, data, and activities to geographical locations. These network models serve as a starting point for designing the communication systems for dis-

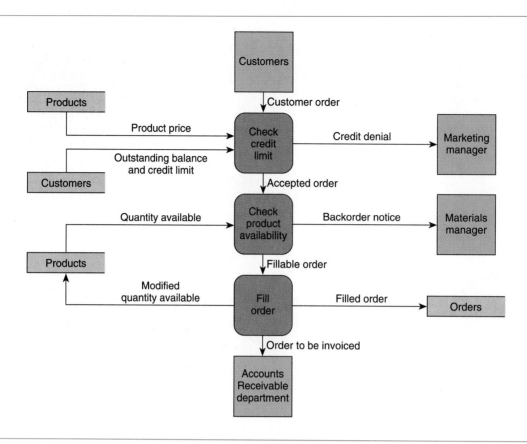

FIGURE 7.17   **Data Flow Diagrams**   Data flow diagrams are a favorite tool for documenting process or activity requirements for a new system. The rounded rectangles are processes or activities that transform inputs into outputs. The open-ended boxes are files or data stores. The squares are system boundaries.

tributing the data, activities, and people to the various geographical locations.

The best systems analysts can develop models that provide no hint of how the system will or might be implemented. There are no references to forms, files, terminals, databases, computer connections, or the like.

Some systems analysts understand how to draw the models, but have great difficulty asking the right questions to obtain the necessary facts to draw the models. For now, experience is the best teacher; however, The Next Generation feature for this chapter offers some hope for an easier future.

All of the system models are developed with cooperation between the systems analyst, system users, and systems owners. In many organizations a data analyst or administrator must be involved with the development of the data model, and a network analyst or manager must aid in the development of the network model. This ensures that databases and networks evolve according to a master strategy or plan.

The business and user requirements identified in the previous activity along with the current system details identified during the study phase are the inputs

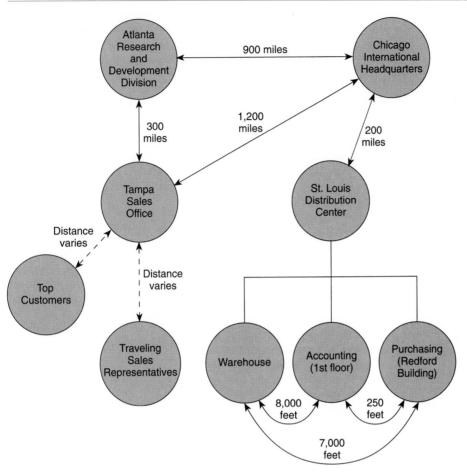

FIGURE 7.18 **Connectivity Diagrams** Connectivity diagrams are useful for depicting geographic locations where data and processing may become important. The circles are locations within the business. The squares are locations outside of the business.

used to develop the <u>proposed system models</u>. If the project was triggered by systems planning, <u>models from strategic planning</u> may also be used.

The following techniques and skills are applicable to this activity:

- Data modeling (see Chapter 9, Data Modeling).
- Process modeling (see Chapter 10, Process Modeling).
- Network modeling (see Chapter 11, Network Modeling).

Activity 3: Build Discovery Prototypes (If Necessary)

An alternative or complementary approach to translating and validating requirements is to build discovery prototypes. You may recall the following definitions:

> **Prototyping** is the act of building a small-scale, representative, or working model of the users' requirements for purposes of discovering or verifying the users' requirements.

The Next Generation

EXPERT TECHNOLOGY FOR SYSTEMS MODELING

Systems modeling is one of the most popular techniques for verifying user requirements. Unfortunately, many systems analysts find it difficult to ask the right questions of system owners and users to build those models. Some analysts have a knack for it, whereas others struggle.

One possible future solution is based on the use of expert systems technology (introduced in Chapter 2) in CASE tools for systems analysis. Two variations on this approach involve *templates* and *expert dialogues.*

Analysis Templates Some CASE market forecasters believe that analysis templates will soon become the next "hot" value-added or after-market products for CASE tools.

    Analysis templates are generic, reusable system models

for a specific industry and/or application.

Normally, system models for each business and application are built from scratch. But just as software vendors have learned to develop and market applications software that can support multiple businesses, CASE vendors and consultants are learning that the system models for different businesses in the same industry, or for similar applications, have much in common.

At a minimum, analysis templates will include data and process models. In the future, network models should also become common.

For example, the data and processing requirements for course scheduling in the average higher-education institution are probably more alike than different. A vendor or consultant could easily develop templates for this application for a

variety of popular CASE tools. The analyst and user start with the template and change the vocabulary and rules to fit their own organization.

Looking further ahead, the template could be packaged with its own computer-based customization procedure. For instance, the data model template for course scheduling may include an entity for SUBJECT. At a specific university, the customizer might allow the analyst to (1) rename the entity DEPARTMENT to reflect the local vocabulary, and (2) establish or change a rule to reflect that at this university, a department teaches one and only one subject.

Templates offer the promise of accelerating the system-modeling tasks through commonalities.

Analysis Expert Dialogues Looking still further ahead, what if we could capture the intelligence and exper-
continued

Discovery prototypes are simple mock-ups of screens and reports that are intended to help systems analysts "discover" requirements. The discovered requirements would normally be added to system models. A synonym is *requirements prototypes.*

Discovery prototyping is done primarily by the systems analyst working closely with system owners and system users. Although discovery prototyping is optional, it is frequently applied to systems development projects — especially in those cases when the parties involved are having difficulty developing or completing system models. The philosophy is that the users will recognize their requirements when they see them.

Prototyping requires at least some understanding of business and user requirements. System models may or may not be complete. The analyst uses powerful screen and report "painting" tools to quickly build sample screens and reports. The system owners and users then react to the prototype to help the analyst discover missing or misunderstood requirements. These discovery prototypes and refined requirements will either be recorded in the repository for

---

The Next Generation

concluded

tise of those analysts who have that "knack"? This is a classic *knowledge engineering* opportunity. Knowledge engineering is the preferred technique for building expert systems. When combined with a *natural language processor* (that understands English), this expert system might be called an *analysis expert dialogue*.

An **analysis expert dialogue** system would pose business-oriented questions to users and analysts. The answers would both: (1) lead to follow-up questions, and (2) generate or refine system models.

Consider the following dialogue (joined in progress):

System: By what term do you call a customer?
User: Customer.
System: Do you have any other customers?
User: No.
System: By any chance, are

there different types of customers?
User: Yes.
System: Please list customer types:
User: Franchise customers. Nonfranchise customers. Employees.
System: Employees were mentioned earlier in our dialog. Are these the same employees?
User: Yes.
System: Are franchise customers, nonfranchise customers, and employees described by exactly the same data attributes?
User: What is an attribute?
System: For example, "employee name" is an attribute of "employee." Are franchise customers, nonfranchise customers, and employees described

by exactly the same data attributes?
User: No.
System: Do you process transactions for customers?
User: Yes.
System: Please list all transactions.

This is how a typical dialogue would proceed. As the expert system analyzes user responses, it builds a data model based on those responses. Once a draft model is complete, the analyst and user can work together to refine the model.

Now combine the two major themes of this box feature. Tomorrow's innovative CASE vendor could combine expert dialogues with templates to further accelerate the model-building process. The expert dialogues would start with the template and customize the template based on the dialogue.

Of course, this technology is all somewhere in the next generation. . . .

---

later use, or immediately incorporated in the system models (note the arrow, new ideas and discovery prototypes, back to Activity 2).

It is important to emphasize that discovery prototypes are rarely intended to simulate the "look" of the final system. In fact, the danger of requirements prototyping is that users may: (1) become overly concerned with format of screens and reports, and/or (2) consider the form of the prototype to be the form of the final system.

The benefits of prototyping are numerous. End-users become more active participants in systems development. They tend to be more excited by working prototypes than design specifications on paper. Requirements definition is simplified through realization that many end-users will not understand or be able to state their requirements until they see a prototype. The likelihood that end-users will approve requirements documented on paper only to reject the subsequent design and implementation of that requirement is greatly reduced. Finally, design by prototyping is said to reduce development time; some experts, however, dispute the claim of savings.

This book does not teach prototyping skills, per se. If your course includes a laboratory component, you may be exposed to discovery-prototyping capabilities built into CASE tools (such as Excelerator), screen painters (such as Dan Bricklin's Demo), or a fourth-generation language/application generator (such as dBASE IV, Rbase, PowerBuilder, or PC/Focus).

Activity 4: Prioritize Business Requirements

We stated earlier that the success of a systems development project can be measured in terms of the degree to which business requirements are met. But actually, all requirements are not equal. If a project gets behind schedule or over budget, it may be useful to recognize which requirements are more important than others. Systems can be built in versions. Early versions should deliver the most important requirements.

Thus, given the proposed system requirements, models, and prototypes defined earlier, the systems owner and systems analyst should evaluate the business requirements according to the following criteria:

- Is the business requirement "mandatory"? In other words, is it something that the system owner *must* have? Careful! There is a temptation to label too many requirements as mandatory. By definition, if a system does not include a mandatory requirement, it cannot fulfill its mission — at all!

  Perform the following test on any suspected mandatory requirements: Rank them. If you can rank them, they are *not* mandatory. Why? Because you should not be able to rank requirements that you absolutely must have! All mandatory requirements are essential to the first version of the system!

- Is the business requirement "desirable" but not essential? Desirable requirements are things the user eventually wants; however, the early versions of the system can provide value without them. Unlike mandatory requirements, desirable requirements can and should be ranked.

- Is the business requirement "optional"? This is a catch-all category for those features and capabilities that you could live without indefinitely. Although these would be "nice to have," they are not really requirements.

The priorities resulting from this analysis will allow the system owner and systems analyst to make intelligent decisions if cost and schedule become constrained. Systems analysts should keep in mind that the system is for the user! Therefore, the actual priorities should be specified jointly by the system owners and users.

There are no tangible skills required for this activity except for experience. For young systems analysts, the lack of experience can be compensated for by working with more experienced analysts.

Activity 5: Modify Project Plan and Scope

Here again, recall that project scope is a moving target. Now that we're approaching the completion of the definition phase, we should step back and redefine our understanding of the project scope and adapt our project plan accordingly. The systems analyst and system owner are the key individuals in this activity.

The analyst and system owner will consider the possibility that the new system may be larger than expected, and they may have to reduce the scope or definitions to meet a deadline. In this case the systems analyst and system owner will consider the prioritized business requirements from the previous activity and adjust the scope accordingly.

To accomplish this activity, the systems analyst and system owners will refer to the project plan and size estimates that were last revised during the study phase and the proposed system requirements, priorities, and system models identified in the previous definition phase activities. The result of this activity is a modified project plan and size estimates for the project. Here again, the overall project schedule may be adjusted and a detailed schedule for completing the next phase would be developed.

The following techniques and skills are applicable to this activity:

- Function point analysis (see Part Five, Module A, Project Management).
- Project management (again, see Module A).

Activity 6: Review Requirements Specifications

As with the survey and study phases, the results of the definition phase should be reviewed. Two levels of review are possible.

- A **quality review** ensures that the phase was completed in accordance with the methodology and that resulting requirements and documentation (in the repository) are complete, consistent, and in compliance with information systems standards.
- A **feasibility review** reassesses the feasibility of continuing the project. Upon completion of the definition phase, you have a much better understanding of what the owners' and users' expectations are. Frequently, scope has increased to the point that schedules and budgets are in jeopardy. A steering committee is rarely involved; however, the system owner or some higher level of management must be made aware of any unusual increase in possible costs or delayed schedules.

In both cases, the systems analyst or system owner assembles the appropriate supporting documentation from the repository. The resulting business requirements statement may be written as a report, presented verbally to management, or inspected by an auditor or peer group during a walkthrough. Normally, the total requirements statement is too large for a single review. Instead, different reviews focus on different subsets of the total documentation. CASE tool repositories are increasingly used to store the models, prototypes, and details; however, most analysts also keep a printed copy of the documentation. This document might be organized as shown in Figure 7.19.

The business requirements statement is the major deliverable of systems analysis. Pending its reviews, one of the following decisions must be made:

- Authorize the project to continue, as is, to the systems design phases.
- Adjust the scope, cost, and/or schedule for the project and then continue to the systems design phases.
- Cancel the project due to lack of resources to further develop the system.

The techniques and skills needed to complete this activity are all cross life cycle skills:

- Feasibility assessment (see Part Five, Module C, Feasibility Analysis).
- Report writing (see Part Five, Module D, Interpersonal Skills).

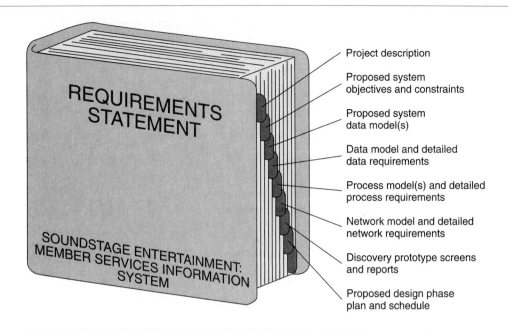

Project description

Proposed system objectives and constraints

Proposed system data model(s)

Data model and detailed data requirements

Process model(s) and detailed process requirements

Network model and detailed network requirements

Discovery prototype screens and reports

Proposed design phase plan and schedule

FIGURE 7.19 **A Hard Copy Requirements Statement**   Requirements statements can be quite large. Reviews and walkthroughs normally focus on subsets of the total documentation. The full documentation is frequently kept in a hard copy binder. (Note: This documentation is frequently produced from a CASE tool repository.)

- Walkthroughs (again, see Part Five, Module D).
- Verbal presentations (again, see Part Five, Module D).

Using JAD Techniques in the Definition Phase

Joint application development techniques are increasingly being used to accelerate the definition phase, even in cases when JAD wasn't used in prior phases. The application of JAD to the definition phase is sometimes called *joint requirements planning (JRP)*. Depending on the size of the system and number of users, the definition phase can be reduced from a couple months to a few days!

The systems analyst facilitates the JAD workshop, keeping it on schedule and mediating conflicts. Conflicts are especially common when trying to identify and prioritize user requirements. The analyst must be careful to balance representation of end-user and management interests in order to properly resolve conflicts.

The workshop would combine Activities 1 through 6. JAD techniques and skills are covered in Part Five, Module B, Fact Finding Techniques.

──────────── *Summary* ────────────

Systems analysis is problem solving. More specifically, it is the study of a current business and information system application and the definition of user requirements and priorities for a new and improved system.

Systems analysis is driven by system users' concerns and business issues—not technical or implementation concerns. It consists of three distinct phases that address the PEOPLE, DATA, ACTIVITIES, and NETWORKS building blocks from the user's perspective.

The three phases are *survey the project feasibility* (or the survey phase), *study the current system* (or the study phase), and *define and prioritize users' requirements* (or the definition phase).

The purpose of the survey phase is to determine the worthiness of a project request. In order to do so, the analyst needs to obtain a general level of understanding about the current system. The objectives will be to: (1) identify problems, opportunities, and/or directives that initiated the project request; and (2) determine whether solving the problems, exploiting the opportunities, and/or satisfying the directives will benefit the business.

To accomplish the survey phase objectives, the systems analyst will work with the system owner, system users, IS manager, and other IS staff to: (1) conduct an initial interview; (2) define the project scope; (3) classify problems, opportunities, and solutions; (4) establish a proposed project plan; and (5) present the survey findings and recommendations. The deliverable for the survey phase is an oral or written project feasibility assessment that must go to a decision-making body, commonly referred to as the *steering committee*. A steering committee is a decision-making body that prioritizes potential information systems projects. Not all projects require this evaluation. For example, projects initiated by formal systems planning have already been justified by the planning team and management. In such cases the project skips directly to the study phase.

The study phase is triggered by a steering committee's approval to continue from the survey phase, or by an applications development project that originated from systems planning. The purpose of the study phase is to answer the questions "Are the problems really worth solving?" and "Is a new system really worth building?" To answer these questions, the objectives of the study phase will be to: (1) understand the business environment of the system; (2) understand the underlying causes and effects of problems; (3) understand the benefits of exploiting opportunities; and (4) understand the implications of noncompliance with directives. James Wetherbe's PIECES framework is particularly useful in identifying and analyzing problems, opportunities, and directives. PIECES is an acronym that represents a class of problems, opportunities, and directives: *P*erformance, *I*nformation, *E*conomics, *C*ontrol (and security), *E*fficiency, and *S*ervice.

To complete the study phase, the analyst will continue to work with the system owner, system users, and other IS management and staff. The systems analyst and appropriate participants will: (1) assign project roles, (2) learn about the current system, (3) model the current system, (4) analyze problems and opportunities, (5) establish the new system's objectives, (6) modify the project scope and plan, and (7) review the findings and recommendations. The end product of the study phase is called the business problem statement. Consistent with the creeping-commitment approach to systems development, this deliverable will be reviewed by a decision-making body (steering committee) to determine whether or not to continue to the definition phase.

The definition phase is critical to the success of any new information system. The purpose of the definition phase is to identify what the new system is to do without the consideration of technology — in other words, to define the business requirements for a new system. The objectives of the definition phase are to: (1) define business requirements that address problems identified within the current system; (2) define business requirements that exploit opportunities identified within the current system; (3) define business requirements that fulfill directives; and (4) offer systems designers absolute flexibility with regard to upcoming design choices.

As in the survey and study phases, the analyst actively works with system users and owners as well as other IS professionals to complete the definition phase. To complete the definition phase, the analyst and appropriate participants will (1) identify requirements; (2) model the system requirements; (3) optionally, build discovery prototypes; (4) prioritize the business requirements; (5) modify the project plan and scope; and (6) review requirements specifications. The end-product of the definition phase is called a business requirements statement. The output is typically a written report that is reviewed by management, auditors, and peer groups. Pending this review, a decision is made as to whether the project should be canceled or should proceed to the design phase.

Key Terms

analysis expert dialogues, p. 286

analysis templates, p. 286

cause/effect analysis, p. 274

connectivity diagrams, p. 283

constraint, p. 275

context diagram, p. 259

control analysis, p. 268

data analysis, p. 266

data flow diagrams, p. 283

data model diagrams, p. 283

data models, p. 272

discovery prototypes, p. 286

economic analysis, p. 267

efficiency analysis, p. 268

entity relationship diagram, p 273

feasibility review, p. 277

function points, p. 260

information analysis, p. 266

joint application development (JAD), pp. 279, 290

model, p. 272

objective, p. 275

performance analysis, p. 264

PIECES, p. 264

process models, p. 272

project repository, p. 253

prototyping, p. 285

quality review, p. 277

response time, p. 265

security analysis, p. 268

service analysis, p. 269

steering committee, p. 262

systems analysis, p. 253

throughput, p. 265

walkthrough, p. 277

Questions, Problems, and Exercises

1. Identify and briefly describe the purpose of the three systems analysis phases.
2. What is a project repository and what role does it play during systems analysis?
3. What important question is addressed during the survey phase? What are the fundamental objectives of the survey phase? How might the information systems building blocks be used to identify the general level of understanding required for fulfilling these objectives?
4. What triggers the survey phase? Do all systems projects go through a survey phase? Explain your answer.
5. Differentiate between problems, opportunities, and directives. Give several examples of each.
6. Characterize each of the following situations as problems, opportunities, or directives. Remember, problems and opportunities are sometimes hard to distinguish. In those cases, describe how the situation might be either a problem or an opportunity.
   a. Management has decided to offer credit to regular customers. Previously, all sales required prepayment.
   b. A bank's manager wonders if bank machine transactions executed at other banks can be posted to customer accounts in fewer than the current average of two days. If so, the bank could save several thousand dollars' worth of interest payments each day.
   c. A baseball manager's competitive edge might be improved by access to information about a hitter's history against specific pitchers.
   d. When a manufacturing workstation breaks down, management finds it difficult to modify the production schedule to reassign work from the broken workstation to workstations that might have unscheduled capacity. Thus, products are not produced at the desired rate.
   e. Management has decided that all computer files should be integrated into databases to improve access to and flexibility of data.

   f. Total cash from customer payments does not equal the sum of payments
       posted against customer accounts.

7. What is a steering committee? Who typically may serve on a steering committee?
    At what points (activities) during systems analysis is this committee involved?
    What decisions might they make at each of these checkpoints?

8. Differentiate between the survey and study phases of the systems analysis process.

9. What are the fundamental objectives of the study phase? Explain how the infor-
    mation systems building blocks aid in identifying the required level of under-
    standing to be obtained in the study phase.

10. Using the PIECES framework, evaluate your local course registration system. Do
    you see problems or opportunities? (*Alternative:* Substitute any system with
    which you are familiar.)

11. Evaluate the following situations according to the PIECES framework. Do not be
    concerned that you are unfamiliar with the application. That isn't unusual for any
    systems analyst. Use the PIECES framework to think of potential problems you
    would ask the end-user about.

    a. The investments officer for ABC Co. has a problem. Currently, its bank con-
        tacts ABC's accounting department each day to relay information regarding
        deposits, check clearing, current and legal bank account balances, and float
        on recently issued checks. Accounting notifies the investments officer, who
        does a cash flow projection. Using pencil and multicolumn ledger sheets,
        this projection takes a couple of hours to complete. These projections are
        used to estimate capital available for investment in stocks, bonds, futures, and
        real estate, among other things. Investments are made on a daily basis. The
        trouble is that the company is too dependent on the bank clerk phoning in
        the information. If the cash projections go too slowly, valuable time is lost
        (the stock market opens early in the morning, but ABC is rarely ready to
        make investments at that time). Considerable capital can be made or lost in
        less than one hour.

    b. National Fund-Raising, Inc., an independent fund-raiser for various nonprofit
        clients, has uncovered a major problem with its data. It keeps considerable
        redundant data on past donors and prospective donors. This results in multi-
        ple contacts for the same donor—and people don't like to be asked to give
        to a single annual fund multiple times! Contacts with donors and follow-ups
        are not well coordinated. Whereas some prospective donors are contacted
        too often, others are overlooked entirely. It is currently impossible to gener-
        ate lists of prospective donors based on specific criteria, despite the fact that
        data on hundreds of criteria has been collected and stored. Donor histories
        are nonexistent, which makes it impossible to establish contribution patterns
        that would help various fund-raising campaigns.

12. Apply the PIECES framework to each of the following situations. Remember, the
    categories of PIECES overlap; therefore, any given situation may have implica-
    tions in more than one category.

    a. During the processing of room assignment applications, much of the same
        data is typed onto different forms at different times.

    b. A sales manager knows total sales for each region but can't tell how well each
        product line is doing in each region.

    c. A new record-keeping system allows students to see their test and project
        scores for any course; however, the system also allows a student to see all of
        the scores for any student in the same course.

    d. The cost of manufacturing a specific product has dramatically reduced that
        product's profit margin.

    e. Warranty claims have increased because defective parts are being used in
        products.

    f. Manufacturing produces too much of some products and not enough of other products.

    g. The chief accountant's office must consolidate the accounting statements for several years, calculating a percentage change (plus or minus) for each item on the statements.

    h. The cost-accounting office needs to determine what percentage of a product's increasing costs are attributed to higher pay scales (which would be out of management's control), and what percentage to lower labor productivity (which management can control).

    i. The marketing department needs to improve the sales staff's ability to make changes to orders that haven't already been shipped.

13. What important question is addressed during the definition phase? What are the fundamental objectives of the definition phase? How might the information systems building blocks be used to identify what requirements must be specified to fulfill these objectives?

14. What is the value of modeling during the systems analysis phases? What types of models might be developed during the survey, study, and definition phases?

15. Explain how the joint application design (JAD) technique can be used during the study and definition phases.

16. What skills are important for a systems analyst to be able to successfully perform systems analysis? How are these skills used in each phase of the systems analysis process?

17. What is the end product of each systems analysis phase? Explain the purpose and content of each of these products.

Projects and Minicases

1. A company is considering awarding your consulting firm a contract to develop a new and improved system. But, at the beginning, they only want to commit to systems analysis. The firm is concerned about your ability to understand its problems and needs. And most of all, it wants to see what kind of computer-based solutions you propose before it contracts you to design and implement a new system. Write a letter of proposal that will address their concerns.

2. You have been a systems analyst with your current employer for the past five years. During this time, the projects that you were assigned came directly from your IS manager. All of the projects were originally submitted to the IS manager upon formal request forms by various users within the company. But now things are different! Several months ago, the company hired a consulting firm to help them develop a strategic plan for the business. How might the resulting strategic plan affect the IS manager in determining which future projects will receive commitment of IS resources? Explain how a systems plan might affect the phases and activities you perform during systems analysis.

3. You have recently completed the survey phase for an assigned systems project. You have become concerned with your ability to complete the study of the current system and the definition of new system requirements within a reasonable time frame. The project is particularly challenging given that it involves numerous users, having differing vocabularies and system perspectives, who are located across several departments. What strategy would you use to reduce the amount of time required to complete the study and definition phases? How would this strategy deal with the issues presented by the diverse user community?

Suggested Reading

Wetherbe, James. *Systems Analysis and Design: Traditional, Structured, and Advanced Concepts and Techniques.* 2nd ed. St. Paul, Minn.: West, 1984. We are indebted to Dr. Wetherbe for the PIECES framework.

Requirements Definition for the New System
(Part I)

Sandra and Bob define the business requirements for a new and improved system: "What are the business and user requirements for the new system?"

Identifying Requirements for the New System

We begin this episode shortly after SoundStage Entertainment Club's Steering Committee approved Sandra and Bob's study phase report. The committee agreed that the projected benefits of solving SoundStage's member services problems exceed the projected costs for continuing the project. We join David Hensley, manager of Customer Services, in a meeting with Sandra and Bob.

"Hi, Bob. I really enjoyed our basketball game yesterday. I hope your ankle is getting better. What are we supposed to accomplish today?" David asked.

Bob took a seat. "Today, I'd like for us to agree on objectives for the new system. The objectives we come up with shouldn't have anything to do with computer solutions or specific reports the system should produce. Instead, we want to establish a set of business objectives that will serve as criteria for evaluating alternative solutions. Keep in mind that we want to be sure to establish objec-

tives that will address the problems and opportunities we uncovered with the current system. An example of such an objective might be that we want to respond to a membership application request within five days. Notice that this objective addresses the problem we identified with the current system's unacceptably slow turnaround time for processing membership requests. In fact, this objective also addresses an opportunity. Though we may consider the current level of customer satisfaction to be acceptable, the fulfillment of this objective will likely have a positive influence on the customer. Do you understand?"

"Sure, I understand. But I really wanted to talk about the computer. I had some good ideas about what we might be able to accomplish using the computer," David replied with disappointment.

"We will, but first we should concentrate on what we want the new system to accomplish, rather than solutions or ways it might be accomplished. Sandra and I believe that the business problem should define the solution, not the other way around. Thus, we'll be sure to end up with a system that best supports your needs, regardless of the way it is actually implemented. Now, how about some business objectives?"

"Well, Bob, I think decreasing the amount of time to process a member's order would be the most important objective."

Bob replied, "I'd like to get a little more specific than that. Let's try to quantify the objective. Currently the system can handle 300 orders a day. I know you want to increase that capacity, but by how much?"

"Okay, I see what you mean. I'd like to be able to process at least 700 orders per day. In fact, management projects that their new promotional campaign will stimulate growth that could soon double the current load."

"That's the kind of information I need," encouraged Bob. "Now we have a precise, measurable target. Any new system we consider or implement can be evaluated on the basis of it being able to process 700 or more orders per day."

"I get it. It's sort of like management by objectives. Our management does this with employees. They establish measurable objectives against which they can evaluate the employees' performance. And of course in establishing the objectives, they defer discussion of how those objectives are to be achieved," explained David as Bob and Sandra smiled and nodded in agreement.

"Okay, how about this one. Some managerial directives have been established that will result in an increase in our membership by 20 percent next year. It currently takes up to three hours for the mail clerk to sort new membership applications into advertisement applications and referral applications. We simply can't afford that kind of time delay. We must be able to process at least 100 membership applications per day."

"That's good!" replied Bob. "I was aware of those management directives being passed down, but they had totally slipped my mind. We'll make sure the new system takes them into account."

"Great! I've got some more. How about . . ."

[We can leave this meeting now. The discussion will continue until all objectives for the new system have been identified.]

Modeling the New System Requirements

Two weeks have passed. Sandra and Bob have been working on the business requirements statement for the new member services system. They have determined the general systems requirements and documented them with diagrams. We join Bob as he is leading a walkthrough to verify his understanding of the new system's data storage requirements with the Member Services staff. Sandra, Bob, David Hensley, the Member Services

manager, Ann Martinelli, the Membership director, Sally Hoover, the director of Order Processing, and Joe Bosley, the Promotions director, are also at the meeting.

Bob dims the lights and places a slide on the overhead projector. "I caution you not to be overwhelmed by this diagram (see Figure E3.1). This diagram is simply a pictorial representation of the data storage requirements for the new system. The purpose of this picture is to illustrate the types of things — we'll call them entities — about which the new system must store data and the relationships that may exist between those entities."

"In other words, this is a picture of the computer files that the new system will maintain," interrupts Joe.

"Not necessarily," Bob answers. "Let's stop thinking of files, both computer files and manual files. Rather, we simply want to concentrate on identifying all entities about which the new system is to store data, independent of how that data is physically stored now or how it might be stored for the new system. In the next phase of the project, we will evaluate different ways to implement the data."

Noting a general consensus of understanding and agreement, Bob points to a rectangular symbol on the slide and begins. "The rectangles on this diagram represent those things or entities about which the new system is to store data. For example, the new system will store data that describes MEMBERS, ORDERS, PROMOTIONS, and so on. Recall that Sandra or I met previously with each of you to help you identify those entities you desired the new system to store data about. Our goal is to confirm your data storage requirements one last time. Now, are there any additional entities you feel we may have omitted or overlooked?"

"What about SUPPLIERS?" asks Sally. "We'd like to be able to determine the supplier for a particular PRODUCT."

"Whoops!" interrupts Sandra. "Bob, I forgot to add it to the diagram."

"No problem; would you add that to the figure? Don't forget to draw any relationships that may exist between it and any other entities." Bob turned his attention back to the end-users. "That's good. Are there any others you feel are missing?"

Ann speaks up, "I can't think of any. You've included all those entities Sandra and I identified last week when we were discussing my people's specific data storage requirements. I assume some

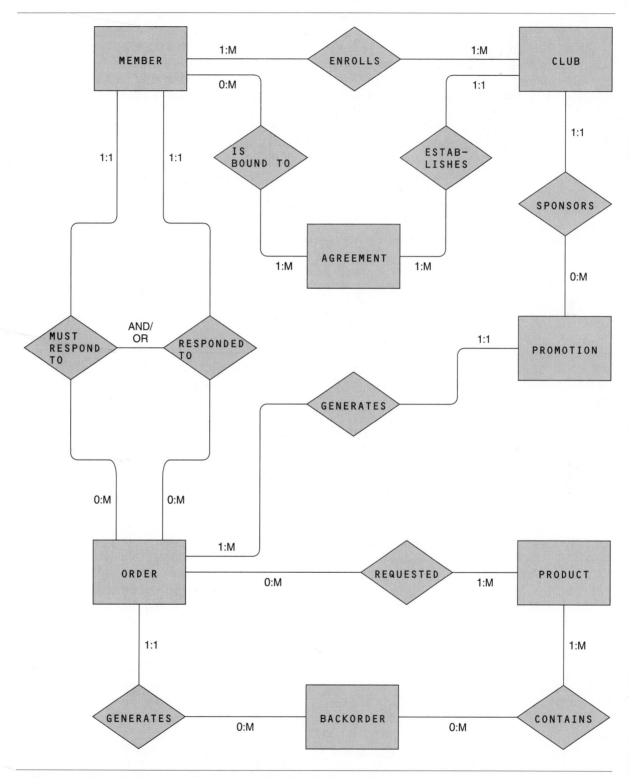

FIGURE E3.1   Data Storage Requirements Model for the New Information System

of these other entities are of interest to the other individuals in this room?"

"That's correct, Ann," replies Bob, "this diagram identifies data storage requirements for the entire Member Services system."

Pausing for a few moments and noting no additional feedback from the end-users, Bob decides to move on. "Ok, let's shift our attention toward identifying and confirming the relationships between these entities. The diamond shapes represent the relationship between groups of entities. For example, a MEMBER is BOUND TO an AGREEMENT and an AGREEMENT BINDS a MEMBER."

"Why are there two relationships between ORDER and MEMBER?" Sally asks.

"Let me answer that question, Bob," Sandra interrupts. "Sally, recall that earlier you explained to me that SoundStage generates automatic orders that members must respond to. The relationship on the left identifies a given MEMBER and the automatic ORDER they MUST RESPOND TO. Or vice versa, it identifies for a particular automatic ORDER which MEMBER the order was placed for."

"Ok. I understand that much," offers Sally.

"Good. Well, essentially the other diamond symbol represents a different type of relationship between an ORDER and MEMBER. For example, it represents ORDERs that the MEMBER MUST RESPOND TO. Of course, that order may have been the original automatic order or some variation."

"Isn't there a relationship between a PROMOTION and a PRODUCT?" asks Joe. "When a particticular CLUB SPONSORS a monthly PROMOTION, that PROMOTION is always for a particular PRODUCT."

"That's very good, Joe. I missed that relationship," Bob admitts as he notes the change.

"What's this 1 : M, 0 : M, and 1 : 1 stuff appearing on the line that connects the rectangles and diamond symbols?" inquires Ann.

"Good question, and if there are no additional questions concerning relationships I will answer that right now," Bob replies. "That portion of the diagram indicates the complexity of the relationships between entities and provides us with a more thorough understanding of their meanings and interpretations. Follow me closely. Let's look at the relationship between PROMOTION and CLUB. A given PROMOTION is sponsored by one and only one CLUB. Conversely, a given CLUB sponsors

zero or more PROMOTIONS. Thus the number on the left represents the minimum, and the number on the right represents the maximum."

"I see. But, if that is how you interpret it, I think I see a mistake with the SUPPLIER entity that Sandra added to the diagram," Ann states with concern. "Though it's true that a given SUPPLIER SUPPLIES one or more PRODUCTS, it's not true that a PRODUCT is supplied by one or more SUPPLIERs. In fact, a particular PRODUCT is SUPPLIED BY one and only one SUPPLIER."

"That's a very good observation, Ann," Sandra offers. Sandra then turns to Bob. "That's great! I was worried about the implications of the complexity of that relationship."

Bob continues the walkthrough of the figure while Sandra takes notes of changes to reflect the additional comments of the end-users. Once the walkthrough verification of the diagram is completed, the walkthrough shifts toward the verification of several additional slides.

"We're not through verifying the data storage requirements for the new system just yet." Bob placed the first of several slides on the overhead projector. "Now we want to verify what data we need to store about each of the entities appearing on that diagram we just finished discussing. We've sampled many of your existing forms and documents to identify specific data that you are currently storing about those entities. We've also talked with you individually to allow you the opportunity to suggest additional data that you would like to start keeping about those entities. Let's confirm our list for each entity. First, let's look at the list of data we would like the new system to store about a MEMBER (see Figure E3.2). Are there any additional items that you'd like to add to this list?"

"Looks complete to me, but why are some of these names underlined?" asks David.

"The underlined items represent the data item or items that uniquely identify a particular occurrence of the entity. In other words, the data item MEMBER NUMBER uniquely identifies a single member. In some cases we use a plus sign to indicate where two or more data items are required to uniquely identify an entity."

"It looks fine to me, but what about SUPPLIERS?" asks Sally.

"That's right! We can't forget about SUPPLIERS," Bob answers. "Let's start with you Sally. What type of data do you need to keep track of for SUPPLIERS?"

CLUB:
CLUB NAME
NUMBER OF MEMBERS ENROLLED
NUMBER CANCELED YTD
CURRENT PROMOTION
TOTAL UNITS SOLD FOR CLUB
MAXIMUM PERIOD OF OBLIGATION

MEMBER:
MEMBER NUMBER or MEMBER NAME
MEMBER ADDRESS consisting of:
    STREET
    P.O.BOX
    CITY
    STATE
    ZIPCODE
MEMBER PHONE
DATE ENROLLED
BALANCE PAST DUE
BONUS CREDITS NOT USED
CLUB GROUP (repeats 1-n times)
consisting of:
    CLUB NAME
    MUSICAL/MOVIE PREFERENCE
    NUMBER OF PURCHASES
    REQUIRED
    NUMBER OF PURCHASES TO
    DATE
    AGREEMENT NUMBER SUFFIX
    AGREEMENT ENROLLMENT DATE
    AGREEMENT EXPIRATION DATE

AGREEMENT:
AGREEMENT NUMBER SUFFIX
CLUB NAME
AGREEMENT EXPIRATION DATE
AGREEMENT PLAN CREATION DATE
MAXIMUM PERIOD OF OBLIGATION
BONUS CREDITS AFTER OBLIGATION
NUMBER OF MEMBERS ENROLLED
NO. MEMBERS WHO HAVE FULFILLED
NO. MEMBERS HAVE NOT FULFILLED

BACKORDER:
ORDER NUMBER + BACKORDER DATE
BACKORDERED ITEM (repeats 1-n times)
  consisting of:
    PRODUCT NUMBER
    MEDIA CODE
    PRODUCT DESCRIPTION
    QUANTITY BACKORDERED

PROMOTION:
CLUB NAME + PROMOTION DATE
PROMOTION TYPE
SELECTION OF MONTH NUMBER
SELECTION OF MONTH TITLE
AUTOMATIC RELEASE DATE
AUTOMATIC FILL DATE

ORDER:
ORDER NUMBER
ORDER DATE
ORDER STATUS
PROMOTION NAME
PROMOTION DATE
AUTOMATIC FILL DATE
MEMBER NUMBER
MEMBER NAME
FORMER MEMBER?
MEMBER ADDRESS consisting of:
    STREET
    P.O.BOX
    CITY
    STATE
    ZIPCODE

ORDERED PRODUCT (repeats 1-n times)
consisting of:
    PRODUCT NUMBER
    MEDIA CODE
    PRODUCT DESCRIPTION
    QUANTITY ORDERED
    ORDERED PRODUCT STATUS
    QUANTITY SHIPPED
    ORDER PRICE
    EXTENDED PRICE
AMOUNT DUE

PRODUCT:
PRODUCT NUMBER + MEDIA CODE
PRODUCT DESCRIPTION
TITLE OF WORK
COPYRIGHT DATE
CURRENT RETAIL PRICE
CURRENT LIST PRICE
SUPPLIER NAME
SUPPLIER ADDRESS consisting of:
    STREET
    P.O.BOX
    CITY
    STATE
    ZIPCODE
QUANTITY ON HAND
UNITS SOLD
VALUE OF UNITS SOLD

FIGURE E3.2 **Typical Data Store Contents**

[We can leave this meeting now. The discussion would continue until all of the objects in the system, the associated data, and their relationships had been discussed and verified.]

Several days have passed since the meeting that verified the new system's data requirements. We join Sandra as she is leading a walkthrough of the order processing requirements with the Member Services staff in order to verify her understanding of the new system's processing requirements.

Sandra places a picture of the proposed system on the overhead projector (see Figure E3.3). "You have already met with Bob and me to confirm the data storage requirements of the new system. Now we want to confirm what processing will be performed on that data. You'll remember seeing pictures similar to this when we were trying to learn about the current member services system. They differ slightly in that these pictures don't show details of how the processing will be performed."

Sandra dims the lights and draws their attention to a data flow appearing on the diagram. "Once a MEMBER ORDER RESPONSE is received from a CLUB MEMBER, the AUTOMATIC DATED ORDER must be revised to reflect any variations. Subsequently, the member's credit status is checked using the CREDIT RATING AND LIMIT from the ACCOUNTS RECEIVABLE. Moving on, if the member's credit status is poor, a REQUEST FOR PREPAYMENT is sent to the CLUB MEMBER using MEMBER ADDRESS & BALANCE details from the MEMBERS file. Note that in this case the PENDING ORDER STATUS is recorded in the ORDERS file. When the MEMBER's PREPAYMENT arrives, the ORDER PAYMENT is sent to the AC-COUNTS RECEIVABLE DEPARTMENT, the pending order status is deleted, and the RELEASED PENDING ORDER is checked for product inventory just like an approved order."

"Excuse me, Sandy," Sally interrupts. "Sometimes the MEMBER ORDER has a notation on it that indicates the member has changed his or her address. When that happens, I have to notify the membership people so that the new address can be recorded."

"That's important, Sally," Sandra responds as she makes the correction to the diagram. "An AP-PROVED ORDER is then checked to determine whether or not the ordered products are available by checking the PRODUCT AVAILABILITY STATUS from the PRODUCTS file. If the order is unfillable, a NEW BACKORDER is recorded in the BACKORDERS file and the CLUB MEMBER is sent a BACKORDER NOTICE. If the order is fillable, the RELEASED ORDER STATUS is stored in the ORDERS file and the ORDER TO BE FILLED is sent to the WAREHOUSE. After the warehouse has filled the order, it returns a SHIPPING NOTICE to us so we can close the order and send a FILLED ORDER FOR BILLING notice to the ACCOUNTS RECEIVABLE DEPARTMENT."

"What about canceling the AUTOMATIC ORDER?" Joe asks. "When we receive a MEMBER ORDER RESPONSE from a CLUB MEMBER we have to cancel the AUTOMATIC DATED ORDER created by the promotion subsystem. Otherwise, the member would receive two orders."

"Isn't that normally done as soon as you determine the member's credit status?" Sandy inquires. "And the AUTOMATIC DATED ORDER is stored in the ORDERS file awaiting the promotion automatic release date, right?"

[We can leave now. This meeting would continue until all of the diagrams illustrating the processing requirements of the new system had been verified.]

Where Do We Go from Here?

This episode introduced the establishment of business objectives for a new system. It also introduced the *entity relationship diagram* for modeling the stored data requirements for the new system. In addition, this episode introduced the use of *data flow diagrams* for modeling the process requirements for the new system. Chapter 8 will discuss data modeling with entity relationship diagrams (ERDs). ERDs are used to document the system's data independent of how that data is or will be used. They also describe relationships between groups of data. Chapter 9 will introduce the use of data flow diagrams, with an emphasis on the use of essential (or logical) DFDs. This is done because, as we define requirements for the new system, we want to concentrate on what it should do, not how it should do it. Therefore, we eliminate the implementation details from the data flow diagrams and look only at the essential business processing steps for defining the general requirements for the new system.

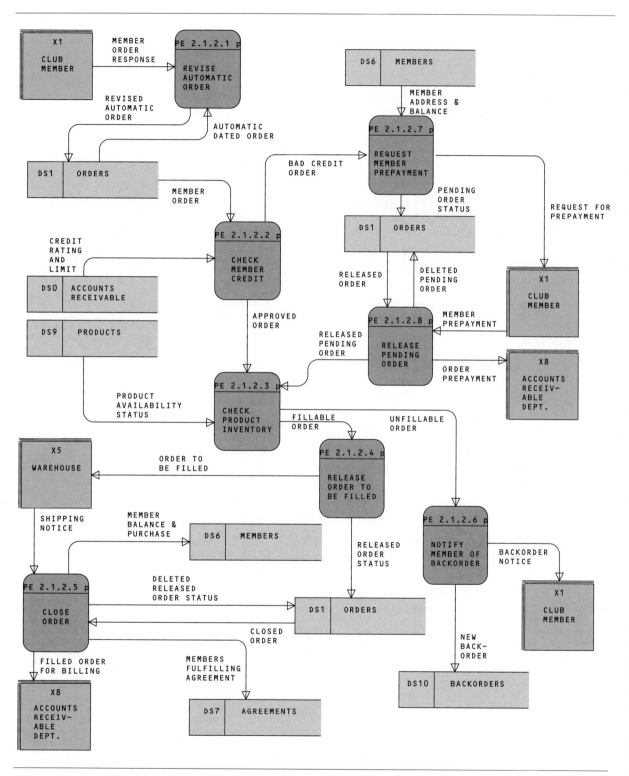

FIGURE E3.3 **Process Requirements Model for the New Information System**

8

Data Modeling

Chapter Preview and Objectives

This is the first of three graphic systems modeling chapters. In this chapter you will learn how to use a popular data modeling tool, entity relationship diagrams, to document a system's data, independently of how that data is or will be used — that is, independently of inputs, outputs, and processing. You will know data modeling as a systems analysis tool when you can

Define
systems modeling and differentiate between essential, implementation, current, and proposed system models.

Define
data modeling and explain why it is important.

Read
and interpret a Peter Chen–style data model.

Identify
fundamental, associative, supertype, and subtype data entities.

Identify
simple and complex relationships and describe their cardinality and ordinality.

Recognize
the similarities between various data modeling notations.

Explain
how data modeling is used in systems planning and analysis.

Draw
an entity relationship data model that depicts the data entities and the natural relationships between those data entities.

Identify
the data attributes that describe each data entity.

BRAGGS & ASSOCIATES

Scene: *Karl Simpson is nervous about his new project assignment for Braggs & Associates, a large manufacturer of fertilizer products for commercial agriculture. Braggs & Associates has been conducting fertilizer and crop research for the past 10 years, during which time a tremendous amount of research data has been collected on magnetic tapes. Now, the research group of Braggs & Associates wants to develop a large database to store this data and wants to make the data available on an ad hoc basis to various researchers, including researchers outside the company. Although Karl is an experienced systems analyst, he feels somewhat uncomfortable about working with research people and scientific terminology. He feels much more comfortable with the traditional business-oriented applications. Let's listen in on a portion of an interview that Karl is conducting with Rich Gazello, a key contact and member of the research group.*

Karl: I have to admit, I'm a little nervous about working with you and your people. Most of my prior projects have been traditional business applications. I'm particularly concerned that I may have difficulty picking up on your terminology. But I think if you can explain the data that you've collected, I can take it from there.

Rich: I find your concerns humorous. My people have always expressed concern with their ability to work with computer people, because of your computer knowledge and lingo.

Karl: I guess I never thought of it that way. Well, I'll try to avoid all the fancy computer lingo. Maybe you and your people can be patient with me.

Rich: It's a deal! Now, what is it you'd like to know about the system?

Karl: You can start by explaining what kind of data has been collected.

Rich: Well, as you probably already know, our group conducts research on various fertilizers and their short- and long-run effects upon various agricultural crops. We've been conducting this research for the past 10 years. This data is on magnetic tapes, but that's inconvenient since we often want to analyze data from different tapes. We'd like you to consolidate the appropriate data into a database that can be accessed and analyzed by our research group and numerous research consultants outside our company.

Karl: Can you be a little more specific as to what that data describes? For example, I assume you keep data about experiments, such as when the experiment was conducted, who performed the experiment, the type of experiment conducted, and the like.

Rich: Yes. I see what you mean. Some of the data pertains to experiments, although we prefer to call the experiment a test. A test can be part of a series of related tests.

Karl: Can I assume that the data about a single test is part of only one series?

Rich: Unfortunately, not! A test may be pertinent to several series. Is that a problem?

Karl: No. It's more difficult, but not really a problem. Let me worry about the implementation concerns. I want the database to match your business situation, not vice versa. Keep telling me about the data.

[Karl appears to be sketching a flowchart-like diagram of the data.]

Rich: Okay. We also obtain data about the crop that we were testing. Among other things, the data tells us the specific location where the crop was planted.

Karl: By location, you mean which farm. I'm aware that your company owns two large farms where you folks grow the crops.

Rich: Yes and no. There is a little more to it than that. Our data tells us the specific location where a batch of crop and fertilizer were tested. For ex-

ample, each farm is divided into a number of designated fields. Each of these farm fields is a specific location. Also, it's important to note that we
have a number of greenhouses. Greenhouses
allow us to simulate field conditions in other geographic regions. As with the farms, each green-
house is divided into four sections that we refer
to as ranges.

Karl: I think I understand. I'm starting to wonder how
            you folks kept track of specific crops and test
            fertilizers.

Rich: The location is very important. The data tells us
            the location of current crops and provides us
            with a history of crops that were planted at a
            particular rotation. As you know, most farmers
            need to know this so that they can properly ro-
            tate their crops.

Karl: What other types of data have been collected?

Rich: Well, there's data about soil types. We were con-
            cerned about picking the crop location based

upon soil types available at those locations. Oh
yes, earlier I mentioned tests. Some of the testing that we do is on a specific location's soil.

Karl: Could you tell me a little about . . .

Challenge

One of the challenges for systems analysts is to communicate their understanding of systems requirements and solutions. To this end, analysts have learned that pictures often
communicate better than words. On a single 8½″ × 11″
sheet of paper, draw a picture that Karl might use to communicate his understanding (or misunderstanding) of the system's data. If you have difficulty, don't be too concerned. This
chapter introduces you to a useful tool.

Hint   If you have ever used a personal computer database
management system, try to draw a picture that could be
implemented in that system. Don't be concerned with keys,
indexes, or name length restrictions.

AN INTRODUCTION TO SYSTEMS MODELING

In the last chapter you were introduced to activities that called for drawing
system models. System models play an important role in systems development.
As a systems analyst or user, you will constantly deal with unstructured problems. One way to structure such problems is to draw models.

A **model** is a representation of reality. Just as a picture is worth a thousand words, most models are graphic representations of reality.

Models can be built for existing systems as a way to better understand those
systems, or for proposed systems as a way to define requirements and designs.
An important concept, in both this chapter and the next, is the distinction
between implementation and essential models.

Implementation models show not only what a system is or does, but
also how the system is physically implemented. Synonyms include *technology model* and *physical model*. (Once preferred, the term physical
model has become less popular in recent years.)

An implementation model of data might describe the format of the data (e.g.,
$$$$9.99, a COBOL PICTURE clause for a data attribute), the media on which
the data is recorded (e.g., preprinted form or computer-generated report), or
how the data is organized (e.g., as an indexed file or a database record).
Implementation models are useful for documenting an existing system's
data. However, analysts have learned that data requirements for proposed systems should ideally be specified using essential models.

Models	Essential System (Describes "What"; Also Known as the Logical System)	Implementation System (Describes "What" plus "How"; Also Known as the Physical System)
Current System (Existing System)	An essential model of the current system depicts those aspects of the current system that are essential to the business and that should be retained—no matter how we choose to implement the system.	An implementation model of the current system depicts how the current system is physically implemented (inclusive of technology). By default, the implementation model includes all of the essential aspects of the current system.
Proposed System (Target System)	An essential model of the proposed system depicts the business and user requirements for the system—regardless of how that system might be implemented.	An implementation model of the proposed system depicts how the proposed system will be physically implemented (inclusive of technology). By default, the implementation model must include all of the essential aspects of the proposed system.

FIGURE 8.1 **Types of System Models** Most system models depict either a current or proposed system. Most system models "should" also depict either the essential or implementation of a system. This chapter deals primarily with "essential, current" and "essential, proposed" system modeling.

Essential models are implementation-independent models—they depict the essence of the system (what the system does or must do), independent of how the system will or could be physically implemented. Essential models are also sometimes called *logical models* or *conceptual models*. (The term logical model has become less popular in recent years.)

There are several reasons why you should use essential models:

- Implementation-independent models remove biases that are the result of the way the existing system is implemented or the way that any one person thinks the system might be implemented. Thus, we overcome the "we've always done it that way" syndrome. Consequently, implementation-independent models encourage creativity.
- Implementation-independent models reduce the risk of missing functional requirements because we are too preoccupied with technical details. Such errors can be costly to correct after the system is implemented. By separating what the system must do from how the system will do it, we can better analyze the requirements for completeness, accuracy, and consistency.
- Implementation-independent models allow us to communicate with end-users in nontechnical or less technical language. We frequently lose requirements in the technical jargon of the computing discipline.

The differences between current versus proposed system models and implementation versus essential system models are summarized in Figure 8.1.

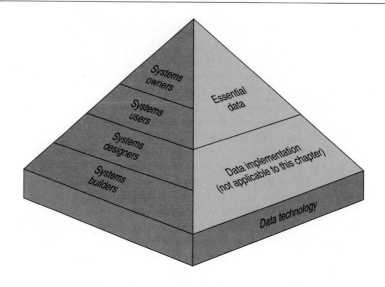

This chapter will present data modeling as a technique for defining data storage requirements. Data modeling was presented in Chapter 4 as a systems development technique.

> **Data modeling** is a technique for organizing and documenting a system's data. Data modeling is sometimes called *database modeling* because a data model is usually implemented as a database. (In fact, this chapter's subject is also taught in the average database or data management course.)

In this chapter we will focus exclusively on essential data modeling. Using your pyramid model (see Figure 8.2), essential data modeling deals with data from the system owners' and system users' points of view. Accordingly, such models are not concerned with implementation details or technology.

Although data in any business is constantly changing, the types of data collected and processed by an average business change very slowly. For this reason, the essential data model of a current system tends to be very similar, if not identical, to the essential data model of any proposed system. For this reason, many systems analysts use data models to simultaneously document current and proposed systems.

Data modeling is also taught in database and data management courses. Database courses tend to teach it from the perspective of data management, independent of applications. Systems analysis courses teach data modeling from the perspective of applications development. Both perspectives are important. In database courses, you may encounter somewhat different terminology and graphical symbology, but these are simply different dialects of the same language. Look closely, and you'll see it's true!

There are numerous data modeling tools; however, we will initially concentrate on one, the entity relationship diagram.

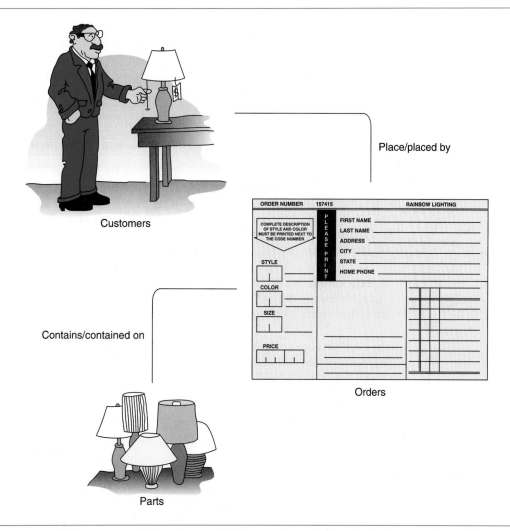

Customers

Place/placed by

Contains/contained on

Orders

Parts

FIGURE 8.3 **An Artistic Data Model** A data model depicts things about which a business stores data and the relationships between those things. Since most of us are not artists, we need a less artistic version of this data model.

ENTITY RELATIONSHIP DIAGRAMS

Let's begin with a formal definition.

> An **entity relationship diagram (ERD)** is a data modeling tool that depicts the associations among different categories of data within a business or information system — it does not imply how data is implemented, created, modified, used, or deleted. Synonyms include *entity model* or *entity diagram,* and *entity relationship attribute (ERA) diagram.*

All systems contain data — usually lots of data! Data describes tangible things, roles, events, or places of interest to the business and its workers. In Figure 8.3, we see three items of interest in a typical order-processing system:

CUSTOMERs (who play a role in the system), ORDERs (which are events that bring new data into the system), and PARTs (which are tangible things in the system). It shouldn't be too hard to imagine data being stored about any of these items. This figure also demonstrates that the things we describe with data are also naturally associated with one another. For example, a customer "places" an order or an order "is placed by" a customer. Figure 8.3 can be thought of as an artistic version of an entity relationship diagram.

Entity Relationship Diagram Conventions and Guidelines

There are various symbolic notations suggested by different authors and experts. Figure 8.4 illustrates the popular Peter Chen entity relationship diagram, the one we will use throughout most of this chapter. Before we teach you how to read the diagram, briefly examine the symbols. The Chen symbols are duplicated in the margin for your reference.

A data entity is the main symbol on an ERD.

> A **data entity** is anything, real or abstract, about which we want to store data. Hereafter, we will refer to the term as **entity.** Synonyms include *entity type, entity class,* and *object.*

Note   As true object modeling techniques become more popular, we expect the term "object" to become less popular in the context of traditional data modeling. In a sense, an object is a superset of the data entity.

An entity is drawn as a rectangular box. This box represents all occurrences of the named entity. For example, the entity AUTHOR represents all authors. In other words, we do not depict specific occurrences of any entity.

In Figure 8.4, we have recorded, inside each entity's rectangle, typical data attributes that might describe all or most occurrences of each entity. The list of attributes is by no means exhaustive. We have also recorded the number of instances of each entity. This emphasizes an important characteristic of data models. They only show types of entities. They do not show individual occurrences of those entities.

A Chen entity relationship diagram also depicts data relationships.

> A **data relationship** is a natural association that exists between one or more entities. Some experts like to think of relationships as business activities or events that link one or more entities. We will refer to a data relationship simply as a **relationship.** Synonyms include *association* and *data association.*

A relationship is depicted by a diamond that is connected to one or more entities. For instance, Figure 8.4 demonstrates that, in addition to storing data about CUSTOMERs and ORDERs, we need to store either data or pointers (or something) that indicates which customer PLACES a specific order or that indicates an order IS PLACED BY a specific customer.

Note   You may argue that you could duplicate certain data, such as CUSTOMER NUMBER, NAME, and ADDRESS in both the CUSTOMER and ORDER entities. After

FIGURE 8.4 An Entity Relationship Diagram as a Data Model One of the most popular data modeling tool's is Peter Chen's entity relationship diagram. Entities (rectangles) are described by data attributes, which in this example are written inside the rectangles. Relationships are depicted by diamonds that are connected to the rectangles.

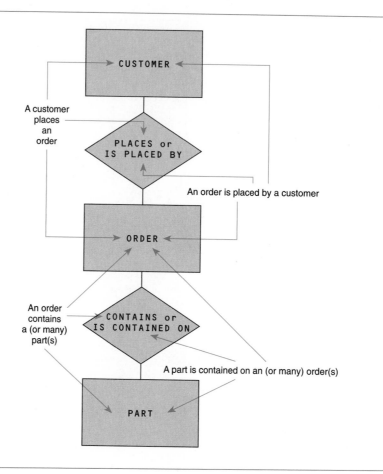

FIGURE 8.5 **How to Read an Entity Relationship Diagram** Reading an ERD is easy. If properly labeled, an ERD reads as a series of simple sentences that describe the realities of the business. Notice that relationships are interpreted in both directions.

all, that customer data is often duplicated on order forms, isn't it? There are two problems with this argument. First, storing redundant data increases the possibility of data inconsistency—for instance, two different customer addresses for the same customer. Second, and perhaps more importantly, we should be dealing with business concerns, not implementation concerns. A business is concerned with reliable data about specific business entities, not where or how that data is stored.

Now, how do you read ERDs? If properly labeled, an ERD should be read as simple sentences. Figure 8.5 duplicates our first ERD to demonstrate this easy-to-learn approach. The sentences describe the natural associations between our entities, which are described by data. Notice that for each relationship there are two sentences, written to the left and right of the relationship. Thus, all relationships have two interpretations—one in each direction.

This bidirectionality is important. For example, we need to know all parts included on an order (ORDER "CONTAINS" PART), but we may also want to find all orders for a specific part (PART "IS CONTAINED ON" ORDER). The ERD forces us to recognize these natural associations in both directions.

Now that you understand the basic symbols and how to read the diagram, let's study entity relationship diagramming conventions in greater detail.

Entities that describe roles played in a system. They usually represent people or organizations.

ACCOUNT	CONTRACTOR	EMPLOYEE	OFFICE
AGENCY	CORRESPONDENT	EMPLOYER	OFFICER
ANIMAL	CLIENT	GROUP	SALESPERSON
APPLICANT	CREDITOR	INSTRUCTOR	STUDENT
BORROWER	CUSTOMER	JOB OPENING	SUPPLIER
CHAPTER	DIVISION	JOB POSITION	TEAM
CHILD	DEPARTMENT	MANAGER	VENDOR
CLASS			

Entities that describe tangible things. Most tangible things are easy to identify because you can see and touch them.

BOOK	EQUIPMENT	METAL	SERVICE
CHEMICAL	ISOTOPE	PART	SUBSTANCE
COURSE	MACHINE	PRODUCT	VEHICLE
DISK	MATERIAL	PROGRAM	

Entities that describe events. Most events are easy to identify because the business records data on forms and in files. Events are characterized by the fact that they happen and have duration.

AGREEMENT	DEPOSIT	PAYMENT	SEMESTER
APPLICATION	DISBURSEMENT	PROJECT	SHIPMENT
APPOINTMENT	FLIGHT	PURCHASE ORDER	STEP
ASSIGNMENT	FORECAST	QUOTE	TASK
BACKORDER	INVOICE	REGISTRATION	TEST
BUDGET	JOB	REQUISITION	TIME BUCKET
CLAIM	LICENSE	RESERVATION	WORK ORDER
CONTRACT	MEETING	RESUME	
DEFECT RETURN			

Entities that describe locations. Again, you can usually see locations.

BRANCH	COUNTRY	ROUTE	STORAGE BIN
BUILDING	COUNTY	SALES REGION	VOTER PRECINCT
CAMPUS	POLICE PRECINCT	SCHOOL ZONE	WAREHOUSE ZONE
CITY	ROOM	STATE	

FIGURE 8.6 Examples of Entities Entities mostly correspond to roles, tangible things, events, and locations. They are named with singular nouns (possibly including descriptive adjectives).

Entities

Once again, an entity is anything about which the user wants to store data. All business systems seek to capture and store data about various entities, regardless of whether computers are used. Entity types tend to fall into four classes: roles, events, locations, or tangible things. Examples are listed in Figure 8.6. The list of entities in any business is finite; therefore, systems analysts and database analysts can eventually build a library of reusable entities—that is, entities that can be used in all subsequent systems development projects for the same business.

Notice that we name entities with nouns that describe the role, event, place, or tangible thing about which we want to store data. Grammatically, names should be singular so as to distinguish the essential concept of the entity from occurrences of the entity. Names may include appropriate adjectives or clauses to better describe the entity—for instance, CUSTOMER INVOICE, to distinguish it from VENDOR INVOICE.

Entities are anything described by data, and data takes the form of data attributes.

Data attributes are characteristics that are common to all or most instances of a particular entity. Hereafter, we will simply refer to them as
attributes. Synonyms include *properties, data elements, descriptors,*
and *fields.*

Any given attribute should describe only one entity; however, you may encounter instances where different entities appear to share the same attribute.
For example, the CUSTOMER and SUPPLIER entities may appear to share
attributes such as NAME and ADDRESS. However, it is best to assign unique
names for each, such as CUSTOMER NAME, SUPPLIER NAME, CUSTOMER
ADDRESS, and SUPPLIER ADDRESS. Other methods allow for global attribute
names that can be shared by multiple entities. For instance, the attribute AD-
DRESS can be standardized for all applications. But analysts and users should
never make such a decision unilaterally. Such decisions should be approved
through an appropriate data administrator who can make sure that the attribute
is standardized with an appropriate length, format, and value ranges that will
serve all future applications.

Attributes take on values for each occurrence of an entity. An attribute must
have more than one legitimate value; otherwise, it is not an attribute—it is a
constant. Attribute values are usually limited by one of the following rules:

- A value range (for instance, {3.65 through 16.85}).
- A finite set of values (for instance, {1, 2, 3, 4, 5} or {Non-Degree, Fresh-
  man, Sophomore, Junior, Senior, Graduate}).
- A binary value (for example, {Yes, No} or {On, Off}).
- A nearly infinite set of values (such as for an attribute such as NAME).

Occasionally you will encounter an attribute that may actually be an entity.
For example, in one instance STATE may be an attribute of the entity CUS-
TOMER or SUPPLIER. However, in another instance—such as the need to store
sales tax information—STATE becomes an entity with attributes such as STATE
NAME, STATE ABBREVIATION, SALES TAX RATE, and STATE TREASURER
ADDRESS. Two simple tests that help us to identify whether an entity exists:

1. How many occurrences exist for the entity. There must be two or more
   occurrences. It doesn't make sense to store a file of one occurrence.
   (*Note:* Some analysts insist that an entity must have a dozen or more occurrences to be considered legitimate.)
2. How many attributes describe the entity. Again, there should be at least
   two attributes describing the entity. If there is only one attribute, that attribute should be assigned to another entity.

When attributes have been identified for entities, the data model is said to be
fully attributed or partially attributed, depending on whether the task is considered complete. In Figure 8.7 a collection of attributes describe a typical entity,
PART, and a simple table demonstrates a common implementation of an entity.
The table shows occurrences of the PART entity (as rows) and values of data
attributes (as columns). Usually, at least one data attribute takes on a unique
value for each occurrence of the entity. In our example PART NUMBER is this
special attribute, called an *identifier.*

An **identifier** is an attribute or combination of attributes that uniquely
identifies one, and only one, occurrence of an entity. Synonyms include
key and *primary key* (although those terms are implementation-oriented).

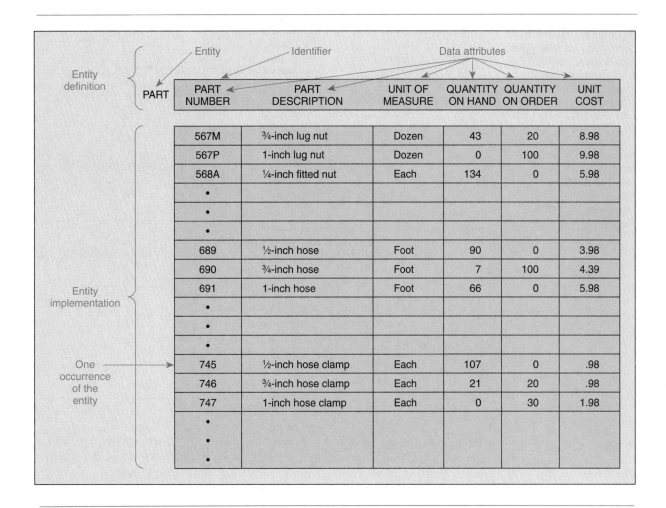

FIGURE 8.7 **Data Attributes that Describe an Entity** Every entity is described by two or more data attributes. When implemented as a simple table, as shown here, the attributes (columns) take on values for each occurrence (row) of the entity. At least one of these attributes should take on a unique value for each occurrence so that the occurrence can be identified.

Sometimes an entity has more than one unique identifier. For instance, a STUDENT might be uniquely identified by either SOCIAL SECURITY NUMBER or STUDENT IDENTIFICATION NUMBER. Sometimes an entity requires combinations of attributes to identify one instance. For example, a STUDENT might also be identified by the combination of STUDENT NAME plus STUDENT ADDRESS (assuming that we are reasonably certain that two students with identical names won't live at the same address).

When identifier attributes have been assigned for all entities in a data model, the model is often said to be a *keyed model.* We'll discuss keys in more detail later in this chapter.

Entity Supertypes and Subtypes

Occasionally, analysts discover that different entities are merely different forms of the same entity. They share some common attributes but also have some

FIGURE 8.8 **Entity Supertypes and Subtypes**
In this ERD, EMPLOYEE is an entity supertype whose occurrences can be grouped into two entity subtypes: SALARIED EMPLOYEE and HOURLY EMPLOYEE. EMPLOYEE contains those attributes common to all employees. The subtypes inherit the common attributes and add attributes common to its unique subtype.

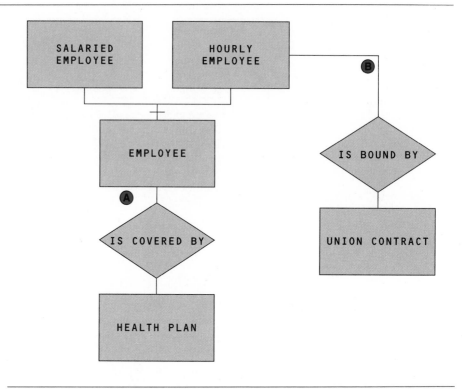

unique attributes. For example, the entities HOURLY EMPLOYEE and SALARIED EMPLOYEE are forms of a general entity, EMPLOYEE. They share attributes such as SOCIAL SECURITY NUMBER, EMPLOYEE NAME, and EMPLOYEE ADDRESS. On the other hand, HOURLY EMPLOYEEs have unique attributes such as HOURLY RATE OF PAY and SENIORITY CODE. Similarly, SALARIED EMPLOYEEs may have their own unique attributes such as MONTHLY SALARY and PROFIT SHARING PLAN. Figure 8.8 demonstrates the Chen modeling notation for this situation.

Notice that an EMPLOYEE entity is still shown on the ERD. It is called an entity supertype.

> An **entity supertype** is an entity whose occurrences can be divided into subtypes that are *not* described by identical data attributes but which share some data attributes. An entity supertype defines those attributes that *are* shared by all groups.

An entity supertype must have two or more entity subtypes.

> An **entity subtype** is an entity whose occurrences inherit some data attributes from an entity supertype and add other data attributes that are unique to occurrences of the subtype.

Entity supertypes and subtypes are depicted by the standard rectangle. The entity subtypes are connected directly to the entity supertype—there is no diamond in the path. Entity subtypes include those data attributes that only describe occurrences of that subtype or group.

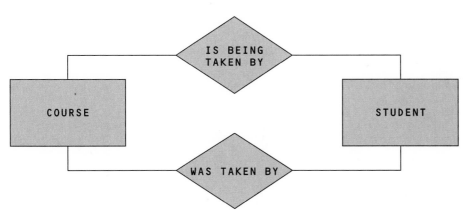

FIGURE 8.9 **Multiple Relationships between the Same Entities** Any pair of entities may participate in multiple relationships, each having its own meaning. Here, for example, we see that one relationship describes a student's current courses, while the other describes courses that have already been taken.

Notice in Figure 8.8 that other entities can:

(A) Be related to all occurrences of an entity (through a connection to the supertype).

(B) Be related only to the instances of one entity subtype (through a connection only to the appropriate subtype(s).

Depicting supertypes and their subtypes allows for clearer communication between systems analysts, users, and database professionals. Frequently, the user may describe policies and procedures that pertain to EMPLOYEEs as a whole or to one particular type of employee. This is an important differentiation. Eventually, these business rules must be implemented in any resulting computer files and databases!

Relationships

Recall that relationships are natural associations between one or more entities. These associations can be determined relatively quickly once the entities are identified. Relationships are important. Whenever we store data in forms, files, and databases, we must understand the potential implications of updating that data. For instance, if we deleted a CUSTOMER for which there are outstanding INVOICEs, we would not be able to collect payments due from that customer. The study of such updating requirements is enhanced through inspection of the data relationships drawn on the data model.

Entities are named with nouns. Relationships should be named with verbs or verb phrases. This enhances the simple sentence interpretation described earlier in the chapter. Although relationships could be named with two verbs to illustrate both directions, this practice can clutter the Chen diagram. For this reason we will name relationships in only one direction on Chen ERDs. The name for the inverse direction should be assumed and implicit.

There are several types of relationships. The most common are *binary relationships*—that is, relationships between two different entities. Such a relationship exists between the entities COURSE and STUDENT (see Figure 8.9). Notice that multiple relationships can be established between the same two entities. In this case, one relationship describes the courses a student is

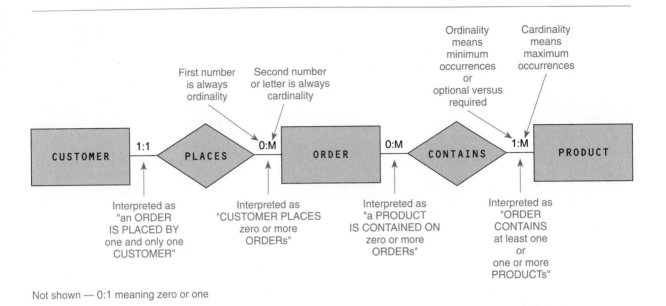

Not shown — 0:1 meaning zero or one

FIGURE 8.10   Ordinality and Cardinality   Ordinality and cardinality define the "business rules" for a relationship between entities. Specifically, ordinality describes the minimum number of occurrences of one entity with respect to another. Cardinality describes the maximum number of occurrences of one entity with respect to another. Study the notation carefully so that you can easily read the ERDs.

currently taking (present tense). The other describes courses that the student has taken (past tense).

In Chapter 2 you learned that a system user's view of data reflects not only knowledge about entities, attributes, and relationships, but also rules. There are two rules that can be defined for every relationship: ordinality and cardinality.

Ordinality defines whether the relationship between the entities is mandatory or optional. In other words, ordinality determines the *minimum* number of occurrences of one entity relative to the other. Ordinality must be defined in both directions. Ordinality is sometimes called *optionality* or *dependency*.

Consider the ERD depicted in Figure 8.10. In each line there are two numbers or letters separated by a colon. Ordinality is recorded to the left of each colon. Thus, for one occurrence of CUSTOMER, there may or may not be a current occurrence of ORDER (in other words, minimum = 0). However, for one instance of ORDER, there must be an occurrence of CUSTOMER (in other words, minimum = 1), because someone placed that order. In almost all cases, ordinality is 0 or 1.

What does this tell us? We can add a new occurrence of CUSTOMER without adding an ORDER for that customer, but we can never add a new occurrence of ORDER without first verifying the existence of its CUSTOMER. If the customer doesn't exist, we must create the CUSTOMER occurrence "before" we add the ORDER.

The second rule that can be defined for every relationship is cardinality.

Cardinality defines the *maximum* number of occurrences of one entity for a single occurrence of the related entity.

Returning to Figure 8.10, cardinality is depicted with the number or character to the right of the colon. Thus, for a CUSTOMER there could be "many" ORDERs. (The letters *n* and *m* are universal data modeling standards meaning "many" or "more." If the specific numbers are constant, those numbers can be substituted—for example, one to twenty or 1:20). However, for an ORDER, there can only be one CUSTOMER.

In this example, both the ordinality and cardinality of an ORDER to a CUSTOMER are equal to one. This is sometimes phrased as "one and only one"—meaning, minimum = 1 and maximum = 1. Thus, for an ORDER, there can be "one and only one" CUSTOMER.

Ordinality and cardinality are typically interpreted in the same sentence. For example, the relationships in Figure 8.10 would be read as follows:

- A CUSTOMER has PLACED zero or more ORDERs (at any given time). On the other hand, an ORDER WAS PLACED BY one and only one CUSTOMER.
- An ORDER CONTAINS one or more PRODUCTs. On the other hand, a PRODUCT IS CONTAINED ON zero or more ORDERs (at any given time).

Note   In some books, methodologies, and businesses, the term *cardinality* is used to describe both minimum and maximum occurrences. For this book we will distinguish between the terms.

Relationships that Can Be Described by Data

For the most part, relationships are not described by data attributes. Instead, occurrences of a relationship serve only to associate or link occurrences of one entity to occurrences of another. But there are exceptions.

When the cardinality between two entities is "many" in both directions, the relationship itself is frequently described by data attributes. This is called a *many-to-many* (sometimes called *m:n*) *relationship*. Consider the relationship between the entities ORDER and PRODUCT in Figure 8.10. An ORDER may request many PRODUCTs. A PRODUCT may be on many ORDERs. This is a classic many-to-many relationship. What data attributes might describe this relationship?

Consider the attributes QUANTITY ORDERED and PRICE AT TIME OF ORDER. Some analysts and users prefer to store these attributes in the entity ORDER, as shown in Figure 8.11A. To many managers or users, they are almost certainly considered attributes of an ORDER. Alternatively, other analysts correctly insist that QUANTITY ORDERED and PRICE (at the time of the order) describe the relationship CONTAINS. In other words, those attributes don't describe ORDER as much as they describe a particular PRODUCT on an ORDER. These analysts might store the attributes as shown in Figure 8.11B.

In reality, this is both a relationship and an entity. It is sometimes called an *associative entity.*

An **associative entity** is a data entity whose attributes describe a relationship or association between two or more fundamental entities. In other data modeling methods it is frequently called a **relational entity** or

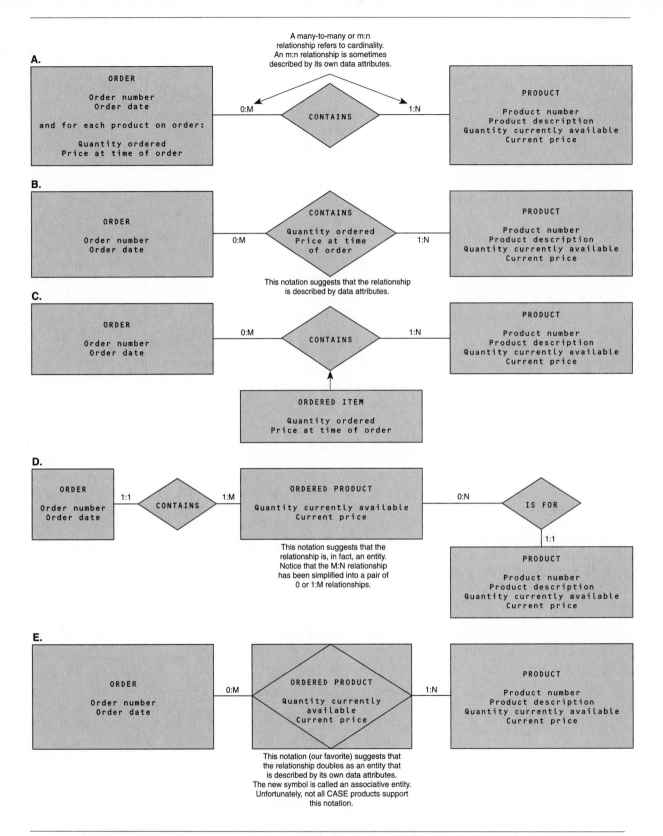

FIGURE 8.11   Alternative Notations for Relationships that Are Described by Data Attributes   Many-to-many relationships are frequently described by their own data attributes. There are several notations for depicting this situation.

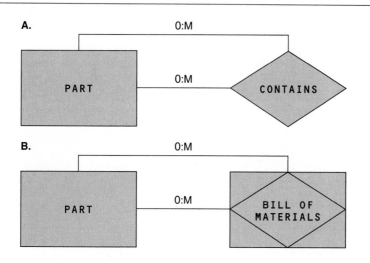

FIGURE 8.12 **Relationships between Occurrences of the Same Entity** Sometimes an entity's occurrences can be related to different occurrences of the same entity. Here, for example, is an ERD that shows the composition of PARTs made from other PARTs. Two notations are shown. The second is preferred when the relationship can be described by data attributes.

intersection entity. An occurrence of an associative entity must be related to one and only one occurrence of its connected fundamental entities.

Figure 8.11C, D, and E show three Chen techniques for depicting associative entities. We prefer Figure 8.11E because: (1) the overlaid rectangle and diamond correctly suggest the dual nature of the associative entity and (2) it uses less clutter while differentiating between fundamental and associative entities. Unfortunately, some CASE tools (including Excelerator, which is used in the SoundStage case study) do not support this notation. Notice in Figure 8.11D and E that the name has been changed from a verb phrase to a noun. It is now considered more than a relationship — it is something about which we can store data.

Suppose you discover a many-to-many relationship, but neither you nor your users can think of any data attributes that describe the relationship. Should you go ahead and decompose the relationship into one-to-many relationships and add the associative entity? Purists would say "yes." They would argue that "Someday, the user may uncover important attributes" or "Database people are going to do it anyway." We consider this "implementation thinking." During systems planning and analysis, data models are supposed to communicate the business problem! We would leave well enough alone and not decompose such many-to-many relationships during systems analysis.

Complex or Extended Relationships

So far, we've only provided examples of classical binary relationships — those that exist between two different entities. Other less common and more complex relationships can also be identified.

Relationships may exist between different occurrences of the same entity. In a manufacturing business, for instance, a bill of materials describes how a product breaks down into parts, which may break down into other parts, which may break down into other parts, and so forth. An ERD for a bill of materials is shown in Figure 8.12A. Because the relationship cardinality is many-to-many, it can be decomposed into the ERD in Figure 8.12B. Attributes of the associative

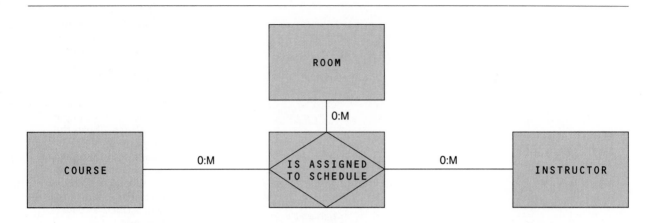

FIGURE 8.13 **N-Ary Relationships** An N-Ary relationship is one defined between N entities. In this example N = 3 — thus, we have what is commonly called a ternary relationship. Most N-Ary relationships can be described by data attributes; therefore, we have used the associative entity symbol to depict the relationship.

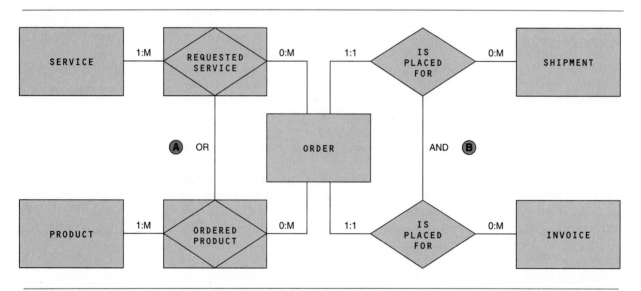

FIGURE 8.14 **Rules Governing Relationships** Some relationships must occur together. They are mutually contingent and denoted by the word "AND" recorded on a line between the relationship. Other relationships must never occur together. They are mutually exclusive and denoted by the word "OR" recorded on a line between the two relationships.

entity describe how many units of one part are required to build one unit of the other part.

Relationships can also exist between more than two different entities. These are sometimes called *N-Ary relationships.* An example of a 3-Ary or ternary relationship is shown in Figure 8.13. This relationship (shown as an associative entity) matches a COURSE, a ROOM, and an INSTRUCTOR to create a SCHEDULED CLASS. The relationship is shown as an associative entity because it also happens to be described by attributes such as DIVISION, DAYS OF WEEK, START TIME, and END TIME.

Finally, relationships may be governed by rules in addition to cardinality and ordinality. They can also be dependent on each another. The straight line labeled OR in Figure 8.14A demonstrates that both relationships cannot exist for a single occurrence of ORDER. The two relationships are said to be *mutually exclusive*. An ORDER may be placed for PARTs or SERVICE, but not both. Thus, there may exist an occurrence of the relationship between an ORDER and PARTs *or* an ORDER and a SERVICE.

Similarly, the straight line labeled AND in Figure 8.14B illustrates a relationship where if a specific ORDER has been shipped (SHIPMENT) it must also be billed (INVOICE). The two relationships in this example are said to be *mutually inclusive* (or *mutually contingent*).

Also, some relationships may exhibit the AND/OR property, meaning either, or, or both. As you can see, relationships demonstrate important associations that must be maintained between different collections of data to be stored!

Note The Chen notation doesn't actually provide standards for modeling mutually exclusive or mutually contingent relationships. Because most non-Chen notations provide this capability, we enhanced the Chen notation accordingly. The proposed notation can be implemented in several CASE tools, including Excelerator.

Other Popular Data Modeling Tools: Martin and Bachman

As mentioned earlier, there are numerous data modeling tools. Both systems analysts and users will likely encounter different tools — sometimes even within the same organization. Therefore, it is important to be familiar with two other commonly used data modeling tools — the Martin and the Bachman data models. These two tools are intended to depict and communicate the same information as the entity relationship diagram; however, they each have a different symbolic notation.

Figure 8.15 presents a sample Chen diagram that exhibits many of the modeling concepts that were presented in the last section. Review these features on the diagram:

Ⓐ A typical entity.

Ⓑ Supertype and subtype entities. Notice that TRANSACTIONs apply to both MEMBERs and NON-MEMBERs; however, DEPOSITs apply only to NON-MEMBERs.

Ⓒ An associative entity — a many-to-many relationship with data attributes.

Ⓓ A typical relationship.

Ⓔ A typical ordinality to cardinality expression.

Ⓕ A many-to-many relationship without data attributes.

Ⓖ An entity whose occurrences may be related to other occurrences of the same entity. Here, a MOVIE may be part of a MOVIE series.

Ⓗ A mutually exclusive relationship.

Now, let's compare the Chen model to the Martin and Bachman notations.

Figure 8.16 shows a sample James Martin – style data model. This data model is popular in information engineering literature. Slight variations are common. We call your attention to the following letters on the diagram:

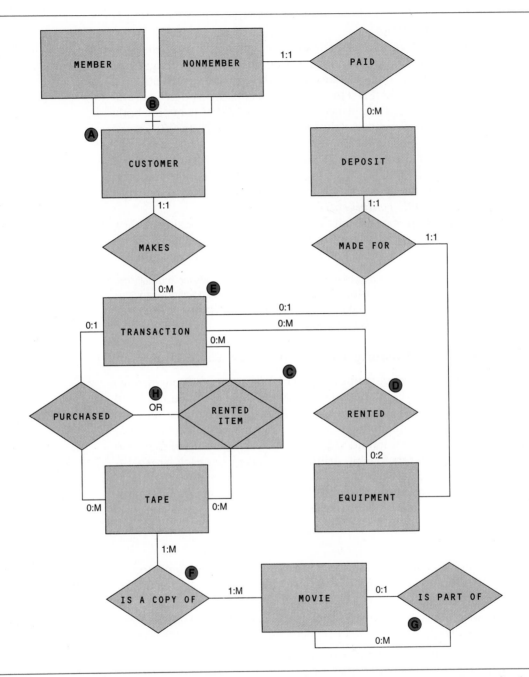

FIGURE 8.15   **A Chen-Style Data Model**   This entity relationship diagram depicts the Chen notation for data modeling.

Ⓐ Like Chen, the Martin notation uses rectangles to represent most entities.

Ⓑ Entity supertypes are depicted with a larger rectangle. Entity subtypes are represented by a partitioned row within the entity CUSTOMER.

Ⓒ An associative entity is depicted exactly as in the Chen notation. As in Chen, the diamond is optional but adds clarity.

Ⓓ Relationships between entities are depicted using a connecting line.

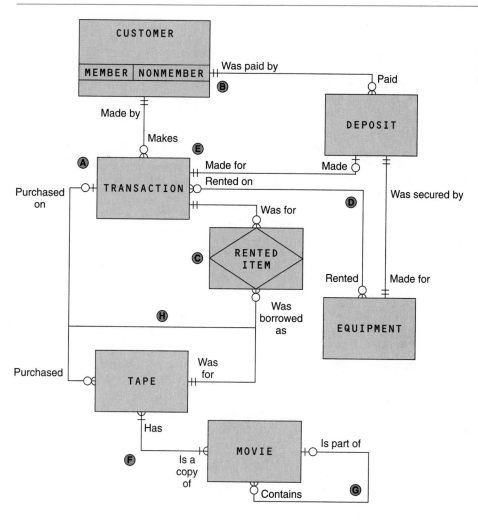

FIGURE 8.16 **A Martin-Style Data Model** This entity relationship diagram depicts the Martin notation for data modeling.

(There is no diamond or alternate shape.) Relationships are usually labeled in both directions.

Ⓔ Ordinality and cardinality are indicated using a pair of symbols at each end of the relationship line. The inside symbols indicate ordinality. The outside symbols indicate cardinality. The symbols are interpreted as follows:

· A circle indicates "zero."

· A bar or crossing line indicates "one."

· A crows-foot indicates "many."

These three symbols allow us to depict all possible ordinality to cardinality combinations. Study the relationship between CUSTOMER and TRANSACTION. We can interpret this portion of the diagram as follows:

A CUSTOMER (rectangle) makes zero (small circle) or more (crows feet) TRANSACTIONs (rectangle).

A TRANSACTION (rectangle) is made by one and only one (small bar) CUSTOMER (rectangle).

FIGURE 8.17   A Bach-man-Style Data Model
This entity relationship diagram depicts the Bach-man notation for data modeling.

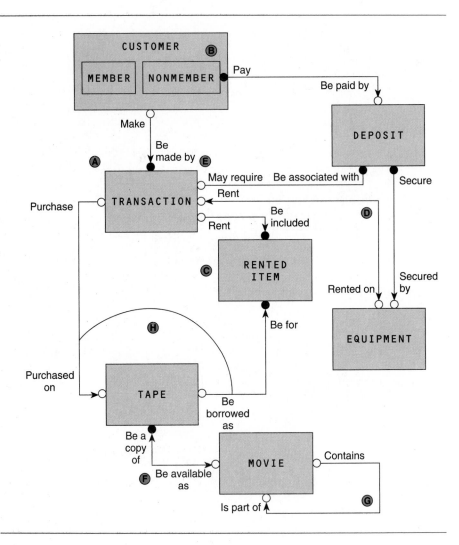

F A many-to-many relationship without data is depicted with crows feet at both ends of the relationship.

G An entity is connected to itself to indicate a 1-Ary relationship. If the re-lationship had its own data attributes, an associative entity would have been introduced with two relationships back to MOVIE.

H A single bar or line between two relations indicates a mutually exclusive relationship. A double line would have indicated a mutually contingent relationship.

There are several variations of the Charles Bachman notation. Figure 8.17 shows the Bachman notation used in the latest Bachman CASE tools. Once again, refer to the following letters on the diagram.

A Like Chen and Martin, the Bachman notation also uses rectangles to rep-resent entities.

B Unlike Chen and Martin, Bachman depicts entity supertypes as rectangles and entity subtypes as smaller rectangles located within the supertype.

C There is no special notation for associative entities.

D Relationships between entities are depicted using a connecting line. Like Martin, both directions are usually labeled. Notice, however, that the labels are reversed from the Martin approach.

E As with the Martin notation, cardinality is indicated using one or more combinations of symbols located near each end of the relationship. The cardinality symbols and their meanings for Bachman models are as follows:

· A small, darkened circle at the end of a relationship means "must." It is an alternative expression for minimum = 1 (not optional).

· A small, clear circle at the end of a relationship means "may." It is an alternative expression for minimum = 0 (optional).

· An arrowhead on the end of a relationship means "one or more."

· The absence of an arrowhead on the end means "one and only one."

Study the relationship between TRANSACTION and DEPOSIT. Using the legend of symbols above, we can interpret this Bachman diagram as follows:

A TRANSACTION (rectangle) may (small, clear circle touching the transaction entity) REQUIRE one and only one (absence of an arrow near the deposit entity) DEPOSIT (rectangle).

A DEPOSIT (rectangle) must (small, black circle touching the deposit entity) be associated with one and only one (absence of an arrow near the transaction entity) TRANSACTION.

F A many-to-many relationship without data attributes has an arrow on both ends of the line.

G The 1-Ary relationship is implemented similar to Martin. Again, if data attributes had described the relationship, a new entity would have to be introduced and two relationships would connect that entity back to MOVIE.

H A single-line arc that connects two relationships indicates that the relationships are mutually exclusive. A double-line arc would have indicated that the relationships were mutually inclusive (equivalent to AND in the Chen notation).

Figure 8.18 depicts an earlier and simpler Bachman notation that still shows up in books and CASE tools.

A Ellipses or rectangles are used to depict entities. We used ellipses here for a change of pace.

B Supertypes and subtypes are shown as in the more advanced Bachman notation.

C Once again, there is no special symbol for associative entities.

D Relationships are depicted by lines as in the Martin and more advanced Bachman notation.

E Arrows are used to depict cardinality. One arrow means "one." Two arrows mean "many." Ordinality is not depicted.

F Double arrows on both ends of a line indicate a many-to-many relationship without data attributes.

FIGURE 8.18 **A Simpler Bachman-Style Data Model** This entity relationship diagram depicts an earlier and simpler Bachman notation for data modeling.

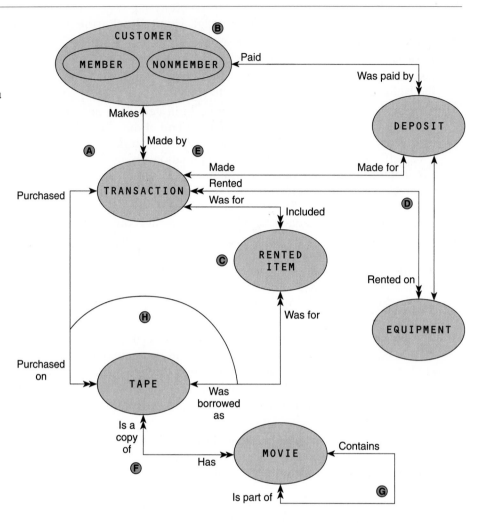

(G) The 1-Ary relationship is depicted with only cardinalities.

(H) Mutually exclusive (and inclusive) relationships are depicted as in the more advanced Bachman notation but are frequently left off this simpler diagram.

There are many data modeling tools. While you are most likely to encounter one or more of the four models we've presented thus far, you should be able to easily adapt to most other models by simply recognizing their symbolic notations. Now let's examine some technology that can help in the development of data models.

Computer-Aided Systems Engineering (CASE) for ERDs

Computer-aided systems engineering (CASE) was introduced in Chapter 5 as an emerging, enabling technology for systems analysis and design methods. Many CASE products support computer-assisted data modeling. CASE takes the drudgery out of drawing and maintaining the diagrams.

Using a CASE product, you can easily create professional, readable ERDs without the use of paper, pencil, erasers, and templates. The ERDs can be easily modified to reflect corrections and changes suggested by end-users; you don't have to start over! Also, most CASE products provide powerful analytical tools that can check your ERDs for mechanical errors, completeness, and consistency. Some CASE products can even help you analyze the data model for consistency, completeness, and flexibility. (Data analysis, an important analytical technique, will be described later in this chapter.) The potential time and quality savings are substantial.

CASE tools do have their limitations. Not all data model conventions are supported by all CASE products. Therefore, it is very likely that any given CASE product may force the company to adapt their methodology's data modeling symbols or approach so that it is workable within the limitations of their CASE tool.

All of the remaining ERDs in this book were created with a popular CASE product called Excelerator. Only the color screens and annotations were added by an artist. All of the entity types on our ERDs were automatically cataloged into a project repository where details about attributes and cardinality are recorded. Figure 8.19 demonstrates a sequence of Excelerator screens for computer-assisted data modeling.

HOW TO CONSTRUCT DATA MODELS

There are many opportunities for and approaches to performing data modeling. In this section, we'll examine when data modeling might be performed during systems development. We'll also demonstrate a step-by-step approach for performing data modeling.

Data Modeling throughout the Life Cycle

Data modeling may be performed during various phases of the systems development life cycle. Data models are progressive — there is no such thing as the "final" data model for a business or application. Instead, a data model should be considered a living document that will change in response to a changing business. Data models should ideally be stored in a repository so that they can be retrieved, expanded, and edited over time. Let's examine how data modeling may come into play during systems planning and analysis.

Data Modeling during Systems Planning

Most systems planning methodologies and techniques require data modeling. Data is the fundamental resource around which all applications are built. Therefore, if you get control of your data, you can better harness it to produce information and competitive advantage. Let's briefly examine how data models might evolve out of the three systems planning phases: study, definition, and analysis (these phases were described in detail in Chapter 6).

During the study phase of systems planning, data modeling usually does not come into play; rather, the focus is entirely upon studying the business and its mission. However, the business focus may call for identification of critical information needs that effectively identify data entities. For example, executive

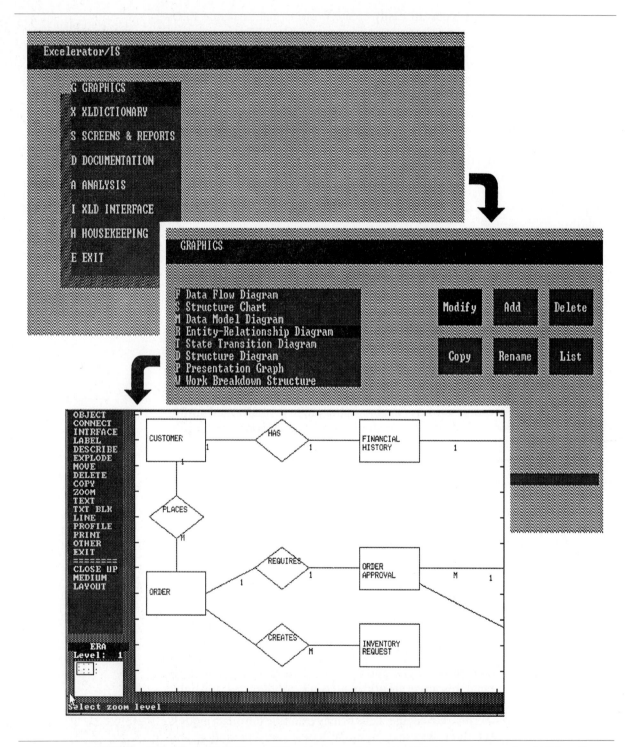

FIGURE 8.19   CASE Tools for Data Modeling   This sequence of screens demonstrates a popular CASE tool's usefulness in drawing data models. (Screens were captured from INTERSOLV's Excelerator/IS.)

managers may determine that summary information about customers, orders, and products is critical to evaluating progress toward achieving business goals, objectives, and critical success factors. A data model is not usually drawn, but the entities CUSTOMER, ORDER, and PRODUCT have been identified.

During the definition phase of systems planning, an **enterprise data model** is usually constructed. This model reflects management's high-level view of its critical information entities. The enterprise data model is also called a **subject data model** because its entities represent high-level subjects about which management seeks information. Entity supertypes are often defined without consideration of their subtypes.

The result is a single, enterprisewide data model. Data attributes are not defined in most methodologies since senior executives frequently don't have the in-depth knowledge to attribute the model. Maintenance of this model is typically assigned to the data administration group. They will expand and refine the model based on the results of any subsequent systems planning and application development.

Note Even in the absence of formal information systems planning, data administrators often take the initiative to create and maintain an enterprise data model.

In the information engineering methods, systems planning proceeds to another phase: analysis of business areas. A business area or collection of similar business functions is identified. The entities and relationships that are pertinent to that business area are grouped into a **business area data** (sub)**model.** Working with the managers and supervisors of the business area, that business area data submodel is thoroughly examined, expanded, and refined. Subtypes, complex relationships, identifiers, attributes, ordinalities, and cardinalities are recorded. All changes recorded on the business area data model must be carried backward to the enterprise data model. Thus, as business areas are analyzed over time, the enterprise data model will become more detailed and complete. Meanwhile, business area data models become the foundation for applications development projects.

Data Modeling during Systems Analysis

Application development begins with systems analysis. Recall that systems analysis consists of three phases: survey, study, and definition. Data modeling plays a significant role in these phases.

During the survey and study phases, a simple **application context data model** might be quickly constructed to define project scope. Attributes, ordinalities, and cardinalities are rarely, if ever, defined. It can be interesting to compare this data model with either an enterprise data model or a business area data model (if either of those exist). In fact, this context data model may help the systems analyst and data administrator carve out the appropriate subset of the enterprise or business area data model for further development.

Data modeling is rarely associated with the study phase of systems analysis. This is unfortunate. Most analysts prefer to draw process models to document the current system. But some experts (such as Steven Mellor and, recently, Ed Yourdon) report that data modeling may be a far superior technique for studying a current system or application. We agree! Consider the following:

- Data models help analysts to quickly identify business vocabulary more completely than process models.

The Next Generation

OBJECT AND BUSINESS RULE MODELING

Today, data-driven and data modeling–based methods, such as information engineering, dominate systems development. They evolved in response to perceived inadequacies in process-driven and process modeling methods, such as structured analysis and design. But the pendulum is always swinging! In this box we examine two emerging methods, object modeling and business rule modeling. Object modeling is almost here today! Business rule modeling is in its formative stages.

Object Modeling The next major paradigm shift is already in progress—object-oriented analysis and design. Many experts view it as the heir apparent to information engineering, today's method of choice. The entire paradigm shift is based on the merging concept of an object. An **object** is the combination of data and all processes (called methods in the object-oriented world) that access and maintain that data. Processes are invoked by messages sent from one object to another.

To most, object-oriented methods represent a radical shift in thinking and techniques. Unfortunately, the perception of radical change is slowing the acceptance of object-oriented analysis and design. Several methodologies have emerged, with none the clear winner at the time this box was written. One technique, however, may hold some promise—James Martin's object-oriented information engineering. While object purists may argue (and rebuke) the trueness of Martin's approach, the approach offers a relatively smooth migration from the comfortable world of information engineering and data models, to object-oriented analysis with object models.

(Continued)

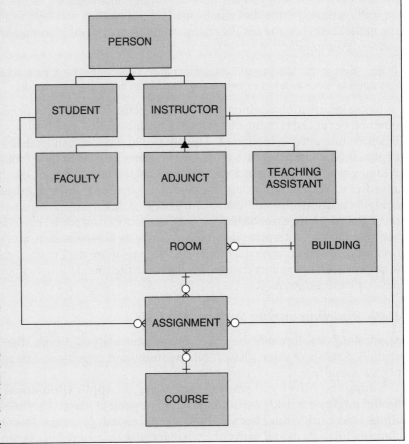

- Data models are almost always built more quickly than process models.
- A complete data model can be fit on a single sheet of paper. Process models often require dozens of sheets of paper.
- Process modelers too easily get hung up on unnecessary detail—which is unnecessary for the study phase.
- Data models for existing and proposed systems are far more similar than process models for existing and proposed systems. Consequently, there is less work to throw away as you move into later phases.

The Next Generation

(*Concluded*)

Martin's object models are very similar to data models. Thus, it should be fairly straightforward to adapt existing CASE tools to draw the models. (The same cannot be said of most object-oriented models.) Information engineering's entity relationship diagram, with extensions, becomes an object relationship diagram.

The diagram illustrates one of the object-oriented paradigm's major concepts—the object hierarchy, a more formalized treatment of entity supertypes and subtypes. The analyst more explicitly looks for hierarchies of objects (entities) that can pass their data attributes on to object subtypes. The small triangle indicates that object subtypes *inherit* the attributes of their supertypes (as well as the processes that access and maintain the supertypes). For example, in a university registration system we define a supertype object called PERSON which has subtypes STUDENT and INSTRUCTOR. The STUDENT and INSTRUCTOR objects inherit the attributes of PERSON. The INSTRUCTOR object might be further refined into three objects—FACULTY, ADJUNCT, and TEACHING ASSISTANT—each of which inherits the common attributes of INSTRUCTOR and PERSON. Otherwise, an ERD's cardinality and ordinality symbols are retained in the object relationship diagram.

This box has, out of necessity, greatly oversimplified object relationship diagrams. Additionally, Martin's approach calls for an object behavior diagram that models the processes or methods to be encapsulated within objects, and the messages they pass to one another to invoke the processes. For more information on object modeling and object-oriented analysis and design, read the book *Object-Oriented Analysis and Design* by James Martin and James O'Dell (Prentice Hall, 1992).

Business Rule Modeling The 1970s brought us structured analysis and design. The 1980s brought us information engineering. The 1990s will bring us object-oriented analysis and design. Why so many modeling paradigms? Because we continue to chase the elusive goal of overcoming the communications gap that persists between information professionals and business professionals. So what's next? Although many would have you believe that objects are the last word—the proverbial silver bullet to all our communications problems. But some already believe that objects, while an improvement, are still too complicated to bridge the gap. And some experts believe that business rules are the next (or "ultimate") answer.

In a sense, business rules have been the goal all along. All of the modeling techniques to date have tried to achieve the goal of modeling reality—modeling the business rules and policies that underlie the business. There are many definitions of business rules. Daniel Appleton defines a **business rule** as "an explicit statement stipulating a condition that must exist in a business information environment for information extracted from the environment to be consistent with business policy." Ideally, business rules are expressed in a language that is natural to business experts and which can be automatically expressed or transformed into a language that is natural to information professionals (such as the models discussed in this chapter).

No popular business rule analysis methods or languages have emerged to deal with this problem. However, considerable research and discussion is occurring to develop methods and languages, especially in systems development and database circles. At least one CASE tool, Sapiens, is sold to model and implement information systems in terms of business rules. The subject of business rule modeling and analysis certainly warrants your attention for the next generation of systems development.

As we move into the definition phase, data modeling is almost always performed. If an enterprise or business area data model already exists, it is expanded or refined to reflect application requirements. Otherwise, the data model from the study or definition phase is expanded or refined. In either case, the result is an **application essential data model** that reflects data requirements for the target system, independent of implementation technology concerns. (Again, this model may merely be an update to the business area and enterprise data models.) Attributes, identifiers, complex relationships, ordinalities, cardinalities, and other aspects covered in this chapter should be defined.

All definition phase data modeling should be a team effort that includes systems analysts, users and managers, and data analysts or administrators. The data analysts or administrators often have final approval of all data models. On the other hand, some data analysts and administrators are too technical. They want to discuss specific database technology and implementation issues such as tables, concatenated keys, foreign keys, indexes, and the like. Such discussion alienates or confuses users and managers. The systems analyst should intercede to keep both the discussion and the data model at the business level of detail as described in this chapter. The Next Generation box examines models that many believe will replace data models in future systems analysis.

Looking Ahead to Systems Design

The essential data model from systems analysis describes business data requirements, not technical solutions. As we proceed to systems design, the data model must become more technical. It must become an **application implementation data model** that will guide the technical implementation of files and databases.

File and database design often begins with a formal **data analysis** or **data normalization** that is performed on the essential data model. At the risk of oversimplification, this data analysis prepares the data model for implementation.

- It ensures that stored data is not stored redundantly. Ideally, attributes should be stored in one and only one entity. Thus, when a value changes, it only has to be changed in one location. This ensures the integrity of the data.
- It ensures that stored data is flexible and adaptable to changing requirements. All too often, data is physically organized to meet current requirements only. As new requirements become known, the data is not properly organized to fulfill those requirements. Physical data organization cannot be easily changed without costly changes to existing programs that use the data. Data analysis can maximize data flexibility prior to file and database design.

Some methodologies insist that data analysis be performed as part of systems analysis. We used to teach it that way ourselves. Today, we consider data analysis to be implementation-oriented—that is, it is more appropriate for systems design. Still, other qualified experts consider it part of systems analysis. Data analysis is covered in Chapter 13.

After data analysis, files and databases are physically designed. Implementation data models for files and databases are covered in Chapter 15.

A Step-by-Step Data Modeling Approach

You already know enough about ERDs to read them. But as a systems analyst or knowledgeable end-user, you must learn how to draw them. We will use the SoundStage Entertainment Club systems project to teach you how to draw ERDs. This example teaches you to draw the data model from scratch. In reality, you should always look for an existing data model. If such models exist, they are usually maintained by the data management or data administration group.

All data models in this section were created with the CASE tool, Excelerator/ IS. For the case study, we did not screen objects in color. Except for bullets that call your attention to specific points, they are printed here as they would appear if a good quality laser printer had been used.

Step 1: Identify Entities

The first task in data modeling is relatively easy. You need to identify those entities in the system that are or might be described by data. You should not restrict your thinking to entities about which the end-users know they want to store data. There are several techniques that may be used to identify entities.

*   During interviews or JAD sessions with system owners and users, pay attention to key words in their discussion. For example, during an interview with an individual discussing SoundStage's membership processing, the individual may state that "We have to keep track of all our members and the many clubs in which they are enrolled." Notice that the key words in this statement are "members" and "clubs." Both MEMBER and CLUB are likely entities!
*   During interviews or JAD sessions, specifically ask the system owners and users to identify things about which they would like to capture, store, and produce information. More often than not, those things represent entities that should be depicted on the data model.

*   Another technique for identifying entities is to study the forms and files sampled during the study phase. Some forms identify event entities. Examples include ORDERs, REQUISITIONs, PAYMENTs, DEPOSITs, and so forth. But most of these same forms also describe other entities. Consider a course request form for your school's course registration system. A COURSE REQUEST is itself an event entity. But the average course request form also contains items that describe other entities, such as STUDENT (a person), COURSE (an event), INSTRUCTOR (a person), ADVISER (a combination), SECTION (a somewhat abstract location), and the like. These same entities would also be derived from other course registration system forms and files.
*   Technology may also help you identify entities. Reverse engineering CASE tools are evolving to read program data structures, file structures, and database structures and to recover first draft essential data models from those structures. The analyst must usually clean up the resulting model by replacing implementation names, codes, and abbreviations with more complete business-oriented and recognizable names.

While these techniques may prove useful in identifying entities, they occasionally play tricks on you. A simple, quick quality check can eliminate false entities. Ask your user to specify the number of occurrences of the entity. A true entity has multiple occurrences—dozens, hundreds, thousands, or more! If not, the entity is false.

Give simple, meaningful, business-oriented names to each entity. Try not to abbreviate or use acronyms. Also, remember that entity names should be singular. To verify your understanding of vocabulary, define each entity in business terms to the project repository. This glossary of terms will serve both you and future analysts and users for years to come. For your convenience, descriptions of each entities identified for the SoundStage project are presented in Figure 8.20.

FIGURE 8.20 **Sound-
Stage Member Services
Data Entities** Data
entity definitions are
recorded in a project
repository. This simple
glossary report could be
an output from that
repository.

AGREEMENT: A contract whereby a member agrees to purchase a certain number of products within a certain time period. After fulfilling that agreement, the member will receive
bonus credits, as specified in the agreement, for each additional purchase.

BACKORDER: An order or partial order created in response to SoundStage's inability to fill
a member's order due to lack of stock.

CLUB: A SoundStage membership group to which members can belong. Examples include
the Compact Disc Club and the VHS Videotape Club.

MEMBER: A member of one or more clubs.

ORDER: An automatic, dated order generated in response to a monthly promotion. The
order may be approved, revised, or canceled via the member's response. If not canceled
or revised by a specified date, the order is normally shipped.

PRODUCT: Audio, video, or other entertainment merchandise sold to members through the
club.

PROMOTION: Monthly or quarterly events whereby automatic, dated orders are created for
all members in a club. The promotion specifies a "Selection of the Month" that will automatically be filled unless it is canceled or revised by the member within a specified time period.

Step 2: Define Identifiers for Each Entity

Next, identify the data attribute(s) that uniquely identifies one and only one
occurrence of each entity. In many cases, these identifiers already exist in either
computer files or printed forms.

Many entities have one and only one identifier. In SoundStage, ORDER
NUMBER is the only attribute that uniquely identifies an ORDER. In other
cases, multiple attributes may each (by themselves) uniquely identify an entity.
For example, in SoundStage, either MEMBER NUMBER or MEMBER NAME
uniquely identifies a MEMBER. (Note: The more MEMBERs that exist, the less
likely this is to be true.)

Some entities cannot be identified by a single attribute. Instead, they require
a combination of attributes to uniquely identify an occurrence of the entity. For
example, in SoundStage, the cassette tape and compact disc version of the same
recording are considered different products. Thus, a specific PRODUCT can
only be identified by the combination of a PRODUCT NUMBER and a MEDIA
CODE. You should consider inventing a simpler key as an alternative to such a
combination key. Simpler keys can be extremely useful, not only for computer
files but also for easier inquiries and forms.

What do you do if you have two or more entities that have the same key? You
may find situations where different entities legitimately share the same key.
Consider the event entity ORDER. Other entities, such as BACKORDER, are
triggered by ORDERs. It is not uncommon for a business to assign the same key,
in this case ORDER NUMBER, to all of the related entities. This makes the
ORDER easier to track. Each entity is a distinct event. You could, however,
create a combination key for each entity to more clearly distinguish between
the entities. For example, we can give BACKORDER the combination key
ORDER NUMBER + BACKORDER DATE.

The keys for our entities are depicted in Figure 8.21. We have adopted the
common practice of underlining primary keys.

If you cannot define keys for an entity, it may be that the entity doesn't really
exist—that is, *multiple* occurrences of the so-called entity do not exist.

AGREEMENT:	AGREEMENTS NUMBER SUFFIX
BACKORDER:	ORDER NUMBER + BACKORDER DATE
CLUB:	CLUB NAME
MEMBER:	MEMBER NUMBER or MEMBER NAME
ORDER:	ORDER NUMBER
PRODUCT:	PRODUCT NUMBER + MEDIA CODE
PROMOTION:	CLUB NAME + PROMOTION DATE

FIGURE 8.21 SoundStage Data Entity Identifiers Data entity occurrences must be uniquely identifiable by an attribute or some combination of attributes. Identifiers are recorded in a project repository.

ENTITY	MEMBER	CLUB	AGREEMENT	PROMOTION	ORDER	PRODUCT	BACKORDER
MEMBER		ENROLLED IN one or more	IS BOUND TO one or more		MUST RESPOND TO zero or more –and– RESPONDED TO zero or more		
CLUB	ENROLLS zero or more		ESTABLISHES one or more	SPONSORS zero or more			
AGREEMENT	BINDS zero or more	ESTABLISHED FOR one and only one					
PROMOTION		SPONSORED BY one and only one			GENERATES one or more	PROMOTES one or more	
ORDER	AWAITS RESPONSE FROM one and only one –and– RESPONDED TO BY one and only one			GENERATED BY one and only one		REQUESTED one or more	GENERATES zero or one
PRODUCT				PROMOTED BY zero or more	REQUESTED BY zero or more		CONTAINED ON zero or more
BACKORDER					GENERATED FOR one and only one	CONTAINS one or more	

FIGURE 8.22 Relationship Matrix Relationships can sometimes be defined using a simple entity-to-entity matrix. Some CASE tools support such a matrix and can generate an ERD from such a matrix.

Step 3: Draw a Rough Draft of the Entity Relationship Data Model

The next task in data modeling is to build a first draft of the entity relationship diagram. We complete the draft model by brainstorming relationships between our entities. A simple matrix such as the one shown in Figure 8.22 can serve as a useful tool when you are brainstorming relationships between entities. Our sample matrix represents relationships that exist between entities that were identified for the definition phase of the SoundStage project. Notice that the name of each entity appears across the top row and down the first column. Using a matrix allows you to more easily address all possible relationships that might exist between entities on our ERD. A matrix template can be created in most word processors, and the matrix can easily be completed by the analyst working with the users.

Once the relationships have been identified, we can now draw our ERD. We have completed this step in Figure 8.23. Don't worry if you don't get it perfect the first time—that would be rare. Once we begin mapping attributes, new entities and relationships may surface. The ERD that was generated using our matrix communicates the following:

(A) A CLUB—for example, compact disc club, VHS video tape club, and so on—ESTABLISHES one or more membership AGREEMENTs. Members will learn about these AGREEMENTs through advertisements in newspapers and magazines.

A specific AGREEMENT is ESTABLISHED FOR one and only one CLUB.

(B) A CLUB ENROLLS zero or more MEMBERs. Why zero? The club may be new and awaiting receipt of its first membership application.

A MEMBER may ENROLL IN one or more CLUBs. Note that this is our first many-to-many relationship.

(C) A MEMBER IS BOUND TO one or more AGREEMENTs, depending on how many clubs that member has joined.

An AGREEMENT BINDS zero or more MEMBERs. Again, the agreement may be awaiting its first member enrollment.

(D) Each month, a CLUB SPONSORs zero or more marketing PROMOTIONs. Each PROMOTION, however, is SPONSORED BY one and only one CLUB.

(E) A club's PROMOTION automatically GENERATES one or more ORDERs for members in that club.

Members have to respond by a specific date to cancel the order or have it filled. Most, but not all, ORDERs are GENERATED BY one and only one PROMOTION. The lone exception are ORDERs placed as a result of a member first joining the club (a bonus order for joining the club).

(F) A PROMOTION PROMOTES one or more PRODUCTs.

A PRODUCT may be PROMOTED BY zero or more PROMOTIONs.

(G) A MEMBER MUST RESPOND TO zero or more outstanding ORDERs. Note that at any given time, a member may have no outstanding ORDERs to which he or she must respond. Alternatively, if a member belongs to more than one club, he or she may have several ORDERs.

An ORDER MUST BE RESPONDED TO by one and only one MEMBER.

(H) MEMBERs will RESPOND TO zero or more ORDERs by canceling, approving, or altering the ORDERs.

An ORDER is RESPONDED TO by one and only one MEMBER.

Note that the relationships marked **(G)** and **(H)** are exclusive—that is, an ORDER may either be awaiting the member's response or it has been responded to, but not both.

(I) If one or more PRODUCTs cannot be filled due to lack of stock, an ORDER GENERATES one and only one BACKORDER.

Similarly, a BACKORDER is GENERATED FOR one and only one ORDER. This is the first one-to-one relationship we've encountered.

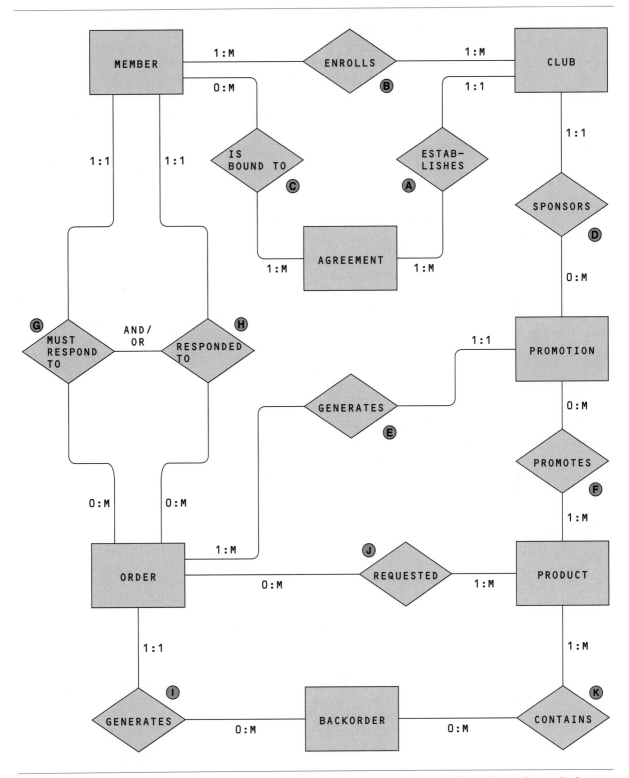

FIGURE 8.23 **The SoundStage Essential Data Model**   These are the entities and relationships about which SoundStage wants to store data.

Ⓙ An ORDER REQUESTED one or more PRODUCT, usually the product being advertised as the "Selection of the Month."

At the same time, a specific PRODUCT may be CONTAINED ON zero or more ORDERs (zero if the product is not currently selling well). Notice how this inverse relationship could help us track sales of all PRODUCTs. Also note that this is a many-to-many relationship.

Ⓚ A BACKORDER CONTAINS one or more PRODUCTs.

Similarly, a PRODUCT is contained on zero or more BACKORDERs.

If you read each of the preceding items carefully, you probably learned a great deal about the Entertainment Club. ERDs have become increasingly popular as a tool for explaining the total business picture for systems projects.

Step 4: Identify Data Attributes

It may seem like a trivial task to identify data attributes; however, analysts not familiar with data modeling frequently encounter problems. To accomplish this task, you must have a thorough understanding of the data attributes for the system. By studying the forms, files, and reports, you identify those data attributes that are essential to the system you are developing. If a business area data model exists, most of the attributes may have already been identified and recorded in a repository. If not, we suggest the following strategy for identifying data attributes, which is best accomplished by working with your system owners and system users:

1. Sample a form, document, printout of a file, report, or other implementation of stored data. (Note: Sample size and sampling are discussed in Part Five, Module B, Fact-Finding Techniques.) We will demonstrate the remaining steps using the sample paper form in Figure 8.24.
2. Circle each unique item on the form. In Figure 8.24, notice that there is not always a one-to-one correspondence between an item on the form and its associated attribute. For instance, checked boxes frequently represent multiple values for a single attribute.
3. Draw an X through circled items that won't be stored by the new system.
4. Draw an X through extraneous items, such as signatures (assuming that you circled them).
5. Draw an X through constant information. In other words, if every occurrence of the form has the same value for a field, that field is not an attribute. For instance, the "Remit Payment To" entry will always have the value "SoundStage Entertainment Club." Therefore, it is not an attribute.
6. Verify your attributes with your end-users, especially if there is any question as to whether some of the attributes are really needed.

Another recommended technique for identifying attributes is **brainstorming.** Pick an entity and ask yourself and your users what characteristics (data attributes) describe that entity.

You must also logically name the data attributes. Although the sample form has names, they are not sufficient. Forms and files usually have abbreviated names called *labels*. On those forms or in those files, the names make sense. Taken out of that context, they don't always make sense. Consider the attribute NAME. On the order form, we know that this is a MEMBER NAME. Taken out of

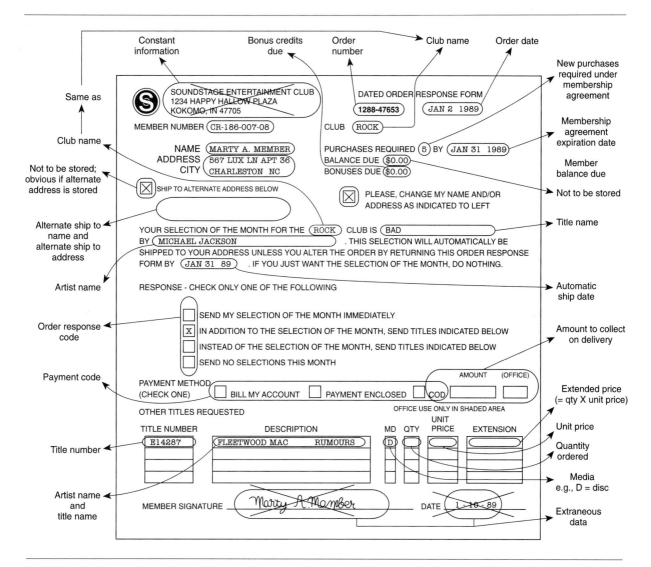

FIGURE 8.24 Forms Sampling Data attributes can be gleaned from sample forms and files. This form has been marked to identify data attributes.

the context of the form, there could be other interpretations: ARTIST NAME, PRODUCT NAME, or CLUB NAME. Naming guidelines for data attributes follow:

- Do not use abbreviations unless absolutely necessary. For example, a label may read COD. After studying completed forms, you find that a dollar amount is actually recorded. The correct logical name is AMOUNT TO COLLECT ON DELIVERY.

- Expand on generic labels. A label may say DATE, but date of what? Try ORDER DATE. This distinguishes the attribute from dates for other event entity types. Notice that this also places the name of the data entity being described first, followed by the actual attribute descriptor. Other examples include CUSTOMER NAME instead of NAME, QUANTITY ORDERED instead of QUANTITY, and ORDER AMOUNT DUE instead of DUE.

- Add a question mark to data attributes whose values are *yes* and *no.* For example, the attribute name CHANGE OF ADDRESS? is more descriptive than CHANGE OF ADDRESS. The first clearly implies that the value of the attribute answers the question. You can't be sure what the values of the second attribute name are.
- Some data attributes on forms are not labeled. Recall the example of a multiple-choice attribute. The form describes the values of the attribute instead of the attribute itself. You must name the attribute (see MEDIA on the sample form).

Step 5: Map Data Attributes to Entities

Next, you need to identify and map all data attributes from Step 4 to the entities. Here again, if a corporatewide data model exists, it is likely that the data attributes describing each entity are documented and readily available. If not, we suggest the following approach:

1. Pick an entity or relationship (associative entity) from your data model.
2. Find the forms, file printouts, reports, and so on whose data describes the entity.
3. Record the attributes for that entity.
4. If you can't find any sample data for an entity, interview end-users to identify appropriate or desired data attributes.
5. Repeat steps 1 to 4 for each entity.
6. If you have attributes left over, you may have missed an entity (and some relationships).

IMPORTANT   A data attribute should be associated with one and only one data entity. Do not redundantly record attributes in different entities. Either give the attributes unique names to differentiate them or determine which entity they "best" describe.

Review these contents of entities with appropriate system owners and users. Brainstorm new attributes wherever possible. In general, attributes need not be deleted since the data model is not a design commitment.

For your information, Figure 8.25 provides the mapping of data attributes to entities for the definition phase of our SoundStage systems project.

──────────────────────── *Summary* ────────────────────────

Data modeling is a technique for defining requirements for data that will be stored in a new system. This chapter presented a popular data modeling tool, entity relationship diagrams (ERDs). ERDs and data models describe the logical—as opposed to the physical—requirements for a system. Essential models do not show the physical implementation of the model, even if that implementation is known or anticipated.

ERDs show entities and relationships. Entities are persons, objects, events, and locations that can be described by data attributes or attributes. Examples include customers, products, orders, and regions. Relationships are natural associations between entities. For instance, customers PLACE orders that CONTAIN products.

ERDs are relatively easy to draw, since the analyst only has to define the entities, relationships, and data attributes needed. Most of this can be gleaned from existing

CLUB:
CLUB NAME
NUMBER OF MEMBERS ENROLLED
NUMBER CANCELED YTD
CURRENT PROMOTION
TOTAL UNITS SOLD FOR CLUB
MAXIMUM PERIOD OF OBLIGATION

MEMBER:
MEMBER NUMBER or MEMBER NAME
MEMBER ADDRESS consisting of:
    STREET
    P.O.BOX
    CITY
    STATE
    ZIPCODE
MEMBER PHONE
DATE ENROLLED
BALANCE PAST DUE
BONUS CREDITS NOT USED
CLUB GROUP (repeats 1-n times)
consisting of:
    CLUB NAME
    MUSICAL/MOVIE PREFERENCE
    NUMBER OF PURCHASES REQUIRED
    NUMBER OF PURCHASES TO DATE
    AGREEMENT NUMBER SUFFIX
    AGREEMENT ENROLLMENT DATE
    AGREEMENT EXPIRATION DATE

AGREEMENT:
AGREEMENT NUMBER SUFFIX
CLUB NAME
AGREEMENT EXPIRATION DATE
AGREEMENT PLAN CREATION DATE
MAXIMUM PERIOD OF OBLIGATION
BONUS CREDITS AFTER OBLIGATION
NUMBER OF MEMBERS ENROLLED
NO. MEMBERS WHO HAVE FULFILLED
NO. MEMBERS HAVE NOT FULFILLED

BACKORDER:
ORDER NUMBER + BACKORDER DATE
BACKORDERED ITEM (repeats 1-n times)
consisting of:
    PRODUCT NUMBER
    MEDIA CODE
    PRODUCT DESCRPTION
    QUANTITY BACKORDERED

PROMOTION:
CLUB NAME + PROMOTION DATE
PROMOTION TYPE
SELECTION OF MONTH NUMBER
SELECTION OF MONTH TITLE
AUTOMATIC RELEASE DATE
AUTOMATIC FILL DATE

ORDER:
ORDER NUMBER
ORDER DATE
ORDER STATUS
PROMOTION NAME
PROMOTION DATE
AUTOMATIC FILL DATE
MEMBER NUMBER
MEMBER NAME
FORMER MEMBER?
MEMBER ADDRESS consisting of:
    STREET
    P.O.BOX
    CITY
    STATE
    ZIPCODE
ORDERED PRODUCT (repeats 1-n times)
consisting of:
    PRODUCT NUMBER
    MEDIA CODE
    PRODUCT DESCRIPTION
    QUANTITY ORDERED
    ORDERED PRODUCT STATUS
    QUANTITY SHIPPED
    ORDER PRICE
    EXTENDED PRICE
AMOUNT DUE

PRODUCT:
PRODUCT NUMBER + MEDIA CODE
PRODUCT DESCRIPTION
TITLE OF WORK
COPYRIGHT DATE
CURRENT RETAIL PRICE
CURRENT LIST PRICE
SUPPLIER NAME
SUPPLIER ADDRESS consisting of:
    STREET
    P.O.BOX
    CITY
    STATE
    ZIPCODE
QUANTITY ON HAND
UNITS SOLD
VALUE OF UNITS SOLD

FIGURE 8.25 **Fully Attributed Entities** Notice that each attribute is recorded in one and only one entity.

forms and files, supplemented by interviews and walkthroughs. However, data models should always be analyzed to structure the data so that it is flexible, nonredundant, and easy to understand.

Data analysis is a technique for structuring data in its simplest, most flexible form. It uses an approach called *normalization* to simplify the data model. Normalization is covered in Chapter 13.

Key Terms

application context data
   model, p. 329
application essential data
   model, p. 331
application
   implementation data
   model, p. 332
associative entity, p. 317
attribute, p. 312
brainstorming, p. 338
business area data
   model, p. 329
business rule, p. 331

cardinality, p. 317
data analysis, p. 332
data attributes, p. 312
data entity, p. 308
data modeling, p. 306
data normalization, p. 332
data relationship, p. 308
enterprise data
   model, p. 329
entity, p. 308
entity relationship
   diagram, p. 307
entity subtype, p. 314

entity supertype, p. 314
essential model, p. 305
identifier, p. 312
implementation
   model, p. 304
intersection entity, p. 319
model, p. 304
object, p. 330
ordinality, p. 316
relational entity, p. 317
relationship, p. 308
subject data model, p. 329

Problems and Exercises

1. Give three reasons why many analysts believe that requirements should be specified in an implementation-independent fashion.

2. Differentiate between an entity relationship diagram (ERD) and a data flow diagram (DFD).

3. Most data entities correspond to persons, objects, events, or locations in the business environment. Give three examples of each data entity class.

4. Obtain three sample business forms from a business, your school, or your instructor. What entities are described by the fields on the forms?

5. What entities are described on your class schedule form or your school's course registration form?

6. Give two examples of each of the following data relationship complexities: one-to-one (1 : 1), one-to-many (1 : M or M : 1), and many-to-many (M : M). Draw an ERD for each of your examples. Be sure to label data entities using nouns and label data relationships using verbs. Annotate the graph to communicate the relationship complexity.

7. During the survey and study phases, an analyst collected numerous samples, including documents, forms, and reports. Explain how these samples will prove useful for data modeling.

8. During the definition phase, the actual data model is drawn, refined, and improved. Identify the data modeling issues that must be specified when alternative solutions are being identified and analyzed in the selection phase.

9. List and explain the seven steps for constructing a data model.

10. Given the following narrative description of entities and their relationships, prepare a draft entity relationship diagram (ERD). Be sure to state any reasonable assumptions that you are making.

Burger World Distribution Center serves as a supplier to 45 Burger World franchises. You are involved with a project to build a database system for distribution. Each franchise submits a day-by-day projection of sales for each of Burger World's menu products (the products listed on the menu at each restaurant) for the coming month. All menu products require ingredients and/or packaging items. Based on projected sales for the store, the system must generate a day-by-day ingredients need and then collapse those needs into one-per-week purchase requisitions and shipments.

11. Write a paragraph or two explaining how you would present and verify the ERD prepared in Exercise 10 to a group of end-users who are not familiar with computer concepts.

12. What is an identifier? Give at least three examples of identifiers and the entities they describe.

13. What is an entity supertype? What is an entity subtype? Give two examples of each.

14. What is an associative entity? How is it different from a fundamental entity? Give at least two examples of associative entities.

15. What is an N-ary relationship? What is a mutually exclusive relationship? What is a mutually inclusive relationship?

16. Identify and briefly explain three different popular data modeling tools.

Projects and Minicases

1. Obtain copies of all of the forms used in your school's course registration system. Your instructor may be able to supply these forms. Using the five steps of data modeling, derive a Chen-style data model to support a course registration system.

2. Convert the data model from Project 1 above into a Martin-style data model.

3. Convert the data model from Project 1 above into a Bachman-style data model.

4. Given the following data attributes and entities, indicate which attributes could be identifiers for each of the entities. Draw a rough draft entity relationship diagram.

Microcomputer Property Accounting and Maintenance System

Entity: Equipment

Serial number
Model number
Manufacturer name
Manufacturer location
Item description
Installed in serial number (if applicable)
Date purchased
Supplier name
Supplier address
Purchase price
Replacement cost
Current end-user location
Current responsible end-user

Entity: Software Package

Serial number
Package name
Package type
Version number
Registration date
Number of end-users allowed
Date purchased
Supplier name
Supplier address
Purchase price
1 or more of:
    Current end-user
    Current location

Entity: Spare Part

Internal part number
Vendor part number
Supplier part number
Part description
Used in one or more of:
    Manufacturer name
    Model number

Entity: Warranty

Serial number
Model number
Manufacturer name
Manufacturer location
Item description
Date purchased
Date warranty expires
Warranty servicer name
Warranty servicer address
Warranty servicer phone

Entity: Maintenance Contract

Serial number
Model number
Manufacturer name
Item description
Maintenance purchase date
Maintenance start date
Maintenance finish date
Labor covered?
Parts covered?
Bill rate for costs not covered

5. Given the following data attributes and entities, indicate which attributes could be identifiers for each of the entities. You may have to combine attributes or even add some attributes that are not listed. Map all of the attributes to their appropriate entity. Remember, each attribute should describe one and only one entity. Draw a rough draft entity relationship diagram.

Green Acres Real Estate System

Entities:

Seller	House	Closing
Buyer	Offer	Showing
Listing	Property	Room

Attributes:

Seller name	Offer amount
Square foot size	Listing date
Seller address	Property description
House style	Offer date
Closing location	Room type
Listing price	Property size
Number of bathrooms	Showing time
Garage size	Room size
Showing date	Elementary school zone
Garage location	Buyer phone number
Buyer name	Closing date
Basement size	Sales terms
House heating method	

Suggested Readings

Flavin, Matt. *Fundamental Concepts of Information Modeling.* New York: Yourdon Press, 1981. A classic book on information modeling and the entity relationship diagram approach.

Martin, James, and Clive Finkelstein. *Information Engineering.* 2 vols. New York: Savant Institute, 1981. Information engineering is a formal, database, and fourth-generation language-oriented methodology. The method is logically equivalent; however, the authors use entity diagrams instead of entity relationship diagrams. ERDs could easily be substituted.

Mellor, Stephen. *Object Oriented Design.* Englewood Cliffs, N.J.: Prentice Hall, 1987. This modern book presents an object (entity)-oriented approach to structured systems analysis and design.

Weaver, Audrey. *Using the Structured Techniques.* Englewood Cliffs, N.J.: Prentice Hall, 1987. This book presents, in case study form, a structured methodology based on the complementary use of information models and process models (which we cover in the next chapter).

9

Process Modeling

Chapter Preview and Objectives

This is the second of three graphic systems modeling chapters. In this chapter you will learn how to use a popular process modeling tool, data flow diagrams, to document a system's processing and data flow, independent of how that processing and data flow is or will be implemented. You will know process modeling as a systems analysis tool when you can:

Define
process modeling and explain why it is important.

Explain
how process modeling is useful for
systems planning and analysis.

Factor
a system into component subsystems, functions, and tasks and
depict its structure using a decomposition diagram.

Document
the interactions between subsystems, functions, and tasks
using data flow diagrams.

Explain
the complementary relationship between process models (DFDs)
and data models (ERDs).

Differentiate
between the explosion and expansion approaches for
drawing data flow diagrams.

THE INTERNAL REVENUE SERVICE[1]

The following case describes how the typical IRS regional center processes your tax return. The adaptations are intended to simplify the challenge that follows the minicase.

Scene: *Wayne Richards, an experienced systems analyst, is interviewing Paul Adams, an Internal Revenue Service (IRS) regional supervisor. Wayne has been given the assignment of streamlining and simplifying the way in which the IRS processes tax returns in order to speed up both taxpayer refunds and payment collections. In order to do this, Wayne is trying to understand the current processing methods. Wayne is not at all familiar with the procedures involved since his background is in the area of manufacturing information systems. Let's join Wayne and Paul during an interview.*

Wayne: I'm not really very comfortable with this application area but if you can explain to me how the tax returns are processed, I will feel much better.

Paul: Initially, postal trucks bring tax returns to the regional center. The envelopes are then sorted by type of return—for example, long form versus short form and whether or not the envelope contains a payment. The sorted envelopes are sent to Receipt and Control, where they are further separated into 27 types falling into three general categories: short forms requesting refunds, long forms requesting refunds, and returns containing tax payments.

Wayne: Why are they sorted twice?

Paul: That's a good question. The main reason is because of the sheer volume of the returns. It's not unusual for us to receive more than 200,000 returns in one day. The first sort divides that total to make the job more manageable.

Wayne: Twenty-seven types seems like an awful lot.

Why so many? I always thought there were only tax payments and refunds.

Paul: It's not quite that simple. Some returns are requests for extensions for filing. Others are quarterly estimated tax payments. There are over 500 official government forms for filing tax returns!

Wayne: Maybe it would be easier if we started with the simplest form; which one is that?

Paul: In order to process short forms requesting refunds, operators submit forms to a machine that scans the returns and stores the data for later processing. The data is read by the main computer. It determines the correct tax, decides whether a refund should be sent, updates taxpayers' files, and prints letters, notices, liens, etc.

Wayne: Do we print the actual refund checks?

Paul: No, the refund information is sent to the National Computing Center, which subsequently triggers the Treasury Department to issue the actual refund checks. Letters, notices, and other communications are sent to local IRS sites around the country, from which appropriate information is sent to taxpayers.

Wayne: That seems fairly straightforward; what about the long forms requesting refunds?

Paul: The processing of long forms requesting refunds is similar, but not identical, to the processing of the short forms.

Wayne: Why are they processed differently?

Paul: Partly because the long forms usually include multiple schedules of information, such as itemized deductions. First, returns are sorted into blocks of batches to be processed as single units. Batches are numbered to ensure that no returns are lost or excessively delayed. The batches are then forwarded to examiners. The

[1] Adapted from "The IRS: How Your Return Is Processed," *USA Today,* January 8, 1986, p. 7A.

Wayne: examiners check for and correct errors and code the returns for processing.

Wayne: What about returns that are incomplete or unreadable?

Paul: The examiners send back to the taxpayers any returns with incomplete or uncorrectable data. Also, clerks stamp a document locator number on each return for additional tracking capability as the return moves through the system. From this point, the processing is similar to the short form. Returns are input to the computer system. Data is stored for subsequent processing. The data is read by the main computer. It determines the correct tax, decides whether a refund should be sent, updates taxpayers' files, selects returns for possible tax audits, and prints letters, notices, liens, etc. Refund information is sent to the National Computing Center, which subsequently triggers the Treasury Department to issue the actual refund checks. Notices and information regarding audits are sent to local IRS sites around the country, from which appropriate information is sent to taxpayers.

Wayne: How are tax payments handled?

Paul: For returns containing tax payments, examiners check for and correct errors, code the returns for processing, and send back to taxpayers any returns with incomplete or uncorrectable data. Returns are entered into the computer. The computer checks taxpayer calculations and amounts, assigns document locator numbers, and stores the data. Then, the preceding steps are repeated using different operators.

Wayne: Why use two different sets of operators to repeat something that's already been done?

Paul: The data from the second operators is checked against the first set for accuracy. Error reports are sent to examiners. Accurate data is stored for subsequent processing. Checks are collected for daily deposit into the Federal Reserve Bank.

Wayne: What do the examiners do with the error reports?

Paul: Examiners check for errors, correcting any errors they can, and write the taxpayers for any missing information. At this point, the returns follow identical processing as described for the long forms requesting refunds.

Challenge

One of the challenges for systems analysts is to communicate their understanding of systems requirements and solutions. To this end, analysts have learned that pictures often communicate better than words. On a single 8½″ × 11″ sheet of paper, draw a picture that Wayne might use to communicate his understanding (or misunderstanding) of the system's processing. If you have difficulty, don't be too concerned. This chapter introduces you to a useful tool.

SYSTEMS MODELING: THE ESSENCE OF A SYSTEM

Note This section is repeated from Chapter 8, Data Modeling. If you have already covered Chapter 8, you may skim or skip this section. Otherwise, this section introduces *system* modeling as a preface to *process* modeling.

In Chapters 6 and/or 7 you were introduced to activities that called for drawing system models. System models play an important role in systems development. As a systems analyst or user, you will constantly deal with unstructured problems. One way to structure such problems is to draw models.

A **model** is a representation of reality. Just as a picture is worth a thousand words, most models are graphic representations of reality.

Models can be built for existing systems as a way to better understand those systems, or for proposed systems as a way to define requirements and designs.

An important concept, in both this chapter and the next, is the distinction between *implementation* and *essential models*.

An **implementation model** shows not only what a system is or does, but also how the system is physically implemented. Synonyms include *tech-*

Models	Essential System (Describes "What", Also Known as the Logical System)	Implementation System (Describes "What" plus "How", Also Known as the Physical System)
Current System (Existing System)	An essential model of the current system depicts those aspects of the current system that are essential to the business and that should be retained—no matter how we choose to implement the system.	An implementation model of the current system depicts how the current system is physically implemented (inclusive of technology). By default, the implementation model includes all of the essential aspects of the current system.
Proposed System (Target System)	An essential model of the proposed system depicts the business and user requirements for the system—regardless of how that system might be implemented.	An implementation model of the proposed system depicts how the proposed system will be physically implemented (inclusive of technology). By default, the implementation model must include all of the essential requirements of the proposed system.

FIGURE 9.1 **Types of System Models**   Most system models depict either a current or proposed system. Most system models "should" also depict either the essential or implementation of a system. This chapter deals primarily with "essential, current" and "essential, proposed" system modeling.

nology model and *physical model.* (Once preferred, the term *physical model* has become less popular in recent years.)

For example, a structure chart or flowchart is an implementation model of a program. It explains how that program is to be structured or coded.

Implementation models are useful for documenting an existing system's data. However, analysts have learned that data requirements for proposed systems should ideally be specified using essential models.

Essential models are implementation-independent models—they depict the essence of the system (what the system does or must do), independent of how the system will or could be physically implemented. Essential models are also sometimes called *logical models* or *conceptual models.* (The term logical model has become less popular in recent years.)

There are several reasons why you should initially use essential models:

- Implementation-independent models remove biases that are the result of the way the existing system is implemented or the way that any one person thinks the system might be implemented. Thus, we overcome the "we've always done it that way" syndrome. Consequently, implementation-independent models encourage creativity.
- Implementation-independent models reduce the risk of missing functional requirements because we are too preoccupied with technical details. Such errors can be costly to correct after the system is implemented. By separating what the system must do from how the system will do it, we can better analyze the requirements for completeness, accuracy, and consistency.

FIGURE 9.2 **Process Modeling Relationship with the IS Building Blocks**   This chapter is about essential process modeling from the perspective of the system owners and users. The focus is on the essential processes required to run the business.

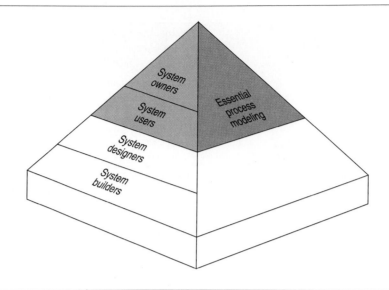

- Implementation-independent models allow us to communicate with end-users in nontechnical or less technical language. We frequently lose requirements in the technical jargon of the computing discipline.

The difference between current versus proposed system models and implementation versus essential system models are summarized in Figure 9.1.

This chapter will present essential process modeling as a technique for defining users' processing requirements. Process modeling was presented earlier in Chapter 4 as a systems development technique.

> **Process modeling** is a technique for organizing and documenting a system's processes, inputs, outputs, and data stores. Process modeling is a software engineering technique; therefore, you may have encountered similar models in software engineering courses. On the other hand, the usefulness of process models extends far beyond describing software processes.

In this chapter we will focus exclusively on **essential process modeling.** Using your pyramid model (see Figure 9.2), essential process modeling deals with *business* processes from the system owners' and system users' points-of-view. Accordingly, such models are not concerned with implementation details or technology.

There are numerous process modeling tools; however, we will concentrate on one, the *data flow diagram.*

DATA FLOW DIAGRAMS

Let's begin with a formal definition.

> A **data flow diagram (DFD)** is a process modeling tool that depicts the flow of data through a system and the work or processing performed by

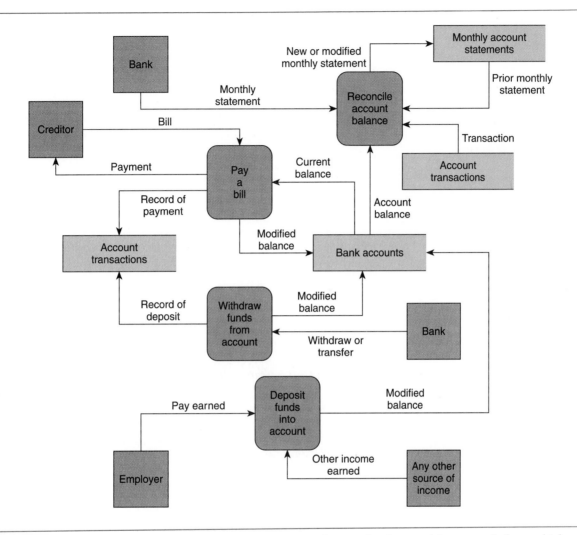

FIGURE 9.3 Gane and Sarson DFD   This DFD demonstrates the popular Gane and Sarson symbol set, which will be used throughout most of this chapter.

that system. Synonyms include *bubble chart, transformation graph,* and *process model.*

Recall from Chapter 2 that one of the fundamental building blocks of information systems is activities. All information systems perform activities — usually lots of them! Activities (or processes) are necessary to transform data (another building block) into useful information. We need a formal way to model and study the processing in an information system.

Data Flow Diagram Conventions and Guidelines

There are various symbolic notations suggested by different authors and experts. Figure 9.3 demonstrates the popular Chris Gane and Trish Sarson data flow diagram symbol set, the one we will use throughout most of this chapter. First, we will briefly examine the symbols and then learn how to read the diagram. The Gane and Sarson symbols are duplicated in the margin for your reference.

A *process* is the main symbol on a DFD.

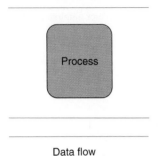

A **process** is work or actions performed on incoming data flows to produce outgoing data flows. Although processes can be performed by people, departments, robots, machines, or computers, we once again want to focus on what work or action is being performed, not on who or what is doing that work or activity. Synonyms include *bubble* and *transform*.

Since the purpose of a process is to transform data flows, all processes must be attached to at least one input and one output data flow.

A **data flow** represents the input of data to a process or the output of data from a process. It can also represent the update of data in a file, database, or other data store. Think of a data flow as a highway down which packets of data of known composition are allowed to travel. The name implies what data may travel down any given highway.

But where do the data flows come from? Most data flows occur inside the system, but some data flows bring new data from outside the system into the system. For example, a new order from a customer is a data flow that brings new data (the new order) into the system. In this case the customer is an internal or external agent to the system.

Internal and external agents define a system's boundaries. They provide net inputs into a system, and net outputs from a system. A very common synonym is **internal and external entity** (not to be confused with the term *data entity*). Agents are also sometimes called *sources* (of net inputs to the system) or *destinations* (of net outputs from the system).

Finally, both are sometimes called *external processors* because they perform processes that have been excluded from the scope of the system being modeled. (Note: We are particularly fond of this synonym since changes in scope determine whether a process is truly external or internal.)

Agents are usually considered external if they are clearly external to your business. Examples include customers, suppliers, and government agencies. Agents are internal if they represent work performed inside your business, but that work is not part of your system's scope, and yet they provide inputs to or receive outputs from your system. This may include other departments, employees, or information systems. Internal agents may also include your system's end-users, who are frequently the source of inputs (= data) and recipients of outputs (= information).

Most information systems capture data for later use. That data is kept in the *data store,* the last symbol on a data flow diagram.

A **data store** is an "inventory" of data. Synonyms include *file* and *database* (although those terms are too implementation-oriented for essential process modeling).

Ideally, essential data stores should describe "things" about which the business wants to store data. This includes:

- Roles (e.g., customers, suppliers, employees, students, instructors, etc.).
- Objects (e.g., products, parts, textbooks, equipment, etc.).

- Locations (e.g., warehouse, sales region, building, room, etc.).
- Events (e.g., order, time card, requisition, course, registration).

Note   A data store may sound similar to what we called a "data entity" in Chapter 8. However, it is different. A data entity represents the concept of something about which we want to store data. A data store represents all occurrences of that something. As such, the data store represents the common link between data and process models.

Now, how do you read a DFD? If properly labeled, a DFD should be easy to read. A process receives a data flow(s) from an agent, a data store, or another process (perhaps all three). It carries out the action described in the process symbol. Then it produces the output data flow that may be received from another process or an agent. It may also update one or more data stores.

Don't confuse DFDs with flowcharts! Most of you are reading this book after some exposure to computer programming. Program design frequently involves the use of flowcharts. But data flow diagrams are very different! Let's summarize the differences.

- Processes on a DFD can operate in parallel. Thus, several processes might be executing or working simultaneously. This is consistent with the way businesses work. On the other hand, processes on classical flowcharts can only execute one at a time.
- DFDs show the flow of data through the system. Their arrows represent data flows. Looping and branching are not shown. On the other hand, flowcharts show the sequence of processes or operations in an algorithm or program. Their arrows represent pointers to the next process or operation. This may include looping and branching.
- DFDs can show processes that have dramatically different timing. For example, a single DFD might include hourly, daily, weekly, yearly, and on-demand processes. This would be rare on classical flowcharts.

The DeMarco and/or Yourdon Notation

There is an alternative but equivalent DFD symbol set. The Tom DeMarco and/or Ed Yourdon symbol set, shown in Figure 9.4, differs as follows:

- Circles or "bubbles" represent processes (hence, the synonym *bubble chart*).
- Rectangles represent internal and external agents.
- Open-ended boxes, on both ends, represent data stores.

Compare Figures 9.3 and 9.4. Convince yourself that both symbol notations communicate "exactly" the same meaning. Don't get caught up in irrelevant arguments about which symbols are best. It really doesn't matter! Just use the symbols preferred by your instructor or business. In some cases your CASE product of choice may dictate the symbol set.

Now that you understand the basic symbols and how to read the diagram, let's study data flow diagram conventions in greater detail.

Essential Processes

Once again, processes are work or actions that are performed on incoming data flows to produce outgoing data flows. Essential processes are work or actions that must be performed no matter how you implement the system. Essential

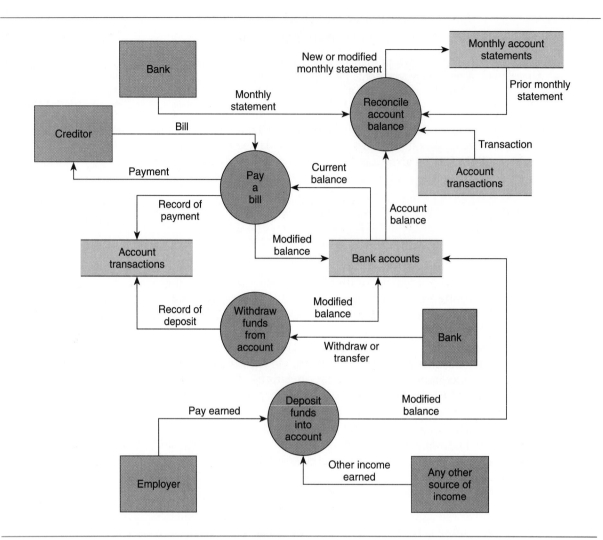

FIGURE 9.4 DeMarco and/or Yourdon DFD   This DFD demonstrates the DeMarco or Yourdon symbol set. Convince yourself that this symbol set is virtually identical in meaning to that in Figure 9.3.

processes can eventually be implemented as (1) work performed by people, (2) work performed by robots or machines, or (3) work performed by computer software. It doesn't matter which implementation is used, however, because an essential DFD's process should only show that there is work that must be done.

Naming conventions for essential processes depends on how general or detailed the DFD is. Whenever possible, all processes should be named with a strong action verb followed by an object clause that describes what the work is performed on or for. Examples of detailed names are listed in the margin on the next page. Such description is not always possible. On general or high-level DFDs, you are trying to communicate the big picture—that is, management's view of the business or system. For such views, a process actually represents a set of detailed processes. Thus, general process names are used—usually a noun that adequately reflects the entire function. Examples are listed in the margin (on the next page).

Essential DFDs omit any processes that do nothing more than move or route data, thus leaving the data unchanged. Business systems frequently implement such processes, but they are not essential. Indeed, they are often inefficient. Thus, you should omit a process that corresponds to a secretary receiving and simply forwarding a variety of documents to their next processing location. In the end, you should be left only with essential processes that:

- Perform computations—for instance, calculating your grade-point average.
- Make decisions—for example, establishing the financial aid status of a student based on rules that include grade-point average, number of semesters enrolled, number of credits earned, and charges past due.
- Split data flows based on content or business rules—for instance, separating approved orders from rejected orders based on credit rules.
- Combine data flows—for instance, combining requested courses with available courses to create a student schedule.
- Filter and/or summarize data flows to produce a new data flow(s)—for instance, filtering invoice data to identify only overdue accounts or summarizing course enrollment data to identify high-demand courses. In both cases, the data isn't changed, but its structure is different.

You may have noticed that a process can have many incoming and outgoing data flows. Some may occur every time the process is executed. Others may occur optionally, under certain conditions. DFDs do not show which data flows must occur and which occur under specific circumstances. Ideally, processes should have as few inputs and outputs as possible. The best way to achieve this goal is to split a complex process into multiple, simpler, cohesive processes. A cohesive process does only one thing. Cohesive DFD processes eventually become cohesive modules in programs. Cohesive modules offer the benefit of being reusable and more maintainable.

Be careful to avoid three common mechanical errors with processes: (1) black holes, (2) miracles, and (3) gray holes. These are illustrated in Figure 9.5.

- What is wrong with process 1 in Figure 9.5? The process has inputs but no outputs. We call this a **black hole** because data enters the process and then disappears. In most cases, the analyst just forgot the output.
- What is wrong with process 2? The process has outputs but no input. Unless you are David Copperfield, this is a **miracle!** In this case, the input flows were likely forgotten.
- What is wrong with process 3? The inputs are insufficient to produce the output. We call this a **gray hole.** There are several possible causes including: (1) a misnamed process, (2) misnamed inputs and/or outputs, or (3) incomplete facts.

Gray holes are the most common errors—and the most embarrassing. Once handed to a programmer, the input data flows to a process (to be implemented as a program) must be sufficient to produce the output data flows. To avoid this problem, the analyst must ultimately define the specific attributes or fields that make up each data flow.

———— ✓ ————

**Detailed Process Names
(Start with a Verb)**

Validate customer account
  exists
Get customer account data
Validate customer credit
  standing
Validate ordered product
  exists
Determine if inventory is
  available
Update inventory data
Create backorder
Create packing order
Create shipping order

———— ✓ ————

**General Process Names
(a Noun Clause)**

Marketing system
Sales subsystem
Order processing function

FIGURE 9.5 Common DFD Errors Process 1 has inputs but produces no outputs. This is called a *black hole*. Process 2 produces outputs but receives no inputs. This is called a *miracle*. Process 3 has inputs and outputs; however, the inputs are not sufficient to produce the outputs. This is called a gray hole.

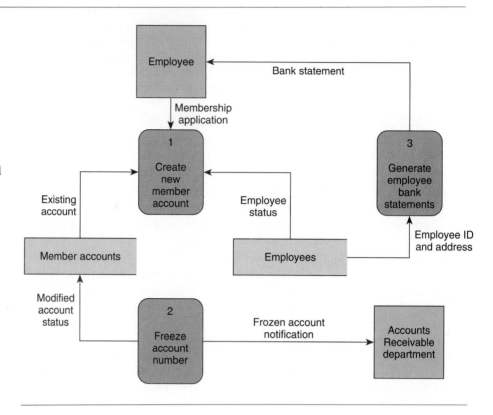

Data Flows

Recall that a data flow represents inputs to and outputs from a process. Clearly, the output of one process may be the input to another process. Essential data flows should represent the minimum essential data needed by the process that receives the data flow. By ensuring that processes only receive as much data as they really need, we reduce the coupling or dependence between processes. This is a very desirable property to carry forward to programs that might eventually implement the process, because it minimizes the effects of changes in one process as they might relate to other processes. Stated more simply, it leads to reduced program maintenance cost and effort!

Data flow names should be descriptive nouns and noun phrases that are singular, as opposed to plural. We don't want to imply that occurrences of the flow must be implemented as batch. Again, nonbatch implementations might be possible. Data flow names also should be unique (except for those to and from data stores). Accordingly, adjectives and adverbs can help to describe how processing has changed a data flow. For example, if an input to a process is named ORDER, the output should not be named ORDER. It might be named VALID ORDER, APPROVED ORDER, ORDER WITH VALID PRODUCTS, ORDER WITH APPROVED CREDIT or any other more descriptive name that reflects what the process does to the order.

If a process does nothing more than route the data flow to the next process, can data flow names be duplicated? Businesses are certainly replete with such

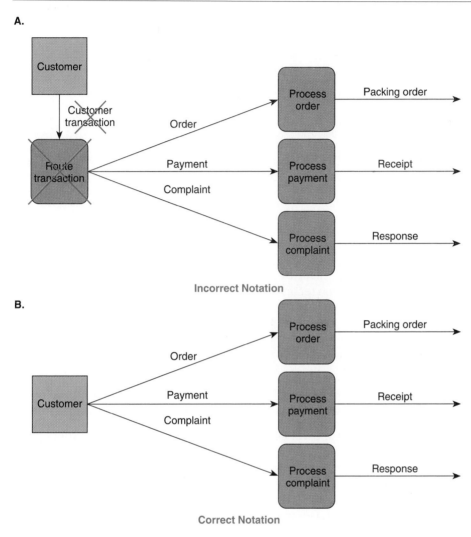

A.

Incorrect Notation

B.

Correct Notation

FIGURE 9.6 **Eliminating Routing Processes** Processes that do not change or make decisions using incoming data should be eliminated. This also eliminates duplicated data flow names.

processes (see Figure 9.6A). But a routing process is nonessential, if not unnecessary. The process should be eliminated from the essential model so that we might at least consider alternative implementations that do not involve routing and paper-shuffling. The duplicated data flow name would automatically disappear once the routing process was deleted from the DFD (see Figure 9.6B).

No data flow should ever go completed unnamed. Unnamed data flows are frequently the result of flowchart thinking (e.g., step 1, step 2, etc.). If you can't give the data flow a reasonable name, it probably does not exist!

Consistent with our goal of essential modeling, data flow names should describe the data flow without describing how the flow is or could be implemented. Suppose, for example, that end-users explain their system as follows: "We fill out a '23' in triplicate and send it to. . . ." The real name of the "23" may actually be COURSE REQUEST FORM. If you use the term COURSE REQUEST, you eliminate two implementation terms — the identification number 23 and the reference to a paper form. This term is much more accurate because a

FIGURE 9.7 **The Data Flow Packet Concept**
If two or more separate data flows always travel together, they should be shown as a single data flow.

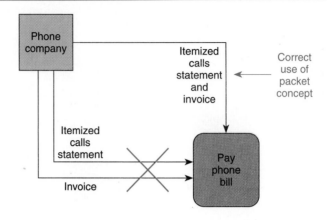

course request may be made by phone, by computer, or, if you are more futuristic, by voice recognition techniques. A paper form is not the only implementation alternative. The name COURSE REQUEST tells us what we need to know.

Analysts draw DFDs at different levels of detail. General-level diagrams often consolidate multiple processes into functions or subsystems. Frequently, analysts also consolidate data flows on general-level DFDs into composite data flows.

> **Composite data flows** are flows that actually consist of multiple primitive data flows. They are used to make general-level DFDs easier to read.

For example, a general-level DFD might consolidate all types of orders into a single data flow, ORDER, input to a single process, PROCESS ORDER. Other, more detailed DFDs might replace the composite data flow with specific primitive data flows such as STANDARD ORDER, STANDING ORDER, EMERGENCY ORDER, and EMPLOYEE ORDER. Why? Perhaps different orders require somewhat different processing or decision making, or perhaps the different flows do not consist of identical data.

Another common use of composite data flows is to consolidate all reports and inquiry responses into two composite flows. There are two reasons for this. First, these outputs can be quite numerous. Second, many modern systems provide extensive user-defined reports and inquiries that cannot be predicted prior to the system's implementation and use.

Ultimately, all data flows should be identified at their primitive level.

> A **primitive data flow** is one that consists of specific data attributes. These attributes always travel together as a single packet.

The packet concept is critical. Data that should travel together should be shown as a single data flow, no matter how many physical documents might be involved. This is necessary because, once again, alternative implementations might split a single data flow into two or more flows, possibly with differing implementations. The packet concept is illustrated in Figure 9.7, which shows the correct and incorrect ways to show how a phone company combines an itemized invoice and payment request.

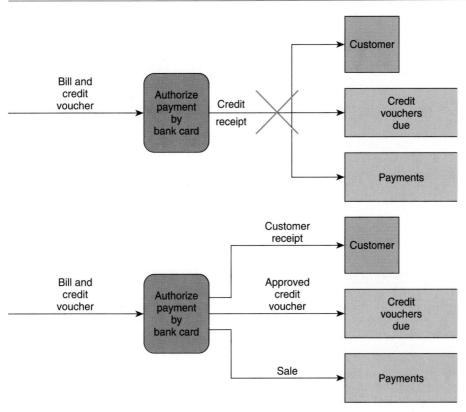

FIGURE 9.8 **Diverging Data Flows** Diverging data flows are implementation-oriented and should be avoided on essential DFDs. Replace each diverging flow with a separate, uniquely named data flow.

Do not use diverging data flows (see Figure 9.8) on essential DFDs. Diverging flows model implementation solutions such as multipart forms and multiple copies of reports. Once again, you should emphasize the continuing concept of depicting the essence of the system. Rarely do all destinations require exactly the same subset of data. Use separate and uniquely named data flows in place of each potential copy of the diverging data flow. There are alternative implementations that wouldn't necessarily require multipart forms or copies.

All data flows must begin and/or end at a process, because data flows either initiate a process or result from a process! Consequently, all of the data flows depicted on the left side of Figure 9.9 are illegal. The corrected diagrams are shown on the right side.

Essential Agents

Remember that every system has a boundary and that agents define the system's boundaries. Essential agents are people, organizations, and other systems with which the system you are modeling must interact. Their inclusion on a DFD means that your system interacts with these agents. They are almost always one of the following:

- An office, department, division, or individual within your company that, while directly using the system you are modeling, either provides inputs

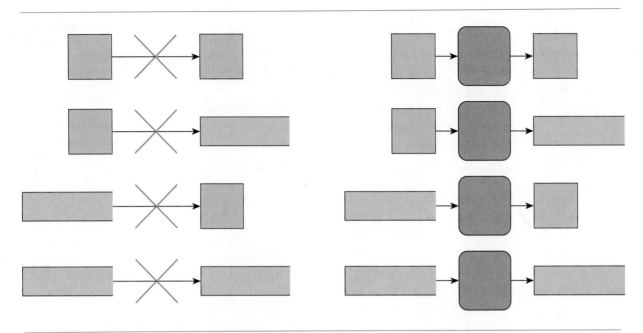

FIGURE 9.9 **Illegal Data Flows** All data flows must begin and/or end at a process. The diagrams on the left side violate this rule. The diagrams on the right side correct the mistakes.

to that system, receives outputs from that system, or both. These are internal agents.

- Organizations, agencies, or individual(s) that are outside your company but that provide inputs to, or receive outputs from, your system. These are external agents. Examples include customers, suppliers, contractors, banks, and government agencies.

- Another system—possibly, though not necessarily, computer-based—that is separate from the system you are modeling but with which your system must interact. In general, this internal agent should only be used if your system communicates directly with the programs in the other system. In other words, best current practice does not use the agent symbol to reflect access to another system's files or database. In that situation use the data store symbols.

- One of your system's end-users or managers. In this case, the user or manager is either a net source of data to be input to your system and/or a net destination of outputs to be produced by your system. This is a special type of internal agent.

It is important to realize that scope can change as one progresses through a project. For instance, entire subsystems that were studied and modeled as processes of the current system may be removed from the scope if they are considered adequate or of lower priority. Thus, these subsystems and the people and machines that perform them can become internal agents when you model the requirements for a new system. If so, you are graphically stating that these agents' current physical implementation of interfaces to your system will *not* change.

Agents should be named with descriptive nouns, such as REGISTRAR, SUP-
PLIER, or FINANCIAL AID INFORMATION SYSTEM. If the agent describes an
individual, use titles rather than the person's name (for example, use AC-
COUNT CLERK, not Mary Jacobs). Plural nouns are discouraged.

To avoid crossing data flow lines on a DFD, it is permissible to duplicate
external and internal agents on DFDs. If external or internal agents are duplicated on or between pages of DFDs, they can be marked as indicated in the
margin. This mark tells readers that they can expect additional occurrences of
this same agent on the same page or other pages. (Note: This practice may be
limited by your CASE tool.)

As a general rule, external and internal agents should be located on the
perimeters of the page, consistent with their definition as a system boundary.
This is not a strict rule, however.

Essential Data Stores

Again, data stores are inventories of data. Essential data stores represent essential, reusable data to be collected and stored by the system. How you identify
data stores depends on your systems development methodology:

- If you do data modeling before process modeling, identification of most
  data stores is simplified by the following rule: There should be one data
  store for each data entity on your entity relationship diagram.
- If you do process modeling *before* data modeling, it can be somewhat
  more difficult. Our best recommendation is to identify existing implementations of files or data stores (e.g., computer files and databases, file
  cabinets, record books, catalogs, etc.) and then rename them to reflect
  the essential data to be stored.

Generally speaking, data stores should be named after the data model or
entities. The word data is unnecessary because it is implicit that a data store
stores data. Additionally, names should not imply implementation media. For
instance, avoid terms such as file, database, file cabinet, file folder, and the like.
Because a data store represents all occurrences of an entity, names should be
plural.

If you have already read or studied Chapter 8, you know that data models
sometimes include relationships that are described by data. They were called
associative or *relational entities*. For example, in Figure 9.10, we see a data
model that includes three fundamental entities (CUSTOMER, ORDER, and
PRODUCT) and two relationships (PLACES and ORDERED PRODUCT). One
of those relationships, ORDERED PRODUCT, is described by data (e.g., Quantity Ordered). But ORDERED PRODUCT data is always used in the context of a
specific ORDER or PRODUCT. Thus, Gane suggests that the relational entity be
added to the data stores for the associated entities as shown in the margin. This
notation is offered for explanation only; we will not use it in the SoundStage
case study.

As was the case with data flows, data stores might be consolidated into
composite data stores on general-level DFDs in order to simplify reading. The
most common example of this practice is to draw a single data store that represents an entire data model (entity relationship diagram) as drawn in Chapter 8.
This composite data store might be named _____ SYSTEM DATA MODEL.

Internal
or
external
agent

CUSTOMERS

ORDERS AND
ORDERED PRODUCTS

PRODUCTS AND
ORDERED PRODUCTS

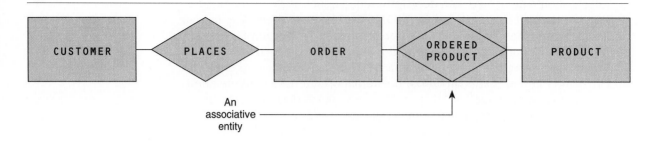

FIGURE 9.10   **Subset of a Data Model**   Data modeling, covered in Chapter 8, is linked to process modeling. The entities in this typical data model correspond to data stores on the process model.

When including data stores in a DFD consider the following guidelines, illustrated in Figure 9.11:

1. Only processes may connect with data stores. Data cannot be used or updated except by way of a process.
2. The direction of arrows is significant. Data flow direction is interpreted as follows:
   a. A data flow from a data store to a process means that the process uses that data. Notice that we said "uses," not "reads." The read is assumed. The data flow name can reflect the specific data used, or it can be the same as, or similar to, the data store name itself.
   b. A data flow to a data store means that the process updates the data store. Updates may include any or all of the following:
      · *Adding or storing new records*—for instance, adding a new customer to the CUSTOMERS data store. The data flow name should not contain verbs. For example, use NEW CUSTOMER, not ADD A CUSTOMER.
      · *Deleting or removing old records*—for instance, deleting inactive customers from a store. Once again, the data flow name should not contain verbs. For example, DELETED CUSTOMER is better than DELETE A CUSTOMER. (Note: Archive data stores should not be shown on essential data flow diagrams.)
      · *Changing existing records*—for instance, changing the credit rating, address, or balance of an existing customer. And once again, the data flow name should not contain verbs. Use MODIFIED CREDIT RATING, not MODIFY CREDIT RATING, for example.
3. Although you know that you can't update a record without reading it, you don't depict this detail! It clutters the diagrams by requiring every interaction with every data store to have two flows. In this case show only the net or final data flow. In most cases this means showing the ADD, DELETE, or CHANGE.
4. Some processes legitimately use and update data stores. The key word is use. For example, calculations or decisions might be necessary before updating the data stores. If so, as a general rule, use separate data flows for the use and the update.

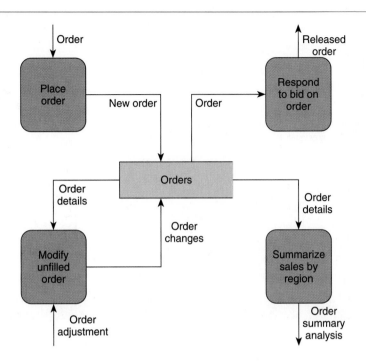

FIGURE 9.11 **Data Store Guidelines** Use these guidelines when connecting data stores to processes.

To prevent crossing data flow lines on a diagram, try to place data stores in the middle of the page (or in the middle of the processes that use and update the data stores). If crossed data flow lines are impossible to avoid, it is permissible to duplicate a data store on a page. (Some CASE products prevent this practice.) If duplicated, it is recommended that all duplicated data stores be marked as indicated in the margin. This tells the reader to look elsewhere on the same page for a duplicate. (Note: Not all CASE products support the margin's special notation.)

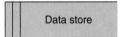

Computer-Aided Systems Engineering (CASE) for DFDs

Computer-aided systems engineering (CASE) was introduced in Chapter 5 as an emerging, enabling technology for systems analysis and design methods. Virtually all CASE products support computer-aided process modeling. Most CASE products specifically support DFDs. CASE takes the drudgery out of drawing and maintaining the diagrams.

Using a CASE product, you can easily create professional, readable DFDs without the use of paper, pencil, erasers, and templates. The DFDs can be easily modified to reflect corrections and changes suggested by end-users; you don't have to start over! Also, most CASE products provide powerful analytical tools that can check your DFDs for completeness as well as mechanical and consistency errors. The potential time and quality savings are substantial.

CASE tools do have their limitations. Not all process modeling conventions are supported by all CASE products. In fact, many organizations establish their own process modeling symbology that may be substantially different from most

FIGURE 9.12 **CASE Tool for DFDs** Data flow diagrams can be drawn with most upper-case tools. This screen was captured from the popular CASE tool, Excelerator/IS.

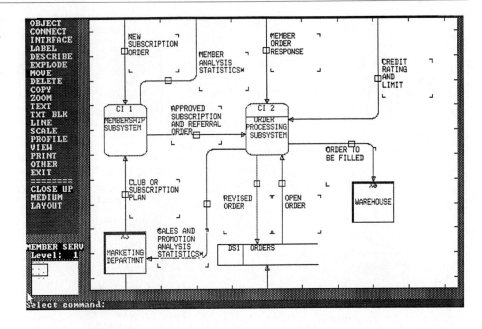

process models commonly supported by most CASE tools. Therefore, it is very likely that any given CASE product may force the company to adapt their methodology's process modeling symbols or approach so that it is workable within the limitations of their CASE tool.

All of the DFDs in this book were created with a popular CASE product called Excelerator/IS. Only the color screens and annotations were added by an artist. All of the objects on our DFDs were automatically cataloged into a project repository so that we can add detailed information, such as data elements, as the project moves through analysis, design, and implementation. Figure 9.12 demonstrates an Excelerator screen for computer-aided process modeling.

HOW TO CONSTRUCT PROCESS MODELS

There are many opportunities to perform process modeling—and many approaches. In this section, we'll examine when process modeling might be performed during systems development. We'll also demonstrate a step-by-step approach for performing process modeling.

When to Build Process Models

Process modeling may be performed during various phases of the systems development life cycle. Process models are progressive. In other words, there is no such thing as the final process model for a business or application. Instead, the model should be considered a living document that will change in response to a changing business. Process models should ideally be stored in a repository so that they can be retrieved, expanded, and edited over time. Let's examine how process modeling may come into play during systems planning and analysis. Also, The Next Generation box examines some new potential uses for process models.

Process Modeling during Systems Planning

Many systems planning methodologies and techniques require some degree of process modeling. Let's briefly examine how process models might evolve out of the three systems planning phases: study, definition, and analysis. (These phases were described in detail in Chapter 6.)

During the study phase of systems planning, process modeling usually does not come into play. Rather, the focus is entirely upon studying the business and its mission.

During the definition phase of systems planning, an **enterprise process-like model** may be constructed. This model reflects management's high-level view of its business functions (= groups of processes). The resulting model is usually called a *decomposition diagram.* This diagram resembles the structure charts that are a part of many programming courses. The model decomposes the business into subsystems and functions, but it rarely gets so detailed as to define specific processes that make up the functions.

In the information engineering methods, systems planning proceeds to another phase: analysis of business areas. A business area or collection of similar business processes is identified. These processes may actually cross the boundaries of existing business functions. A **business area process model** or data flow diagram is drawn. The level of detail varies according to specific methodologies; however, these DFDs are usually not detailed enough to support the design of specific computer applications. Instead, they are used to identify and prioritize specific computer applications for subsequent analysis and design.

Data Modeling during Systems Analysis

Application development begins with systems analysis. Recall that systems analysis consists of three phases: survey, study, and definition. Process modeling plays a significant role in most of these phases.

During the study phase, a simple **application context process model** might be quickly constructed to define project scope. This model is a simple, one-process DFD that shows how the system fits into the business. The one process is the system. Everything else is an agent or data store with which the system interacts. If a business area process model exists, it should serve as a point of departure for this context model.

Process modeling used to be a major part of many study phase techniques and methodologies. Analysts would draw implementation DFDs of current systems to gain insight into the systems and their problems. Some analysts would convert these implementation DFDs into essential DFDs to eliminate the bias of the existing implementation when considering alternatives. Today, such DFDs are largely discouraged because of the excessive time they take. On the other hand, the current interest in business process reengineering (see The Next Generation box in Chapter 6) may ultimately generate renewed interest in essential DFDs of the current system. Such DFDs might help analysts and managers identify areas in which business processes could be eliminated or streamlined.

The Next Generation

BUSINESS PROCESS REDESIGN AND WORKFLOW SYSTEMS

Data flow diagrams (DFDs) have had staying power for systems developers. They were born of the structured analysis and design methods. And they have survived the transition from structured analysis and design to information engineering. What does the future hold for them? Historically, DFDs have been used primarily to model application system processes as a prelude to software design. Tomorrow, we may be using DFDs to redesign fundamental business processes and workflows.

Business process redesign has been a recurring theme in The Next Generation boxes. The idea makes sense. Throughout the history of data processing and information systems, we have focused our efforts on automating and supporting existing business processes with computers. Today, businesses are questioning the value of that approach. There is concern that we may have been automating outdated and inefficient business processes—doing the same old wrong things, only faster.

Consider the following simple, but real-life example. In examining its professional hiring business processes, a company learned that it could hire a new worker with four signatures on a contract: the employee, the department head, the

section manager, and the division vice president. But to get that new employee on payroll, a payroll authorization form requiring 14 signatures was needed—three the same as those on the contract. No new information is transferred from the contract to the payroll authorization. In many documented cases employees wait two, and sometimes three, months to get their first check. Payroll changes can take a similar amount of time. Several attempts to improve computer processing did not solve the problem because the "red tape" of the payroll authorization was the true bottleneck.

The business response is the

emerging emphasis on business process redesign. Instead of automating business processes, we analyze business processes, eliminate those which contribute little or no value, and streamline process and data flow through systems. After business processes are redesigned, the supporting computer processes can also be redesigned.

Business process redesign requires a fresh, cross-functional analysis of existing business processes. By cross-functional, we mean that users from various affected areas get together to study process and data flows, and ask

(Continued)

Process modeling is almost always performed in the definition phase. If an enterprise or business area process model already exists, it is expanded or refined to reflect application requirements. Otherwise, the process model for the application is built from scratch, using the context model from the survey phase as a point of departure. In either case, the result is an **application essential process model** that reflects inputs, outputs, data storage, and processing requirements for the target system, independent of implementation technology concerns. Most methods also require a certain amount of detail

---

The Next Generation

(*Concluded*)

questions about "why" we do things the way we do them and "why not" try other, better approaches. Information systems professionals become involved because the computer is usually involved in existing processes.

Data flow diagrams can be used to document existing business processes and design new ones. A variation on DFDs called *process dependency diagrams* (PDDs) may emerge as a improved modeling technique for business process redesign. They more clearly document the dependencies between processes—in a manner similar to entity relationship diagrams.

A sample is shown here. Unlike DFDs, not all arrows depict data flows. Connections that terminate in an arrow indicate a data flow. All other connections indicate sequence or flows of control, which are labeled with logical constructs such as "if . . ." and "for . . .".

But PDDs go much further than DFDs by showing process dependencies in the form of ordinality and cardinality. If you've read Chapter 8, Data Modeling, these concepts are already familiar to you. The terminating mark indicates cardinality—the maximum number of times the second process will execute for a single execution of the first process. The cardinality symbol is preceded by an ordinality symbol that indicates whether the second process "must" execute after the first process or whether it "may" execute after the first process.

Unlike DFDs, PDDs also show which data flows into a process are optional or conditional. Arrows are labeled with the specific conditions. Thus, they more clearly communicate the logical paths through the process model. This promises to simplify the transformation of the model into program code. DFDs, by contrast, are dependent on underlying pseudocode-like logic specifications to accomplish the same goal. Since many analysts don't take the time to write those underlying pseudocode-like specifications, DFDs frequently do not provide a smooth path from model to code.

Business process redesign is paving the way for a new technology solution, *workflow systems.* Workflow systems are designed to automate business processes in a manner that more closely approximates business rules and policies. Workflow systems pass electronic documents, information, and messages through a system to improve productivity. Workflow system technology attempts to break down the barriers of physical paper flow as well as disparate computer technologies and networks that have evolved in most businesses.

In one company, the development of workflow systems to replace paper documentation is already forcing the business to reexamine the efficiency and effectiveness of existing business processes. Approval sequences and information consolidations that took days or weeks now take hours or minutes. Workflow systems automatically forward electronic documents and information to appropriate workers and managers according to the rules and policies of the business. For example, high-cost transactions may follow one path, while low-cost transactions may follow a different path. Workflow systems can also provide expertise or decision support assistance to appropriate business processes.

In an ironic twist, business process redesign and workflow systems are triggering somewhat of a renaissance of an original data flow diagramming practice of drawing physical and logical models of the current system. That practice had come to be disfavored as a time-consuming bottleneck in the structured systems analysis methodology. But as businesses increasingly reexamine their underlying business processes, DFDs for cross-functional business areas may reemerge as a next generation technique.

---

beyond that depicted on the DFDs. These details, describing objects on the DFDs, are recorded in the project repository.

Looking Ahead to Systems Design

The essential process model from systems analysis describes business processing requirements, not technical solutions. As we proceed to systems design, the process model must become more technical. It must become an **application**

implementation process model that will guide the technical implementation of programs.

Thus, DFDs can also be used in the design and implementation phases. Such DFDs will include the implementation details that we discouraged earlier in this chapter. Other types of process models, such as program structure charts, will also emerge during design. Implementation data flow diagrams and structure charts will be discussed in greater detail in Part Three of this textbook.

A Step-by-Step Process Modeling Approach

As a systems analyst or knowledgeable end-user, you must learn how to draw DFDs to model process requirements. We will use the SoundStage Entertainment Club project to teach you how to draw DFDs. We have completed the survey and study phases of the systems development life cycle and fully understand the current system's strengths, weaknesses, limitations, problems, opportunities, and constraints. We have already defined data requirements for an improved system, which resulted in the data model presented in Chapter 8. We will now model the corresponding process requirements. We will use a step-by-step approach to provide a complete set of DFDs that you can use in the future.

Step 1: Draw a Context Data Flow Diagram

All projects have scope. A project's scope defines what aspect of the business a system or application is supposed to support. A project's scope also defines how the system or application being modeled must interact with other systems and the business as a whole. The definition of project scope is an important first step in process modeling.

A **context data flow diagram** defines the scope and boundary for the system and project. The scope of any project is always subject to change; therefore, the context data flow diagram is always subject to change. Synonyms include *context diagram, context model,* and *environmental model.*

Context diagrams tend to be difficult to draw because they must define scope. We suggest the following strategy for determining the systems boundary and scope:

1. Think of the system as a container in order to distinguish the inside from the outside.
2. Ignore the inner workings of the container. This is the classic black box concept of systems theory.
3. Ask your end-users what events or transactions a system must respond to. Business events simply happen and bring new data into the system. An example might be an order.
4. For each event, ask your end-users what responses must be produced by the system. Examples for the order event might include backorders, picking orders, and invoices. On the other hand, some systems are intended to produce reports as their response.
5. Ask your users what fixed-format reports must be produced by the system. (Do not include ad hoc reports and inquiries; they tend to clutter the diagram. If your users insist, show them as a single composite data flow.)

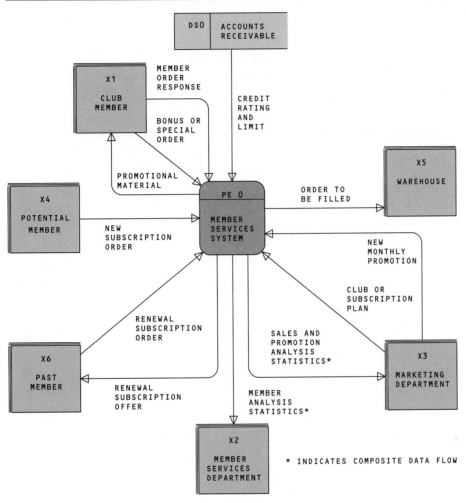

FIGURE 9.13 **A Context Diagram** The context (data flow) diagram is the most general process model you can draw for a system or application. The system or application is drawn as a single process. Its interactions with the business, other systems, and the outside world are drawn as input and output data flows.

6. Identify the net sources of data about each event. These sources will be external or internal agents on the DFDs.

7. Identify the net recipients of each response or output that the system should generate. These destinations will also be external or internal agents.

8. Identify any external data stores. Many systems require access to the files or databases of other systems. They may use the data in those files or databases. Sometimes they may update the certain data in those files and databases. But generally, they are not permitted to change the structure of those files and databases.

9. Draw your context diagram from all of the preceding information.

The context diagram contains one and only one process (see Figure 9.13). Internal and external agents are drawn around the perimeter. External data stores are added to the perimeter. Data flows define the interactions of your system with the internal and external agents and with the external data stores.

Note First-generation context diagramming did not permit data stores on the context diagram. Those data stores were drawn as external entities. Not only was this confusing to many analysts and users (who knew they were data stores), but it prevented the analyst from defining the external data store's structure (separate from the system's data model) in the CASE tool. Most CASE tools do not allow a data structure or model to be defined for an agent—you must use a data store! Consequently, best current practice now permits external data stores to be included on the context diagram.

If you try to include all of the inputs and outputs between a system and the rest of the business and outside world, a typical context diagram might show as many as 50 or more data flows. Such a diagram would have little, if any, communication value. Therefore, we suggest the following strategy:

- Only show those data flows that represent the main objective or most common inputs and outputs of the system. Defer less common data flows to more detailed DFDs to be drawn later.
- Use composite data flows for reports, inquiries, and similar transactions.

This strategy is applied in Figure 9.13. The main purpose of our system is to respond to subscription orders (an initial order and request for membership), monthly membership promotions, and member orders. Management has also placed great emphasis on the need for various sales and member analysis reports (shown as composite flows). Other specific reports will be identified on lower-level DFDs.

Notice that we used a composite data flow to consolidate regular orders for records and tapes with special merchandise orders. Later, we can factor the consolidated flow into separate, primitive flows on more detailed DFDs.

Finally, notice that the Accounts Receivable data store appears on our context diagram. Accounts Receivable (A/R) has agreed to provide read-only access to their database to facilitate credit checking for member orders. The store is external because we cannot change its structure. Any structural changes to the A/R database might cause the need to modify several programs in the A/R information system.

Step 2: Draw a Decomposition Diagram to Outline DFDs

There are two basic approaches to drawing DFDs for a complete system or application:

- Draw two data flow diagrams, each on its own sheet of paper. As demonstrated in Figure 9.14, the first diagram, called the *overview* or *level-zero DFD,* serves as a general-level diagram, usually consisting of 12 or fewer processes. The second diagram, called a *system* or *level-one DFD,* is an expansion of the first diagram, providing a more detailed view of the system. It may contain 10 to 30 processes. Both diagrams focus

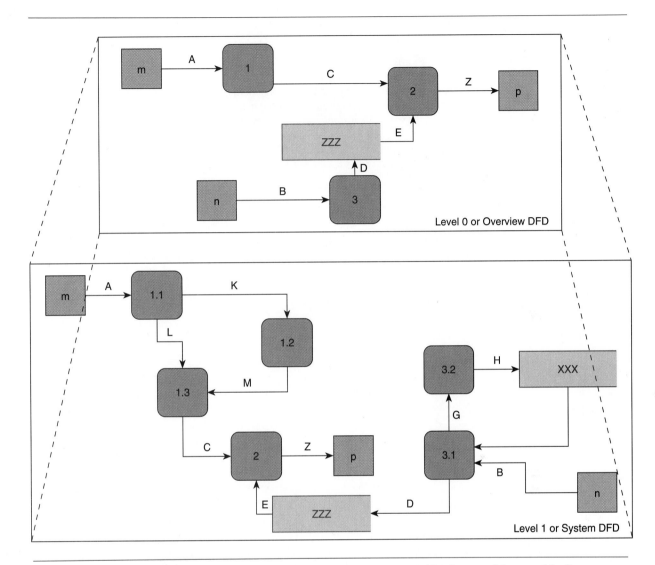

FIGURE 9.14 **The Expansion Approach to Drawing DFDs** As suggested by Gane and Sarson, this diagramming technique requires the analyst to draw two DFDs—one overview DFD and a detailed DFD. Both diagrams model the entire system of interest.

on user-level requirements, not computer-related or implementation details. This approach was first suggested by Gane and Sarson and is popular in EDS's STRADIS methodology. The approach may be used with either the Gane and Sarson or the DeMarco or Yourdon DFD symbols.

- Draw a leveled set of data flow diagrams. This approach uses an explosion technique instead of the expansion approach. As Figure 9.15 shows, the process on the context data flow diagram is exploded into its own data flow diagram that reveals the underlying subsystems, which are shown

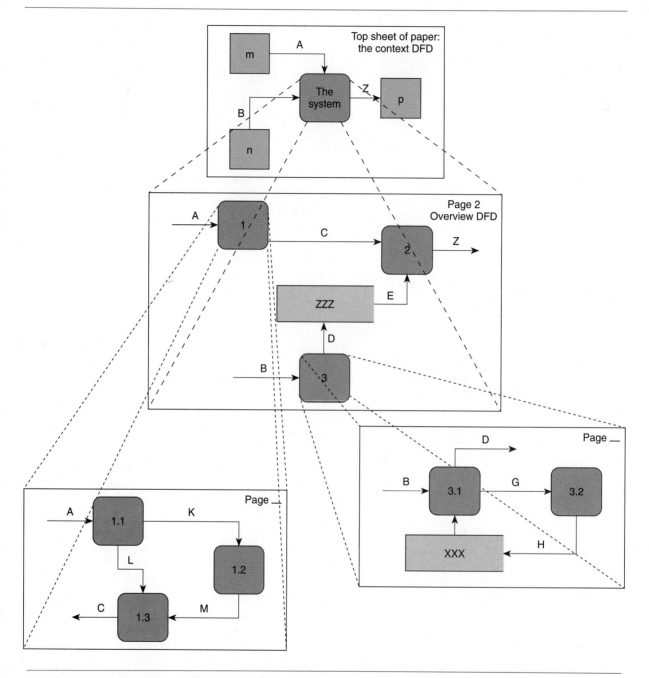

FIGURE 9.15   The Explosion Approach to Drawing DFDs   As suggested by DeMarco and others, this diagramming technique requires the analyst to draw multiple DFDs, each one exploding from a single process on another diagram, until the system is completely modeled.

as subprocesses. Each of these subprocesses may, in turn, be exploded into its own data flow diagram to show more detailed processes. Process explosions continue until the user requirements are completely detailed. Once again, computer implementations or ideas are not shown. This approach was first suggested by DeMarco and is widely practiced in many methodologies. Once again, the approach may be used with either the Gane and Sarson or DeMarco or Yourdon DFD structures.

In recent months, we have developed greater appreciation for the Gane and Sarson approach, but we still recognize that the DeMarco approach is more widely taught and practiced. Therefore, our samples will be based on DeMarco's explosion approach, leading to a leveled set of data flow diagrams. Each level reveals the system, a subsystem, or a function in somewhat greater detail. (Note: The Gane and Sarson technique's equivalent DFDs are presented in the next section for comparison.)

Exploding or leveling a system into subsystems and functions is a time-proven technique — divide and conquer! Unfortunately, we tend to do this without any real plan. The result is often a collection of illogical subsystems and DFDs that, while somewhat correct, don't seem to make much sense. Consequently, we will teach you to plan your explosions by using a pictorial outline called a decomposition diagram to draw your DFDs.

A **decomposition diagram,** also called a *hierarchy chart,* shows the top-down functional decomposition or structure of a system. It also provides us with an outline for drawing our DFDs.

The process symbol is the only symbol used on the decomposition diagram; it is the same process symbol used in DFDs. The processes are connected to form a treelike structure. Process names should conform to the naming guidelines described for DFDs. The top process, also called the *root,* represents the entire system for which you are defining requirements. The root process is exploded or factored into subsystems, functions, and tasks — the number of levels of which is entirely dependent on the size of your project.

Figures 9.16 and 9.17 are the decomposition diagrams for the SoundStage Entertainment Club project. Let's study these diagrams. The numbering scheme used for processes in the diagram will help you keep track of the subsystems and functions as you draw DFDs. It is a good idea to establish a standard numbering scheme before you draw decomposition diagrams and data flow diagrams. We follow these popular guidelines:

1. The root process is numbered 0.
2. The root process is factored into processes that are numbered consecutively, 1, 2, 3, and so on.
3. With subsequent factoring — also called leveling — of any process into subprocesses, each subprocess is numbered as a decimal of the parent process. For instance, process 1 is factored into processes 1.1, 1.2, 1.3, and so forth. Process 1.2 is factored into processes 1.2.1, 1.2.2, and so on.

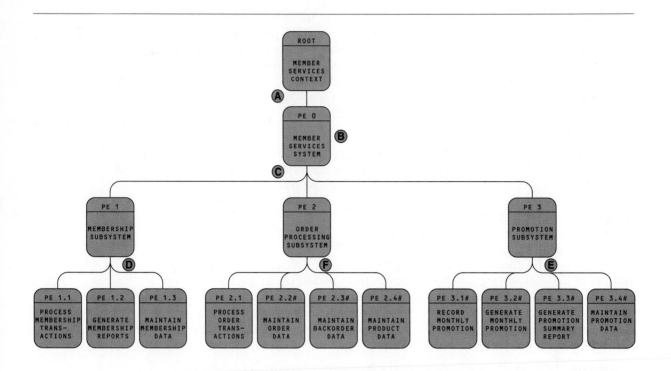

FIGURE 9.16   Decomposition Diagram   This is the first page of the decomposition diagram for our SoundStage project. Again, our DFDs will ultimately be exploded according to this outline.

This strategy is repeated throughout the decomposition diagram and the subsequent leveled set of DFDs.

The following is an item-by-item discussion of the sample decomposition diagrams. The circled letters correspond to specific locations on Figure 9.16.

(A) We used Excelerator's Presentation Graph facility to draw our decomposition diagram. The double root was required because of our CASE tool. It links the decomposition diagram directly to the DFDs, allowing us to examine the DFD for any given process on the decomposition diagram.

   The first root explodes to the context diagram that contains the process PE 0. (The letters PE stand for proposed essential system.) The second root actually starts the decomposition. It explodes to a DFD that contains the processes PE 1, PE 2, and PE 3.

(B) The root PE 0 corresponds to the entire system.

(C) The system is initially factored into subsystems or functions. (Caution: Subsystems and functions do not necessarily correspond to organization units on organization charts.)

How many subsystems or functions can you have? Technically, there is no limit. On the other hand, our goal is eventually to draw DFDs. Advocates of

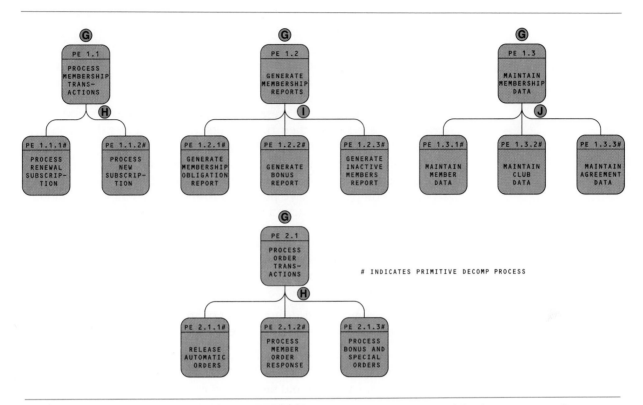

FIGURE 9.17 **Decomposition Diagram** *(concluded)* This is the second page of the decomposition diagram started in Figure 9.16. Again, our DFDs will ultimately be exploded according to this outline.

DeMarco's DFDs argue that no single DFD should contain more than seven processes. This enhances readability by keeping each DFD relatively simple. If we apply this rule to the decomposition diagram, no parent process should have more than seven child processes. This is a good rule of thumb; however, you must temper the rule with common sense — you want your groupings of processes to make business sense.

What about a minimum number of child processes for any given parent process? Certainly factoring a parent process into a single child process provides no additional detail; therefore, if you plan to factor a process, it should be factored into two to seven child processes.

How do we further factor our functions in the decomposition diagram? There are several reasonable strategies. We illustrate three different approaches (but they are not exhaustive).

(D) Membership is a large subsystem that consists of more than seven total processes. We were able to group these processes into three functions: processing various membership transactions, generating various membership reports, and maintaining demographic membership data. We factored the membership subsystem accordingly. Each process is further factored on the second page of our diagram (see Figure 9.17).

(E) The promotion subsystem had only two transactions, one report, and one data store to be maintained. We felt no need to factor this subsystem into functions. Instead, we factored the subsystem directly into its transactions, report, and data maintenance.

To indicate that none of these processes will be further factored, we have placed a # sign by their ID numbers. (Note: You may define any convention supported by your CASE tool.)

(F) The order processing subsystem is a cross between D and E. It has several transactions, a few types of data maintenance, but no fixed-format reporting requirements. (We prefer not to show ad hoc reporting and inquiries.) We created a function for all transactions. Meanwhile, we added the three processes that maintain order data stores.

Notice that the # sign tell us that three of the processes are not further factored. On the other hand, the absence of a # sign on PE 2.1 indicates that we should look for further factoring of that process. That factoring is shown in Figure 9.17.

Our entire decomposition diagram did not fit on a single page. Thus, we continue our diagram in Figure 9.17. Note the following:

(G) The roots on this page were duplicated from the first page of the decomposition diagram in order to provide backward reference.

(H) Transaction functions should be factored into one process per business event. Virtually all systems respond to transaction events. These events just happen. As they happen, they bring new data into the system. Examples include orders, payments, requisitions, shipments, receipts, returns, work—recorded on time slips—and the like. Systems must respond to transaction events with appropriate outputs. Sometimes the outputs simply store the input data for later processing.

(I) Reporting functions should be factored into one process per fixed-format report to be produced (unless multiple reports absolutely must be produced simultaneously).

(J) Data maintenance functions should be divided into one process for each data store to be maintained (or, if you've already completed your data model, each fundamental data entity to be maintained). This provides the custodial activities needed to ensure that accurate data is available for transaction processing and reporting. Examples include routine maintenance of data stores concerning parts, customers, employees, and the like. Most of you have probably written file maintenance programs to add, change, and delete records in master files. These programs are part of this maintenance activity.

Notice the common systems pattern: transaction and maintenance activities bring new data into the system. Reporting activities generate useful information from that data. Virtually all information systems will benefit from this factoring strategy!

Continuing with Figure 9.17, only one other process must be factored from the first page of our decomposition diagram, "PE 2.1." Once again, we factored this transaction system into one process per business event.

There is no need to factor beyond the transactions, maintained stores, and reports. That would be like outlining down to the final paragraphs or sentences in a paper. However, this diagram will serve as a good outline for an integrated, leveled set of DFDs.

Step 3: Identify Data Stores

Before we draw our data flow diagrams, it can be useful to identify candidate data stores that will be used throughout our diagrams. Since a data model was drawn in Chapter 8 for the SoundStage project, we can use that data model to identify our data stores. In Figure 9.18, we use a decomposition diagram to identify our data stores. First, we create a composite data store that represents all system data. This store explodes to our data model. Next, we identify the primitive data stores, one per entity or associative entity on the data model.

Step 4: Draw an Overview Data Flow Diagram

Using our decomposition diagram as an outline, we can now explode the process on the context diagram into a more detailed picture of the system. This second DFD is usually called the **overview data flow diagram.** It shows the

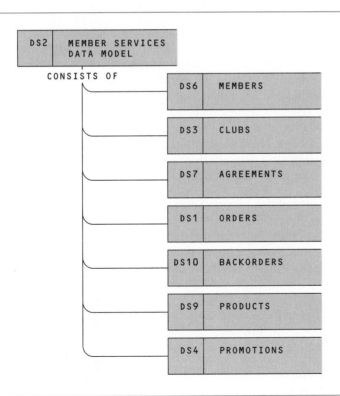

FIGURE 9.18 Data Store Decomposition This decomposition diagram depicts the data stores to be used in our diagrams.

major subsystems or functions and how they interact with one another. It is useful for communicating the big picture of the system.

Whenever you explode a process, the resulting DFD is a more detailed look at that same process. Accordingly, exploding should add detail while retaining the essence of the details from the more general diagram. Let's demonstrate by exploding the lone process from the context diagram (see Figure 9.15).

That context diagram process is exploded into the systems diagram shown in Figure 9.19. Notice that the three processes in this figure correspond precisely to the three processes on the second level of our decomposition diagram (see Figure 9.16). In other words, we are following our original graphic outline. This does not mean that you never deviate from that outline—any outline can be changed.

The following is an item-by-item description of the sample systems diagram:

(A) The data flows that appeared on the context diagram also appear on this diagram. This is done to maintain consistency with the previous diagram. Most books call this *balancing the diagrams*.

(B) Note that we also duplicated the external and internal agents from the context diagram. Some experts prefer not to duplicate these agents. Instead, they draw the data flow without the connecting object (as in Figure 9.20). The data flows that appear to go nowhere or come from nowhere are supposed to direct the reader's attention back to the parent DFD, in this case the context diagram. We think that this detracts from communication; therefore, we duplicate the agents.

(C) The subsystems or functions on most systems diagrams share data stores, usually several data stores. To simplify the systems diagram, we generally consolidate all data stores into a composite data store that represents all of the system's shared data. Using a single data store keeps this diagram readable. Our CASE tool allows us to link this data store directly to our data model. In other words, the data store explodes to our system entity relationship diagram from Chapter 8. (Note: Even if you haven't drawn a data model, this strategy should work.)

(D) The DFD should show significant data flows that occur between the subsystems or functions. This is the principal benefit of high- and middle-level DFDs: they show the interfaces between groups of processes.

Notice that the subsystems mostly communicate through the data store. This is a highly desirable characteristic of modern information systems.

(E) We still haven't exploded composite data flows into primitive data flows. Why? Composite flows are best left unexploded until specific processes have been identified to deal with the primitive flows.

(F) We are using composite data flows to consolidate the many types of uses and updates to the data store.

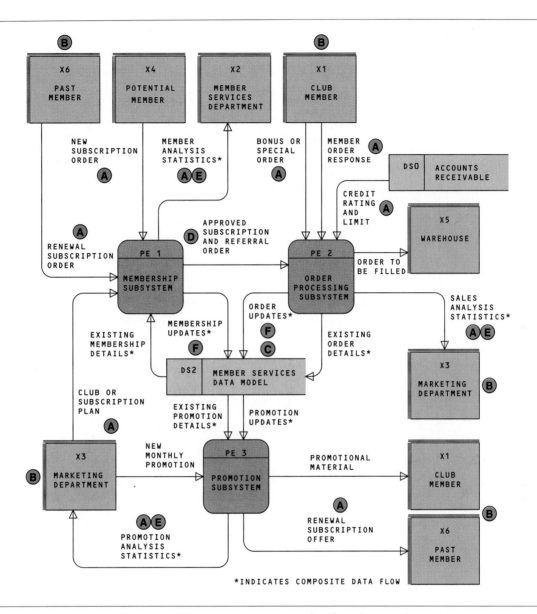

FIGURE 9.19 An Overview Data Flow Diagram   An overview data flow diagram shows the interaction between key subsystems and/or functions.

If you study Figures 9.15 and 9.19 carefully, you'll see that all of the inputs and outputs to process PE 0 on the context diagram (Figure 9.15) also appear on the overview DFD. These two diagrams are said to be balanced.

Balancing is the task of ensuring that no details are lost when a process on one DFD is exploded to a more detailed DFD. Balancing ensures consistency between different levels.

FIGURE 9.19 concluded

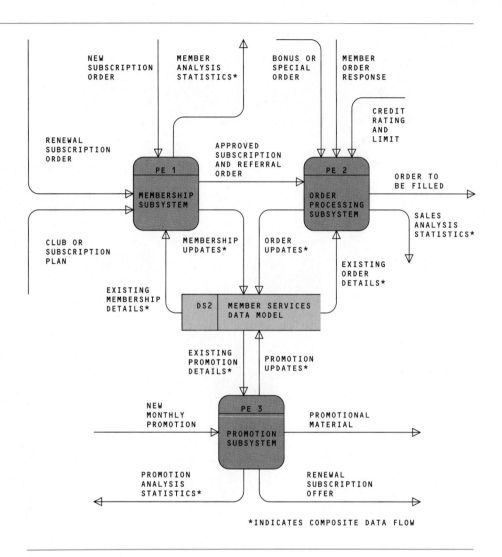

*INDICATES COMPOSITE DATA FLOW

Most upper-CASE tools check balancing. Some automatically carry down data flows when a process is exploded in order to ensure balancing.

If we had introduced some new agents or flows to or from agents on our overview diagram, should those flows have been added to the context diagram also? Some experts argue yes. They want to keep the different levels of DFDs perfectly in balance. We respectfully disagree. The new flows may have been left off the context diagram to keep it readable (as suggested in Step 1). Adding them may undo that readability. Our rule of thumb is that flows from an exploded process *must* be carried down to the more detailed DFD; however, new flows on that detailed DFD need not be carried back up to the original parent process. This is clearly a matter of opinion or preference.

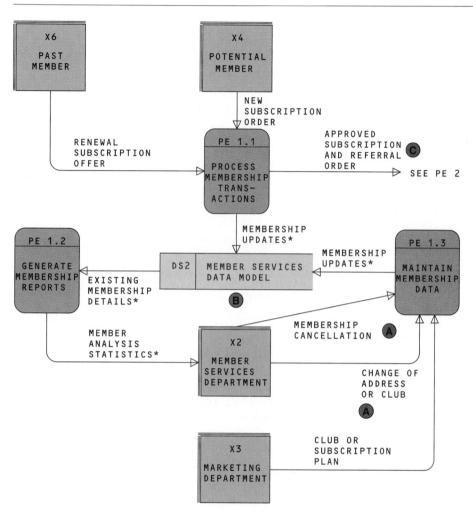

FIGURE 9.20 A Middle-level Data Flow Diagram A middle-level diagram consists of processes that will still be exploded to reveal more detail.

Step 5: Draw Middle-Level Data Flow Diagrams

After drawing the systems diagram, we can explode each of the processes on that DFD to reveal still greater detail about the subsystems. Any process on any DFD can be exploded to reveal a more detailed DFD for that process. Explosion is continued until we have depicted a sufficient level of detail. All but the lowest-level DFDs are frequently called middle-levels.

In the decomposition diagram (Figure 9.16), notice that PE 1, the MEMBER-SHIP SUBSYSTEM, should be exploded into a DFD that includes three processes. That DFD is shown in Figure 9.20. All three processes are still high-level functions, each of which must be further exploded. All the flows to and from the parent process (PE 1 in Figure 9.19) have been carried down to this diagram. We also call your attention to the following:

FIGURE 9.21 Another Middle-Level DFD for Membership

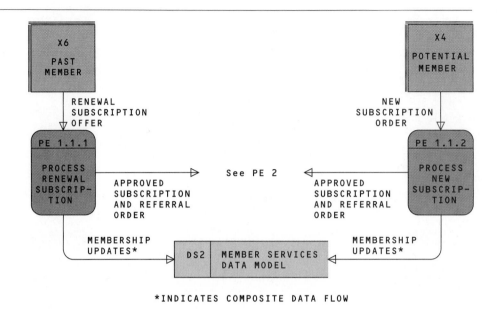

*INDICATES COMPOSITE DATA FLOW

(A) Some new, less common data flows (MEMBERSHIP CANCELLATION and CHANGE OF ADDRESS OR CLUB) have been introduced. Recall that our guideline does not require that these new flows be added to the parent process (seen in Figure 9.18). The diagram is still considered balanced with its parent process.

(B) We still haven't expanded the composite data store into its primitive data stores since our processes are still fairly high level.

(C) The data flow APPROVED SUBSCRIPTION AND REFERRAL ORDER appears to be going nowhere. This is called an *interface data flow.* It is required to balance the DFD back to its parent process and diagram.

We added the note "SEE PE 2" to describe the destination. The note indicates that the flow is going to process PE 2 on the parent diagram (see Figure 9.18).

If we continue to use our decomposition diagram as our model, process PE 1.1 should be exploded to the diagram shown in Figure 9.21. No new concepts are introduced here.

There are two more middle-level diagrams. Figure 9.22 depicts the order processing subsystem in greater detail. Please note the following:

(A) Processes PE 2.2, PE 2.3, and PE 2.4 are primitive processes—that is, they won't be further exploded for user requirements modeling purposes. The letter "p" after the ID number indicates primitive processes.[1]

(B) Because we have introduced primitive processes on this diagram, we must finally expand the composite data store from Figure 9.18 into its

---

[1] Note: The letter "p" denotes primitive processes in DFDs. This is *not* the same notation as "#" used in decomposition diagrams (Figure 9.17). A "#" designated process in the decomposition diagram will usually be further decomposed into "p" designated processes in the DFDs.

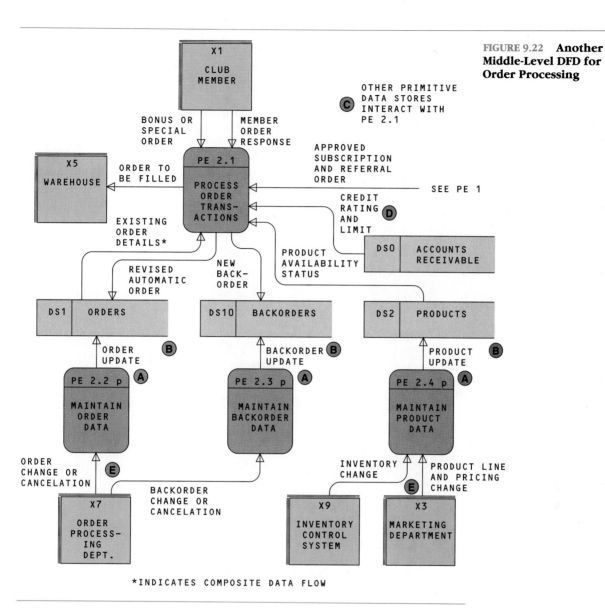

FIGURE 9.22 **Another Middle-Level DFD for Order Processing**

*INDICATES COMPOSITE DATA FLOW

primitive data stores. Our primitive stores correspond to the data model we built in Chapter 8.

C Common sense suggests that PE 2.1, PROCESS ORDER TRANSAC-TIONS, uses and updates other data stores such as MEMBERS. These data stores do not appear on the diagram because for nonprimitive processes, data stores should only be shown if they are shared by at least two processes on the DFD. Otherwise, their introduction should be delayed until explosion DFDs are drawn. (Of course, primitive data stores must be shown for primitive processes!)

D The ACCOUNTS RECEIVABLE data store appears to violate the preceding rule. However, remember that data store was first introduced in the

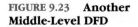

FIGURE 9.23 Another Middle-Level DFD

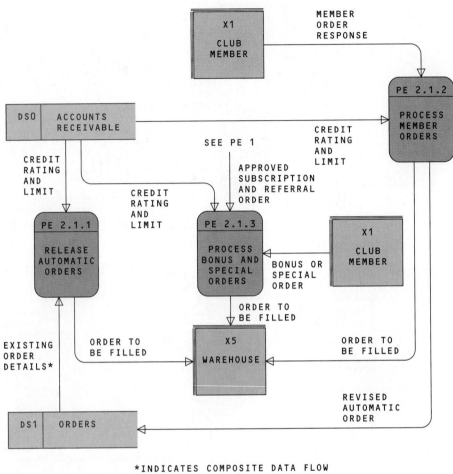

*INDICATES COMPOSITE DATA FLOW

context diagram. Once introduced, a data store must be carried down to appropriate explosion DFDs in order to preserve balancing.

Ⓔ We added new input data flows to expand detail for the data maintenance processes.

Figure 9.23 is the explosion DFD for process PE 2.1 on Figure 9.22. This DFD is self-explanatory.

Step 6: Draw Primitive-Level Data Flow Diagrams

Let's now complete our leveled set of DFDs by drawing those diagrams that show detailed processing requirements for the system. These are called lower- and primitive-level DFDs. You should periodically review the decomposition diagrams to assure yourself that the original outline is being followed.

Let's start with a fairly simple primitive-level DFD. Figure 9.24 is the explosion DFD for process PE 3 on Figure 9.19. This diagram contains one example of each type of primitive process:

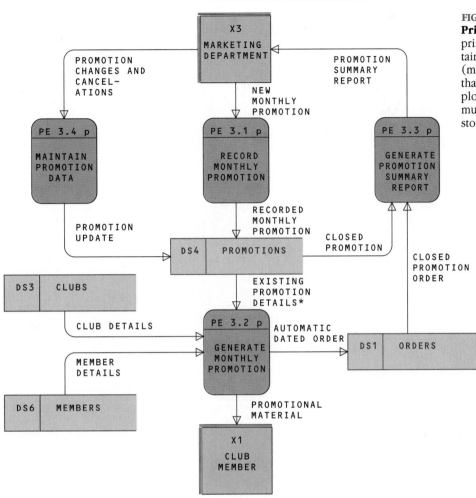

FIGURE 9.24 A Simple Primitive-Level DFD A primitive-level DFD contains some processes (marked by letter "p") that do not further explode. These processes must show primitive data stores and data flows.

- A simple input transaction process (PE 3.1 — more complex transactions would require further explosion).
- An output transaction process (PE 3.2).
- A reporting process (PE 3.3).
- A data maintenance process (PE 3.4).

Notice that primitive DFDs must show all appropriate primitive data stores and data flows.

Now let's examine a more typical transaction process. Figure 9.25 is the explosion DFD for process PE 2.1.1 on Figure 9.23. The processes show the detailed business processing required for a single transaction. Once again, all composite data flows and data stores must be shown on this diagram since it will not be further exploded. Figures 9.26 and 9.27 are similar explosion DFDs for PE 2.1.2 and PE 2.1.3 on Figure 9.23.

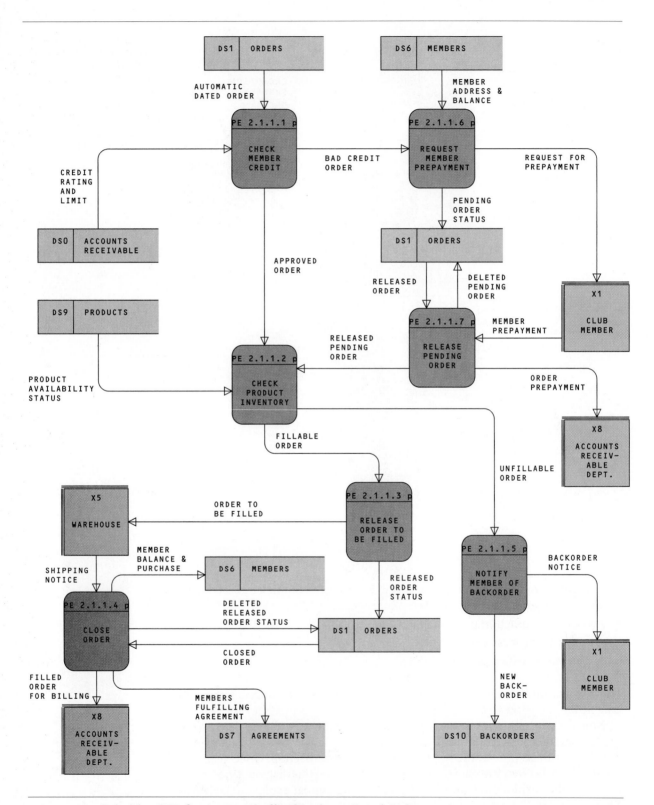

FIGURE 9.25   Primitive DFD for Automatically Shipping a Dated Order

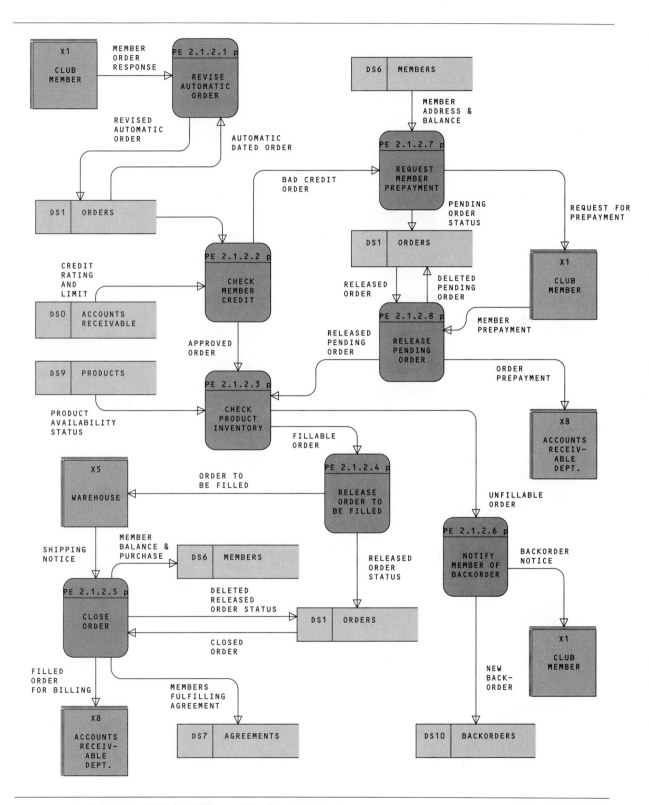

FIGURE 9.26   Primitive DFD for Filling a Member's Order Response

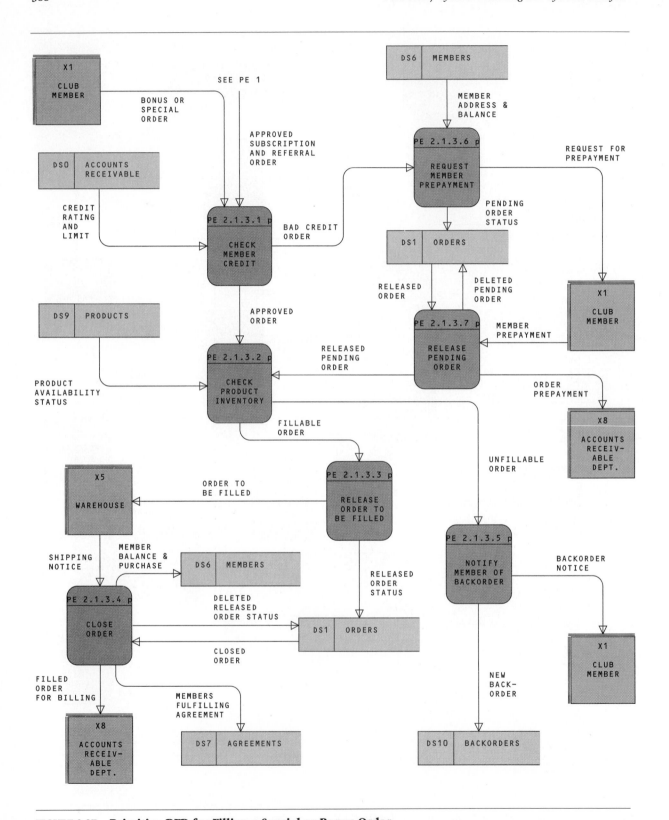

FIGURE 9.27   Primitive DFD for Filling a Special or Bonus Order

All that remains to complete our leveled set of DFDs are the explosion DFDs for the membership subsystem. These diagrams are shown in Figure 9.28. And there you have it—a complete process model for the SoundStage Entertainment Club project. Although we could have drawn the set with somewhat fewer diagrams, we felt it was important to demonstrate a variety of decomposition and explosion approaches.

THE EXPANSION APPROACH TO DRAWING DFDs

Gane and Sarson suggest an alternative approach to drawing DFDs. This approach is a blueprintlike approach that usually results in only two DFDs. Whereas DeMarco advocates the explosion approach, Gane and Sarson suggests an **expansion approach.**

The first data flow diagram is called the *overview* or *level-zero DFD.* It serves as a general-level diagram, usually consisting of 12 or fewer processes. The advantage, according to Gane and Sarson, is improved readability. They argue

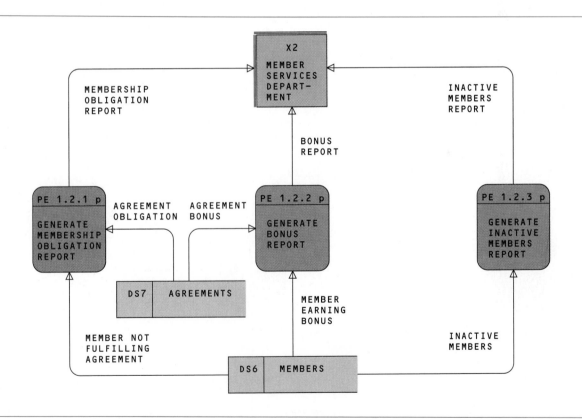

FIGURE 9.28 **Primitive DFD for the Membership Reporting Subsystem**

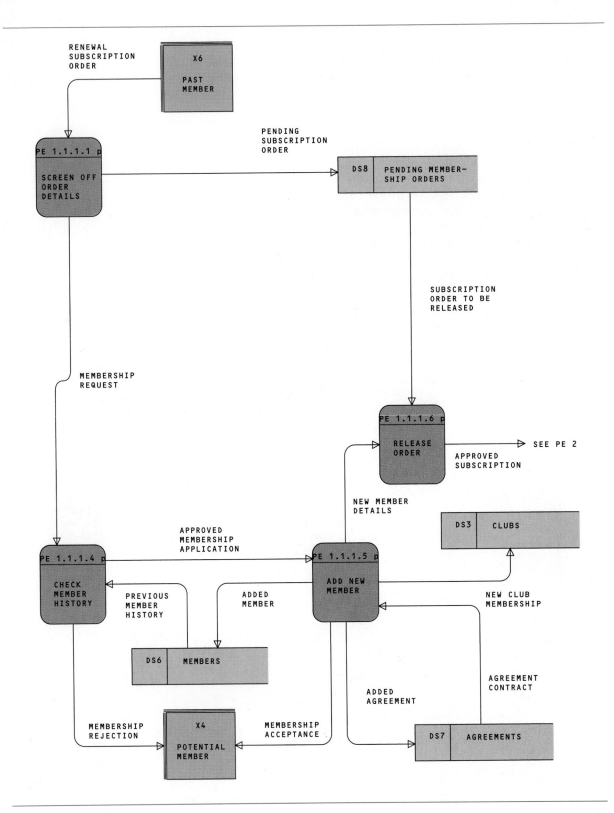

FIGURE 9.28 continued

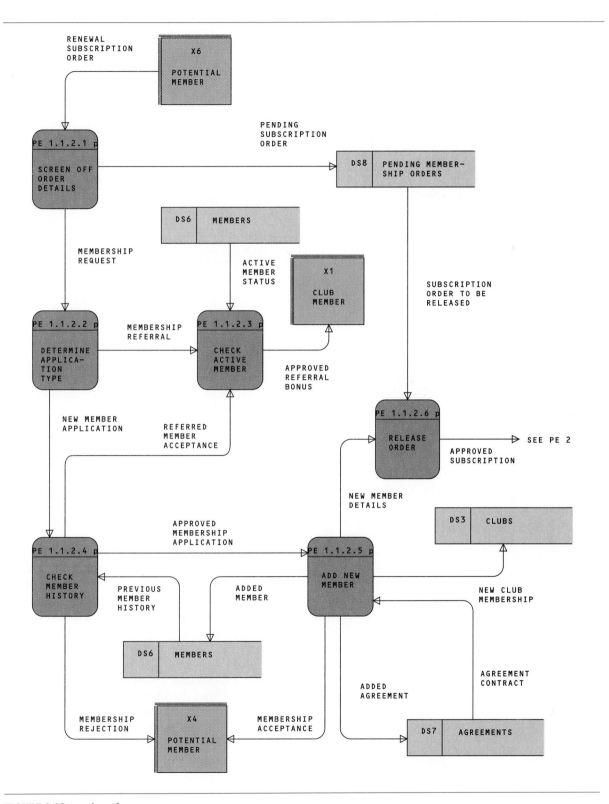

FIGURE 9.28 continued

FIGURE 9.28 concluded

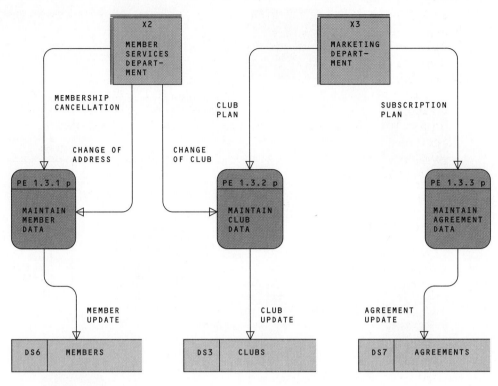

that users get lost in a leveled set of DFDs. As demonstrated in Figure 9.29, we tend to agree.

The level-zero diagram typically includes one process per transaction and key report. Processes for data maintenance and less important reports and ad hoc reports are not shown.

The second diagram, called a *system* or *level-one DFD,* is an expansion of the first diagram. It provides a more detailed view of the system and may contain as many as 30 processes. The level-one diagram is not exploded from any process or processes on the level-zero diagram. Instead, the level-one diagram is merely an expanded, more detailed version of the entire level-zero diagram. As shown in Figure 9.30, it typically: (1) expands the transaction processes from the level-zero diagram and (2) adds the data maintenance processes.

The two-diagram rule is not rigid. Many Gane and Sarson advocates begin with a DeMarco-style context DFD. And for larger systems, the level-zero and level-one diagrams might require two or maybe three DFDs.

In conclusion, the simplicity and user friendliness of the Gane and Sarson approach is preferable. We've discovered additional advantages in the transition to systems design. Thus, we actually practice Gane and Sarson whenever possible. But we acknowledge that you are more likely to encounter and practice the DeMarco leveling approach.

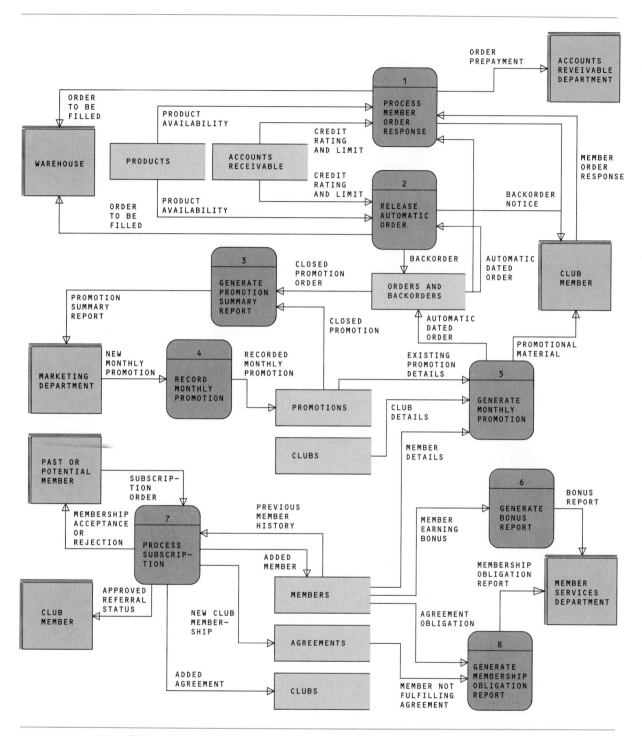

FIGURE 9.29 A Level-Zero DFD Using the Gane and Sarson Expansion Approach The expansion approach begins with a single DFD that models the entire system at a fairly high level of detail.

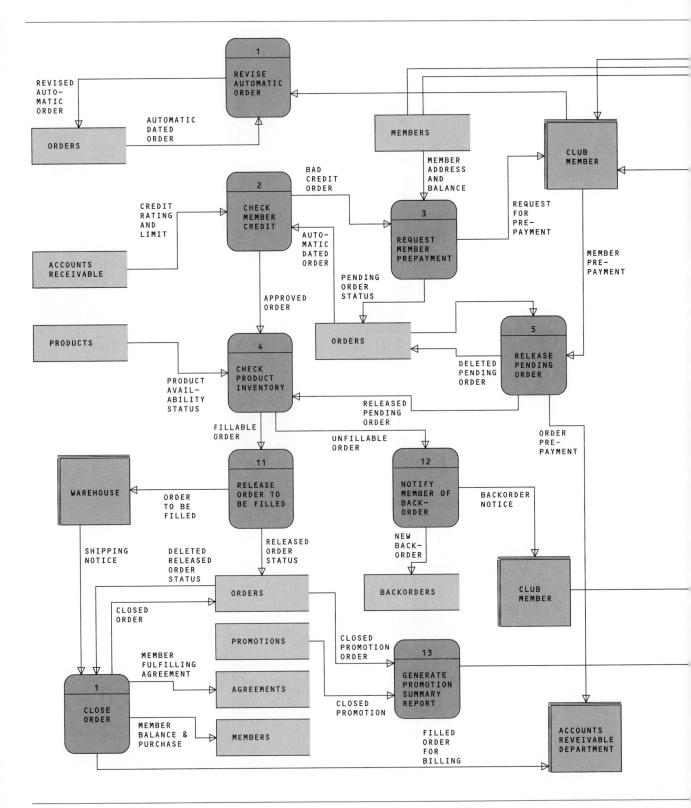

FIGURE 9.30   A Level-One DFD Using the Gane and Sarson Expansion Approach   The expansion approach concludes with an expanded view of the level-zero DFD—once again, a single DFD.

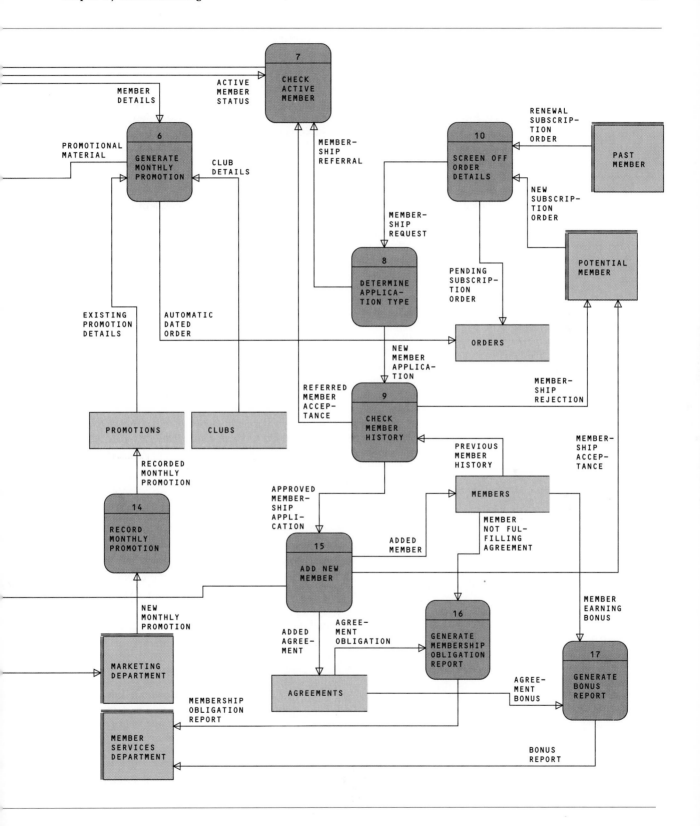

Summary

An essential data flow diagram (DFD) is a tool for drawing a model or picture of an information system's processing requirements. This is called process modeling. DFDs illustrate the flow of data and work through a system. There are only four symbols that can appear on a DFD: the process, the external or internal agent, the data store, and the data flow. With these symbols, you can model processing for virtually any information system. Essential DFDs differ from implementation DFDs in that they model the essential system requirements independently of any technology or methods that might be used to implement those requirements. This opens the door for more creative solutions. Data flow diagrams can effectively model all transaction processing, management reporting, and decision support functions in a system.

Data flow diagrams have many uses in systems planning, analysis, and design. This is a tribute to their versatility in modeling different levels of detail. Today, CASE technology is used extensively to draw and maintain DFDs.

Most analysts draw a series of DFDs that depict the system in various levels of detail, from general to explicit. This is called the explosion approach. Processes on one diagram explode to a more detailed diagram until the combined diagrams document the entire system.

In the explosion approach the most general DFD is the context DFD, which shows the scope or boundary of the system. The next level of detail, the overview DFD, shows key subsystems or functions and their interfaces. Middle-level DFDs show still greater detail about subsystems and/or subfunctions. Finally, primitive DFD levels show explicit data flows and processes for a small, manageable piece of the system.

To intelligently plan the levels of DFDs, a decomposition diagram is used to functionally decompose a system. This chapter presented useful strategies that are applicable to virtually all information systems.

Some analysts prefer to draw DFDs using expansion rather than explosion. A single high-level DFD is drawn. It is then expanded into a single, more detailed, but equivalent DFD.

Key Terms

application context
    process model, p. 365
application essential
    process model, p. 366
application
    implementation
    process model, p. 367
balancing, p. 379
black hole, p. 355
business area process
    model, p. 365
composite data flow, p. 358
context data flow diagram,
    p. 368

data flow, p. 352
data flow diagram, p. 350
data store, p. 352
decomposition diagram,
    p. 373
enterprise process-like
    model, p. 365
essential model, p. 349
essential process
    modeling, p. 350
expansion approach, p.
    389
external agent, p. 352
external entity, p. 352

gray hole, p. 355
implementation model, p.
    348
internal agent, p. 352
internal entity, p. 352
miracle, p. 355
model, p. 348
overview data flow
    diagram, p. 377
primitive data flow, p. 358
process, p. 352
process modeling, p. 350

Problems and Exercises

1. Explain to an end-user or manager the difference between essential and implementation modeling. When and why would you use each?

2. Compare process models and data models. What does each model show? Should you choose between the two modeling strategies? Why or why not?

3. A manager who has noted your use of essential data flow diagrams (DFDs) to document a proposed system's requirements has expressed some concern because of the lack of details that demonstrate the computer's role in the system. Defend your use of essential DFDs. Concisely explain the symbolism and how to read a DFD. (*Note:* The answer to this exercise should be a standard component in any report that will include DFDs. You cannot be certain that the person who reads this report will be familiar with the tool.)

4. Explain why you should exclude implementation details when drawing an essential DFD. Can you think of any circumstances in which implementation details might be useful?

5. Explain why a systems analyst might want to draw essential models of an automated portion of an existing information system rather than simply accepting the existing technical information systems documentation, such as systems flowcharts and program flowcharts.

6. Draw an essential DFD to document the flow of data in your school's course registration and scheduling system.

7. Draw an essential DFD for some day-to-day "system" that you use or observe in use — for instance, your morning routine; making your favorite meal, including appetizer, entree, side dishes, and dessert; constructing something from scratch.

Projects and Minicases
---

1. Given the following narrative description, draw a context DFD for the portion of the activities described.

   The purpose of the TEXTBOOK INVENTORY SYSTEM at a campus bookstore is to supply textbooks to students for classes at a local university. The university's academic departments submit initial data about courses, instructors, textbooks, and projected enrollments to the bookstore on a TEXTBOOK MASTER LIST. The bookstore generates a PURCHASE ORDER, which is sent to publishing companies supplying textbooks. Book orders arrive at the bookstore accompanied by a PACKING SLIP, which is checked and verified by the receiving department. Students fill out a BOOK REQUEST that includes course information. When they pay for their books, the students are given a SALES RECEIPT.

2. Given the following narrative description, draw a context DFD for the portion of the activities described.

   The purpose of the PLANT SCIENCE INFORMATION SYSTEM is to document the study results from a wide variety of experiments performed on selected plants. A study is initiated by a researcher who submits a STUDY PROPOSAL. After a panel review by a group of scientists, the researcher is required to submit a DETAILED STUDY DESCRIPTION. An FDA PERMIT REQUEST is sent to the Food and Drug Administration, which sends back a PERMIT. As the experiment progresses, the researcher fills out and submits DETAILED EXPERIMENT NOTES. At the conclusion of the project, the researcher's results are reported on a PLOT ANALYSIS DIAGRAM.

3. Given the following narrative description of a system, draw a context and system-level DFD.

   The purpose of the production scheduling system is to respond to a PRODUCTION REQUEST (submitted by the SALES DEPARTMENT) by generating a daily PRODUCTION SCHEDULE, generating MATERIAL REQUESTS (sent to the STORES DEPARTMENT) for all production orders scheduled for the next day, and generating JOB TICKETS for the work to be completed at each workstation during the next day (sent to the SHOP LINE SUPERVISOR). The work is described in the following paragraphs.

   The production scheduling problem can be conveniently broken down into three functions: routing, loading, and releasing. For each product on a PRODUCTION REQUEST, we must determine which workstations are needed, in what sequence the work must be done, and how much time should be necessary at each workstation to complete the work. This data is available from the PRODUCT ROUTE SHEETS. This process, which is referred to as ROUTING THE ORDER, results in a ROUTE TICKET.

Given a ROUTE TICKET (for a single product on the original PRODUCTION RE-QUEST), we then LOAD THE REQUEST. Loading is nothing more than reserving dates and times at specific workstations. The reservations that have already been made are recorded in the WORKSTATION LOAD SHEETS. Loading requires us to look for the earliest available time slot for each task, being careful to preserve the required sequence of tasks (determined from the ROUTE TICKET).

At the end of each day, the WORKSTATION LOAD SHEETS for each workstation are used to produce a PRODUCTION SCHEDULE. JOB TICKETS are prepared for each task at each workstation. The materials needed are determined from the BILL OF MATERIALS data store, and MATERIAL REQUESTS are generated for appropriate quantities.

4. Health Care Plus is a supplemental health insurance company that pays claims after its policyholders' primary insurance benefits through their employer or another policy have been exhausted. The following narrative partially describes its claims-processing system. Draw a context and system-level DFD for the portion of the systems activities described.

Policyholders must submit an Explanation of Health Care Benefits (EOHCB) along with proof that their primary health policy claim has been paid. All claims are mailed to the claims-processing department.

All claims are initially sorted by the claims screening clerk. This clerk returns all requests that do not include the EOHCB or EOHCB reference number. For those requests returned, a PENDING CLAIM is created, dated, and stored by date. Once each week, the clerk deletes all tickets that are more than 45 days old and sends a letter to the policyholders notifying them that their case has been closed. Requests that include the EOHCB are then sorted according to type of claim. Requests that include an EOHCB reference number are matched up with an EOHCB form, which is pulled from the OPEN CLAIMS data store. At the end of each day, all these claims are forwarded to the preprocessing department.

In the preprocessing department, clerks screen the EOHCB for missing data. They complete the form if possible. Otherwise, a copy of the claim is returned to the policyholder with a letter requesting the missing data. The original EOHCB is placed in the OPEN CLAIMS data store, and a PENDING CLAIM is sent to the claims screening clerk. Completed claims are assigned a claim number, and the claim is microfilmed and filed for archival purposes.

A different clerk checks to see if the proof of primary health care policy payment was included or is on file in the PRIMARY PAYMENT data store. If it is not available, the policyholder is sent a letter requesting the proof. The EOHCB is placed in a PENDING PROOF data store. Claims are automatically purged if they remain in this file for more than 14 days (a letter is sent to policyholders whose claims have been purged).

If proof is available, another clerk pulls the policyholder's policy record from the POL-ICY data store, records policy and action codes on the EOHCB, and refiles the policy. At the end of the day, all preprocessed claims are forwarded to Information Systems.

5. Given the following narrative description, draw a context DFD and the system-level DFD for the portion of the activities described.

The purpose of the GREEN ACRES REAL ESTATE INFORMATION SYSTEM is to assist agents as they sell houses. Sellers contact the agency, and an agent is assigned to help the seller complete a SALES REQUEST. Information about the house and lot taken from that request is stored. Personal information about the sellers is copied by the agent onto a SELLERS PERSONAL INFO SHEET and stored.

When a buyer contacts the agency, he or she fills out a BUYER REQUEST. Every two weeks, the agency send prospective buyers a HOMES FOR SALE MAGAZINE and an AD-DRESS KEY for the magazine containing street addresses. Periodically, the agent will find a particular house that satisfies most or all of a specific buyer's requirements, as indicated in the BUYERS REQUIREMENTS REPORT distributed weekly to all agents. The agent will occasionally photocopy a picture of the house along with vital data and send the MULTIPLE LISTING SERVICE (MLS) SHEET to the potential buyer.

When the buyer selects a house, he or she fills out a PURCHASE AGREEMENT that is forwarded through the real estate agency to the seller, who responds with either OFFER ACCEPTANCE or a COUNTEROFFER. After a PURCHASE AGREEMENT has been accepted, the agency sends an APPRAISAL REQUEST to an appraiser, who appraises the value of the house and lot. The agency also notifies its finance company with a FINANCE REQUEST.

6. Given the following narrative description, draw a context DFD and the system-level DFD for the portion of the activities described.

The purpose of the OPEN ROAD INSURANCE SYSTEM is to provide automotive insurance to car owners. Initially, customers are required to fill out an INSURANCE APPLICATION REQUEST. A DRIVER'S RECORD REQUEST is sent to the local police department, which sends back a DRIVER'S RECORD REPORT. Also, a VEHICLE REGISTRATION REQUEST is sent to the Department of Motor Vehicles, which supplies a VEHICLE REGISTRATION. POLICY CONTRACTS are sent in by various insurance companies. The agent determines the best policy for the type and level of coverage desired and gives the customer a copy of the INSURANCE POLICY along with an INSURANCE COVERAGE CARD. The customer information is now stored. Periodically, a FEE STATEMENT is generated, which—along with ADDENDUMS TO POLICY—is sent to the customer, who responds by sending in a PAYMENT with the FEE STUB.

Suggested Readings

DeMarco, Tom. *Structured Analysis and System Specification.* Englewood Cliffs, N.J.: Prentice Hall, 1978. This is the classic book on the structured systems analysis methodology, which is built heavily around the use of data flow diagrams.

Gane, Chris, and Trish Sarson. *Structured Systems Analysis: Tools and Techniques.* Englewood Cliffs, N.J.: Prentice Hall, 1979. This is an early structured analysis methodology book. Although not as thorough as DeMarco, it is relatively easy to read and grasp. Their DFD symbolism is identical to ours. A methodology called STRADIS, which is marketed through Electronic Data Systems (EDS), has been developed around this approach.

Keller, Robert. *The Practice of Structured Analysis: Exploding Myths.* New York: Yourdon Press, 1983. This is a concise overview of the first-generation structured analysis methodology.

Martin, James, and Carma McClure. *Structured Techniques: The Basis for CASE.* Englewood Cliffs, N.J.: Prentice Hall, 1988. This book provides a critical analysis of DFDs and their modern cousins, dependency diagrams.

McMenamin, Stephen M., and John F. Palmer. *Essential Systems Analysis.* New York: Yourdon Press, 1984. This is the most thorough reference to date on essential DFDs. The book also hints at ways to integrate process and information modeling. We recommend that you read DeMarco before you read McMenamin and Palmer.

Ward, Paul. *Systems Development without Pain.* New York: Yourdon Press, 1985. This book provided our first insight into the possibilities of merging the process modeling and information modeling approaches.

Yourdon, Edward. *Modern Structured Analysis.* Englewood Cliffs, N.J.: Yourdon Press, 1989. This is the long-awaited update to DeMarco's classic book. It includes all the suggested updates to the structured analysis methodology.

10

Network Modeling

Chapter Preview and Objectives

This is the third of three graphic systems modeling chapters. In this chapter you will learn how to use a unique network modeling tool, location connectivity diagrams, to document a system's network locations, independent of how data and processes will be distributed to those locations. You will know network modeling as a systems analysis tool when you can:

Describe
why network modeling may become an important skill for applications developers in the next several years.

Define
network modeling and explain why it is important.

Explain
how network modeling is useful for systems planning and analysis.

Factor
a system's or application's locations into component locations using a special location decomposition diagram.

Document
the connections and essential data flows between locations using location connectivity diagrams (LCDs).

Explain
the complementary relationship between network models (LCDs), process models (DFDs), and data models (ERDs).

MINICASE

REMINGTON STEEL

Remington Steel is a northwest Indiana headquartered steel manufacturer with decentralized manufacturing, finishing, and distribution plants located throughout the country. Historically, Remington's information systems have been decentralized but autonomous. Leased line networks connect the locations, but only for purposes of data consolidation and electronic mail.

Scene: *Mike Ironsides, a young systems analyst, has been called into the office of his immediate supervisor, Bill Riker. Bill has been a key player in Remington's Information Strategy Planning Project (ISPP), an effort to define a long-term architecture for future information system applications and technology deployment.*

Mike: You wanted to see me, Bill?

Bill: Yes, Mike. Have a seat. I assume you've heard about the new just-in-time manufacturing project?

Mike: Yes, that project got a fairly high priority in the information strategy plan. But I thought that the project wasn't scheduled to begin for eight or nine months.

Bill: You're right. But I've got a special preproject assignment for you. Have you read the information strategy plan's system architecture?

Mike: I skimmed it. I don't remember too much.

Bill: Let me amplify on it. The plan has some unique elements, most notably its call for application rightsizing through client/server distribution. I don't know how familiar you are with this new client/server trend, but let me tell you how I see it. To date, we've focused on decentralizing data and processes—but keeping all data and processes on a few computers, usually mainframes or minis. In the future, we will further distribute processes. The processes for a single application might be split among multiple computers of different sizes. We'll use a client/server architecture to split these processes.

Mike: I've been reading a lot about this client/server phenomenon. I've seen so many different definitions, I'm not sure what to believe. Some say it's the end of the mainframe. Others say it may include the mainframe.

Bill: Let me give you the planning committee's perspective on client/server. Client/server is cooperative processing between computers. The client is usually a personal computer. The server may be a mainframe, minicomputer, or superserver. In a client/server application, nonshared processes or programs are located on the clients. Examples include input and most output functions. Shared processes, programs, and technologies are located on servers. Examples include database processes and certain expensive technologies, such as fax and high-speed printing.

Mike: So we might have separate servers for database, fax, and high-speed laser output?

Bill: That's right. Clients will send requests and data to those servers as needed and get responses as appropriate. This will all be transparent to the users of the client workstations.

Mike: You seem to imply the mainframe is not dead because of client/server.

Bill: Certainly not because of client/server. First, we cannot convert everything to client/server overnight. Second, a mainframe can be a server in a client/server platform. The key, today, is rightsizing—using the right-sized computer for a specific task.

Now that I've said that, I must qualify the remarks. I don't believe the mainframe will play a long-term role in our scheme of things. Clearly, today's minicomputers and tomorrow's superservers will continue to grow in power, approaching or surpassing today's mainframe. I believe that the mainframe, as we know it, will either play a different role in our business, or give way to smaller but equally powerful superservers. Just think of how far the personal computer has come in a very short time. Parallel processing microcomputers are on the horizon. That's my view, and I think it is shared by all of the ISPP team and IS managers.

Mike: What's the possibility that client/server is just a fad?

Bill: Like all new technologies, it's getting a lot of press, mostly good. Publications tend to overglamorize new technologies, like they did to CASE. After a while, the press turns bad as early users discover that the technology has problems. In reality, the technology is usually good, but we haven't yet learned how to best apply it. Then, over time, the application techniques catch up and the technology is taken for granted—we wonder why we didn't always do it that way.

I think client/server makes sense. Put data and processes where they can best meet requirements—some call that rightsizing. Then make the various processes cooperate with one another in ways that are transparent to users. The real problem, as I see it, is how do we make intelligent rightsizing decisions in the best interest of the company?

Mike: So, what's my assignment?

Bill: Well that's just it! I've been thinking about this distribution problem. I don't think our new methodology has any provisions to handle distribution, rightsizing, and client/server architecture. We both know that our methodology is sound in process and data modeling. But it seems to me that those models are no longer sufficient. Don't we need to extend our modeling to consider distribution issues?

Mike: I agree. Is that my assignment?

Bill: Yes, Mike. I want you to find or create extensions to our methodology to help us deal with these new technology alternatives.

Mike: We are probably going to have to create something. This may be one of those situations where the technology is truly evolving faster than the techniques needed to deal with the technology. It seems to me that we need a model of a system's geography—the locations in which we do business. Then we need to be able to attach portions of our data and process models to the locations. And, of course, we need to keep all three models consistent. This sounds real interesting. I'll get started right away. Is there anything else?

Bill: Yes. Congratulations, Mike. I've decided to let you head up the just-in-time inventory project. You've earned it. Besides, it'll give you a chance to field-test whatever you come up with in this project. Keep me informed. I've got an academic interest in this project.

Discussion Questions

1. Think of a typical mainframe computer application. How might it be different in the world of client/server technology? How would you expect application development (e.g., programming) to differ?

2. What are some of the problems or issues that you expect Mike to encounter in his new assignment?

3. How will the concepts of essential versus implementation modeling come into play for network modeling? In other words, what would an essential network model depict? What would an implementation network model depict?

NETWORK MODELING—NOT JUST FOR COMPUTER NETWORKS

We begin this chapter with a famous saying:

Necessity is the mother of invention.

Nowhere is that old saying more applicable than in this chapter. What "necessity"? Many new technologies evolve faster than our ability to properly apply them. That is the case with client/server computing. Its use is called *application downsizing* or *rightsizing*. (We'll define these terms soon enough.) Appli-

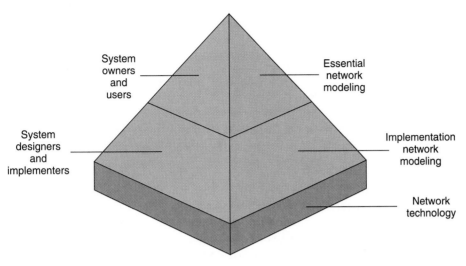

FIGURE 10.1 **Network Modeling Relationship with the IS Building Blocks** This chapter is about essential network modeling from the perspective of the system owners and users. The focus is on the essential locations required to run the business. The need to understand this aspect of systems is being driven by the evolution of networking technology.

cation downsizing and rightsizing require analysts to make intelligent decisions about the distribution of data and processes to multiple locations. But how does the systems analyst make intelligent rightsizing and client/server technology decisions?

The analyst needs a model of application's shape or geography. This need is reinforced by John Zachman's information systems framework, the conceptual basis for your pyramid model. Figure 10.1 illustrates network modeling in the pyramid.

Unlike process modeling (with DFDs) and data modeling (with ERDs), no generally accepted network modeling standards have emerged. Thus, we had to invent a tool, location connectivity diagrams (LCDs). Since this entire chapter represents future thinking, we purposefully elected not to include The Next Generation box.

Why study a topic or technique that is not in the mainstream of current practice? Simple! Analysts are trying to build network-based, client/server systems and applications *today!* The trend is toward application downsizing and rightsizing.

> **Application downsizing** is the deliberate effort to redevelop or reengineer information systems applications to run on smaller, cheaper computers or networks of computers.
>
> **Application rightsizing** is the deliberate effort to design and deliver new information systems using appropriate-sized computers or networks. In other words, applications are not automatically designed for mainframe computers as they have been in the past.

We expect these trends to continue or escalate. And we expect that today's analysts' yearning for tools and techniques will help them better deal with the data and process distribution issues of the 1990s. Today, we can use any advantage we can get our hands on — even if better tools are forthcoming tomorrow.

THE HISTORY AND CONCEPTS OF SYSTEMS DISTRIBUTION

This chapter examines the evolution of system distribution and its impact on the modern systems analyst.

> **Network modeling** is a diagrammatic technique used to document the shape of a business or information system in terms of its user, data, and processing locations.

In this chapter we will focus exclusively on **essential network modeling**— that is, the modeling of business network requirements independent of their implementation. Consistent with data and process modeling, essential models eventually give way to implementation models that describe computer networks to be implemented or used by specific information systems.

To better understand the need for essential network models, it helps to examine the evolution of the corresponding implementation models—the technology of distribution and networks. Today's systems analyst is faced with an increasing variety of distribution decisions based on evolving technologies.

Centralized Computing and Timesharing

Much of today's existing base of information systems and applications uses an old, reasonably successful, but increasingly expensive, computing technology—central computers and timesharing.

> **Central computing** is an application architecture that uses a single computer, usually located in a central data processing center or departmental data processing location. The computer is usually a mainframe or mini-computer that provides access to multiple simultaneous users. It is also called *central processing*.

> **Timesharing** is the method by which users share a central computer. In a timesharing system, the user interface (screens), input and output, data storage and retrieval processes, and business logic are all performed on the single, central processor.

Central computing and timesharing are illustrated in Figure 10.2. Notice that users communicate with the computer through terminals. Also notice that personal computers running terminal emulation software can also access the central computer, but in this case the central computer is still doing all the work (except for the terminal emulation).

It is interesting to note that **decentralized computing** is appreciably different than centralized computing. A decentralized environment merely duplicates central computers of different sizes at different locations. Each central computer still provides timesharing to its own users, and the central computer still does all of the work.

Despite the growing use of more sophisticated alternatives, central computing and timesharing still account for the bulk of today's information systems and applications. Many companies still develop systems using this formula, although many are finding that many smaller, decentralized computers (e.g., minicomputers and superservers) can handle the timesharing workload just as well as one or a few traditional mainframes.

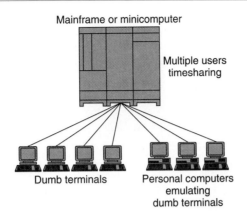

Mainframe or minicomputer

Multiple users
timesharing

Dumb terminals

Personal computers
emulating
dumb terminals

FIGURE 10.2 **Centralized Computing with Time-sharing** Today, most existing information systems applications use centralized computing with timesharing. All processing is performed on the central computer. Some businesses duplicate this model to create decentralized computing with timesharing.

Stand-Alone Personal Computing

No phenomenon is more responsible for the changing landscape of computing alternatives than is the personal computer.

> The **personal computer,** also called an *intelligent workstation* or *programmable workstation,* brings decentralized computing power to the desktop and on the road.

The PC is similar to the timesharing computer in that the user interface (screens), input and output, data storage and retrieval processes, and business logic are all still processed on a single computer. The difference is that all the computer's resources are dedicated to a single user.

While the PC presented a new option for applications, it did not represent the best way to share resources, especially data storage. Still, PC end-users have yet to scratch the surface of the potential of the PC to provide personal information systems. That potential will remain largely untapped until end-users are better educated to develop more sophisticated applications with today's available tools. (Note: The techniques of systems analysis and design can contribute to development of improved personal information systems.)

Distributed Computing

The centralized data processing shop became less important as the numbers of departmental minicomputers and personal computers grew. But as these decentralized computers grew in number, users and information professionals alike saw great potential in connecting them. Out of this grew the technology of networking and data communications, and many new choices became available for systems analysts developing modern information systems and applications.

Computer networks allowed mainframe and minicomputer applications to exchange data and information. More importantly, computer networks allowed analysts to consider a new application alternative, distributed processing.

> **Distributed processing** allowed data processing shops to reallocate data storage, input/output (I/O), and processing to multiple computers and to exchange data and information between those computers.

FIGURE 10.3 **Distributed Computing** Distributed computing reallocates application processes to a network of timesharing computers. Notice also that PCs are grouped into local area networks, which are connected to form wide area networks. Also notice that PCs can be connected to timesharing computers to enhance the network.

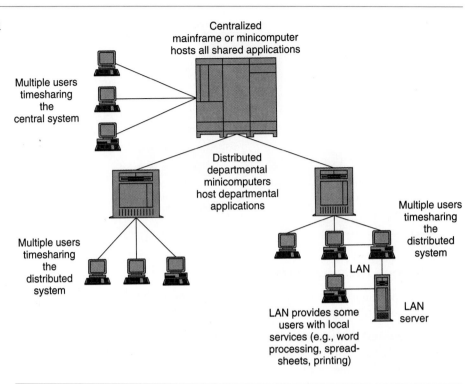

The computers in a distributed computing architecture still provide timesharing service to their users. All processing, I/O, user interfaces, and data storage and retrieval are still performed by each individual computer in the network. The primary difference is that applications could be distributed to appropriate locations but still be integrated through data and information exchanges across the network. This is illustrated in Figure 10.3.

Distributed processing has grown slowly over the years as the problems with network security and database integrity were discovered and solved. A key result of distributed processing is distributed data storage. Data can now be located closer to departments and users while still preserving the ability to share some central data and exchange any data between locations. The majority of businesses that operate in multiple locations today use some form of distributed computing to support information systems.

At the same time as distributed processing between decentralized computers grew, the growth in numbers of personal computers led to the linking of those personal computers into **local area networks** (LANs) to share data, software, and technology. The growth of local area networks eventually gave rise to **internetworking** that created **wide area networks** (WANs). Additionally, technology to allow PCs, LANs, and WANs to communicate and exchange data with central computers, such as mainframes and minicomputers, became widely available.

Eventually, all the technology was in place to enable sophisticated, new application development alternatives to be established, such as cooperative processing and client/server computing.

Cooperative Computing

Cooperative computing is an extension of distributed computing.

> In **cooperative computing,** two or more computers cooperate to perform a given task. This is also called *cooperative processing.*

When applied to information system applications, the processes of the application are divided between different computers. Both computers are required to execute the application; however, the user perceives that a single computer is doing all the work. This relatively new application alternative is sometimes called *seamless* or *transparent cooperation.*

The most common example of cooperative computing is sometimes called *screen scraping.* The popularity of graphical user interfaces (e.g., Windows, Presentation Manager, Motif, NextStep) on personal computers makes it possible to move the user interface for an application off the mainframe and onto the user's own PC. Additionally, some or all data editing can be moved to the PC. The new user interface on the PC seamlessly communicates with a host computer (e.g., mainframe or minicomputer), which does the rest of the processing, including data storage and retrieval.

In the CASE industry, cooperative processing is being used by some tools to seamlessly integrate workstation functions with host-based repositories and code libraries. Examples include IBM's OS/2 interface to its host-based Cross System Product, and ViaSoft's Existing System Workbench's OS/2 cooperation with host-based code libraries and software reengineering tools.

A classic application is the automatic teller machine (ATM). Each ATM handles the user interface, input and card verification, and output (cash and receipts). However, it must cooperate with the bank's branch or central computer to perform certain transactions, such as account balance lookup and inter-account transfers.

In still another example, a point-of-sale PC cash register can process sales transactions and cooperatively provide that data to host-based inventory and sales analysis applications.

At the very least, modern analysts should consider the value of cooperative processing for putting a graphical user interface on mainframe applications (instead of the classic, character-based, 3270 terminal, CICS interface of the past). This can be accomplished through CASE tools such as Easel.

Client/Server Computing

Client/server computing is an extension of cooperative processing made possible by the evolution of PCs, LANs, WANs, host connectivity, graphical user interfaces, and distributed database management systems.

> In **client/server computing** an application's processing is distributed among multiple computers in an LAN or WAN. Server computers will provide shared or common services to the application or system. The classic example is a database server whose sole job is to process reads and writes to a database(s). A print server does nothing but respond to printing requests. A fax server does nothing but send or receive faxes. Meanwhile, client computers perform local processing to the end-user. The local processing usually includes the user interface (usually graphical), input edit-

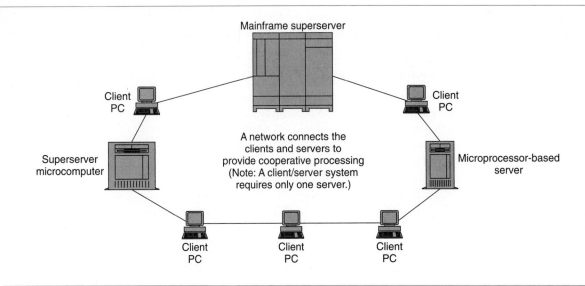

FIGURE 10.4 **Client/Server Computing** In a client/server network, processing workloads are shared between client workstations (PCs) and task-specific servers. Notice that mainframes and minicomputers can be servers in the client/server network.

ing, output, and some or all of the business-specific logic. Clients and servers interact through cooperative processing to form a complete application.

Client/server computing is illustrated in Figure 10.4. There are several reasons for the modern interest in client/server computing:

- Clients are increasingly powerful and cheaper than mainframes and minicomputers.
- Server computers are becoming powerful enough to handle the workload of many clients, again at a lower cost than mainframes and minicomputers.
- Data storage can be moved ever closer to the end-user where it becomes a more valuable business resource.
- Graphical client interfaces are said to make applications easier to learn and use.
- Client/server applications are said to be cheaper and easier to construct and maintain. (Only time will prove or disprove this claim.)

There is no reason that either the mainframe or the minicomputer could not become a server in a client/server environment. This is especially true of minicomputers. On the other hand, if superservers (e.g., IBM RS/6000s, Compaq SystemPros, NCR 3400 series and above) continue to provide increased processing power at a declining cost (compared to minis and mainframes), the host computer, as we know it, could and should become extinct.

To the end user, a typical client/server application resembles a very powerful PC application. Behind the scenes, each user's PC is sending work requests to appropriate servers that do the actual work. The big question in client/server application design is "how should the processing be distributed?" Most agree that processing for the user interface, input and editing, and private or local databases should be on client workstations. Most agree that processing for shared databases should be on a server(s). But experts disagree on how to distribute the business logic that sits between input, data storage and retrieval, and output. In the near future these issues will likely be resolved.

Implications for the Systems Analyst

The choices confronting today's systems analyst are confusing. Today's analyst must seek answers to new questions:

- What locations are applicable to this information system or application?
- How many users are at each location?
- Do any users travel while using (or potentially using) the system?
- Are any of our suppliers, customers, contractors, or other external agents to be considered locations for using the system?
- What are the user's data and processing requirements at each location?
- How much of a location's data must be available to other locations? What data is unique to a location?
- How might data and processes be distributed between locations?
- How might data and processes be distributed within a location?

These questions and others are driving the need for the analyst to fully understand the geography of each information system.

LOCATION CONNECTIVITY DIAGRAMS

The need to understand the geography of each information system has driven us to develop a new tool—the location connectivity diagram (LCD). This tool models system geography independent of any possible implementation.

> A **location connectivity diagram** (LCD) is a network modeling tool that depicts the shape of a system in terms of its user, process, and data locations and the necessary interconnections between those locations.

Recall from Chapter 2 that one of the fundamental building blocks of information systems is networks. All information systems have a geography—some more complex than others! Geography provides a basis for eventually distributing data and processes to specific locations and computers. We need a formal way to model and study the processing in an information system.

Location Connectivity Diagram Conventions and Guidelines

Figure 10.5 demonstrates a symbol set for documenting a business network (the geography of a single business information system or application). The symbols are duplicated in the margin for your reference.

A location is the main symbol on a location connectivity diagram.

> A **location** is any place at which users exist to use or interact with the information system or application.

Location

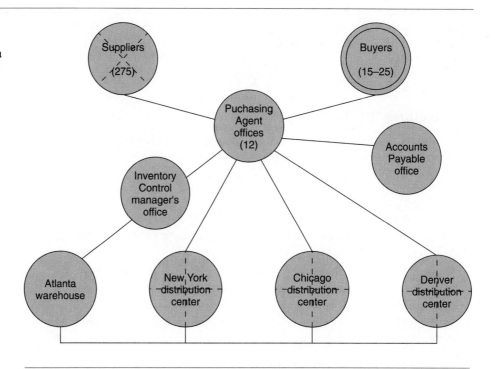

A location can mean different things depending on whether you are modeling the essential system (business requirements) or implementation system (computer and networking requirements). Examples include:

Essential Locations	**Implementation Locations**
• City.	• Computer or server site.
• Campus.	• Terminal site.
• Building.	• Terminal cluster site.
• Office.	• Local area network cluster.
• Cluster of offices.	• Wide area network connection.
• Area (e.g., warehouse).	• Peripheral site.
• Subsidiary.	• Communications peripheral site.
• Customer.	
• Supplier.	
• Contractor.	

For the remainder of this chapter, we are concerned exclusively with essential locations of the business network.

Locations can:

• Be scattered throughout an information system.
• Be on the move (e.g., traveling sales representatives).

- Represent clusters of similar locations.
- Represent organizations and agents outside of the company but which interact with or use the information system.
- Represent all or part of any information system.
- Require similar business data and processes.

Ideally, we need a more precise symbolic notation for locations.

Some locations have sublocations. For example, a college may include many campuses. A campus may include many buildings. A building may include many offices (or office clusters), classrooms (or classroom clusters), laboratories, equipment rooms, and so on. A circle superimposed with a plus sign indicates that sublocations are documented on another page.

Some locations are not stationary. Sales representatives and purchasing agents may be on the road, but nonetheless they use your information system. They should be considered a special part of the network. They might interact via dial-up access through modems or cellular modems. We represent moving locations with a pair of concentric circles (which resemble a wheel).

Some locations represent external organizations and agents (e.g., customers, suppliers, taxpayers, contractors, etc.). Many organizations are trying to link directly with the systems of external organizations to reduce response time and improve throughput of transactions. For example, Sears Roebuck can electronically order new inventory directly from their suppliers (the companies that manufacture and/or distribute their products). A circle superimposed with an ''X'' indicates an external location.

The simple circle represents all other primitive locations that are internal to the business. A primitive location is one that will not be divided further into sublocations. A primitive location does not necessarily represent a single user. To keep a diagram from being cluttered, a circle may represent a cluster of similar locations and users.

That takes care of our basic shapes. Connections between locations represent the possibility of data flows between locations. The interpretation might be as follows (see Figure 10.5):

> For this application, the NEW YORK DISTRIBUTION CENTER (a cluster) needs to communicate with or interact with the PURCHASING AGENT OFFICES.

Connections are drawn without arrows because each connection is a conceptual two-way highway that may support numerous business data flows that must pass between locations. Also, until we know how data and processes will be distributed to locations (a design and implementation decision) we can't possibly know which business data flows will travel each connection.

As an LCD progresses from essential requirements to implementation requirements, specific data flows will need to be associated with connections, and the volume of data traffic for each connection will have to be summed. This will help the network designer or manager determine network capacity requirements and technical options. Data flows and their individual volumes would normally be recorded in a project repository entry for each connection (see Chapter 11). Once they are calculated, the sum of the data flow volumes should probably be recorded directly on the implementation-level LCD.

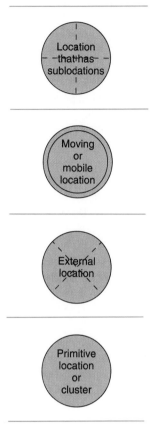

——————— ✓ ———————
Location Names

Paris, France
Indianapolis, Indiana
Grissom Hall
Building 105
Grant Street building
Room 222
Rooms 230–250
Warehouse
Shipping dock
———————————————

——————— ✓ ———————
User Names (as locations)

Order Entry Department
Order clerk
Order clerks (a cluster)
Customers
Suppliers
Students
———————————————

Essential Locations (All Types)

Essential locations are places where we may eventually distribute data and processes. Initially, we should not make data or process distribution decisions on an LCD.

Location names should describe the location and/or users. Try to use proper nouns for locations, but use titles for users. Use singular and plural nouns where appropriate. Locations that have sublocations tend to have more general names. Examples of naming conventions are provided in the margin.

Clustering allows a single symbol to represent "like locations." This simplifies the diagram. But you don't want to oversimplify the business model. Use common sense to determine when to cluster.

A group of locations or users should be represented as a single cluster if it is expected that they will likely share the same data and processes (to be assigned from the data and process models). For example, most order clerks share the same data and processes. A single location labeled ORDER CLERKS is appropriate. In some systems, you might even toss the SALES MANAGER in the same symbol. On the other hand, if the sales manager will be assigned unique data or processes, it might be best to show separate locations—even if the locations are in close proximity.

Note   This doesn't prevent the network designer from putting the clerks and manager on the same computer network. But it may suggest the need to assign security levels such that the manager has access to data and processes that aren't available to the clerks.

Essential Connections

Essential connections are not named on the LCD. It is useful, however, to label each connection by noting the distance between locations. A range of distances should be indicated for mobile locations. A range of distances is also appropriate for geographically scattered locations (e.g., CUSTOMERS).

Do not let current or proposed computer network thinking guide your choice of connections between locations. Like essential data and process models, essential network models are supposed to stimulate your creativity when you eventually make computer networking decisions. Always ask yourself, might this location-to-location be useful? If so, include it. If it doesn't make business sense, exclude it. Also, don't get caught up in routing. In other words, don't eliminate a possible connection just because you can get there by following a route of other connections. If a direct connection makes business sense, put the direct connection on the diagram. The network designer will determine the ultimate route later.

Clearly, the best way to identify possible connections is to discuss the business possibilities directly with system owners and users.

Miscellaneous Symbols

In appropriate situations it is permissible to annotate LCDs with symbols from existing essential DFDs. For example, external agents (squares) may be included to represent external connections that absolutely will not be computerized or otherwise automated. For example, suppose we want to directly place orders for stock with our largest suppliers only. We'll use mail for all other suppliers. This can be represented on our LCD with two symbols: an external location circle labeled SELECT SUPPLIERS and a DFD external agent labeled OTHER SUPPLIERS. In this case both symbols should be drawn in the same general area.

Be very careful of constraining your creativity by overusing external agents. With evolving technology, it is becoming increasingly possible to at least consider connecting any two locations. For example, the telephone has become a modern-day terminal and keyboard through creative use of touch-tone technology.

Another useful DFD symbol is the data store. When used on an LCD it distributes or attaches specific data storage to that location. Once again, be careful not to constrain your thinking. Are there other ways to distribute the data and achieve the same, if not better, results?

The Relationships between Network, Data, and Processing Models

Model relationships are usually established by linking objects on one model with objects on another model. For example, in Chapters 8 and 9 you learned that entity objects on an entity relationship diagram can and should be linked to data stores on a DFD. This keeps the two models consistent with one another.

Similar linking is possible (and recommended) between network models and their corresponding data and process models. This linking is not always appropriate to systems analysis; however, it makes sense to discuss it so you understand how the LCDs will evolve as part of a more complete set of models.

The purpose of network modeling is to gain an understanding of the business network such that data and processes might be intelligently distributed to those locations. As a project moves closer to systems design, distribution decisions become important. Some locations will become computer processor sites (e.g., for hosts, clients, or servers). These locations will be replaced by the appropriate number of processor location symbols.

Each of these processor locations will be linked (or exploded) to a process model that represents the processes to be distributed to that location. Note that a single process model might be duplicated on several processors if that makes more sense than a central processing option.

We must also make data distribution decisions. These decisions can be added to the LCD in the form of data stores attached to specific processors. Note that data may be either partitioned between processors or, in carefully controlled situations, duplicated at multiple processors. For example, consider the entity CUSTOMER in a data model. That entity might be partitioned into subsets appropriate for each location (processor) that performs order processing. Thus, the data store CUSTOMERS may be connected to several different processors on the implementation LCD.

The essential network models will link to components of the data and process models after we make system design decisions concerning data and process distribution.

Computer-Aided Systems Engineering (CASE) for Network Modeling

Computer-aided systems engineering (CASE) was introduced in Chapter 5 as an emerging, enabling technology for systems analysis and design methods. Very few CASE products explicitly support computer-aided network modeling. Some CASE tools do provide generic modeling facilities for unique or emerging modeling tools.

For example, INTERSOLV's Excelerator/IS provides a generic modeling facility called Presentation Graph that can be used to draw network models. It doesn't support all the variations of circles we have introduced; however, you could easily annotate the diagrams by hand to reflect this chapter's recommended standards. Presentation Graph objects can be described to Excelerator's local repository, which means you could easily describe details such as security, distance, and the like.

If your CASE tool doesn't have a generic modeling capability, try using DFDs to model network requirements. Use processes as locations — just give them location-oriented names to keep them separated from real processes. Use bidirectional or double arrows to indicate connections — only use distances to keep the network connections easily distinguished from true data flows.

As the demand for network modeling tools grows, and certain tools emerge as widely used de facto standards, we expect explicit and sophisticated network modeling CASE facilities to emerge in existing and new products.

All of the LCDs in this book were created with a popular CASE product called Excelerator/IS. Only the color screens and annotations were added by an artist. All of the objects on our LCDs were automatically cataloged into a project repository so that we can add detailed information, such as assigned data and processes, as the project moves through analysis, design, and implementation.

How to Construct Network Models

There are many opportunities to perform network modeling. Some methodologies formally recommend network modeling techniques. For example, STRADIS by Structured Solutions (formally marketed by EDS, McDonnell-Douglas, and Improved System Technologies) includes network modeling as part of application development. (Note: They call their LCD a *generalized architecture schematic* or *GAS*.) There are also many approaches to performing network modeling. In this section we'll examine when network modeling might be performed during systems development. We'll also demonstrate a step-by-step approach for performing network modeling.

When to Build Network Models

Network modeling may be performed during various phases of the systems development life cycle. Network models are progressive. In other words, there is no such thing as the final network model for a business or application. Instead, the model should be considered a living document that will change in response to a changing business. Network models should ideally be stored in a repository so that they can be retrieved, expanded, and edited over time. Let's examine how network modeling may come into play during systems planning and analysis.

Network Modeling during Systems Planning

Many systems planning methodologies and techniques result in a network architecture to guide the design of all future computer networks and applications that use those networks. Consequently, network modeling is an appropriate technique for systems planning. Let's briefly examine how network models might evolve out of the three systems planning phases: study, definition, and analysis. (These phases were described in detail in Chapter 6.)

During the study phase of systems planning, network modeling usually does not come into play. Rather, during this phase the focus is entirely upon studying the business and its mission.

During the definition phase of systems planning, an **enterprise network model** may be constructed. This model reflects management's high-level view of its business locations. The resulting model may not take the form of an LCD. Instead, it may take the form of a map or a top-down decomposition diagram that logically groups locations.

In the information engineering methods, systems planning proceeds to another phase: analysis of business areas. A business area or collection of similar business processes is identified. These processes may actually cross the boundaries of existing business functions. A **business area network model** may be drawn to establish moderate-level network requirements for the business area.

Network Modeling during Systems Analysis

Application development begins with systems analysis. Recall that system analysis consists of three phases: survey, study, and definition. Network modeling should play a significant role in most of these phases.

During the study phase, the analyst should review any existing network models, essential or implementation. It is probably not worthwhile to draw a network model for an existing system.

As we move into the definition phase, network modeling becomes more important. If an enterprise or business area network model already exists, it is expanded or refined to reflect application requirements. Otherwise, a model should be built from scratch. In either case, the result is a detailed **essential application network model** that defines locations and connectivity requirements for the application.

Looking Ahead to Systems Design

The essential application network model from systems analysis describes business networking requirements, not technical solutions. As we proceed to systems design, network models must become more technical. They must become **application implementation network model** that will guide the technical implementation of programs.

Thus, LCDs can also be used in the design and implementation phases. As you might guess, design- and implementation-phase LCDs will include the physical implementation details that we discouraged in this chapter. They will describe host processors, servers, and clients. They will describe connectivity protocols and will document expected data traffic.

A Step-by-Step Network Modeling Approach

As a systems analyst or knowledgeable end-user, you must learn how to draw LCDs to model network requirements. We will use the SoundStage Entertainment Club project to teach you how to draw LCDs. We have completed the survey and study phases of the systems development life cycle. We fully understand the current system's strengths, weaknesses, limitations, problems, opportunities, and constraints. We will now model the network requirements. The process, as you will soon see, is fairly straightforward by comparison to data and process models.

Step 1: Identify Locations

Make a list of your locations. If you've completed your DFDs already, study the external agents to identify possible external locations. Add any moving locations to the list. Finally, think of as many internal locations as you can and add them to the list.

Step 2: Draw a Decomposition Diagram to Outline and Group Locations

You learned how to draw process decomposition diagrams in Chapter 9. In this chapter we use decomposition diagrams to logically decompose and group locations.

Figure 10.6 is the location decomposition diagram for SoundStage. There is only one symbol used on the location decomposition diagram—the location—and it is the same location symbol used in LCDs. The locations are connected to form a top-down, treelike structure. A parent location may consist of those child locations beneath it.

To group locations in the decomposition diagram, keep similar locations on the same level or within the same branch of the tree. For example, don't combine cities with buildings or buildings with rooms. Instead, keep cities with cities, buildings with buildings, and rooms with rooms. It makes the diagrams easier to read. More importantly, it makes it possible to produce a leveled set of LCDs (much in the manner that process decomposition diagrams made it possible to produce a sensibly leveled set of DFDs).

Clustering reduces clutter through simplification; however, there is a danger of oversimplifying the model. Once again, common sense should guide the decision to cluster. Cluster a location or its users if the data and processing requirements for all users are expected to be the same. For example, rather than show each ORDER ENTRY CLERK, it probably makes more sense to show one location labeled ORDER ENTRY CLERKS (plural).

Suppose these clerks have order entry supervisors who perform many of the same tasks as the clerks, but who perform special tasks reserved for, or restricted to, the supervisors. It is probably best to show the ORDER ENTRY MANAGERs

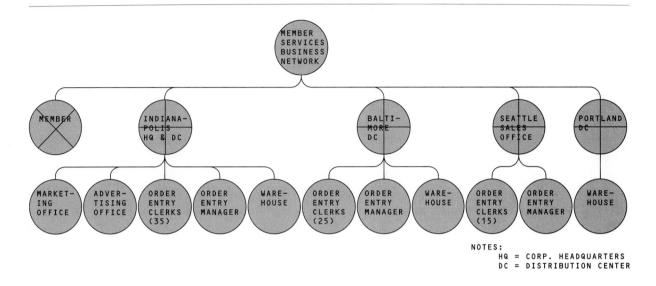

FIGURE 10.6 **A Location Decomposition Diagram** This decomposition diagram illustrates the hierarchy of locations for the Member Services system.

as a separate cluster. This doesn't preclude a single computer network from being implemented for both clerks and supervisors. However, the separation sends a strong signal to the network designer to define different security levels and access privileges for the two groups.

Step 3: Draw a System Location Connectivity Diagram

The first LCD we draw will be a systemwide model. It will include any external locations and locations that have sublocations. The SoundStage system model is shown in Figure 10.7.

Notice that we included an external location called MEMBER. This external location was selected to fulfill a system goal to permit members to make account inquiries and place orders as Member Services System users. We will probably implement this node as a terminal (e.g., a touch-tone telephone response system or a dial-up account for members with their own PCs). The implementation is not yet relevant, but the location is.

Notice that we also included sublocation symbols for each city. These will be exploded in Step 4 to reveal the sublocations and their interactions. Finally, notice that each connection's distance is recorded.

Step 4: Draw Explosion Location Connectivity Diagrams

For any sublocations node on the system diagram, an explosion diagram such as Figure 10.8 is drawn.

Notice that the connections from the parent location were brought down from the system diagram. This maintains consistency between levels. (It was called *balancing* in Chapter 9). The new nodes correspond to the parent's child nodes on the decomposition diagram. It's all very straightforward. Once again the connections are labeled to reflect distances.

FIGURE 10.7 **A System Location Connectivity Diagram** This LCD reflects the highest-level view of business locations for the Member Services subsystem. The diagram was created using INTERSOLV's Excelerator/IS Presentation Graph facility.

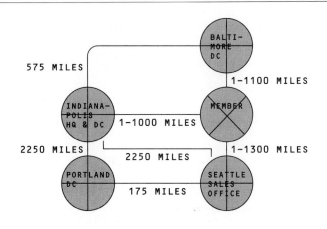

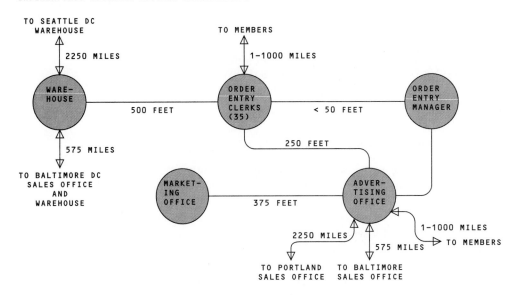

INDIANAPOLIS BUSINESS NETWORK REQUIREMENTS

FIGURE 10.8 **An Explosion Location Connectivity Diagram** This LCD reflects the detailed view of business locations that are part of the Indianapolis corporate headquarters. Only locations relevant to the Member Services system are shown.

Although our example doesn't show it, this diagram could have contained additional sublocation nodes. If so, those nodes would have to be exploded to their own diagram. Once again, the parent's connections would be carried down to the more detailed level to preserve balancing.

That's all there is to essential network modeling. Later, processor nodes will be defined, and essential data and processes will be distributed to the processors. But for systems analysis, the network model is somewhat easier to build than the data and process models.

Summary

The rapid growth of network technologies such as distributed processing, cooperative processing, and client/server computing has created a need for tools that help systems analysts downsize and rightsize information systems applications. The analyst needs a tool to help make intelligent choices about which data and processes (from the data and process models) to distribute to various business and processor locations.

An essential location connectivity diagram (LCD) is a tool for drawing a model or picture of an information system's networking requirements. This is called network modeling. LCDs illustrate work locations and potential for communication between those locations. There is only one basic symbol that can appear on an LCD—the location. Annotations of location symbols indicate whether the location is (1) external or internal, (2) fixed or moving, or (3) primitive or clustered. With these symbols, you can model locations and connections for virtually any information system. Essential LCDs only model business locations and connections, not their implementation.

Location connectivity diagrams have many uses in systems planning, analysis, and design. During systems design, LCDs link with components of data and process models to document distribution decisions. Unfortunately, few CASE products explicitly support network models of any type; however, many CASE tools can be adapted to model such requirements.

Key Terms

application downsizing, p. 403

application implementation network models, p. 415

application rightsizing, p. 403

business area network model, p. 415

central computing, p. 404

client/server computing, p. 407

cooperative computing, p. 407

decentralized computing, p. 404

distributed processing, p. 405

enterprise network model, p. 415

essential application network model, p. 415

essential connection, p. 412

essential location, p. 412

essential network modeling, p. 404

implementation location, p. 410

internetworking, p. 406

local area network, p. 406

location, p. 409

location connectivity diagram, p. 409

network modeling, p. 404

personal computer, p. 405

timesharing, p. 404

wide area network, p. 406

Problems and Exercises

1. Differentiate between application downsizing and application rightsizing. Give an example of each.
2. Explain why downsizing and rightsizing, in part, substantiate the need for network modeling.
3. Differentiate between essential and implementation networking. Explain to a manager why we shouldn't restrict network modeling to *computer* network design only.
4. Differentiate between centralized and decentralized computing.
5. Differentiate between distributed and cooperative processing.
6. Differentiate between cooperative processing and client/server computing.
7. Explain why client/server computing has become so popular.

8. Consider a typical mainframe application that you've either used or programmed. In layperson's terms, explain how that application might be redesigned as a client/server application.

9. Compare network, data, and process models. What does each model show? Should you choose between the three modeling strategies? Why or why not?

10. A manager who has noted your use of essential LCDs to document a proposed system's business network requirements has expressed some concern because of the lack of nodes that depict processor locations. Defend your use of essential LCDs. Concisely explain the symbolism and how to read an LCD.

11. Draw an essential LCD to document the locations and connections in either your school's admissions system or course registration and scheduling system.

12. Draw an essential LCD for some day-to-day "system" that you use or observe in use—for instance, your morning routine; making your favorite meal, including appetizer, entree, side dishes, and dessert; constructing something from scratch.

Projects and Minicases

1. Formally research the subject of network modeling, both essential and implementation. Compare and contrast techniques uncovered and make recommendations for expanding or complementing the guidelines provided in this chapter.

2. Make an appointment to interview a network manager in your local area. Find out how he or she documents computer networks, then, teach him or her to use essential LCDs as an applications development tool. Together, analyze the usefulness of the LCD as a means of communicating essential requirements to the network designer. How could it be improved without making the diagram implementation-dependent?

3. Make an appointment with an application development manager in your local community. Find out to what degree the shop is doing downsizing, rightsizing, cooperative processing, and client/server computing. If they aren't moving in these directions, find out why. If they are moving in these directions, how have they adapted their systems development methodology? Explain the purpose of LCDs to the manager, and teach the manager how to use them. Together, analyze the tool's applicability to the shop. Make suggestions for improvement.

4. Make an appointment with a systems analyst in your local community. Discuss an application that he or she is working on or is familiar with. Interview the analyst to discover the locations and potential business connections. Draw the LCD and present it to the analyst. Together, analyze the tool's value for modeling network requirements. Make suggestions for improvements.

Suggested Readings

"Client/Server Computing." Dayton, Ohio: NCR Corporation, 1992. This white paper does an excellent job of introducing the evolution and technology of client/server computing as contrasted to centralized, decentralized, distributed, and cooperative computing alternatives.

Zachman, John. "A Framework for Information Systems Architecture," *IBM Systems Journal* 26, no. 3, March 1987. This paper is the basis for creating network modeling as a complementary strategy with data and process modeling. You may recall that Zachman's framework is also the basis for your pyramid model.

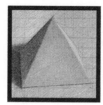

Requirements Definition for the New System
(Part II)

Sandra and Bob prioritize business requirements and store important system documentation in the repository: "What are the detailed business and user requirements for the new system?"

Recording Detailed Requirements in the Repository

Two weeks have passed. Sandra and Bob have completed documenting the general requirements for the new system with entity relationship diagrams and data flow diagrams. Sandra was finishing up on defining network requirements for the new system. Meanwhile, Bob was working on the detailed systems requirements. We join Bob as he discusses some of those requirements with Ann Martinelli, the Membership director.

"Well, that should finish the membership subsystems processing requirements. Can we work on the automatic orders now? If you need to get back to work we can do this later."

"That's okay," Ann responded. "I have a few more minutes."

Bob continued, "First, let's define these data flows. Where did I put those definitions? Ah, here they are (see Figure E4.1). We just defined AUTO-MATIC DATED ORDER, an input to this CHECK MEMBER CREDIT process. We need to define the CREDIT RATING AND LIMIT data flow."

"That's easy, CREDIT RATING AND LIMIT from the ACCOUNTS RECEIVABLE file is just the member's CREDIT RATING and CREDIT LIMIT attributes," Ann offered.

Bob continued, "You can see from this diagram (Figure E4.2) that when the AUTOMATIC DATED ORDER is combined with CREDIT RATING AND LIMIT, the output of this procedure is either an APPROVED ORDER or a BAD CREDIT ORDER. Now can we discuss the procedure itself?"

Ann replied, "I think I'm beginning to understand. First we calculate and verify the TOTAL ORDER PRICE. Then, if the PREPAID AMOUNT is less, we check the member's CREDIT RATING AND LIMIT from the ACCOUNTS RECEIVABLE file. For automatic orders, the amount of prepayment is usually zero."

"Is that all?" Bob asked as he wrote a CHECK MEMBER CREDIT procedure description (see Figure E4.3).

"Well, not really," Ann answered. "If the member's credit is bad, we mail him or her a prepayment request and hold the order until we receive payment. If the order was approved, we check the product inventory."

"But both of those are separate processes that we will define later," Bob added. He then showed Ann the procedure description and she verified its correctness.

"I think we should stop here for today. I need to get back to work," Ann said.

"Thanks for your help," Bob responded.

[The above process would continue until the contents of all data flows on all the data flow diagrams had been defined and verified and the procedures for each of the processes had been documented using Structured English.]

Where Do We Go from Here?

This episode reinforced the use of data flow diagrams as a process modeling tool (discussed in Chapter 9) for outlining the *general* requirements of a new information system. The case also introduced two new tools used to specify the *detailed* requirements of a system—a project repository and Structured English. Both of these tools are very

---

FIGURE E4.1 Repository Entry for Sample Data Flow Content

An AUTOMATIC DATED ORDER is a dated order created by a promotion that is automatically filled and shipped to the club member.

An occurrence of an AUTOMATIC DATED ORDER is uniquely identified by the data attribute ORDER NUMBER.

An AUTOMATIC DATED ORDER consists of the following data attributes:
    ORDER NUMBER
    ORDER DATE
    ORDER STATUS
    PREPAID AMOUNT
    TOTAL ORDER PRICE
    1 or more of the following attributes:
        PRODUCT NUMBER
        MEDIA CODE
        QUANTITY ORDERED
        QUANTITY SHIPPED
        ORDER PRICE

FIGURE E4.2 Sample Data Flow Diagram

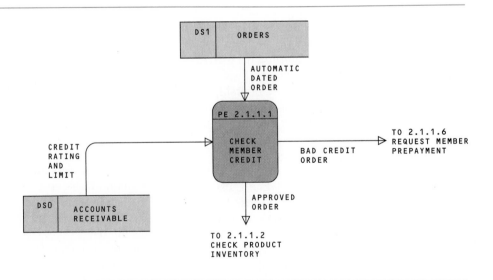

popular for specifying detailed systems requirements.

Chapter 11 introduces you to the concept and use of a project repository. The project repository is used to document the contents of data flows and data stores depicted on data flow diagrams. Chapter 11 also introduces you to tools and techniques for documenting policies and procedures, specifically Structured English and decision tables. These tools specify how processes on a data flow diagram accomplish their work.

This episode also emphasized the need for analysts to develop sound communications skills. Sandra and Bob are spending considerable amounts of time talking with the system's end-users. Systems analysts don't spend all their time behind a desk. Once again, we direct you to Part Five, Module D, to learn more about making presentations and conducting walkthroughs.

```
For each AUTOMATIC DATED ORDER, do the following:
    For each ordered product do the following:
        Calculate the TOTAL ORDER PRICE using the following formula:
            TOTAL ORDER PRICE = TOTAL ORDER PRICE +
            (PRODUCT PRICE × QUANTITY ORDERED)
        If the PREPAID AMOUNT is less than the TOTAL ORDER PRICE then:
        If the CREDIT RATING is "GOOD" and LIMIT is >
            TOTAL ORDER PRICE − PREPAID AMOUNT:
            Approve the order and forward the APPROVED ORDER
            to CHECK PRODUCT INVENTORY
        Otherwise:
            Disapprove the order and forward the BAD CREDIT ORDER
            to REQUEST MEMBER PREPAYMENT.
```

FIGURE E4.3 **Sample Structured English for a Typical Process**

11

Using a Project Repository

Chapter Preview and Objectives

This chapter focuses upon the project repository as a tool for documenting systems that are being developed. In this chapter you will learn how to record details about the data, process, and network models you learned to draw in Chapters 8, 9, and 10. You will have mastered the use of the project repository as a systems analysis tool when you can:

Describe
the need for a project repository, its contents, and its value as a documentation tool.

Define
the contents of data entities, data flows, and data stores in terms of restricted data structures that consist of data attributes.

Create
complete project repository entries for data entities, data flows, data stores, and locations. These entries should include pertinent facts about terminology, properties, and content.

Create
complete project repository entries for data attributes and codes for data attributes. These entries should include pertinent facts about terminology, properties, and values (ranges).

Differentiate
between policy and procedure.

Describe
some of the typical problems encountered in documenting procedures, and explain the ambiguities of ordinary English as a policy and procedure specification tool.

Construct
a decision table to describe a policy in terms of conditions and actions to be taken under various combinations of those conditions.

Use
Structured English to write procedure specifications.

AMERICANA PLASTICS

Americana Plastics is a manufacturer of custom-engineered plastic parts for sale to original equipment manufacturers. As the lead analyst on the cost accounting information systems project, Kay had designed a system to serve two groups of users: the manufacturing group and the accounting group.

Scene: *Jill Latin, a key contact with the manufacturing group on Kay's cost accounting project, placed a phone call to Kay. Jill had just received a number of reports from the new cost accounting system and is not pleased with them. Jill is trying to explain to Kay why the report is useful to the accounting group, but confusing and worthless to her manufacturing group.*

Jill: No! What they are calling a part number, although unique, is actually a mold number.

Kay: Okay, you folks call it a mold number, they call it a part number. In reality, it's the same thing.

Jill: Hold on a minute, I'm not finished. Mold numbers identify the basic molds used to manufacture parts.

Kay: So what you are telling me is that it doesn't identify a part.

Jill: Yes and no. To those guys it does; to us it doesn't. We identify the actual part by a combination of the mold number and two process codes — the manufacturing method code and the insert codes. The part process method code is a one-letter code that tells us whether the plastic parts are to be formed using heat (H), cold (C), or pressure (P). The insert codes identify slugs that can be inserted into the basic mold to form slightly modified versions of the plastic part to be manufactured.

Kay: Is this really a problem? After all, the six-character basic code is unique. Can't we go ahead and use that code for all reports? I'll even design two sets of reports, one set using the heading "Part Number" and the other set using "Mold Number."

Jill: No, we need those additional codes to determine which production methods and insert requirements are responsible for cost overruns or inefficiencies. How will we be able to use the information to effect changes to our manufacturing methods without the extra codes?

Kay: Okay, I see your point.

Jill: Good, then maybe we can suggest alternative structures for the manufacturing reports to make them more useful for pinpointing detailed problems. We'd like to see alternative report structures organized around the different processing methods and inserts as well as the basic molds.

Kay: You know I will work with you to provide whatever reports you need. I'm sorry about the confusion. I'm still a little perplexed. I don't know how I'm going to keep track of the details when you two groups use a different terminology.

Discussion Questions

1. How would you suggest that an analyst keep track of the special terminology characteristics of most business applications?

2. Could Kay have done something to avoid wasting the time she spent in designing the reports?

3. What would you have done differently? If you had tried to establish terminology and content before format, how would you have communicated your understanding (or lack thereof) back to the two end-user groups?

WHAT IS A PROJECT REPOSITORY?

Suppose you are reviewing the entity relationship diagrams (ERDs) or data flow diagrams (DFDs) for a new information system. These diagrams model the essential requirements for the new system, but only in general terms. Details still must be defined. Questions such as these may arise:

What data are we putting in that invoices data store?

Just exactly what information is needed on that overtime analysis report?

Did we remember to include the new disputed amount field in the customer statement?

What steps are involved in that process that schedules work orders?

These are questions that can easily be answered in a project repository. In this chapter we examine the project repository and its purpose, content, and use.

A **project repository** is a place where documents, documentation, and programs that are associated with the application and project are kept. Common synonyms include *data dictionary, project dictionary,* and *encyclopedia.*

A sample CASE-based repository entry, in this case for a data flow, is shown in Figure 11.1. The repository helps the systems analyst keep track of the enormous volume of details that is part of every system, even small ones. By using a project repository, the analyst minimizes the chance of becoming overwhelmed by these details. The Next Generation box describes trends toward corporatewide repositories where analysts will share data from a central repository containing data for the business as a whole.

We have purposely avoided using the popular CASE-derived synonyms: data dictionary, project dictionary, and encyclopedia. In particular, the term *data dictionary* has a database management system (DBMS) connotation generally accepted in the computing industry. A DBMS's data dictionary stores facts about the files and databases for all systems that use the DBMS.

The Purpose and Content of the Project Repository

Why do you need a project repository? During the definition phase, the analyst defines the essential requirements for a new information system. No matter what your methods are, you will likely generate the following:

- Data to be captured and stored in the system files and databases.
- Inputs to the system.
- Outputs to be generated by the system.
- Processes and procedures to be supported by the system.
- Locations to be supported by the system networks.

If you drew data, process, and network models of the system, all of these objects would appear on that diagram as either shapes or connections.

A project repository stores these models and expands on the objects appearing on them. For instance, given a data flow object, such as MEMBERSHIP SUBSCRIPTION, we need to define the content of a typical occurrence of that data flow. Or, given a data store or data entity type (recall that they both describe

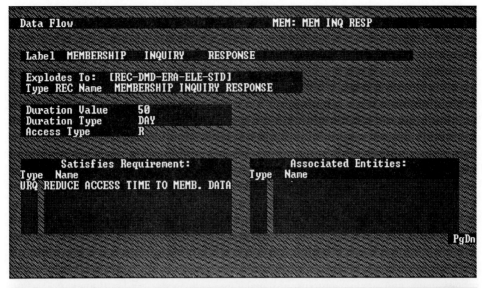

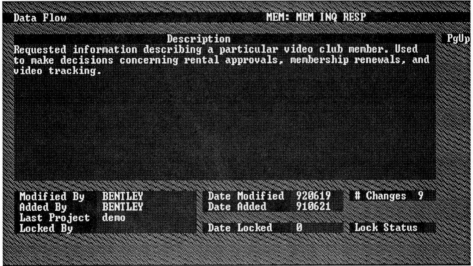

FIGURE 11.1 **Sample Project Repository Entry** This sample entry in a CASE-based project repository describes a data flow. These screens came from INTERSOLV's Excelerator/IS.

stored data) such as PROMOTION(S), we need to define the contents of a single record in that data store or entity type. The project repository provides a vehicle for recording such definitions.

Consistent with our ongoing systems analysis theme, we want to focus on the essential data, process, and network requirements, independent of its current or future implementation. Once this description has been completed, you can sit down with the end-users and discuss that content objectively.

The Next Generation

INTEGRATED REPOSITORIES

In many businesses, data has become the most precious resource—and the most uncontrolled resource! As data files and databases proliferate, data administration (or the lack thereof) becomes a serious problem. Just how serious can the problem get?

Consider a law that mandates a nine-digit zip code (or larger). Do you have any idea how difficult it is to find every computer program and file that has a zip code field in it? Over the years different programmers and analysts introduced a bit of uniqueness into their businesses' systems and computer program libraries. How many unique field names exist for zip code? Let's see—there's ZIPCODE, ZIP.CODE, ZIP, ZP, Z, ZC, ZCODE, ZCOD. Where and how do you look for all the fields? There must be a better way of keeping track.

And there is! In some businesses a formal data administration func-

tion has developed—that's *data* administration, not *database* administration. The idea is to create and maintain a data repository for the business as a whole. That repository keeps track of where data attributes are used and stored—every file, database, computer program, report, document, and so forth. The aliases, descriptions, formats, limitations, and properties of every data attribute are recorded in a centralized data repository. As new systems are developed and old systems are maintained, analysts and programmers are urged to use the repository to curb the proliferation of aliases and to minimize the creation of new data attributes. In other words, the project data repository concept has been expanded to include the business as a whole!

The only way to implement an enterprisewide data repository is to use one of the commercially available data repository software packages. Automated data repositories

aren't new. They've been with us almost since the advent of database management systems software. In fact, you usually get a repository package when you buy a database management system (database-independent products are also available). However, those automated data repositories were originally conceived to help the database professionals keep track of the data attributes and records in databases. They are being increasingly used to catalog all the data flows, stores, and attributes in a business.

What can an automated data repository do for you? Data repositories may allow you to do the following:

- Get full listings of all known facts about all or specific subsets (e.g., a group of related applications, a single related application, and/or a single program) of data flows, stores, attributes, and so forth.

(Continued)

Organization of a Project Repository

A project repository should be organized to allow convenient reference to definitions that you or the end-user need. The repository can be keyed to the names in the systems models. The ERDs, DFDs, and LCDs from Chapters 8 through 10 provide the framework for your project repository. They are normally included in the repository to serve as a graphic table of contents into the more detailed specifications (Figure 11.2).

As the figure suggests, detailed requirements and specifications would be placed after the models. Ideally, they would be organized by type—for instance, data flow, data store, data attribute, process, or the like. Consequently, you can easily examine details pertaining to all data flows, data stores, and so on.

A repository for any project, even a small one, can get very large. However, size is only a disadvantage if you make the mistake of dumping the entire project repository into the laps of the end-users and requesting their verification. No-

---

The Next Generation

(Concluded)

- Get partial listings (e.g., name and descriptions only) for data flows, stores, attributes, and so forth.
- Reorganize facts into a convenient format. For example, give full facts for all attributes that are part of specific data flows.
- Find all data flows, data stores, or data attributes that share a certain property (e.g., the same unit of measure or the same length).
- Answer queries about data. For instance, LIST ALL KNOWN ALIASES FOR ZIPCODE. And then, LIST ALL FILES IN WHICH ZIPCODE (including aliases) IS STORED. And finally, LIST ALL PROGRAMS THAT USE THE FIELD ZIP CODE (again, including all its aliases).

Where does the systems analyst fit into this picture? The analyst's

project repository can be a subset of the business data repository.

Virtually all CASE products include a proprietary project repository. However, the trend is clearly to have a central CASE repository that can be accessed by multiple CASE and DBMS tools from a wide variety of vendors. This implies that some standard information model must exist in the central repository and that all vendors must agree to import and export their proprietary repository contents from and to the central repository.

Two such open, central repositories are currently being developed. IBM is advocating its AD/Cycle repository called Repository Manager, which runs on mainframe computers only. (*Note:* IBM is in the process of downsizing its central repository to run on UNIX, AIX, and OS/2 database servers.) Similarly, DEC's Cohesion includes an open, central repository for participating CASE tool vendors.

CASE vendor KnowledgeWare offers a mainframe-based central repository for users of its own CASE tool family. Similarly, INTERSOLV is building a LAN-based central repository for its own CASE tools. (*Note:* Both KnowledgeWare and INTERSOLV's CASE tools write, or will soon write, to IBM's Repository Manager as well!).

One issue that will have to be addressed is the interface between CASE repositories (for systems being developed) and production data repositories. Few corporate data administrators are willing to allow that automatic transfer from CASE to occur without a quality check occurring.

Thus, the systems analyst of the future may not have to worry about proliferating the creation of duplicate data attributes and data stores. Indeed, the future systems analyst may not be allowed to operate in isolation from other analysts and information systems!

---

body will read such an imposing document. When you want the end-user to verify a project repository, you should present it in pieces. Extract the desired repository definitions and organize them so the reader doesn't have to shuffle back and forth through the pages.

Increasingly, the models and descriptions are stored in a CASE-based repository. CASE technology allows the data to be easily output in a wide variety of useful formats.

SPECIFYING DATA AND INFORMATION REQUIREMENTS IN A PROJECT REPOSITORY

There are several notations for documenting essential data and information requirements. In this section, we will examine two popular notations—the Boolean algebraic notation and an English interpretation of the algebraic notation.

FIGURE 11.2 **The Orga-nization of a Project Repository** Although many project repositories are maintained on computers, the contents are usually printed for reference. This book demonstrates sample contents and organization of a typical printed project repository.

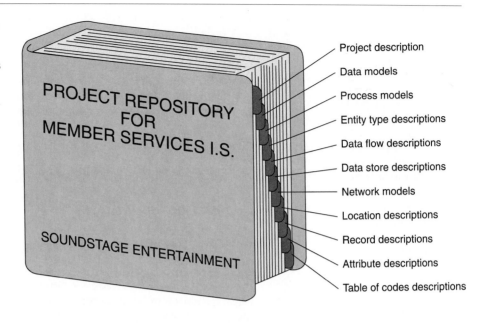

Project description
Data models
Process models
Entity type descriptions
Data flow descriptions
Data store descriptions
Network models
Location descriptions
Record descriptions
Attribute descriptions
Table of codes descriptions

Defining Data and Information Structure

We are defining data and information requirements for inputs, outputs, and data stores (or entity types stored in data stores). These components consist of data attributes. By themselves, data attributes have little value; however, they combine to form data structures that do have meaning.

> **Data structures** are a specific arrangement of data attributes that define one occurrence of an input, output, or data store (data entity).

Inputs, outputs, and data stores can be defined in terms of the following types of data structures:

- A sequence of data attributes or group of data attributes that occur one after the other.
- The selection of one or more data attributes from a set of data attributes.
- The repetition of an individual data attribute or a group of data attributes.

Two notations are presented. The most common is the Boolean algebraic notation suggested by Tom DeMarco and supported by numerous authors. We also present the form that we prefer—an English interpretation of the algebraic notation that is especially useful for verification by end-users who may be intimidated by the algebraic notation. In our examples we will present the English notation first since it helps explain the algebraic notation.

The definition of the data structure for any data flow or data store should begin with this English statement:

An occurrence of _____ consists of the following data attributes:

The Boolean algebraic equivalent of this statement could also be used:

_____ =

English Notation	Algebraic Notation
An occurrence of WAGE AND TAX STATEMENT consists of the following data attributes:	**WAGE AND TAX STATEMENT =**
SOCIAL SECURITY NUMBER	**SOCIAL SECURITY NUMBER +**
EMPLOYEE NAME	**EMPLOYEE NAME +**
EMPLOYEE ADDRESS	**EMPLOYEE ADDRESS +**
EMPLOYER NAME	**EMPLOYER NAME +**
EMPLOYER ADDRESS	**EMPLOYER ADDRESS +**
WAGES, TIPS, COMPENSATION	**WAGES, TIPS, COMPENSATION +**
FEDERAL TAX WITHHELD	**FEDERAL TAX WITHHELD +**
STATE TAX WITHHELD	**STATE TAX WITHHELD +**
FICA TAX WITHHELD	**FICA TAX WITHHELD**

FIGURE 11.3 The Sequence Data Structure The sequence data structure consists of a group of serial data attributes and/or groups.

where _____ is the name of the data flow, data store, or data entity. Note that the equal sign is not used in its arithmetic sense. It means "consists of," "is composed of," or "contains."

The Sequence Data Structure

The sequence of data attributes that define a data flow, data store, or data entity would be documented as illustrated in Figure 11.3. The sequence of attributes is boldfaced. Notice in both notations that we indented the data attribute names to improve readability. Each of the data attributes is required to assume a value for every occurrence of WAGE AND TAX STATEMENT. In the algebraic notation the plus sign is interpreted as the word "and."

The Selection Data Structure

The selection construct allows you to show situations where, given any single occurrence of a data flow, data store, or data entity, one of the following is true:

- One and only one data attribute from a list of data attributes will assume a value (often called *exclusive-or*).
- One or more data attributes in a list will assume a value (often called *inclusive-or*).

There must be at least two data attributes in the list from which you choose. Some examples will make the selection construct clear.

Figure 11.4 demonstrates the exclusive-or construct. Notice how the sequence and selection constructs combine to define the data structure. An order can be placed either by an individual or by a company, but not both. Thus, any given occurrence of an order will consist of either an ORDER DATE and a SOCIAL SECURITY NUMBER, or an ORDER DATE and a CUSTOMER ACCOUNT NUMBER. In the algebraic notation the selection attributes are set within the square brackets and separated by commas.

FIGURE 11.4 **The Exclusive-or Selection Data Structure** The exclusive-or selection data structure describes a group of data attributes from which one and only one may assume a value for the overall data structure.

English Notation	Algebraic Notation

An ORDER consists of the following data attributes:
    ORDER DATE
    Only one of the following attributes:
        SOCIAL SECURITY NUMBER
        CUSTOMER ACCOUNT NUMBER
	ORDER =
	    ORDER DATE +
	    [SOCIAL SECURITY NUMBER,
	        CUSTOMER ACCOUNT NUMBER]

FIGURE 11.5 **The Inclusive-or Selection Data Structure** The inclusive-or selection data structure describes a group of data attributes from which one or more may assume a value for the overall data structure.

English Notation	Algebraic Notation

A TRAVEL EXPENSE VOUCHER consists of the following data attributes:
    EMPLOYEE ID NUMBER
    EMPLOYEE NAME
    DATE TRIP STARTED
    DATE TRIP COMPLETED
    PURPOSE OF TRIP
    MILES TRAVELED
    MILEAGE CHARGE
    One or more of the following attributes:
        AIR TRAVEL EXPENSE
        TAXI FARE EXPENSE
        REGISTRATION FEES
        LODGING EXPENSES
        MEAL EXPENSES
    TOTAL EXPENSES
	TRAVEL EXPENSE VOUCHER =
	    EMPLOYEE ID NUMBER +
	    EMPLOYEE NAME +
	    DATE TRIP STARTED +
	    DATE TRIP COMPLETED +
	    PURPOSE OF TRIP +
	    MILES TRAVELED +
	    MILEAGE CHARGE +
	    < **AIR TRAVEL EXPENSE,**
	        TAXI FARE EXPENSE,
	        REGISTRATION EXPENSE,
	        LODGING EXPENSES,
	        MEAL EXPENSES > +
	    TOTAL EXPENSES

Figure 11.5 demonstrates the inclusive-or selection construct. Both notations suggest that a TRAVEL EXPENSE VOUCHER contains one or more of the listed expenses. TOTAL EXPENSES always occurs. In the algebraic notation the attributes from which you select are placed between the angled brackets and separated by commas.

The data structures in Figure 11.6**A** depict a common error: FRESHMAN, SOPHOMORE, JUNIOR, SENIOR are values of the data attribute STUDENT CLASSIFICATION, not data attributes themselves. The correct data structures are shown in Figure 11.6**B**. Later in this chapter you'll learn where to document values and value ranges for data attributes in the project repository.

The Repetition Data Structure

The repetition construct is used to set off a data attribute or group of data attributes that will repeat a specified number of times for a single occurrence of the data flow, data store, or data entity. The English and algebraic notations are

FIGURE 11.6 A Common Misuse of the Selection Data Structure These data structures demonstrate a common mistake: using the selection structure to document values of data attributes instead of data attributes themselves. Part B fixes the mistake.

FIGURE 11.7 The Repetition Data Structure The repetition data structure consists of a group of data attributes that repeat as a group for each occurrence of the main data structure.

---

documented as

  M to *N* occurrences of each of the following attributes:

and

  M {list of attributes or groups of attributes} *N*

where *M* is the minimum number of occurrences and *N* is the maximum number of occurrences of the data structure. If *M* = 0, the entire group occurs optionally. In any case, the entire group of data attributes repeats as a group. If the number of repetitions is indefinite, you may write

English Notation	Algebraic Notation

A TIME WORKED RECORD consists of the following data attributes:
   EMPLOYEE IDENTIFICATION NUMBER
   EMPLOYEE NAME
   1 to 4 occurrences of the following:
      START DATE FOR PAY PERIOD
      END DATE FOR PAY PERIOD
      1 to 7 occurrences of the following:
         HOURS WORKED
	TIME WORKED RECORD =
	   EMPLOYEE IDENTIFICATION NUMBER +
	   EMPLOYEE NAME +
	   1{ START DATE FOR PAY PERIOD +
	         END DATE FOR PAY PERIOD +
	       1{ HOURS WORKED }7 }4

FIGURE 11.8 A Nested Repeating Data Structure A nested repeating data structure contains a group of data attributes that repeat inside another group, which itself repeats.

> One or more occurrences of:

for the English notation or simply not assign a numeric value to N in the algebraic notation.

The contents of a repeating group may contain a sequence of data attributes, a selection construct or constructs, or even additional (sometimes called nested) repetition constructs. Let's expand on our ORDER example to demonstrate the repetition construct. Figure 11.7 shows both notations. Note that a sequence of related data attributes was defined as the repeating group. Figure 11.8 provides another example. There are four weeks in a pay period. That explains the first repeating group. For any or all of those weeks, an employee will have from one to seven values of hours worked—one value for each day worked. This is a nested repeating group—one repeating group inside another. For the algebraic notation, the inner bracket occurs one to seven times (for days of the week) for each occurrence of pay period, which also happens one to four times.

A Notation for Optional Data Attributes

Sometimes certain data attributes or groups of data attributes optionally take on values for occurrences of data flows. In Figure 11.9, for any occurrence of CLAIM, the attribute SPOUSE NAME may or may not take on a value, depending on the policyholder's marital status. In the algebraic notation, parentheses set off the optional data attribute.

Note that we still used the repeating construct for the optional group of data attributes. Grouping the attributes suggests that both attributes (DEPENDENT NAME and RELATIONSHIP) occur together or neither attribute occurs. It's all or nothing.

Groups of Attributes

Groups of data attributes that "always" occur together can be given a group name that can be documented as a separate data structure. This approach, demonstrated in Figure 11.10, allows many data structures to refer to a common substructure. Thus, instead of having to separately define data attributes for employee address, customer address, supplier address, and the like, we need only set each of those addresses equal to the common definition of ADDRESS.

English Notation	Algebraic Notation

A CLAIM consists of the following data attributes:
  POLICY NUMBER
  POLICYHOLDER NAME
  POLICYHOLDER ADDRESS
  SPOUSE NAME (optional)
  0 to 15 occurrences of the following:
    DEPENDENT NAME
    RELATIONSHIP
  CLAIMANT NAME
  1 or more of the following:
    EXPENSE DESCRIPTION
    NAME OR FIRM PROVIDING SERVICE
    TOTAL CHARGE FOR SERVICE
	CLAIM =
	  POLICY NUMBER +
	  POLICYHOLDER NAME +
	  POLICYHOLDER ADDRESS +
	  (**SPOUSE NAME**)+
	  0{ **DEPENDENT NAME** +
	      RELATIONSHIP }15 +
	  CLAIMANT NAME +
	  1{ EXPENSE DESCRIPTION +
	      NAME OR FIRM PROVIDING SERVICE +
	      TOTAL CHARGE FOR SERVICE }M

FIGURE 11.9   **Optional Data Attributes in a Data Structure**   This figure demonstrates how to indicate that an attribute does not have to take on a value. The example also demonstrates how to show that a repeating group does not have to occur for an occurrence of the entire data structure.

English Notation	Algebraic Notation

A WAGE AND TAX STATEMENT consists of the following data attributes:
  SOCIAL SECURITY NUMBER
  EMPLOYEE NAME
  EMPLOYEE ADDRESS which is defined separately as ADDRESS
  EMPLOYER NAME
  EMPLOYER ADDRESS which is defined separately as ADDRESS
  WAGES, TIPS, COMPENSATION
  FEDERAL TAX WITHHELD
  STATE TAX WITHHELD
  FICA TAX WITHHELD
	WAGE AND TAX STATEMENT =
	  SOCIAL SECURITY NUMBER +
	  EMPLOYEE NAME +
	  EMPLOYEE ADDRESS = ADDRESS +
	  EMPLOYER NAME +
	  EMPLOYER ADDRESS = ADDRESS +
	  WAGES, TIPS, COMPENSATION +
	  FEDERAL TAX WITHHELD +
	  STATE TAX WITHHELD +
	  FICA TAX WITHHELD +

ADDRESS consists of the following data attributes:
  One or both of the following:
    STREET ADDRESS
    POST OFFICE BOX NUMBER
  CITY
  STATE
  ZIPCODE
	ADDRESS =
	  < STREET ADDRESS,
	      POST OFFICE BOX NUMBER > +
	  CITY +
	  STATE +
	  ZIPCODE

FIGURE 11.10   **Common Data Structures**   It is useful to define reusable data structures that can be referenced from within other data structures. Thus, we avoid reinventing the wheel when it comes to common structures such as ADDRESS.

SPECIFYING ESSENTIAL PROCESS REQUIREMENTS IN THE PROJECT REPOSITORY

In addition to specifying data and information requirements in a project repository, process requirements must also be specified. In this section we will examine the differences between business policies and procedures and will present specification tools for each.

What Are Business Policies or Procedures?

Processes on DFDs represent required tasks performed by the system. These tasks are performed according to business policies and procedures.

A policy is a set of rules that govern some task or function in the business.

In most firms, policies are the basis for decision making. For instance, most companies have a credit policy for determining whether to accept or reject an order. A credit card company must bill cardholders according to policies that adhere to restrictions imposed by state and federal governments (maximum interest rates and minimum payments, for instance). Policies consist of rules that can often be translated into computer programs if the systems analyst can accurately convey those rules to the computer programmer.

What are procedures? To the programmer, procedures may represent the executable instructions in a computer program. But to the end-user, and for purposes of defining process requirements during systems analysis,

A **procedure** is a step-by-step set of instructions for accomplishing a task or tasks.

In many businesses, well-defined procedures are sorely missed by the end-users and management. Procedures put those policies into action. For instance, most companies have a policy on vacations, leaves, sick days, and the like. Those policies are implemented by procedures that define how to call in sick, request and approve vacations, and so forth.

Many of you have come from a programming course or background. As computer programmers, your job is to translate business requirements into syntactically correct code. Unfortunately, the language and idiosyncrasies of the business world and the programming world are vastly different. The language of computer programming is extremely precise, much more so than natural English. For this reason, many programmers spend more time debugging the business requirements than they do their computer programs! We would like to suggest that the programmer should not have to interpret or clarify business requirements. That is the analyst's job!

The Problems with Procedures

When was the last time you received a programming assignment that you didn't spend time clarifying? In defense of analysts and teachers, the complexity of English and other natural languages can make it difficult to convert business policies and procedures into computer programs. Manual procedures are no easier to write. Anyone who has experienced the frustration of completing a tax return knows that manual procedures are often difficult to communicate.

General Criticisms of Procedures

Leslie Matthies has some unique insight into the often-ignored art of procedure writing. In *The New Playscript Procedure* (1977), Matthies describes several problems encountered with typical procedures, three of which are pertinent to policy and procedure specification.

- Simply writing a procedure may not be enough. Not only do many of us not write well, but we also tend not to question our writing abilities. It's important to write clearly and accurately.

- Most of us are too educated! It's often difficult for a highly educated person to communicate with an audience that may not have had the same educational opportunities. For example, the average college graduate (including most analysts) has a working vocabulary of 10,000 to 20,000 words; on the other hand, the average non–college graduate has a working vocabulary of around 5,000 words. This underscores the importance of the first law of good writing—know your audience. A procedure has little value if it cannot be interpreted by those who will perform it.

- A related problem deals with jargon. Often, we allow the jargon of computing to dominate our procedures. The computer industry constantly invents terms and acronyms to describe its products and discipline.

Problems with Ordinary English

The English language itself can be a source of problems when specifying procedures.

- Statements tend to have an excessive or confusing scope. How would you carry out this procedure: "If customers walk in the door and they do not want to withdraw money from their account or deposit money to their account or make a loan payment, send them to the trust department." Does this mean that the only time you should not send the customer to the trust department is when he or she wishes to do all three of the transactions? Or does it mean that if a customer does not wish to perform at least one of the three transactions, that customer should not be sent to the trust department? The scope of the procedure is not clear.

- Compound sentences (two complete sentences connected by and, but, or other conjunctions) can be another serious problem. Consider the following procedure, which describes how to replace an electrical outlet: "Remove the screws that hold the outlet cover to the wall. Remove the outlet cover. Disconnect each wire from the plug, but first make sure the power to the outlet has been turned off." Did you catch the compound sentence structure in the last instruction? An unwary person might try to disconnect the wires prior to turning off the power!

- Multiple definitions associated with many words are another problem. An example of this problem was seen in the chapter opening minicase, which explained how part numbers did not mean the same thing to the end-user groups at Americana Plastics.

- Undefined adjectives confuse readers. Each semester, we receive several "Good Student Driver" discount forms from insurance companies. They want to know if a student is in good standing. We need to know their definition of "good" before we can respond. They sometimes define good as "the upper 10 percent of his/her class." Ten percent of what class? The entire university's class? The class of a specific major?

- Conditional instructions sometimes present difficult problems as well. These instructions determine whether or not certain steps are performed. The statement of the conditions can be part of the problem. If we state that "all applicants under the age of 19 must secure parental permission," do we really mean less than 19 or less than or equal to 19? Although programmers are familiar with this problem, end-users and analysts frequently forget to carefully specify value ranges.

B.

Check-Cashing Policy

	Check-Cashing Policy	Rules				
		1	2	3	4	5
C o n d i t i o n s	Type of check	1	2	1	2	2
	Check amount less than or equal to $75	Y	–	N	N	–
	Company accredited by store	–	Y	–	Y	N
A c t i o n s	Cash check	X	X			
	Refuse check			X	X	X

A.

CHECK-CASHING IDENTIFICATION CARD

Upon presentation person named hereon is entitled to cash personal checks up to $75.00 and payroll checks of accredited companies at Save Super Markets. Card is issued in accordance with terms and conditions of application, remains property of Save Super Markets, Inc., and shall be returned upon request.

Charles C. Parker, Jr

SIGNATURE
ISSUED BY
SAVE SUPER MARKETS, INC.

FIGURE 11.11   A Decision Table for Specifying a Store's Check-Cashing Policy   Decision tables offer a number of advantages over ordinary English. Although the narrative equivalent of the decision table is also short and concise, the decision table clearly ensures that the policy is complete and without contradictions.

This problem is further complicated by combinations of conditions. For example, credit approval may be a function of several conditions: credit rating, credit ceiling, annual dollar sales for that customer, and payment history. Different combinations of these factors result in different decisions, such as accept order on credit, reject order on credit and require full prepayment, or reject order until a down payment is received. As the number of conditions and possible combinations increases, the procedure becomes more and more tedious and difficult to write.

Decision Tables: A Policy Specification Tool

Fortunately, there are ways to formalize the specification of policies and procedures. For instance, a popular tool for specifying business policies is a decision table.

A **decision table** is a tabular form of presentation that specifies a set of conditions and their corresponding actions.

Decision tables, unfortunately, don't get enough respect! People who are unfamiliar with them tend to avoid them. But decision tables are very useful for specifying complex policies and decision-making rules. Figure 11.11 illustrates the three components of a standard decision table.

- **Condition stubs** describe the conditions or factors that will affect the decision or policy.
- **Action stubs** describe, in the form of statements, the possible policy actions or decisions.
- **Rules** describe which actions are to be taken under a specific combination of conditions.

Figure 11.11**A** depicts a check-cashing policy that appears on the back of a check-cashing card for a grocery store. In Figure 11.11**B**, this same policy has been defined with a decision table. Three conditions affect the check-cashing decision: the type of check (1 = Personal, 2 = Payroll), whether the amount of the check exceeds the maximum limit (Y = Yes, N = No), and whether the company is accredited by the store (Y = Yes, N = No). The actions (decisions) are either to cash the check or to refuse to cash the check. Notice that each combination of conditions defines a rule that results in an action, denoted by an X. Finally, note that rules 1, 3, and 5 contain a "—" entry for certain conditions. This means that the condition is irrelevant for these rules.

Decision tables offer a number of advantages over ordinary English instructions. They use a standard format and handle combinations of conditions in a very concise manner. The English equivalent of a decision table would be much more difficult to write and read, because each combination of conditions would have to be described. Decision tables also provide techniques for identifying policy incompleteness and contradictions. In other words, if a combination of conditions has no actions, the policy is incomplete.

Let's learn how to construct decision tables. But first, we have a challenge for you. Let's see if you can solve the dilemma presented in the following story.

The Poker Chip Challenge

Joe, Gordon, and Susan own Granger's Restaurant Supply. They are in dire financial straits. They need $250,000 to meet their debts, and cannot get a bank loan because of their poor credit rating. Among them, they can only collect $50,000.

They have decided on a drastic and risky solution to their problem. They will go to Atlantic City and try to gamble their $50,000 into enough money to cover their debts. There is one problem, however. They are lousy gamblers! Within one short hour, they lose the entire $50,000. As they leave the casino, they run into the president of Premier Restaurant & Supply, Inc., their fiercest competitor. He has been trying, unsuccessfully, to buy Granger's for some time. The unlucky trio offer Granger's to the greedy competitor for a bargain basement price. However, their rival, sensing an opportunity to get business for absolutely nothing, offers the following proposition:

"I have five poker chips in my pocket — three black and two white. I propose to blindfold each of you and then give you each a chip. One by one, I will remove your blindfolds. You will be permitted to see the chip in your colleagues' hands; however, you must keep your own chip concealed in your closed palm. If any one of you can tell me the color of your own chip, then I will give you $1 million, more than enough to ensure the financial future of your business. Each of you has the option of guessing or not guessing. However, if any one of you guesses wrong, you must give me your company, free and clear: Is it a deal?"

The partners have little choice and no other reasonable hope, so they accept the challenge. The competitor then shows them the five chips — three black and two white — and chuckles as he places the blindfolds in place and gives each person one chip. He returns the two unused chips to his pocket.

The blindfold is removed from Joe, the eldest businessman and a world-class chess master. He looks at his partners' chips but, despite his logical mind, cannot determine the color of his own chip. He responds, "I just cannot give an

FIGURE 11.12 **A Decision Table for Solving the Poker Chip Problem** Through the process of elimination, Susan was able to deduce the color of her chip.

Process Name		Rules							
		1	2	3	4	5	6	7	8
C o n d i t i o n s	Joe	W	W	W	W	B	B	B	B
	Gordon	W	W	B	B	W	W	B	B
	Susan	W	Ⓑ	W	Ⓑ	W	Ⓑ	W	Ⓑ
A c t i o n s	Impossible—only two white chips	X							
	Joe would have guessed					X			
	Gordon would have guessed			X					
	Susan knows her chip is black		↑		↑		↑		↑

answer: It's too risky. I'm better off giving my partners the opportunity for a better guess.''

The blindfold is removed from Gordon, a graduate of a prestigious business school. After looking at the chips of his two partners, he too is unable to guess the color of his own chip. He passes the opportunity to Susan.

The competitor grins as he starts to remove the blindfold from Susan. He doesn't give her any more of a chance than he gave Joe or Gordon.

Susan interrupts confidently, "You can leave my blindfold on. How about double or nothing!" The competitor laughs aloud, "It's your funeral!"

Susan replies, "I'll take that $2 million in cash! I know from the answers of my colleagues that my chip is _____." She is correct, and the winnings save Granger's from financial ruin.

What color was Susan's chip? Can you prove it without any doubt? You should make the following assumptions: (1) none of the characters in the story cheated, and (2) all of the characters are reasonably intelligent—that is, they would have guessed any obvious answer. Give up? Read on.

Solving the Poker Chip Problem

Figure 11.12 sets up a decision table for solving our poker chip problem. All possible combinations of poker chips are recorded as rules. The actions represent possible outcomes of the problem. Note that the first rule results in the action "Impossible" because we know that there were only two white chips. We can also eliminate rules 3 and 5 because, if either Joe or Gordon had seen two white chips, he would have known his own chip was black. But that didn't happen. We cannot eliminate rule 2 because Susan requested that her blindfold not be removed.

Examine the remaining rules for the third condition stub carefully. Do you see that only one of the remaining rules (rule 7) results in Susan having a white chip? She solved the problem by concluding from her colleagues' responses that she didn't have a white chip. Here's how to eliminate rule 7.

If rule 7 had been true, Gordon would have known that his chip was black, and the business would have been saved *before* Susan was asked to guess. Why? Because Joe couldn't answer, Gordon knew that Joe did not see two white chips. Joe had to see either two black chips or a white and a black chip. Now, remove the blindfold from Gordon and assume that rule 7 is true. If rule 7 is true, then Gordon is looking at a white chip in Susan's hand. Therefore, his own chip could not possibly be white because, if Joe had seen a white chip in both other people's hands, he would have known that his was black. Thus, we can eliminate rule 7 because Gordon, as an intelligent person, would have known his chip was black.

We now see that all the remaining rules result in Susan's chip being black. Therefore, she didn't need to see her colleagues' chips. All she was required to do was recognize that her chip was black—all rules resulting in her having a white chip had been eliminated by her colleagues' answers!

Building a Decision Table

Let's begin with a policy statement.

> A local credit union offers two types of savings accounts, regular rate and split rate. The regular-rate account pays dividends on the account balance at the end of each quarter—funds withdrawn during the quarter earn no dividends. There is no minimum balance on the regular-rate account. Regular-rate accounts may be insured. Insured accounts pay 5.75 percent annual interest. Uninsured regular-rate accounts pay 6.00 percent annual interest.
>
> For split-rate accounts, dividends are paid monthly on the average daily balance for that month. Daily balances go up and down according to deposits and withdrawals. The average daily balance is determined by adding each day's closing balance and dividing this sum by the number of days in the month. If the average daily balance is less than \$25, then no dividend is paid. If the average daily balance is \$25 or more, 6 percent per annum is paid on the first \$500, 6.5 percent on the next \$1,500, and 7 percent on funds over \$2,000. There is no insurance on split-rate accounts.

Let's construct a decision table for this policy statement example. Recall that the decision table presented in Figure 11.11 contained simple condition stubs that could only assume one of two values, yes and no. Many conditions, such as those presented in the preceding policy, may assume more than two possible values. However, the rules for constructing all decision tables are identical.

1. *Identify the conditions and values.* Identify the data attribute each condition tests and all of the values that these data attributes can assume. For our example:

Data Attributes or Conditions	Values
Account type	R = Regular rate
	S = Split rate
Insurance?	Y = Yes
	N = No
Balance dropped below \$25 during month?	Y = Yes
	N = No
Average daily balance	1 = \$0.00 – \$24.99
	2 = \$25.00 – \$500.00
	3 = \$500.01 – \$2,000.00
	4 = more than \$2,000.00

Note that we have yes and no conditions as well as multivalue conditions.

Process Name/Dividend Rate		Rules																															
		1	2	3	4	5	6	7	8	9	10	11	12	13	14	15	16	17	18	19	20	21	22	23	24	25	26	27	28	29	30	31	32
C o n d i t i o n s Account type	(a)	R	S	R	S	R	S	R	S	R	S	R	S	R	S	R	S	R	S	R	S	R	S	R	S	R	S	R	S	R	S	R	S
Insurance	(b)	Y	Y	N	N	Y	Y	N	N	Y	Y	N	N	Y	Y	N	N	Y	Y	N	N	Y	Y	N	N	Y	Y	N	N	Y	Y	N	N
Balance dropped below $25 during month	(c)	Y	Y	Y	Y	N	N	N	N	Y	Y	Y	Y	N	N	N	N	Y	Y	Y	Y	N	N	N	N	Y	Y	Y	Y	N	N	N	N
Average daily balance	(d)	1	1	1	1	1	1	1	1	2	2	2	2	2	2	2	2	3	3	3	3	3	3	3	3	4	4	4	4	4	4	4	4
A c t i o n s Pay no dividend																																	
5.750% ÷ 4 as the quarterly dividend on entire balance																																	
6.000% ÷ 4 as the quarterly dividend on entire balance																																	
6.000% ÷ 12 as the monthly dividend on balance up to $500																																	
6.500% ÷ 12 as the monthly dividend on balance between $500.01 and $2,000																																	
7.000% ÷ 12 as the monthly dividend on balance over $2,000																																	

FIGURE 11.13   Entering All Possible Rules into a Decision Table   This decision table identifies all possible combinations of conditions. Each condition defines a rule that must be verified or eliminated as impossible. The circled letters correspond to steps described in the text to determine all possible conditions.

2. *Determine the maximum number of rules.* Multiply the number of values for each condition data attribute by each other to determine the maximum number of rules in a decision table. For example:

Condition 1 offers two values	2
Condition 2 offers two values	× 2
Condition 3 offers two values	× 2
Condition 4 offers four values	× 4
Number of rules in decision table	32

3. *Identify the possible actions.* Identify each independent action to be taken for the decision or policy. For our example:

   Pay no dividend

   Pay 5.750% ÷ 4 as the quarterly dividend on entire balance

   Pay 6.000% ÷ 4 as the quarterly dividend on entire balance

   Pay 6.000% ÷ 12 as the monthly dividend on balance up to $500

   Pay 6.500% ÷ 12 as the monthly dividend on balance between $500.01 and $2,000

   Pay 7.000% ÷ 12 as the monthly dividend on balance over $2,000

4. *Enter all possible rules.* Record the conditions and actions in their respective places in the decision table (see Figure 11.13). All possible rules can easily be identified by completing the following steps:

Process Name/Dividend Rate		Rules																															
		1	2	3	4	5	6	7	8	9	10	11	12	13	14	15	16	17	18	19	20	21	22	23	24	25	26	27	28	29	30	31	32
Conditions	Account type	R	S	R	S	R	S	R	S	R	S	R	S	R	S	R	S	R	S	R	S	R	S	R	S	R	S	R	S	R	S	R	S
	Insurance	Y	Y	N	N	Y	Y	N	N	Y	Y	N	N	Y	Y	N	N	Y	Y	N	N	Y	Y	N	N	Y	Y	N	N	Y	Y	N	N
	Balance dropped below $25 during month	Y	Y	Y	Y	N	N	N	N	Y	Y	Y	Y	N	N	N	N	Y	Y	Y	Y	N	N	N	N	Y	Y	Y	Y	N	N	N	N
	Average daily balance	1	1	1	1	1	1	1	1	2	2	2	2	2	2	2	2	3	3	3	3	3	3	3	3	4	4	4	4	4	4	4	4
Actions	Pay no dividend		?		X						X								X								X						
	5.750% ÷ 4 as the quarterly dividend on entire balance	X	?						X				X				X				X				X				X				
	6.000% ÷ 4 as the quarterly dividend on entire balance		?	X						X				X				X				X				X				X			
	6.000% ÷ 12 as the monthly dividend on balance up to $500		?													X								X									X
	6.500% ÷ 12 as the monthly dividend on balance between $500.01 and $2,000		?																					X									X
	7.000% ÷ 12 as the monthly dividend on balance over $2,000		?																														X
	Impossible		?			X	X	X	X			X			X					X			X					X			X		

FIGURE 11.14 Defining the Actions for Each Rule in a Decision Table   This decision resulted in some rules (combinations of conditions) that are impossible. For other rules, we are uncertain as to what actions should be performed.

(a) For the first condition, alternate its possible values.

(b) Note the size of the pattern that repeats in step a (in our example, two values). Cover each pattern of two values (R S) in the previous condition with the values of the second condition, repeating as necessary until the row is filled.

(c) Again, note the size of the pattern that repeats in the second condition (this time, four values: Y Y N N). Cover each pattern of four values with the values of the next condition, repeating as necessary until the row is filled.

(d) Once again, note the size of the pattern that repeats in the third condition (this time, eight values: Y Y Y Y N N N N). Cover each pattern of eight values with the values of the next condition, repeating as necessary until the row is filled.

This simple process defines all possible combinations of conditions for any decision table. We could have continued the process for any number of conditions.

5. *Define the actions for each rule.* Determine which actions are appropriate for each rule and mark them with an X. In the event that certain rules are impossible (cannot happen), add an action stub "Impossible" and mark the rules with an X. A question mark denotes that an action for a rule is unknown. It reminds you to check with your end-users to learn how this rule should be handled. Figure 11.14 illustrates the actions for our rules.

FIGURE 11.15 Simplifying a Decision Table
Most decision tables can be simplified by collapsing combinations of certain rules into single rules. For teaching purposes, we placed original rule numbers above the collapsed rules. In practice, the rules would be renumbered from 1 to 9 (for the nine remaining rules).

A.

Conditions		1	2	3	4	5	6	7	8	9	10	11	12
	Condition 1	Y	N	Y	N	Y	N	Y	N	Y	N	Y	N
	Condition 2	Y	Y	N	N	Y	Y	N	N	Y	Y	N	N
	Condition 3	L	L	L	L	M	M	M	M	H	H	H	H
Actions	Action 1	X				X				X			
	Action 2		X					X			X	X	X
	Action 3			X	X								
	Action 4				X			X					

B.

Conditions		1, 5, 9	2	3	4	6	7	8	10	11, 12			
	Condition 1	Y	N	Y	N	N	Y	N	N	–			
	Condition 2	Y	Y	N	N	Y	N	N	Y	N			
	Condition 3	–	L	L	L	M	M	M	H	H			
Actions	Action 1	X											
	Action 2		X					X	X	X			
	Action 3				X	X							
	Action 4			X			X						

6. *Verify the policy.* Your completed decision table should be reviewed with your end-users. Resolve any rules for which the actions are not specified. Verify that rules specified as impossible cannot occur. Resolve apparent contradictions, such as one rule with two possible interest rates covering a single balance. Finally, verify that each rule's actions are correct.

7. *Simplify the decision table.* At this point, our decision table is both complete and correct. Still, 32 rules can be a bit overwhelming. We can simplify the decision table by eliminating and consolidating certain rules. (*Note:* This step should never be done until step 6 is complete!) The technique is described as follows:

   a. Eliminate impossible rules.

   b. Look for indifferent conditions. An **indifferent condition** is a condition whose values do not affect the decision and always result in the same action. These rules can be consolidated into a single rule. The technique is described as follows:

Process Name/Dividend Rate		Rules					
		1	2	3	4	5	6
C o n d i t i o n s	Account type	R	R	S	S	S	S
	Insurance	Y	N	–	N	N	N
	Balance dropped below $25 during month	–	–	Y	N	N	N
	Average daily balance	–	–	–	2	3	4
A c t i o n s	Pay no dividend			X			
	5.750% ÷ 4 as the quarterly dividend on entire balance	X					
	6.000% ÷ 4 as the quarterly dividend on entire balance		X				
	6.000% ÷ 12 as the monthly dividend on balance up to $500				X	X	X
	6.500% ÷ 12 as the monthly dividend on balance between $500.01 and $2,000					X	X
	7.000% ÷ 12 as the monthly dividend on balance over $2,000						X

Account type: R = Regular rate
             S = Split rate

Insurance: Y = Yes
         N = No

Balance dropped below
  $25 during month: Y = Yes
                      N = No

Average daily balance: 1 = $0.00–$24.99
                       2 = $25.00–$500.00
                       3 = $500.01–$2,000.00
                       4 = More than $2,000.00

FIGURE 11.16 **Simplified Decision Table for the Credit Union Dividend Policy** Notice that conditions 3 and 4 are indifferent to regular accounts. Thus, more than one condition can be indifferent to a single action.

1. Find a set (pair, trio, and so on) of rules for which
   · The actions are identical.
   · The condition values are the same except for one and only one condition or factor.
   · All possible values of an entry for a given condition must become indifferent before the rules can be collapsed. (This is important, especially in extended entry tables.)
2. Consolidate that set of rules into a single rule, replacing the value of the indifferent condition with a minus sign—the indifference symbol.

This technique should be repeated as often as sets of rules satisfy the criteria in step b. But be careful! Never consolidate rules based on conditions that have already been identified as indifferent!

Before we simplify our dividend table, let's look at an easier example. In Figure 11.15**A**, notice that rules 1, 5, and 9 result in the same action. Also note

FIGURE 11.17   **A Sample Structured English Description of a Business Procedure**

> For each LOAN ACCOUNT NUMBER in the LOAN ACCOUNT FILE do the following steps:
>   If the AMOUNT PAST DUE is greater than $0.00 then:
>   While there are LOAN ACCOUNT NUMBERs for the CUSTOMER NAME do the
>   following steps:
>   Sum the OUTSTANDING LOAN BALANCES.
>   Sum the MINIMUM PAYMENTs.
>   Sum the PAST DUE AMOUNTs.
>   Report the CUSTOMER NAME, LOAN ACCOUNTs on OVERDUE CUSTOMER LOAN
>   ANALYSIS.

that all condition values for the three rules are the same except for the third condition. In addition, observe that the three rules cover all possible values for the third condition. Do you see that the third condition's value does not affect the action? That condition is indifferent. We can consolidate the three rules into a single rule, as in Figure 11.15**B**. We recorded the original rule numbers above the consolidated rules for your convenience. Also note that we consolidated rules 11 and 12 based on indifference for the first condition stub.

You may have been tempted to consolidate rules 3 and 7. But notice that the third condition hasn't been satisfied: only two (L and M) of the three values (L, M, and H) of the alleged indifferent condition are covered! Therefore, that condition is not indifferent.

Figure 11.16 depicts the final simplified decision table for the credit union's dividend policy. Convince yourself that our simplifications are valid. This condensed decision table is much easier to read than its equivalent in Figure 11.14.

Structured English: A Procedure Specification Tool

Whereas decision tables are particularly effective for describing policies, Structured English is a tool for describing procedures. Structured English is based on the principles of structured programming. An example of Structured English is illustrated in Figure 11.17.

An understanding of computer program pseudocode is important to using the Structured English tool effectively.

Pseudocode is a tool to define detailed computer program algorithms or logic prior to coding.

There are similarities between Structured English and pseudocode; however, pseudocode often tends to take on a programming accent that makes it unsuitable for nonprogrammers. For instance, array and variable initialization, opening and closing files, and read/write operations are often included in pseudocode. Also, most programmers tend to accent their pseudocode with the syntax of a computer programming language (for example, BASIC, COBOL, FORTRAN), usually the first such language they learned.

Structured English borrows the logical constructs of structured pseudocode but restricts the use of nouns, verbs, adjectives, adverbs, and computer jargon to make the specification easier to read for end-users. Two other popular proce-

For each LOAN ACCOUNT NUMBER in the LOAN ACCOUNT FILE
 If the AMOUNT PAST DUE is greater than $0.00
  Do while there are LOAN ACCOUNT NUMBERs for CUSTOMER NAME
   Sum the OUTSTANDING LOAN BALANCE
   Sum the MINIMUM PAYMENTs
   Sum the PAST DUE AMOUNTs
 Report the CUSTOMER NAME and LOAN ACCOUNTs on OVERDUE CUSTOMER
 LOAN ANALYSIS

FIGURE 11.18 **A Sample Action Diagram Description of a Business Procedure** This action diagram is equivalent to the Structured English in Figure 11.17. It was drawn with KnowledgeWare's ADW Analysis CASE tool.

To complete an overdue customer loan analysis for each loan account:
 Step 1: Identify loan accounts that are past due.
 Step 2: Summarize past due loan accounts by customer as follows:
  Step 2.1 Sum the outstanding loan balances for each overdue loan account belonging to a given customer.
  Step 2.2 Sum the minimum required payment balances for each overdue loan account belonging to a given customer.
  Step 2.3 Sum the current past due amounts for each overdue loan account belonging to a given customer.
 Step 3: Report the name of each customer, the loan account, and loan account sums.

FIGURE 11.19 **A Sample Tight English Description of a Business Procedure** This Tight English sample is equivalent to the Structured English in Figure 11.17.

dure specification tools are action diagrams and tight English. Essentially, an **action diagram** (Figure 11.18) uses brackets to group instructions to visually represent different control structures. **Tight English** (Figure 11.19) differs from Structured English because it groups instructions into specific steps, avoids using keywords and capitalization, and references decision tables.

Structured English Guidelines and Syntax

Structured English is the marriage of the English language with the syntax of structured programming. The following restrictions are placed on the use of English within Structured English:

1. Only strong, imperative verbs may be used.
2. Only nouns and terms defined in the project repository may be used. These nouns may include names of data flows, data stores, entity types, records, data attributes, and tables of code (and also decision tables).
3. Compound sentences should be avoided.
4. Undefined adjectives and adverbs (the word *good,* for instance) are not permitted unless clearly defined in the project repository as value ranges for data attribute descriptions.
5. Avoid language that destroys the natural flow. Examples include the terms *go to . . ., do . . .,* and *perform,* all of which are programming verbs that have no meaning in Structured English.

Find the MATERIAL NUMBER in the INVENTORY FILE.
Select the appropriate case:
    Case 1: MATERIAL CLASS = 'stock,' then:
        If the QUANTITY ON HAND is greater than or equal to the QUANTITY REQUISI-
        TIONED then:
            Calculate new QUANTITY ON HAND using the formula:
                QUANTITY ON HAND − QUANTITY REQUISITIONED
            Record QUANTITY ON HAND in the INVENTORY FILE.
            Issue a STORES TICKET.
        Otherwise (QUANTITY ON HAND is not greater then the QUANTITY REQUISI-
        TIONED) then:
            Issue a STORES STOCKOUT TICKET.
    Case 2: MATERIAL CLASS = 'seasonal,' then:
        Calculate QUANTITY NEEDED using the formula:
            REQUISITIONED QUANTITY × SEASONAL ADJUST RATE
        Issue a PURCHASE REQUISITION.
    Case 3: MATERIAL CLASS = 'requisition,' then:
        Issue a PURCHASE REQUISITION.

6. A limited set of logic or flow constructs must be used. These constructs are familiar to those who practice structured programming. The three valid constructs are
   a. A sequence of single declarative statements.
   b. The selection of one or more declarative statements based on a decision (the if-then-otherwise decision construct).
   c. The repetition of one or more declarative statements (the looping construct).

The best way to learn Structured English is to study some examples. We try to write simple but complete sentences. To as great an extent as possible, we want to eliminate the rigid style of computer programming while maintaining the syntax.

Sequential constructs are simple, declarative statements that follow one another. Declarative statements should begin with a strong action verb that describes exactly what should be done in that step of the procedure. Avoid vague, meaningless verbs such as *process, handle,* and *perform.* Also avoid computer programming language verbs such as *move, open,* or *close.* Many declarative statements are arithmetic, specifying how to calculate such data attributes as GROSS PAY, FEDERAL TAX WITHHELD, and NET PAY. The statement should always specify the formula to be used. For example:

    Calculate <insert data attribute> using the formula
    <insert formula>

All data attributes should be defined in the project repository. The following statements are valid sequential instructions:

    Find the MEMBER ACCOUNT using the MEMBER ACCOUNT NUMBER.
    Compute the new ACCOUNT BALANCE using the formula:
        ACCOUNT BALANCE = ACCOUNT BALANCE + ADJUSTMENT
        AMOUNT.

Record the new ACCOUNT BALANCE in the MEMBER ACCOUNT data store.

Write the new ACCOUNT BALANCE on the CUSTOMER STATEMENT.

Note that we capitalized the names of data flows, data stores, data attributes, and other nouns recorded elsewhere in the project repository.

The **decision construct** of Structured English allows you to place branching instructions within a sequence of instructions. The following formats for the decision construct are permitted.

- If <insert condition> then:
    <insert instruction(s)>
  Otherwise: (not condition)
    <insert instruction(s)>.
- Select the appropriate case:
  Case 1: <insert condition value 1>
          <insert instruction(s)>
  Case 2: <insert condition value 2>
          <insert instruction(s)>
  .
  .
  .

  Case *n*: <insert condition value *n*>
          <insert instruction(s)>

Use the if-then-otherwise format for conditions that can only assume two values (for example, yes and no, or male and female). Use the case format whenever the condition can assume more than two values (for example, freshman, sophomore, junior, and senior). The instructions within the decision construct include one or more statements.

Figure 11.20 demonstrates the case and if-then-otherwise constructs. The instructions for each case are indented and blocked for readability. The constructs are nested, one within another. Also, the negative of the condition is recorded in parentheses to enhance readability.

The **repetition construct,** or *looping construct,* allows us to specify that a sequence of instructions is to be repeated until some condition or desired result is satisfied. The following formats are permitted:

- Do the following <insert some number> times:
    <insert instruction(s)>
- Repeat the following steps:
    <insert instruction(s)>
  Until <insert condition> is satisfied.
- While <insert condition>, do the following steps:
    <insert condition(s)>
- For each <insert condition>, do the following steps:
    <insert instruction(s)>

The first format allows you to specify that certain steps be performed a specific number of times. Do you see the difference between the second and third formats? The second format requires that the instructions be executed at least one time. The third format specifies that the instructions might not be executed at all, if the condition is not initially satisfied. The fourth format functions like

FIGURE 11.21   **The Repetition Construct of Structured English**

> **For each CUSTOMER NUMBER in the CUSTOMER ACCOUNT file, do the following:**
> **Repeat the following steps for each ACCOUNT NUMBER:**
> **For each ACCOUNT TRANSACTION for the ACCOUNT NUMBER, do the following:**
> Report each ACCOUNT TRANSACTION.
> Sum the following account totals:
> NUMBER OF DEBIT TRANSACTIONS
> NUMBER OF CREDIT TRANSACTIONS
> TOTAL OF DEBIT TRANSACTIONS
> TOTAL OF CREDIT TRANSACTIONS
> ACCOUNT EXPENSES
> Report the account totals for the ACCOUNT NUMBER.
> **Until there are no more ACCOUNT NUMBERs for the CUSTOMER NUMBER.**

the third, specifying that a condition determines whether the instructions will be executed.

Figure 11.21 demonstrates the repetition construct for the procedure that produces a consolidated bank statement for each customer. The first repetition construct, "For each . . .," drives the entire procedure. Because a customer can have numerous accounts but must have at least one account, we use the repeat-until construct for an inner loop of the procedure.

Decision Table or Structured English?

This is the first chapter in which we've offered two tools for specifying one object, a process. Both decision tables and Structured English can describe a single process. How then do we determine which tool to use? Recall that procedures often implement policies. In the same way, we can use Structured English to describe all procedures, and decision tables to describe any policies implemented by the procedures.

COMPUTER-ASSISTED SYSTEMS ENGINEERING (CASE) FOR SPECIFYING DATA AND PROCESS REQUIREMENTS

Computer-Assisted Systems Engineering (CASE) was introduced in Chapter 5 as an enabling, emerging technology for systems analysis and design. All CASE products are built around a project repository, which is usually fragmented into project repository subsets. The project repository catalogs all systems models and details about the objects contained on those models. The analyst can easily call up a model, such as a data flow diagram, select a data flow from that model, and instantly call up the data structure for that data flow.

CASE products also make it possible to look at subsets of the project repository, such as all data flows for a particular subsystem or end-user. In addition, CASE products provide numerous analytical tools to check for consistency and completeness in the project repository. Many CASE tools can import data from and export data to other project dictionaries. CASE is becoming the tool of choice for managing the overwhelming detail that makes all but the smallest systems projects difficult.

Alternatively, if CASE is not available, there are several ways to implement a project repository. One of the more popular options is to print special forms (recall Figure 11.1) for documenting entity types, data flows, data stores, data attributes, and so forth. The advantage of forms is that the analyst is provided with standards for what to include in the repository.

Another technique is to implement a simple computerized project repository using a word processor or line editor on a computer. This technique offers the advantage of easier editing; however, the disadvantage is that it can be difficult to get listings of subsets of the repository—it's all or nothing.

In the remaining section of this chapter, we will demonstrate how to enter complete descriptions of our data and process models into the project repository. All of the repository entries in this chapter were created with a popular CASE tool, Excelerator. Only the color screens and annotations were added by an artist. Each CASE tool has its own way of dealing with data structures. Excelerator is no different. We will explain its idiosyncrasies in the examples.

HOW TO SPECIFY DATA AND INFORMATION REQUIREMENTS IN A REPOSITORY

We've learned how to define data structures for essential data flows, data stores, and entity types. Our goal, however, is to define all essential facts about these components. In this section, you will learn how to write complete essential project repository entries.

These entries have been gleaned from the essential models for the Sound-Stage Entertainment Club case study. We are not presenting this section step by step, since you would normally begin recording the repository entries as you drew the information and process models. In other words, the step "Record details in the project repository" would immediately follow the last step of information modeling (Chapter 8), process modeling (Chapter 9), and network modeling (Chapter 10).

Specifying Data Entities in a Project Repository

Data entities and relationships were the objects on a data model (entity relationship diagram, or ERD). Either object may be "described" or "exploded" to specify data attributes that describe the object. A sample repository screen or form for an entity type is shown in Figure 11.22. The following descriptions correspond to the circled letters in that figure.

(A) Every entity type should have a name and label, which are normally the same thing. Naming conventions were suggested in Chapter 8. The label will correspond to the name of a data entity appearing on an ERD.

(B) The content and data structure for every entity type can be described by way of a record or a data model. Regardless, as a general rule, we give the record or data model the same name as the entity type. In this example the data structure of the entity MEMBER is further described by a record.

(C) As they become known, user requirements that may be satisfied (or addressed) by the data entity, and other related items in the repository may be identified.

(D) Every data entity should contain a brief definition to establish vocabulary.

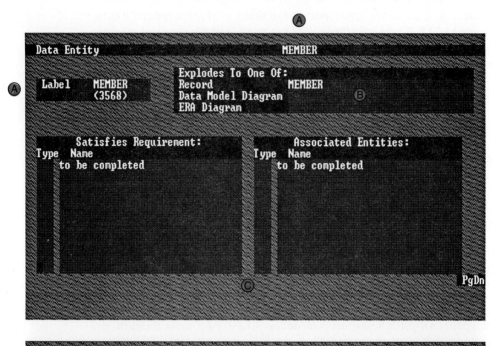

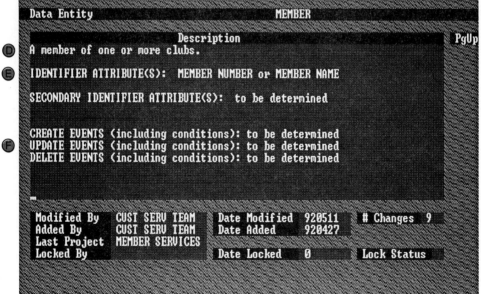

FIGURE 11.22 **Project Repository Entry for a Data Entity from a Data Model**   This is a typical repository entry for a data entity. The explosion attribute points to a separate description for a record.

E Every data entity's unique identifier attribute(s) should be recorded in the repository. Optionally, you want to record non-unique subsetting attributes. These attributes divide all occurrences of the data entity types into useful subsets.

F The data entity repository entry may also contain information about how the occurrences of the data entity are maintained (discussed in Chapter 13).

Specifying Data Flows in a Project Repository

Data flows are objects on a data flow diagram (DFD). Figure 11.23 represents a sample screen specification describing a data flow to the repository. The following descriptions correspond to the circled letters in that figure.

A Every data flow should be named. If desired, labels can be used to give longer, more descriptive names that will appear on the DFDs.

B The contents of data flows should be described as records. A record description, which will be explained shortly, documents the data structure that was covered earlier in the chapter. Why record the data structure separately from the data flow description? Because by separating the record, that record can be reused for other data flows that are identical, though perhaps with different names. As a general rule, we give the record the same name as the data flow except when its contents correspond to an entity type (a data flow from the data store for that entity type).

C We've indicated the duration value (or average volume) of occurrences for the data flow, as well as the duration type (or unit of time).

D The access type specifies what the data flow does. In this case, the "A" means that the data flow causes an A(dd). Data flows can also trigger updates, reads, or deletions.

E As they become known, user requirements that may be satisfied by (or addressed by) the data flow will be identified, as will other related items in the repository.

F Every data flow should contain a brief definition to establish vocabulary.

Specifying Data Stores in a Project Repository

Data stores are also easy to describe using a project repository. The screen for a typical data store entry is illustrated in Figure 11.24. The following annotation will help to clarify the figure:

A Every data store should have a unique name. (Naming conventions were covered in Chapter 9.) The label appears on any DFD that depicts this data store.

B Like data entity types and data flows, the content or data structure will normally be described separately as a record. In some cases, however, our data store's data structure is defined by a data model that contains data entities, each of which are further described separately as a record.

C The data repository entry for a data store should include such characteristics as location, physical implementation, and volumes as they become known.

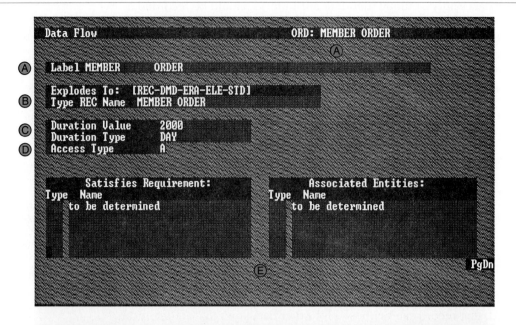

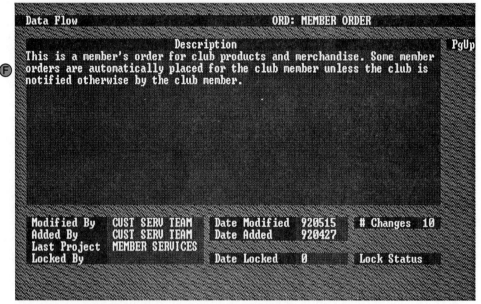

FIGURE 11.23   **Project Repository Entry for a Data Flow from a Process Model**   This is a typical repository entry for a data flow. The explosion attribute points to a separate description for a record.

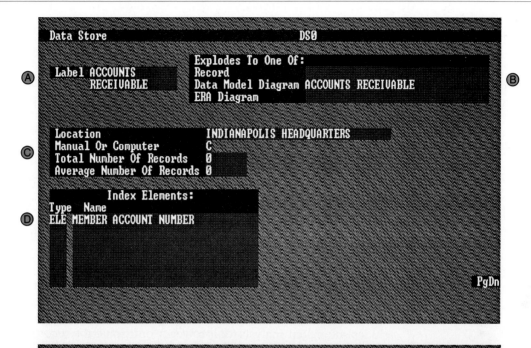

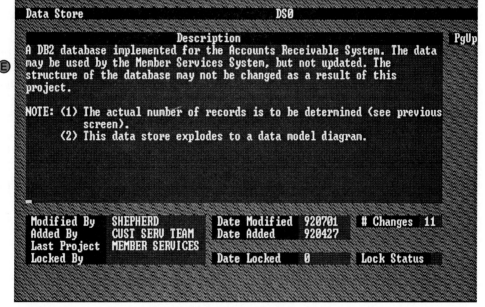

FIGURE 11.24 **Project Repository Entry for a Data Store from a Process Model** This is a typical repository entry for a data store. The explosion attribute points to a data model diagram.

(D) Unique identifier attributes should be specified.

(E) Every data store should include a brief definition to establish vocabulary.

Specifying Records in a Project Repository

Recall that repository entries for data entities, data flows, and data stores may be further described as a record. Records describe the content and data structures that were covered earlier in the chapter. Figure 11.25**A** illustrates a simple CASE product's record description in a repository. This record corresponds to an entity type or data store. The following descriptions correspond to the circled letters in Figure 11.25**A**.

(A) Every record has a unique name. Recall that the named record (data structure) may be shared by many data flows, data stores, and data entity types.

(B) Some records may be known to end-users by other or alternative names. These aliases should be recorded to provide reference.

(C) The Normalized attribute happens to be specific to the CASE product we used to demonstrate this entity. It is marked Y for yes or N for no to indicate whether the record is in third normal form (normalization and third normal form are discussed in Chapter 13). The CASE product includes automated analysis tools to verify or dispute the recorded claim.

(D) The data structure is recorded in terms of data attribute (referred to as "element" by the CASE product) names and/or data group names. A data group is called a *subrecord*. A subrecord would be described in a separate record description screen or form. For example, Figure 11.25**B** represents the subrecord MEMBER ORDERED PRODUCT. This practice of defining a separate record description allows common subrecords to also be referenced by other data flows, data stores, and data entities.

(E) The Occ attribute means maximum number of occurrences of the attribute or group for a single occurrence of the record. Notice that the subrecord MEMBER ORDERED PRODUCT can occur up to 99 times for a single occurrence of the record MEMBER ORDER.

(F) The Seq attribute is pertinent to the CASE product but is not important to this discussion.

(G) The Type attribute describes the type of attribute or group that is being described. For attributes, the type can be

    K   meaning   identifier attribute (key)

    E   meaning   nonidentifier attribute (element)

For groups or subrecords, the type can be

    R   meaning   subrecord

If type is R, there must be another record description in the repository to describe that subrecord's data structure.

Specifying Data Attributes in a Project Repository

Preparing repository entries to describe data attributes is just as important as preparing repository entries to describe entity types, data flows, data stores, and records. Data attributes are not composed of data structures or other data attributes. However, data attributes are assigned values, and it is very important that

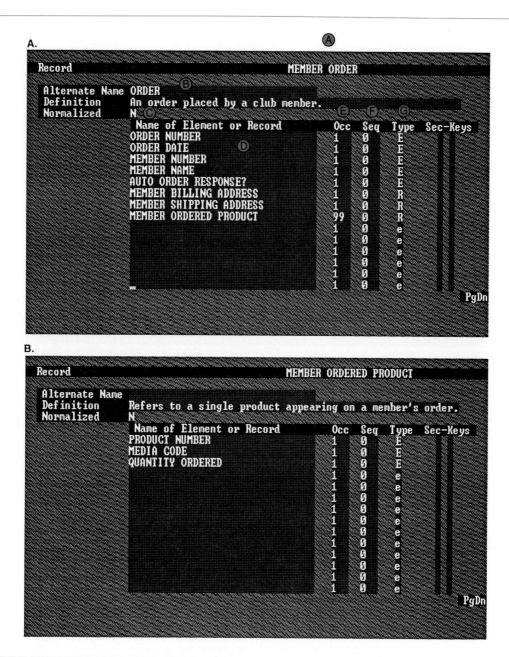

FIGURE 11.25 **Project Repository Entry for a Record** This is a typical repository entry for a record. A record describes a data structure. The record in the top figure contains a repeating group that is described as a separate record description in the bottom figure.

we learn the legitimate values for data attributes so we can design controls in information systems to guarantee that occurrences of data attributes are valid. We can use dictionaries to describe the data attributes and to verify our understanding of the data attribute with the end-user. A complete data attribute entry in the repository is illustrated in Figure 11.26. It is more difficult to separate essential and implementation attributes for data attributes; therefore, they are

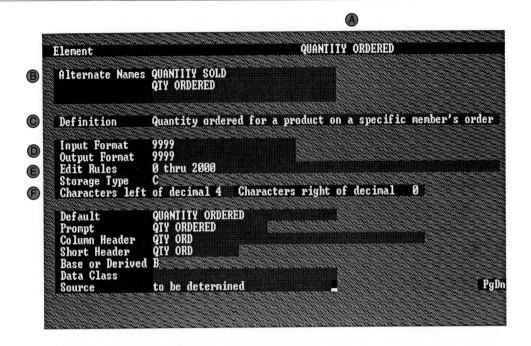

FIGURE 11.26   **Project Repository Entry for a Data Attribute**

combined on this form/screen. We will discuss only the pertinent attributes at this time:

(A) Every data attribute should have a logical, descriptive name. Naming conventions for data attributes were described in Chapter 8.

(B) Data attributes frequently have business aliases, synonyms, or acronyms. They should be recorded as a frame of reference for end-users and analysts.

(C) Every data attribute should include a brief definition to establish vocabulary.

(D) Data attributes may be constrained by certain business formatting restrictions. The CASE product we used required that the format be specified as a COBOL-like PICTURE clause.

(E) Legitimate data attribute values must be specified. Values fall into three categories:

· *Indefinite values.* In this case the attribute can assume a virtually limitless number of values. The Value Range attribute is left blank.

· *Value ranges.* In this case the attribute can assume any value in the range of values—for example, 3.35 through 6.50. Such a range is recorded in the Value Range attribute.

· *Finite values.* In this case the attribute may assume any one of a limited set of well-defined values. These values can also be listed in the Value Range attribute. Sometimes the set of values is a set of codes, each of which carries a meaning. In this case, the attribute Value Range should be specified as FROM name of a table of codes. Tables of codes are described separately in the repository.

(F) Decimal positioning should be specified for all numeric data attributes.

Specifying Tables of Codes in a Project Repository

If our goal is to define all facts about the data structures for essential data flows, data stores, and entity types, we must not forget about codes.

A **code** is a group of characters and/or digits that identifies and describes something in the business system.

Codes are frequently used to describe customers, products, materials, or events. The use of codes is popular for several reasons. First, codes can often be used for quick and easy identification of people, objects, and events. For instance, a product coded A-57-G may mean "product number 57, gallon can, in warehouse zone A." Second, codes usually condense numerous facts into concise format. And finally, the concise format usually reduces data storage space requirements.

Systems analysts are frequently charged with analyzing and defining coding schemes for information systems. First, we'll examine some of the more common coding schemes, and then we'll show you how to describe them to the project repository.

Sequential codes and **serial codes** are quite similar. Both number items with consecutive numbers—for example, 1,2,3 . . . n. Sequential numbers are typically assigned to a set of items, such as customers, that have been previously ordered (alphabetically, for example). Although the scheme is simple, new items cannot easily be inserted without disrupting the original ordering or changing many assigned numbers.

Serial numbers are assigned according to *when* new items are first identified. The first item identified is numbered 1, the second item is numbered 2, and so forth. Although serial coding also offers simplicity and a nearly infinite number of occurrences, the code has little information value.

Sequential and serial coding are frequently used as components in the more complicated coding schemes described in the following sections.

Block codes are a variation on sequential coding. A set of sequential or serial codes is divided into blocks that classify items into specific classes. For instance, a block code could be defined for customers as follows:

1000 through 4999	Record/Tape Club Customers
5000 through 6999	Compact Disc Club Members
7000 through 7999	Video Club Members

Every customer would be assigned a serial or sequential number within its proper block classification. The number for any given customer identifies the customer type.

Alphabetic codes use different combinations of letters and/or numbers to describe items. Alphabetic codes have greater information value than any of the previously described codes. Many such codes have been standardized—for example, the two-letter system of abbreviations for states. Most alphabetic codes are abbreviations; therefore, with a little practice, such codes can easily be interpreted.

Group codes are the most powerful of the coding schemes because they convey much more information content than the other coding schemes. Each position or group of positions in the code describes some pertinent characteristic of the item being coded. Thus, the code number tells the reader a great deal about the item itself. There are two common types of group codes: significant position codes and hierarchical codes.

For **significant position codes,** each digit or group of digits describes a measurable or identifiable characteristic of the item. Significant digit codes are frequently used to code inventory items. For example, the following code might be defined for a music/video product:

First two digits	Product classification:
	RK = Rock
	CL = Classical
	JZ = Jazz
	CW = Country/western
	EL = Easy listening
	MU = Music
	CH = Children
	DR = Drama
	CC = Classics
	SC = Science fiction
	AD = Adventure
	HR = Horror
	FY = Fantasy
	CM = Comedy
	DC = Documentary
Third and fourth digits	Year of issue
Fifth through eighth digits	Sequential assigned no.
Ninth and tenth digits	Media code:
	L8 = 8″ laser disc
	LD = Standard laser disc
	LL = Letterbox laser disc
	CV = CDV laser disc
	VH = VHS videotape
	8M = 8MM videotape
	DC = Digital cassette
	CA = Cassette
	CD = Compact disc
	CM = Compact mini disc
	CS = Compact disc single

As was the case with alphabetic codes, significant position codes are frequently standardized for some industries (zip codes, light bulb codes, tire codes).

Hierarchical codes provide a top-down interpretation for an item. Every item coded is factored into groups, subgroups, and so forth. Each group and subgroup can be coded such that the codes identify specific groups and subgroups. For instance, we could code—or partially code—all inventory items by warehouse location as follows:

First digit	Warehouse zone (A, B, C, D, or E)
Second digit	Section in zone (1–5)
Third and fourth digits	Aisle in section (1–20)
Fifth digit	Shelf number (A–M)

All the coding schemes discussed here can be combined in any arrangement desired to achieve business goals. Codes are intended to make the handling of data easier. Whenever you encounter a data attribute, which can only assume a

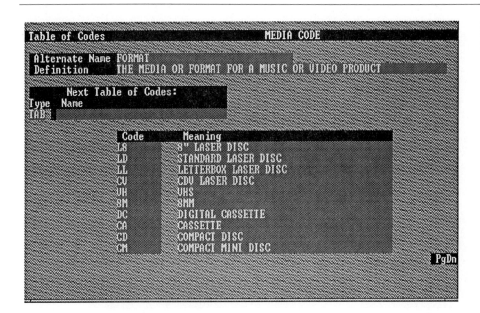

FIGURE 11.27 **Project Repository Entry for a Table of Codes**

finite set of values, a coding scheme may have to be studied, defined, or expanded. Codes are, first and foremost, a business tool. Therefore, you should consider the following business issues when analyzing or proposing codes:

1. *Codes should be expandable.* To accommodate natural growth, a code should allow for additional entries.
2. *Codes should be unique.* Each occurrence should define one, and only one, occurrence of data.
3. *Size of codes is important.* A code should be large enough to describe relevant characteristics but small enough to be easily read and interpreted by people.
4. *Codes should be convenient.* A new occurrence of a code should be easy to construct and interpret. A computer should not be required.

Once a code has been defined, it can be entered into the project repository. Figure 11.27 demonstrates one such entry. It is self-explanatory.

HOW TO DESCRIBE PROCESS REQUIREMENTS IN A REPOSITORY

Process models (data flow diagrams) were introduced in Chapter 9 as an effective requirements modeling tool. The models show the essential processes, their inputs, outputs, and data store interactions. In the previous section we learned how to further document data flows and data stores. In this section we'll describe how to complete the specification of primitive processes.

Primitive processes — those that are not exploded into more detailed DFDs — should be exploded into project repository descriptions that include Structured English and, where applicable, decision tables.

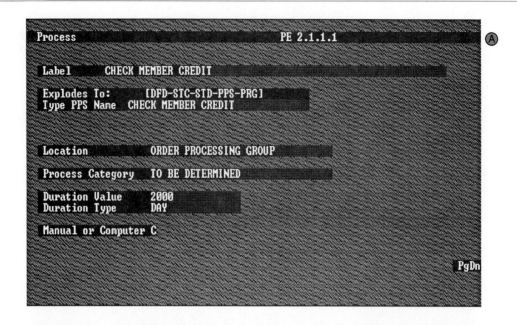

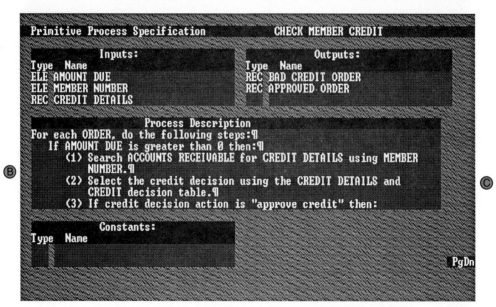

FIGURE 11.28 **Project Repository Entry for a Process Appearing on a Process Model** Note that this sample screen includes a reference to a decision table and the use of a numbering convention within the procedure.

Specifying Processes in a Repository Using Structured English

A sample repository screen or form for a primitive process is demonstrated in Figure 11.28. These sample entries were created using a CASE product. The following annotation will help to clarify the figure:

(A) The primitive process specification has a name that corresponds to the name of the process appearing on the DFD (see Chapter 9, DFDs for Entertainment Club case study).

Cond/ Act	Description	1	2	3
Ⓐ CREDIT DECISION TABLE				
Ⓑ This decision table defines the rules for approving or rejecting member credit for an order.				
C	AMOUNT DUE > MEMBER CREDIT LIMIT Ⓒ	N	Y	Y
C	CREDIT RATING = "A" OR "B"	–	Y	N
			Ⓔ	
A	APPROVE CREDIT	X		
A	APPROVE CREDIT BUT SEND REMINDER Ⓓ		X	
A	REJECT CREDIT			X

FIGURE 11.29 **Project Repository Entry for a Decision Table**

Ⓑ The procedure statement is expressed in Structured English.

Ⓒ A primitive process may be further described by a decision table that is referenced within the procedure.

Specifying Processes in a Repository Using Decision Tables

Decision tables are always referenced from within a primitive process specification (see Figure 11.28). The decision table itself is described separately in the repository, as shown in Figure 11.29.

Ⓐ Every decision table should have a descriptive, logical name.

Ⓑ Every decision table should have a definition that suggests its purpose, if the name is not self-explanatory.

Ⓒ Conditions should be expressed in terms of data attributes that are described elsewhere in the repository (discussed in the previous section).

Ⓓ Actions are recorded in the usual location.

Ⓔ Only rules that result from simplification of the decision table are entered.

HOW TO DESCRIBE NETWORK REQUIREMENTS IN A REPOSITORY

Network models (location connectivity diagrams) were introduced in Chapter 10 as an effective tool for communicating geographical requirements for a business. The model depicted connectivity requirements between different locations. In this section we'll describe how to complete the specification of locations. The following annotations will help to clarify the sample repository entry in Figure 11.30.

Ⓐ Every location should have a name. The label corresponds to the name of a location appearing on a network model.

Ⓑ A location may be further described by separate repository entries. In this instance, the location explodes to a more detailed network model.

Ⓒ A network location should contain a brief description.

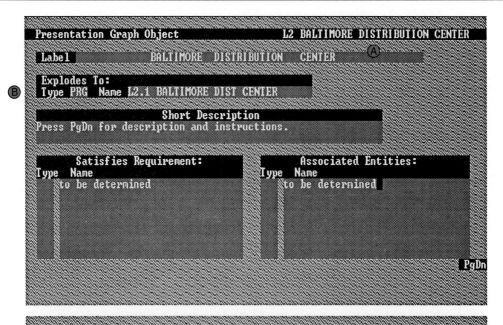

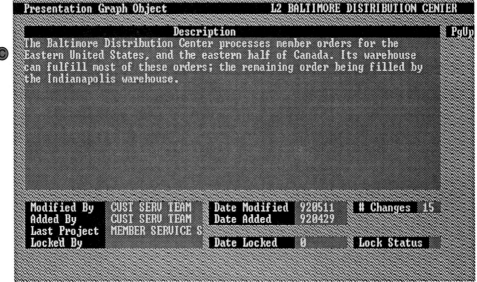

FIGURE 11.30   **Project Repository Entry for a Location**

Summary

A project repository is a catalog of details for a new information system's data, process, and network requirements. The principal purpose of the repository is to record details about objects appearing on essential data, process, and network models.

Data objects (i.e., data entity, data flow, data store) consist of data attributes—that is, the smallest unit of data or information that has any meaning to an end-user. Data attributes are arranged in specific patterns, called data structures, that describe a single occurrence of the object. These data structures, along with other logical attributes, are recorded in the repository.

The content of any occurrence of a data object can be documented as combinations of three data structures: a sequence of data attributes, the selection from a list of data attributes, and the repetition of one or more elements.

Process objects are documented in the repository as policies and procedures. Policies specify management decisions and rules. Procedures execute those policies as well as the tasks that support day-to-day business operations and management. Although most existing procedures are specified in ordinary English, we have learned that most such procedures are often unclear and incomplete.

Fortunately, two tools can help us specify policies and procedures more effectively. Decision tables are particularly effective for policies and for procedures that contain complex combinations of decisions. Structured English is a procedure specification tool that overcomes many of the limitations and ambiguities of ordinary English. Together, these two tools complement process models. Furthermore, these tools help analysts to communicate business requirements to computer programmers and end-users alike!

A project repository is normally initiated during the study phase of the systems development life cycle; however, it expands most rapidly during the definition phase, when end-user requirements are the central focus. It remains the principal source of facts throughout the life cycle.

Project repositories can be implemented manually, with forms, or automatically, using CASE tools. Repository entries should be defined for all objects appearing on data, process, and network models.

Key Terms

action diagram, p. 447

action stub, p. 438

alphabetic code, p. 459

block code, p. 459

code, p. 459

condition stub, p. 438

data structure, p. 430

decision construct, p. 449

decision table, p. 438

group code, p. 459

hierarchical code, p. 460

indifferent condition, p. 444

policy, p. 436

procedure, p. 436

project repository, p. 426

pseudocode, p. 446

repetition construct, p. 449

rules, p. 438

sequential code, p. 459

sequential construct, p. 448

serial code, p. 459

significant position code, p. 460

Tight English, p. 447

Questions, Problems, and Exercises

1. Why is a project repository a valuable systems analysis tool? What are the possible consequences of not creating a project repository during systems analysis?

2. Can you think of any specific times that a project repository might have been helpful when you were writing a computer program? Can you think of a situation in which you misinterpreted a computer program requirement because you didn't know something that could have been recorded in a project repository?

3. Dig out your last computer program. Prepare project repository entries for the following:
   a. Inputs (data flows).
   b. Outputs (data flows).
   c. Files or database (data store).
   d. All variables or fields (data attributes).

4. Describe the relationship between a project repository and data and process models.

5. You have compiled a complete project repository for your new inventory control system. It is time to verify the contents of three summary reports specified for your end-users. Each data flow (report), record (data structure), and data attribute should be reviewed. Unfortunately, the entire repository is 373 pages long. You can't mark the relevant pages and "thumb" back and forth between pages during your review. How should the report specifications be presented?

6. Why shouldn't a project repository be organized alphabetically independent of type—for example, data flow, data store, data attribute, and so on—like a traditional repository such as a dictionary?

7. Using the English data structure notation in this chapter, create a project repository entry for the following:
    a. Your driver's license.
    b. Your course registration form.
    c. Your class schedule.
    d. IRS Form 1040 (any version).
    e. An account statement and invoice for a credit card.
    f. Your telephone, electric, or gas bill.
    g. An order form in a catalog.
    h. An application for anything (e.g., insurance, housing).
    i. A retail store catalog.
    j. A typical real estate listing.
    k. A computer printout from a business office or computer course.
    l. A catalog that describes the classes to be offered next semester.
    m. Your checkbook.
    n. Your bank statement.

8. Repeat Exercise 7 using the algebraic data structure notation.

9. Select one of the data structures you developed in Exercise 7 or 8 and complete a set of data attribute repository entries for each attribute appearing in the project repository.

10. During the study phase of systems analysis, the analyst must gather facts concerning both the manual and automated portions of the system. Why would it be desirable for a systems analyst to obtain samples of the existing computer files and computer-generated outputs? What value would project repository entries for computer files and computer-generated outputs have during systems analysis?

11. Visit a local business or school office. Ask for samples of five business forms, or regular reports. Prepare complete project repository entries for each sample. If possible, review your entries with the end-users. Did they find your entries easy to read and understand? Can they think of additional data attributes that would make their job easier? Add these attributes to your project repository entries.

12. Find an example of a business code, possibly on the forms from Problem 11. Make a data attribute project repository entry for that code. Analyze that code according to the guidelines presented in this chapter. Can you suggest a better coding scheme?

13. Explain the difference between a policy and a procedure. Give an example of a policy and the procedure to implement or administer that policy.

14. Obtain a formal statement of a policy and procedure, such as a policy for a credit card. Evaluate the policy and procedure statement in terms of the common specification problems identified earlier in this chapter.

15. Reconstruct the policy and procedure used in Exercise 14 with the tools you learned in this chapter. Simplify the decision table.

16. Simplify the following decision table:

	Rules											
	1	**2**	**3**	**4**	**5**	**6**	**7**	**8**	**9**	**10**	**11**	**12**
Condition 1	Y	N	Y	N	Y	N	Y	N	Y	N	Y	N
Condition 2	Y	N	Y	N	Y	N	Y	N	Y	N	Y	N
Condition 3	A	A	A	A	B	B	B	B	C	C	C	C
Action 1			X				X				X	
Action 2	X				X				X			
Action 3		X				X						
Action 4				X				X		X		X

17. Were you able to combine rules 2 and 6 in Problem 16? Why?

18. Write a mini spec (Structured English) for balancing your checkbook.

19. Produce a mini spec to describe how to prepare your favorite recipe, tune a car, or perform some other familiar task. Ask a novice to perform the task, working from your specification.

Projects and Minicases

1. Create a simple project repository using your local word processor or line editor. Try to implement standard forms that can be read into the repository (to initiate new entity types, data flows, data stores, records, data attributes, and table of codes entries).

2. Through information systems trade journals, research a commercial CASE product. Evaluate that package's project repository. Can you define both information and process models? Can you describe data structures to the repository? Can you describe individual data attributes to the repository? What types of analytical reports can be generated from the repository?

3. The district sales manager for Grayson Industries, Inc. has requested an improved sales analysis report that is based on the following information:

   Grayson Industries' sales regions cover four territories, each divided into two to four districts. There are 45 salespersons, and each salesperson is assigned to one and only one district.

   Prepare a project repository entry to describe the content of the requested sales analysis report. Why would developing a project repository be beneficial even though the manager already has an idea of what he wants? What additional information concerning this report should be included in the project repository before the layout of the final report is designed?

4. Identify and select a system from which you can gather three to five related forms. Discuss the forms with someone who is knowledgeable about them. Then, prepare an essential project repository for the consolidated forms. The forms correspond to data flows. Define contents as records, and be sure to define all data attributes. No flow, record, or attribute should be redundantly stored, even if it exists redundantly on the forms under separate names. Be sure to use logical, implementation-independent naming conventions. Review your essential repository with your initial contact person. Try to evaluate how well the repository describes the essence of the forms.

5. Obtain a copy of your last programming assignment. Using Structured English, write procedure specifications to clearly communicate the procedures that you were asked to implement. Did you avoid unnecessary details? Does your Structured English take on an unnecessary programming accent? Does your Structured English accurately communicate the processing procedures that are to be implemented as a computer program? Did you avoid programming-dependent statements, such as OPEN and PERFORM? Would you have appreciated such a description when you were originally given the assignment?

6. Prepare a decision table that accurately reflects the following course grading policy:
   A student may receive a final course grade of A, B, C, D, or F. In deriving the student's final course grade, the instructor first determines an initial or tentative grade for the student, which is determined in the following manner:

   A student who has scored a total of no lower than 90 percent on the first three assignments and exams and received a score no lower than 70 percent on the fourth assignment will receive an initial grade of A for the course. A student who has scored a total lower than 90 percent but no lower than 80 percent on the first three assignments and exams and received a score no lower than 70 percent on the fourth assignment will receive an initial grade of B for the course. A student who has scored a total lower than 80 percent but no lower than 70 percent on the first three assignments and exams and received a score no lower than 70 percent on the fourth assignment will receive an initial grade of C for the course. A student who has scored a total lower than 70 percent but no lower than 60 percent on the first three assignments and exams and received a score no lower than 70 percent on the fourth assignment will receive an initial grade of D for the course. A student who has scored a total lower than 60 percent on the first three assignments and exams, or received a score lower than 70 percent on the fourth assignment, will receive an initial and final grade of F for the course. Once the instructor has determined the initial course grade for the student, the final course grade will be determined. The student's final course grade will be the same as his or her initial course grade if no more than three class periods during the semester were missed. Otherwise, the student's final course grade will be one letter grade lower than his or her initial course grade (for example, an A will become a B).

   Are there any conditions for which there was no action specified for the instructor to take? If so, what would you do to correct the problem? Can your decision table be simplified? If so, simplify it.

Suggested Readings

Copi, I. R. *Introduction to Logic.* New York: Macmillan, 1972. Copi provides a number of problem-solving illustrations and exercises that aid in the study of logic. The poker chip problem in our chapter case was adapted from one of Copi's reasoning exercises.

DeMarco, Tom. *Structured Analysis and System Specification.* Englewood Cliffs, N.J.: Prentice Hall, 1978. This classic book introduced Structured English. Chapters 11 through 14 present the most comprehensive treatment of the project repository that we've seen thus far. DeMarco uses an algebraic notation for specifying data structures.

Gane, Chris, and Trish Sarson. *Structured Systems Analysis: Tools and Techniques.* Englewood Cliffs, N.J.: Prentice Hall, 1979. Gane and Sarson include an entire chapter on defining process logic. In addition to discussing decision tables and Structured English, this chapter explains how decision trees and pseudocode can be used to define process logic.

Gildersleeve, T. R. *Successful Data Processing Systems Analysis.* Englewood Cliffs, N.J.: Prentice Hall, 1978. The first edition of this book includes an entire chapter on the construction of decision tables. Gildersleeve does an excellent job of demonstrating how narrative process descriptions can be translated into condition and action entries in decision tables. Unfortunately, the chapter was deleted from the second edition.

Martin, James, and Carma McClure. *Action Diagrams: Towards Clearly Specified Programs.* Englewood Cliffs, N.J.: Prentice Hall, 1986. This book describes a formal grammar of Structured

English that encourages the natural progression of a process (program) from Structured English to code. Action diagrams are supported directly through CASE tools sold by KnowledgeWare, Inc.

Matthies, Leslie H. *The New Playscript Procedure.* Stamford, Conn.: Office Publications, 1977. This book provides a thorough explanation and examples of the weaknesses of the English language as a tool for specifying business procedures.

art Three introduces you to the systems design phases of systems development. Seven chapters make up this unit. First, Chapter 12 expands on the three systems design phases of systems development: selection, acquisition, and design and integration. Systems design is presented as consisting of the first phases of systems development to begin to address the implementation aspects of the target system. The chapter also introduces several tools and techniques that will be covered in greater detail in subsequent chapters and cross life cycle modules.

Chapter 13 teaches a procedure that prepares the data model, developed during systems planning and systems analysis, for implementation by ensuring that the model is simple, flexible, and nonredundant. This procedure, called data analysis, results in data entities that are normalized into third normal form. The chapter also presents event analysis, a technique to identify business events which cause data to be added, modified, or deleted. This is important because it helps to ensure that the model is kept accurate and current at all times and that all necessary maintenance processing for our data entities has been specified.

Chapter 14 presents the process analysis and design to transform essential DFD's into implementation data flow diagrams (or system flowcharts). This high-level, general design serves as a blueprint for detailed design in the form of design units. The design units can be assigned to different teams for detailed design, construction, and unit testing. This process ensures that the appropriate data and processing will be distributed throughout the business as needed.

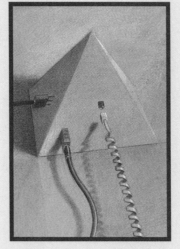

Chapter 15 introduces the design of physical data stores from the data model developed earlier. This overview of file and database design teaches both traditional file based methods and the modern database approach to data storage design. Internal controls such as access, retention, and backup and recovery are presented. Database management systems are introduced. The chapter also includes a discussion on the new client/server approach.

Chapter 16 teaches the physical design and implementation of both computer inputs and computer outputs. Important components of input and output design such as formats, methods, media, human factors, and internal controls are stressed. The chapter also emphasizes prototyping as a way of finding, documenting, and communicating design requirements.

Chapter 17 presents user interface design, one of the most important and visible components of the system. The design of the user interface is crucial because user acceptance of the system is frequently dependent upon a friendly, easy-to-use interface. Several techniques for controlling dialogue are presented, as well as some useful human engineering guidelines.

Finally, Chapter 18 introduces program design as two activities: modular design and packaging. Several popular tools and strategies for modular design are presented, including the authors' common sense approach. Finally, packaging is presented to ensure that complete specifications are passed on to computer programmers in the implementation phase.

SYSTEMS DESIGN

12

Systems Design

Chapter Preview and Objectives

This chapter expands on the systems design phases that were introduced in Chapter 3, A Systems Development Life Cycle. This process consists of three phases: selection, acquisition, and design and integration. You will have mastered the complete systems design process when you can:

Define
the systems design process in terms of the selection, acquisition, and design and integration phases of the life cycle.

Describe
selection, acquisition, and design and integration phases in terms of your information building blocks.

Describe
the selection, acquisition, and design and integration phases in terms of objectives, activities, roles, inputs and outputs, and tools and techniques.

Identify
those chapters and modules in this textbook that can help you actually perform the activities of systems design.

Describe
traditional and prototyping approaches to systems design.

SCHUSTER AND PETRIE, INC.

Scene: *Keith Stallard is a relatively new programmer/ analyst at Schuster and Petrie, Inc. Having spent two years as a programmer, he was promoted one year ago. The Information Services division of S & P requires job performance reviews twice a year. Tim Hayes, associate director of Financial Systems, has scheduled a job performance review with Keith.*

Tim: Well, Keith, do you still want this programmer/analyst job? You've had about six months to get used to your new responsibilities.

Keith: More than ever! Now that I've had a taste of systems work, I know it's right for me. I assume this meeting will determine if I'm making progress. How am I doing?

Tim: You're right. I've discussed your performance on the job cost accounting system project with both your supervisor and your key user contact. Your technical design statement was quite impressive. But I have to ask you, where did you learn to complete such thorough design specifications? The implementation appears to be moving along more smoothly than expected, largely because of your specifications.

Keith: I did a lot of reading in systems analysis and design textbooks at the college library. I also queried both users and programmers about problems with typical specifications. My own experience as a programmer has influenced my specifications. But to be honest, I was really embarrassed by my performance on the account aging project. That's why I did all those things!

Tim: I don't understand. We gave you acceptable ratings. The account aging project was a little off schedule, but that's the only problem I recall.

Keith: It was a little more complicated than that. Bill had done the systems analysis and was supervising me since it was my first experience with systems design. But he had to be called off the project to fix a major flaw in another system. I kept working on the design and passed the design document to Rita [a programmer]. Then lightning struck for a second time. Rita had to go into the hospital, and I had to assume her programming responsibilities. It was the first time I ever had to cut code from my own specifications. There were so many details, and I hadn't documented all of them. Surprisingly, I couldn't even remember all of the thought process that went into my own specifications. Now I know how the maintenance programmers feel!

Eventually I got the system up and running— only to find out that some of the reports were not acceptable to my users. The content was there, but the format was wrong. I had to take certain liberties with the format. In my school days, that was what we did with all programming assignments. I just didn't appreciate the importance of user involvement in the design process. I assumed that systems analysis took care of all the user issues. And then, the Internal Audit department got ahold of my design specifications. They didn't like them at all! There weren't enough internal controls to satisfy their standards. By that time, I had half the programs written and tested. I had to redesign many system components and rewrite several affected programs. To make a long story short, I never want to go through that kind of design experience again. So I learned about systems design.

Tim: And you still want to be an analyst? After all that?

Keith: Yes. Despite the problems, I found the work to be so much more satisfying than programming. I knew it wasn't going to be easy. But it was enjoyable.

Tim: Well, we've discussed your strengths. But we do need to work on a few things. First, as you know, most of our older systems are being converted to on-line systems using databases. I'm sending you to a one-week intensive course at

IBM. There you will learn about their DB2 database management system. I also want you to go through our user interface course the next time it's offered. I don't know if you've heard that the on-line interface on your accounts aging system hasn't lived up to expectations. You also need to work on your writing and speaking skills. The report you did to sell the new job costing system wasn't well organized, was too wordy, and contained numerous grammatical errors and typos. You were lucky that Bill got it before your users. You might have lost the sale. And your presentation of that system to management could have gone a little smoother. Public speaking is tough, I realize that. But you did not seem confident. You made a good recommendation! But if you don't seem confident and comfortable with your own recommendation, how will management feel about it? We did get it through, though. But your communications skills need improvement . . . especially if you want my job in the future. You have that potential, Keith. Don't waste it!

Keith: I understand. And I appreciate your honesty. I've suspected the problem. I guess I never took those English and communications courses seri-

ously. I'll get enrolled in some evening continuing education courses for the next term. I'm not going to let poor communications skills get in the way of my future.

Tim: Let's get to the bottom line, Keith. You've shown better than average progress in your new assignment. That's why, effective next month, you'll see a little increase in your paycheck. If you keep up the good work and improve in the areas we've outlined, I'm certain you'll be promoted to systems analyst within two years. Now, let's talk about design specification some more. Do you think we could teach our other analysts to do that?

Discussion Questions

1. Think back to your programming courses (or experiences). What are some of the problems you've had responding to programming assignments?

2. What did Keith learn about working from his own specifications?

3. As a systems analyst working on the design phase of a project, what types of people did Keith have to communicate with? Why does communication become tougher during systems design than during systems analysis?

WHAT IS SYSTEMS DESIGN?

Systems design was first defined in Chapter 3:

> **Systems design** is the evaluation of alternative solutions and the specification of a detailed computer-based solution. It is also called *physical design.*

The key term here is *design.* Whereas systems analysis primarily focused on the logical, implementation-*independent* aspects of a system (the requirements), systems design deals with the physical or implementation-*dependent* aspects of a system (the system's technical specifications).

Systems design is driven by various system designers, including the systems analysts. Hence, in the pyramid model, it addresses PEOPLE, DATA, ACTIVITIES, and NETWORKS from the system designer's perspective. What about the TECHNOLOGY building block? Often, that technology is in place, or specified by a predefined technology architecture (from systems planning). In other cases, the analyst must select or supplement the technology. In all cases, systems design builds on the knowledge derived from systems planning and systems analysis.

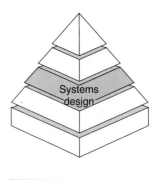

Most of us place too restrictive a definition on the process of design. We envision ourselves drawing blueprints of the computer-based systems to be programmed and developed by ourselves or our own programmers. Thus, we

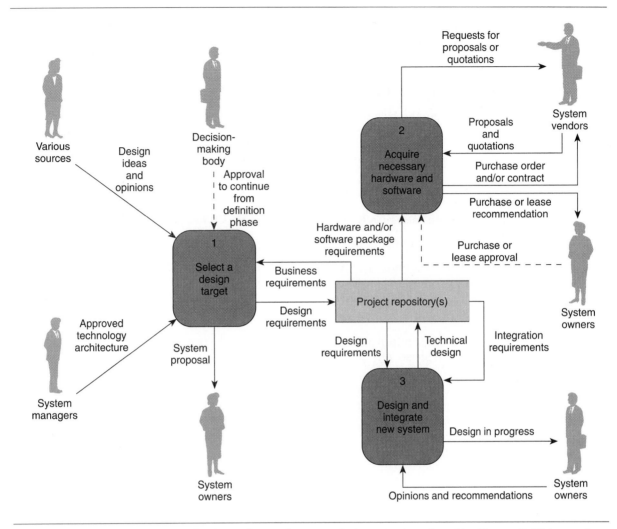

FIGURE 12.1 Systems Design The three phases of systems design were introduced in Chapter 3. We've made some revisions here. The deliverables don't really pass from phase to phase. Instead, they are recorded into various project repositories for use in later phases.

design inputs, outputs, files, databases, and other computer components. Recruiters of computer-educated graduates refer to this restrictive definition as the "not-invented-here syndrome." In reality, many companies purchase more software than they write in-house. That shouldn't surprise you. Why reinvent the wheel? Many systems are sufficiently generic that computer vendors have written adequate — but rarely, if ever, perfect — software packages that can be bought and possibly modified to fulfill end-user requirements.

Consequently, we must expand our definition of the design process to include the evaluation and selection of alternative solutions and the acquisition or purchase of computer software and hardware — as well as the more traditional physical design and integration of computer-based components. Figure 12.1 depicts these three phases of systems design. This chapter will examine each of the phases in greater detail. To give you the feel of a true systems development methodology, each phase is described in terms of the: (1) purpose of the phase, (2) activities that should be performed, (3) roles played by various people in each activity, (4) inputs and outputs for each activity, and (5) techniques and skills that can be used to complete each activity.

THE SELECTION PHASE OF SYSTEMS DESIGN

Given the business requirements for an improved information system, we can finally address how the new system — including computer-based alternatives — might operate. You should never automatically go with your first hunch. During the selection phase, it is imperative that you identify options, analyze options, and then sell feasible solutions based on the analysis.

Building Blocks for the Selection Phase

There are two fundamental objectives in the selection phase:

1. To identify and research alternative manual and computer-based solutions to support our target information system.
2. To evaluate the feasibility of alternative solutions and recommend the best overall alternative solution.

Your information system building blocks provide a framework for the selection phase. Alternative solutions to be considered should be those that address the business requirements of the information system. Recall that in the definition phase, we established PEOPLE, DATA, ACTIVITIES, and NETWORK requirements for the target system. The selection phase marks the first point in the systems development process that we have placed emphasis on how the new system might operate. Thus, we will address how TECHNOLOGY may be used to support the target system.

Selection Phase Activities, Roles, and Techniques

Figure 12.2 illustrates the typical activities of the selection phase. This section details each activity in the selection phase, the roles that users and other individuals play, and some popular techniques for selecting a design target.

Activity 1: Specify Alternative Solutions

As is shown in Figure 12.2, the selection phase is triggered by an approval to continue from the definition phase. This is consistent with our "creeping commitment" approach to systems development.

Given the business requirements established in the definition phase, we must identify alternative candidate solutions. Some candidate solutions will be posed by design ideas and opinions from system owners and users. Others may come from various sources including: systems analysts, systems designers, technical consultants, and other IS professionals. Some technical choices may be limited by a predefined, approved technology architecture provided by system managers.

The amount of information describing the characteristics of any one candidate solution may be overwhelming. A matrix is a useful tool for effectively capturing, organizing, and communicating the characteristics for candidate

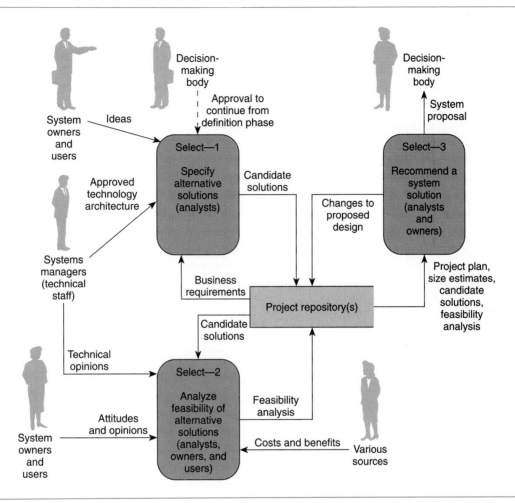

FIGURE 12.2 **Selection Phase Activities** These activities will lead to completion of the selection phase. The activities are generally completed in a counterclockwise sequence from the top; however, the project repository allows you to overlap activities or return to any previous activity.

solutions. A partially completed candidate matrix is depicted in Figure 12.3. This allows for a side-by-side comparison of the different characteristics for a number of candidates.

Activity 2: Analyze Feasibility of Alternative Solutions

Once alternative candidate design solutions have been identified, each candidate must be analyzed for feasibility. Feasibility analysis should not be limited to costs and benefits. Most analysts evaluate solutions against four sets of criteria:

1. **Technical feasibility.** Is the solution technically practical? Does our staff have the technical expertise to design and build this solution?
2. **Operational feasibility.** Will the solution fulfill the user's requirements? To what degree? How will the solution change the user's work environment? How do users feel about such a solution?

Characteristics	Candidate 1	Candidate 2	Candidate 3
Portion of System Computerized Brief description of that portion of the system that would be computerized in this candidate.	The scheduling and reporting subsystems would both be computerized.	Same as Candidate 1.	Same as Candidate 1.
Benefits Brief description of that portion of the system that would be computerized in this candidate.	Scheduling: This candidate will allow the schedules of all social workers to be consolidated. This will allow for easy identification of available meeting times. Schedules could be consolidated based on a number of options including, by day, week, or month. Reporting: Case/meeting information would be made readily available for each social worker. Thus, government and internal reporting requirements would be more easily fulfilled.	Scheduling: Same as Candidate 1. However, this candidate will also allow adhoc social worker schedule inquiries based upon a number of "subjects." Reporting: Same as Candidate 1.	Scheduling: Same as Candidate 2. Reporting: Same as Candidate 1.
Software Tools/Applications Needed Software tools needed to design or build the candidate (e.g., database management system, spreadsheet, word processor, terminal emulators, programming languages, etc.). Also, a brief description of software to be purchased, built, accessed, or some combination of these techniques.	This candidate would require that the scheduling subsystem be "purchased" in-house. The reporting subsystem would be built using spreadsheet template(s).	Same as Candidate 1 except: the scheduling subsystem would also be "built" in-house. The scheduling subsystem would be built using a database management system.	Both the scheduling and reporting subsystems would be "purchased."

FIGURE 12.3   **Partially Completed Candidate Matrix**   A matrix is a very useful tool for specifying characteristics for alternative candidate solutions. Notice that the first column is used to identify the type of characteristics.

3. **Economic feasibility.** Is the solution cost-effective?
4. **Schedule feasibility.** Can the solution be designed and implemented within an acceptable time period?

Systems analysts must work very closely with system owners and users when evaluating candidate solutions. The attitudes and opinions of system owners and users are particularly vital inputs toward measuring the operational feasibility of candidates. The resulting feasibility analysis often reflects the input from a number of additional individuals. To aid in measuring the technical feasibility of candidates, the technical opinions of system managers and technical staff are often sought. Also, costs and benefit information may be provided by a number of various sources.

Once again, a matrix can be used to communicate the large volume of information about candidate solutions. The matrix in Figure 12.4 allows for a side-by-side unveiling of the different feasibility analyses for a number of candidates. When completing the feasibility matrix, the analyst and users are careful not to make comparisons between the candidates. The feasibility analysis is performed upon each individual candidate without regard to the feasibility of other candidates. This approach discourages the analyst and users from prematurely making a decision concerning which candidate is the best.

The ability to perform a feasibility assessment is an extremely important skill requirement. Feasibility assessment techniques and skills are more fully covered in Part Five, Module C, Feasibility Analysis.

Feasibility Criteria	Candidate 1	Candidate 2	Candidate 3
Operational Feasibility Brief description of the functionality: to what degree the candidate would benefit the organization and how well the system will work. Also, a brief description of the political feasibility: how well-received the solution would be from the owner's and user's perspectives.	A brief survey of scheduling packages revealed that such packages can provide the users with improved accessibility to information concerning social workers and cases/meetings. This solution should decrease the amount of time needed to schedule social workers. It is felt that management would be satisfied with this candidate only if the direct system users find the packaged application to their satisfaction.	Same as Candidate 1, except a few users will find the capability to do adhoc inquiries according to "subjects' of particular benefit.	Same as Candidate 2.
Technical Feasibility Brief assessment of the maturity, availability, and desirability of the computer technology needed to support the candidate. Also, an assessment of the technical expertise needed to develop, operate, and maintain the candidate.	There are numerous, highly rated scheduling packages available to date. Once the system users have been properly trained in the application, expertise requirements would be minimal. The same is true for spreadsheet reporting software and application.	The technology and expertise to build the scheduling and reporting subsystems are readily available.	Same as Candidate 1. There is also the added concern that no existing packages provides needed support for the reporting subsystem.
Economic Feasibility Cost to develop: Payback period (discounted): Net present value: Detailed calculations:	Approximately $1,000. Approximately 6 months. Approximately $8,300. See Attachment A.	Approximately $2,700. Approximately 2.5 years. Approximately $5,500. See Attachment B.	Approximately $1,500. Approximately 7 months. Approximately $9,000. See Attachment C.
Schedule Feasibility An assessment of how long the solution will take to design and implement.	Approximately 3 months.	Approximately 9 months.	Approximately 4 months.

FIGURE 12.4   Partially Completed Feasibility Matrix   A matrix is a very useful tool for specifying feasibility analyses for alternative candidate solutions. Notice that the first column is used to identify the feasibility criteria.

Activity 3: Recommend a System Solution

Once the feasibility analysis has been completed for each candidate solution, we can now select a candidate solution to recommend. First, any infeasible candidates are usually eliminated from further consideration. Since we are looking for the most feasible solution of those remaining, we will identify and recommend the candidate that offers the best overall combination of technical, operational, economic, and schedule feasibilities.

Once again, a matrix tool can be used to aid in decision making. A matrix may simply record the feasibility score (e.g., 85 percent on a 1 percent to 100 percent scale) or ranking (first, second, third, etc.) for each candidate. A sample rankings matrix is depicted in Figure 12.5. The analyst and users can determine a proper weight that should be given to the feasibility criteria. Once that has been established, the matrix provides for side-by-side comparison of each candidate to more easily identify the best overall choice.

The key deliverable for the selection phase is a formal written or verbal system proposal. This proposal is usually intended for the system owners who will normally make the final decision. The proposal will contain the project plans, size estimates, candidate solutions, and feasibility analysis. Based on the outcome of the proposal, changes to proposed design requirements are established for the new systems components we will "buy" or "make."

Feasibility Criteria	Candidate 1	Candidate 2	Candidate 3
Operational Feasibility	85	90	87
Technical Feasibility	90	90	87
Economic Feasibility	86	75	92
Scheduling Feasibility	90	82	90

FIGURE 12.5 **Sample Rankings Matrix**   A matrix is a very useful tool for specifying the candidate that offers the "best" overall combination of feasibilities.

Finally, the techniques and skills needed to complete this activity are all cross life cycle skills:

- Feasibility assessment (see Part Five, Module C, Feasibility Analysis).
- Report writing (see Part Five, Module D, Interpersonal Skills).
- Verbal presentations (see Part Five, Module D).

THE ACQUISITION PHASE OF SYSTEMS DESIGN

The acquisition of software and hardware (computer equipment) is not necessary for all new systems. On the other hand, when new software or hardware is needed, the selection of appropriate products is often difficult. Decisions are complicated by technical, economic, and political considerations. A poor decision can ruin an otherwise successful analysis and design. The systems analyst is becoming increasingly involved in the acquisition of software packages, peripherals, and computers to support specific applications being developed by that analyst.

Building Blocks for the Acquisition Phase

There are four fundamental objectives of the acquisition phase:

1. To identify and research specific products that could support our recommended solution for the target information system.
2. To solicit, evaluate, and rank vendor proposals.
3. To select and recommend the best vendor proposal.
4. To establish requirements for integrating the awarded vendor's products.

Here again, your information system building blocks provide a framework for understanding the acquisition phase. Alternative vendor hardware or software products to be considered should be those that provide the best overall support for the target information system. The acquisition phase will address specific build/buy TECHNOLOGY products that support the PEOPLE, DATA, ACTIVITIES, and NETWORK requirements for the target system.

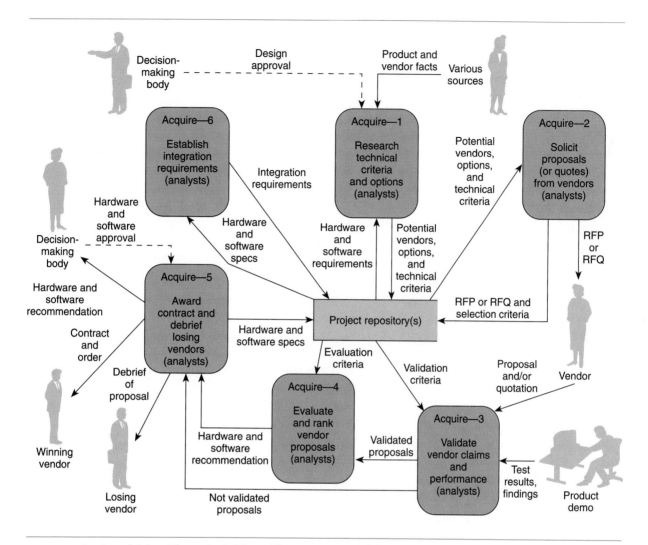

FIGURE 12.6   Acquisition Phase Activities   These activities will lead to completion of the acquisition phase. Once again, the activities are generally completed in a clockwise sequence from the top; however, the project repository allows you to overlap activities or return to any previous activity.

Acquisition Phase Activities, Roles, and Techniques

Figure 12.6 illustrates the typical activities of the acquisition phase. The acquisition phase is optional — that is, it is triggered by a design approval requiring new hardware or software. Let's examine each acquisition phase activity and the roles that users and other individuals play. In addition, let's learn about some popular techniques for selecting a design target.

Activity 1: Research Technical Criteria and Options

The first activity is to research technical alternatives. This task identifies specifications that are important to the hardware and/or software that is to be selected. The task responds to the hardware and/or software requirements established in the selection phase. These requirements specify the functionality, features, and critical performance parameters. Examples of criteria for hardware and software are listed in the margin.

To complete this task analysts must do their homework. They obtain product and vendor facts from various sources. They are careful not to get their information solely from a salesperson — not that sales representatives are dishonest, but the number one rule of salesmanship is to emphasize your product's strengths and deemphasize its weaknesses.

Most analysts read appropriate magazines and journals to help them identify those technical and business issues and specifications that will become important to the selection decision. Other sources of information for conducting research include the following:

- *Internal standards* may exist for hardware and software selection. Some companies insist that certain technology will be bought from specific vendors if those vendors offer it. For instance, some companies have standardized on specific brands of microcomputers, terminals, printers, database management systems, network managers, data communications software, spreadsheets, and programming languages. A little homework here can save you a lot of unnecessary research.

- *Information services* are primarily intended to constantly survey the marketplace for new products and advise prospective buyers on what specifications to consider. They also provide information such as the number of installations and general customer satisfaction with the products. Some information services are listed in the margin.

- *Trade newspapers and periodicals* offer articles and experiences on various types of hardware and software that you may be considering. Some examples of these are listed in the margin. Many can be found in school and company libraries. Subscriptions (sometimes free) are also available.

The research should also identify potential vendors that supply the products to be considered. After the analysts have completed their homework, they will initiate contact with these vendors. Thus, the analysts will be better equipped to deal with vendor sales pitches after doing their research!

Activity 2: Solicit Proposals (or Quotes) from Vendors

Given the potential vendors, options, and technical criteria, your next task is to solicit proposals from vendors. If your company is committed to buying from a single source (IBM, for example), the task is quite informal. You simply contact the supplier and request price quotations and terms. On the other hand, most decisions offer numerous alternatives. In this situation, good business sense dictates that you use the competitive marketplace to your advantage.

The solicitation task requires the preparation of one of two documents: a request for quotations (RFQ) or a request for proposals (RFP). The **request for quotations** is used when you have already decided on the specific product, but that product can be acquired from several distributors. Its primary intent is to solicit specific configurations, prices, maintenance agreements, conditions regarding changes made by buyers, and servicing. The **request for proposals** is used when several different vendors and/or products are candidates and you

---

✓

Technical Criteria

Quality of documentation
Ease of learning
Ease of use
Response time
Throughput
Number of installed copies
Number of improved
  versions over time
  (maturity)
Licensing arrangements
Training
Maximum file database size
Internal controls

✓

Information Services

Data Pro
EDP Auerbach
International Computer
  Programs
The Source

✓

Trade Publications

Computerworld (weekly)
InfoWorld (weekly)
Information Week (weekly)
Datamation (monthly)
Computer Decisions
  (monthly)
Infosystems (monthly)
Mini-Micro Systems
  (monthly)
Communications Week
  (weekly)
PC (biweekly)
PC Week (weekly)
PC World (monthly)

Request for Proposals

I. Introduction.
   A. Background.
   B. Brief summary of needs.
   C. Explanation of RFP document.
   D. Call for action on part of vendor.

II. Standards and instructions.
   A. Schedule of events leading to contract.
   B. Ground rules that will govern selection decision.
      1. Who may talk with whom and when.
      2. Who pays for what.
      3. Required format for a proposal.
      4. Demonstration expectations.
      5. Contractual expectations.
      6. References expected.
      7. Documentation expectations.

III. Requirements and features.
   A. Hardware.
      1. Mandatory requirements, features, and criteria.
      2. Essential requirements, features, and criteria.
      3. Desirable requirements, features, and criteria.
   B. Software.
      1. Mandatory requirements, features, and criteria.
      2. Essential requirements, features, and criteria.
      3. Desirable requirements, features, and criteria.
   C. Service.
      1. Mandatory requirements.
      2. Essential requirements.
      3. Desirable requirements.

IV. Technical questionnaires.

V. Conclusion.

FIGURE 12.7 Request for Proposals (RFP) This is an outline for a typical request for proposals. The outline for a request for quotations would be similar; however, the RFQ's requirements are more technical and don't allow the vendor as much flexibility to tailor alternatives to the customer's business.

want to solicit competitive proposals and quotes. RFPs can be thought of as a superset of RFQs. We'll address the RFP for the remainder of this task description. Both serve to define <u>selection criteria</u>.

The quality of an RFP has a significant impact on the quality and completeness of the resulting proposals. A suggested outline for an RFP is presented in Figure 12.7, since an actual RFP is too lengthy to include in this book. Obviously, your ability to write clearly will affect the quality of proposals you get in response to your RFP. Furthermore, you can expect that any RFP will raise additional questions that you will address in meetings and other communications with prospective vendors. Therefore, verbal communication skills will also be tested in this task.

The primary purpose of the RFP is to communicate requirements and desired features to prospective vendors. Requirements and desired features must be categorized as *mandatory* (must be provided by the vendor), *extremely important* (desired from the vendor but can be obtained in-house or from a third-party vendor), or *desirable* (can be done without). Requirements might also be classified by two alternate criteria: those that satisfy the needs of the systems and those that satisfy our needs from the vendor (for example, service).

Entire books could be written on the flow and processing of RFPs and RFQs (see Suggested Readings at the end of the chapter). Mechanisms must be implemented to answer vendor questions and control the format of the vendor's subsequent proposals. Often, vendors are invited to a bidder's meeting where common questions and issues can be addressed. Ultimately, interested vendors will submit proposals. The remaining tasks address the analysis of proposals.

Many of the skills you developed in Part Two, such as process and data modeling, can be very useful for communicating requirements in the RFP. We have found that vendors are very receptive to these tools because they find it easier to match products and options and package a proposal that is directed toward your needs. Other important skills include report writing (covered in Part Five, Module D, Interpersonal Skills) and developing questionnaires (covered in Part Five, Module B, Fact-Finding Techniques). These skills are extremely important since everyone benefits from a clear and complete statement of requirements.

Activity 3: Validate Vendor Claims and Performance

Soon after the RFPs or RFQs are sent to prospective vendors, you will begin receiving proposal(s) and/or quotation(s). Because proposals cannot and should not be taken at face value, claims and performance must be validated. This task is performed independently for each proposal; proposals are not compared with one another.

Eliminate any proposal that does not meet all of your mandatory requirements. If you clearly specified your requirements, no vendor should have submitted such a proposal. For proposals that cannot meet one or more extremely important requirements, verify that the requirements or features can be fulfilled by some other means. Finally, validate vendor claims and promises against validation criteria.

Claims about mandatory, extremely important, and desirable requirements and features can be validated by completed questionnaires and checklists (included in the RFP) with appropriate vendor-supplied references to user's and technical manuals. Promises can only be validated by ensuring that they are written into the contract. Performance is best validated by a demonstration, which is particularly important when you are evaluating software packages. Demonstrations allow you to obtain test results and findings that confirm capabilities, features, and ease of use.

Activity 4: Evaluate and Rank Vendor Proposals

The validated proposals can now be evaluated and ranked. The evaluation and ranking task is, in reality, another cost-benefit analysis performed during systems development. The evaluation criteria and scoring system should be established before the actual evaluation takes place so as not to bias the criteria and scoring to subconsciously favor any one proposal.

Some methods suggest that requirements be weighted on a point scale. Better approaches use dollars and cents! Monetary systems are easier to defend to management than points. One such technique is to evaluate the proposals on the basis of hard and soft dollars. Hard-dollar costs are the costs you will have to pay to the selected vendor for the equipment or software. Soft-dollar costs are additional costs you will incur if you select a particular vendor (for instance, if you select vendor A, you may incur an additional expense to vendor B to overcome a shortcoming of vendor A's proposed system). This approach awards the contract to the vendor who fulfills all essential requirements while offering the lowest total hard-dollar cost plus soft-dollar penalties for desired features not provided (for a detailed explanation of this method see Isshiki, 1982 or Joslin, 1977).

Once again the ability to perform a feasibility assessment is an extremely important skill requirement. Feasibility assessment techniques and skills are covered more fully in Part Five, Module C, Feasibility Analysis.

Activity 5: Award (or Let) Contract and Debrief Losing Vendors

Having ranked the proposals, the analyst usually presents a hardware and software recommendation for final approval. Once again, communication skills, especially salesmanship, are important if the analyst is to persuade management to follow the recommendations.

Once the final hardware and software approval decision is made, a contract must be negotiated with the winning vendor. Certain special conditions and terms may have to be written into the standard contract and order. Ideally, no computer contract should be signed without the advice of a lawyer. For microcomputers and software, legal advice can be prohibitively expensive (compared to the cost of the products themselves). In this case the analyst must be careful to read and clarify all licensing agreements. No final decision should be approved without the consent of a qualified accountant or management. Purchasing, leasing, and leasing with a purchase option involve complex tax considerations.

Out of common courtesy, and to maintain good relationships, the analyst may provide a debriefing of proposals for losing vendors. The purpose of this meeting is not to allow the vendors a second-chance opportunity to be awarded the contract; rather, the briefing is strictly intended to inform the losing vendors of precise weaknesses in their proposal and/or product(s).

Activity 6: Establish Integration Requirements

Finally, given the hardware and software specifications of the awarded vendor's products, the analysts must determine how they can be integrated with other existing information systems. It is not merely enough to purchase or build systems that fulfill the target system requirements. The analyst must integrate or interface the new system to the myriad of other existing systems that are essential to the business. Many of these systems may use dramatically different technology, techniques, and file structures.

The analyst must consider how the target system fits into the federation of systems of which it is a part. The integration requirements that are specified are vital to ensuring that the target system will work in harmony with those systems.

THE DESIGN AND INTEGRATION PHASE OF SYSTEMS DESIGN

Now we come to a more traditional phase of the systems design, the design and integration phase (see Figure 12.1). Given design and integration requirements for the target system, this phase involves developing technical design specifications.

Building Blocks for Completing the Design and Integration Phase

The goal of the design and integration phase is twofold. First and foremost, the analyst seeks to design a system that both fulfills requirements and will be friendly to its end-users. Human engineering will play a pivotal role during design. Second, and still very important, the analyst seeks to present clear and complete specifications to the computer programmers and technicians. Our information systems building blocks serve as a framework for completing the design phase:

- NETWORKS. During the systems analysis phase, we established the network requirements for the target system. Now we need to analyze and distribute the systems data and processes.
- DATA. We specified the content of each data and information flow during the definition phase. We specified the media during the selection phase. Now we need to design the style, organization, and format of all inputs and outputs. We also must specify format, organization, and access methods for all files and databases to be used in the computer-based system.
- ACTIVITIES. During design, the sequence of steps and flow of control through the new system must be specified. The processing methods and intermediate manual procedures must also be clearly documented.
- PEOPLE. The roles people play in the new system must be specified. For instance, who will capture and input data? Who will receive outputs?
- TECHNOLOGY. Although hardware is not selected or designed during the physical design phase, the hardware does constrain the system. The specific hardware configuration specified during the selection and acquisition phases must be considered as various other components are designed.

Clearly, the physical design phase gets into considerably greater detail than any of the previous phases of the life cycle.

Design and Integration Phase Activities, Roles, and Techniques

Figure 12.8 depicts the specific activities for the design and integration phase. Notice that the activities appearing on the figure have been divided into two parts—general design and detailed design. **General design** includes those activities that serve to develop an outline of the overall design of the target system. The **detailed design** activities are those activities that focus on developing the detailed design specifications for components in the outline. This distinction is important from a vocabulary standpoint since many methodologies view general and detailed design as separate phases.

Activity 1: Analyze and Distribute Data

Before the analyst can design computer files and/or databases for the target system, the analyst must perform some additional analysis and address distribution issues of the data. If you recall, a data model of the target system already exists. That model was created during systems analysis as a tool for communicating and documenting the data requirements for the target system. However, that data model does not usually represent a good file or database design. In fact, it may contain structural characteristics that may lead to numerous problems.

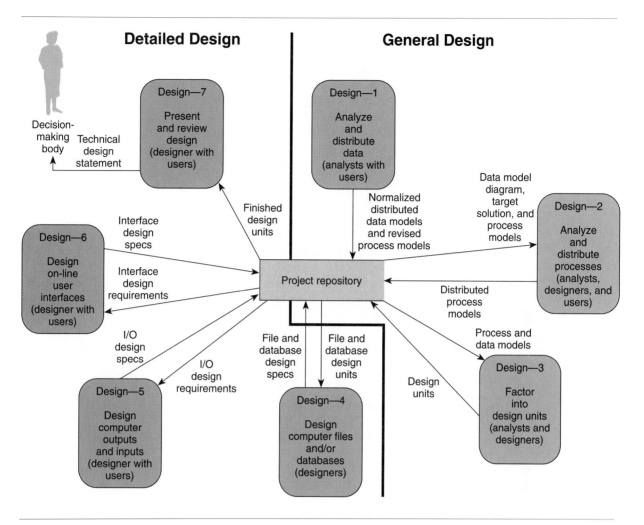

Detailed Design | **General Design**

Decision-making body

Technical design statement

Design—7: Present and review design (designer with users)

Design—6: Design on-line user interfaces (designer with users)

Interface design specs

Interface design requirements

I/O design specs

I/O design requirements

Design—5: Design computer outputs and inputs (designer with users)

Finished design units

File and database design specs

File and database design units

Design—4: Design computer files and/or databases (designers)

Project repository

Design—1: Analyze and distribute data (analysts with users)

Normalized distributed data models and revised process models

Data model diagram, target solution, and process models

Design—2: Analyze and distribute processes (analysts, designers, and users)

Distributed process models

Process and data models

Design units

Design—3: Factor into design units (analysts and designers)

FIGURE 12.8 Design and Integration Phase Activities These activities will lead to completion of the design and integration phase. The activities are generally completed in a clockwise sequence from the top—beginning with general design. The project repository allows you to overlap activities or return to any previous activity.

During this activity, the analyst will work closely with users to develop a good data model—that is, a data model that will allow the development of ideal file and database solutions. Data analysis is the technique used to derive a good data model.

Data analysis is a procedure that prepares a data model for implementation as a nonredundant, flexible, and adaptable file/database.

Normalization is the procedure that is used to simplify entities, eliminate redundancy, and build flexibility and adaptability into the data model.

Normalization of data refers to the way data attributes are grouped together to form stable, flexible, and adaptive entities.

Once data analysis has been completed, event analysis will be performed to address the analyst's obligations to ensure that the end-users' data will be kept accurate and up to date.

Event analysis is a technique that studies the entities of a fully normalized data model to identify business events and conditions that cause data to be created, deleted, or modified.

Since data and event analysis will likely have an impact on the process models for the target system, the target system data flow diagrams (DFDs) may need to be revised. Data analysis, event analysis, and their impact upon DFDs are covered in Chapter 13. The end products of this first activity are the normalized distributed data models and revised process models.

Activity 2: Analyze and Distribute Processes

We can now shift our focus away from data and toward processes. The revised data and process models can now be used to address distribution issues for the target system data. In Chapter 10 we detailed network requirements for the new system. In Chapter 14 you'll learn how those requirements can be addressed and reflected within the design of the target system.

Given the data model diagram, target solution, and process models the analyst will develop distributed process models. To complete this activity, the analyst may involve a number of systems designers and users.

Activity 3: Factor into Design Units

The remaining activity in general design is to factor the target system into separate design units. Using the process and data models, the analyst and system designers will determine how the target system could best be factored into smaller pieces whereby an individual can easily complete the design details. The resulting design units will serve as an outline that will guide the detailed design activities.

Tools and techniques for factoring a target system into design units will be fully covered in Chapter 14.

Activity 4: Design Computer Files and/or Databases

Given the file and database design units from general design, systems designers must develop the corresponding file and database design specifications. The design of data goes far beyond the simple layout of records. Files/databases are a shared resource. Many programs will typically use them. Future programs may use files and databases in ways not originally envisioned. Consequently, the designer must be especially attentive to designing files and databases that are adaptable to future requirements and expansion.

The designer must also analyze how programs will access the data in order to improve performance. You may already be somewhat familiar with various programming data structures (e.g., sequential, ISAM, VSAM, relative, and linked lists) and their impact on performance and flexibility. These issues affect file/database organization decisions. Other issues to be addressed during both file and database design include record size and storage volume requirements. Finally, because files and databases are shared resources, the designer must also design internal controls to ensure proper security and disaster recovery techniques, in case data is lost or destroyed.

Chapter 15 provides complete coverage on designing computer files and databases.

Activity 5: Design Computer Outputs and Inputs

Given the input and output design requirements for the target system, the systems designer must work closely with system users to develop input and output specifications. Because end-users and managers will have to work with inputs and outputs, the designer must be careful to solicit their ideas and suggestions, especially regarding format.

Transaction outputs will frequently be designed as preprinted forms onto which transaction details will be printed. Reports and other outputs are usually printed directly onto paper or displayed on a terminal screen. In any event, the precise format and layout of the outputs must be specified. Management involvement is also necessary because management must approve expenditures on all outputs. Finally, internal controls must be specified to ensure that the outputs are not lost, misrouted, misused, or incomplete.

For inputs, it is crucial to design the data capture method to be used. For instance, you may design a form on which data to be input will be initially recorded. You want to make it easy for the data to be recorded on the form, but you also want to simplify the entry of the data from the form into the computer or onto a computer-readable medium. This is particularly true if the data is to be input by people who are not familiar with the business application (keypunch operators, for example). The layout of the input record as it will be presented to the computer must also be designed.

Any time you input data to the system, you can make mistakes. We need to define editing controls to ensure the accuracy of input data. Normally, we will be forced to define additional outputs (called *edit reports* or *edit screens*) to identify input errors. As an additional control to prevent lost or erroneous inputs, you usually design historical reports that commit input transaction processing to paper, where they can be audited and confirmed.

Computer output and input design will be covered in Chapter 16.

Activity 6: Design On-Line User Interfaces

This activity is omitted from many designs. However, for on-line systems, the development of interface (the dialogue between the end-user and the computer) design specifications from interface design requirements may well be the most critical design activity. Too many on-line systems are difficult to learn and use because they exhibit poor human engineering.

The idea behind user interface design is to build an easy-to-learn and easy-to-use dialogue around the on-line input and output screens that were designed in earlier tasks. This dialogue must take into consideration such factors as terminal familiarity, possible errors and misunderstandings that the end-user may have or may encounter, the need for additional instructions or help at certain points in time, and screen content and layout. Essentially, you are trying to anticipate every little error or keystroke that an end-user might make — no matter how improbable. Furthermore, you are trying to make it easy for the end-user to understand what the screen is displaying at any given time.

In Chapter 17 you'll learn how to design on-line user interfaces.

Activity 7: Present and Review Design

This final detailed design activity packages all of the specifications from the previous tasks into computer program specifications that will guide the computer programmer's activities during the construction phase of the systems development life cycle.

There is more to this task than packaging, however. How much more depends on where you draw the line between the systems designer's and computer programmer's responsibilities (this issue is moot if the analyst and programmer are the same person). In addition to packaging, you need to determine the overall program structure. There are numerous strategies for top-down, modular decomposition, which will be surveyed in Chapter 18.

Given the finished design units, you need to prepare two more components:

1. An implementation plan that presents a proposed schedule for the construction and delivery phases (detailed in Chapter 19)
2. A final cost-benefit analysis that determines if the design is still feasible

The final technical design statement specifications are typically organized into a workbook or technical report. Technical design specifications evolve from the essential requirements specifications that were prepared during systems analysis. Thus, the project repository that was started during systems analysis will eventually become the design specifications document.

The systems design should be reviewed with all appropriate audiences, which may include the following:

- *System users.* End-users have already seen and approved the outputs, inputs, and terminal dialogue. The overall work and data flow for the new system should get a final walkthrough and approval.

- *System owners.* Management should get a final chance to question the project's feasibility, given the latest cost-benefit estimates.

- *Technical support staff.* Computer center operations management and staff should get a final chance to review the technical specifications to be sure that nothing has been forgotten and so that they can commit computer time to the construction and delivery phases of the project.

- *Audit staff.* Many firms have full-time audit staffs whose job it is to pass judgment on the internal controls in a new system.

As you probably guessed, the results of any of these reviews may necessitate a return to previous tasks in the design phase.

DESIGN BY PROTOTYPING

Traditionally, physical design has been a paper-and-pencil process. Analysts drew pictures that depicted the layout or structure of outputs, inputs, and files and the flow of dialogue and procedures. This is a time-consuming process that is prone to considerable error and omissions. Frequently, the resulting paper specifications did not prove themselves inadequate, incomplete, or inaccurate until programming started.

Today, many analysts are turning to prototyping, a modern engineering-based approach to design. A **prototype,** according to Webster's dictionary, is "an original or model on which something is patterned" and/or "a first full-scale and usually functional form of a new type or design of a construction (as an airplane)." Engineers build prototypes of engines, machines, automobiles, and the like, prior to building the actual products. **Prototyping** allows engineers to isolate problems in both requirements and designs.

Systems analysts are using powerful prototyping tools and languages to implement this concept. In this section we examine how prototyping is being used to improve the design and integration phase of our life cycle.

The Prototyping Approach: Advantages and Disadvantages

The prototyping approach has several advantages.

- Prototyping encourages and requires active end-user participation. This increases end-user morale and support for the project. End-user morale is enhanced because the system appears real to them.

- Iteration and change are a natural consequence of systems development — that is, end-users tend to change their minds. Prototyping better fits this natural situation since it assumes that a prototype evolves, through iteration, into the required system.

- It has often been said that end-users don't fully know their requirements until they see them implemented. If so, prototyping endorses this philosophy.

- Prototypes are an active, not passive, model that end-users can see, touch, feel, and experience. Indeed, if a picture such as a DFD is worth a thousand words, then a working model of a system is worth a thousand pictures.

- An approved prototype is a working equivalent to a paper design specification, with one exception — errors can be detected much earlier.

- Prototyping can increase creativity because it allows for quicker user feedback which can lead to better solutions. (*Note:* See the list of disadvantages for ways creativity can be stifled by prototyping.)

- Prototyping accelerates several phases of the life cycle, possibly bypassing the programmer. In fact, prototyping consolidates parts of phases that normally occur one after the other. These phases include the following:

  · *Definition.* Prototyping can be used to quickly experiment with different requirements. Each prototype can change not only the design, but the actual requirements — until the end-users accept the requirements. In many cases, requirements can be defined more quickly with this approach.

  · *Design.* Screen and report layouts can be very quickly changed until end-users accept their design. Terminal dialogue can be user-tested for friendliness and completeness. Even if the prototype is reconstructed using a traditional language such as COBOL, the prototype serves as a model of how the system must work. In most cases the design, as developed through prototyping, can be completed faster than one developed with paper and pencil. Also, the working prototype has been seen by end-users; therefore, it is less likely to be redesigned after it has been implemented in final form.

· *Construction.* The very act of prototyping requires construction, which is also known as programming. The analyst programs — if it's even possible to use that term — the prototype. Although many prototypes are eventually discarded in favor of a final system implemented in a traditional language such as COBOL, many prototypes are being implemented in the prototyping language. This can significantly reduce implementation time and effort.

There are also disadvantages or pitfalls to using the prototyping approach. Prototyping is not without disadvantages. Most of these can be summed up in one statement: Prototyping encourages ill-advised shortcuts through the life cycle. Fortunately, the following pitfalls can all be avoided through proper discipline.

- Prototyping encourages a return to the "code, implement, and repair" life cycle that used to dominate information systems. As many companies have learned, systems developed in prototyping languages can present the same maintenance problems that have plagued systems developed in languages such as COBOL.
- Prototyping does not negate the need for the survey and study phases. A prototype can just as easily solve the wrong problems and opportunities as a conventionally developed system.
- You cannot completely substitute any prototype for a paper specification. No engineer would prototype an engine without some paper design. Yet many information systems professionals try to prototype without a specification. Prototyping should be used to complement, not replace, other methodologies. The level of detail required of the paper design may be reduced, but it is most certainly not eliminated. (In the next section, we'll discuss just how much paper design is needed.)
- There are numerous design issues not addressed by prototyping. These issues can inadvertently be forgotten if you are not careful.
- Prototyping often leads to premature commitment to a design. In other words, the selection and acquisition phases gets shortchanged.
- When prototyping, the scope and complexity of the system can quickly expand beyond original plans. This can easily get out of control.
- Prototyping can reduce creativity in designs. The very nature of any implementation — for instance, a prototype of a report — can prevent analysts and end-users from looking for better solutions.
- Prototypes often suffer from slower performance than their third-generation language counterparts.

All of these disadvantages can be overcome through discipline. You need only remind yourself that prototyping does not replace any phase of the life cycle; it merely improves your productivity and quality in several of the phases.

The Technology and Strategy of Prototyping

Given the advantages of prototyping, and realizing that all disadvantages can be overcome, let's examine the technology that makes prototyping possible and the overall strategy for prototyping.

Prototyping Languages and Tools

Building prototypes makes so much sense that you may wonder why we didn't always do it. The reason is simple: the technology wasn't available. Traditional languages such as COBOL, FORTRAN, BASIC, Pascal, and C (often called **third-generation languages**) don't lend themselves to prototyping. Prototypes must be developed and modified quickly, neither of which is possible with third-generation languages. Consider the prospects of continually modifying the DATA and PROCEDURE divisions of a COBOL program as end-users try to make up their mind what they want and how it should look.

Fourth-generation languages (4GLs) and applications generators (AGs) are software tools that make building systems a simpler task. At the risk of oversimplifying 4GLs and AGs, they are less procedural than traditional languages. This means that the tools specify more of what the system is or what it should do, and less of how to do it. In other words, they are not as dependent on specification of logic.

The syntax of a 4GL or AG tends to be more concise and English-like. Many 4GLs and AGs substitute menus and question-and-answer dialogues for most of the procedural specification common to traditional languages. As a result, 4GLs and AGs allow analysts and programmers to define and load databases, develop input records, define terminal screens, develop terminal dialogues, and write reports—all within a matter of hours or days instead of the usual weeks and months associated with languages such as BASIC, COBOL, and PL/1! Such advantages account for the increased popularity of prototyping as a technique.

Virtually all 4GLs and AGs are built around the technology of a database management system (DBMS). A DBMS helps you organize and store different, but related, collections of data (for instance, CUSTOMERS, ORDERS, and PARTS are different sets of data, but they are related since CUSTOMERS "place" ORDERS "for" PARTS). It is becoming increasingly difficult to distinguish between DBMSs, 4GLs, and AGs, since they are inclusive to most database products. In fact, it is increasingly difficult to differentiate between 4GLs and AGs, since one rarely occurs without the other today.

Prototyping technology is widely available in mainframe computer, minicomputer, and microcomputer environments. As just noted, most mainframe DBMSs include a 4GL/AG. Examples include Cullinet's IDMS database with its ADS/ON-LINE applications generator and ADR's DATACOM with its IDEAL applications generator. Traditional mainframe 4GLs/AGs include Information Builder's FOCUS (arguably the most widely used 4GL) and Mathematica's RAMIS, both of which include their own proprietary DBMS. Other popular prototyping tools are listed in the margin.

Many mainframe prototyping tools have been released as PC versions. For instance, both FOCUS and RAMIS are now available for microcomputers (with compatibility between mainframe- and microcomputer-developed systems). It may actually surprise you that traditional microcomputer database packages are also suitable for prototyping. Most people, for example, think of dBASE IV and R:BASE System V as database packages. In reality, they have evolved into sophisticated 4GLs and AGs that are, in most respects, the equals of their mainframe counterparts.

It should also be noted that many computer-assisted systems engineering (CASE) products now contain limited prototyping tools for designing screens

——————— ✓ ———————

Sample Prototyping Tools

ADS/ON-LINE
APPLICATION FACTORY
DATATRIEVE
dBASE IV
FOCUS
IDEAL
NATURAL
INTELLECT
MANTIS
NOMAD2
ObjectView
POWERBUILDER
RAMIS
SAS
SPSS
TELON
VISUAL BASIC

and reports. Some experts believe that CASE will ultimately become the embodiment of both 4GLs and AGs.

Prototyping Strategy

The design-by-prototyping strategy is being used by an increasing number of businesses. Unfortunately, prototyping is also misused by a large number of computer specialists. The method requires a somewhat different approach to the traditional life cycle, but not a radically different one. Although prototyping is being presented here as a design strategy, its impact on the entire life cycle should be fully understood.

We've already emphasized that the study phase is still crucial, no matter what design approach is forthcoming. Problems and opportunities must be identified, analyzed, and understood so that objectives may be established for the new system.

Although the definition phase can be simplified in those cases where prototyping will be used, some general requirements should be specified prior to the generation of prototypes. Specifically, since prototyping languages utilize database technology, it makes sense to use a well-conceived data model in building databases. You learned how to build a data model in Chapter 8. You'll learn how to improve upon it in Chapter 13, and how to convert it into a database in Chapter 15.

In addition to a data model, process models can be useful for defining those processes that will be designed through prototypes. You can save some time in the definition phase by not defining detailed requirements — for example, records, data elements, policies, and procedures — for the process model. Instead, prototypes will be used to zero in on the detailed requirements for those components.

It is extremely important to complete the selection phase prior to prototyping so as to ensure that the target system is the most feasible solution to the end-users' problems and requirements. Perhaps a software package would be better, or perhaps a microcomputer-based solution would be better than a mainframe solution. The selection phase deals with such issues.

Given the most feasible solution, prototyping can begin. Prototypes can be quickly developed using the prototyping technology just described. Prototypes can be built for simple outputs, computer dialogues, key functions, entire subsystems, or even the entire system. Each prototype system is reviewed by end-users and management, who make recommendations about requirements, methods, and formats. The prototype is then corrected, enhanced, or refined to reflect the new requirements. Prototyping technology makes such revisions in a relatively straightforward manner. The revision and review process continues until the prototype is accepted. At that point, the end-users are accepting both the requirements and the design that fulfills those requirements.

Design by prototyping doesn't necessarily fulfill all design requirements. For instance, prototypes don't always address important performance issues and storage constraints. Prototypes rarely incorporate internal controls. These must still be specified by the analyst.

Design is usually followed by the construction and delivery phases of the life cycle. But haven't we, simply by virtue of the fact that we used prototyping, constructed the system? Not necessarily! A decision must be made on whether to reprogram the system into a more traditional language in order to improve performance or perhaps to standardize for purposes of maintenance. Many prototypes perform satisfactorily with their small test database; however, performance declines as the database size grows to more realistic and expected levels. Also, a prototype may be developed on a microcomputer for eventual implementation on a mainframe computer. Thus, there are several reasons to possibly reconstruct the system.

Even if the system is implemented using the prototype — in other words, if we let the prototype evolve into the final system — the acceptance of the prototype does not signal the completion of the construction phase. Prototype programs must usually be modified to include internal controls. For example, in the interest of speed, prototypes may not include many or any edits for inputs; the end-users can input garbage data. These edits must be added to programs. Also, prototypes rarely create audit trails to track updates to the database or files. This could lead to an inability to recover data that is accidentally lost. Such audit trails would have to be added. Thus, we see that the construction phase must still follow design by prototyping.

The delivery phase is clearly not affected by the choice of the traditional or prototyping design strategy. End-users must still be trained. The old system must still be converted to the new system. The final system should still be audited after it is placed into operation.

Rapid versus Systems Prototyping

There are two distinctly different types or styles of prototyping — rapid prototyping and systems prototyping.

Rapid prototyping, our term, is the simplest type of prototyping, albeit a very powerful type. Rapid prototyping allows you to create and test input designs, output designs, terminal dialogues, and simple procedures. You are not building a prototype system. Instead, you are building prototypes of selected components of a system.

The technology of rapid prototyping is unique in that it is not built around a complete AG or 4GL. And instead of being built around a DBMS, it is built around a computerized data or project dictionary. This dictionary may be part of a CASE product such as Index Technology's Excelerator, or it may be self-contained as part of a dedicated rapid prototyping product such as Pansophic's TELON.

Rapid prototyping proceeds as follows. The prototyping tool is used to create screens or reports. For example, in a screen prototype the analyst can place headings, comments, instructions, and the like anywhere on the screen. The analyst can also define fields (or variables) to appear on the screen. As fields are positioned, the analyst can call up previously described data element specifications from the dictionary to be applied to the field or variable. Such specifications would define attributes such as size, format, and value ranges. The analyst can alter these attributes and even add additional design attributes such as display attributes (for example, reverse video and blinking), additional editing rules, help messages, and error diagnostics. The screen can be chained to other screens to define a sequence of screens.

Once the screens are completed, the end-users can sit at the terminal or microcomputer and use them. The screens use the underlying dictionary to simulate normal operation. Thus, the end-user can receive and interpret help messages, react to error diagnostics, enter data, and even save that data, which can then be passed to prototype reports created via rapid prototyping. The end-users can tell the analyst what they don't like and what needs to be added, deleted, or changed. Once the design is approved, the analyst can usually generate code for several alternative technical environments.

This entire process can be completed in a relatively small number of hours! Other aspects of systems design would be handled by traditional physical design or systems prototyping methods.

Systems prototyping requires the use of a true 4GL/AG. The physical design tasks that were described earlier in the chapter are still valid; the only difference is that they are completed via prototyping. The process is as follows:

1. A prototype database would be designed, using whatever constraints are imposed by the 4GL/AG's underlying database. The prototype database would be loaded with a sufficient collection of test data.
2. The following tasks could occur in parallel or in any sequence:
   a. Prototype outputs can be created using the report generator of the 4GL/AG. Report generators allow new reports to be quickly defined. Report fields will be filled with test data from the prototype database.
   b. Prototype inputs can be created and generated using the screen or report generator of the 4GL/AG. Screens can be chained to form a dialogue. Normally, the input screens would be designed with minimum data editing and no security features.
3. Once the inputs and outputs are completed, they would be integrated around some sort of user-friendly shell. The most common shell consists of menus and submenus.

Once the completed prototype system has been accepted, the analyst can add data editing and security features to the system, unless the prototype will be discarded in favor of an implementation using a more traditional language. The analyst can also experiment with the database structure to improve systems efficiency.

It is difficult to demonstrate prototyping in a book. Prototyping is a "live" computing technique. You need the prototyping technology to fully appreciate it. Still, as you progress through the design chapters, we will show you some sample prototyping screens that will help you contrast prototyping with traditional approaches.

Summary

Systems design is the process whereby the end-users' requirements are transformed into a software package and/or a specification for a computer-based information system. Systems design consists of three phases that can be successfully completed through a series of well-defined activities that are common to all projects.

The purpose of the selection phase is to identify alternative manual and computer-based solutions. It is during this phase that the make-versus-buy decision is made. Each alternative solution is analyzed for feasibility. Specifically, each candidate is analyzed for operational, technical, economic, and schedule feasibility. The candidate offering the best overall combination of feasibility is normally recommended.

The purpose of the acquisition phase is to evaluate and select specific software packages and/or computer equipment that fulfills the target system requirements. The most important document of the phase is the request for proposal (RFP) or request for quotation (RFQ). These documents communicate our needs to prospective vendors, who will respond with formal proposals.

The most detailed phase of systems development is the design and integration phase. The purpose of this phase is to generate detailed specifications for the computer elements of the new information system (or for modifications and enhancements to a software package). These design specifications will be passed on to the computer programmers for implementation. Obviously, the degree to which computer programmers will be able to construct the system without further assistance is dependent on the completeness and clarity of the design specifications. Although the ultimate goal of systems design is to communicate specifications to programmers for implementation, the importance of end-user participation cannot be overemphasized. Systems design is the phase in which the outputs, inputs, and on-line dialogues take form. An understanding of the importance of human engineering and end-user acceptance is crucial to overall project success.

Prototyping has emerged as a preferred strategy for physical design. Prototypes are working models of a system. Analysts can quickly build prototypes using modern fourth-generation languages and applications generators. The prototyping strategy is not a substitute for the life cycle. Each phase of the life cycle is still essential to successful systems development. However, prototyping does consolidate portions of the definition, physical design, and construction phases of the traditional life cycle. Consequently, prototyping accelerates productivity. There are two types of prototyping—rapid and systems. Rapid prototyping builds models of distinct systems components, whereas systems prototyping builds models of an entire working system.

--- *Key Terms* ---

data analysis, p. 485

detailed design, p. 484

economic feasibility,
   p. 476

event analysis, p. 486

fourth-generation
   languages, p. 491

general design, p. 484

normalization, p. 485

operational feasibility,
   p. 475

prototype, p. 488

prototyping, p. 489

rapid prototyping, p. 493

request for proposals
   (RFP), p. 480

request for quotations
   (RFQ), p. 480

schedule feasibility, p. 476

systems design, p. 472

systems prototyping, p. 494

technical feasibility, p. 475

third-generation
   languages, p. 491

--- *Questions, Problems, and Exercises* ---

1. How can a successful and thorough systems analysis be ruined by a poor systems design? Answer the question relative to these factors:
   a. The impact on the subsequent implementation (in other words, the systems implementation phases, which you studied in Chapter 3).
   b. The lifetime of the system after it is placed into operation.
   c. The impact on future projects.

2. What skills are important during systems design? Create an itemized list of these skills. Identify other computer, business, and general education courses that would help you develop or improve your skills. Prepare a plan and schedule for taking the courses. (If you are not in school, prepare a plan for using available corporate training resources, reading appropriate books, enrolling in seminars or continuing education courses, etc.) Review your plan with your counselor, advisor, or instructor.

3. How does your information systems pyramid model aid in systems design?

4. What by-products of the systems analysis phases are used in the systems design

phases? Why are they important? How are they used? What would happen if they were incomplete or inaccurate?

5. What are the end products of the selection, acquisition, and design and integration phases? What is the content of each end product?

6. United Films Cinemas has asked you to help them acquire microcomputer systems for their theaters and main office. Write a letter that proposes a disciplined approach to acquiring an appropriate system. Assume that your end-user is inclined to ignore a disciplined approach and would prefer to go to the local computer store and just buy something. In other words, defend your approach.

7. Distinguish between the terms validation and evaluation as they apply to the selection of computer equipment and software.

8. Explain the difference between a request for proposal and a request for quotation.

9. Explain what you would do if a vendor said the following in response to an RFP or RFQ.

> This thing is not useful to you or me. It rarely tells me what you really want or need. I can do a better job by visiting your business and configuring a system to meet your needs. Also, it takes too long for me to answer all the questions in this RFP. And even if I do, you may not fully understand or appreciate the answers and their implications.

10. A programming assignment in the classroom is a subset of a systems design. Obtain a copy of a programming assignment from a current course. Evaluate the design from the perspective of the systems design phase tasks and the completeness of the design specification.

11. Distinguish between general design and detailed design.

12. Obtain a copy of a computer programming assignment. Assume that the assignment is to be implemented on a microcomputer that has not been acquired. Estimate the costs necessary to complete the project (hardware, programming, etc.). State your assumptions about salaries, supplies, and other relevant factors.

Projects and Minicases

1. Make an appointment with or write to a hardware and software vendor. Tell them you would like to see and discuss a typical RFP. Ask the vendor how they feel about RFPs. If they don't like them, find out why. How could RFPs be improved from the vendor's point of view? Do the vendor's attitudes about RFPs help the vendor, the end-user, or both?

2. Make an appointment to discuss physical design standards of a local information systems operation. Does it have standards? Does it follow them? Why or why not? Does the company use 4GLs or AGs to prototype systems? Why or why not? If it does prototype systems, has the approach been successful?

3. The city of Granada's art museum recently purchased an IBM PS/2 Model 50 microcomputer. They read an article about an art collection inventory system software package that they want to put on that computer. You, having experienced end-users who too hastily purchased software that didn't fulfill promises and expectations, are concerned that they are jumping the gun and should approach the software selection decision with great care. Write a letter to the museum's board of trustees that expresses your concerns and proposes a better approach.

4. Write a letter to your last (or favorite) programming instructor. Suggest a disciplined approach to developing a systems specification to guide the programming assignments for the next term. Your goal should be a system (of programming) specification that will eliminate or drastically reduce the need for students to request clarification from the systems analyst, played by the instructor. Defend your approach.

─────────────── *Suggested Readings* ───────────────

Boar, Benard. *Application Prototyping: A Requirements Definition Strategy for the 80s.* New York: Wiley, 1984. This is one of the first books to appear on the subject of systems prototyping. It provides a good discussion of when and how to do prototyping, as well as thorough coverage of the benefits that may be realized through this approach.

Connor, Denis. *Information System Specification and Design Road Map.* Englewood Cliffs, N.J.: Prentice Hall, 1985. This book compares prototyping with other popular analysis and design methodologies. It makes a good case for not prototyping without a specification.

Isshiki, Koichiro R. *Small Business Computers: A Guide to Evaluation and Selection.* Englewood Cliffs, N.J.: Prentice Hall, 1982. Although it is oriented toward small computers, this book surveys most of the better-known strategies for evaluating vendor proposals. It also surveys most of the steps of the selection process, although they are not put in the perspective of the entire systems development life cycle.

Joslin, Edward O. *Computer Selection.* Rev. ed. Fairfax Station, Va.: Technology Press, 1977. Although somewhat dated, the concepts and selection methodology originally suggested in this classic book are still applicable. The book provides keen insights into vendor, customer, and end-user relations.

Lantz, Kenneth E. *The Prototyping Methodology.* Englewood Cliffs, N.J.: Prentice Hall, 1986. This book provides excellent coverage of the prototyping methodology.

Martin, James. *Fourth Generation Languages.* 2 vols. Englewood Cliffs, N.J.: Prentice Hall, 1985 and 1986. Volume 1 covers principles that underlie mainframe 4GLs and AGs. It is highly recommended reading for anyone planning a career in systems analysis. Volume 2 surveys the 4GL/AG marketplace. It can help you better appreciate the capabilities of 4GLs and AGs. You may find that your school or business owns one of the packages discussed in this book.

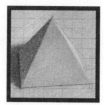

Data and Process Analysis and Design for the New System

When we left Sandra and Bob they had just completed defining the detailed requirements for the new Member Services system.

Data and Process Analysis for the New System

Sandra and Bob had completed documenting the detailed requirements for the new system. They have now begun to complete a general design for the new system. Bob is working on a data analysis for the new system. Recall that Sandra and Bob created a data model during the definition phase to serve as a vehicle for communicating business data requirements for the new system. Bob has been busy revising the model for Sandra and for his own use to more accurately reflect many of the details that they purposefully omitted from the earlier data model. Many of the details were omitted to avoid confusing or overwhelming the users. Let's join Bob as he discusses the revised data model with Sandra.

"Here it is Sandra! The final data model (see Figure E5.1). It sure looks different from the data model that we showed the users."

"It sure does," responded Sandra. "But it is obvious that this model doesn't reflect their view or understanding of their system's data."

"As you can see, we've gone from the seven data entities that our users identified for us to a grand total of thirteen data entities," Bob offered. "Only a few of the new data entities are associative data entities. I discovered most of those new entities when I was trying to put the entities in first normal form. Some were identified by simply examining the original data model for many-to-many relationships between related entities."

"It's sometimes amazing what one can uncover by performing data analysis. Did you document the data structures for all the new data entities?" asked Sandra.

"Yes I did," answered Bob. "I also went ahead and got started with event analysis to identify business events that would affect the data entities. I've identified business events that would require the creation, modification, and deletion of occurrences for each data entity. I also specified any conditions that would govern these actions."

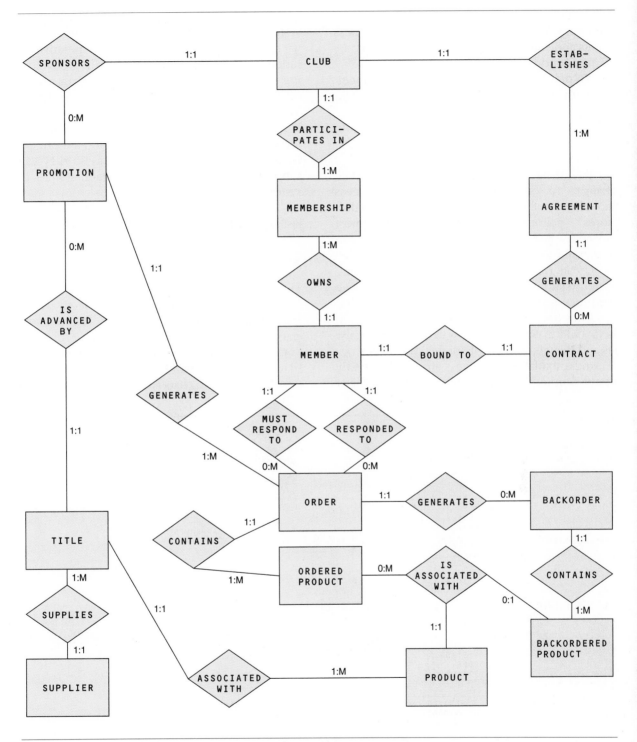

FIGURE E5.1 **Typical Data Model upon Completion of Data Analysis** Compare this model to the data model in Episode 3.

"Don't forget to check this out with the users. They may know of other more subtle business events and conditions you may have overlooked," warned Sandra.

"No problem. I already conducted a few meetings with appropriate users to verify the events and their effects and conditions," replied Bob.

"Wow! You really worked hard while I was out with that nasty cold," said Sandra. "So, where exactly are you in the project?"

"Well, I had to go back and make appropriate changes to our data flow diagrams (process model) to reflect additional data entities (data stores), actions, and conditions. I'm almost finished—just one or two more new diagrams to be revised," explained Bob.

"Great! How about I take some of the completed data flow diagrams and begin completing the process analysis and design?" offered Sandra.

"I was hoping you'd offer to do that. I'd like to help, but I'm not too sure how to do process analysis and design," Bob explained with concern.

Sandra explained, "Process analysis is actually pretty simple. There are a few techniques that I plan to apply for determining how to distribute the system's data and processes to locations since we're dealing with a distributed or cooperative solution to our system. Essentially, I will be developing some implementation data flow diagrams to document these decisions (see Figure E5.2). These diagrams will represent a design unit which is a collection of data and processes specific to a location. You and I will then develop a more detailed design for the design unit(s) so they can subsequently be implemented as stand-alone subsystems."

"Let me finish making those revisions to those last few diagrams and then we can get started."

Where Do We Go from Here?

This episode introduced data and process analysis and design. These activities are collectively referred to as *general systems design.* The next two chapters cover general systems design.

Chapter 13 will cover data analysis and design. In Chapter 13, you will learn a step-by-step procedure for completing data analysis. The chapter will

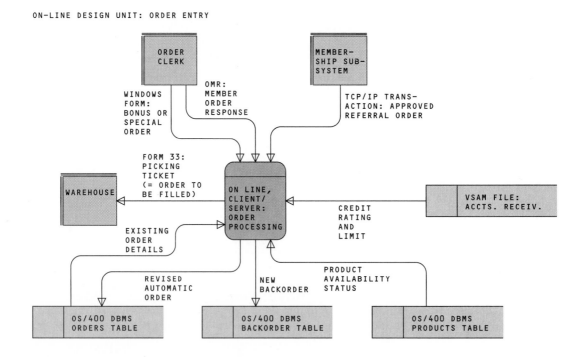

FIGURE E5.2 **Typical Implementation Data Flow Diagram**

also introduce you to *event analysis*—a technique for identifying business events that cause data to be added, modified, or deleted. This chapter will also discuss the impact that data analysis has on process models that must be revised to be more accurate and current.

Chapter 14 introduces process analysis and design as a process modeling technique for systems design. This chapter will teach you techniques for analyzing processes to distribute a system's data and processes into convenient, smaller design units for implementation.

13

Data Analysis

Chapter Preview and Objectives

This chapter teaches you techniques for analyzing data to derive a data model that is simple, nonredundant, flexible, and adaptable. The chapter builds on the data modeling techniques taught in Chapter 8. Normalization will be introduced as a procedure for completing data analysis. The chapter will also introduce event analysis as a technique for ensuring that a system's data is kept accurate and current at all times.

You will know that you understand data and event analysis when you can:

Explain
the need to analyze a data model for
simplicity, redundancy, flexibility, and adaptability.

Describe
a step-by-step approach to performing data analysis.

Explain
the need to analyze business events and their
impact on a system's data.

Describe
a step-by-step approach to performing event analysis.

Recognize
the impact that data and event analysis has
on process models.

PRECIOUS JEWELS DIAMOND CENTERS

Precious Jewels Diamond Centers is a franchised jewelry store that specializes in diamonds and other gems, custom selected by and for customers. Gems are custom set into rings, pendants, and the like. Precious Jewels also serves as a diamond broker, providing gems to other franchises and jewelry stores. These gems are sent out on approval. The stores have the option of purchasing the gems or returning them.

Scene: *Frank Burnside, a systems consultant to Precious Jewels, is meeting with Jeff Kassels, vice president. They are discussing the possibility of hiring Frank to develop improved information systems for their IBM and Compaq microcomputers.*

Jeff: Where do I start? About two years ago, we decided to purchase two microcomputers. On the recommendation of the computer store, we also bought some software packages.

Frank: What packages?

Jeff: Excel, Paradox, and Word for Windows. Unfortunately, we just didn't have the training to exploit these packages, especially the Paradox database package. Therefore, I hired some young students who were into micros; you'd probably call them hackers. They didn't have much experience.

Frank: And what happened?

Jeff: They wrote some Paradox programs and Excel macros for inventory control and sales. The programs seemed to work. We entered lots of data into the system, and it generated several reports. On the other hand, we later realized the need for new reports and inquiries. We tried to generate them ourselves, but I guess we just didn't understand the report writer in Paradox. The original students were unavailable, so we hired a woman who does Paradox programming on the side. She couldn't seem to generate the reports from the data. I know the data is in there because we put it there and it does come out on the original reports. I don't see how you can put data in and not be able to get it out.

Frank: It's not an unusual phenomenon. Actually, it's quite common. Inexperienced analysts—no, even many experienced analysts—tend to design systems to meet today's needs. They fail to recognize that the structure of stored data can make it difficult or impossible to adapt to changing needs and situations. Your data is probably not well organized. Poorly designed computer files are no different from poorly organized manual files.

Jeff: That's not all. There are growing problems with data already in the system. As I scan the original reports, I note that records exist that should have been deleted a long time ago. For instance, I find records of purchase orders for gems that were sold and paid for a year ago. To make matters worse, I find records of gems for which I can't find a purchase order. I need that purchase order to compare my valuations and prices to the valuation and prices I was charged by my suppliers.

Frank: This situation is typical of poorly designed databases. Before a database is built, I put considerable effort into understanding its data and the complex business relationships between different sets of data. I try to understand when data needs to be created, when it should be changed, and when to delete or retain it. For instance, I suspect that you keep customer data. You probably wouldn't want me to delete a customer for which you have outstanding invoices, would you? By studying your data, I come to understand your values, needs, requirements, and policy constraints.

Jeff: Can I add new fields to existing files? The second consultant told me that she'd have to rewrite many of the existing programs.

Frank: Unfortunately, that's probably true. You see, the data files are too closely associated with the programs that use them. The programs expect the data in specific files. If you now realize that some of that data belongs in different files, the original programs must be modified to reflect the new location.

Jeff: Are you telling me that this is an unavoidable by-product of using computers? If so, I may want to scrap this whole idea.

Frank: Absolutely not! Data can be structured independently of the programs that will use it. You can easily minimize the likelihood of extensive modifications to programs.

Jeff: Okay. But can you solve this problem? I'm getting reports from the system that show conflicting data. I don't see how that can happen.

Frank: It's another database design flaw. I suspect that many data attributes are stored redundantly in different files. When you store data in more than one location, you increase the chances that you will modify data in one location and not in the other locations. Consequently, you find data conflicts in the reports generated from the different files. Data redundancy should be minimized as much as possible.

Jeff: This is interesting. When our system was designed, the students just sat down and drew some sort of flow diagram based on the reports I needed. I never considered the strategy of focusing on the raw data.

Frank: If you think about it, it makes more sense. If I can understand your data, help you get control of and capture that data, and organize it in a way that is flexible and adaptable, you'll realize two important advantages. First, you'll still be able to generate the reports you need now. But second, you'll be better able to create new reports as the need arises. After all, the data has been captured and stored in a flexible format. We can fix this system. Give me the chance to show you how!

Discussion Questions

1. How does Frank's approach to identifying requirements differ from the traditional "tell me what outputs you want" approach?

2. What benefits do you think can be derived from studying data before you study output needs and processing requirements?

3. Why do you think that consultants — and experienced analysts — so frequently ignore or do not adequately consider the future implications of the systems they design?

DATA ANALYSIS FOR DESIGN DECISIONS

In Chapter 8 you learned how to model data. We created a data model including entity relationship diagrams (ERDs) and a list of data attributes that describe each entity. These data modeling specifications provided a vehicle for communication with end-users during systems analysis. However, in this chapter our emphasis shifts away from using the data model as a vehicle for communicating business requirements (to system users). Rather, we move toward preparing that data model to communicate database *design* requirements (to system builders).

While a systems analysis data model effectively identifies data requirements, it does not usually represent a good database design. It may contain structural characteristics that may lead to the very problems demonstrated in the chapter opening minicase. Therefore, we must "prepare" the data model for implementation.

This section will discuss the characteristics of a good data model — a data model that will allow us to develop ideal file and database solutions. We'll also present the process used to convert a poor data model to a good one.

What Is a Good Data Model?

Our goal is to define a good data model for use during systems design.

- *A good data model is simple.* As a general rule, the data attributes that comprise any entity should describe only that entity. In Figure 13.1 we have reproduced the list of the data attributes that were mapped to the data entities for our SoundStage Entertainment Club case. (This list was created in Chapter 8.) Examine the contents of the ORDER entity. Does MEMBER NUMBER describe the entity ORDER? No, it really describes the entity MEMBER. Our model is not as simple as it could be.

    Also, an entity is considered simple if, for one occurrence of the entity, all its data attributes assume one and only one value. Reexamine the contents of the ORDER entity. For a single order, all the data attributes that make up the group ORDERED PRODUCT will assume many values, one value for each product on the order. Thus, we again see that our model can be simplified.

- *A good data model is nonredundant.* This means that no data attribute, other than keys, describes more than one entity. We definitely have some problems here. Referring again to Figure 13.1, MEMBER ADDRESS has been mapped to the MEMBER and ORDER entities. Such redundancy, if physically implemented by our eventual design, can wreak havoc with data consistency. Suppose we get a change of address for a member. We must redundantly change two records or risk having different packages or letters sent to different addresses.

    We also have some subtle redundancy in our model. PRODUCT DESCRIPTION (in the ORDER entity) may be logically equivalent to TITLE OF WORK (in the PRODUCT entity).

- *A good data model should be flexible and adaptable to future needs.* We tend to design files and databases to fulfill today's requirements. Such a data model is sometimes called an **applications data model** because it is conceived to support today's application needs. Then, when new needs arise, we can't change the files or databases without rewriting many or all of the programs that used those files and databases. We can't change the fact that we tend to work on applications; however, we can make our data model as application-independent as possible to encourage the eventual design of files and databases that are not dependent on changing data and information requirements. We call this a **subject data model.**

What Is Data Analysis?

The technique used to derive a good data model is called *data analysis.*

> **Data analysis** is a procedure that prepares a data model for implementation as a nonredundant, flexible, and adaptable database.

Data analysis typically uses a procedure called *normalization* to simplify entities, eliminate redundancy, and to build flexibility and adaptability into the data model.

> **Normalization** is a data analysis method that organizes data attributes such that they are grouped together to form stable, flexible, and adaptive entities.

CLUB:
    <u>CLUB NAME</u>
    NUMBER OF MEMBERS ENROLLED
    NUMBER CANCELED YTD
    CURRENT PROMOTION
    TOTAL UNITS SOLD FOR CLUB
    MAXIMUM PERIOD OF OBLIGATION

MEMBER:
    <u>MEMBER NUMBER</u> or <u>MEMBER NAME</u>
    MEMBER ADDRESS consisting of:
        STREET
        P.O. BOX
        CITY
        STATE
        ZIP CODE
    MEMBER PHONE
    DATE ENROLLED
    BALANCE PAST DUE
    BONUS CREDITS NOT USED
    CLUB GROUP (repeats 1-n times)
    consisting of:
        CLUB NAME
        MUSICAL/MOVIE PREFERENCE
        NUMBER OF PURCHASES REQUIRED
        NUMBER OF PURCHASES TO DATE
        AGREEMENT NUMBER SUFFIX
        AGREEMENT ENROLLMENT DATE
        AGREEMENT EXPIRATION DATE

PROMOTION:
    <u>CLUB NAME + PROMOTION DATE</u>
    PROMOTION TYPE
    SELECTION OF MONTH NUMBER
    SELECTION OF MONTH TITLE
    AUTOMATIC RELEASE DATE
    AUTOMATIC FILL DATE

ORDER:
    <u>ORDER NUMBER</u>
    ORDER DATE
    ORDER STATUS
    PROMOTION NAME
    PROMOTION DATE
    AUTOMATIC FILL DATE
    MEMBER NUMBER
    MEMBER NAME
    FORMER MEMBER?
    MEMBER ADDRESS consisting of:
        STREET
        P.O. BOX
        CITY
        STATE
        ZIP CODE
    ORDERED PRODUCT (repeats 1-n times)
    consisting of:
        PRODUCT NUMBER
        MEDIA CODE
        PRODUCT DESCRIPTION
        QUANTITY ORDERED
        ORDERED PRODUCT STATUS
        QUANTITY SHIPPED
        ORDER PRICE
        EXTENDED PRICE
    AMOUNT DUE:

AGREEMENT:
    <u>AGREEMENT NUMBER SUFFIX</u>
    CLUB NAME
    AGREEMENT EXPIRATION DATE
    AGREEMENT PLAN CREATION DATE
    MAXIMUM PERIOD OF OBLIGATION
    BONUS CREDITS AFTER OBLIGATION
    NUMBER OF MEMBERS ENROLLED
    NO. MEMBERS WHO HAVE FULFILLED
    NO. MEMBERS HAVE NOT FULFILLED

PRODUCT:
    <u>PRODUCT NUMBER + MEDIA CODE</u>
    PRODUCT DESCRIPTION
    TITLE OF WORK
    COPYRIGHT DATE
    CURRENT RETAIL PRICE
    CURRENT LIST PRICE
    SUPPLIER NAME
    SUPPLIER ADDRESS consisting of:
        STREET
        P.O. BOX
        CITY
        STATE
        ZIP CODE
    QUANTITY ON HAND
    UNITS SOLD
    VALUE OF UNITS SOLD

BACKORDER:
    <u>ORDERED NUMBER + BACKORDER DATE</u>
    BACKORDERED ITEM (repeats 1-n times)
    consisting of:
        PRODUCT NUMBER
        MEDIA CODE
        PRODUCT DESCRIPTION
        QUANTITY BACKORDERED

FIGURE 13.1   **Data Entities and Attributes for SoundStage Entertainment Club**   This figure was reproduced from an earlier figure in Chapter 8, Data Modeling.

Normalization is a three-step procedure that places the data model into first normal form, second normal form, and third normal form. Don't get hung up on the terminology—it's easier than it sounds. For right now, let's establish an initial understanding of these three forms.

- Simply stated, an entity is in **first normal form (1NF)** if there are no attributes (or groups of attributes) that repeat for a single occurrence of the entity. Any attributes that repeat actually describe an occurrence of a separate entity, possibly an entity that we haven't yet defined in our data model.

- An entity is in **second normal form (2NF)** if all non-key attributes are dependent on the full key attribute(s)—not just part of it. In 2NF it is assumed that you have previously placed all entities into 1NF—that is, you have removed all repeating groups of attributes.

- An entity is in **third normal form (3NF)** if the values of its non-key attributes are not dependent on any other non-key attributes. In 3NF it is assumed that you have previously placed all entities into 2NF.

WHEN TO PERFORM DATA ANALYSIS

In most organizations, data analysis is performed by the systems analyst and/or database administrator. The end-user is also involved, but primarily as a reviewer of the final data model. In this book we present data analysis as a systems design chapter and technique. As you'll learn in the next few subsections, however, data analysis is not restricted to systems design.

Data Analysis during Systems Planning

While most planning methodologies include data modeling (see Chapter 6), most do not advocate data analysis and normalization. On the other hand, the most popular current planning technique, information engineering, frequently requires data analysis and normalization of business area data models. Thus, the applicability of data analysis is clearly dependent on your choice of methodology.

Data Analysis during Systems Analysis

Many methodologies require data analysis during systems analysis. Although we did not present data analysis during our coverage of systems analysis, we acknowledge that there are some good reasons to do it during systems analysis. First, data analysis deals with the essential data requirements. Essential models are most typically associated with systems analysis, specifically the "definition of the users' data requirements." Also, a normalized essential data model represents the true essence of those requirements. Consequently, we would not be surprised to learn that your instructor has you reading this chapter immediately after Chapter 8, Data Modeling.

Data Analysis during Systems Design

There are equally compelling reasons to delay data analysis and normalization until systems design. First, normalized data models are less user-friendly—that is, they are harder to read. Although analysts and database professionals thoroughly understand them, they don't always make sense to system owners and users. If the purpose of modeling requirements is verification, then you could argue that data models should not be normalized during systems analysis.

Second, normalization is historically rooted in database design. It prepares a database for implementation. Accordingly, it introduces technology-driven requirements such as foreign keys. During database implementation, the data model is often denormalized for performance tuning. This is dangerous! The database administrator must be careful to balance the advantages (such as improved performance) and disadvantages (such as introducing redundancy, decreasing flexibility and adaptability) of denormalizing a data model.

A STEP-BY-STEP APPROACH TO DATA ANALYSIS

There are numerous approaches to data analysis. We have chosen to present normalization as a technique for completing data analysis. We'll draw upon the SoundStage case study to demonstrate the steps.

Step 1: Verify or Add Keys to Entities

Recall that Chapter 8 introduced the concept of attributes serving as **identifiers.** Data analysis is performed by systems analysts and/or database administrators who frequently use the term **key,** rather than *identifier,* when communicating to fellow IS professionals. Therefore, in this chapter, we will switch our vocabulary to the use of the term key. The remaining steps for completing data analysis are very much tied to the concept of keys for data entities. Before proceeding, we must ensure that we have identified keys for each data entity.

Systems analysts and database administrators use certain buzzwords to differentiate between specific types of keys.

Primary key refers to an attribute or attributes that uniquely identify one and only one occurrence of an entity.

Candidate key refers to *alternative* primary keys used to uniquely identify one and only one occurrence of the entity.

Concatenated key refers to a primary key that is composed of more than one data attribute. A common synonym is *combination key.*

Keys for SoundStage entities are underlined in Figure 13.1.

Step 2: Place Entities into 1NF

The first step in data analysis is to place each entity into 1NF. Reexamine Figure 13.1. Which entities are not in 1NF? You should find three: MEMBER, BACKORDER, and ORDER. Each contains a group of attributes that repeat $1 - n$ times for a single occurrence of the entity. For example, a MEMBER can contain data about membership in more than one CLUB. An ORDER or BACKORDER can contain data about more than one ORDERED PRODUCT(s).

Figures 13.2 through 13.4 demonstrate how to place these three entities into 1NF. The unnormalized entity is on the left side of the page. The entities in 1NF are on the right side of the page. Also, since normalization changes the graphic data model, all of these figures show not only the redistribution of attributes, but also the portion of the entity relationship data model that has changed.

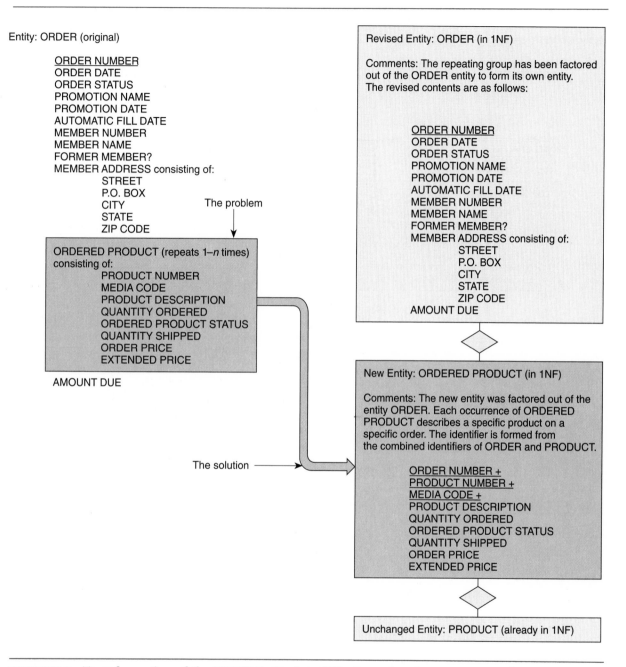

FIGURE 13.2 **Transformation of the ORDER Entity into 1NF** Repeating elements are moved to their own entity types to improve data flexibility and adaptability.

Let's examine the ORDER entity. First, we take the repeating group of attributes out of the ORDER entity. That alone places ORDER in 1NF. But what do we do with the removed group? We create a new entity, ORDERED PRODUCT. Each occurrence of the attributes describes one PRODUCT on a single ORDER.

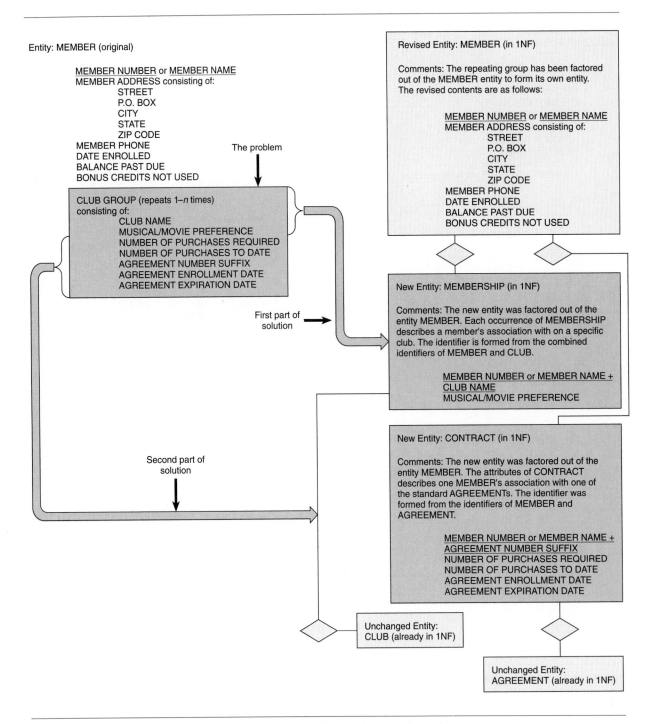

FIGURE 13.3   Transformation of the MEMBER Entity into 1NF

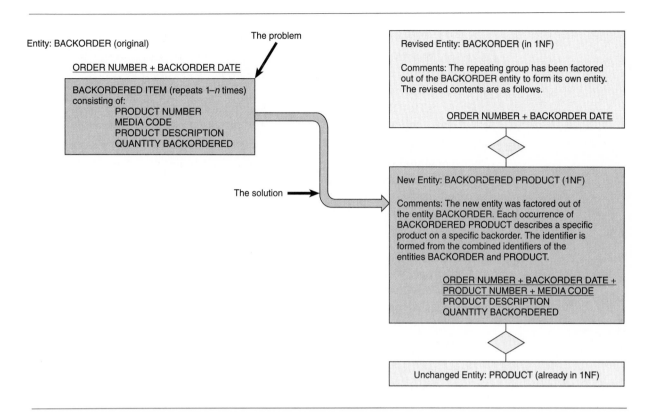

FIGURE 13.4 **Transformation of the BACKORDER Entity into 1NF**

Thus, if an order contains five products, there will be five occurrences of the ORDERED PRODUCT entity. The primary key of ORDERED PRODUCT is created in a similar way—that is, by combining the primary key of the original entity, ORDER NUMBER, with the primary key attribute of the group, PRODUCT NUMBER + MEDIA CODE (a concatenated key).

It may help to think about what we've done from a graphic data modeling viewpoint. As you can see in Figure 13.2, we have added the new entity ORDERED PRODUCT between the entities ORDER and PRODUCT. This structure replaces the original, direct relationship between ORDER and PRODUCT. ORDERED PRODUCT is sometimes called an **associative entity** (discussed in Chapter 8) since it describes data about the association between two other entities—in this case one PRODUCT's association with various ORDERs (or one ORDER's association with various PRODUCTs). Most associative entities are usually derived from many-to-many association (such as the one that originally existed between ORDER and PRODUCT). That many-to-many relationship is no longer needed now that the associative entity and relationships are added to the data model. After we complete our discussion of data analysis, we'll redraw the entire data model.

A somewhat more challenging example of 1NF is shown in Figure 13.3 for the MEMBER entity. The repeating group, CLUB GROUP, is easy to spot, but let's study the attributes in the group. The attribute MUSICAL/MOVIE PREFERENCE describes one MEMBER's association with one CLUB. The other attributes, however, describe one MEMBER's association with a specific AGREE-MENT. Thus, we create two new entities:

1. MEMBERSHIP, whose data describes one MEMBER in one CLUB (thus, the primary key is a concatenated key composed of MEMBER NUMBER + CLUB NAME).
2. CONTRACT, whose data attributes describe one MEMBER as bound by one general AGREEMENT (thus, the primary key is a concatenated key consisting of MEMBER NUMBER + AGREEMENT NUMBER SUFFIX).

Convince yourself that BACKORDER has also been placed into 1NF (Figure 13.4). The procedure is the same as was applied to ORDER and MEMBER. All other entities that did not have repeating groups of attributes are, by default, already in 1NF.

Step 3: Place Entities into 2NF

The next step of data analysis is to place the entities into 2NF. It is assumed that you have already placed all entities into 1NF. We only need to check those entities that have a concatenated key (PRODUCT, MEMBERSHIP, CONTRACT, PROMOTION, ORDERED PRODUCT, and BACKORDER). All other entities are automatically in 2NF.

First, let's check the associative entities that were created when we did our 1NF analysis. Reexamine the 1NF for the MEMBERSHIP entity in Figure 13.3. The concatenated key is the combination of MEMBER NUMBER and CLUB NAME. There is only one nonidentifier attribute, MUSICAL/MOVIE PREFER-ENCE. That attribute's value cannot be determined if you have only a MEMBER NUMBER because a member may have different preferences in different clubs to which he or she belongs. Similarly, the attribute's value cannot be deter-mined if you only have CLUB NAME since a club has many members. Truly, MUSICAL/MOVIE PREFERENCE describes a specific member in a specific club—it requires the full concatenated key. Thus, the entity MEMBERSHIP is already in 2NF.

While you are still studying Figure 13.3, you can also analyze CONTRACT. The 1NF concatenated key is MEMBER NUMBER + AGREEMENT NUMBER SUFFIX. None of the nonidentifier attributes can be described if you have only MEMBER NUMBER or only AGREEMENT NUMBER SUFFIX. Thus, CON-TRACT is also already in 2NF since all attributes describe one MEMBER's promise to fulfill one specific AGREEMENT. In other words, all the values of all the nonidentifier attributes (AGREEMENT ENROLLMENT DATE, AGREE-MENT EXPIRATION DATE, NUMBER OF PURCHASES REQUIRED, and NUMBER OF PURCHASES TO DATE) require MEMBER NUMBER + AGREEMENT NUMBER as a unique key.

Those are the cases where 1NF entities require no changes for 2NF. Now let's examine an entity that must be placed into 2NF—namely, the ORDERED PRODUCT entity in Figure 13.5. The concatenated key is ORDER NUMBER + PRODUCT NUMBER + MEDIA CODE. Certainly, several of the non-key attrib-utes describe one PRODUCT on one ORDER. Examples include QUANTITY ORDERED, ORDERED PRODUCT STATUS, QUANTITY SHIPPED, ORDER PRICE, and EXTENDED PRICE (which is price times quantity). But there are non-key attributes that describe PRODUCT, independent of any specific ORDER that might contain that product. For instance, you don't need an

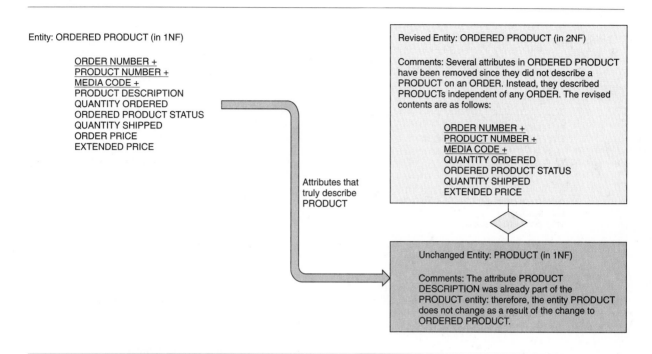

Entity: ORDERED PRODUCT (in 1NF)

ORDER NUMBER +
PRODUCT NUMBER +
MEDIA CODE +
PRODUCT DESCRIPTION
QUANTITY ORDERED
ORDERED PRODUCT STATUS
QUANTITY SHIPPED
ORDER PRICE
EXTENDED PRICE

Attributes that
truly describe
PRODUCT

Revised Entity: ORDERED PRODUCT (in 2NF)

Comments: Several attributes in ORDERED PRODUCT
have been removed since they did not describe a
PRODUCT on an ORDER. Instead, they described
PRODUCTs independent of any ORDER. The revised
contents are as follows:

ORDER NUMBER +
PRODUCT NUMBER +
MEDIA CODE +
QUANTITY ORDERED
ORDERED PRODUCT STATUS
QUANTITY SHIPPED
EXTENDED PRICE

Unchanged Entity: PRODUCT (in 1NF)

Comments: The attribute PRODUCT
DESCRIPTION was already part of the
PRODUCT entity; therefore, the entity PRODUCT
does not change as a result of the change to
ORDERED PRODUCT.

FIGURE 13.5 **Analysis and Transformation of the ORDERED PRODUCT Entity into 2NF** This revision of our initial mapping shows how certain elements were moved from this concatenated key entity to the entity that they really describe.

ORDER NUMBER to define values for PRODUCT DESCRIPTION. PRODUCT DESCRIPTION truly describes a PRODUCT, not an ORDERED PRODUCT. What about ORDER PRICE? At first glance, it appears to describe PRODUCT, not ORDERED PRODUCT; however, if you consider it more carefully, you discover that it describes the price of a product at the time of order, not necessarily the current price.

Figure 13.5 demonstrates how we get ORDERED PRODUCT into 2NF. The non-key attributes that don't describe ORDERED PRODUCT were removed from ORDERED PRODUCT and merged into the PRODUCT entity. As it turned out, the attribute PRODUCT DESCRIPTION was already contained in PRODUCT; therefore, that entity was unchanged. The entity BACKORDERED PRODUCT would undergo the same basic transformation to 2NF.

Another example is shown in Figure 13.6. Although the entity PRODUCT was one of our original entities (not created for 1NF), it does have a concatenated key, PRODUCT NUMBER + MEDIA CODE. Recall that the MEDIA CODE helps differentiate between the cassette, record, and compact disc versions of the same audio PRODUCT NUMBER, as well as the VHS tape, Beta tape, 8mm tape, and videodisc versions of the same video PRODUCT NUMBER. Therein lies our problem. Some of the values of non-key attributes do not depend on the value MEDIA CODE. For instance, the PRODUCT DESCRIPTION for PRODUCT NUMBER V34566 is *Beverly Hills Cop II,* no matter what the MEDIA CODE value is. On the other hand, QUANTITY ON HAND requires the full key since we probably have different quantities of *Terminator II* in the various video formats. Those attributes that depend on the full key will remain in the PRODUCT entity. Those that depend only on PRODUCT NUMBER will

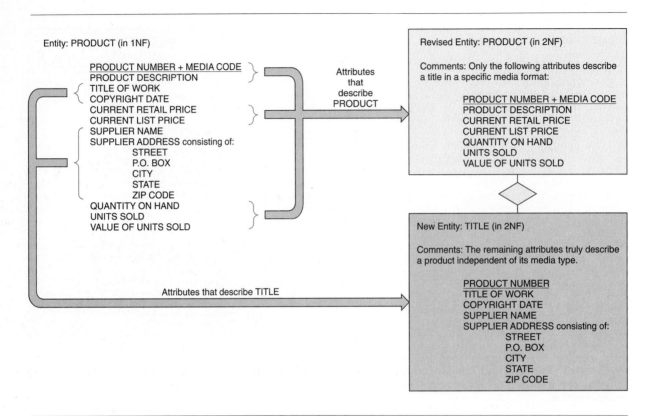

Entity: PRODUCT (in 1NF)

PRODUCT NUMBER + MEDIA CODE
PRODUCT DESCRIPTION
TITLE OF WORK
COPYRIGHT DATE
CURRENT RETAIL PRICE
CURRENT LIST PRICE
SUPPLIER NAME
SUPPLIER ADDRESS consisting of:
    STREET
    P.O. BOX
    CITY
    STATE
    ZIP CODE
QUANTITY ON HAND
UNITS SOLD
VALUE OF UNITS SOLD

Attributes
that
describe
PRODUCT

Attributes that describe TITLE

Revised Entity: PRODUCT (in 2NF)

Comments: Only the following attributes describe
a title in a specific media format:

PRODUCT NUMBER + MEDIA CODE
PRODUCT DESCRIPTION
CURRENT RETAIL PRICE
CURRENT LIST PRICE
QUANTITY ON HAND
UNITS SOLD
VALUE OF UNITS SOLD

New Entity: TITLE (in 2NF)

Comments: The remaining attributes truly describe
a product independent of its media type.

PRODUCT NUMBER
TITLE OF WORK
COPYRIGHT DATE
SUPPLIER NAME
SUPPLIER ADDRESS consisting of:
    STREET
    P.O. BOX
    CITY
    STATE
    ZIP CODE

FIGURE 13.6   **Transformation of the PRODUCT Entity into 2NF**

be moved to a new entity (which we call TITLE) whose key is PRODUCT
NUMBER. This new entity does not have a combination of attributes that serve
as its key; therefore, it is automatically in 2NF.

In Figure 13.5 and 13.6 attributes were moved so that they truly describe the
entity in which they were placed. In some cases, this resulted in the identification of new, simpler entities (with fewer attributes). Eventually, we will redraw
our ERD to depict the new entities and some new relationships.

Step 4: Place Entities into 3NF

We can further simplify our entities by placing them into 3NF. Entities are
assumed to be in 2NF before beginning 3NF analysis. The first type of 3NF
analysis is easy: examine each entity and delete any data attributes whose values
can be calculated or derived (through logic or formula) from other data attributes *in that entity*. For example, look at the ORDERED PRODUCT entity in
Figure 13.7. EXTENDED PRICE is calculated by multiplying QUANTITY
SHIPPED by ORDER PRICE. Therefore, EXTENDED PRICE (a non-key attribute) is dependent on two other non-key attributes. Thus, we simplify the entity
by deleting EXTENDED PRICE. Figure 13.7 also shows the attribute NO. MEM-
BERS HAVE NOT FULFILLED being deleted from the AGREEMENT entity
since it can be derived from other attributes within AGREEMENT. Be careful
not to delete attributes that can be derived from the values of attributes con-

Entity: ORDERED PRODUCT (in 2NF)

      ORDER NUMBER +
      PRODUCT NUMBER +
      MEDIA CODE +
      QUANTITY ORDERED
      ORDERED PRODUCT STATUS
      QUANTITY SHIPPED
      ORDER PRICE
      EXTENDED PRICE

Revised Entity: ORDERED PRODUCT (in 3NF)

Comments: The attribute EXTENDED PRICE was deleted since it can be calculated from the attributes QUANTITY ORDERED and ORDER PRICE.

      ORDER NUMBER +
      PRODUCT NUMBER +
      MEDIA CODE +
      QUANTITY ORDERED
      ORDERED PRODUCT STATUS
      QUANTITY SHIPPED
      ORDER PRICE

Entity: AGREEMENT (in 2NF)

      AGREEMENT NUMBER SUFFIX
      CLUB NAME
      AGREEMENT EXPIRATION DATE
      AGREEMENT PLAN CREATION DATE
      MAXIMUM PERIOD OF OBLIGATION
      BONUS CREDITS AFTER OBLIGATION
      NUMBER OF MEMBERS ENROLLED
      NO. MEMBERS WHO HAVE FULFILLED
      NO. MEMBERS HAVE NOT FULFILLED

Entity: AGREEMENT (in 3NF)

Comments: NO. MEMBERS HAVE NOT FULFILLED was deleted since it could be calculated from NO. MEMBERS WHO HAVE FULFILLED and NUMBER OF MEMBERS ENROLLED.

      AGREEMENT NUMBER SUFFIX
      CLUB NAME
      AGREEMENT EXPIRATION DATE
      AGREEMENT PLAN CREATION DATE
      MAXIMUM PERIOD OF OBLIGATION
      BONUS CREDITS AFTER OBLIGATION
      NUMBER OF MEMBERS ENROLLED
      NO. MEMBERS WHO HAVE FULFILLED

FIGURE 13.7   **Analysis and Transformation of Entities into 3NF**   Attributes that can be derived from other attributes in the same entity have either been deleted or removed (or merged) into a separate entity that they really describe.

tained in more than one entity (at least not as part of 3NF analysis). We did not depict relationships on Figure 13.7 since relationships have no effect on this analysis.

Another 3NF analysis checks non-key attributes to see if they really describe a separate entity. This analysis is only performed on those entities that do not have a concatenated key. In our example, this includes ORDER, TITLE, AGREEMENT, MEMBER (which has candidate keys — not a concatonated key), and CLUB. Figure 13.8 demonstrates only those entities that change for this analysis. The other entities were already in 3NF.

For example, examine the TITLE entity in Figure 13.8**A.** Most of the attributes are dependent on the value of the key attribute, namely PRODUCT NUMBER. However, the attribute SUPPLIER ADDRESS (a group of attributes) is not dependent on the entity's key attribute — that is, its PRODUCT NUMBER. Instead, it is dependent on the non-key, SUPPLIER NAME. This suggests that the attribute does not truly describe the entity TITLE. It also suggests that SUPPLIER NAME is a foreign key. A **foreign key** is an attribute that points to another entity. In this case, SUPPLIER NAME points to the entity SUPPLIER. To correct this problem, we factor out SUPPLIER NAME and SUPPLIER ADDRESS and place them in a new entity, SUPPLIER. When we revise our data model (ERD), we will also add a relationship between TITLE and SUPPLIER to retain the association between the data attributes that describe each.

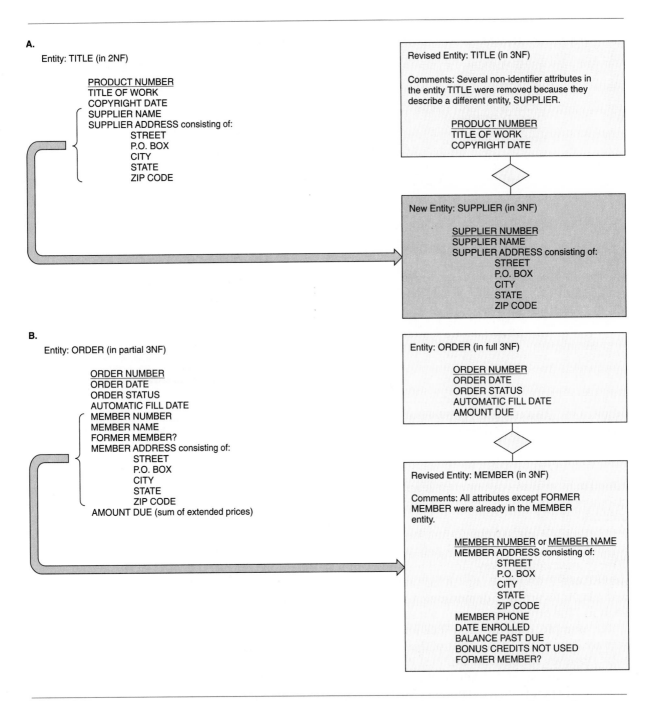

A.

Entity: TITLE (in 2NF)

> PRODUCT NUMBER
> TITLE OF WORK
> COPYRIGHT DATE
> SUPPLIER NAME
> SUPPLIER ADDRESS consisting of:
> > STREET
> > P.O. BOX
> > CITY
> > STATE
> > ZIP CODE

Revised Entity: TITLE (in 3NF)

Comments: Several non-identifier attributes in the entity TITLE were removed because they describe a different entity, SUPPLIER.

> PRODUCT NUMBER
> TITLE OF WORK
> COPYRIGHT DATE

New Entity: SUPPLIER (in 3NF)

> SUPPLIER NUMBER
> SUPPLIER NAME
> SUPPLIER ADDRESS consisting of:
> > STREET
> > P.O. BOX
> > CITY
> > STATE
> > ZIP CODE

B.

Entity: ORDER (in partial 3NF)

> ORDER NUMBER
> ORDER DATE
> ORDER STATUS
> AUTOMATIC FILL DATE
> MEMBER NUMBER
> MEMBER NAME
> FORMER MEMBER?
> MEMBER ADDRESS consisting of:
> > STREET
> > P.O. BOX
> > CITY
> > STATE
> > ZIP CODE
> AMOUNT DUE (sum of extended prices)

Entity: ORDER (in full 3NF)

> ORDER NUMBER
> ORDER DATE
> ORDER STATUS
> AUTOMATIC FILL DATE
> AMOUNT DUE

Revised Entity: MEMBER (in 3NF)

Comments: All attributes except FORMER MEMBER were already in the MEMBER entity.

> MEMBER NUMBER or MEMBER NAME
> MEMBER ADDRESS consisting of:
> > STREET
> > P.O. BOX
> > CITY
> > STATE
> > ZIP CODE
> MEMBER PHONE
> DATE ENROLLED
> BALANCE PAST DUE
> BONUS CREDITS NOT USED
> FORMER MEMBER?

FIGURE 13.8 **Analysis and Transformation of Entities into 3NF**   Non-key attributes that truly describe a separate entity are removed. This approach is only performed upon those entities that do not have a concatenated key.

In Figure 13.8**B** the analysis is the same except that we don't have to create the entity MEMBER. Furthermore, with the exception of FORMER MEMBER?, the other attributes removed from ORDER were already in the MEMBER entity.

Before we leave the subject of normalization, we should acknowledge that several normal forms beyond 3NF exist. Each successive normal form makes the data model simpler, less redundant, and more flexible. However, systems analysts rarely take data models for their applications beyond 3NF. Database experts tend to step beyond 3NF when they merge data models from various applications into a single, corporate data model (or database). Consequently, we will leave any further discussion of normalization to database textbooks.

The first few times you normalize a data model, the process will appear slow and tedious. However, with time and practice, it becomes quick and routine.

Step 5: Further Simplify through Inspection

Normalization is a fairly mechanical process. It is dependent on naming consistencies in the original data model (prior to normalization). When several analysts work on a common application, it is not unusual to create problems that won't be taken care of by normalization. These problems are best solved through **simplification by inspection,** a process wherein a data entity in 3NF is further simplified by such efforts as addressing implicit data redundancy.

Examine all 3NF non-key attributes that also participate as identifiers in other entities. For instance, in the AGREEMENT entity in Figure 13.7, we see CLUB NAME as a non-key attribute. On our ERD, we have already established a relationship between the entities AGREEMENT and CLUB; therefore, we can delete CLUB NAME from the AGREEMENT entity (*Note:* Those who are familiar with database technology will recognize that CLUB NAME is a foreign key; however, foreign keys are physical solutions to linking entities and should not be included in our essential model.)

Carefully check every entity for synonym attributes—attributes that have different names but are really the same thing. They can actually be created through normalization. For example, in our PROMOTION entity, we have a SELECTION OF MONTH NUMBER and SELECTION OF MONTH TITLE. These are synonyms for the TITLE entity's attributes, PRODUCT NUMBER and TITLE OF WORK, respectively. We delete the synonyms from the PROMOTION entity and create a relationship on the ERD to the TITLE entity (Figure 13.9).

Also, check attribute names to ensure that they won't cause confusion. In the CONTRACT entity (see Figure 13.3), we have two attributes named AGREEMENT ENROLLMENT DATE and AGREEMENT EXPIRATION DATE. At the time, the names seemed appropriate. After normalization, they ended up in the CONTRACT entity. We will change them to CONTRACT ENROLLMENT DATE and CONTRACT EXPIRATION DATE to reduce confusion.

You may also check entities for attributes that can be derived from attributes in different entities. For instance, AMOUNT DUE (in the entity ORDER) can be calculated as the sum of QUANTITY SHIPPED times ORDER PRICE (in the entity ORDERED PRODUCT) for each product on the order. Thus, AMOUNT DUE can be deleted.

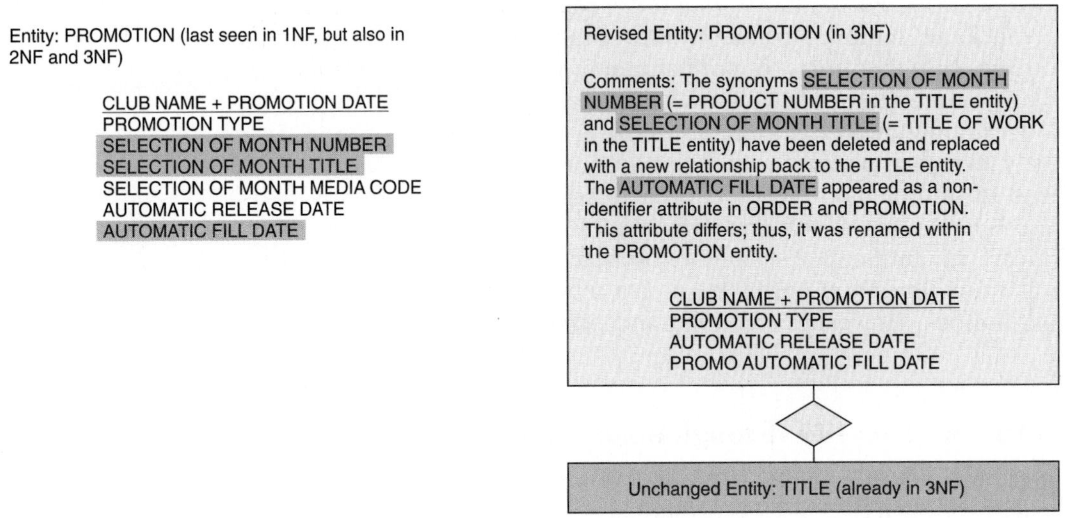

Entity: PROMOTION (last seen in 1NF, but also in 2NF and 3NF)

    CLUB NAME + PROMOTION DATE
    PROMOTION TYPE
    SELECTION OF MONTH NUMBER
    SELECTION OF MONTH TITLE
    SELECTION OF MONTH MEDIA CODE
    AUTOMATIC RELEASE DATE
    AUTOMATIC FILL DATE

Revised Entity: PROMOTION (in 3NF)

Comments: The synonyms SELECTION OF MONTH NUMBER (= PRODUCT NUMBER in the TITLE entity) and SELECTION OF MONTH TITLE (= TITLE OF WORK in the TITLE entity) have been deleted and replaced with a new relationship back to the TITLE entity. The AUTOMATIC FILL DATE appeared as a non-identifier attribute in ORDER and PROMOTION. This attribute differs; thus, it was renamed within the PROMOTION entity.

    CLUB NAME + PROMOTION DATE
    PROMOTION TYPE
    AUTOMATIC RELEASE DATE
    PROMO AUTOMATIC FILL DATE

Unchanged Entity: TITLE (already in 3NF)

FIGURE 13.9   **Further Simplification through Inspection**   Attributes that are synonyms represent implied redundancy and must be removed.

Step 6: Redraw Your Refined ERD

Now we can redraw our ERD to reflect the changes inspired through data analysis. First, quickly review the ERD created in Chapter 8. Next, examine the new ERD shown in Figure 13.10. Note the following changes:

(A) We have added the associative entities that were created when we went to 1NF. In each case, we replaced the many-to-many relationship that used to exist between the associated entities.

(B) We have added the TITLE entity that was created when we placed PRODUCT into 2NF. Because it was factored out of PRODUCT, we added a relationship between TITLE and PRODUCT.

(C) We have added the SUPPLIER entity that was created when we placed TITLE in 3NF.

(D) We have added the PROMOTION-to-TITLE relationship that was identified during simplification by inspection.

The final mapping of the data attributes to these entities is shown in Figure 13.11. Each entity's attributes describe only that entity. Attributes (with the exception of identifiers) are not redundantly stored within or between entities. For one occurrence of each entity, each attribute occurs, at most, one time. The model is simple. It is flexible. It is nonredundant.

We may now determine attributes that may serve as secondary keys.

A **secondary key** is any attribute whose values, while not unique to each occurrence of the entity, effectively divide occurrences of the entities into groups or subsets. The classic example is an attribute such as SEX, which divides occurrences of an entity—for example, STUDENT or EMPLOYEE

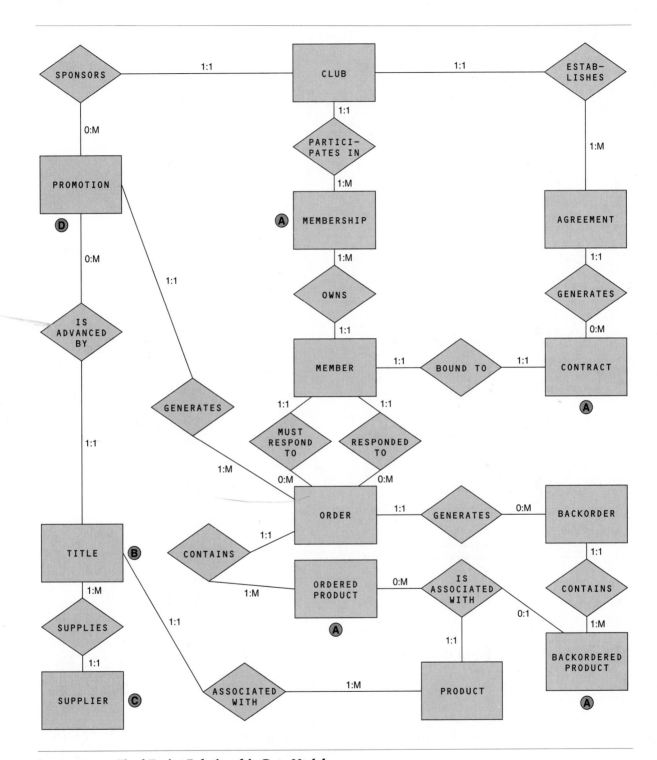

FIGURE 13.10   **Final Entity Relationship Data Model**

Entity: MEMBER

    <u>MEMBER NUMBER</u> or <u>MEMBER NAME</u>
    MEMBER ADDRESS consisting of:
        STREET
        P.O. BOX
        CITY
        STATE
        ZIP CODE
    MEMBER PHONE
    DATE ENROLLED
    BALANCE PAST DUE
    BONUS CREDITS NOT USED
    FORMER MEMBER?

Entity: ORDER

    <u>ORDER NUMBER</u>
    ORDER DATE
    ORDER STATUS
    AUTOMATIC FILL DATE
    AMOUNT DUE

Entity: ORDERED PRODUCT

    <u>ORDER NUMBER +</u>
    <u>PRODUCT NUMBER +</u>
    <u>MEDIA CODE</u>
    QUANTITY ORDERED
    ORDERED PRODUCT STATUS
    QUANTITY SHIPPED
    ORDER PRICE

Entity: PRODUCT

    <u>PRODUCT NUMBER + MEDIA CODE</u>
    PRODUCT DESCRIPTION
    CURRENT RETAIL PRICE
    CURRENT LIST PRICE
    QUANTITY ON HAND
    UNITS SOLD
    VALUE OF UNITS SOLD

Entity: BACKORDER

    <u>ORDER NUMBER + BACKORDER DATE</u>

Entity: BACKORDERED PRODUCT

    <u>ORDER NUMBER + BACKORDER DATE +</u>
    <u>PRODUCT NUMBER + MEDIA CODE</u>
    QUANTITY BACKORDERED

Entity: PROMOTION

    <u>CLUB NAME + PROMOTION DATE</u>
    PROMOTION TYPE
    AUTOMATIC RELEASE DATE
    PROMO AUTOMATIC FILL DATE

Entity: CLUB

    <u>CLUB NAME</u>
    NUMBER OF MEMBERS ENROLLED
    NUMBER CANCELED YTD
    CURRENT PROMOTION
    TOTAL UNITS SOLD FOR CLUB
    MAXIMUM PERIOD OF OBLIGATION

Entity: MEMBERSHIP

    <u>MEMBER NUMBER</u> or <u>MEMBER NAME +</u>
    <u>CLUB NAME</u>
    MUSICAL/MOVIE PREFERENCE

Entity: AGREEMENT

    <u>AGREEMENT NUMBER SUFFIX</u>
    AGREEMENT EXPIRATION DATE
    AGREEMENT PLAN CREATION DATE
    MAXIMUM PERIOD OF OBLIGATION
    BONUS CREDITS AFTER OBLIGATION
    NUMBER OF MEMBERS ENROLLED
    NO. MEMBERS WHO HAVE FULFILLED

Entity: CONTRACT

    <u>MEMBER NUMBER</u> or <u>MEMBER NAME +</u>
    <u>AGREEMENT NUMBER SUFFIX</u>
    NUMBER OF PURCHASES REQUIRED
    NUMBER OF PURCHASES TO DATE
    CONTRACT ENROLLMENT DATE
    CONTRACT EXPIRATION DATE

Entity: SUPPLIER

    <u>SUPPLIER NUMBER</u>
    SUPPLIER NAME
    SUPPLIER ADDRESS consisting of:
        STREET
        P.O. BOX
        CITY
        STATE
        ZIP CODE

Entity: BACKORDERED PRODUCT

    <u>PRODUCT NUMBER</u>
    TITLE OF WORK
    COPYRIGHT DATE

FIGURE 13.11   **Final Mapping of Data Attributes to Data Entities**

— into male and female subsets. Common synonyms include *secondary ID* and *subsetting criteria.*

Step 7: Review and Refine Your Data Model

The data model (ERD) and the contents of the entities should be reviewed with appropriate end-users. Our experience with user acceptance of data models is generally good, especially when you explain that the data model is a picture of the things about which the system must capture and store data and the natural business relationships between those things. Users frequently respond by adding new and potentially useful entities, relationships, and data attributes.

Some users react negatively to such terms as "candidate keys" and "concatenated keys." Avoid those terms; simply use "alternative keys" or "parts of key," respectively.

Many users have difficulty with the idea of an associative entity. They might argue, "Why have you split out ORDERED PRODUCT from ORDER? Aren't ORDERED PRODUCTs part of an ORDER?" Of course they are — at least on the form most people call an order. The best response to such a question is: "This is something database people do to prepare the data for implementation. Don't worry. The data that you will see on the screen will be consolidated."

While the final data model may be reviewed by end-users, you must be careful that the model does not overwhelm the users. Remember, during the analysis phase we chose to only depict fundamental (nonassociative) entities on our data model. This was done because those entities represented those things that the end-user could recognize and relate to. With this in mind, it would be more appropriate to provide a version of the final model that omits the associative entities than to review the final data model in its current form. Think about it. We did identify some new, nonassociative entities, such as SUPPLIER and AGREEMENT. These are likely entities that the end-user will readily identify with. You can expect them to be able to give feedback concerning their relationships to other fundamental entities and to the attribute list. (*Note:* Now that these entities have been discovered and brought to the attention of the end-user, the end-user may identify new attributes he or she wishes to start keeping for a supplier and for an agreement).

Finally, if database administration was not involved in the data analysis, a second review is appropriate. The review involving the database administrator should involve the final data model — including the depiction of associative entities. It should be pointed out that for most organizations, database administration is actively involved throughout data analysis.

EVENT ANALYSIS FOR DESIGN DECISIONS

We now have a data model that is simple, flexible, and nonredundant, but we are not yet ready to begin designing the new system. We still have some issues and potential problems that need to be addressed for our data. Thus far we have focused on what data is to be stored for the new system. We must not forget our obligation to the end-users to ensure that that data is kept accurate and current at all times.

The Need to Perform Event Analysis

We have completed our modeling of the data storage requirements for our new system; however, what have we done as far as defining processing requirements that will ensure that our data will be accurate and current? The answer is, very little. Yes, we have created process models that document what processes use and maintain what data stores (data entities), but those process models (developed in Chapter 9) depicted only prenormalized entities—those that were known prior to performing data analysis. Thus, our process models are now incomplete. Also, those process models depicted the processing of only those inputs that the end-users could identify as input requirements. Did the end-users or analyst miss any inputs?

The implications are substantial. As with our data model (ERD), our process models (DFDs) will serve as a blueprint for completing the designs. As we did with our data model, we must "prepare" our process model for implementation.

We need to ensure that our process model depicts all the maintenance processing needed for all entities. Once more, we must ensure that our process models depict the processing requirements for all inputs—not just those mentioned by the end-users. Not identifying all system input requirements may result in the failure to capture needed data for creating, modifying, or deleting occurrences of the same data entities. This results in incomplete or inaccurate data.

What Is Event Analysis?

The technique that ensures the integrity of a system's data is called *event analysis.*

> **Event analysis** is a technique that studies the entities of a fully normalized data model to identify business events and conditions that cause data to be created, deleted, or modified.
>
> A **business event** is something that "happens" and that causes business data to change. Business events require processes to utilize or maintain the data. These processes may include:
> · Creating a new occurrence of an entity (sometimes called *add*).
> · Reading an occurrence of an entity (sometimes called *use* or *access*).
> · Updating an occurrence of an entity (sometimes called *change* or *modify*).
> · Deleting or archiving an occurrence of an entity.

Business events are also commonly referred to as *inputs, transactions,* or *triggers.* Typical business events may include a "customer places an order," "customer makes a payment," "customer application is approved," "product is approved for sale," "product is deleted from catalog," or "in-stock inventory item falls below a certain level." Each of these represents a business event that triggers processes that will create, delete, or modify occurrences of entities.

When a business event occurs, the action to be taken is often subject to a condition. A **condition** represents a policy that governs the action(s) to be taken when a business event occurs. For example, the business event whereby a "customer places an order" causes the need to create a new occurrence of the data entity ORDER. The action for this business event could be based on the condition that "we must first ensure that we have a corresponding record for the CUSTOMER who placed the order." Taking things even further, the business event may be subject to the additional condition that "the order was preapproved for credit." Thus, there can be multiple conditions, and conditions may be statements concerning how data is related to other data (such as the first condition) or a statement concerning a general business practice (such as the second condition).

When to Perform Event Analysis

In most organizations, event analysis is performed by the systems analyst and end-users. In this book, we present event analysis as a systems design chapter. We do this for the same reasons that we presented data analysis in the design chapter and because event analysis is commonly viewed as a natural extension of data analysis. Whereas data analysis is concerned with preparing our data model implementation, event analysis is concerned with making sure that we identify those processing requirements and rules for ensuring that we implement necessary processes to maintain that data.

A STEP-BY-STEP APPROACH TO EVENT ANALYSIS

This section details a step-by-step approach to event analysis. We'll draw upon the SoundStage case study to demonstrate the steps.

Step 1: Identify Events for Fundamental Entities

The first step in event analysis is to identify business events that affect changes to fundamental entities. The fundamental entities for the SoundStage project included MEMBER, ORDER, TITLE, PRODUCT, SUPPLIER, CLUB, AGREEMENT, and PROMOTION (see Figure 13.11). For each of these entities, we need to identify all business events in terms of actions and corresponding conditions.

Figure 13.12 provides a table to help organize our analysis. The table has the following entries:

(A) A column used to identify the entity for which we will perform the event analysis.

(B) A column for providing a brief narrative description of the business event that affects the entity.

(C) A column for assigning a name to the business event. We will want to eventually revise our DFDs to include all input requirements. This business event name will appear as a data flow on DFDs containing the same name.

Since a business event may appear repeatedly throughout our table as an event that affects different entities, we need to be consistent by using the same name. For example, we don't want to call an event "ORDER" in one place and call it a "CUSTOMER SALE" in another. Decide on a name and consistently use that name. The danger is that we may make the silly mistake of assuming these are two separate and distinct events that will both need to be depicted on our DFDs as separate requirements.

| Ⓐ | Ⓑ | Ⓒ | Ⓓ | Ⓔ |
Entity Name	Event Description	Event Name	CRUD	Condition(s)

FIGURE 13.12   **Sample Table for Recording Event Analysis**

Ⓓ A column to indicate the type of action that is caused to that entity by the business event. Note that the acronym CRUD corresponds to the first letter of each possible action (C = create, R = read, U = update, D = delete).

Ⓔ A column to describe any condition that applies to a given business event and entity.

This table provides an excellent tool for communicating with end-users which is essential in event analysis.

Figure 13.13 presents a partially completed table for a few fundamental entities in our Member Services system for the SoundStage Entertainment Club.

Ⓐ Notice that a single business event may affect more than one entity.

Ⓑ Some business events may be subject to more than one condition. Identifying conditions requires input from end-users. Recall that there are two types of conditions. The end-users will play a very important role in identifying conditions concerning general business practices that will govern business events. But the systems analysts can assume much of the responsibility for identifying conditions that represent policies concerning how data is related to other data.

If the systems analyst is not the business expert, how can he or she identify conditions? The fully normalized data model (ERD) is the answer. The data model provides a visual clue that helps the analyst to identify conditions concerning how data must be related to other data. For example, let's refer back to the ERD in Figure 13.10. Conditions can be identified as follows:

· Examine the relationship between any two data entities. Identify which, if either, of the two entities would represent the parent entity and which would represent the child entity. The **parent entity** is an entity that, given one occurrence of the related entity, may assume at

Entity Name	Event Description	Event Name	CRUD	Condition(s)
PROMOTION	1. The Marketing Department submits promotion cancellations.	PROMOTION CANCELLATION	D	1. There must be no existing, corresponding occurrences of the entity ORDER.
	2. The Marketing Department provides information about new promotions for the month.	NEW MONTHLY PROMOTION	A	1. A corresponding occurrence of the entity TITLE must already exist. 2. A corresponding occurrence of the entity CLUB must already exist.
	3. The Marketing Department notifies Customer Services of changes to existing promotions.	PROMOTION CHANGES	U	1. None Note: Modifications/updates can affect either of the following attributes: AUTOMATIC RELEASE DATE PROMO AUTOMATIC FILL DATE
ORDERS	1. The Marketing Department notifies Customer Services of promotions for the next month.	NEW MONTHLY PROMOTION	A	1. A corresponding occurrence of the entity MEMBER must already exist. 2. A corresponding occurrence of the entity PROMOTION must already exist.

(Page 1 of 12)

FIGURE 13.13   Partially Completed Event Analysis Table   This table contains the event analysis of the data entity PROMOTION and a portion of the event analysis for the entity ORDERS. Event analysis would be performed upon all data entities once data analysis has been completed.

most one occurrence. A **child entity** is an entity that, given one occurrence of the related entity, may assume a variable number of occurrences (for example, zero or one, zero or more, one or more, etc.).

Study the relationship between PROMOTION and TITLE. The TITLE entity would represent the parent entity, and PROMOTION would represent the child entity in this relationship.

Now examine the relationship between PROMOTION and ORDER. In this relationship, PROMOTION would represent the parent entity and ORDER would represent the child entity. What about the relationship between PROMOTION and CLUB? CLUB is the parent, PROMOTION is the child. Which is the parent and which is the child in the relationship between the entities ORDER and BACKORDER?

Now that we have studied our ERD to identify parent versus child in relationships between data entities, we can recognize some absolute conditions governing business events that affect two data entities.

For an entity that participates as a child entity, there must be a condition statement for any business event that would require an occurrence of that entity to be created. That condition would state that before adding the new occurrence, we must make sure that a corresponding occurrence of the parent entity exists. For example, PROMOTION appeared as a child entity in the relationship between PROMO-

TION and TITLE, and as a child in the relationship between PROMOTION and CLUB. Thus, for any business event that would require a new occurrence to be created for PROMOTION, there must be two conditions—one for each relationship. One condition must state that "a corresponding occurrence for TITLE must first exist," and another condition must state that "a corresponding occurrence of CLUB must first exist." Both of these conditions must be met.

For an entity that participates as a parent, there must be a condition statement for any business event that would require an occurrence of that entity to be deleted. That condition would state that before deleting the occurrence, we must make sure that there are no corresponding occurrences of the child entity existing. For example, PROMOTION participates as a parent entity in its relationship with the entity ORDER. Thus, for any business event that would require an occurrence of PROMOTION to be deleted, there must be a condition statement that "no corresponding occurrences of ORDER may exist."

Notice how important these conditions are to ensuring the integrity of the business data. The first two conditions ensure that we will never have a PROMOTION that isn't sponsored by a particular club or that doesn't promote a particular TITLE. Likewise, the later condition ensures that for each ORDER, we'll always be able to associate the ORDER with the PROMOTION that generated it.

ⓒ For business events that cause an update, the data entity's attributes that might be updated are indicated. Since business events that cause updates may only affect one data attribute, identifying these business events is best accomplished by examining each data attribute on an individual basis. For each attribute, we need to ask "what business event might cause the value of this data attribute to be changed (updated)?"

Our table worked well, but can CASE tools help? Unfortunately, most CASE products do not provide the capabilities for creating a table that records event analysis for all entities. Rather, most typically the event analysis is recorded separately for each individual data entity. Figure 13.14 illustrates how the event analysis for the PROMOTION entity would be documented for a popular CASE product.

Step 2: Identify Events for Associative Entities

You might be asking, "Why did we ignore associative entities in step 1? Is event analysis different for associative entities?" The answer is no. Simply stated, it is best to delay addressing associative entities until the event analysis has been completed for all fundamental entities. Here's why.

Locate the associative entity ORDERED PRODUCT appearing on our ERD in Figure 13.10. Notice that ORDERED PRODUCT is an associative entity that describes the association between the fundamental entities ORDER and PRODUCT. What business events might you identify that would affect the entity ORDERED PRODUCT? Which of those business events, if any, do not also represent business events that affect either of ORDERED PRODUCT's related

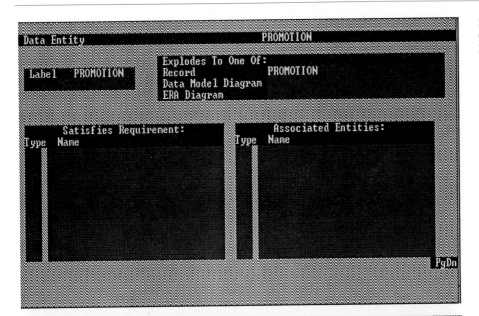

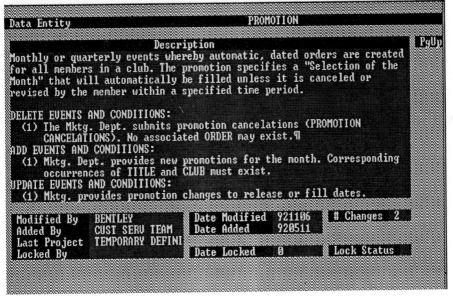

FIGURE 13.14 **Sample CASE Tool Entry for Event Analysis**

fundamental entities ORDER or PRODUCT? Most often, the second answer is none. This tells us that if we first perform event analysis upon our fundamental entities, later we can easily identify the majority of business events that may affect their related associative entities. Let's be careful, however. While this approach will help to automatically identify many business events that affect an associative entity, the actual action and any conditions may vary.

Step 3: Consolidate Common Events

After identifying events for associative entries, we should cross-check the names of our business events. Do the names represent good description names that are appropriate for the DFDs that we will be revising? Have we been consistent with our business event names? For example, if that business event was identified as affecting two or more entities, did we call it by the same name in all locations? Any inconsistencies should be corrected at this point.

Finally, we need to develop a list that will represent a reorganized version of the table we created earlier. This table should be organized by business events. It should identify each business event identified during event analysis, data entities that would be affected by the business event, the action to be taken (e.g., create, update, or delete), and the names of those entities that would need to be used (accessed) to check for any applicable conditions.

IMPACT OF DATA AND EVENT ANALYSIS UPON DFDS

While the DFDs created in Chapter 9 served our purpose of documenting the process requirements for the target system, they were not accurate and complete. Data analysis and event analysis reveal new data entities, actions, and conditions. Since, during analysis, we were only concerned with identifying processing requirements from the standpoint of the end-users, many of these data entities, actions, and conditions were overlooked. Once data and event analysis have been completed, we must revise those DFDs. As you'll learn in Chapter 14, an accurate and complete set of DFDs serves as an essential vehicle for many of the remaining design phases.

Let's examine the impact that data and event analysis may have upon a typical DFD. Figure 13.15 represents the revised DFD for the PROMOTION subsystem in our case study. Let's examine some of the specific impacts that data and event analysis had upon our original DFD:

(A) Notice that we added the data store ORDERS as an input to the MAINTAIN PROMOTION DATA process. This data store was added to reflect the need to ensure that we do not delete an occurrence of the entity PROMOTION as long as there exists a corresponding occurrence of the data entity ORDER. If you refer back to Figure 13.13 (row 1), you will see that this condition was established for the event PROMOTION CANCELLATION.

(B) Our event analysis also established the need to ensure that corresponding occurrences already exist for TITLE and CLUB before a new occurrence of the data entity PROMOTION is created (see Figure 13.13, row 2). Checks for these conditions are included by adding the data stores CLUBS and TITLES as inputs to the RECORD MONTHLY PROMOTION process on our DFD.

The data stores CLUBS and TITLES were also added as inputs to the GENERATE PROMOTION SUMMARY REPORT. After reviewing the contents of the data flow PROMOTION SUMMARY REPORT, it was

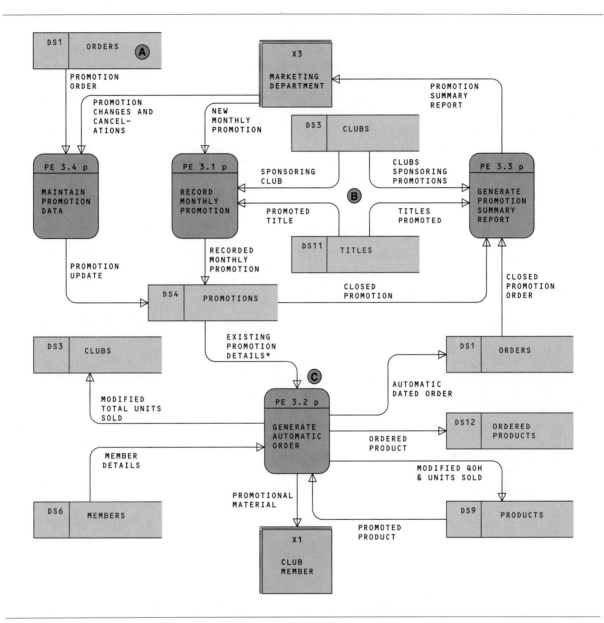

FIGURE 13.15 Revised DFD for PROMOTION Subsystem This DFD, initially developed in the survey phase, has been revised to reflect implications of data and event analysis.

discovered that some of the attributes needed to create the output were attributes that describe these data entities.

Ⓒ The GENERATE AUTOMATIC ORDER process is responsible for generating orders at some predetermined time, once NEW MONTHLY PROMOTIONS have been created. If Figure 13.13 was expanded to include all event analysis, you'd find numerous data entities that were affected by this event. Those entities and any conditions governing the actions taken upon them have been added to our DFD. In addition, appropriate input data flows and data stores have been added in those cases where the processes needed data from additional entities in order to generate the output PROMOTIONAL MATERIAL.

Notice that we do not show checks for all conditions as incoming data flows from data stores. Checks for some conditions are implied. For instance, we did not show an incoming data flow from ORDER that represents a conditional check to first make sure we had an occurrence of the data entity ORDER prior to adding an occurrence of ORDERED PRODUCT. This condition would be specified as part of the internal logic for the process GENERATE AUTOMATIC ORDER (first add ORDER, then add ORDERED PRODUCT).

As mentioned earlier, your instructor may have chosen to cover this chapter immediately after Chapter 7 — during systems analysis. If so, the implications of data and event analysis are the same; you are simply addressing these same implications as you create the DFDs presented in Chapter 9.

COMPUTER-AIDED SYSTEMS ENGINEERING (CASE) FOR DATA AND EVENT ANALYSIS

In Chapter 8 you learned how CASE may be used to support the drudgery of drawing ERDs. But many CASE tools go well beyond merely providing drawing capabilities. Most powerful CASE tools also provide analytical tools to assist in data analysis. These analytical tools can be used to check your ERDs for mechanical errors, completeness, and consistency errors. The potential time and quality savings are substantial!

What about CASE support for event analysis? The event analysis is normally recorded as additional supporting documentation in the data entity repository descriptions. Additionally, data flow diagramming tools can be used to make changes to existing DFDs to reflect any impacts that your event analysis uncovered. The ability to quickly examine an ERD, and query data entities and their content, makes it much easier and less time-consuming to complete event analysis.

Summary

Data analysis is a technique for structuring data in its simplest, most flexible form. It uses normalization to simplify the data model. Through normalization, entities are placed in first, second, and third normal form, in that sequence.

- First normal form entities contain no repeating group of attributes (for a single occurrence of the data entity).
- Second normal form entities contain no attributes that are only partially dependent on a key that is made up from the combination of more than one attribute.
- Third normal form entities contain no attributes that can be derived from the values of other attributes in the same entity.

As you normalize the data model, the ERD may be modified to include new entities and/or relationships. Additionally, the process of normalization can delete unneeded entities or shift attributes to new or different entities that they truly describe.

Once data analysis is completed, event analysis is performed to ensure the integrity of the system data. Event analysis studies the entities of a fully normalized data model to identify business events and conditions that cause data to be created, deleted, or modified.

Both data and event analysis may have a significant impact upon process models (DFDs). For this reason and others, some methodologies perform data and event analysis during the analysis phase—as a natural extension of data modeling—prior to process modeling.

Key Terms

applications data model, p. 507

associative entity, p. 513

business event, p. 524

candidate key, p. 510

child entity, p. 527

concatenated key, p. 510

condition, p. 524

data analysis, p. 507

event analysis, p. 524

first normal form (1NF), p. 509

foreign key, p. 517

identifiers, p. 510

key, p. 510

normalization, p. 507

parent entity, p. 526

primary key, p. 510

second normal form (2NF), p. 509

secondary key, p. 520

simplification by inspection, p. 519

subject data model, p. 507

third normal form (3NF), p. 509

Problems and Exercises

1. Give three characteristics of a good data model.

2. Differentiate between an applications data model and a subject data model.

3. List and briefly describe the three steps of normalization.

4. When can data analysis be performed? Give two reasons why data analysis might be performed during systems analysis.

5. List and explain the seven steps for performing data analysis.

6. What is the difference between a data entity in first normal form (1NF) and second normal form (2NF)? Give an example of an entity in 1NF and show its conversion to 2NF.

7. What is the difference between a data entity in second normal form (2NF) and third normal form (3NF)? Give an example of an entity in 2NF and show its conversion to 3NF.

8. Explain the differences between each of the following: primary key, candidate key, concatenated key, secondary key.

9. List the four possible actions that may be caused by a business event.

10. List and explain the three steps for performing event analysis.

Projects and Minicases

1. Using the case study ERD appearing in Figure 13.10, study the parent and child relationships between associated data entities. Prepare a table with the following column headings: Entity Name, Action, and Condition. For each entity on the

ERD, identify the data entity and specify any conditions implied by its parent/child relationship that would govern adding, modifying, and deleting occurrences of the entity.

2. Given the sample form that follows, prepare a list of entities and their associated data attributes as determined from the document. Then, completely normalize the entities to 3NF and draw a hypothetical ERD. Your instructor should be the final interpreter for the form.

PURCHASING REQUISITION Form 12 Rev. 1988

INSTRUCTIONS — INCLUDE IN EACH REQUISITION ONLY SUCH ARTICLES AS MAY BE PURCHASED FROM ONE FIRM. IF SPECIAL HANDLING IS DESIRED, NOTE. SEE REVERSE SIDE FOR SPECIAL COMMENTS BY REQUESTOR.

DEPARTMENT COMPLETES UNSHADED AREA — PURCHASING COMPLETES SHADED AREA — ORDER NO.

DEPT. OR FUNCTION: Computer Information Systems

COMMITMENT NO. | COMMODITY CODE | ORDER TYPE

M F C	RES. CODE	ACCOUNT NUMBER				DEPT. REFERENCE	AMOUNT	FUND EXPIRATION DATE
		FUND	CENTER DEPT. — PROJ.	OBJECT				
1				5-6207			8,736.00	
2				5-6106			399.00	
3				5-6107			84.00	

SHIP TO STAFF MEMBER: Jonathan Doe
DEPT. 242
BUILDING & ROOM: Administration

ORDER DATE

FOLLOW UP

PRICING METHOD
- [ ] RQ #
- [ ] 1 Phone/Verbal Quote
- [ ] 2 Agreement/Contract
- [ ] 3 Price List on File
- [ ] 4 Repair Negotiation
- [ ] 5 None of the above
BUYER

REQUISITIONER'S PHONE NO.: 555-4545

MATERIAL WILL BE USED FOR

VENDOR SUGGESTED:
IBM
Main Street
Somewhere, IN 47906

VENDOR NAME | VENDOR NUMBER

FOB: [ ] 1 DESTINATION [ ] 2 DESTINATION PREPAY & ADD [ ] 3 SHIPPING POINT [ ] 4 SHIPPING POINT FREIGHT ALLOWED [ ] 5 SEE BELOW | VIA | | TERMS

ITEM #	ITEM DESCRIPTION	MFC	QUANTITY	UNIT	UNIT PRICE	EXTENDED PRICE	DELIVER ON	EST.	COMM.
	IBM PS/2 Model 70 386 8570-121	1	1		7,995.00	4,797.00			
	IBM PS/2 2-8 MB Memory Module Expansion Option #5211	1	1		1,695.00	1,017.00			
	IBM PS/2 2MB Memory Module Kit #5213	1	3		1,395.00	2,511.00			
	IBM 8513 PS/2 Color Display	1	1		685.00	411.00			
	IBM 8770 PS/2 Mouse	2	1		95.00	57.00			
	IBM PC Network Adapter II/A #150122	2	1		570.00	342.00			
	IBM DOS 3.3	3	1		120.00	84.00			

REQUESTED — HEAD OF DEPT.: *Thomas J. Mathieu* | DATE: 6-3-89 | APPROVED — FOR THE COMPTROLLER | DATE | BYPASS APPROVAL REQUESTED [ ] | PURCHASING APPROVALS: PA | AD | DIR

RECOMMENDED — DEAN OR ADMINISTRATOR | DATE | APPROVED — FOR THE EXECUTIVE VICE PRESIDENT AND TREASURER | DATE | APPROVAL SIGNATURE/DATE

OCGBA PREAUDIT
BY: DATE:

Suggested Readings

Flavin, Matt. *Fundamental Concepts of Information Modeling.* New York: Yourdon Press, 1981. This book remains a classic on the topic of data modeling (including data analysis).

Martin, James, and Clive Finkelstein. *Information Engineering.* 2 vols. New York: Savant Institute, 1981. Information engineering is a formal, database, and fourth-generation language-oriented methodology. The method is logically equivalent; however, the authors use entity diagrams instead of entity relationship diagrams. ERDs could easily be substituted.

Mellor, Stephen. *Object Oriented Design.* Englewood Cliffs, N.J.: Prentice Hall, 1987. This modern book presents an object-oriented (entity) approach to systems analysis and design.

Weaver, Audrey. *Using the Structured Techniques.* Englewood Cliffs, N.J.: Prentice Hall, 1987. This book presents, in case study form, a structured methodology based on the complementary use of information models and process models.

14

Process Analysis and Design

Chapter Preview and Objectives

This chapter teaches you techniques for analyzing processes in order to derive an application's general design. Process analysis includes techniques for distributing data and processes to locations in a distributed or cooperative solution. Implementation data flow diagrams are used to document this general design in terms of design units—cohesive collections of data and processes at specific locations—that can be designed in greater detail and subsequently implemented as stand-alone subsystems. You will know that you understand process and data analysis when you can:

Define
centralized, distributed, and cooperative processing as
design alternatives.

Define
client/server computing as a cooperative processing design alternative,
and describe the role of connectivity
and interoperability in client/server computing.

Describe
various networking topologies and the importance
of internetworking.

Describe
various modern design alternatives available
to system designers.

Differentiate
between essential and implementation data flow diagrams.

Draw
implementation data flow diagrams.

Distribute
data and processes in a system using implementation
data flow diagrams.

PACIFIC IMPORTS

For this chapter's minicase, we will describe a process design solution and diagramming challenge instead of the usual playscript narrative.

Pacific Imports, a wholesale distributor of a wide variety of imported products, is located in Los Angeles, California. Pacific has just completed the automation of its mail order system. This minicase reviews the design and implementation.

At approximately 8:30 each morning, the Order Entry department receives all new sales order forms. These forms include orders received both by mail and over the phone. Order Entry enters and edits all orders on personal computers, creating a batch of orders on a network file server.

The batch of orders is transmitted from the server's disk to an AS/400 mid-range computer via a network gateway over a TCP/IP point-to-point network connection. There, the order-processing program checks the inventory master file (actually, a database table) to validate products ordered and to determine the availability and price of the products ordered. For any products that are out of stock, a backorder is created in the database and printed as a notice for the customer. Fillable orders (and partial orders) are then processed as follows.

Another program checks the customer accounts master file to check credit. For customers who have a poor credit rating, a prepayment request is printed. The order is placed on hold in the order master file. Also, a report of these notices is prepared for management. The program prints an order confirmation letter for those orders that will be filled.

Another program produces a four-part warehouse order form that includes a picking copy, a packing copy, a shipping copy, and an invoicing copy. The processed order is moved to a sales master file, a copy of which is archived on magnetic tape for backup and later use.

The system also includes an on-line program that allows the sales manager both to query the inventory master file to obtain prices and to query the customer accounts master file to obtain credit information needed to manually override a credit rejection. This program is available to the sales manager from 8 A.M. until noon each working day.

Customer order cancellations are processed immediately upon receipt (by mail or phone). When the request is received by a clerk, he or she enters the order cancellation using the PC as a terminal into the AS/400. An on-line program reads the sales master file to determine the order's status. If the order has not been filled, the clerk phones the warehouse to have the order terminated. At the end of the day, the program generates cancellation notices for customers and a cancellation report for the sales manager.

This narrative description of the system is becoming lengthy. We haven't yet talked about how the customer billing and payment operations are performed. But let's stop here — we have a challenge for you.

Discussion Questions

1. What are the limitations of the English language for describing a design to programmers who must implement that design?

2. Did you spot any errors of omission in the narrative description — that is, important questions or processes that were not covered?

3. Remember, we have stated in several chapters that a picture is worth a thousand words. Draw a picture of your choice to describe this technical solution to a programmer.

4. In Chapter 9 you used DFDs to model user requirements — a nontechnical picture of those requirements. Do you think DFDs could serve to model the technical design of a system? Why or why not?

GENERAL DESIGN DECISIONS

In Chapter 12 you learned about general and detailed systems design. General design provides the blueprint for detailed design and implementation. It is during general systems design that basic technological decisions are made. These decisions include:

- Will the system use centralized, distributed, or cooperative processes?
- Will the system's data stores be centralized or distributed?
- How will data be input?
- How will outputs be generated?

Let's briefly examine these design decisions.

Centralized, Distributed, and Cooperative Processing

At one time, all data processing was centralized, with data recorded, input, and processed on a central computer.

> In **centralized processing** applications, a host computer (usually a mainframe) handles all activities including input, output, data storage and retrieval, and business logic.

The cost of placing computers closer to the end-user was prohibitive. Over time, the cost of computers went down, and processors were distributed to multiple sites; however, even today, many (if not most) applications remain centralized on large mainframe computers. But while some businesses continue to develop centralized processing solutions, most are now moving toward other alternatives based on distributed processing.

> In **distributed processing** applications, multiple computers (hosts, minicomputers, and sometimes personal computers) handle activities. Each computer in the network handles its own input, output, data storage and retrieval, and business logic.

Computerized processes (and their data stores) were distributed to processor sites. Sites exchange data, but each site generally serves as home to specific processes and data. Distributed processing is not all that new, but its popularity has increased with the availability of lower-cost, mid-range computers, personal computers, and networks.

Today, the cost of computers continues its downward spiral. Personal computers and workstations possess a power and capacity that rivals yesterday's minicomputers. With advances in parallel processors in today's superservers, those superservers possess a power and capacity that rivals today's mainframes.

More importantly, advances in networks and connectivity have allowed computers to interoperate.

> **Connectivity** defines how computers are connected to "talk" to one another.

> **Interoperability** is a state in which connected computers cooperate with one another in a manner that is transparent to their users.

These computer and networking advances have led to yet another processing alternative, cooperative processing in the form of client/server computing.

In **cooperative processing** applications, multiple computers share activities. These computers cooperate in a manner that is transparent to the users — each user perceives that a single computer (possibly their own PC) is doing all the work.

One computer may handle all the input and/or output. Another may handle data storage and retrieval. Business logic for the application may be split between multiple processors (again, transparent to the user).

Client/server computing is the most popular form of cooperative processing.

Client/server computing involves a network of client computers (usually PCs and workstations on the users' desktops) and server computers (usually LAN servers, minicomputers, mainframes — but increasingly, superservers) that interoperate to support applications.

Today, companies are rushing to **downsize** or **rightsize** mission-critical applications from traditional mainframe and mid-range computers to local area networked client/server architectures. The most typical client/server application approach is as follows:

- Input and output functions are placed on the clients. (This includes input editing.) Because PCs and workstations are used as clients, the user interface in client/server applications is usually graphical (e.g., Microsoft Windows, IBM Presentation Manager, or OSF/Motif).
- Data storage and retrieval functions are placed on the server(s). Typically, the server provides a relational, client/server database management system such as Oracle, Sybase, Ingress, or DB2/2. Clients issue SQL (an ANSI standard relational database language) and stored procedure requests to the server databases.
- Business functions and processes are distributed as follows:
  - Business rules, such as those that govern data, are enforced on the server, often as *stored procedures.*
  - Other business logic is programmed to run on the clients.

Of course, all of this is transparent to the end-users, thanks to the power of local and wide area network operating systems and protocols.

Why the popularity of client/server computing? In a nutshell, it's economics! Companies are speculating that they can downsize applications to much cheaper platforms that, through cooperative processing, can achieve equal or better throughput and response time. Furthermore, users find the graphical user interfaces of clients much more consistent, easier to learn, and easier to use!

Note Recent evidence suggests that while client/server hardware costs are cheaper, development costs, maintenance costs, and support costs may be just as expensive as centralized computing. Regardless, the distribution of data and decentralization of processes continues to drive the client/server trend.

Regardless of the choice of mainframes and mid-range computers (now called *hosts*), personal computers and workstations, and LAN servers and superservers, the key to modern computing is networking. Let's briefly survey some of the options. These options are covered more extensively in data communications, telecommunications, and networking courses.

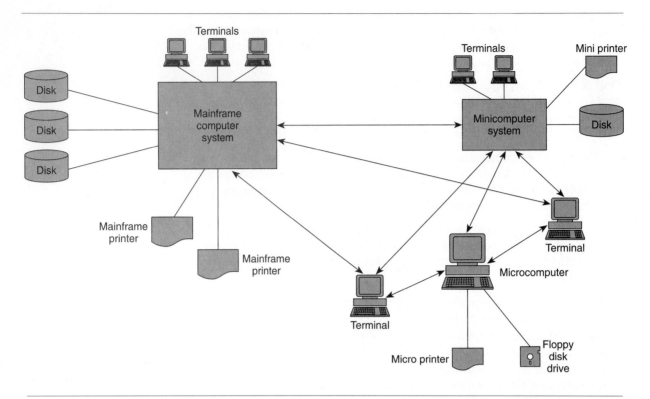

FIGURE 14.1   Point-to-Point Network   The simplest distributed processing network architecture is a point-to-point network, whereby a dedicated data path is placed between two devices. That data path has only to concern itself with understanding the devices on each end.

Point-to-Point and Bus Networks

The simplest networking strategy is to provide a direct point-to-point link between any two computer systems. This **point-to-point networking** concept is illustrated in Figure 14.1. Notice that the network can contain mainframes, minicomputers (or mid-range computers), personal computers, and dumb and intelligent terminals. To completely connect all points between N computers, you would need $N(N\text{-}1)/2$ direct paths. Unless each path is heavily utilized, the cost will prove prohibitive.

What can you do if the direct paths will not be heavily utilized? You could have several computers share a single point-to-point data path (see Figure 14.2). The data path in this case is called a *bus*. Only one computer can send data through the bus at any given time. Incidentally, the computer systems (and other devices) are said to be **multidropped** off the bus.

Ethernet, a product developed jointly by Xerox, Intel, and Digital Equipment Corporation, is an example of a popular **bus networking** strategy. It is most frequently used to implement local area networks (or LANs), a critical component in client/server architectures.

A **local area network** is a collection of computers (usually personal computers and LAN servers), printers, and other devices that are connected through cable over relatively short distances—for instance, in a single department or in a single building.

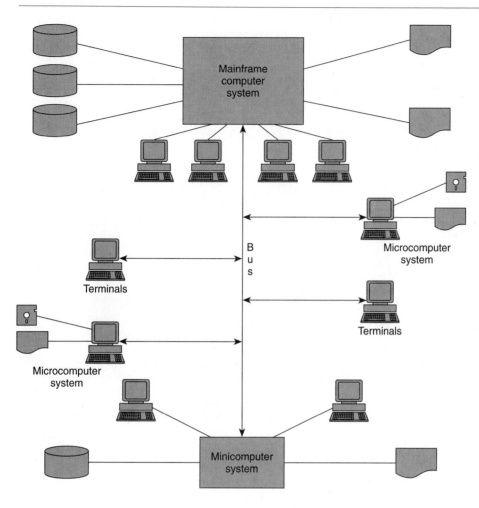

FIGURE 14.2 **Bus Network** A bus network is similar to point-to-point networks except that multiple devices share a single point-to-point pathway. Only two devices, a sender and receiver, may use the pathway at any given time.

Ethernet is one hardware architecture for LANs. LANs also require a software architecture—a network operating system—that manages the sharing of data and peripherals across the LAN's nodes. Examples include Novell's Netware, IBM's OS/2 LAN Manager, UNIX, Microsoft's LAN Manager and Windows/NT Advanced Server, and Banyon's Vines.

Ethernet's bus network manages point-to-point communication between computers and devices on the bus, resolving contention that occurs when more than one computer or device attempts to send a message, instruction, or data at the same time. Ethernet is currently the most dominant LAN architecture for client/server computing.

Star and Hierarchical Networks

A **star network** links multiple computer systems through a central computer (see Figure 14.3). Some would argue that this is a throwback to centralized computing. However, the central computer does not have to be a mainframe or even a mid-range. The central computer, whatever the size, is merely a traffic

FIGURE 14.3   Star Network   In a star network, a central computer plays traffic cop to satellite processors and devices that are trying to communicate with each other and with the central computer.

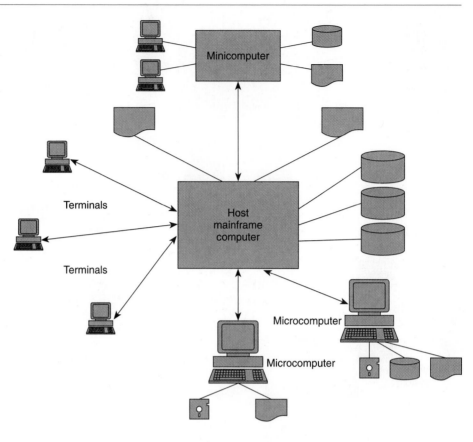

cop used to control the transmission of data and messages between the other computers.

A **hierarchical network** can be thought of as a multiple star network, where the communications processors are arranged in a hierarchy (see Figure 14.4). The top computer system (usually a mainframe) controls the entire network. Its satellite processors (in this case, mid-range computers) have their own satellites (in this case, personal computers and terminals). Notice that each satellite may have its own complement of peripherals. IBM's proprietary Systems Network Architecture (SNA) is essentially a hierarchical network for IBM-compatible computers of all sizes.

Ring Networks

A **ring network** connects multiple computers and some peripherals into a ringlike structure (see Figure 14.5). Each computer can transmit messages, instructions, and data (called *packets*) to only one other computer (or node on the network). Every transmission includes an address, similar to an address you print on an envelope. When a computer receives a packet, it checks the address. If the packet's address is different than the computer's address, it passes it on to the next computer or node. Eventually, the packet arrives at its destination. Ring networks generally transmit packets in one direction; therefore, many computers can transmit at the same time to increase network throughput.

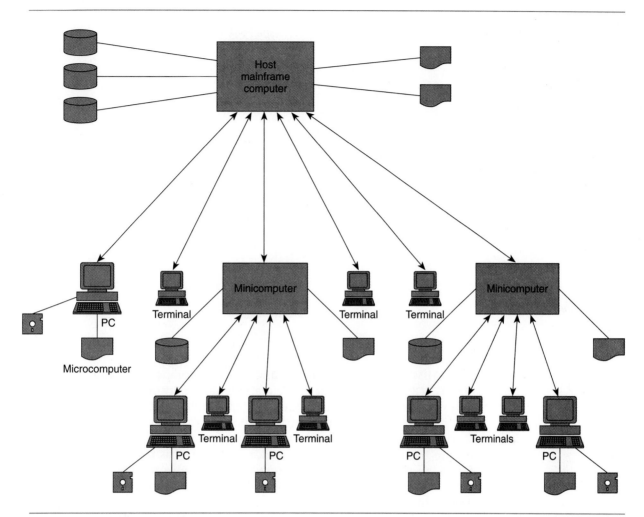

FIGURE 14.4   Hierarchical Network   Hierarchical networks, such as IBM's SNA, use a host computer to control satellite processors and devices, which in turn may control other satellite processors and devices, and so forth. The host computer supervises the entire network.

IBM's Token Ring Network is an example of a ring network. It has become a successful alternative for LANs and wide area networks (WANs) that may include personal computers and workstations, LAN servers and superservers, mid-range computers, and even mainframes.

Internetworking

LANs have dramatically changed the landscape of personal computing. The **interconnection** of LANs into WANs has similarly changed the landscape of enterprisewide computing. Today, servers of various shapes and capacities can be connected to clients (PCs and workstations) to provide the aforementioned client/server architecture that is becoming increasingly popular for applications design.

Interconnection is accomplished through **network protocols** that permit different types of computers to communicate and interoperate. Examples in-

FIGURE 14.5 Ring Network Ring networks link computers, one to another, but only in one direction. Data transmissions are passed around the ring until they arrive at their intended destination.

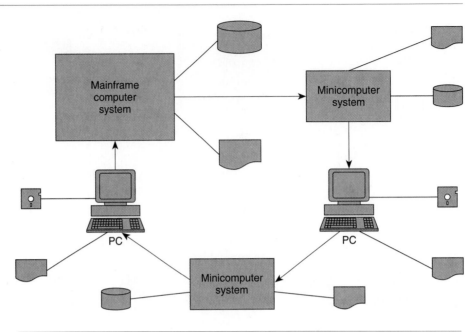

clude IBM's SNA (proprietary to IBM and IBM-compatible computers of all sizes), TCP/IP (the current de facto standard for networks that contain computers of various architecture), and the forthcoming OSI (a standard protocol that promises open system interconnection for a wider variety of systems).

The network architecture in most businesses has evolved by chance. In more mature businesses, a formal network architecture has been planned for the future. Specific LAN, WAN, and telecommunications topologies have been agreed to, and all applications must be developed within the constraints of that architecture.

The many complexities of data communications are beyond the scope of this book. However, as a systems analyst, you will often enlist data communications and networking specialists to help answer questions that arise. You'll find it almost essential to have at least a working knowledge of the vocabulary and technology of data communications. Enroll in your local introductory data communications and networking course; you can't go wrong.

Centralized and Distributed Data Storage

Process distribution is not the only design issue. The location of data stores is also an issue. Historically, the need to control data centrally has been considered essential. Until recently, the only practical way to accomplish this goal was to store all data on a central computer(s), with absolute control by a data administration group. Some ''local'' only data might be distributed to midrange computers, while remaining under the control of a data administration group.

To support client/server computing, we need to be able to distribute data to multiple servers while maintaining control over its integrity (meaning who can update the data and under what conditions and rules). This is being accomplished through distributed database management systems technology.

> **Distributed database management systems (DDBMSs),** available from a number of vendors (e.g., Oracle, IBM, and Sybase), allow data to be distributed or duplicated on multiple hosts and servers while maintaining control of data integrity.

In a distributed database environment, transactions are (or can be) designed such that they update data in more than one physical location, possibly on different database management systems.

As with networking, all of the modern database distribution and management issues and techniques are beyond the scope of this book. Systems analysts work with data professionals to apply the database technology. Thus, a comprehensive data and/or database management course has become another essential component in the aspiring systems analyst's curriculum.

Inputs and Outputs

Other fundamental design decisions must be made regarding inputs and outputs. The decision used to be simple — *batch* versus *on-line.* Today, we must also consider modern alternatives such as remote batch, keyless data entry, pen data entry, graphical user interfaces, electronic data interchange, imaging and document interchange, among others. Let's briefly examine these alternatives.

Batch Input/Output

In **batch processing,** transactions are accumulated into batches for periodic processing. The batch inputs are processed against master files or databases. Transaction files or databases may also be created or updated by the transactions. Most outputs tend to be generated to paper or microfiche on a scheduled basis. Others might be produced on demand or within a specified period of time (e.g., 24 hours).

Contrary to popular belief, batch-based applications are not obsolete. Some application requirements lend themselves nicely to batch processing. Perhaps the inputs arrive in natural batches (e.g., mail), or perhaps outputs are generated in natural batches (e.g., invoices). There is, however, the definite trend away from batch input and output to on-line approaches. As the cost of computer processing decreases, and network technology enables paperless document flow, we expect batch inputs and outputs to decline. In any case, as older batch-based systems become candidates for replacement, other alternatives should at least be explored.

On-Line Input/Output

The majority of systems have slowly evolved from batch I/O to on-line I/O. On-line systems provide for a conversational dialogue between user and computer. Business transactions and inquiries are often best processed when they occur. Errors are identified and corrected more quickly. Transactions tend to be processed earlier since on-line systems eliminate the need for batch data file preparation. Furthermore, on-line methods permit greater human interaction in decision making, even if the data arrives in natural batches. Inquiries and reports can, for the most part, be processed immediately. And as PCs have replaced dumb terminals, PC software can both support more creative formatting of I/O and print to local (and network) printers.

The lower response time requirements of most applications and the desire for human interaction during processing has driven systems development to on-line alternatives. Other contributing factors include faster computers, increased capacity to handle more simultaneous users, and better on-line development and control software.

Client/server applications are simply a new form of on-line processing. Input editing and output formatting occur on client computers in an on-line mode. Input transactions and information requests are transmitted on-line to several computers for processing.

Remote Batch

Remote batch combines the best aspects of batch and on-line I/O. Distributed on-line computers handle data input and editing. Edited transactions are collected into a batch file for later transmission to host computers that process the file as a batch. Results are usually transmitted as a batch back to the original computers.

Remote batch is hardly a new alternative, but personal computers have made the option increasingly more attractive. PCs provide low-cost, on-line data capture and editing power with adequate storage for resultant batches. PCs with graphical user interfaces (e.g., Windows) provide for simpler data entry and user assistance. Finally, PC-to-host communications technology makes it easy to transmit the batch to another computer for processing and to receive the results of that processing.

Remote batch using PCs should get another boost with the advances in laptop and palmtop computer technology. These three- to seven-pound computers can be used to collect batches of everything from mortgage applications to university and housing applications. Don't forget this technology of remote batch I/O as you redesign old batch and on-line systems.

Keyless Data Entry

Keying errors have always been a major source of errors in computer inputs (and inquiries). Any technology that reduces or eliminates the possibility of keying errors should be considered for system design.

In batch systems, keying errors can be eliminated through **optical character reading (OCR)** and **optical mark reading (OMR)** technology. Both are still viable options for input design.

The real advances in keyless data entry are coming for on-line systems. **Bar coding** systems (similar to universal product code systems that are commonplace in the grocery and retail industries) are widely available for many modern applications. For example, Federal Express creates a bar code–based label for all packages when you take the package to a center for delivery. The bar codes can be read and traced as the package moves across the country to its final destination.

Keyless data entry should be considered for appropriate high-volume transaction-based systems as they become candidates for redesign.

Pen Input

Pen-based computing is starting to evolve. As pen-based operating systems (e.g., Microsoft's Pen Windows) become more widely used and tools for building pen-based applications become available, we expect to see more system designs that exploit this technology. Some businesses already use this technology for remote data collection. For example, UPS uses pen-based notebook systems to communicate deliveries to drivers and to collect delivery confirmation signatures and data from customers and drivers. When a driver returns to their distribution center, the data is transmitted from the pen-based notebook computer to host computers.

Graphical User Interfaces

Graphical user interfaces (GUIs) were popularized by the success of Apple's Macintosh and Microsoft's Windows. While the commercial success has been driven by applications such as word processing and spreadsheets, the popularity of the interface is driving all applications to the interface.

Technology exists to create GUI-like applications for dumb terminals. Technology also exists to create true PC-based GUIs that work with host applications via cooperative processing. And most importantly, GUI technology has become the user interface of choice for client/server applications.

GUIs do not automatically make an application better. Poorly designed GUIs can negate the alleged advantages of consistent user interfaces. Fortunately, GUI standards are evolving to guide system designers to create consistent interfaces. For example, DOS/Windows and OS/2 Presentation Manager are based on a standard called *Common User Access (CUA)*. Properly designed GUIs simplify input, reduce keystrokes required, and provide interesting and useful formatting options for outputs. Many businesses are mandating their use for all new systems.

Electronic Data Interchange

Businesses that operate in many locations and businesses that seek more efficient exchange of transactions with their suppliers and/or customers often utilize electronic data interchange.

> **Electronic data interchange (EDI)** is the electronic flow of business transactions between customers and suppliers.

With EDI, a business can eliminate its dependence on paper documents and mail. For example, Sears uses EDI to directly submit their purchase orders from their own computers to their suppliers' computers. The competitive advantages of reduced response time should be obvious. It is expected that larger businesses will convert most of their customer and supplier transactions to EDI by the end of the decade.

Imaging and Document Interchange

Another emerging I/O technology is based on image and document interchange. This is similar to EDI except that the actual images of forms and data are transmitted and received. It is particularly useful in applications in which the form images or graphics are required. For example, the insurance industry has made great strides in electronically transmitting, storing, and using claims images. Other imaging applications combine data with pictures or graphs. For example, a law enforcement application can store, transmit, and receive photographic images and fingerprints.

IMPLEMENTATION MODELS

Just as we modeled user requirements during systems analysis, we should model implementation requirements during systems design. The models serve as blueprints for prototyping and/or detailed systems design. We'll cover two popular tools for modeling the design decisions discussed in the first section of this chapter: implementation data flow diagrams and system flowcharts. The former is the modern tool of choice. The latter used to be the tool of choice and is still encountered when studying documentation of older systems.

Implementation Data Flow Diagrams

Data flow diagrams were introduced in Chapter 9 as a tool for modeling the essential (sometimes called *logical*) requirements of a system. With just a few extensions of the graphical language, DFDs can also be used to model the implementation (sometimes called *physical design*) of the system.

> **Implementation data flow diagrams** model the physical and technical design decisions made for a new or improved system. They communicate technical and other design constraints to those who will actually implement the system — in other words, they serve as a blueprint for the implementation. Implementation DFDs are also said to model the general (as opposed to the detailed) design of the system.

> *Note*   This chapter deals with the transition of essential models to implementation models. You may find it useful to briefly review Chapter 9, the section titled "Systems Modeling: The Essence of a System" for a review of the "essential versus implementation" modeling concept.

Let's examine the graphical conventions for implementation DFDs. Implementation DFDs use the same symbols as essential DFDs (Chapter 9): processes, external agents, data stores, and data flows. Only the object naming standards (and a few rules) are changed to extend the language to document technology and design decisions. In this section, we'll focus only on the new naming standards and rules.

Implementation Processes

A process is the main symbol on any DFD.

> An **implementation process** is work or actions to be performed on incoming data flows to produce outgoing data flows. An implementation process clearly indicates which person(s) or what technology is to be used to perform the named work.

Some essential processes must be split into multiple processes on an implementation DFD because the essential process may require more than one different implementation. For example, part of a process may be manual and part may be computerized. Or parts of an essential process might be implemented with different technologies. Or, perhaps, different parts of the essential process are to be distributed to client and server processors.

Some design decisions may require that we add new processes that do not directly support user requirements. Instead, they are necessary only because of the way we choose to implement the system. Examples include utility processes such as sorting, merging, and extracting data from files.

In all cases, if you split an essential process into multiple implementation processes or add new implementation processes, you have to add any necessary data flows to preserve the essence of the essential process. In other words, the implementation processes must still meet the essential requirements.

Process names use the same *action* verb + object clause convention as before; however, the name is preceded by an implementation method. The format is:

IMPLEMENTATION METHOD : PROCESS NAME

An example (see margin samples) is ON-LINE COBOL : CHECK CUSTOMER CREDIT. We could have been more specific; for example, CICS+COBOL : CHECK CUSTOMER CREDIT. Let's further examine implementation methods.

First, all implementation processes are either manual or computerized. If a process is to be part manual and part computerized, it must be split into two processes and associated data flows must be added between the processes.

For manual processes, the implementation method should indicate who will perform that process. We recommend titles, not proper names. An example is SUPERVISOR: or ORDER CLERK:. Of course, the colon would be followed by the essential name of the process. Why not just change the manual process to an external agent symbol? Because, as analysts, a system design is not complete until all business processes for the new system are also described. This includes (standard) operating procedures for all manual processes.

For computerized processes, the implementation method is, in part, chosen from one of the processing methods listed in the margin. The remainder of the implementation method depends on whether the process will be one of the following:

- A package (possibly to be selected).
- A utility program (possibly to be selected).
- An existing program (from a program library).
- A program to be written.

For a package to be selected, the remainder of the method should be PACKAGE. For instance, ON-LINE PACKAGE : ENCUMBER FUNDS specifies that an on-line application package, to be chosen, must be able to encumber funds. For an existing or preselected package, the method would be ON-LINE FMIS: ENCUMBER FUNDS where FMIS (Financial Management Information System) is a purchased software package.

For a utility to be selected, the implementation method should be UTILITY. If the specific utility already exists or has been determined, the implementation name should be the utility's name. For another example, BATCH QUICKSORT: SORT INVOICES indicates that a batch (the method) sort utility called Quick-Sort (the name) will be used to sort invoices.

For an existing program to be reused, the implementation method should be LIBRARY. In this case, the essential name should be replaced with the specific library program to be used. For example, the process name ON-LINE, LIBRARY : EDIT-STU.COB indicates that the essential process to edit student data will be implemented using an existing on-line COBOL program in the library called EDIT-STU.COB.

1

ON-LINE COBOL: CHECK CUSTOMER CREDIT

2

ORDER CLERK: TAKE ORDER OVER PHONE

Computer Processing Methods

Batch
On-line
Client
Server

Finally, for a program to be written, the implementation method should indicate the language or tool to be used to write the program. A few examples include:

COBOL2

CSP (a fourth-generation language that generates COBOL code)

C

C++

Focus (a fourth-generation language)

SAS (another fourth-generation language)

DB2-SQL (a database programming language)

dBASE (another database programming language)

Excel (a spreadsheet and macro language)

Powerbuilder (a client/server development tool/language)

ObjectView (another client/server development tool/language)

Note that specifying a language does not eliminate the possibility of using a code generator. Also note that, today, it is not uncommon to use different languages for different parts of a system. Thus, the following process names are representative of alternative designs for the same essential process:

- COBOL+CICS : RESERVE SEAT IN COURSE.
- COBOL+DB2: RESERVE SEAT IN COURSE.
- Powerbuilder+Oracle SQL: RESERVE SEAT IN COURSE.

The number of processes on an implementation DFD will usually be greater than the number of processes on its equivalent essential DFD. For one thing, processes may be added to reflect data flow collection, filtering, forwarding, preparation, business controls—all in response to the implementation target that has been selected. Also, some essential processes may be split into multiple processes to reflect portions of a process to be done manually versus by a computer, to be implemented with different technology, or to be distributed to clients, servers, or different host computers. It is important that the final implementation DFDs reflect all manual and computer processes required for the chosen implementation strategy.

Implementation Data Flows

Recall that all processes have at least one input and one output data flow.

BARCODE:
TOOL
CHECKOUT

$\longrightarrow$

GUI SCREEN:
CUSTOMER
ORDER FORM

$\longrightarrow$

EDI: PURCHASE
ORDER

$\longrightarrow$

An **implementation data flow** represents the implementation of an input to or output from an implementation process. It can also indicate an access to or update of a file or database. It can represent the import of data to or the export of data from another system across a network. Finally, it can represent the internal transfer of data between two processes implemented within the same program.

Implementation data flow names use the following general format (see margin samples):

IMPLEMENTATION MEDIUM: ESSENTIAL DATA FLOW NAME

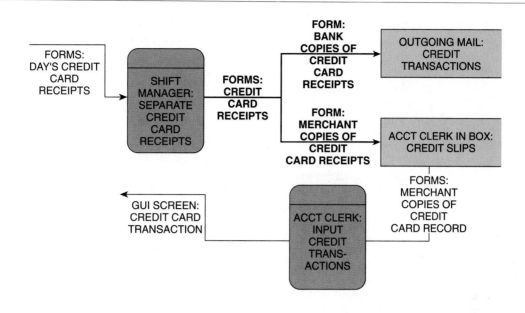

FIGURE 14.6 **Diverging Data Flows** Diverging data flows on implementation DFDs indicate the common design decision to use multipart forms.

Representative input media include PUNCHED CARDS, KEY-TO-TAPE FILE, KEY-TO-DISK FILE, OMR (optical mark reader), OCR (optical character reader), BARCODE, EDI (electronic data interchange), CUI SCREEN (character-based user interface screen), and GUI (graphical user interface screen). For GUIs, you could use the specific GUI name such as WINDOWS, PM, or MOTIF.

Representative output media include PRINTOUT, MICROFICHE, GUI SCREEN, EDI, and IMAGE. Once again, specific GUI names could be used.

For data transmitted across a network, the implementation media should indicate the file transfer protocol to be used. Examples include SNA, TCP/IP, and OSI. This notation will be especially important for communication between cooperative processes that are to be implemented on clients and servers.

For data transmitted between processes that are to be part of the same program, do not specify an implementation media. Such data flows are implemented within COBOL data divisions or by passing variable names to and from subroutines.

Implementation DFDs must also indicate any business forms used in the system. Use the medium FORM. If form numbers are already known, they can be included as part of the name (for instance, FORM 23: COURSE REQUEST FORM). Business forms frequently use a multiple (carbon or carbonless) copy implementation. At some point in processing, the different copies are split and travel to different manual processes. This is shown on an implementation DFD as a **diverging data flow** (see Figure 14.6). Notice that specific copy information is appended parenthetically to the essential data flow name.

Most essential data flows are carried forward to the implementation DFDs. Some are consolidated in single data flows that represent business forms.

Others are split into multiple flows as a result of having split essential processes into multiple implementation processes. For example, if an essential process is to be implemented partially on client workstations and partially on a server or host, the process is split into a client portion and a server portion. Data flows between the client and server indicate what data (and commands) are passed from client to server and back again.

External Agents

External agents are carried over from the essential DFD to the implementation DFD unchanged. Why? By definition, external agents were classified during systems analysis as outside the scope of the systems and, therefore, not subject to change. Only a change in requirements can initiate a change in external agents.

Implementation Data Stores

From Chapter 13 you know that each data store on the essential DFD now represents a *data entity* on a normalized **entity relationship diagram.** Accordingly,

> An **implementation data store** represents all instances of one or more named data entities to be stored as a file or a database record/table. Additional data stores may represent temporary data files necessitated by implementation processes.

The name of an implementation data store uses the following format (see margin samples):

FILE OR DATABASE ORGANIZATION: DATA STORE NAME

Representative file and database organizations are listed in the margin.

One special notation is important. As you may recall from a programming or file organization class, conventional files may be either fixed length (based on a single data entity, such as CUSTOMER) or variable length (based on two data entities, such as ORDER and ORDER ITEM). In the latter case, one record in the ORDER with ORDER ITEMs file describes one order plus one or more ordered items for that order. Since different orders have different numbers of order items, the records in the file are said to be of variable length. The implementation name of the variable length data store is ORDER + ORDER ITEMS.

Some designs require that temporary files be created to act as a queue or buffer between processes. Such files are documented in the same manner except that their name should include some indication of their temporary status.

A general design may also include noncomputerized files. If this is the case, the storage mechanism name replaces the file organization. For example, FILE CABINETS: RADIOACTIVE ISOTOPE RESEARCH FILE indicates that records describing research involving radioactive isotopes are stored in file cabinets. Despite futurist predictions about the demise of paper files, they will remain a part of many systems well into the foreseeable future — if for no other reasons than (1) there is psychological comfort in paper and (2) government frequently requires it!

ORACLE TABLE:
ORDERS

VSAM FILE:
PRODUCTS

SEQ FILE:
SORTED
ORDERS (temp)

FILE CABINET:
PERSONNEL RECORDS

✓

Representative File and Database Organizations

Sequential File
Direct File
ISAM File
VSAM File
Inverted File
IDMS Set
IMS Data Set
DB2 Table
Oracle Table
Paradox Table
Focus Database

Drawing Implementation DFD

The mechanics for drawing implementation DFDs are identical to those of essential DFDs. The rules of correctness are also identical. We'll not repeat those rules here. (Review Chapter 9 for those rules.)

Implementation DFDs document the high-level design of the new system. An acceptable design results in

- A system that works.
- A system that fulfills user requirements (specified in the derivative essential DFDs).
- A system that provides adequate performance (throughput and response time).
- A system that includes sufficient internal controls (to eliminate human and computer errors, ensure data integrity and security, and satisfy auditing constraints).
- A system that is adaptable to ever-changing requirements and enhancements.

We could develop a single implementation DFD or leveled set of implementation DFDs for the target system. Alternatively, we could develop a lot of small, self-contained implementation DFDs that could be assigned to different implementation teams. These are called design units.

> A **design unit** is a self-contained collection of processes, data stores, and data flows that share similar design attributes. A design unit serves as a subset of the total system whose inputs, outputs, files and databases, and programs can be designed, constructed, and unit tested as a single subsystem. (Note: The concept of design units was first proposed by McDonnell Douglas in the STRADIS methodology.)

A design unit may be documented using either implementation data flow diagrams or system flowcharts.

An example would be a set of processes (one or more) to be designed as a single COBOL, CICS, DB2 program. The design unit could then be assigned to a single programmer (or team) who (which) can work independently of other programmers and teams without adversely affecting the work of the other programmers. The implemented units would then be assembled into the final application system. Design units can also be prioritized for purposes of implementing versions of a system.

Computer-Aided Systems Engineering (CASE) for Implementation DFD

CASE was introduced in Chapter 4 as a technology for systems development. Most CASE products support DFDs, but don't distinguish between essential (logical) and implementation (physical) models. Some CASE users simply modify the DFDs that they created during systems analysis. The problem with that approach is that you are overwriting the essential DFDs — the requirements themselves! A better approach is to copy the essential models and then modify them.

Many CASE tools cannot completely support the graphical language extensions described in this chapter. Limitations on name lengths is the most common problem. You can overcome the limitations by agreeing on a standard set of implementation technology abbreviations. For example, BC can mean "Bar Code"; therefore, BC: TOOL ASSIGNMENT might mean that the data flow that checks out a tool to an employee will be implemented using bar-code technology. Alternatively, implementation details can be recorded in the repository description of objects on the DFDs.

The ultimate goal of CASE technology is to be able to automatically generate application programs (code) from these models and supporting detailed specifications (all stored in the CASE tools' repositories).

System Flowcharts

Before we teach you how to draw implementation (and design unit) DFDs, we should introduce an alternative. Before the popularity of DFDs, systems flowcharts were a popular tool for modeling design decisions.

> **System flowcharts** are diagrams that show the flow of control through a system while specifying all programs, inputs, outputs, and file/database accesses and retrievals.

While the popularity of system flowcharts is clearly on the decline, you may encounter them (especially as existing system documentation for older information systems). You need to learn how to read them.

The American National Standards Institute (ANSI) has established certain symbols that have been widely used in the computer industry to describe the flow of process control in systems. Although the symbols have been standardized, their use has not. Thus, many system flowcharts look incomprehensible to those who would like to use them.

System flowcharts are supposed to be the basis for communication between systems analysts, end-users, applications programmers, and computer operators. Think of system flowcharts as a chance to prove or disprove that a specific technical solution to the end-users' requirements will work.

In this section, you'll learn how to read system flowcharts.

System Flowchart Symbols

✓

System Flowchart Symbol Subsets

Processes
Batch input
Batch output
Files and databases
On-line inputs and outputs
Miscellaneous

Figure 14.7 shows the system flowchart symbols and their meanings. The symbols can be conveniently classified into six subsets listed in the margin.

There are four symbols for processing: the computer program (to be written), the library program (that already exists—possibly a utility program), the manual operation (indicating who and/or what—the symbol is also used as a start or finish symbol in a flowchart), and the auxiliary operation (used to indicate operations performed by other office equipment). The name (or identification) of the process is recorded in all the symbols.

There are four symbols for batch input: the source document (from which data will be keyed into the system), the key-to-punched-card (a dated and increasingly rare batch input medium), the key-to-disk (KTD) or key-to-tape (KTT), and the optical character (or mark) document (which could also be used for EDI and imaged documents). The name or identification of the document, or resultant batch input file, is recorded in the symbol.

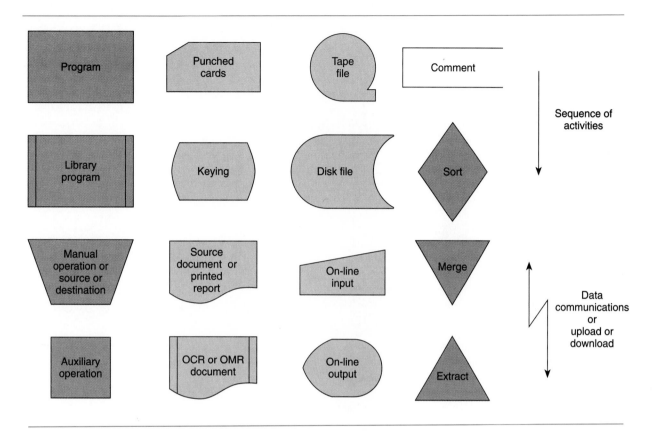

FIGURE 14.7 **Systems Flowchart Symbols and Their Meanings**

Batch output forms or files are represented by a single printed output symbol. The symbol could also be used for microfiche output. The name or identification of the output is recorded in the symbol.

System flowcharts show only those files and databases stored on computers. There are only two symbols: the tape file, and the disk file or database. Because tape files are always sequential, updates always occur in pairs — an original file is input to a program and a new file is produced by the program. Disk files need not be duplicated since reads and writes are processed against the same copy of the file or database. On the other hand, applications sometimes create distinct subsets or reorganizations of files for a program's exclusive use. These versions are shown as separate symbols.

Databases serve to integrate many files. They can be depicted as one integrated symbol or as one symbol per file, record type, or table — depending on what level of detail you are trying to depict. For all tape and disk symbols, the label indicates the name or identification of the file, database, record type, or table.

The symbolism for on-line inputs and outputs can be somewhat tricky. If we had to show every possible screen, the systems flowchart would get very cluttered. Therefore, most designers show only the net inputs and outputs and exclude the on-line dialogue that gets you to those inputs and outputs. There are two on-line symbols: the on-line input and the on-line output. As usual, the name of the input or output is recorded in the symbol.

FIGURE 14.8 System Flowchart (Page 1)
This system flowchart depicts the methods and procedures typically associated with batch processing of a transaction. Particularly common for batch processing is the keying/edit loop. Not all batch systems require the sorting steps shown in this example.

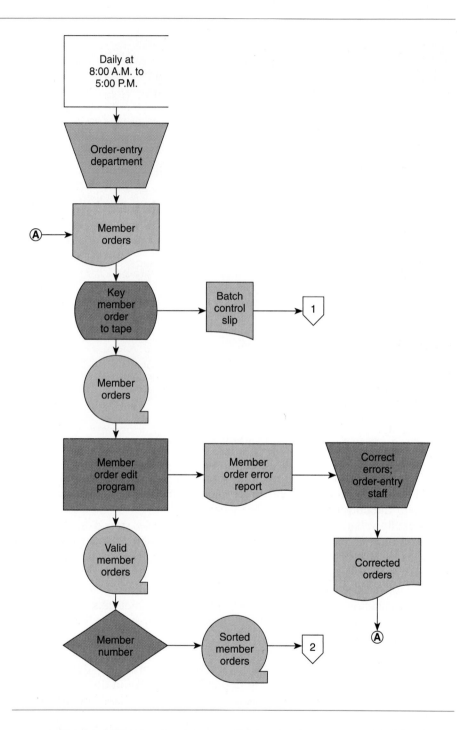

There are a number of miscellaneous symbols in the ANSI standard. They are used to document aspects of methods not covered by the other symbols. The most important of these symbols is the comment. It may be connected via a dashed line to any other symbol to add information about such things as timing, security, or instructions. Other miscellaneous symbols include sort, merge, and extract.

Reading System Flowcharts

The symbols are connected by one of three lines: a single-ended arrow (indicating either the sequence of activities or processing in the system or a read-only or write-only access to a file or database), a double-ended arrow (indicating read-write operations against files and databases), or a jagged-double-ended arrow (indicating on-line dialogue or data flow). Unlike DFDs, connections on system flowcharts are not labeled or named.

Symbols and connections are combined in classic input-process-output patterns to document the design of a system. Figures 14.8 and 14.9 demonstrate sample system flowcharts for part of a system. Notice the small circle (labeled A) and small pentagons (labeled 1 and 2). The circles are on-page connectors that show transfer of flow of control on the same page (to avoid crossed lines). The pentagons are off-page connectors that show transfer of flow of control to the matching symbols on another page. The rest of the diagram should be fairly self-explanatory.

GENERAL SYSTEM DESIGN FOR PROCESSES

The use of DFDs to model process requirements is a fairly accepted practice. However, the transition from analysis-oriented essential DFDs to design-oriented implementation DFDs has historically been somewhat mysterious and elusive. What we need is a high-level general design that can serve as a framework for detailed design. Implementation DFDs provide such a framework. But how does one transform the essential DFDs into the implementation DFDs? Let's set the table by describing the prerequisites to creating implementation DFDs. They include:

- Essential data model (an entity relationship diagram created in Chapter 8 and normalized in Chapter 13).
- Essential process model (data flow diagrams created in Chapter 9 and synchronized to the normalized data model in Chapter 13).
- Essential network model (optional — location connectivity diagrams created in Chapter 10; data flow diagrams could be adapted for the same purpose).
- Essential details of data, process, and network models as described in a project repository (Chapter 11).

Given these models and details, we can distribute data and processes to create a general design. Your general design will normally be constrained by one or more of the following:

- Standards or an approved architecture that has predetermined the choice of database management systems, network topology and technology, and/or processing methods.
- Project objectives that were defined at the beginning of systems analysis and refined throughout systems analysis.
- Feasibility of chosen technology and methods. Feasibility analysis methods are covered in Part Five, Module C).

Within any restrictions of those constraints, the ensuing techniques can be applied.

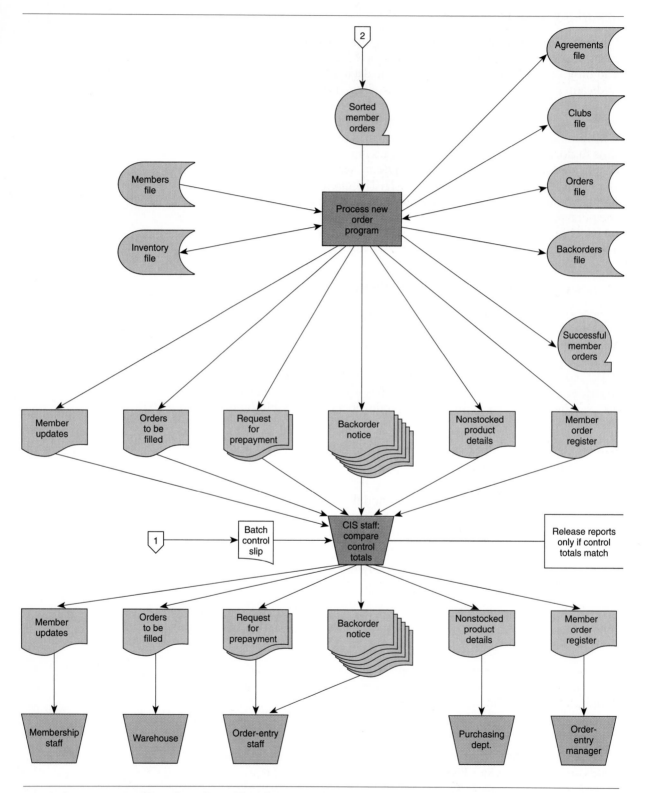

FIGURE 14.9 System Flowchart (Page 2)   This system flowchart is a continuation from Figure 14.8. This solution assumes a computer that allows a large number of files to be open at the same time. Notice the connection symbols back to the first page.

Process Analysis and Design

Process analysis and design is a general systems design technique.

> **Process analysis and design** is an analytical technique for transforming essential data stores and processes into implementation locations, processes, and data stores.

We begin with the essential DFDs created in Chapter 9 and revised in Chapter 13.

Note   In the event that you haven't read Chapter 13, the principle changes made to the DFDs were: (1) ensuring that each data entity on the normalized data model has a corresponding data store(s) in the process models, and (2) ensuring that the processes included in the DFDs are sufficient to maintain (create, delete, and modify) each data entity in the data model.

Step 1: Establish Computer and Manual Processes

The first step of process analysis is to determine which essential primitive processes will be implemented as computer processes (hardware, firmware, and software) and which ones will be implemented as manual processes (performed by people).

Note   Recall from Chapter 9 that primitive processes are those that were not further factored into more detailed data flow diagrams.

Sometimes called establishing a **person/machine boundary,** this step establishes initial design units: manual and computer. Establishing a person/machine boundary is not difficult, but it is not as simple as you might first think. The difficulty arises when the person/machine boundary cuts through an essential process—in other words, part of the process is to be manual and part is to be computerized (see Figure 14.10). This situation is common on essential DFDs since they are drawn at a user level of interest and without regard to implementation alternatives.

Figure 14.11 indicates the solution. Notice that **manual design units** are factored out and drawn as separate DFDs. Note that, in a manual design unit, all processes are completely manual. All computerized processes are depicted as external agents (which will be, in subsequent steps, factored into their own design unit DFDs).

Manual design unit DFDs may be subsequently fleshed out into more detailed manual DFDs that include internal controls required to record and prepare data for input to computer processes. Detailed procedures for each manual process should be designed and documented in a form suitable to train and guide the persons performing those processes.

Eliminate the manual processes on the resultant DFD to make that DFD correspond to the first-draft computer application design unit. External agents are added to replace the manual design units and to show how they interface to the computer design unit.

Step 2: Establish Batch and On-Line Computer Processes

Response-time requirements help to determine how much of a system must be on-line versus how much could be batch. Note the word *could.* A constraint that requires the designer to implement all processes on-line is not uncommon.

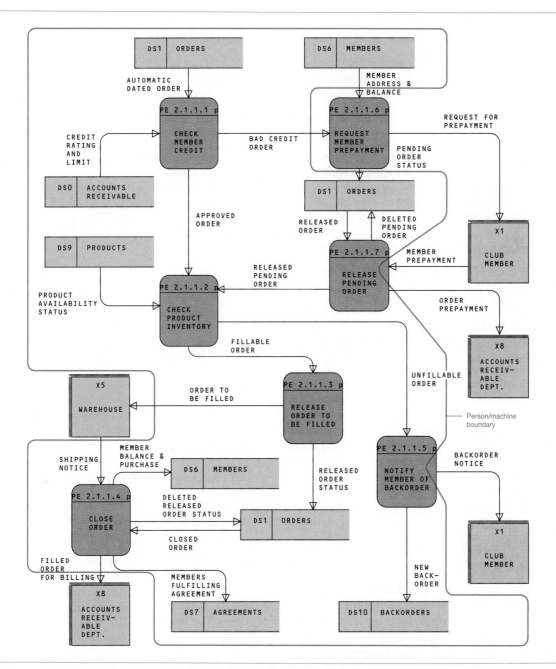

FIGURE 14.10 **A Person/Machine Boundary** A person/machine boundary establishes what parts of a system
will be automated and what parts will be manual. A boundary can cross through a process to indicate that part of a
process is automated and part is manual.

Still, on-line systems are more difficult to design and implement. They also take
longer to build and are more expensive. Therefore, a case *could* be made to
always choose the simplest design alternative, even if it doesn't seem modern.
As engineers are fond of saying, "great engineering is simple engineering."
(Information systems professionals are frequently guilty of designing overly
complex systems using the latest fad technology.)

MANUAL DESIGN UNIT: REQUEST MEMBER PREPAYMENT

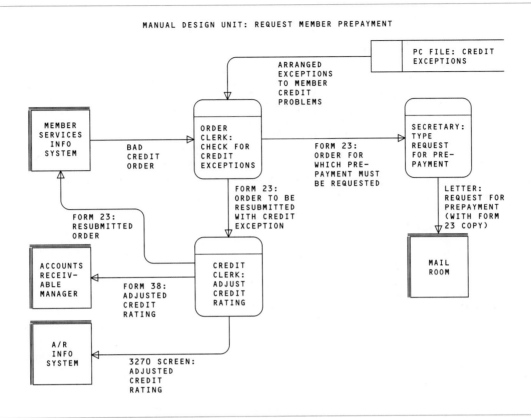

FIGURE 14.11 **A Manual Design Unit** This implementation DFD documents a manual design unit factored out of an essential DFD (Figure 14.10).

Response-time analysis is applied to all essential data flows to and from all primitive computer processes on the first-draft computer application design unit (established in Step 1). The procedure is as follows:

2a. The analyst must determine a suitable threshold (e.g., 60 seconds) for differentiating between immediate and nonimmediate data flows. We used a 30-second threshold for the SoundStage case.

2b. Identify all data flows "to" external agents. These are the net outputs of the system. Representative SoundStage net outputs are shown in Figure 14.12. The annotations are described in steps 2c through 2f.

2c. For each net output of the system, determine if the output is needed immediately on demand (IOD) or as soon as possible (ASAP) but not necessarily immediately (see Figure 14.12). Use your threshold from step 2a to make this determination. The required response times of these net outputs will be used to analyze and identify which processes must be implemented on-line and which could be implemented as batch or scheduled processes.

2d. For processes that have only IOD outputs, mark all input data flows as IOD as well. This should make sense to you since the process cannot produce immediate results unless the inputs are also available on demand. Working backward through the DFDs, mark the inputs and out-

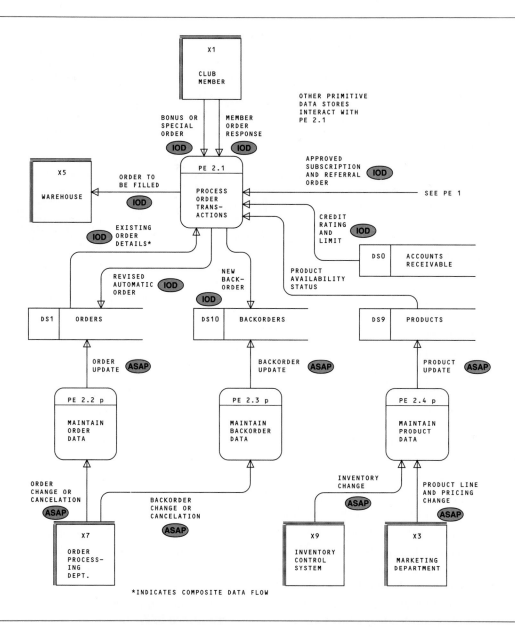

FIGURE 14.12 **Analyzing Response-Time Requirements** This essential DFD has been annotated to indicate which data flows must be made available immediately on demand (IOD) and which may be made available as soon as possible (ASAP). The timing of these data flows can assist in defining batch versus on-line processes.

puts for any prerequisite process as IOD until you reach either an external agent or a data store (see Figure 14.12).

2e. For processes that have only ASAP outputs, mark all input data flows as ASAP as well. Working backward through the DFDs, mark the inputs and outputs for any prerequisite process as IOD until you reach either an external agent or a data store (see Figure 14.12).

2f. For processes that have a mixture of IOD and ASAP net outputs, we must do some factoring. Split the process into two or more processes

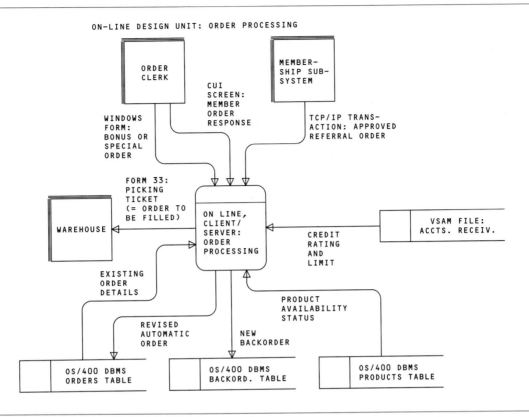

FIGURE 14.13 **An On-Line Design Unit** This on-line design unit (implementation DFD) was factored out of the annotated DFD in Figure 14.12.

that have only IOD output data flows or only ASAP output data flows. Rename the processes to clearly represent the data they are processing. Some input data flows may be required for both processes—they should be given unique names if possible. Next, repeat steps 2d and 2e for these new processes (see Figure 14.12).

2g. Redraw each stream or collection of IOD- or ASAP-based processes into its own design unit DFD. Use all of the implementation DFD conventions described earlier in the chapter. Each design unit should be clearly labeled as batch, on-line, cooperative, client/server, or other technology (possibly future technology) as described earlier in the chapter (see Figure 14.13 for one sample).

Step 3: Determine Process Cycles

All businesses operate on business cycles. Examples of business cycles are listed in the margin. Business cycles are triggered by transactions or events. Transactions have already been documented in essential DFDs. Events have not been documented on DFDs. Examples include time (e.g., an end-of-month process that produces reports) and requests for reports. Some analysts like to include such events on essential DFDs using dashed arrows labeled to describe

----- ✓ -----

Representative Processing Cycles

Daily
Weekly
Monthly
Quarterly
Annually

the event. Timed data flows can originate from an external agent labeled SYS-TEM CLOCK.

This step is straightforward. All processes on a single implementation DFD should be documented as separate design units. This step requires you to examine all triggers in each design unit DFD. For each trigger, determine its business cycle. If all business cycles in the design unit are the same, make no changes. Otherwise, factor the design unit DFD into separate design unit DFDs for each business cycle.

Step 4: Establish Processing Locations

The concept of system geography has been stressed throughout the book when discussing the building blocks of information systems. If you read Chapter 10, you know that system geography is defined in terms of locations.

> **Essential locations** are places where people create, modify, delete, and use data.

In a business system, these locations may be defined in terms of countries, regions, states, cities, campuses, buildings, departments, and so forth.

> **Implementation locations** are places where users interact with systems (e.g., terminals and bar code readers) or with computer processors (e.g., mainframes, minis, servers, workstations, desktop PCs, laptops, and note-pads).

Locations become nodes in business and computer networks. In Chapter 10 we introduced location connectivity diagrams as a tool for modeling the geography or shape of a business system. We could use the same tool to model the geography or shape of a computer network (commonly called a **network topology**). Alternatively, we could use implementation DFDs to accomplish the same purpose. In many shops the network professionals may have adopted their own standard for modeling network topology.

Since all readers now know DFDs, we'll use them in this chapter. One advantage of DFDs is that they can be drawn with most CASE tools and, therefore, can be integrated with other models and specifications.

Network topology DFDs (see Figure 14.14) need to be annotated to show somewhat different specifications than normal DFDs. They don't show specific data flows per se. Instead, they show highways over which data flows may travel in either direction. Also, network topology DFDs indicate the following:

- User locations and numbers.
- Processor locations, types, and numbers.
- Storage devices and data capacities.
- Connection protocols and traffic volumes.

The network topology DFD can be used to either design a computer network or to document the design of a computer network. Finally, if you use DFDs to document a network topology, do not mix the network topology diagrams with the actual DFDs.

Use the following steps to build an implementation network model:

4a. Use external agents to identify users and other points of data entry. An external agent may represent groups of users. A parenthetical number

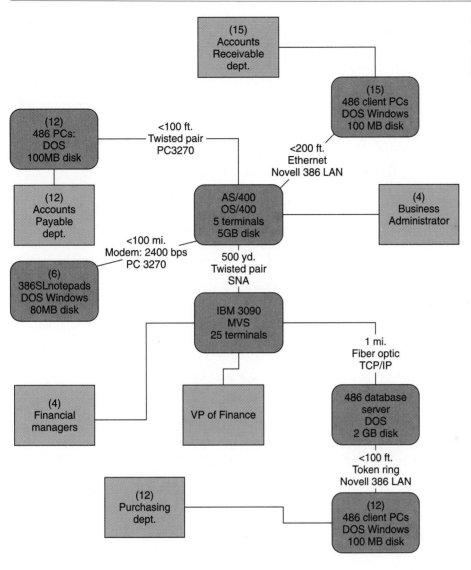

FIGURE 14.14 **A Network Topology DFD**
This figure demonstrates the use of a DFD to document a network topology.

indicates a number of similar users. Most external agents should be placed on the perimeter of the model (see Figure 14.15).

4b. Use processes to represent processors. Central processors should be in the middle of the diagram. Other processors can be layered around the diagram in logical groups (again, see Figure 14.15). The processor should be labeled to indicate its location and type, quantity (only for clients in an LAN), storage capacity, and number of I/O devices.

4c. Connect the processors and users to represent the shape of the network. Ideally, the connections should not have arrows or should have double-ended arrows. On each connection, indicate the distance (possibly a range), and, if known, the network architecture and protocols.

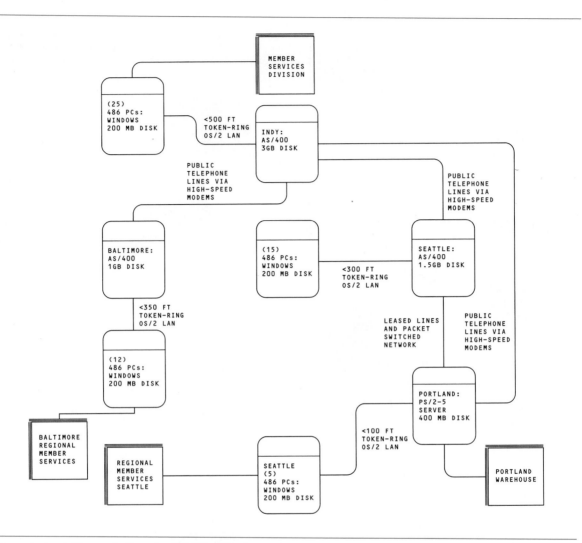

FIGURE 14.15   A Network Topology DFD   This figure documents the topology of the SoundStage network.

Step 5: Distribute Data Stores to Locations

The next step is to distribute data stores to the locations. Essential data stores are already known. We need only determine where each will be physically stored and how.

One possibility is to store all data on a single processor. Until recently, this was the most common solution. A single data store labeled ALL DATA would be connected to the appropriate processor.

Increasingly, data is distributed. Distributed data may be partitioned into subsets of specific data entities (from the data model) for locations and/or duplicated for locations. Duplicated data presents special implementation requirements that ensure simultaneous updates of the same instance of a data entity that is stored in more than one location. (Note: Some distributed database management systems automatically perform such updates.)

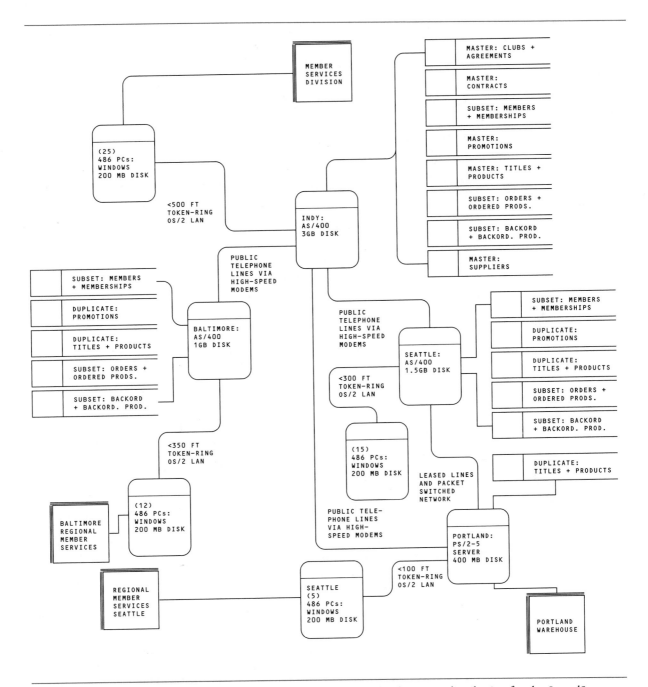

FIGURE 14.16 **Distributed Data Stores**   This figure documents the data store distribution for the SoundStage application.

Why distribute data storage? There are many possible reasons. First, some data instances are of local interest only. Second, performance can often be improved by subsetting data to multiple locations. Finally, some data needs to be localized to assign custodianship of that data.

Data distribution decisions can be very complex—normally the decisions are guided by data and database professionals and taught in data management

FIGURE 14.17 **Distrib-
uted Processes** This figure documents the pro-
cess distribution for the
SoundStage application.

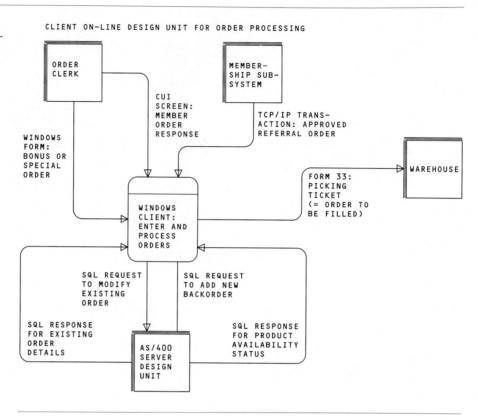

FIGURE 14.17 **Distributed Processes** This figure documents the process distribution for the SoundStage application.

courses and textbooks. In this book we want to consider only how to document the partition and duplication decisions.

Data stores should be connected in clusters to the appropriate processors (see Figure 14.16). Each data store label should be annotated to indicate whether it is a master data set (all instances of the entity), a duplicated data set, or a partitioned data set (subset of all instances of the entity). We have used the codes M, D, and P for these options.

Step 6: Distribute Processes to Locations

How are processes then distributed to processors? You need to analyze processes on design unit DFDs with respect to processing locations on your network topology DFD.

For each design unit DFD, examine the processes. Determine which process location will handle the process. For any process that will be split across multiple processors (possible in client/server solutions), factor the process into separate implementation processes. If all processes on the design unit DFD are to be implemented on one processor, or duplicated on many processors, make no changes to the design unit DFD. But if processes on the DFD are to be distributed, factor the design unit into separate design units for each processor.

The SoundStage solution will be client/server with an AS/400 database server and PC clients. Notice in Figure 14.17 that the data stores disappeared from the design unit for the PC clients. They all appeared on the network topology DFD for the AS/400 server (recall Figure 14.16).

CLIENT ON-LINE DESIGN UNIT FOR ORDER PROCESSING

FIGURE 14.18 **A Final Design Unit DFD** This figure documents one of many final design unit DFDs.

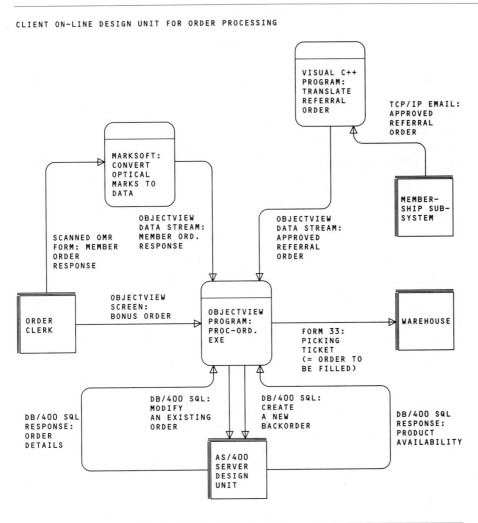

Step 7: Assign Technology

The last step is to complete the design unit DFDs by assigning technology choices. Technology choices may be dictated by an approved technology architecture or list of approved technologies. If not, alternative choices need to be analyzed for technical, operational, and economic feasibility as described in Part Five, Module C.

Using the implementation DFD conventions for processes, data flows, and data stores (as described earlier in the chapter), the analyst simply records technology choices on each design unit DFD.

We now have design units that can be assigned to specific analysts or teams for subsequent design and to programmers for implementation. In summary, we have taken our essential DFDs that represent user requirements and have factored them into several self-contained design unit DFDs. Each design unit is a blueprint for detailed design of files, inputs, outputs, interface, and programs. Alternatively, one or more design units may become the basis for selecting and purchasing an application package or for outsourcing development to a third-party contractor.

One complete design unit DFD for SoundStage is shown in Figure 14.18.

Summary

Today's designer is faced with many more design alternatives than in the past. The most pervasive of these choices deals with processes. Whereas most existing processes were designed and implemented on centralized processors, today's processes are increasingly decentralized or cooperative. In the latter case, many applications are now being redesigned and implemented with a cooperative processing technique called client/server computing. Distributed and cooperative solutions require multiple computers that are networked to provide connectivity and interoperability.

Analysts are faced with many networking options including point-to-point, bus, star, hierarchical, and ring networks. Local area networks are bridged to create wide area networks. Telecommunications technology further expands internetworking.

Analysts are also faced with centralized versus distributed database alternatives and a wide variety of input and output alternatives including batch, on-line, keyless data entry, pen computing, electronic data interchange, and others. With all of these design alternatives, an analyst is well advised to precede any detailed design effort with a higher-level general design. The general design serves as a blueprint for the detailed design. General design results in implementation models to outline the design effort.

This chapter introduced two implementation model alternatives: implementation data flow diagrams and system flowcharts. The former is the modern tool of choice, but the latter often documents the existing implementation of many current systems. Implementation DFDs are very similar to essential DFDs except that they document specific design decisions.

A technique called process analysis and design provides for the systematic transformation of essential DFDs into implementation DFDs. These implementation DFDs take the form of design units that are so self-contained that they can be independently assigned to different teams for detailed design, construction, and unit testing. Design units are factored out of the original essential DFDs by analyzing (1) person/machine boundaries, (2) response-time requirements, (3) business cycles, (4) processing locations, (5) data distribution, and (6) process distribution. As a final step, technology is assigned to all processes, data flows, and data stores in each design unit.

Key Terms

bar coding, p. 546

batch processing, p. 545

bus networking, p. 540

centralized processing, p. 538

client/server computing, p. 539

connectivity, p. 538

cooperative processing, p. 539

design unit, p. 553

distributed database management systems (DDBMSs), p. 545

distributed processing, p. 538

diverging data flow, p. 551

downsize, p. 539

electronic data interchange (EDI), p. 547

entity relationship diagram, p. 552

essential locations, p. 564

graphical user interfaces (GUIs), p. 547

hierarchical network, p. 542

implementation data flow, p. 550

implementation data flow diagrams, p. 548

implementation data store, p. 552

implementation locations, p. 564

implementation process, p. 548

interconnection, p. 543

interoperability, p. 538

local area network, p. 540

manual design units, p. 559

Problems and Exercises

1. Differentiate between centralized, distributed, and cooperative processing alternatives. Why is new application development seriously considering the cooperative processing alternative?

2. Differentiate between connectivity and interoperability. Why can you have connectivity and lack interoperability? Which is most important?

3. What type of processing drives client/server computing?

4. Respond to the following editorial:

   Client/server computing is too big a step for us at this time. I realize that PCs and networks may—and I emphasize "may"—be cheaper to acquire, operate, and maintain, but we can't just get rid of our mainframe and all its legacy applications just like that. We can't rush into this mainframe versus server issue just yet.

5. Recent business reports and editorials have concluded that client/server computing is not the economic panacea that most initially believed. Why might cost "not" be a factor in the decision to go client/server for all new applications?

6. Downsizing, rightsizing, and client/server computing involve distribution decisions. What are the three components that are distributed?

7. Why is networking and data communications an essential subject for future systems designers who will build rightsized, client/server applications?

8. Compare and contrast five different strategies for distributed processing.

9. Make an appointment to visit a local business computing facility. How has the growth in numbers of microcomputers affected the networking strategy of the business? What is the networking strategy? What topology was chosen and why?

10. Explain the difference between batch, on-line, and remote batch input methods. Define an input and conceive a situation that might call for each of the three methods to be used.

11. Why are keyless data entry alternatives attractive to business? Describe three keyless data entry technologies.

12. Differentiate between essential and implementation data flow diagrams. When is each relevant?

13. Prepare a physical data flow diagram (or, if your instructor prefers, a systems flowchart) to describe the backup and recovery procedures for a lost or damaged master file or database.

14. What symbols can be connected to the following symbols on a physical data flow diagram?
    a. External agent.
    b. Process.
    c. Data store.

15. Your business has many old-style systems flowcharts that you want converted to physical data flow diagrams since all of your staff are familiar with PDFDs. Create a translation legend that suggests how symbols on a system flowchart correspond

to their equivalent symbols on data flow diagrams. A small team should be able to use this as a conversion guide for the project.

16. What is process analysis and design?

17. Acquire a network topology model for your school (or a local business). Convert it to a network topology DFD as described in this chapter.

18. Acquire a systems flowchart from a local business' application, an older systems analysis and design textbook, or your instructor (who probably has access to many in his or her course archives). Convert that systems flowchart to a physical data flow diagram.

Projects and Minicases

1. Research the evolving subject of client/server applications development. Study the costs and benefits and the issues and problems to be addressed. Be sure to find pro and con examples—both exist! Present your findings in a recommendations report to information systems management.

2. Client/server technology has yet to stabilize. Research and prepare a technology update report on one of the following subjects:
   a. Database management systems for servers in a client/server architecture.
   b. Network topology and operating systems for a client/server environment.
   c. Distributed computing environment (DCE), an evolving standard for open systems interoperability for client/server computing. (Note: Open systems is a goal whereby dissimilar computers and software can communicate and cooperate in a manner that is entirely transparent to users.)
   d. Client programming languages versus downsized COBOL/CICS. Examples of the former include Powerbuilder, ObjectView, and Visual BASIC.
   e. Middleware for easy, transparent access to existing data stored using dissimilar structure and different servers and hosts.

3. Chapter 9 provided numerous essential data flow diagrams for the SoundStage project. Chapter 13 provided a normalized data model. Using your school or business's technical environment as a constraint (or one provided by your instructor), use the process analysis and design technique in this chapter to derive a design. The design should be documented using physical data flow diagrams. (Note: Your technical environment should at least specify processors, operating systems, languages, network topology, network operating systems and protocols, and database management systems.)

Suggested Readings

See Chapter 9 for more data flow diagramming references.

Fitzgerald, Jerry. *Business Data Communications.* New York: Wiley, 1990. If you want to learn more about data communications without taking a course, this popular textbook is representative of those that have a business emphasis.

Gane, Chris, and Trish Sarson. *Structured Systems Analysis: Tools and Techniques.* Englewood Cliffs, N.J.: Prentice Hall, 1979. This classic on process modeling became the basis for the McDonnell Douglas' STRADIS methodology (now owned by Structured Solutions) and the derived design techniques taught in this book.

Gelber, Stan. *Introduction to Data Communications: A Practical Approach.* Horsham, Pa., 1991. This professional-market book is an excellent primer for working professionals seeking a relatively concise primer.

Theby, Stephen E. "Derived Design: Bridging Analysis and Design." McDonnell Douglas Professional Services: Improved System Technologies, 1987. The techniques described in this paper are the basis for a phase in STRADIS (Structured Analysis, Design, and Implementation of Information Systems), a systems development methodology. The technique was altered and simplified to make it suitable to the level of this textbook. As authors, we were quite impressed with the full derived design technique as advocated in the STRADIS methodology.

Whitten, Jeffrey L.; Lonnie D. Bentley; and Victor M. Barlow. *Systems Analysis and Design Methods.* Second ed. Homewood, Ill.: Richard D. Irwin, 1989. If you are interested in a more comprehensive tutorial on systems flowcharts, this older edition of our book covered that material in Chapter 18.

File and Database Design for the New System

When we left Sandra and Bob, they had completed the general design for the new system. Now they must design the files and/or databases in the new system.

Database Specifications for the New System

Sandra and Bob have completed the general design for the new system. They have been able to work independently during much of the detailed design effort. Bob has been working on the design of computer files and databases while Sandra focused her efforts on the design of computer outputs. They will cross-check their final specifications to maintain consistency.

Bob has finished the first draft design specifications for the database required by the new system. Recall that he and Sandra made a decision during general design to implement all the new system's data using the DB/400 relational database that is built into the AS/400 operating system. Because of this decision, Bob had no computer files to design.

Bob did not directly involve the end-users because of the highly technical nature of file and database design. We join Bob as he is finalizing the design documentation.

Sandra enters Bob's office. "Hi, Bob. Is that the DB/400 database schema you are working on?"

Sandra was referring to the diagram on Bob's desk (see Figure E6.1).

"Yes it is. I'm just checking the final design. Are you finished with outputs?"

Sandra sighed and said, "No, I still have to conduct a walkthrough with the end-users to verify my report designs."

"Be sure and let me know if you identify any new data storage requirements in your walkthrough," Bob responded. "Maybe you could help me review this schema?"

"No problem," offered Sandra. "Why don't you hand me the essential data model so I can compare it to your database schema? Oh yes, I'll need the data structures for each of the entities so I can check the keys for each entity."

"I've got everything right here in my binder. Let me take them out. I'll spread them out on this table so you can easily reference everything," responded Bob.

"I see a problem right away. The join between title and supplier needs to be by supplier number, not product number," said Sandra.

"That's right! The supplier entity was the parent entity, so I needed to duplicate its key for joining the two tables," replied Bob. "This is great. I really appreciate you reviewing my work. Sometimes you work so intensely on something that you become blinded to the obvious. It's getting late. How about I buy you lunch and we can look over the rest of these specifications?"

Where Do We Go from Here?

This episode introduced an activity that must be accomplished during detailed systems design—the design of computer files/databases. In Chapter 15 you will learn about the design issues and tools used to design computer files and databases.

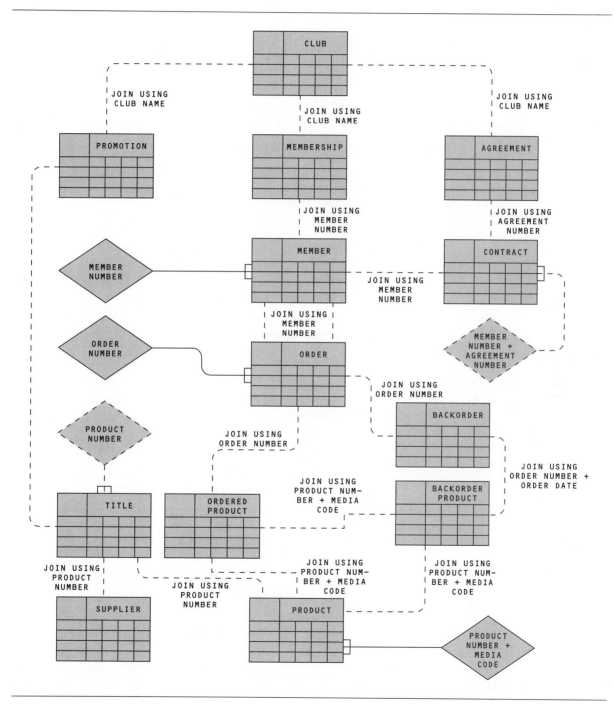

FIGURE E6.1 Typical Database Schema

15

File and Database Design

───────────── *Chapter Preview and Objectives* ─────────────

This chapter teaches the design of physical data stores. It teaches you how to design conventional computer-based files, and, as a modern alternative to files, the database. You will know that you have mastered the tools and techniques of file and database design when you can.

Compare and contrast
conventional files and modern databases.

Differentiate
between applications and end-user databases.

Describe
the architecture of a database management system and how this
architecture supports the prototyping approach to
systems analysis and design.

Design
fixed- and variable-length logical records and internal controls
for files in a computer-based information system and
record those requirements in a project repository.

Explain
why records are blocked, and determine the blocking factor for a file.

Identify and differentiate
between several types of files.

Determine
the best file organization for a given file by studying that file's required
access methods and usage.

Describe
a relational database implementation model.

Transform
a simple, logical data model into a simple version of a
relational model or schema.

MINICASE

MARTIN AND SHORT CONSULTING COMPANY

Scene: *Tim Robbins, a programmer/analyst for Martin and Short Consulting, is sitting at his desk reviewing the design specifications for the new computer-based requisition processing system. Martin and Short Consulting is a small computer consulting company in Asheville, North Carolina that develops information systems for other businesses. Tim recently graduated from a local university and joined the company. He has been responsible for this project from the very beginning. Most of the programs for the system have already been implemented, but the final two programs have been returned by the programmer with a note explaining that they are impossible to code according to the specifications. Because Tim has been in a hurry to implement this new system, design specifications have been passed along to the programmer as soon as they were completed. Now, Tim is frustrated. His boss, Mike Foster, enters the office.*

Mike: Hi, Tim. What's the matter? You look like you've just lost your best friend.

Tim: You're not going to be happy. I may have really messed up this time. I've worked so hard implementing that darn requisition system. But I've encountered a problem with the last two programs.

Mike: Does this mean that the system might not meet its deadline? You know how important this system is. What type of problem? What do the programs involve?

Tim: One of them produces a management report and the other an on-line inquiry response.

Mike: That hardly seems like an overwhelming problem. Why can't you just generate the programs to produce the reports? I don't understand, we do that all the time. You know that. We did capture the necessary data, right?

Tim: Well, it's not all that simple. The inquiry program requires quick response. That part of the data model was implemented as a sequential file to tie into some of the client's other existing systems. But, I didn't anticipate this query requirement. It wasn't in the end-user's initial request. The report requires data to be retrieved and printed by product number. That presents a problem because the file is organized as an indexed file with requisition number as the primary key. I didn't set this file up with any secondary indexes because none of the previous programs needed to access the file by anything other than the requisition number. You see . . .

Mike: This is a very serious problem. You mean you have already designed and implemented the file? Why?

Tim: Yes, we were behind schedule. I figured that the highest volume of transactions against the file would be executed by the programs that maintained the file. I organized the file to make those programs run efficiently. The rest of the data model was implemented as a relational database, but I also deliberately denormalized it to improve performance. Then I passed the specifications for the files and those particular programs on to the programmers so we could begin writing those programs.

Mike: That's not very smart, Tim. Now you've found that the file and database organization you chose is not best for the way the new programs need to access the file. First, I'm a bit surprised that you took such a shortcut. It's not like you. Second, if you were to have requested such a shortcut, my staff would have strongly encouraged that you implement the file as indexed sequential structure with secondary keys to minimize the impact of just such problems. And you obviously did not carefully consider the impact of denormalization when you designed the database. There are no shortcuts. You're just going to have to choose between the lesser of two evils. First, we could write programs to extract data from the

Tim: existing file and database and place it in temporary files whose organization suits the new programs.

Tim: But then we'll have the same data stored redundantly, and that could cause data integrity problems!

Mike: I know, I'm not crazy about that option either. It looks to me like you're going to have to temporarily halt implementation so that you can redesign the file and database. But, I must have this system on time, Tim. It looks like you'll be working some overtime. And tell the programmers that they will likely be working some overtime, too —once you have redesigned the file and database. I'm sorry, but you brought this on yourself.

Let's just try to pick up the pieces and not let it happen again. Don't be too discouraged. I can accept mistakes. Just learn from them, okay?

Tim: Yes, sir. I'm sorry. It won't happen again. The project will come in on time, even if I have to work on some of the code myself.

Discussion Questions

1. Tim made a big mistake. Where did he go wrong?

2. What should Tim do now?

3. Why did Tim combine both conventional files and a database?

4. Can you describe how the files and database should have been implemented?

CONVENTIONAL FILES VERSUS THE DATABASE

All information systems create, maintain, and use data. This data is stored in files and databases.

Files are collections of similar records. Examples include CUSTOMERS, ORDERS, and PRODUCTS.

Databases are collections of interrelated files. For example, a SALES database might contain ORDER records that are "linked" to their associated CUSTOMER and PRODUCT RECORDS.

Let's compare the file and database alternatives. Figure 15.1 illustrates the fundamental difference between the file and database environments. In the file environment, data storage is built around the applications that will use the files. In the database environment, applications will be built around the integrated database. Ideally, the database is not dependent on the applications that will use it. Each environment has advantages and disadvantages. Let's examine these differences.

Pros and Cons of Conventional Files

Conventional files are relatively easy to design and implement because they are normally based on a single application or information system, such as accounts receivable or payroll. If you understand the end-user's output needs for that system, you can determine the data that will have to be captured and stored to fulfill those needs and define the best file organization for those requirements.

Another advantage of conventional files is processing speed. Database technology uses complex indices, linked lists, trees, and other data structures. Such data structures are sometimes too slow to handle large volumes of transactions with an adequate throughput. This limitation of database technology is rapidly disappearing, however.

Still, even if database technology matches or exceeds the performance of file management systems, numerous file systems are already in place. Given the shortage of information systems personnel and growth in demand for new information systems, many firms cannot afford the time or money required to

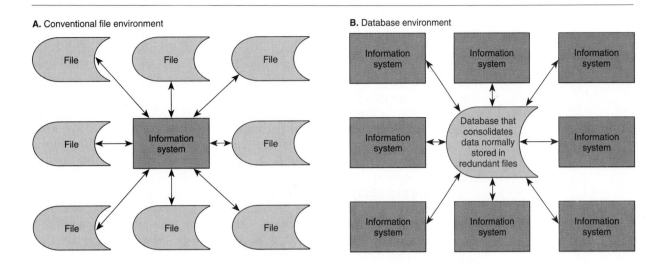

A. Conventional file environment

B. Database environment

FIGURE 15.1 **Conventional Files versus the Database Approach** File-based environments emphasize the system or application. As applications are developed, customized files are built around them. These files frequently do not meet future needs. Database environments emphasize the data, independent of the applications that will use the data. The applications evolve around a database designed such that it can adapt to changing needs.

redesign all of their current file-based systems as database systems. To accelerate change, makers of CASE tools now provide technology to assist with the database design and conversion processes.

It is likely that database technology eventually will replace file technology. Conventional files have numerous disadvantages. Duplication of data between files is normally cited as the principal disadvantage of file-based systems. Files tend to be built around single applications without regard to other applications. Over time, the development of multiple applications results in duplicated data in files.

Duplicate data results in duplicate input, duplicate maintenance, duplicate storage, and possibly data integrity errors. And what happens if the data format changes? Consider the problem faced by many firms if all systems must support a nine-digit zip code. Do you have any idea how many redundant files would have to be located and changed in a typical organization? Add to this the enormous volume of programs that use the field, and you have a nightmare of a program maintenance project.

Another disadvantage of files is inflexibility. Files are typically designed to support the end-user's current requirements and programs. New needs — such as reports and queries — often require files to be redesigned because existing files cannot easily support the new needs. If those files were to be restructured, all programs using those files would have to be rewritten. In other words, the current programs have become dependent on the files, and vice versa. Since reorganization is impractical, redundant files are often created to meet the new requirements. Thus, we escalate the redundancy and the associated disadvantages of redundancy.

Most of these problems are the result of file design inadequacies. These problems can be overcome by basing conventional file designs on sensible data models developed during systems design. But in the final analysis, true database makes fewer compromises than conventional files.

Pros and Cons of Database

The database approach is not a solution to every problem. We've already stated the principal advantage of database—the sharing of data. A common misconception about the database approach is that you can build a single database that contains all data items of interest to an organization. This notion, however desirable, is not currently practical. The reality of such a solution is that it would take forever to build such a complex database. Realistically, most organizations build several databases, each one sharing data with many information systems. Yes, there will be some redundancy between databases. However, this redundancy is both reduced and controlled—even more so than in well-designed conventional files.

Database technology also offers the advantage of storing data in more flexible formats. This is made possible because databases are defined externally from the programs that will use them. Theoretically, this allows us to use the data in ways not originally specified by the end-users. Care must be taken to achieve this data independence. Different combinations of the same data can be easily accessed to fulfill new report and query needs. When fully realized, data independence permits data formats and structure to change (recall our zip-code example) without having to change any of the computer programs that currently use that data. Thus, new fields and record types can be added to the database without affecting current programs.

On the other hand, database technology is more complex than file technology. Special software, called a *database management system (DBMS),* is required. The flexibility provided by a DBMS usually makes it slower (again, the speed difference between conventional files and databases is decreasing). Thus, many organizations must buy faster computers.

The cost of developing databases is higher because analysts and programmers must learn how to use the DBMS. Also, in order to achieve data independence, analysts and database specialists must adhere to rigorous design principles. The DBMS technology itself can be expensive to acquire and maintain.

Another problem with the database approach is the increased vulnerability inherent in the use of databases. Because you are using a shared data resource, you are literally placing all your eggs in one basket. Therefore, backup and recovery procedures and security issues become more complex and costly.

Despite the problems discussed, database usage is growing by leaps and bounds. The technology will get better, and performance limitations will disappear. Design methods and tools will also improve. But for the time being, we'll have to adjust to a world that uses both files and database systems. In the next two sections we will explore some of the fundamental concepts of file and database design. We will start with a review of the more traditional file design concepts.

FILE AND DATABASE DESIGN CONCEPTS

Conventional files and modern databases are the lifeblood of many information systems. Almost every computer program will require access to files and/or databases. The design of computer files and databases can be difficult because the storage and organization of data on computer media require the analyst to consider complex and often conflicting issues, such as storage capacity and performance.

Many of the technical issues important to file and database design are taught in different courses. We will review those technical issues that are pertinent to the systems analyst's responsibilities.

Fields

Fields are common to both conventional files and modern databases.

> A **field** is the implementation of a *data attribute* (introduced in Chapter 8). Fields are the smallest unit of data to be stored in a file or database.

There are four types of fields that can be stored: primary keys, secondary keys, foreign keys, and descriptors.

> **Primary keys** are fields whose values identify one and only one record in a file. They implement *identifiers* of entities in an essential data model (Chapters 8 and 13).

For instance, CUSTOMER NUMBER uniquely identifies a customer record, and ORDER NUMBER uniquely identifies an order record.

> **Secondary keys** are alternate indexes for identifying an entity. A secondary key's values may identify single entity occurrences or a subset of all entity occurrences.

For example, GENDER defines two subsets of the EMPLOYEE entity, male employees and female employees. A file may have several secondary keys.

> **Foreign keys** are pointers or links to occurrences of a different file. A foreign key in one entity must be a primary key in another entity (or logically equivalent). Foreign keys physically implement relationships on an ERD.

For example, an ORDER record contains the foreign key CUSTOMER NUMBER to "point" to the CUSTOMER record that is associated with the order.

> **Descriptors** are any other fields that describe business entities. For example, given the business entity EMPLOYEE, some descriptor fields include EMPLOYEE NAME, DATE HIRED, PAY RATE, and YEAR-TO-DATE WAGES. Both primary and secondary keys are descriptors.

Keys (identifiers) and descriptors (attributes) were defined when you performed data modeling in systems analysis (Chapter 8).

Field Storage Formats

There are three basic field storage formats: binary codes, fixed-point numbers, and floating-point numbers.

> A **binary code** is a combination of ones and zeros that represent a character, number, or symbol. There are three common binary codes: EBCDIC, ASCII, and packed decimal.

During file design, your choice of storage formats for fields will determine how much disk or tape capacity will be needed for your files.

EBCDIC (Extended Binary Coded Decimal Interchange Code) is a popular code used on IBM mainframe computers and compatibles. Each letter, number, and special symbol is represented by an eight-bit code. A field value "Bill" would be stored as "11100010" (B), "10001001" (i), "10010011" (l), and "10010011" (l). This field requires four bytes of storage (a byte is eight bits long on most business-oriented machines).

ASCII (American Standard Code for Information Interchange) is another popular code used by many computer manufacturers. ASCII represents each letter, number, or symbol with a unique seven-bit code. An eighth bit is added, but the last bit is called a *parity bit* and is used only to check a stream of ASCII characters for correct transmission from one place to another, such as from computer to printer. A field value "Bill" would be stored as "01000010" (B), "10010110" (i), "00110110" (l), and "00110110" (l). This field also requires four bytes of storage. Standard tables for EBCDIC and ASCII codes are widely available.

Most computers do not use EBCDIC or ASCII codes in arithmetic operations. Some computers perform arithmetic only on fixed- and floating-point binary numbers. These are pure binary (base 2) formats that significantly reduce the required storage space. On the other hand, IBM mainframe computers and others do arithmetic on packed-decimal numbers.

> **Packed decimal** is equivalent to EBCDIC, except that only the last four bits are used. By storing numeric fields in packed-decimal format, the field requires only half as much storage space.

Field storage decisions should be recorded in the project repository. Notice that field design decisions are recorded in the attribute screens (storage type is equivalent to storage format). The CASE product used to create the attribute repository description in Figure 15.2 permits the following storage types:

C   Character, either ASCII or EBCDIC
B   Binary
P   Packed
F   Floating Point
D   Date

Most business information systems exclusively use character and packed formats for storage.

Records

Fields are organized into records. Like fields, records are common to both files and databases.

> **A record** is a collection of fields arranged in a predefined format. It is also the smallest unit of data storage that is operated on by most computer programs (most computers can read records, but not individual fields from a record—in other words, all or nothing).

A Record for CUSTOMER may be described by the fields listed in the margin. The primary key is underlined.

Record Storage Formats

During systems design, records will be classified as either fixed-length or variable-length records.

————— ✓ —————

Customer Record

Customer number
Customer name
Customer address
Date initiated
Customer credit limit
Customer balance
Balance past due

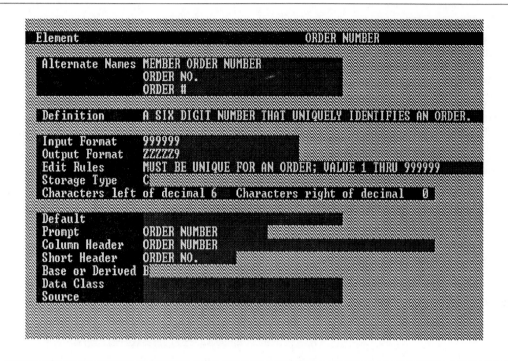

FIGURE 15.2 **Sample Attribute Repository Description** As attributes are defined, their associated descriptions in the project repository must be updated.

Occurrences of **fixed-length records** will be equal in length. Normally, every record occurrence will also be defined by the same fields. The fixed-length record will contain no repeating elements or groups. In a database, fixed-length records are indicative of *first normal form* entities (see Chapter 13).

While fixed-length records may be more common, conventional files often require variable-length records.

Occurrences of **variable-length records** have different lengths because certain fields or groups of fields repeat a different number of times for each record occurrence.

For example, an ORDER variable-length record will contain some fields that occur once for any given order. Examples include ORDER NUMBER, ORDER DATE, and CUSTOMER NUMBER. On the other hand, a single order may contain multiple occurrences of the following fields: ORDERED PART NUMBER, UNIT PRICE, and QUANTITY ORDERED. Note that variable-length records normally contain a fixed-length group of fields and a repeating group of fields. Figure 15.3 demonstrates the difference between occurrences of fixed- and variable-length records. The length of any given occurrence of the record will depend on how many products are on the order.

For the analyst, the specification of fixed- or variable-length records will affect file or database size and performance in ways that we'll study later in this

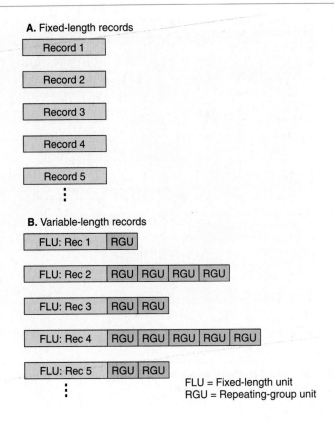

FIGURE 15.3 Fixed-Length versus Variable-Length Records All occurrences of a fixed-length record will be the same size and contain the same fields. Variable-length records are usually composed of a fixed-size root segment followed by a variable number of occurrences of the other fields.

chapter. Variable-length records have an adverse effect on flexibility of data, which is not surprising since a variable-length record is not, by definition, in first normal form, a concept that was introduced in Chapter 13.

As records are designed, design decisions must be documented in the project repository. Figure 15.4**A** demonstrates a typical record layout. Notice that the occurs (Occ) column suggests that the fields composing the subrecord ORDERED PRODUCT repeat. Thus, this repository entry describes a variable-length record. If all the occurs entries were 1, the record would be fixed length. Each field on the record explodes to an attribute description as previously illustrated (see Figure 15.2). The subrecord ORDERED PRODUCT explodes to a separate record layout description (see Figure 15.4**B**). Note that the TYPE entry does not refer to storage type. Its values are K = key, E = element, and R = subrecord. (Notice that this repository does not distinguish between types of keys.)

Record Blocking

When you execute a read or write instruction to a file from your computer program, what are you actually reading or writing? One record? Think about it! Wouldn't it be more efficient to read groups of records at a time? That is what most computers do. The operating system reads and writes *blocks* of records to

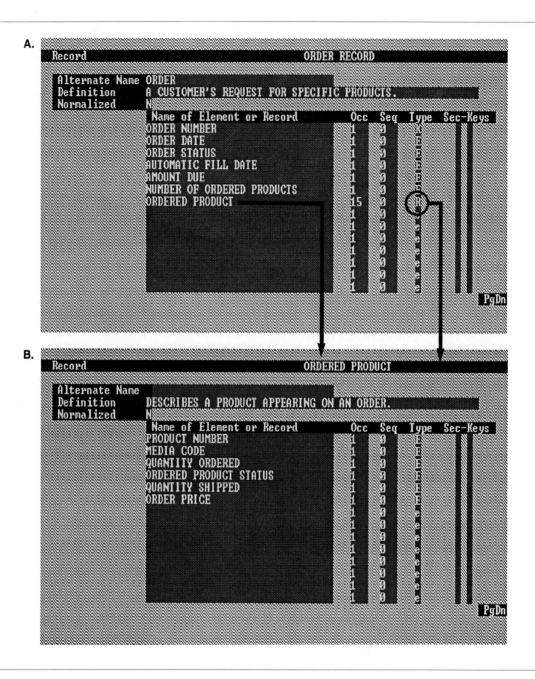

A.

Record	ORDER RECORD

Alternate Name ORDER
Definition   A CUSTOMER'S REQUEST FOR SPECIFIC PRODUCTS.
Normalized   N

Name of Element or Record	Occ	Seq	Type	Sec-Keys
ORDER NUMBER	1	0	K	
ORDER DATE	1	0	E	
ORDER STATUS	1	0	E	
AUTOMATIC FILL DATE	1	0	E	
AMOUNT DUE	1	0	E	
NUMBER OF ORDERED PRODUCTS	1	0	E	
ORDERED PRODUCT	15	0	R	
	1	0		
	1	0		
	1	0		
	1	0		
	1	0		
	1	0		
	1	0		

PgDn

B.

Record	ORDERED PRODUCT

Alternate Name
Definition   DESCRIBES A PRODUCT APPEARING ON AN ORDER.
Normalized   N

Name of Element or Record	Occ	Seq	Type	Sec-Keys
PRODUCT NUMBER	1	0	E	
MEDIA CODE	1	0	E	
QUANTITY ORDERED	1	0	E	
ORDERED PRODUCT STATUS	1	0	E	
QUANTITY SHIPPED	1	0	E	
ORDER PRICE	1	0	E	
	1	0		
	1	0		
	1	0		
	1	0		
	1	0		
	1	0		
	1	0		
	1	0		

PgDn

FIGURE 15.4 Sample Repository Entry for the ORDERS FILE   Fields (attributes) must be organized into records for a computer file.

and from a workspace in your program. When the workspace has been processed, a revised block of records can be written back to the file or a new block of records can be read from the file. The operating system handles this feature for you.

A **blocking factor** is the number of logical records in a block for a file. The block is sometimes called a *physical record*.

On some computer systems, the blocking factor is determined or optimized by the computer's operating system or database management system. Other times, a systems analyst or database professional must calculate the ideal blocking factor.

Obviously, we would like to make the blocking factor as large as possible in order to increase efficiency, but you can't make the blocking factor equal to the number of records in the file or database. Why not? The computer's memory (also called RAM) is nearly always only a fraction of the size of most files and databases. Also, the storage device itself — for example, a magnetic disk — may impose limitations on the maximum size for a block. For instance, many systems cannot read or write more than one disk track at a time. This would limit the blocking factor to the number of records that can be placed on a single track or part thereof. Blocking factors are normally a compromise between the above constraints.

Files

Similar records are organized into groups called *files*. In conventional file-based systems, these files are the final storage unit to be designed. In database systems, a file corresponds to one and only one set of similar records; sometimes called a **table.**

> A **file** is the set of all occurrences of a designed record. Typically, several types of files are encountered in information systems.

Some of the types of files include:

- **Master files** contain records that are relatively permanent. Once a record has been added to a master file, it remains in the system indefinitely. The values of fields for the record will change over its lifetime, but the record occurrence normally remains active for a long period of time. Examples of master file records include CUSTOMER, PRODUCT, and VENDOR. Most master files contain fixed-length records.

- **Transaction files** contain records that describe business events. The data describing these events normally has a limited useful lifetime. For instance, an INVOICE record is ordinarily useful until the invoice has been paid or written off as uncollectible. In information systems, transaction records are frequently retained on-line for some period of time. The inactive, off-line records are retained as an archive file (see below). Examples of transaction files include ORDER, INVOICE, and MATERIAL REQUISITION. Many transaction files contain variable-length records.

- **Archive files** contain the off-line records of master and transaction files. Records are rarely deleted; they are archived when they are not accessed as frequently. Archives would be used when data must be recalled for subsequent audit or analysis.

- **Scratch files** (also called *work files* or *temporary files*) are special files that contain temporary duplicates or subsets or alternate sequencing of a master or transaction file. A scratch file is created, used by the appropriate computer program, and then erased. In other words, it is created for a single task and must be recreated each time that task is performed. Scratch files are not found in databases.

- **Log files** are special records of updates to other files, especially master and transaction files. They are used in conjunction with archive files to recover "lost" data.

When designing files and databases, one of the major responsibilities of the systems analyst is to determine file access and organization. Files are a shared data resource. Because several programs will likely use and maintain the same file, the records should be organized so each program can easily access them.

File Access

The method by which a computer program will read records from a file is called **file access.** Every computer program will access a file in one of two ways: sequentially or directly.

The **sequential access method** starts reading or writing with a particular record in the file (normally the first) and proceeds—one record after another—until the entire file has been processed.

Sequential access is used when a program needs to process a relatively high percentage of all records in a file. For instance, a monthly invoice program may need to access virtually every record in the ACCOUNT file. Sequential access is normally required for batch transaction processing as well as for management reporting.

Some applications may not need to access most of the records in a file. In this case, the direct access method may be more appropriate.

The **direct access method**—also called *random access*—permits access to any record in a file without reading all previous records in that file.

Direct access is appropriate for programs that need to access only one or a few records in a file (e.g., queries). This approach is normally needed for on-line transaction processing and decision support.

File Organization

File organization defines how records in a single file are related to one another. Different file organizations are designed to optimize performance for one or both file access methods. There are several common file organizations:

- *Sequential* file records are usually arranged in a sequence determined by the value of the primary key. For instance, customer records could be organized sequentially according to CUSTOMER NUMBER (a primary key) or CUSTOMER NAME (a secondary key). Sequential records can be stored adjacent to one another as in Figure 15.5**A**, or they can be arranged sequentially with a linked list, as shown in Figure 15.5**B**. The latter is common in database systems athough users and programmers are isolated from such details. The figure illustrates the possible use of multiple linked lists that support different ways to sequentially process the file. Sequential files are optimized for situations in which high percentages of all the records in a file must be read by all programs that use the file.
- *Relative* or *direct* file records are physically located at an address that is calculated from its primary key field. The calculation of the address from

FIGURE 15.5 Alternative Sequential File Organizations of Records
There are two ways to store records in a sequential file. First, you can store the records one after the other, as in Figure 15.5**A**. Second, you can store the records using a linked list—called a *logically sequential file*—as in Figure 15.5**B**.

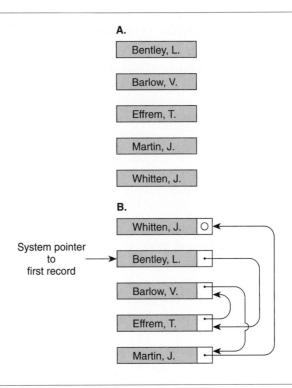

the key field is called *hashing*. Thus, record occurrences are scattered on the storage device instead of being arranged next to one another or linked via pointer fields. Records can be retrieved rapidly by applying the hashing formula to the primary key field of the desired record. Unfortunately, records cannot be processed sequentially. Although records are scattered, they are generally stored in preallocated areas that, when full, must be reallocated.

Relative files are optimized for situations in which rapid access to a small percentage of the records in a file is needed by all programs. Thus, relative files work well for on-line programs, whereas sequential files work better for high-volume batch programs.

- An *indexed* file organization is illustrated in Figure 15.6. Records are pointed to by indexes. The advantage of an index is that it is relatively small and easy to update. At a minimum, the file will have one index per primary key. However, Figure 15.6 also shows the use of indexes to implement secondary keys, with each value pointing to multiple records. A file that has several indexes—for primary and secondary keys—is said to have an *inverted file organization*. For all indexed files, there is an inherent cost of maintaining the indexes. If not properly maintained, the indexes can become corrupt and require rebuilding.

What if some programs need sequential access while others need direct access? The indexed sequential file organization is a special variation on the indexed file concept. Records are physically arranged in sequence to allow sequential access, in contiguous locations or using a linked list or both. However, the file

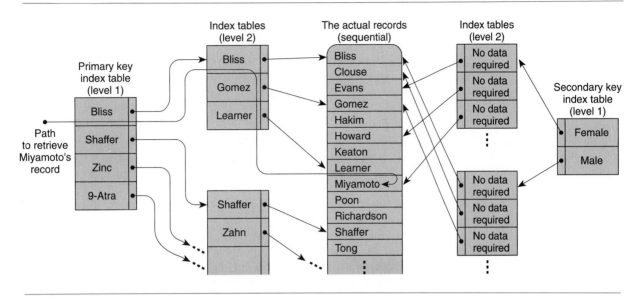

FIGURE 15.6 **Indexed File Organization** This employee file is organized as an indexed file. In indexed files, the actual records—shown in the middle of the figure—are usually stored sequentially. As records are added to the sequential file, index tables—shown on the left and right of the figure—are built that allow any specific record to be retrieved without reading sequentially through the whole file. The path for retrieving Miyamoto's record is indicated.

also contains an index of record keys and their physical addresses that can be used to provide semidirect access to records within the sequential file. Each program determines how it needs to access the records, sequentially or directly (via the index).

Indexed sequential file organization is supported by the operating system software or program compilers so that when records are added to or deleted from the sequential file the index is also maintained automatically. Two of the more common environments are VSAM (Virtual Storage Access Method) and ISAM (Indexed Sequential Access Method). VSAM and ISAM differ primarily in the manner in which the physical files are organized. This has a greater impact on the programmer than on the analyst.

The selection of the best type of organization for a file is a function of the file access methods required by programs that will use that file. Today, it is frequently mandated that all master and transaction files be organized as indexed sequential files, because in the future a combination of sequential and direct access might be required even if that combination is not needed today. ISAM and VSAM files offer such flexibility.

In database systems, the database software often predetermines or limits the file organizations for all files that are contained in the database. If file organization must be determined, a database professional makes that determination.

As was the case with fields and records, file design decisions must also be recorded in the project repository. A file is one physical implementation of a data store; therefore, design specifications are recorded in a data store entry. Design decisions such as blocking factors and file organization, to name only two, should be recorded in the repository. We'll demonstrate how to complete this repository entry later in the chapter.

Databases

Conventional files may be the lifeblood of many information systems; however, they are slowly but surely being replaced with databases. Recall that a **database** may loosely be thought of as a set of interrelated files. By *interrelated,* we mean that records in one file may be associated with the records in a different file.

For example, a CUSTOMER record may be linked to all of that customer's ORDER records. In turn, each of those ORDER records may be linked to relevant PRODUCT records. This linking allows us to eliminate most of the need to redundantly store fields in the various files. Indeed, in a very real sense, conventional files are consolidated into a single file, the database.

The idea of relationships between different collections of data was introduced in Chapter 8. In that chapter, you learned to analyze and model data as entities and relationships. Database is the modern implementation of those entities and relationships.

So many applications are now being built around database technology that database design has become an important skill for the analyst. Indeed, database technology, once considered important only to the largest corporations with the largest computers, is now common for applications developed on microcomputers.

The history of information systems has led to one inescapable conclusion: data is a resource that must be managed. Very few experienced information systems staffs have avoided the frustration of uncontrolled growth and duplication of data stored in their computer files and systems. As systems were developed, implemented, and maintained, the common data needed by the different systems was duplicated in multiple, conventional files. This duplication carried with it a number of costs: extra storage space required, duplicated input to maintain redundantly stored data and files, and data integrity problems (e.g., the ADDRESS for a customer not matching in the various files that contain customer ADDRESSes).

Out of necessity, database technology was created so an organization could maintain and use its data as an integrated whole instead of as separate data files. We can now develop a shared data resource that can be used by several information systems.

Database Environments

Data becomes the central resource in a database environment. Information systems are built around this central resource to give both computer programmers and end-users flexible access to data. Let's examine this environment more closely by studying two different types of database situations.

Figure 15.7 illustrates the way many companies have evolved into a database environment. Many companies have numerous conventional file-based information systems, most of which were developed prior to the popularity of database. In many cases, the processing efficiency of files or the projected cost to redesign files has slowed conversion of these systems to database.

Production databases are used to support transaction processing for major information systems. Access to these databases is limited to computer programs that use the DBMS to process transactions, maintain the data, and generate regularly scheduled management reports. Some query access may also be provided.

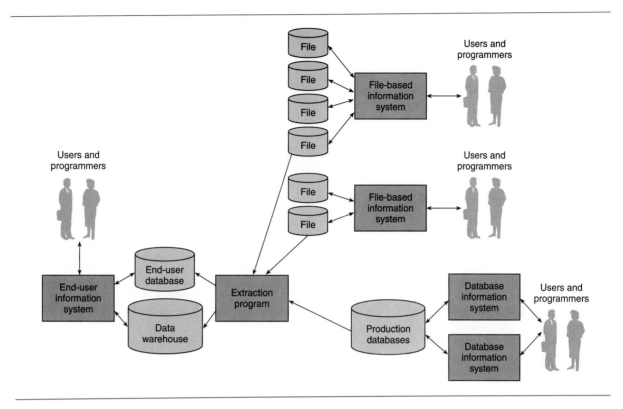

FIGURE 15.7   A Typical Database Environment   A typical database environment may still contain numerous files. Additionally, it may contain both production and end-user databases. Databases are frequently loaded and updated from conventional files.

Many information systems shops hesitate to give end-users access to production databases for queries and reports. The volume of unscheduled reports and queries could overload the computers. Instead data warehouses are developed, possibly on separate computers.

Data warehouses store data that is extracted from the production databases and conventional files. Fourth-generation programming languages, query tools, and decision support tools are used to generate reports and analyses off these data warehouses.

Admittedly, this second scenario is advanced, but many firms are currently using variations of it.

Additionally, PC database technology has matured to allow end-users to develop **personal databases** that may include data downloaded from files, production databases, and/or data warehouses. Personal databases are built using PC database software such as dBASE, Paradox, FoxPro, and Access.

Database Architecture

In a database environment, the notion of conventional files disappears. In its place, we have a structure where records of one type, such as ORDERs, can be related to records of different types, such as CUSTOMERs. If this all sounds similar to the data modeling concepts you learned in Chapters 8 and 13, that is

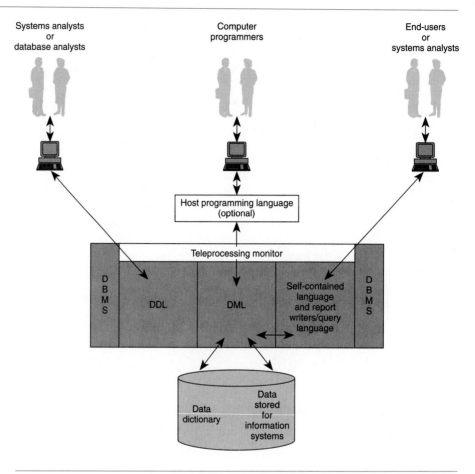

FIGURE 15.8 Architecture of an Ideal DBMS
Not all DBMSs contain all the components shown in this diagram; however, the diagram is representative of a typical DBMS. The key components are the data definition language and the data manipulation language, components common to all DBMSs.

precisely correct! Database physically implements essential entities and relationships.

A **database management system (DBMS)** is specialized computer software available from computer vendors that is used to create, access, control, and manage the database.

Figure 15.8 depicts the database technical environment. A systems analyst, or database analyst, designs the structure of the data in terms of record types, fields contained in those record types, and relationships that exist between record types.

Data definition language (DDL) is used by the DBMS to physically establish those record types, fields, and relationships. Additionally, the DDL defines views of the database. Views restrict the portion of a database that may be used or accessed by different users and programs. DDLs record the definitions in a permanent data repository.

Some data dictionaries include formal, elaborate software that helps database specialists keep track of everything stored in the database — including generation of data repository reports, analyses, and inquiry responses.

Computer programs are then written to load, maintain, and use actual data. These programs may be written in a host programming language—such as COBOL, PL/1, or BASIC—that is supported by the DBMS.

A **data manipulation language (DML)** is used to retrieve, create, delete, and modify records and to navigate between record types—for example, from CUSTOMER to ORDERs for that customer. The programmers don't have to understand how the data is physically stored or accessed. The DBMS hides such details. The DML refers to the DDL during execution.

Alternatively, many DBMSs don't require a host programming language. They provide their own self-contained programming language that includes a DDL and a DML. Generally, these self-contained languages greatly simplify applications prototyping and development. These languages and features are typically designed to be simple to learn and use—so much so that experienced programmers can be replaced by analysts and end-users. This alternative is also depicted in Figure 15.8.

Many mainframe DBMSs greatly simplify internal controls by automatically logging all updates and enforcing security as defined by the database analysts. Eventually, this should even be true of microcomputer DBMSs.

A **teleprocessing** (or *TP monitor*) is specialized software, frequently included in a multiple-user DBMS environment, that supervises and controls access to the database via terminals in on-line environments. Most DBMSs can also interface with industry TP monitors, such as IBM's CICS.

Relational Database Management Systems

There are several types of database management systems. They can be classified according to the way they structure records. Early database management systems organized records in hierarchies or networks implemented with indexes and linked lists. You may study these further in a database course. But today, most successful database management systems are based on *relational* technology.

Relational databases use a model intended to greatly simplify the end-users' and programmers' view of a database.

In a **relational database,** as shown in Figure 15.9, files are seen as simple tables, also known as **relations.** The rows are record occurrences—also called **tuples**—and the columns are fields, also called **domains.** True relational databases do not store pointers to other tables. Instead, foreign keys infer relationships between different tables.

To write reports and answer inquiries, the DML (usually *structured query language* or *SQL*) lets the programmer or end-user perform simple table operations to create temporary tables. These operations include:

- **SELECTing** specific records from a table and creating a new, but temporary, table that contains only those occurrences. Criteria can be set to determine which records to select from the initial table.

- **PROJECTing** out specific fields from a table, creating a temporary table that has fewer fields.

Customer Table			
Cust. No.	Cust. Name	Cust. Address	•••
10112	Lucky Star	Ann Arbor	
10113	Pemrose	Grand Rapids	
⋮			

Order Table		
Order No.	Cust. No.	•••
A6334	10112	
A6335	10113	
⋮		

Part Table		
Part No.	Part Desc.	Quant. on Hand
77B12	Widget	8000
77B13	Widget	0
⋮		

Part records describe data about parts in general. Ordered Part describes data about specific parts on specific orders.

Ordered Part Table			
Order No.	Part No.	Quantity Ordered	•••
A6334	77B12	50	
A6334	77B13	100	
A6335	77B13	25	
A6335	77B12	4	
⋮			

FIGURE 15.9   Relational Data Structures A relational DBMS stores records as simple tables, similar to those seen in some spreadsheets that exhibit relational-like qualities. Tables are related to one another via intentionally redundant fields, usually keys. Relational DBMSs provide operators that build temporary working tables from the permanent tables that are physically stored.

- **JOINing** two or more tables across a common field (this is the same as navigating relationship paths in a hierarchical or network database). Again, a temporary table is created.

In all cases, these three relational operators—which, unfortunately, have different names in the various relational DBMSs on the market—create temporary, working tables that will disappear when you exit the program. If changes are made to the data in those tables, those changes will be updated into the permanently stored tables before the working tables are discarded (assuming the end-user has update authority).

Also, note that the relational commands can be combined. For instance, assume we need a table of orders for certain customers, perhaps to produce a report. First, we can use SELECT on the CUSTOMER table to create a working table of those customers needed. Second, we can PROJECT out only those fields needed in both the CUSTOMER and ORDER working tables to reduce their size. Finally, we JOIN the two working tables to give us the final working table needed for the report.

Query languages and report writers that are easy to learn and use have been built around relational databases. Most relational databases are currently converging to supersets of a de facto standard DML **Structured Query Language**

(SQL), which is utilized to create, update, and use tables. SQL includes the SELECT, PROJECT, and JOIN operators.

Examples of relational DBMSs include IBM's DB2 and SQL/DL, ADR's DATACOM, Relational Technology's Ingress, Oracle Corporation's Oracle, and Information Builder's FOCUS (which is also hierarchical). Additionally, most microcomputer DBMSs are relational. Examples include Ashton-Tate's dBASE IV (which recently embraced true SQL relational standards) and Microrim's R:BASE for DOS. Many relational DBMSs are offered in both mainframe and micro versions. FOCUS is one such DBMS.

Database textbooks and courses offer entire chapters and units on relational databases. This is only an introduction to database — we encourage you to learn more!

The Database Administrator

Systems analysts and database analysts are becoming increasingly involved in database design. However, in most cases, they rarely make the final recommendations. Who manages the database environment?

A **database administrator (DBA)** oversees a staff of database specialists. These specialists make the final recommendations on all database designs because they have the global, application-independent view that many systems analysts lack. DBAs also load and maintain databases, establish security controls, perform backup and recovery, and maintain the DBMS software. In addition, they plan and control database definition to minimize redundancy and keep track of where all data is stored and how various systems use that data. In smaller shops, a systems analyst may perform some or most of these duties.

In some organizations, the role of the DBA is split between two administrators. The **database administrator** manages the technical environment of database. The **data administrator** manages the data itself, keeping track of where all data is stored and what programs and end-users require access to what data.

HOW TO DESIGN AND DOCUMENT CONVENTIONAL FILES

In this section, we discuss how files are designed and documented. Remember that some physical design decisions (from earlier chapters) have already been recorded in the project repository. We'll design a file to demonstrate this important systems design task.

File Design Technique

Two important and related issues will affect file design: redundancy and internal controls. These issues influence the value and usefulness of the files you will design.

Minimizing Redundancy in Files

The minimization of redundant data in files has long been a suggested philosophy of file design. Ideally, no single field would be included in more than one record. This is not just because redundantly stored fields use considerably more storage space. The real problem is even more significant. If the same field is stored in more than one record, the data must be updated in more than one place when it changes. Otherwise the same field could take on different values in different records.

So why do systems frequently store redundant data? The reasoning is not unsound. First, there are natural relationships between files. If we need to use the data in one file (e.g., CUSTOMER) to cross-reference the data in another file (e.g., INVOICE), we redundantly store some customer fields in both files. Second, to improve performance of some programs, we can reduce the number of files needed by a program — thus reducing the number of files opened, read from, and written to — by redundantly storing some common data in more than one file.

Redundancy, as a term, has an undeserved reputation. Redundancy is not necessarily wrong. Uncontrolled redundancy is the more common problem! We suggest you restrict duplication of fields in different records to primary key fields (needed to cross-reference files) and relatively stable descriptor fields (those that don't change values often). For instance, CUSTOMER ADDRESS is relatively stable, but CUSTOMER BALANCE is not. The more volatile the field, the greater the danger in storing that field redundantly in different files.

The problems of data integrity and redundancy can be addressed by basing your file design on a third normal form data model as presented in Chapter 13. Conventional files, as you will soon learn, tend to compromise that ideal data model; however, the compromises can be more carefully thought out using the data model as a foundation.

Internal Controls for Conventional Files

Internal controls are a requirement in all computer-based systems. They ensure that the computer-based system is protected against both accidental and intentional errors and use, including fraud. (We will cover controls separately for the design components in their respective chapters.)

Internal controls are designed into files to ensure the integrity and security of the data in those files. Internal control issues for files include the following:

- **Access controls** should be specified for all files. Access controls determine who will be able to read from and write to the file. Some end-users will have read- and write-access. Other end-users will have only read-access. If corrections must be made to the file's data, who will authorize those changes? Remember, the data belongs to the end-user; therefore, authorization should not come from the analyst.

- **Length of retention** of files is another control issue. Master file records are retained indefinitely. However, transaction records should be retained in the file only so long as the data is immediately useful. Otherwise, valuable storage capacity will be wasted. When should such records be archived? How long should archived records be retained? In some cases, government regulations may dictate a long retention cycle.

- **Backup methods and procedures** should be defined for all data files. Backups are used to recover data when files have been lost or destroyed. All master and transaction files should be periodically copied onto tape or diskette.

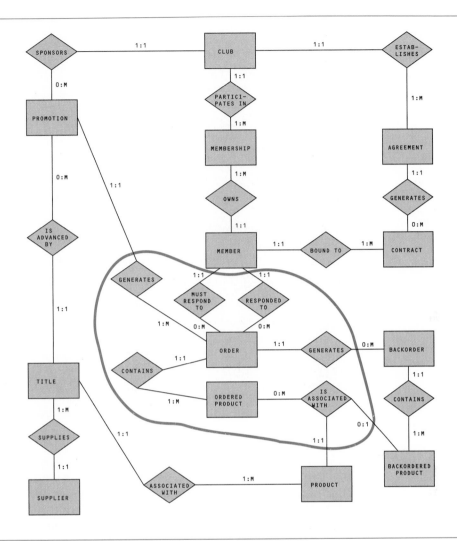

FIGURE 15.10   Partitioned Entity Relationship Data Model for SoundStage   The partitioned area represents the scope of implementation for the ORDERS FILE.

- **Audit trails** should be established for all files. An audit trail makes a permanent record of any update—an added record, deleted record, or changed record—that is made by a program. If such changes are processed only in batches, the program needs to print a report of the updates. Today, on-line processing of updates is more common. On-line programs need to post updates to a special audit file that can be printed at convenient times. Many file management systems automatically create audit trails, relieving the analyst and programmer of that responsibility.

  Computer files that need to be designed are easily identified by studying the most recent entity relationship diagram. (You learned to define data storage requirements with these tools in Chapters 9 and 13.)

  An entity relationship diagram (or data model) depicts the data to be stored (see Figure 15.10). The actual files may be constructed from combinations of

FIGURE 15.11 **Preliminary Attribute Content of the ORDERS FILE**
The attribute content for the two entities contained in the ORDERS FILE is identified in Figure 15.11**A**. These third normal form attributes were compiled during data analysis (normalization) in Chapter 13. The consolidated (denormalized) attribute contents for a record in the ORDERS FILE is shown in Figure 15.11**B**.

A. ORDER is composed of the following attributes:
    <u>ORDER NUMBER</u>
    ORDER DATE
    ORDER STATUS (added by end-users)
    AUTOMATIC FILL DATE
    ORDER AMOUNT DUE

ORDERED PRODUCT is composed of the following attributes:
    <u>ORDER NUMBER</u>
    <u>PRODUCT NUMBER</u>
    <u>MEDIA CODE</u>
    QUANTITY ORDERED
    QUANTITY SHIPPED
    ORDER PRICE
    ORDERED PRODUCT STATUS

B. An ORDERS FILE record contains the following attributes:
    ORDER NUMBER
    ORDER DATE
    ORDER STATUS (added by end-users)
    AUTOMATIC FILL DATE
    ORDER AMOUNT DUE
    NUMBER OF ORDERED PRODUCTS (added to indicate number of occurrences of order product)
        ~~ORDER NUMBER~~
        PRODUCT NUMBER
        MEDIA CODE
        QUANTITY ORDERED
        QUANTITY SHIPPED
        ORDER PRICE
        ORDERED PRODUCT STATUS

entities. The entity relationship diagram also describes mandatory relationships that must be implemented between different types of data to be stored.

On the data flow diagram, files appear as data stores. If, for example, the stores ORDER and ORDERED PRODUCT were combined (from Chapter 14), the data store could be implemented as a file containing variable-length ORDER records.

Let's study a file design example to see how this procedure is applied. We'll demonstrate the complete design for only one file. We will continue to use our CASE product, Excelerator, to demonstrate one way of recording design decisions into the project repository.

Step 1: Determine the Data Content of the File

The first step in file design is to determine the actual data to be stored in the files. For demonstration purposes, the remaining steps will focus on designing the ORDERS FILE. Let's examine the substeps required to define the contents of the ORDER record.

Step 1(a): Identify Attributes to Be Stored in the File

According to our partitioned entity relationship diagram, the ORDERS FILE is to contain data describing the ORDER and ORDERED PRODUCT entities, as well as a number of relationships. There's no need to start from scratch and brainstorm the attributes of the file.

Think back. Repositories were initiated in systems analysis to define the logical, or implementation-independent, contents of each entity. The contents were described in a logical record description in the project repository. For your convenience, the attributes for each entity are listed in Figure 15.11**A**. Keys are underlined.

Step 1(b): Consolidate the Attributes into a Single Record

A single occurrence of the final ORDERS record should contain attributes describing one occurrence of the entity ORDER plus one or more occurrences of the entity ORDERED PRODUCT. Figure 15.11**B** shows the integrated attributes from the two entities. Notice that the attributes from the ORDERED PRODUCT entity appear in the repeating portion of the record. The attribute NUMBER OF ORDERED PRODUCTS was added to the list to indicate the actual number of occurrences of the repetition of attributes for any given ORDER. Also, recall that attributes appearing in the repeating portion of a variable-length record cannot be keys in most conventional files. Therefore, ORDER NUMBER, PRODUCT NUMBER, and MEDIA CODE are no longer underlined.

We also eliminate any redundancy in the resulting record. Thus, ORDER NUMBER is deleted from the group of repeating elements because it is the key of the fixed-length portion of the record.

Step 1(c): Add Attributes to Implement Relationships to Other Entities

Review our partitioned entity relationship diagram in Figure 15.10. The ORDERS FILE is to store additional attributes used to implement relationships. Figure 15.12 revises our ORDERS FILE record to describe the required relationships. The relationships are to be implemented as follows:

- The relationship between ORDER and PROMOTION is implemented by adding the key of PROMOTION (CLUB NAME and PROMOTION DATE) to the record.
- The relationship between ORDER and MEMBER is implemented by adding two attributes. First, the attribute MEMBER NUMBER is used to associate an ORDER to the MEMBER that placed the order. We also brought in the elements MEMBER NAME, P.O. BOX, STREET, CITY, STATE, and ZIP CODE. These are relatively stable attributes; therefore, the redundancy is not very risky. The benefit is that we may not have to open the MEMBERS FILE to print most reports about orders, because the data we need about members is also in the ORDERS FILE. This is planned redundancy, which, of course, must be documented if it is to be controlled.

  Second, the attribute MEMBER RESPONSE STATUS will distinguish the relationships between ORDER and MEMBER. In other words, MEMBER RESPONSE STATUS describes whether the order is awaiting the member's response or released to be filled after receiving the member's response.
- The attribute BACKORDER DATE will implement the relationship between ORDER and BACKORDER. If it takes on a value, then we know that a backorder record exists in the BACKORDERS FILE.

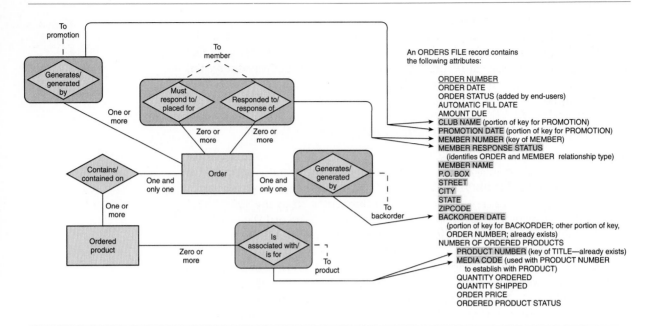

FIGURE 15.12 Implementing Relationships through Attributes   Relationships between the ORDERS FILE and other entities appearing in separate files are implemented by redundantly storing the keys of those other entities.

- The attribute PRODUCT NUMBER will be used to establish the relationship between ORDERED PRODUCT (in the ORDERS FILE) and TITLE. Specific occurrences of the entity PRODUCT will then be obtained by using the MEDIA CODE attribute common to ORDERED PRODUCT and PRODUCT. If you wish, you could also duplicate some relatively stable elements from TITLE into this record.

Step 1(d): Add Audit Fields

Finally, we may choose to add audit and control fields to the content of our record. This is not always necessary because some operating systems and file management software automatically provides audit and control mechanisms. Figure 15.13 represents the final contents for the ORDERS record.

Now we can describe our record to the project repository as a record containing attributes. (Typical screens were shown earlier in the chapter.) The ORDER record must specify the maximum number of occurrences for each field for one occurrence of the record. Each field should be described as a attribute, including specifications for storage format and length (which we learned is dependent on storage format). After entering the record and describing each attribute in the record, we can print out or prepare a detailed record layout (see Figure 15.14).

Step 2: Update the Record to Reflect Physical Requirements

Before we can design a file, there are a number of implementation-independent file attributes that must be reviewed and, if necessary, updated. These attributes will have an impact on any design solution. Figure 15.15 shows the

An ORDERS FILE record contains the following attributes:
    <u>ORDER NUMBER</u>
    ORDER DATE
    ORDER STATUS (added by end-users)
    CLUB NAME (portion of key for PROMOTION)
    PROMOTION DATE (portion of key for PROMOTION)
    MEMBER NUMBER (key of MEMBER)
    MEMBER NAME
    MEMBER ADDRESS consisting of:
        P.O. BOX
        STREET
        CITY
        STATE
        ZIP CODE
BACKORDER DATE (portion of key for BACKORDER; other portion of key, ORDER
NUMBER; already exists
    AUTOMATIC FILL DATE
    ORDER AMOUNT DUE
    NUMBER OF ORDERED PRODUCTS
        PRODUCT NUMBER (key of TITLE—already exists)
        MEDIA CODE (used with PRODUCT NUMBER as key of PRODUCT)
        QUANTITY ORDERED
        ORDERED PRODUCT STATUS
        QUANTITY SHIPPED
        ORDER PRICE

FIGURE 15.13 Final Attribute Contents of the ORDERS FILE The final contents of the ORDERS FILE contains attributes describing entities and relationships.

logical requirements for our ORDERS FILE. Pay particular attention to the following:

- The "Explodes to" entry points to the record description we just finished.
- The Volume entries describe the number of occurrences of ORDER that will be active at any given time. Volume data should have been recorded when the entity ORDER was defined in the systems analysis phase of the project. We'll soon need this data to calculate disk space requirements for our ORDERS FILE.

Step 3: Design the File

We can now complete the design of our ORDERS FILE. Figure 15.15 also presents the remaining physical specifications for the ORDERS FILE. Skim through these repository entries, and then direct your attention to the following:

Ⓐ This file is to be implemented as a variable-length VSAM record. VSAM — an indexed file organization — maximizes file access flexibility providing both sequential and direct access opportunities.

Ⓑ The record size is specified as a minimum and maximum number of bytes. Minimum record size is easy to calculate. Because every ORDER must have at least one ORDERED PRODUCT, we simply sum the sizes — specified in bytes — of the individual attributes appearing in the ORDER record.

FIGURE 15.14 Typical CASE Repository Report This report describes the ORDERS FILE record structure (including subrecords, which are indented) and key field attributes.

```
DATE: 15-DEC-92              RECORD - EXPLOSION
TIME: 00:08                  NAME: ORDER RECORD

NAME:            ORDER RECORD
ALIAS:           ORDER

ELEMENT/RECORD                          OFF  OCC  TYPE  LEN
-----------------------------           ----  ---- ----- ----

ORDER NUMBER                            000  001   K    006

ORDER DATE                              006  001   E    002

ORDER STATUS                            008  001   E    001

CLUB NAME                               009  001   E    015

PROMOTION DATE                          024  001   E    002

MEMBER NUMBER                           026  001   E    006

MEMBER NAME                             032  001   E    025

P.O.BOX                                 057  001   E    010

STREET                                  067  001   E    015

CITY                                    082  001   E    015

STATE                                   097  001   E    002

ZIPCODE                                 099  001   E    009

BACKORDER DATE                          108  001   E    002

AUTOMATIC FILL DATE                     110  001   E    002

ORDER AMOUNT DUE                        112  001   E    006

DATE ORDER CREATED                      118  001   E    002

ORDER CREATED BY ID                     120  001   E    004

DATE ORDER LAST MODIFIED                124  001   E    002

ORDER LAST MODIFIED BY ID               126  001   E    004

NUMBER OF ORDERED PRODUCTS              130  001   E    002

ORDERED PRODUCT                         132  015   R
  PRODUCT NUMBER                        132  001   E    008
  MEDIA CODE                            140  001   E    001
  QUANTITY ORDERED                      141  001   E    005
  ORDERED PRODUCT STATUS                146  001   E    001
  QUANTITY SHIPPED                      147  001   E    005
  ORDER PRICE                           152  001   E    005

Record length is 507.
```

Maximum field size is a little more difficult to determine than minimum. Maximum record size is based on the maximum number of times that the repeating group can occur (in our case, 15). The maximum record size equals the sum of the field sizes in the fixed-length portion of the record plus the sum of 15 times the sum of the repeating elements.

FIGURE 15.15 **Detailed Logical Requirements and File Design Specifications for the ORDERS FILE**

```
DATA STORE - OUTPUT
NAME: ORDERS FILE

TYPE Data Store              NAME ORDERS FILE

                             Explodes To One Of:
Label ORDERS FILE     Record      ORDER RECORD
                             Data Model Diagram
                             ERA Diagram

Location                     ORDER PROCESSING
Manual Or Computer           C
Total Number Of Records      20034
Average Number Of Records    17300

Index Elements:      ORDER NUMBER
                     ORDER STATUS
                     CLUB NAME + PROMOTION DATE
                     MEMBER NUMBER

        Description
AN AUTOMATIC, DATED ORDER GENERATED IN RESPONSE TO A PROMOTION. IT MAY
BE REVISED, APPROVED, OR CANCELLED VIA A MEMBER RESPONSE.

LOGICAL, IMPLEMENTATION-INDEPENDENT ATTRIBUTES OF THIS DATA STORE

   VOLUME
     AVERAGE: 17,300   GROWTH RATE (PER TIME PERIOD): 5% PER YEAR
     PEAK:    20,034   WHEN: NOVEMBER THROUGH CHRISTMAS (DECEMBER)

   USER REFERENCES: ORDER PROCESSING STAFF

   CONSTRAINTS: NONE

PHYSICAL, IMPLEMENTATION-DEPENDENT ATTRIBUTES OF THIS DATA STORE

Ⓐ IMPLEMENTED AS: VARIABLE-LENGTH, VSAM
   COMPUTER NAMES OR IDs: ORDER.DAT
   MAXIMUM PHYSICAL RECORD SIZE: 507 Ⓑ
   BLOCKING CONSTRAINTS: 1/3 TRACK BLOCKING
   BLOCKING FACTOR: 5 Ⓒ BLOCK SIZE: 2603
   FILE SIZE (IN BYTES): 10,430,221 Ⓓ
   FILE SIZE (IN TRACKS OR INCHES): 1336 Ⓔ
   FILE SIZE (IN CYLINDERS): 149 Ⓕ
   AVERAGE RECORD LIFETIME UNTIL ARCHIVE: 2 MONTHS
   BACKUP AND RECOVERY
     BACKUP TIMING: DAILY 8:00 PM      BACKUP MEDIA: TAPE
     BACKUP RETENTION: 1 MONTH
     AUDIT TRAIL METHOD: UPDATES LOGGED TO DISK, PRINTED AT END OF DAY
   SECURITY: ONLY ORDER PROCESSING STAFF MAY UPDATE
```

132 bytes per record + (25 bytes per record × 15 groups)
= 507 bytes per record

Ⓒ There is usually a limit to the size of a block—for example, half of a track on the disk. Some information systems departments even establish a standard for block size. For any given disk drive, there are tables that provide blocking reference data for that disk drive. Different tables are used for files to be stored with and without keys. The blocking factor table tells you how big the block can be.

The Next Generation

REVERSE DATABASE ENGINEERING

There is a growing crisis in the commercial database applications arena. For many years, most databases were implemented with hierarchical DBMSs, particularly IBM's IMS. Thousands of databases were built. These hierarchical databases have survived mostly because they provided much better transaction performance than other databases.

Network database management systems like Cincom's TOTAL and Cullinet's IDMS offered non-IMS shops the opportunity to take the database plunge without the complex overhead of hierarchical systems. Hierarchical systems still provided better transaction throughput; however, thousands of network databases have been built.

All the while, researchers were touting a better way, the relational data model. But the relational database seemed more of a promise for the future since performance could not approach that of either hierarchical or network systems.

But the technology of relational database kept improving. It still is. Today, IBM's relational DBMS,

DB2, is approaching the throughput of IBM's kingpin, the hierarchical IMS. Other relational DBMSs have also improved. The writing is on the wall. Relational databases, as once predicted, should replace both hierarchical and network databases.

But there is one final roadblock —the thousands of nonrelational databases that have been written and the programs that were written to update and use those databases! It would take years, perhaps decades, to convert those hierarchical and network databases to relational equivalents.

Enter the next generation of CASE, reverse database engineering. Reverse database engineering, in its simplest form, will read the data definition language (DDL) for one database and automatically convert it to the DDL for another database—for example, IMS hierarchical to DB2 relational.

That sounds impressive, but to work, the reverse engineering must also automatically rewrite the original computer programs to re-

place all data manipulation language (DML) instructions with the DML instructions for the new database.

We are very close to seeing CASE tools with these capabilities. But it will get even better. With the improvement of artificial intelligence, CASE products can eventually incorporate expert systems capabilities that will help systems analysts and database administrators correct analysis and design errors in the original databases. For example, many of those databases were conceived before the practice of normalization was common. We might be able to undo those errors.

Thus, CASE reverse engineering may soon provide a migration path from one DBMS's first generation to its second, as well as providing the opportunity for correcting old design errors. In fact, reverse engineering may be the only hope for converting those old hierarchical and network databases. It probably can't be done with manual methods!

For our example, the Entertainment Club's Information Systems Department adopted a ⅓-track blocking standard; therefore, a block size consists of 2,603 bytes. Blocking factor (BF) is calculated as follows:

(2,603 bytes per block)/(507 bytes per record)
$$= 5.13 \text{ records per block} = 5 \text{ records per block}$$

Blocking factors are always truncated at the decimal point because a fraction of a record cannot be stored in a block.

Ⓓ File size is an important calculation because we can't store data for which we don't have capacity. To calculate file size, we first determined, by consulting the end users, that there are currently 17,300 ORDERS and that orders are increasing by 5 percent per year. To calculate the file size in bytes, tracks, and cylinders, the following calculations were used:

17,300 records $\times$ 1.158/5 records per block

$\times$ 2,603 bytes per block = 10,430,221 bytes of storage

Note that the blocking factor had to be considered because the wasted storage in any block cannot be reclaimed, since block size is fixed at 2,603 bytes per physical record. Also note that we included anticipated growth (5 percent = 1.05) over three years ($1.05^3 = 1.158$) in our formula. Finally, after dividing by the blocking factor, that result was increased to the next greatest integer value, because a file and a block must both contain complete records.

(E) It is also useful to express file size in terms of tracks and cylinders, because those tracks and cylinders might have to be dedicated to the file.

(17,300 records per volume $\times$ 1.158/5 records per block)

/3 blocks per track = 1,336 tracks

Again, after performing the operations, we increased the result to the next greatest integer value.

(F) To determine the number of cylinders required to store the file, you need to understand another characteristic of the disk packs used by Entertainment Club. The disk packs currently in use have 9 tracks per cylinder (this may vary for other disk packs).

1,336 tracks per file/9 tracks per cylinder = 149 cylinders per file

You've seen how to design a conventional computer-based file containing variable-length records. Design of a file containing fixed-length records is accomplished in the same manner. The blocking factors and file-size calculations would be simplified since all records would be fixed length.

HOW TO DESIGN AND DOCUMENT DATABASES

Computer-assisted systems engineering (CASE) has been a continuing theme throughout this book. There are specific CASE products that address database analysis and design. Examples include Chen & Associates' ER Modeler and Bachman's Database Administrator. Database design facilities are also finding their way into the more popular CASE tools for systems analysis and design. For a look at future directions of CASE and database design, see The Next Generation box for this chapter.

The design of any database will usually involve the DBA and database staff. They will handle the technical details and cross-application issues. Still, it is useful for the systems analyst to understand the basic design principles for relational databases. The designs presented here are simplified; we will not cover the technical idiosyncrasies of any specific database management system. Database courses and textbooks usually detail the technical side of the subject in greater detail.

Step 1: Review Database Requirements

As was the case for conventional file design, the system's data model—in our case, an entity relationship diagram (ERD)—serves as a starting point. Figure

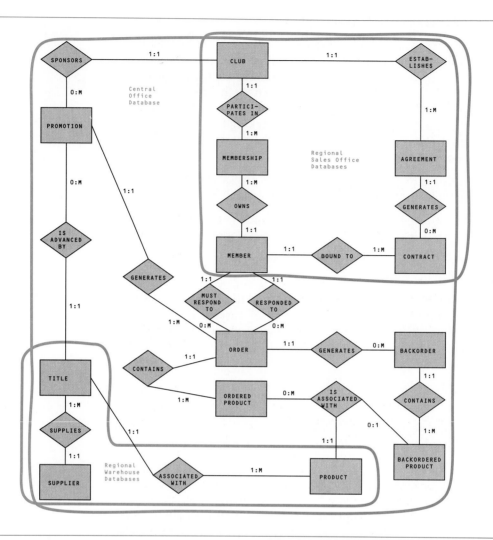

FIGURE 15.16 **Distributed Data Model for SoundStage** This version of the original data model reflects the approved design decisions.

15.16 identifies the entities and associations that must be designed. These requirements were determined during the definition phase of systems analysis in our life cycle. Recall that each entity is further described in the project repository by a record that defines the attributes describing the entity. Also recall that the data model and repository were subjected to a rigorous data analysis called *normalization* (see Chapter 13, Data Analysis).

Notice that our data model has been "bounded and annotated" to indicate **database distribution** decisions. There is a trend towards distribution of data to multiple servers in a client/server architecture (Chapter 14). Some data is replicated at multiple sites. Distributed DBMS technology (e.g., Oracle and Sybase) provides control features to properly manage distributed data.

Step 2: Design the Logical Schema for the Database

The proposed implementation of a database is depicted in a special model called a schema.

A **schema** is the structural model for a database. Any given DBMS supports two schemas, a logical schema and a physical schema.

These two schemas specify both the physical and logical structure of the records in a database.

The **physical schema** defines data structures, access methods, file organizations, indexes, blocking, pointers, and other physical attributes.

Thus, DBMSs don't replace file details — they just hide them from the programmers and end-users. This aspect of the database is not of concern to most systems analysts.

The **logical schema** defines the database in simpler terms as seen by end-users and programmers. This schema is very similar, but not identical, to the data models introduced in Chapter 8. It defines simple records and associations, just like the entity relationship data models. However, the logical schema is constrained by the chosen DBMS's supported structures.

The analyst's role in physical database design is usually restricted to design of the *logical schema*. Recall that the logical schema is a picture or a map of the records and relationships to be implemented by the database. Although it is similar to the ERD, the logical schema reflects the DBMS chosen. The DBA or DB staff will evaluate that schema, make appropriate modifications, generate or modify the DDL, and generate the test database for subsequent prototyping and development. The exception to this procedure is microcomputer databases, which are frequently designed and implemented by the systems analyst.

The schema design and associated repository specifications differ for different classes of DBMS technology. Analysts rarely select a DBMS since the selection of a DBMS involves factors that extend beyond any one application on which the analyst is currently working.

Initially, we record our decision to implement the logical model as a relational database. Recall that a database is a physical implementation of a data store; therefore, we record our implementation decision in a data store description in our project repository.

Schema design is relatively easy. Ideally, each entity on the Distributed ERD (Figure 15.16) is implemented as a relational table. This is demonstrated in Figure 15.17. (Note that the smaller, distributed database schemas would have been subsets of this schema.) Some entities might be split into master tables and secondary tables to separate the more frequently accessed fields from other fields and thereby improve performance. Why follow the ERD's entities so closely? Recall that the ERD was carefully normalized to minimize redundancy and maximize flexibility. The practice of normalization was derived from early research into relational database design. Ideally, we want to retain the advantages of normalization in our database design!

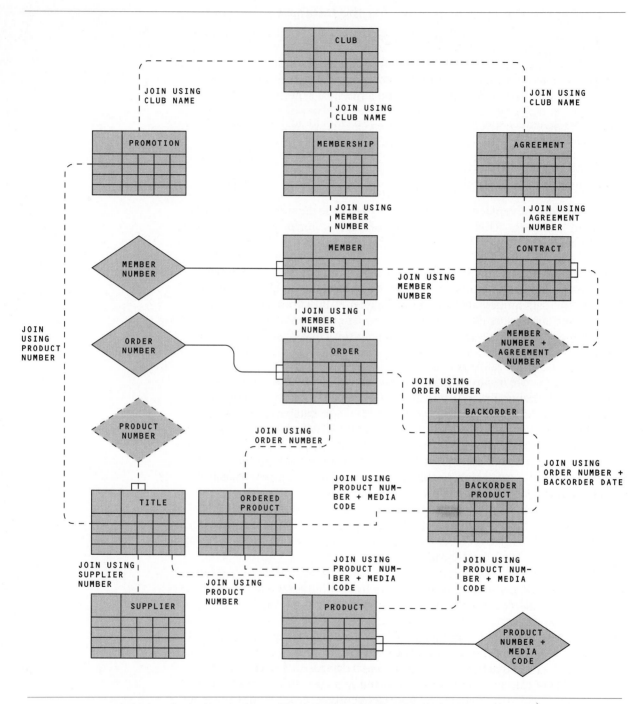

FIGURE 15.17 A Relational Database Schema for SoundStage This database schema illustrates how the SoundStage data model would be implemented as a relational database.

Would you ever want to compromise the third normal form entities when designing the database? For example, would you ever want to combine two third normal form entities into a single table (that would, by default, no longer be in third normal form)? Usually not! Although a DBA may create such a compromise to improve database performance, he or she should carefully weigh the advantages and disadvantages. Although such compromises may mean greater convenience through fewer tables or better overall performance, such combinations may also lead to the possible loss of data independence — should future, new fields necessitate resplitting the table into two tables, programs will have to be rewritten. As a general rule, combining entities into tables is not recommended.

How are relationships in the ERD implemented by a relational database? Recall that relational databases do not, as a rule, store physical pointers between tables. Instead, associations are established at run time by using combinations of SELECT, PROJECT, and JOIN commands to build working tables. Specifically, relationships are implemented by JOINing tables using one or more key fields that the tables have in common. We have represented these virtual relationships on our relational database schema in Figure 15.17 as dashed lines. The lines are labeled to name the foreign keys that would be needed to JOIN the tables.

Most relational databases improve performance by allowing (or requiring) that indexes be established on any table from which you might want to SELECT specific records. At a minimum, such an index would be built on primary and secondary keys. We have indicated indexes by diamonds, labeled with the names of the key fields. A solid diamond represents a permanently stored index. A dashed diamond represents a dynamic index, one that must be built each time you want to run a program that will use that index. The "crow's feet" at the end of the line from the index to the table represent that the index points to many (in fact, most or all) of the records in the table. Purists will rightfully argue that true relational systems do not require indexes. We loosely based our design on the IBM AS/400's relational database. We established permanent indexes for those tables most frequently used and dynamic indexes for medium-volume tables. We did not place indexes on tables with small numbers of records (e.g., CLUB) or on tables best accessed through JOINs with other tables (e.g., ORDERED PART would normally be accessed by JOINing it with either ORDER or PRODUCT, or possibly both).

Each table in the schema explodes to a repository description. Each table is the physical implementation of a data store. A typical data store description is shown in Figure 15.18. The contents of the table (data store) are described as records to the repository. A printout of the AGREEMENT table's record is shown in Figure 15.19. This record is a modified version of the entity contents defined during systems analysis. You can eliminate elements that you have decided not to store and add elements as needed (e.g., the audit fields added at the end of the example). Also note that we added the foreign key, CLUB NAME, to this AGREEMENT table record.

Step 3: Prototype the Database (if Necessary)

Prototyping is not an alternative to carefully thought out database schemas. On the other hand, once the schema is completed, a prototype database can usually be generated very quickly. Most modern DBMSs (mainframe and microcom-

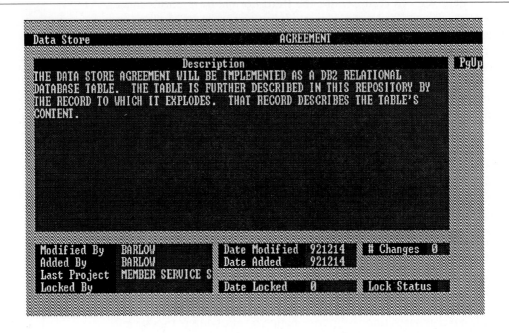

FIGURE 15.18 Project Repository Entry for a Table (Part 1) Each table in a relational database represents its own data store that should be described in the project repository.

FIGURE 15.19 Project Repository Entry for a Table (Part 2) The contents of a typical record in a relational table should be described in the project repository.

```
                                                      RECORD - EXPLOSION
                                                      NAME: AGREEMENT RECORD

       NAME:                AGREEMENT RECORD
       ALIAS:               AGREEMENT

       ELEMENT/RECORD                  OFF  OCC  TYPE  LENGTH
       -------------------------       ---- ---- ----- --------

       AGREEMENT NUMBER SUFFIX         000  001   K    003

       CLUB NAME                       000  001   E    015

       AGREEMENT EXPIRATION DATE       003  001   E    002

       AGREEMENT PLAN CREATION DATE    005  001   E    002

       MAXIMUM PERIOD OF OBLIGATION    007  001   E    004

       BONUS CREDITS AFTER OBLIGATION  011  001   E    002

       NUMBER OF MEMBERS ENROLLED      013  001   E    006

       NO. MEMBERS WHO HAVE FULFILLED  019  001   E    006

       Record length is 25.
```

puter alike) include powerful, menu-driven database generators that automatically create a DDL and generate a prototype database from that DDL. A database can then be loaded with test data that will prove useful for prototyping and testing outputs, inputs, screens, and other systems components.

Even in the absence of a mainframe DBMS, a prototype database built with a relatively inexpensive micro-DBMS, such as dBASE IV, can prove indispensable as the design proceeds, even if you have no intention of implementing that database with the micro-DBMS. Throwaway prototypes have long been used in engineering circles.

Summary

Computer files consist of physical records, records, and fields. A field or attribute is the smallest unit of data to be stored. There are three types of fields: primary keys, secondary keys, and descriptors. Most business-oriented systems store fields as binary codes. There are three common binary codes: EBCDIC, ASCII, and packed decimal. An understanding of storage formats is important during file design because storage format affects file size.

A record is a collection of fields arranged in a predefined format to describe a single entity. A record is also the unit of data storage that is operated on by a computer program. Records for a system are specified by the systems analyst. Depending on the data structure to be stored, the record will have fixed or variable length. The record for a file is normally documented using a record layout chart.

Occurrences of records are blocked to improve input and output efficiency. A block is the smallest number of occurrences of records that can be read or written at one time. The analyst often determines the number of records in the block because this can affect performance.

Finally, occurrences of records make up files. There are several types of files, but the two most important classes are: (1) master files, whose records—usually fixed length—describe basic business entities (e.g., parts, customers, and employees), and (2) transaction files, whose records—which are frequently variable length—describe business events (e.g., orders and invoices).

One of the most important performance decisions made by the systems analyst is file organization. File organization defines how records in a file are related to one another. Three common file organizations are sequential, direct, and indexed. The best way to determine an optimal file organization is to study the file access methods required by the programs that will use and update the file.

To ensure the integrity and security of the data, internal controls should be designed for files. Internal controls for files include access, retention, backup, and recovery controls.

File designs should be based on the data model defined during systems analysis. This helps the analyst ensure that the resulting files implement most of the business data entities and relationships. Initially, the data model should be partitioned to identify the files that must be designed.

For each file to be designed, the analyst must define the record's contents. The essential requirements are already known since the entities were defined during systems analysis. However, the contents of a physical file often require consolidation of entities. Furthermore, the analyst must usually add some new fields from other entities to implement relationships and install internal controls.

Once a file's record has been established, the analyst can review logical requirements (e.g., the number of occurrences that must be stored) and complete the physical design. The physical design includes specification of file organization, blocking factors, and disk space requirements.

A database is an alternative to multiple conventional files. It can be thought of as a single file of dissimilar records. These dissimilar records, such as CUSTOMERs, ORDERs, and PARTs, can be linked or associated with one another such that programmers and end-users can navigate the records—for example, find all ORDERs for a CUSTOMER or all PARTs on an ORDER. Navigation is bidirectional, so that, for instance, we can also find all ORDERs for a PART. Both databases and conventional files have their relative advantages and disadvantages; however, there is a consistent trend toward the use of databases in information systems environments.

Databases are made possible through the use of special software called database management systems (DBMSs). A DBMS uses a data definition language (DDL) to define a database and gives programmers and/or end-users a data manipulation language (DML) to navigate, use, and maintain the database. Some DMLs must be used in conjunction with a host language such as COBOL. The result is to provide programmers and end-users with tools to structure, maintain, and use their data with greater ease and flexibility.

There are three classes of DBMSs. Hierarchical databases organize dissimilar records into easy-to-understand hierarchies or trees that depict the relationships between the records. Network or linked databases organize dissimilar records using linked lists to implement relationships between the records. The current trend is toward relational databases, which store each record as a simple table. Dissimilar records are not linked. Instead, associations are formed when they are needed by joining dissimilar records using fields that they have in common.

Database design, as performed by analysts, is usually restricted to recommending the logical schema or structure of the database needed. A database administrator, or DBA, will fine-tune that design to consider performance and technical idiosyncrasies. Database design is greatly simplified when it is based on data models developed in the systems analysis phase. Such a data model (e.g., an entity relationship diagram) is transformed into a database schema that considers the constraints of the DBMS that will be used to implement the database.

Key Terms

access controls, p. 596
archive files, p. 586
ASCII, p. 582
audit trails, p. 597
backup methods and procedures, p. 596
binary code, p. 581
blocking factor, p. 585
data administrator, p. 595
database, pp. 578, 590
database administrator (DBA), p. 595
database distribution, p. 606
database management system (DBMS), p. 592
data definition language (DDL), p. 592

data manipulation language (DML), p. 593
data warehouses, p. 591
descriptors, p. 581
direct access method, p. 587
domains, p. 593
EBCDIC, p. 581
field, p. 581
file, pp. 578, 586
file access, p. 587
file organization, p. 587
fixed-length records, p. 583
foreign key, p. 581
JOINing, p. 594
length of retention, p. 596

log files, p. 587
logical schema, p. 607
master files, p. 586
packed decimal, p. 582
personal databases, p. 591
physical schema, p. 607
primary keys, p. 581
production databases, p. 590
PROJECTing, p. 593
record, p. 582
relational database, p. 593
relations, p. 593
schema, p. 607
scratch files, p. 586
secondary keys, p. 581
SELECTing, p. 593

Problems and Exercises

1. Define the terms *field, record,* and *file.*
2. What are the advantages of record blocking? How is the blocking factor for a file derived?
3. Identify three types of fields, and give several examples of each.
4. Identify five types of files, and give several examples of each.
5. Define file access, and identify and describe two file access methods.
6. The selection of the best type of organization for a file is a function of the file access methods that will be used for that file. Identify three types of file organizations and how these files organizations are accessed.
7. Differentiate between fixed- and variable-length records. What impact does a record storage format have on a file design?
8. Explain the implications of data modeling for file design. How might file designs be affected by a failure to do data modeling?
9. Calculate the blocking factor for a fixed-length record file, given the following factors:
   ⅓-track blocking (⅓ track = 2,603 bytes)
   Record size = 211 bytes
   How much storage space is wasted in each physical record? How much storage space is required to store 3,750 records? Specify your answer in bytes.
10. Calculate the blocking factor for a variable-length file record, given the following factors:
    ¼-track blocking (¼ track = 1,891 bytes)
    Minimum record size = 75 bytes
    Maximum record size = 313 bytes
    How much storage space is wasted in each physical record? Specify minimum and maximum.
11. Describe the relationship between file access and file organization.
12. Kevin, an inventory manager, is considering a DBMS for his microcomputer. He's not certain that he really understands what a database is. In college, he took an introductory computer course and learned about files. He assumed database is the current buzzword for a collection of files. Write him a memo explaining the difference between a file and database environment. What are the advantages and disadvantages of each environment?
13. Explain the advantages and disadvantages of conventional files versus databases.
14. What is a database? What is the difference between a production database and an end-user database?
15. Briefly explain the differences between a data definition language, a host programming language, and a data manipulation language.
16. What is the difference between a DBMS's physical structure and its logical structure? Prepare a simple schema for a relational database.
17. List and briefly describe the three table operations used to manipulate relational tables.

18. Briefly describe and explain the differences between a primary key, a secondary key, and a foreign key. Give an example of each.

19. Visit a local information systems shop that uses a DBMS. Describe the existing database environment. Do they have production-oriented databases or end-user databases? What host programming language(s) is utilized to load, maintain, and use the data? Ask the systems analyst or database administrator to give you a brief orientation on the physical and logical structures supported by the DBMS. To what extent are conventional files used?

20. Visit a local information systems shop that operates in a strictly conventional file environment. Ask the systems analyst for information describing several of the master and transaction files. Do some of the files contain duplicate data? Is this data input several times? What impact has the duplicated data had on maintenance? Do they experience problems with data integrity? Have the analyst explain the impact of changing the format of one of the files.

21. If databases were created with the ability to solve many of the problems characteristic of conventional file-based systems, why aren't all information systems shops operating in a database environment?

22. For the database environment at a local business, discuss the implications of the database on systems development. Do the analysts do anything differently than when using conventional files? How do the analysts interact with the database administrator?

23. What is the relationship between the systems analyst and the database administrator?

24. Explain how a logical data model aids in the design of databases.

25. How are relationships appearing on an entity relationship diagram implemented by a relational database management system?

Projects and Minicases

1. Robert Williams, a systems analyst for Future Ventures, is facing a perplexing file design problem. He is attempting to design a VENDORS file. This file is to contain variable-length records. The fixed portion of the record contains data describing a vendor. The repeating portion contains data describing a part supplied by the vendor. Future Ventures uses a standard of ⅓-track blocking for all file designs. However, when Robert calculated the record size, he found that its 2,973 bytes exceeded the 2,603-byte limit of ⅓-track blocking. In other words, one logical record would not fit within standard block size. What can he do? Explain at least two alternatives Robert might choose from. Explain the implications of each alternative.

2. Design a file for the following data store structure. State all assumptions, and prepare complete documentation in a data repository.

EMPLOYEE consists of the following elements:	**FIELD SIZE** (in characters)
SOCIAL SECURITY NUMBER	9
EMPLOYEE NAME	30
EMPLOYEE HOME ADDRESS, which consists of	
STREET ADDRESS	30
CITY	15
STATE	2
ZIPCODE	9
EMPLOYEE HOME PHONE	10
DEPARTMENT CODE	2

EMPLOYEE **consists of the following elements:**	**FIELD SIZE** **(in characters)**
DATE EMPLOYED	6
DATE OF BIRTH	6
MARITAL STATUS	1
One of the following:	
MONTHLY SALARY	5
HOURLY RATE	3.2
VACATION DAYS DUE	2
SICK DAYS DUE	2
GROSS PAY YEAR-TO-DATE	6.2
FEDERAL TAX YEAR-TO-DATE	6.2
STATE TAX YEAR-TO-DATE	6.2
FICA TAX YEAR-TO-DATE	6.2

3. Design a file for the following data store structure. State all assumptions, and prepare complete documentation in a data repository.

VENDOR PART **which consists of**	**FIELD SIZE** **(in characters)**
VENDOR NUMBER	8
VENDOR NAME	15
VENDOR ADDRESS, which consists of	
P.O. BOX NUMBER (optional)	2
STREET ADDRESS	25
CITY	15
STATE	2
ZIPCODE	9
VENDOR PHONE	10
VENDOR TERMS, which consists of	
DISCOUNT RATE FOR EARLY PAYMENT	2.1
EARLY PAYMENT PERIOD	2
NET PAYMENT PERIOD	2
1 to 15 occurrences of the following:	
MATERIAL NUMBER	8
MATERIAL DESCRIPTION	25
UNIT PRICE	5.2
QUANTITY REQUIRED FOR DISCOUNT	5
QUANTITY DISCOUNT RATE	2.1

4. Sunset Valley Distributors recently completed a major conversion project. Several months ago, Sunset made the decision to move into the database era. Many of its computer-based files had become unreliable, difficult to maintain, and too inflexible to be used to fulfill many end-user reporting and inquiry requests. A DBMS seemed to be the obvious solution. Two systems analysts were primarily responsible for the conversion project, which took several months to complete. The systems analysts had decided to simply implement each of the computer-based files as a separate table in their relational database. Once the conversion was completed, the same problems that existed with the file-based system reappeared in the database system. Reports contained inaccurate data, report and inquiry requests could not easily be obtained, and data maintenance was still difficult. A consultant was hired to investigate their problems. The consultant acknowledged that many of the problems resulted because the analysts failed to do data modeling. Explain the importance of doing data modeling ahead of time when designing databases.

5. Design the logical schema for a relational database using the entity relationship diagram that follows. The primary keys of the entities are as follows:

CUSTOMER (CUSTOMER NUMBER)

RENTAL (RENTAL NUMBER)

RENTAL ITEM (RENTAL NUMBER and either TITLE, TAPE NUMBER, OR RECORDER SERIAL NUMBER)

MOVIE TITLE (TITLE)

VIDEOTAPE (TAPE NUMBER)

VCR (RECORDER SERIAL NUMBER)

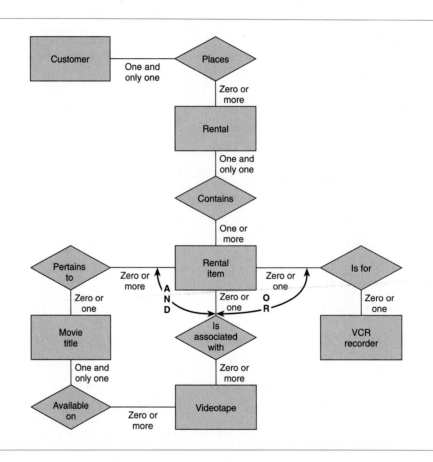

Suggested Readings

Bradley, James. *Introduction to DataBase Management in Business.* 2nd ed. New York: Holt, Rinehart and Winston, 1987. Includes conventional files and distributed databases.

Kroenke, David, and Kathleen Dolan. *Database Processing.* 3rd ed. Chicago: Science Research Associates, 1988. This updated, popular introductory database textbook surveys database concepts, technology, design, and implementation.

Martin, James. *Managing the Database Environment.* Englewood Cliffs, N.J.: Prentice Hall, 1983. Martin is one of the most noted authorities, writers, and lecturers in the database field. No database list would be complete without one of his many titles. We chose this title because of its management orientation and readability.

Pratt, Philip, and Joseph Adamski. *Database Systems Management and Design.* 2nd ed. Boston: Boyd & Fraser Publishing Company, 1991.

Ricardo, Catherine. *Database Systems: Principles, Design, & Implementation.* New York: Macmillan, 1990.

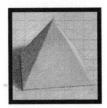

Output, Input, and User Interface Design for the New System

When we left Sandra and Bob, they had completed the database design for the new system. Now they must design the outputs, inputs, and user interface for the system.

Output Design Specifications for the New System

Recall that while Bob was working on the database design, Sandra had been busy designing the outputs to be produced by the new system. Sandra has since completed the output designs and has scheduled a meeting to allow the users the opportunity to review the designs.

Sandra, Joe Bosley, Ann Snyder, and Sally Hoover are reviewing the proposed reports in Sally's office. "You look tired, Joe. What's the matter?" asked Sandra.

"I stayed up late last night watching the baseball game. Four extra innings and then my team loses. I'm still upset."

"Well, I will try to keep this meeting as short as possible. I'd like to review some proposed reports with all of you."

Sally was obviously enthusiastic. "I can hardly wait to hear about the new system and exactly what it will do."

Sandra smiled. "Let's begin with the Order To Be Filled form. Based on the requirements statement we created earlier, I've drawn a sketch of what the form might look like (see Figure E7.1). If it's OK, I'd like to confirm a few facts before we review this form."

Since there weren't any objections, Sandra continued. "The DP department will print these forms each day, but I need to know how many could be printed in a single day."

"Currently, we never process more than 4,000 a day," Sally answered.

"But what about growth?" asked Ann. "We expect to continue adding 100 new members a month. Maybe we had better plan on 5,000 orders a day."

Joe watched as Sandra made some notations on a form (see Figure E7.2). "Sandra, what's the form you're writing on?"

"This is a list of output specifications I'm making for the information systems people. It helps them to anticipate the impact your outputs will have on their facilities. It will also help us to choose the best printer to meet your needs,"

Sandra responded. "Which brings us to my next question. Do you really need four copies of each form?"

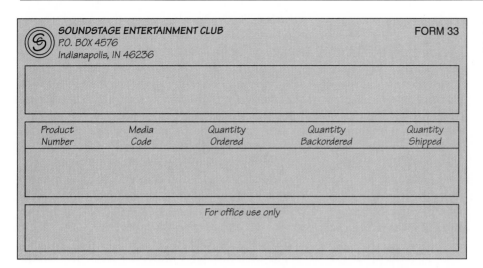

FIGURE E7.1 **Sketch of the New ORDER TO BE FILLED**

SOUNDSTAGE ENTERTAINMENT CLUB FORM 33
P.O. BOX 4576
Indianapolis, IN 46236

| Product Number | Media Code | Quantity Ordered | Quantity Backordered | Quantity Shipped |

For office use only

FIGURE E7.2 **Typical Data Repository Output Specification**

OUTPUT DATA DICTIONARY FORM

NAME OF OUTPUT: Order to be filled
PREPARTED BY Sandy Shepard   DATE:
DESCRIPTION: Four part document describing a sales order

MEDIUM: Paper   OUTPUT TYPE: External/Form
OUTPUT CHARACTERISTICS: Preprinted, 8.5" X 8.5", designed for mailing

FREQUENCY PREPARED: Twice daily at 8 AM and 1 PM
VOLUME: 5,000/day   NUMBER OF COPIES: 3
COPYING METHOD: Chemical carbon
OUTPUT RECIPIENTS: Order entry (all copies)

SPECIAL INSTRUCTIONS: Quantity shipped is hand entered. Quantity backordered may be modified by hand.

OUTPUT COMPOSITION:
    MEMBER NUMBER
    ORDER NUMBER
    MEMBER NAME
    ORDER DATE
    MEMBER ADDRESS

    1 to 15 occurences of: MEMBER ADDRESS is composed of:
                                    P.O. BOX
    PRODUCT NUMBER   STREET
    MEDIA CODE   CITY
    QUANTITY ORDERED   STATE
    QUANTITY SHIPPED   ZIPCODE
    QUANTITY BACKORDERED

REF. NO.	DATA ELEMENT NAME	DATA TYPE	EDIT SIZE	EDIT MASK	SOURCE
1	MEMBER NUMBER	A/N	6	X(6)	
2	ORDER NUMBER	A/N	6	X(6)	
3	MEMBER NAME	A/N	30	X(30)	
4	ORDER DATE	A/N	8	MM/DD/YY	
5	P.O. BOX	A/N	10	X(10)	
6	STREET	A/N	15	X(15)	
7	CITY	A/N	15	X(15)	
8	STATE	A	2	AA	
9	ZIP CODE	A/N	9	X(9)	
10	PRODUCT NUMBER	A/N	7	X(7)	
11	MEDIA CODE	A/N	2	XX	
12	QUANTITY ORDERED	N	4.0	Z,ZZ9	
13	QUANTITY SHIPPED	N	4.0	hand entered	
14	QUANTITY BACKORDERED	N	4.0	Z,ZZ9	

"Absolutely!" Sally answered. "But separating each of the copies from the carbon paper is a real nuisance."

Sandra smiled and said, "There are lots of different ways to produce copies. Maybe chemical carbon paper that is treated with a chemical that darkens under pressure would work best. Since no actual carbon paper is used, you only have to separate the four copies. Okay, let's look at the sketch. What do you think about this design?"

"It looks real good," Ann replied. "But why is FORM 33 in the upper right-hand corner?"

"That is the number of this new form we designed. It's used for information systems control purposes and uniquely identifies this form from all other forms."

"Do we have to fill in each product order line by hand?" Sally asked.

"No," answered Sandra. "The system will do it. The only hand-entered field on the form is the quantity shipped completed by the shipping clerks."

"The QUANTITY BACKORDERED and the QUANTITY SHIPPED are backwards. The form we use now is the other way around. That could create some confusion," observed Joe.

"Oops! I accidentally reversed them. Now you see how important it is for you to verify these sketches." Sandra made corrections to a different form (see Figure E7.3).

Ann noticed the form Sandra was writing on. "That looks awfully complicated—what is it?"

"This is a printer spacing chart," Sandra responded. "It is a technical tool I use to communicate the final design of your outputs to the computer programmers. Getting back to the form—are there any other corrections? Let's look at the product order lines."

The walkthrough continued. Eventually, Sandra completed the specifications for all the outputs. After reviewing the first set of sketches with the users, Sandra asked them to draw other outputs. This created user interest and increased the users' sense of involvement and ownership in the sys-

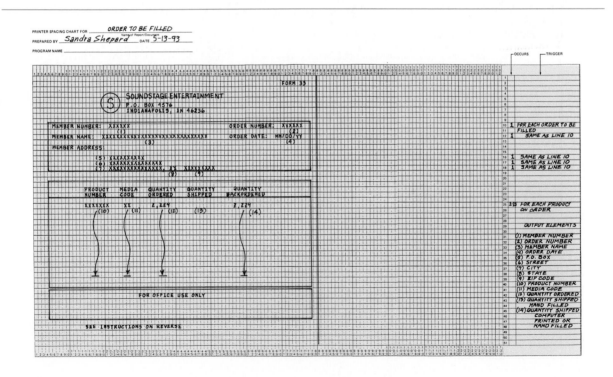

FIGURE E7.3 **Typical Printer Spacing Chart**

tem. Sandra checked their designs carefully and reconciled any differences with the original requirements statement. As the sketches were finalized, Sandra prepared the more technical design specifications for the programmers.

Input Specifications for the New System

Several days have passed since Sandra's meeting for users to review output designs. Since that time, Sandra and Bob worked on completing detailed designs for the new system's inputs. We are now in the conference room where Sandra and Bob scheduled a Saturday morning meeting to review input design specifications with the order-processing staff. Sally Hoover suggested a Saturday morning overtime meeting because she wanted her entire staff to become familiar with the new input and on-line methods that would be used in the new system.

Sandra called the meeting to order. "I think we should go ahead and get started. I know all of you are anxious to enjoy your weekend, so we'll make this as brief as possible. First, Bob and I would like to thank you for your cooperation so far. We realize that we have spent a great deal of time defining what you need in order to do your jobs. We have been deliberately avoiding specific detail of how

to do things because we wanted to build a system that would make your jobs easier. It is your system, and it is shaping up very well."

"Today, we want to review system inputs. Let's begin with a member order." Sandra placed a transparency on the overhead projector and continued. (See Figure E7.4.) "This is the proposed Member Order Form 40. It is similar to the old Form 40. You won't have to enter the order date anymore, the system will do that for you using the current date. Also notice . . ."

Sally Hoover interrupted, "What if the order date is different from the date when the order is entered?"

"Why would they be different?" Sandra responded.

"Sometimes orders are sent to the wrong department, or they don't get entered on the day they are received due to a high volume of orders. Since we generate automatic orders to members based on the date, it's very important that the date on a member's order be correct."

Bob interrupted. "That's my fault. Sally and I discussed this but I forgot to make the change to the specifications. If we have the system fill the date field with the current date but allow the order-processing staff to override that value we can

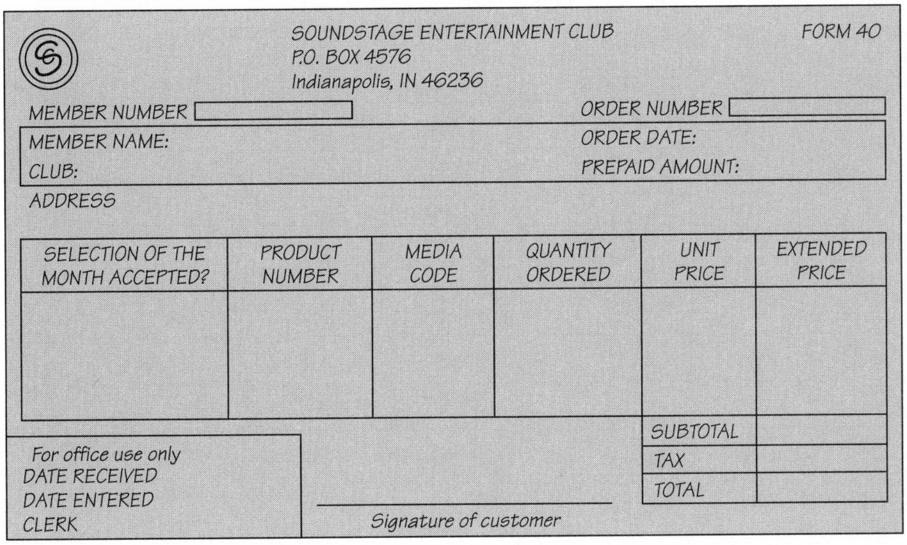

FIGURE E7.4 **Sketch of a Source Document**

621

solve that problem and save them a lot of typing at the same time."

"I have another question, Sandra. Do we have to enter all of those fields to begin processing an order?" Sally asked.

"The order number is automatically generated by the system. Also, when you enter the member number, the member's name will appear for you to verify. In addition to the order date we just talked about, the selection of the month accepted field will be sent to 'YES' and the product number inserted to save you some typing."

"But 70 percent of our members reject the selection of the month," objected Joe Bosley. "That means they will have to change those fields most of the time."

"That's interesting," Sandra admitted. "We will change the default value to 'NO' and leave the product number blank." Noting that there were not any more questions, Sandra continued. "What

we need to do now is to verify the size of each field and get an idea of the range of values each field can assume. This will help us write programs to check the accuracy of the data before it's processed. Each of the fields you'll enter is recorded on this transparency (see Figure E7.5). Don't be intimidated by this form. Most of this information is intended for the computer programmers. All we have to do is verify the size and editing requirements for each field. Each of you will have a computer terminal to enter member orders directly into the computer."

"Oh no!" groaned some unidentified voice.

"Don't worry!" Sandra said. "I remember my first experience with a computer terminal. I thought if I did something wrong, I'd break the machine or something, but that can't happen. First, we are going to develop a pleasant conversational dialog between you and the computer. As much as possible, we will try to make you forget that you are talking with a machine. We will be walking you

Data Element Dictionary: Input

Name of Input: MEMBER ORDER

Ref.	Field Name	Type	Size	Memo	Edit Mask	Editing/Validation
1	MEMBER NUMBER	A/N	7		9(7)	NUMERIC
2	ORDER NUMBER	A/N	6		9(6)	NUMERIC, UNIQUE
3	MEMBER NAME	A/N	30		X(30)	MUST CONTAIN VALUES
4	ORDER DATE	A/N	8		MM/DD/YY XX/XX/XX	VALID DATE
5	CLUB NAME	A/N	10		X(10)	MUST CONTAIN VALUES
6	PREPAID AMOUNT	N	3.2		999.99	NUMERIC ≥ 0
7	P.O. BOX	A/N	10		X(10)	OPTIONAL
8	STREET	A/N	15		X(15)	MUST CONTAIN VALUES
9	CITY	A/N	15		X(15)	" " "
10	STATE	A	2		AA	STANDARD CODES
11	ZIPCODE	A/N	9		9(9)	MUST BE NUMERIC

– CONTINUED

FIGURE E7.5 Typical Specifications for Input Attributes

through that dialog next week. Right now, I'd like to discuss the final input from that dialog, the member order. May I continue?"

Noting that there were no more questions, Sandra continued. "In order to make your job easier, we made the input very similar to the Member Order Form 40 we discussed earlier. All you'll have to do is fill in the blanks."

Pam Turner, a clerk in the Order-Processing department, interrupted. "That sounds good, but I don't think we will all know what to do. I have worked on systems like this at another company and we were always getting stuck. The computer people would tell us to press this key or do that, but we were usually confused, especially if we didn't have our reference guides."

"I know what you mean, Pat," Sandra responded. "I've seen some of those poorly designed systems. What would you think if there wasn't any reference guide and everything you needed to know was right on the screen?"

"That would be fantastic," replied Pam. "But you would need an awfully large screen."

"Not really," Sandra countered. "All we have to do is make sure that simple key assignments and instruction always appear on the screen. Look at this transparency (Figure E7.6). Note that the instructions appear in lines 2 through 4. They will

always appear there. As you enter each field, you press the tab or arrow keys to move to the next field. After you enter the last field, press the enter key and the data will be input to the computer for processing. After you enter all the fields, the system will automatically check the data against the value ranges we discussed earlier. If the value is incorrect, the system will give you an appropriate error message and ask you to reenter the field."

"What happens if I enter something several times and the computer won't take it but I don't understand what I'm doing wrong?" asked Sally.

"That's a good question, Sally," Bob answered. "Notice the 'PRESS F2 FOR HELP' message at the bottom of the screen. Anytime you need more information about a particular field, you can press the F2 key. The system remembers what field you were on and displays a help screen for that field. If, for example, you were trying to enter a prepayment amount and requested help, you'd see something like this. (See Figure E7.7.) That way you can enter the data so that the system will accept it."

"That's great!" Sally exclaimed. "If the system rejects the data, at least we can find out why."

Sandra concluded her discussion of the on-line member order input. Similar discussions were held for each of the on-line inputs.

```
                        ** MEMBER ORDER **
          ENTER THE FOLLOWING ITEMS FROM A MEMBER ORDER (FORM 40). USE TAB OR
          ARROW KEYS TO MOVE FROM ITEM TO ITEM. PRESS ENTER KEY WHEN DONE.

    MEMBER NUMBER:                                ORDER NUMBER:
    MEMBER NAME:                                  ORDER DATE:
    CLUB NAME:                                    PREPAID AMOUNT:
    MEMBER ADDRESS:
       P.O. BOX:
       STREET:
       CITY:
       STATE:        ZIPCODE:

          SELECTION OF        PRODUCT     MEDIA     QUANTITY
          MONTH ACCEPTED?     NUMBER      CODE      ORDERED




    ** PRESS F2 FOR HELP. PRESS F3 TO RETURN TO ORDER PROCESSING OPTIONS MENU **
```

FIGURE E7.6 **Typical Prototype of On-Line Input**

Terminal Dialog Specifications for the New System

One week has passed since the input design walkthrough meeting. Sandra and Bob have prepared some sketches of sample screens of real terminal dialog situations. Various end-users are being walked through typical terminal sessions. Sandra conducts these sessions while Bob makes any necessary modifications to the design specifications. Let's listen in on one walkthrough session.

Sandra began, "Sally, this is a sample of what the screen will look like after you correctly enter your password. (See Figure E7.8.) If you wanted to enter a member order, what would you do next?"

"Well," Sally began, "normally I'd enter 2, but what would happen if I accidentally choose 4 instead?"

```
                    ** MEMBER ORDER **
                       HELP SCREEN



       The prepaid amount is a right-justified, six character,
       positive field in the form of 999.99. The decimal point
       must be entered. For example, a prepaid amount of $21.89
       would be entered as 21.89.



       PRESS ANY KEY TO RETURN TO MEMBER ORDER ==>
```

FIGURE E7.7   **Typical Prototype of On-Line Help**

```
                - MEMBER SERVICES SYSTEM MENU -


       [1] INQUIRE ON PRODUCTS, ORDERS, AND MEMBERS
       [2] PROCESS MEMBER ORDERS
       [3] REPORTS

       SELECT DESIRED OPTION ==>








       PRESS "F10" TO TERMINATE SESSION
```

FIGURE E7.8   **Typical Menu Screen Prototype**

"That's a good question," Sandra answered, as Bob showed Sally a new sample terminal dialog screen. (See Figure E7.9.) "As you can see, the system has identified your response as invalid and is waiting for you to enter another response. Notice the error message at the bottom of the screen. Now what are you going to do?" asked Sandra.

"Okay," Sally conceded, acting somewhat surprised that Sandra and Bob had anticipated incorrect responses. "I'll select option 2."

"That will take you to the member order-processing menu, which looks like this," said Bob, as he showed Sally another sample dialog screen.

"Then I'll choose option 1 for Member Order," added Sally.

"That will take you to the Member Order input screen (refer back to Figure E7.6), which you may remember from last week when we looked at input screens," Sandra said. "You can see that we incorporated several of the changes you suggested in that meeting. You already know how to use that input screen. Let's move on to another sample terminal session."

The walkthrough continued in this manner until all dialog sessions had been simulated with the promotions, membership, and order-processing staff.

Where Do We Go from Here?

This episode introduced three very important activities of systems design: the design of computer outputs, inputs, and user interface. Several important issues must be faced when designing outputs, inputs, and user interface. In Chapter 16 you will learn about important design issues and tools for designing computer inputs and outputs. Important user interface design issues and tools will be discussed in Chapter 17.

Communication skills are very important aspects of designing computer outputs, inputs, and user interface because of the extensive end-user interaction involved in their design and approval. In addition to obtaining an understanding of the issues and tools used for output, input, and terminal dialog design, we strongly encourage you to read Part Four, Module D, "Interpersonal Skills."

```
            - MEMBER SERVICES SYSTEM MENU -


    [1] INQUIRE ON PRODUCTS, ORDERS, AND MEMBERS
    [2] PROCESS MEMBER ORDERS
    [3] REPORTS

    SELECT DESIRED OPTION ==> 4






    INVALID OPTION - CHOOSE 1, 2, OR 3

    PRESS "F10" TO TERMINATE SESSION
```

FIGURE E7.9   **Typical Menu Screen Prototype with Error Message**

16

Input and Output Design

Chapter Preview and Objectives

This chapter teaches you how to design and prototype computer outputs. It also teaches you how to design and prototype computer inputs, both batch and on-line. You will know that you have mastered input and output design tools and techniques when you can

Define
the appropriate format and media for a computer output and the appropriate method and medium for a computer input.

Differentiate
between internal, external, and turnaround outputs.

Apply
human factors to the design of computer outputs and computer inputs.

Design
internal controls for computer outputs and computer inputs.

Identify
data flows on a DFD that must be designed as computer outputs and computer inputs.

Define
output and input design requirements, and record those requirements in a project repository.

Describe
various techniques for prototyping outputs and inputs.

Prepare
layouts to communicate both paper-, screen-, and record-oriented inputs and outputs to programmers who must implement those inputs and outputs.

Explain
the difference between data capture, data entry, and data input.

Design
a source document for data capture.

THE WHOLESALE COST-PLUS CLUB

Scene: *Kathleen Smathers, a programmer/analyst for the Wholesale Cost-Plus Club, had an early afternoon appointment with Linda Pratney, the Accounts Payable assistant manager. Wholesale Cost-Plus Club is a large, citywide warehouse outlet that sells virtually any type of merchandise to club members for a cost very close to wholesale price (significantly below the retail prices charged by grocery stores, drug stores, department stores, and other retail stores). Kathleen had been largely responsible for the Accounts Payable system implemented last fall. Accounts Payable pays off invoices from the suppliers of the club's merchandise. The meeting's purpose was a mandatory post-implementation review of the new system. (Post-implementation reviews occur one month after a new system replaces an old system.) It was no company secret that Linda was displeased with some aspects of the new system.*

Linda: Kathleen, I guess you heard that I'm having a little trouble with this new payables system. You told me that these computer reports would make my job much easier, and that just hasn't happened.

Kathleen: I'm really sorry it hasn't worked for you. But I'm willing to work overtime to make this system work the way you need it to work. What isn't working? I thought I included all the fields you requested. Is it that you're not getting the reports on time?

Linda: That's not the problem. The information is there. It's just that I find the report difficult to use. For example, this report doesn't tell me what I need to know. I spend most of my time trying to interpret it.

Kathleen: Why don't you tell me how you use the report?

Linda: Well, first I read down the report, line by line, and attempt to count and classify the number of invoices that are less than 10 days old, between 10 and 30 days old, and between 31 and 60 days old. You see, I have this graph here that shows the totals for each category over the past 12 months. I compare the new totals from this report with the totals on the graph to identify any significant trends in payment activities.

Kathleen: Why?

Linda: Because we pay some invoices off early to gain a 2 percent discount. Others we defer to the final due date of the supplier, which is usually 30 or 60 days after receipt of the invoice.

Kathleen: I see. Go on.

Linda: I also use the report to identify the costs of discounts that we did not take, choosing instead to defer payment until close to the final due date. To do that, I locate the invoice on the report and look up that same invoice on the previous copy of the report. Then I log the amount as a lost discount to a particular supplier. At the end of the week, I sum the lost discounts by supplier and send the report to my superiors. It really helps. I can think of several occasions when my superiors authorized a change in payment policy for a particular supplier in order to take better advantage of discounts.

Kathleen: When I wrote down specifications for this report, I assumed I understood what information you were requesting. It sounds as if there is some additional data that should have been included on the report. You're wasting a lot of time looking for information that could be automatically generated by the system.

Linda: I'm glad that you are sympathetic to my problem. I hope you can do something about it. I'm pretty frustrated with that report.

Kathleen: I'm sure I can have it fixed in no time. I'm really sorry. I guess I just didn't understand what information you wanted and how you would be using it. Let's see, how should we proceed? I don't want to spend a lot of time designing a new report that isn't precisely what you need. I've got an idea. I just got a new package on my office microcomputer. It's called Lotus 1-2-3. I'm pretty sure that I can quickly mock up and, if necessary, change a sample of the report you want. I can even simulate this pie chart graph you want. Once we get the reports and graphs looking the way you want, we'll use them as models for the programmers.

Linda: That's a good idea.

Kathleen: Okay, what about this other report?

Linda: I'm getting some flack about this report I asked you to produce for my clerks. It was supposed to list those supplier accounts that we owe payment on. Anyhow, my clerks claim the report is too cluttered and is difficult to use. All they really need to know is the supplier's account number, current balance due, discount date, and the final due date. As you can see, there are a number of unnecessary fields, and the report lists two supplier accounts on the same line. Could you clean this report up too?

Kathleen: No problem. Again, I'm sorry I was so off target on those reports. I can probably use Lotus to mock up these reports as well.

Linda: The Accounts Payable department is also dissatisfied with some of the input functions of the system. They claim the system is not easy to use — that it takes them longer than it should to enter a vendor invoice because the dialogue is confusing. They also claim that they're experiencing an increase in vendor complaints.

Kathleen: I don't see how that can be. The program is very easy to use. You should see it. I used a lot of fancy terminal functions — blinking screens, reverse video, and things like that — to make their job of entering sales orders more interesting.

Linda: Did you approve those screen design prototypes with the clerks?

Kathleen: No. It wasn't really necessary as long as all the important information was on the screen. Besides, I didn't want to confuse them with a lot of technical design details.

Linda: This memo states the new system isn't recording the vendor invoices accurately. They say . . .

Kathleen: Impossible! That can't be. The program asks the data entry clerk to enter all the information that's on the form. If they filled the form out right, there wouldn't be any problems.

Linda: Okay. That makes sense. Let's see if we can get to the bottom of this. How did the clerks feel about that form?

Kathleen: I'm not sure. They said they wanted to do vendor invoices on the computer. I sampled the form and then re-created it on their terminal screen.

Linda: I think I'm beginning to understand. Sit down. I think there are a few things we need to talk about.

Discussion Questions

1. What type of reports was Linda receiving?

2. What type of reports does Linda really need?

3. What did Kathleen do wrong? What erroneous assumption did Kathleen make regarding data requirements for the report?

4. What did you think of Kathleen's new strategy of using Lotus 1-2-3 for mocking up new reports? Does the approach sound similar to a strategy that has been frequently described in this book?

5. Why do you suppose the report for Linda's subordinates was inappropriate? Who's to blame, Linda or Kathleen? Why?

6. What mistakes did Kathleen make during input design? What should she have done?

PRINCIPLES AND GUIDELINES FOR INPUT AND OUTPUT DESIGN

Outputs present information to system users. Outputs, the most visible component of a working information system, are the justification for the system. During systems analysis, you defined output needs and requirements, but you didn't design those outputs. In this section, you will learn how to design effective outputs for system users.

There are two basic types of computer outputs. The first type is external outputs.

> **External outputs** leave the system to trigger actions on the part of their recipients or confirm actions to their recipients.

Examples of external outputs are listed in the margin. Most external outputs are created as preprinted forms that are designed and duplicated by forms manufacturers for use on computer printers. Some are designed as turnaround documents.

> **Turnaround outputs** are those which are typically implemented as a form eventually reenters the system as an input.

The revolving charge account invoice depicted in Figure 16.1 is a typical external turnaround document.

Some outputs do not leave the information system. These outputs are called internal outputs.

> **Internal outputs** stay inside the system to support the system's users and managers.

These outputs include detailed reports, summary reports, exception reports, and decision support inquiries. Sample internal outputs are shown in Figure 16.2. All of these output types can be designed using the principles and techniques covered in this chapter.

Outputs are produced from data that is either input or retrieved from files and databases, and data in files must have been input to those files and databases. Input design serves an important goal: capture and get the data into a format suitable for the computer to use.

Designing inputs and outputs is more than working up a few layout charts for reports and screens. Although you may have designed reports for computer programming class assignments, you probably never had to justify your design to system users. There are many input and output issues that must be addressed *before* the input or output is physically designed. And most of them are people-oriented issues.

√

External Outputs

Invoices
Paychecks
Course schedules
Airline tickets
Boarding passes
Travel itineraries
Telephone bills

Data Capture, Data Entry, and Data Input

When you think of "input," you usually think of input devices, such as keyboards and terminals. But input begins long before the data arrives at the device. To actually input business data into a computer, the analyst may have to design an input form, design input screens or records, and design methods and procedures for getting the data into the computer (from customer to form to data entry clerk—if necessary—to disk to tape to computer).

What is the difference between data capture, data entry, and data input? Data happens! It accompanies business events called **transactions**. Examples include orders, time cards, reservations, and the like. We must determine when and how to capture the data.

> **Data capture** is the identification of new data to be input.

FIGURE 16.1 Typical External Turnaround Document This output is external because it leaves the system and goes to the customer to initiate a transaction (a payment). It is also a turnaround output since a portion of the output is returned, with payment, as an input.

When is easy! It's always best to capture the data as soon as possible after it is originated. How is another story!

Where do you capture data? Data is frequently captured from a source document.

A **source document** is a form used to record data that will eventually be input to a computer. Source documents should be easy for the system user to complete and should facilitate rapid data entry into a machine-readable format.

```
                                                                                    PAGE 01
                                    SUMMARY OF MONTHLY
                                 BANK MACHINE TRANSACTIONS
                              FOR THE PERIOD 03/01/89 TO 03/31/89

  MACHINE    BRANCH   NUMBER OF   AMOUNT OF   NUMBER OF    AMOUNT OF   NUMBER OF   AMOUNT OF    NUMBER OF      AMOUNT OF
  NUMBER     NUMBER   DEPOSITS    DEPOSITS    WITHDRAWALS  WITHDRAWALS TRANSFERS   TRANSFERS    LOAN PAYMENTS  LOAN PAYMENTS

    01         01       192      57,600.32       672      31,213.50      140     14,025.33         23          1,725.86
    02         03       134      43,756.45       478      23,144.75       63      6,192.88         11          1,545.38
    03         05       112      47,650.44       462      24,897.26       43      5,023.61         13          1,195.76
    05         07       155      49,864.04       567      27,875.00       97     11,729.58         15          2,304.42
    06         08       234      61,768.34       748      37,563.73      153     17,688.93         26          2,112.45

  TOTAL DOLLARS DEPOSITED        $260,639.59
  TOTAL DOLLARS WITHDRAWN        $144,694.24
  TOTAL DOLLARS PAID ON LOANS    $  8,883.87
  TOTAL DOLLARS TRANSFERRED      $ 54,660.33
```

```
  10/02/89                        INTERNATIONAL MANUFACTURING COMPANY              PAGE  1
                                          BONUS REPORT

       CLOCK                   CLOCK       INCENTIVE      DOWN        BONUS        BONUS
       NUMBER      SHIFT       HOURS        HOURS         TIME        PERCENT      HOURS

        1000         1         35.5         30.0          05.5         150         15.0
        1010         1         40.0         32.0          08.0         110         03.2
        1020         2         40.0         31.5          08.5         142         13.2
        1030         2         36.5         20.3          16.2         113         02.6
        1040         3         09.4         08.2          01.2         144         03.6
        1050         3         10.2         02.8          07.4         107         00.2
        2000         1         55.0         45.3          09.7         134         15.4
        2010         1         50.0         33.2          16.8         139         12.9
        2020         3         12.1         03.4          08.7         132         01.1
        2030         2         20.4         17.9          02.5         125         04.5
        3000         1         16.8         12.6          04.2         127         03.4
        3010         2         40.5         30.1          10.4         104         01.2
        3020         3         40.0         29.0          11.0         143         12.5
        3040         3         32.0         29.5          02.5         147         13.9
        3050         1         07.0         03.8          03.2         141         01.6
        4000         2         60.2         47.8          12.4         150         23.9
        4010         1         61.4         50.3          11.1         117         08.6
        4020         3         14.7         08.5          06.2         121         01.8
        5000         1         50.0         44.1          05.9         100         00.0
        5010         1         52.5         40.0          12.5         133         13.2

       TOTALS:                684.2        520.3         163.9                     151.8
```

FIGURE 16.2 **Sample Internal Outputs** Internal outputs include detailed, summary, and exception reports. They are intended primarily for use by system users.

The top portion of Figure 16.1 is designed to be returned by the customer with their payment. It will become the source document for the input of payment data.

Be careful! Data entry is not the same as data capture.

Data entry is the process of translating the source document into a machine-readable format. That format may be a punched card, an optical-mark form, a magnetic tape, or a floppy diskette, to name a few.

Once data entry has been performed, we are ready for data input.

Data input is the actual entry of data in a machine-readable format into the computer.

Let's examine some of the data capture and data entry issues you should consider during systems design.

Input and Output Methods, Media, and Formats

The analyst usually recommends the method, medium, and format for all inputs and outputs. Let's compare the different input methods and the media alternatives available for modern information systems.

Batch Input Methods and Media

One method of processing input is known as batch input.

> **Batch input** is the oldest and most traditional input method. Source documents are collected and then periodically forwarded to data entry operators, who key the data using a data entry device that translates the data into a machine-readable format.

For many years, punched cards were the most common medium for batch input data. For the most part, batches of punched cards have been replaced by magnetic media. **Key-to-disk (KTD)** and **key-to-tape (KTT)** workstations transcribe data to magnetic disks and magnetic tape, respectively. These workstations are much quieter, which makes life as a data entry clerk more bearable. As each input record is keyed, it is displayed on a screen. The data can be corrected, because it is initially placed into a buffer. The final input file, possible merged from several KTD or KTT workstations, permits much faster data input rates to the computer than those achieved with punched cards.

Figures 16.3**A** and 16.3**B** illustrate the key-to-tape and key-to-disk input procedures, respectively. Notice that we have distinguished the data capture activities, the data entry activities, and the data input activities.

Figure 16.3**C** illustrates another batch input medium, the *optical-mark form*. You may have encountered this medium in machine-scored tests. Optical-mark forms eliminate most or all of the need for data entry. Essentially, the source document becomes the input medium, or so it seems. As the figure illustrates, the source document is directly read by an **optical-mark reader (OMR)** or **optical-character reader (OCR)**. The computer records the data to magnetic tape, which is then input to the computer. OCR and OMR input are generally suitable only for high-volume input activities. By having data directly recorded on a machine-readable document, the cost of data entry is eliminated.

For all these batch media, there is a significant possibility of error when moving from the source document to the input medium. Before data can be processed, it must be edited. If you have written edit programs in your programming courses, you know that edit programs frequently require as much or more effort than the processing of the transaction itself. We'll discuss this issue further when we present internal controls for inputs.

On-Line Input Methods and Media

An increasingly popular alternative method of processing computer inputs is called on-line input.

> **On-line input** is the capture of data at its point of origin in the business and the direct inputting of that data to the computer, preferably as soon as

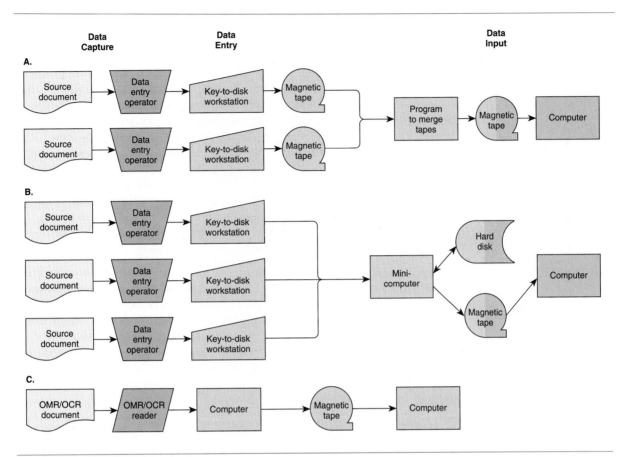

FIGURE 16.3 **Alternative Input Procedures for Batch Input Media** These illustrations show the similarities and differences between the most typical batch input methods.

possible after the data originates. Today, most, but not all, systems have been converted or are being converted to on-line methods.

The most common on-line medium cannot really be classified as a medium; it is the display terminal, or microcomputer display. The on-line system includes a monitor screen and keyboard that are directly connected to a computer system. The system user directly enters the data when — or soon after — that data originates. No data entry clerks are needed! There is no need to record data onto a medium that is later input to the computer; this input is direct! If data is entered incorrectly, the computer's edit program detects the error and immediately requests that the cathode-ray-tube (CRT) operator make a correction.

On-line data input can become even more sophisticated. With today's technology, we can completely eliminate much (and sometimes all) human intervention. Point-of-sale terminals in retail and grocery stores frequently include bar-code and optical-character readers. Everyone has seen the bar codes recorded on today's grocery products (see the example printed in the margin). These **bar codes** eliminate the need for keying data, either by data entry clerks or end-users. Instead, sophisticated laser readers read the bar code and send the data represented by that code directly to the computer for processing.

Universal Product Code

Batch versus On-Line Inputs

Should all systems be designed for on-line input? The technology is certainly cheaper than it used to be. So why bother with batch input?

No matter how cheap and fast on-line processing gets, an on-line program cannot be nearly as fast as its batch equivalent. Many (but not all) on-line programs require some human interaction, and people are slow, relative to computers. Also, for large-volume transactions, too many CRT terminals and operators may be needed to meet demand. As the number of on-line CRTs grows, the overall performance of the computer declines. Futhermore, many inputs naturally occur in batches. For instance, our mail may include a large batch of customer payments on any given day. Postal delivery is, at least today, a batch operation. Additionally, some input data may not require immediate attention. Finally, batch processing may be preferable because internal controls (discussed shortly) are simpler. So you see, batch inputs can still be justified.

But there is a compromise solution, the remote batch.

Remote batch offers on-line advantages for data that is best processed in batches. The data is input on-line with on-line editing. Microcomputers or minicomputer systems can be used to handle this on-line input and editing. The data is not immediately processed. Instead, it is batched, usually to some type of magnetic media. At an appropriate time, the data is uploaded to the main computer, merged, and subsequently processed as a batch. Remote batch is also called *deferred batch* or *deferred processing*.

Output Media and Formats

We assume you are familiar with different output devices, such as printers, plotters, computer output on microfilm (COM), and CRT display terminals. These are standard topics in most introductory information systems courses. In this chapter, we are more concerned with the actual output than with the device. A good systems analyst will consider all available options for implementing an output, especially output medium and output format.

A **medium** is what the output information is recorded on, such as paper or video display device.

Now that we have defined what the information will be stored on, we need to determine exactly how the information will appear.

Format is the way the information is displayed on a medium—for instance, columns of numbers.

The selection of an appropriate medium and format for an output depends on how the output will be used and when it is needed.

Alternative Media for Presenting Information

The most common medium for computer outputs is paper; such outputs are called *printed output*. Currently, paper is the cheapest medium we will survey. Although the paperless office (and business) has been predicted for several years, it has not yet become a reality. Perhaps there is an irreversible psychological dependence on paper as a medium. In any case, paper output will be with us for a long time.

However, paper is bulky and requires considerable storage space. To overcome the storage problem, many businesses have turned to the use of film as an output medium. The first film format is microfilm.

> **Microfilm** is a roll of photographic film that is used to record information in a reduced size.

Another similar medium is microfiche.

> **Microfiche** is a single sheet of film that is capable of storing many pages of reduced output.

The use of film does present its own problems — microfiche and microfilm can only be produced and read by special equipment. Therefore, other than paper, the most common output medium is video.

> **Video** is the fastest-growing medium for computer outputs, the on-line display of information on a visual display device, such as a CRT terminal or microcomputer display.

Although this medium provides the system user with convenient access to information, the information is only temporary. When the image leaves the screen, that information is lost unless it is redisplayed. If a permanent copy of the information is required, paper and film are superior media.

Alternative Formats for Presenting Information

There are several different formats you can choose for communicating information on a medium.

> **Tabular output** using columns of text and numbers is the oldest and most common format for computer outputs. This format presents information as columns. Most of the computer programs you've written probably generated tabular reports.

Another similar output format is zoned output.

> **Zoned output** places text and numbers into designated areas of a form or screen. Zoned output is often used in conjunction with tabular output.

For example, an order output contains zones for customer and order data in addition to tables (or rows of columns) for ordered items.

An increasingly popular alternative format for information is graphic output.

> **Graphic output** is the use of a graph or chart to convey information. To the system user, a picture can be more valuable than words. Bar charts, pie charts, line charts, step charts, histograms, and other graphs can help system users grasp trends and data relationships that cannot be easily seen in columns of numbers.

The popularity of graphic output has been stimulated by the availability of low-cost, easy-to-use graphics printers and software, especially in the microcomputer industry. Later in this chapter we will show you an alternate graphic design for a SoundStage input.

Another increasingly popular output format is the narrative format.

> In the **narrative format,** sentences and paragraphs replace or supplement standard text, numbers, and pictures.

Word-processing technology has exploited the narrative format for reports, business letters, and personalized form letters. For example, an accounts receivable system might interface with a word processor to provide names, addresses, and past due data for personalized credit-reminder letters.

System User Issues for Input and Output Design

Because inputs originate with system users and outputs are used by system users, human factors play a significant role in both input and output design. Inputs should be as simple as possible and designed to reduce the possibility of incorrect data being entered. System users must find computer outputs easy to use and helpful to their jobs. Furthermore, if batch input methods are used, the needs of data entry clerks must also be considered. With this in mind, several human factors should be evaluated.

First, the volume of data to be input should be minimized. The more data that is input, the greater the potential number of input errors and the longer it takes to input that data. These general principles should be followed for input design:

- *Enter only variable data.* Do not enter constant data. For instance, when deciding what elements to include in a SALES ORDER input, we need PART NUMBERs for all parts ordered. However, do we need to input PART DESCRIPTIONs for those parts? PART DESCRIPTION is probably stored in a computer file. If we input PART NUMBER, we can look up PART DESCRIPTION. Permanent (or semipermanent) data should be stored in files. Of course, inputs must be designed for maintaining those files.
- *Do not input data that can be calculated or stored in computer programs.* For example, if you input QUANTITY ORDERED and PRICE, you don't need to input EXTENDED PRICE, which is equal to QUANTITY ORDERED $\times$ PRICE. Another example is incorporating FEDERAL TAX WITHHOLDING data in tables (arrays) instead of keying in that data every time.
- *Use codes for appropriate attributes.* Codes were introduced earlier. Codes can be translated in computer programs by using tables.

Second, source documents should be easy for system users to complete. The following suggestions may help:

- *Include instructions for completing the form.* Also, remember that people don't like to have to read instructions printed on the back side of a form.
- *Minimize the amount of handwriting.* Many people suffer from poor penmanship. The data entry clerk or CRT operator may misread the data and input incorrect data. Use check boxes wherever possible so the system user only needs to check the appropriate values.

Third, design documents so they can be easily and quickly entered into the system. We suggest the following:

- *Data to be entered (keyed) should be sequenced so it can be read like this book, top to bottom and left to right* (see Figure 16.4**A**). The data entry clerk should not have to move from right to left on a line or jump around on the form (see Figure 16.4**B**) to find data items to be entered.

A.

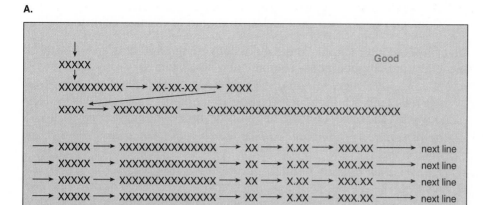

B.

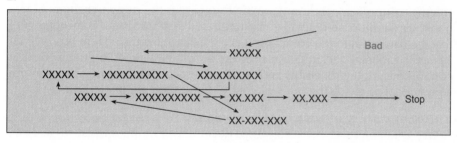

FIGURE 16.4 **Keying from Source Documents** Source documents should be designed to aid in rapid data entry. The source document (part **A**) was designed to allow the data entry clerk to locate and key data in a more natural top-to-bottom, left-to-right sequence. The source document (part **B**) will more likely negatively affect the data entry clerk's ability to quickly locate and enter data.

- *Ideally, portions of the form that are not to be input are placed in or about the lower right portion of the source document* (the last portion encountered when reading top to bottom and left to right). Alternatively, this information can be placed on the back of the form.

These are only guidelines. System users should have the final say on source document design!

Many of these same system user issues also apply to output design. The following general principles are important for output design:

1. *Computer outputs should be simple to read and interpret.* These guidelines may enhance readability:

   · Every report or output screen should have a title.
   · Reports and screens should include section headings to segment large amounts of information.
   · Information in columns should have column headings.
   · Because section headings and column headings are frequently abbreviated to conserve space, reports should include legends to interpret those headings.
   · Legends should also be used to formally define all fields on a report. You never know whose hands a report might end up in! (*Note:* Legends

can be built into on-line outputs using function keys to temporarily interrupt the output to display legends and help.)

· Computer jargon and error messages should be omitted from all outputs.

On many computer outputs, these guidelines are ignored or overlooked; consequently, the outputs appear cluttered and disorganized.

2. *The timing of computer outputs is important.* Outputs must be received by their recipients while the information is pertinent to transactions or decisions. This can affect how the output is designed and implemented.

3. *The distribution of computer outputs must be sufficient to assist all relevant system users.*

4. *The computer outputs must be acceptable to the system users who will receive them.* An output design may contain the required information and still not be acceptable to the system user. To avoid this problem, the systems analyst must understand how the recipient plans to use the output.

Internal Controls for Inputs and Outputs

Internal controls, a continuing theme throughout the design chapters of this book, are a requirement in all computer-based systems. Input controls ensure that the data input to the computer is accurate and that the system is protected against accidental and intentional errors and abuse, including fraud. Output controls ensure the reliability and distribution of the outputs generated by the computer. The following internal control guidelines are offered for inputs:

1. *The number of inputs should be monitored.* This is especially true with the batch method, because source documents may be misplaced, lost, or skipped.

   · In batch systems, data about each batch should be recorded on a batch control slip. Data includes BATCH NUMBER, NUMBER OF DOCUMENTS, and CONTROL TOTALS (e.g., total number of line items on the documents). These totals can be compared with the output totals on a report after processing has been completed. If the totals are not equal, the cause of the discrepancy must be determined.

   · In batch systems, an alternative control would be one-for-one checks. Each source document would be matched against the corresponding historical report detail line that confirms that the document has been processed. This control check may only be necessary when the batch control totals don't match.

   · In on-line systems, each input transaction should be logged to a separate audit file so it can be recovered and reprocessed in the event of a processing error or if data is lost.

2. *Care must also be taken to ensure that the data is valid.* Two types of errors can infiltrate the data: data entry errors and invalid data recorded by system users. Data entry errors include copying errors, transpositions (typing *132* as *123*), and slides (keying *345.36* as *3453.6*). The following techniques are widely used to validate data:

   · **Completeness checks** determine whether all required fields on the input have actually been entered.

   · **Limit and range checks** determine whether the input data for each field falls within the legitimate set or range of values defined for that

MODULUS 11

The following procedure is used to assign a check digit to a key field:

STEP 1: Determine the size of the key field in digits.

$$2\ 4\ 1\ 3\ 5 = 5\ \text{digits}$$

STEP 2: Number each digit location from *right* or *left* beginning with the number "2."

$$2\ 4\ 1\ 3\ 5$$
$$6\ 5\ 4\ 3\ 2$$

STEP 3: Multiply each digit in the key field by its assigned location number.

$$2 \times 6 = 12$$
$$4 \times 5 = 20$$
$$1 \times 4 = \ \ 4$$
$$3 \times 3 = \ \ 9$$
$$5 \times 2 = 10$$

STEP 4: Sum the products from step 3.

$$12 + 20 + 4 + 9 + 10 = 55$$

STEP 5: Divide the sum from step 4 by 11.

$$55/11 = 5\ \text{Remainder}\ 0$$

STEP 6: If the remainder is less than 10, append the remainder digit to the key field. If the remainder is equal to 10, append the character "X" to the key field.

$$2\ 4\ 1\ 3\ 5\ 0$$

FIGURE 16.5 Modulus 11 Self-Checking-Digit Technique Modulus 11 is a very common self-checking-digit technique used to verify that the original/source data has been correctly transcribed into machine-processable form. For example, if a system user read the key field value "24135" and mistakenly keyed in the value "24315," the incorrect data could have been detected by applying the Modulus 11 formula to the key values.

field. For instance, an upper-limit range may be put on PAY RATE to ensure that no employee is paid at a higher rate.

· **Combination checks** determine whether a known relationship between two fields is valid. For instance, if the VEHICLE MAKE is Pontiac, then the VEHICLE MODEL must be one of a limited set of values that comprises cars manufactured by Pontiac (Firebird, Grand Prix, and Bonneville to name a few).

· **Self-checking digits** determine data entry errors on primary keys. A *check digit* is a number or character that is appended to a primary key field. The check digit is calculated by applying a formula, such as Modulus 11, to the actual key (see Figure 16.5). The check digit verifies correct data entry in one of two ways. Some data entry devices can automatically validate data by applying the same formula to the data as it is entered by the data entry clerk. If the check digit entered doesn't match the check digit calculated, an error is displayed. Alternatively, computer programs can also validate check digits by using readily available subroutines.

· **Picture checks** compare data entered against the known COBOL picture or other language format defined for that data. For instance, the input field may have a picture clause XX999AA (where X can be a letter or number, 9 must be a number, and A must be a letter). The field "A4898DH" would pass the picture check, but the field "A4891D8" would not.

Data validation requires that special edit programs be written to perform checks. However, the input validation requirements should be designed when the inputs themselves are designed.

Internal controls must also be specified for outputs. The following guidelines are important output control issues:

1. *The timing and volume of each output must be precisely specified.* You cannot simply state that a report is needed daily. When daily? 8:00 A.M.? 10:30 A.M.? 2:00 P.M.? Computer facilities have limited resources, and the systems analyst must frequently negotiate an appropriate schedule with the computer operations staff.

2. *The distribution of all outputs must be specified.* For each output, the recipients of all copies must be determined. A distribution log, which provides an audit trail for the outputs, is frequently required.

3. *Access controls are used to control accessibility of video (on-line) outputs.* For example, a password may be required to display a certain output on a CRT terminal.

4. *Control totals should be incorporated into all reports.* These controls can be compared with the input controls that will be discussed later in the chapter. The number of records input should equal the number of records output. These control totals are compared before the outputs are distributed. If a discrepancy is found, the outputs are retained until the cause has been determined and corrected.

HOW TO PROTOTYPE AND DESIGN COMPUTER INPUTS

How do you design on-line and batch inputs? In this section, we'll discuss and demonstrate the process of input design, and we'll apply the concepts you learned earlier in the chapter. We'll continue to use our SoundStage case study to demonstrate the process and CASE product to record our specifications into the project repository. We'll also demonstrate how CASE and other tools can be used to prototype inputs.

Step 1: Review Input Requirements

Input requirements, like output requirements, should have been defined during systems analysis. And as was the case for outputs, a good starting point for input design is the design unit data flow diagrams (DFDs) for the new system. The DFD in Figure 16.6 identifies an input to be designed.

Given an input to be designed, we should review the required attributes. The basic content of these inputs should have been recorded in the project repository during systems analysis. If the content has not been recorded, we can define input requirements by studying the output and file designs. An output attribute that can't be retrieved from files or calculated from attributes that are retrieved from files must be input! Additionally, inputs must be designed to maintain the files in the system.

MEMBER ORDER RESPONSE is an input data flow to be designed for the SoundStage Member Services system. The input will initially be designed as a turnaround source document. This means that it will initially be output and mailed to the member, who will respond by completing the order and returning it for input to the order-processing subsystem.

The design specifications for MEMBER ORDER RESPONSE are presented in Figure 16.7. Notice that the physical implementation-oriented attributes have

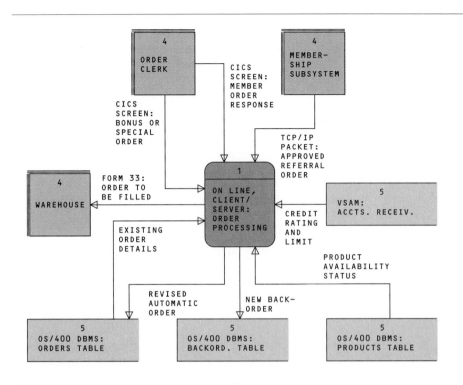

FIGURE 16.6 **A Design Unit Data Flow Diagram for SoundStage Inputs** Those data flows entering the design unit data flow diagram shows inputs that must be designed. The data flows MEMBER ORDER RESPONSE, BONUS OR SPECIAL ORDER, and APPROVED REFERRAL ORDER would need to be designed as computer inputs.

FIGURE 16.7 **Description of a Typical Input** The input design specifications are recorded in the project repository.

DATA FLOW - INPUT
NAME: MEMBER ORDER RESPONSE

TYPE Data Flow   NAME MEMBER ORDER RESPONSE

Label MEMBER ORDER RESPONSE

Explodes To:
Type REC Name   MEMBER ORDER RESPONSE

Duration Value   5,000
Duration Type   DAY
Access Type   U

Description
SOURCE DOCUMENT: MEMBER ORDER (FORM 40), A TURNAROUND DOCUMENT
INPUT METHOD: ON-LINE
INPUT MEDIUM: DISPLAY
TIMING (frequency prepared):   ON DEMAND WEEKDAYS FROM 8 A.M. TO 5 P.M.
AVERAGE VOLUME (in records):   5,000 PER DAY
PEAK VOLUME AND TIMING: 5,800/DAY FROM NOVEMBER THROUGH DECEMBER
CONTROLS AND SPECIAL INSTRUCTIONS:
    THE INPUT WILL INITIALLY BE DESIGNED AS A TURNAROUND SOURCE
    DOCUMENT. IT WILL BE OUTPUT AND MAILED TO THE MEMBER WHO WILL
    RESPOND BY COMPLETING THE ORDER AND RETURNING IT FOR INPUT BY
    ORDER PROCESSING. AN AUDIT TRAIL FILE SHOULD BE CREATED AND ALL
    RETURNED SOURCE DOCUMENTS MUST BE RETAINED.

REPORT OF ATTRIBUTES CONTAINED IN
MEMBER ORDER RESPONSE RECORD

ATTRIBUTE NAME	TYPE	OCC	CHARACTERS LEFT	CHARACTERS RIGHT	INPUT PICTURE	VALUE RANGE
MEMBER NUMBER	C	1	7	0	9999999	MUST BE UNIQUE FOR A MEMBER; VALUE 000001 THRU 9999999
ORDER NUMBER	C	1	6	0	ZZZZZ9	MUST BE UNIQUE FOR AN ORDER; VALUE 1 THRU 999999
MEMBER NAME	C	1	30	0	X(30)	OPTIONAL
ORDER DATE	C	1	8	0	MM/DD/YY	MUST BE VALID MONTH, DAY, AND YEAR (DEFAULT IS TODAY'S DATE)
CLUB NAME	C	1	15	0	X(15)	REQUIRED; MEMBER MUST HOLD AT LEAST 1 CLUB MEMBERSHIP
P.O.BOX	C	1	10	0	X(10)	OPTIONAL
STREET	C	1	15	0	X(15)	
CITY	C	1	15	0	X(15)	
STATE	C	1	2	0	AA	STANDARD STATE CODES
ZIPCODE	C	1	9	0	X(9)	
SELECTION OF MONTH ACCEPTED?	C	1	1	0	A	Y=YES (default) N=NO
PRODUCT NUMBER	C	1-15	7	0	9999999	MUST BE UNIQUE; VALUE 00000000 THRU 99999999
MEDIA CODE	C	1-15	1	0	X	R=RECORD, C=CASSETTE, D=COMPACT DISC, 8=8 TRACK
QUANTITY ORDERED	C	1-15	4	0	ZZZ9	OPTIONAL; IF PRESENT MUST BE > 0
	Ⓐ		Ⓑ		Ⓒ	Ⓓ

FIGURE 16.8   Attribute Repository Entry for the Contents of MEMBER ORDER RESPONSE   This repository printout shows details about the attributes to be input from a MEMBER ORDER RESPONSE.

simply been appended to the logical, requirements-oriented attributes that were defined during systems analysis. This figure is a printout from our CASE product, Excelerator.

The entries are fairly self-explanatory, addressing the issues presented earlier in this chapter. All the entries were developed with the approval of the system users.

There are a number of design considerations for attributes to be included in the input. We were careful to delete any attributes that could be read from files and databases (except keys, which are needed to read files and databases). Physical attributes of each element are also described to the repository. Figure 16.8 is a printout of pertinent attributes of the input. Note the following:

Ⓐ For inputs, all fields—numeric, alphabetic, and alphanumeric—are usually input as character strings (ASCII or EBCDIC). Therefore, TYPE is set to C (for "character") in all but the most unusual of circumstances.

Ⓑ We also specified field size in terms of characters (or positions) to the left and right of the decimal point. For elements with type "character"—meaning "alphanumeric"—the CHARS. RIGHT attribute is not used.

Ⓒ An input PICTURE should not include such special editing symbols as dollar signs and dashes because, as we have already noted, they are not included in the field. However, the PICTURE should indicate which positions of a field can be numbers and characters. The PICTURE also indicates the decimal point position in some numeric fields.

Ⓓ The last column describes the legitimate values or VALUE RANGEs that the attribute can assume. This column will be helpful to programmers who must write the input edit routines. Additionally, this column describes special validation checks to be performed. For instance, key fields may require check digits. Other fields may require limit checks (record the value ranges).

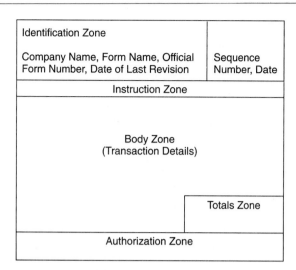

FIGURE 16.9 **Source Document Design Zones** A source document for input may be designed in zones such as those indicated on this template. The locations are fairly typical with the exception of the instruction zone, which may be located almost anywhere on the form.

Step 2: Design or Prototype the Source Document

If a source document will be used to capture data, we should design that document first. The source document is for the system user. In its simplest form, the prototype may be a simple sketch or an industrial artist's rendition.

A well-designed source document will be divided into zones. Some zones are used for identification; these include company name, form name, official form number, date of last revision (an important attribute that is often omitted), and logos. Other zones contain data that identifies a specific occurrence of the form, such as form sequence number (possibly preprinted) and date. The largest portion of the document is used to record transaction data. Data that occurs once and data that repeats should be logically separated. Totals should be relegated to the lower portion of the form because they are usually calculated and, therefore, not input. Many forms include an authorization zone for signatures. Instructions should be placed in a convenient location, preferably not on the back of the form. One possible template for source document design is provided in Figure 16.9.

Prototyping tools have become more advanced in recent years. Spreadsheet programs such as Microsoft's Excel can make very realistic models of forms. These tools give you outstanding control over font styles and sizes, graphics for logos, and the like. Laser printers can produce excellent-quality printouts of the prototypes.

Another way to prototype source documents is to develop a rough model using a word processor. Next, pass the model to one of the growing number of desktop publishing systems that can transform the rough model into impressive looking forms (so impressive, in fact, that some companies now develop forms this way instead of subcontracting their design to a forms manufacturer).

Finally, forms-processing software packages are starting to become popular. This exciting new prototyping and implementation technology is featured in this chapter's The Next Generation box.

The Next Generation

FORMS-PROCESSING SOFTWARE:
SIMPLIFIED ON-LINE INPUTS FOR TRADITIONAL PAPER FORMS

The most popular of on-line input processing is *forms filling*. The form appears on the screen and the end-user simply enters data in the appropriate blanks. Implementing forms filling has involved extensive programming, especially to provide adequate editing and instructions. But that is changing.

One of the newest software categories is *forms-processing software*. Forms-processing software eliminates the need for tedious programming and, even better, permits forms prototyping. The resulting forms can be every bit as detailed and complex as their paper equivalents (they can even be printed and duplicated). Think of it as a tool that permits duplicating actual paper forms on a computer screen. Here's how it works.

Forms-processing software packages generally provide numerous, modifiable templates for commonly encountered business forms. The software also includes powerful, easy-to-use facilities (compared to programming) for

changing the forms. Generally, you are given control over text styles, font sizes, shading, borders, and the like. You can also sometimes import scanned logos and graphics. The tools allow you to customize a form (from template or scratch). Virtually any type of form can be created or recreated.

So far, it sounds just like desktop publishing . . . but there's more. Items on any form can be designated as input fields. These fields can be described to the form processor's internal dictionary. In this dictionary, you can specify whether the field is mandatory or optional, any default values, legal value ranges, date and time stamps, data verification rules, and so on. You can also attach help messages, error messages, and sometimes even detailed instructions.

The forms processor becomes the actual input sub-system. End-users can initially study and try out the screens. Data will automatically be edited and verified, which is useful for creating test data files for

subsequent programming. And there's more!

Depending on the forms processor, the edited data will be copied to a file (similar to the concept of remote batch files). This file, of known structure, may be imported to one or more technical programming environments. For example, the file may automatically conform to VSAM, IMS, or dBASE standards — or, at the very least, an ASCII file that can be read by any program you care to write.

You still write the programs to use the data, but you've just eliminated the need to write complex input, editing, and verification routines.

Many of these packages are starting to appear for microcomputers. But their data files (if ASCII) could easily be uploaded to a mainframe system, thus creating a micro-mainframe version of remote batch processing. The prospects for forms processing as an input prototyping and development environment look promising.

Step 3: Prototype the Layout for System Users

After design decisions have been recorded in the repository and any source documents have been designed, you can now prototype input screens. This is only appropriate, of course, for on-line or remote batch inputs. The best way to design screens is to sketch, or better still, prototype those screens. The sketches or prototypes can be shown to system users and modified based on their feedback.

Modern prototyping tools have made screen design an infinitely easier task. Let's study some of the tools and techniques of prototyping as they apply to inputs.

Prototyping Inputs with CASE Tools

Many CASE products include facilities for rapid prototyping of input screens. They are especially useful since they can use the project repository data recorded during systems analysis. Some products, such as Pansophic's Telon, can ultimately transform approved prototypes into COBOL and PL/1 code.

Products such as Telon assume you have already defined your input requirements (logical, implementation-independent requirements). Throughout this book, we've used the CASE product Excelerator to capture those requirements. Excelerator can pass the requirements to a product such as Telon or provide some simple prototyping capabilities of its own. Approved Excelerator screen prototypes can be used to generate equivalent PICTUREs in several programming languages.

Prototyping Inputs with Database Management Systems or Fourth-Generation Languages

A continuing theme in this book has been the use of database management systems and fourth-generation languages (4GLs) to prototype systems. Virtually all such tools include powerful screen design facilities that make it possible to quickly lay out prototype screens, map screen fields to a repository that defines edit rules, and define help messages and screens. These screens can be directly tested by system users and modified to reflect their opinions. Most 4GL-developed prototypes can eventually evolve into finished production systems, although some must be reprogrammed in traditional languages to improve processing efficiency or security. These 4GL prototyping capabilities can be found in both mainframe and microcomputer databases.

Step 4: If Necessary, Convert User-Oriented Input Layouts into Programmer-Oriented Layouts

In some cases, sketches and prototypes are sufficient for programmers to implement the design; however, in many cases, they are inadequate as final specifications. For example, the prototyping tools may not have been able to implement some of the features required in the final design. Or perhaps the prototype doesn't convey the full range of possibilities to be implemented. In these cases, you must use more formal specification techniques.

Using Display Layout Charts to Document the Format of Input Screens

Visual inputs are frequently documented on display layout charts.

> Terminal screen **display layout charts** (see Figure 16.10) are used to document typical screens (80-column limit). For screens wider than 80 columns, printer spacing charts can be used.

The display grid is used to indicate the exact row and column position of information on the screen. You may have noticed that the number of lines (43) is greater than the number of lines that can be displayed on most screens (24). Why? The additional space allows you to indicate additional information that can be scrolled up as the system user pages through the input. To format the screen input, use COBOL picture clauses and conventions.

> **Function key assignments,** available in most CRT terminals and microcomputers, can be described at the bottom of the chart. Function keys can be assigned to perform a number of tasks at the stroke of a single key. For

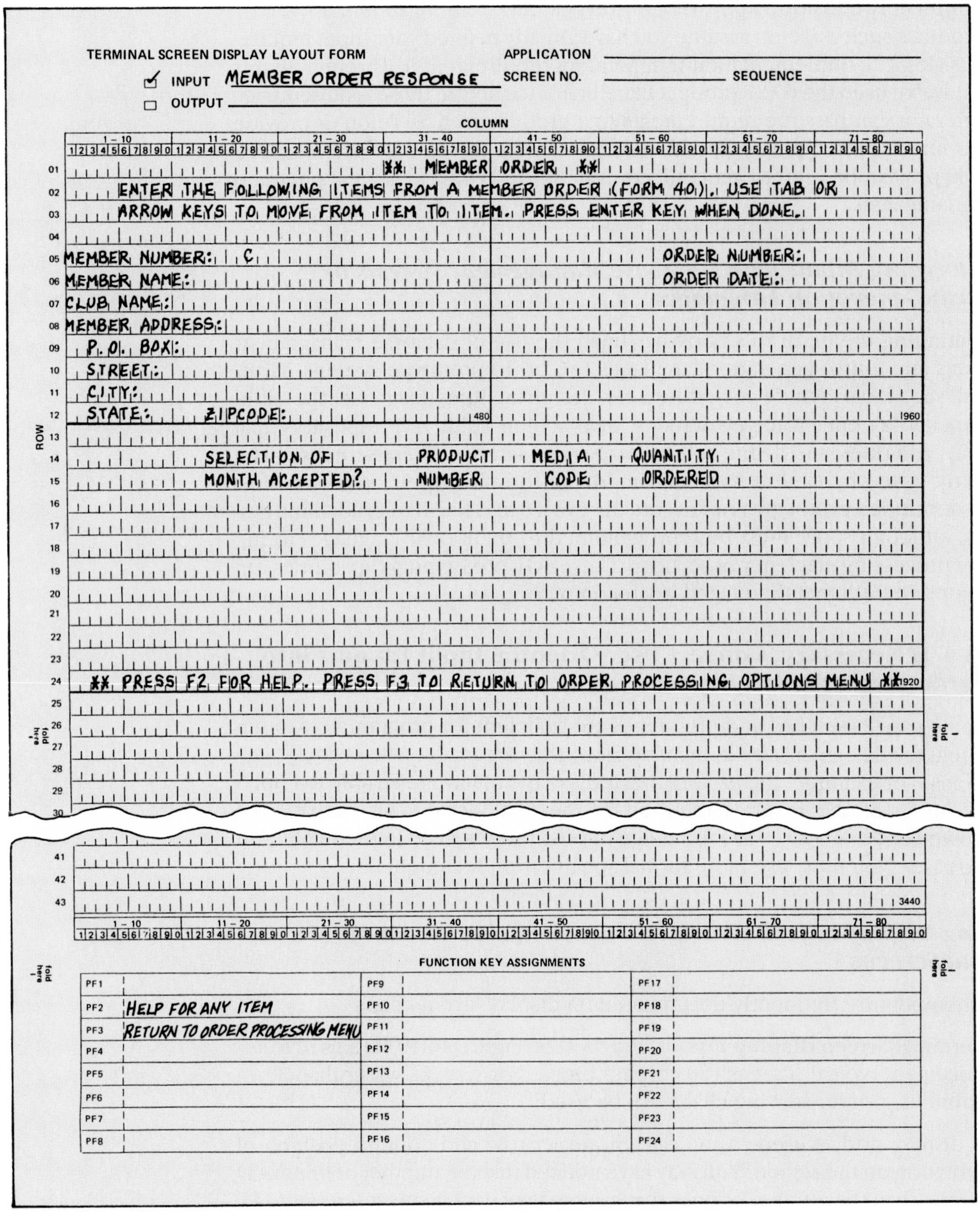

TERMINAL SCREEN DISPLAY LAYOUT FORM   APPLICATION _____

☑ INPUT   MEMBER ORDER RESPONSE   SCREEN NO. _____   SEQUENCE _____

☐ OUTPUT _____

COLUMN

```
01                                    ** MEMBER ORDER **
02      ENTER THE FOLLOWING ITEMS FROM A MEMBER ORDER (FORM 40). USE TAB OR
03      ARROW KEYS TO MOVE FROM ITEM TO ITEM. PRESS ENTER KEY WHEN DONE.
04
05  MEMBER NUMBER:   C                              ORDER NUMBER:
06  MEMBER NAME:                                    ORDER DATE:
07  CLUB NAME:
08  MEMBER ADDRESS:
09    P. O. BOX:
10    STREET:
11    CITY:
12    STATE:    ZIPCODE:                   480                             1960
13
14         SELECTION OF        PRODUCT    MEDIA    QUANTITY
15         MONTH ACCEPTED?     NUMBER     CODE     ORDERED
16
...
24  ** PRESS F2 FOR HELP. PRESS F3 TO RETURN TO ORDER PROCESSING OPTIONS MENU ** 1920
...
43                                                                         3440
```

FUNCTION KEY ASSIGNMENTS

PF1		PF9		PF17	
PF2	HELP FOR ANY ITEM	PF10		PF18	
PF3	RETURN TO ORDER PROCESSING MENU	PF11		PF19	
PF4		PF12		PF20	
PF5		PF13		PF21	
PF6		PF14		PF22	
PF7		PF15		PF23	
PF8		PF16		PF24	

FIGURE 16.10   **Display Layout Chart for On-Line MEMBER ORDER RESPONSE**   The initial screen seen by the system user is a blank order form to be filled in.

instance, the analyst can specify a help key or a save data key to be programmed into the system.

Figure 16.10 provides the layout for the on-line input of MEMBER ORDER RESPONSE. This form may be familiar to you. Display layout charts are intended for the programmer, not the system user. Keep in mind that this chapter deals specifically with the final input screens, not the menus or dialogue that gets the system user to those screens (screen dialogue is covered in the next chapter).

There are two generally accepted ways of designing on-line inputs: question-answer dialogues and form filling.

In the **question-answer mode** the system asks several questions and the operator responds with the appropriate answer. Default answers are often provided in brackets, as in <default = 0>. Care must be taken to design input dialogues to include responses that must occur when improper data is entered.

Another common on-line input technique is called form filling.

Form filling is another on-line input technique. A blank form is painted on the screen. The screen cursor moves to the first field to be entered on the form. When the operator enters the field, the return key sends the cursor to the next field to be entered. The automatic movement of the cursor from field to field is often assisted by dialogue software (e.g., IBM's CICS, a teleprocessing monitor).

Dialogue software and teleprocessing monitors provide functions and subroutines, callable from application programs, that simplify screen and dialogue design. Fields are edited as they are entered, and appropriate error messages are displayed. After all fields have been entered, the operator presses a function key to release the data to the computer program. Most of the screen must be used for the form; however, small windows should be set aside for MESSAGES and ESCAPE. Instructions and expanded messages can be placed in a separate file that can be displayed when a function key assigned to HELP is pressed.

As to the layout of the MEMBER ORDER RESPONSE, let's walk through some of the design. We chose the form-filling method for input. Figure 16.10 represents what the system user will initially see on the CRT screen. The screen cursor will be positioned at the location marked "C." As the instructions indicate, the system user will be permitted to move around the screen from one item to the next or back to the previous item. In the escape area, we show the system user how to discontinue the input of customer orders.

But what about the actual entry of attributes? A programmer would need specific input requirements. An appropriate display layout chart is provided in Figure 16.11. This chart shows the programmer how data is actually input. We call your attention to the following:

(A) This portion of the screen will scroll upward, thus allowing the system user to enter the data for a large number of ordered items. Downward arrows were drawn under repeating attributes to indicate that the system user may enter several values for these attributes.

(B) If at any time the system user enters incorrect data values for an element, a descriptive error message will appear. We explain these error

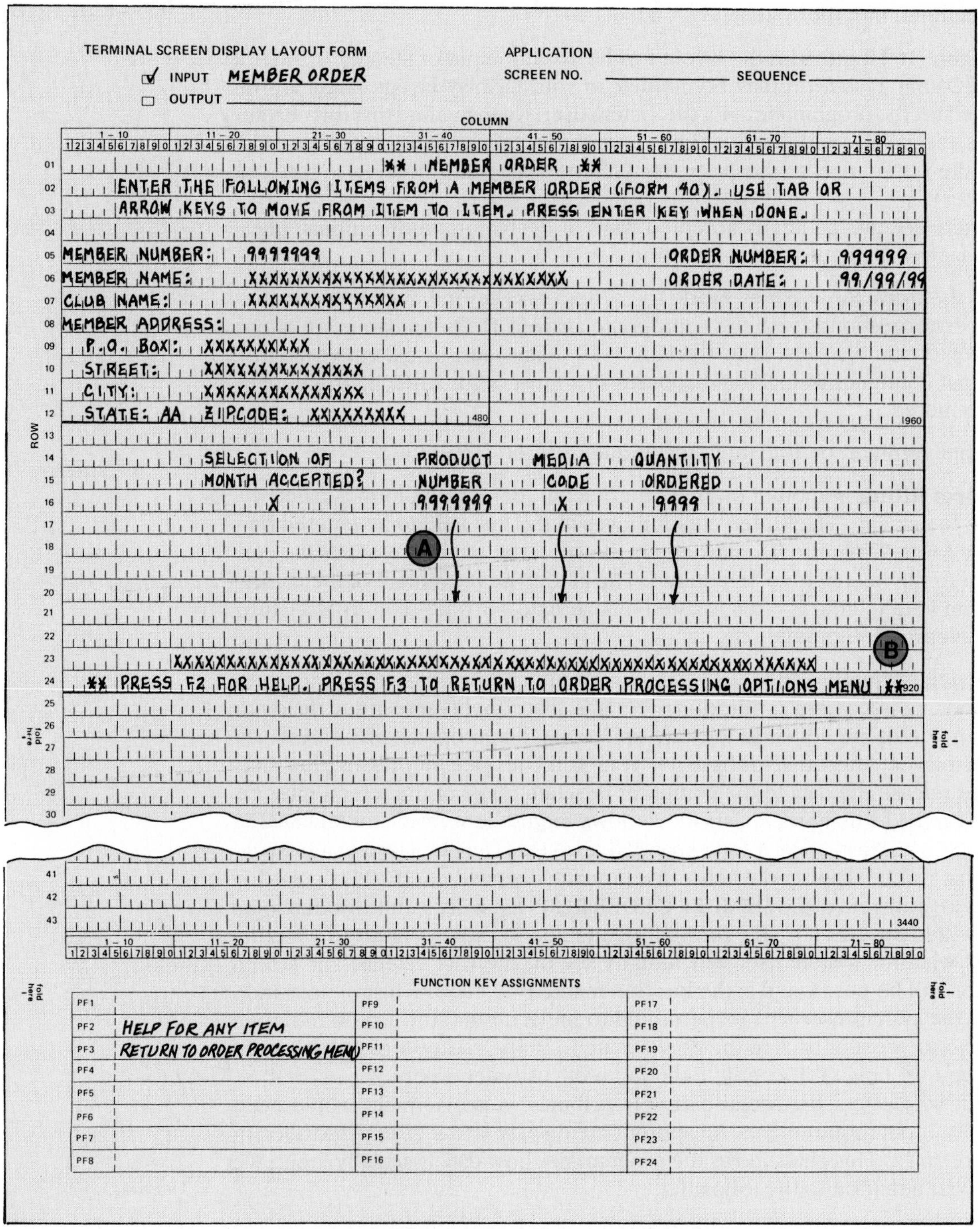

TERMINAL SCREEN DISPLAY LAYOUT FORM   APPLICATION _____

  ☑ INPUT **MEMBER ORDER**   SCREEN NO. _____ SEQUENCE _____

  ☐ OUTPUT _____

```
COLUMN
              1-10        11-20       21-30       31-40       41-50       51-60       61-70       71-80
01                                          ** MEMBER ORDER **
02      ENTER THE FOLLOWING ITEMS FROM A MEMBER ORDER (FORM 40). USE TAB OR
03      ARROW KEYS TO MOVE FROM ITEM TO ITEM. PRESS ENTER KEY WHEN DONE.
04
05  MEMBER NUMBER:   9999999                                   ORDER NUMBER:  999999
06  MEMBER NAME:     XXXXXXXXXXXXXXXXXXXXXXXXXXX               ORDER DATE:    99/99/99
07  CLUB NAME:       XXXXXXXXXXXXXXX
08  MEMBER ADDRESS:
09    P.O. BOX:      XXXXXXXXX
10    STREET:        XXXXXXXXXXXXXXX
11    CITY:          XXXXXXXXXXXXXXX
12    STATE: AA  ZIPCODE: XXXXXXXXX           |480                                          |960
13
14          SELECTION OF      PRODUCT    MEDIA    QUANTITY
15          MONTH ACCEPTED?   NUMBER     CODE     ORDERED
16               X            9999999     X       9999
17
18
19
20
21
22
23      XXXXXXXXXXXXXXXXXXXXXXXXXXXXXXXXXXXXXXXXXXXXXXXXXXXXXXXXXXXX
24  ** PRESS F2 FOR HELP. PRESS F3 TO RETURN TO ORDER PROCESSING OPTIONS MENU **920
25
26
27
28
29
30

41
42
43                                                                                        |3440
```

FUNCTION KEY ASSIGNMENTS

PF1	PF9	PF17
PF2 *HELP FOR ANY ITEM*	PF10	PF18
PF3 *RETURN TO ORDER PROCESSING MENU*	PF11	PF19
PF4	PF12	PF20
PF5	PF13	PF21
PF6	PF14	PF22
PF7	PF15	PF23
PF8	PF16	PF24

FIGURE 16.11 Display Layout Chart for On-Line MEMBER ORDER RESPONSE Fields to be Input This screen design shows the proper edit masks for the fields to be entered by the system user.

MEMO FOR MEMBER ORDER INPUT SCREEN

Ref. No.   Message

a   This field is used to provide the end-user with a descriptive error message when invalid commands or data
             have been entered otherwise the field is not printed. When the message is displayed, it should "blink" to
             grab the end-user's attention. As a reminder, the editing criteria were explained in the report titled "Report of
             Data Attributes Contained in Member Order Record." The following specifies the conditions and types of
             messages to be displayed to the end-user.

Select the appropriate case:

Case 1:   MEMBER NUMBER is invalid
             If the MEMBER NUMBER is not equivalent to a MEMBER NUMBER of an existing MEMBER then:
                 error message = "MEMBER does not exist, please reenter."

Case 2:   ORDER NUMBER is invalid
             If the ORDER NUMBER is equivalent to the ORDER NUMBER of any previous MEMBER ORDER then:
                 error message = "ORDER NUMBER was assigned to previously entered MEMBER ORDER, please reenter."

Case 3:   ORDER DATE is invalid
             Select appropriate case:
                 Case 2.1   ORDER DATE contains no values, then:
                            error message = "The order date must be provided on all orders, please enter."
                 Case 2.2   ORDER DATE contains invalid values for month, day, year, then:
                            error message = "The order date is not valid, please reenter."

Case 4:   PRODUCT NUMBER is not valid product number, then:
             error message = "Entered an incorrect part number (does not exist), please reenter."

Case 5:   MEDIA CODE is invalid, then:
             error message = "media code is invalid, please reenter or press F2 key for list of valid codes and meanings."

Case 6:   QUANTITY ORDERED is not greater than 0, then:
             error message = "Quantity ordered must be greater than 0, please reenter."

**FIGURE 16.12   Specifications for Display Attributes and Error Messages for MEMBER ORDER RESPONSE
Input Screen**   These notes are used to describe attributes—such as blinking fields—and error messages.

messages in Figure 16.12. Notice that we've used Structured English to document when specific error messages may occur. No, we don't consider this documentation to be doing the programmer's job. By conveying these editing requirements in Structured English, we are reducing the time we have to spend conveying the requirements to the programmer.

Earlier, we introduced the concept of graphic output. For the SoundStage case, we have decided to implement visual inputs and outputs as windows-based graphic inputs and outputs. Compare the alternative graphic design presented in Figure 16.13 with the more traditional character-based design in Figure 16.11.

Using Input Record Layout Charts to Document the Structure of Batch and Remote Batch Inputs

The classic **input record layout chart** is used to lay out the format of batch input files—the files that result from keying operations. In conjunction with display layouts, input record layouts also lay out the final format of remote batch

FIGURE 16.13 Alternative Graphic Design for On-Line MEMBER ORDER RESPONSE
Compare this figure with Figure 16.11. Figure 16.13**A** shows the initial MEMBER ORDER RESPONSE input screen. Figure 16.13**B** shows two of the pop-up windows activated when the user clicks on either of the clubs or media codes buttons. Other buttons for common functions are shown on the bottom of the screen.

A.

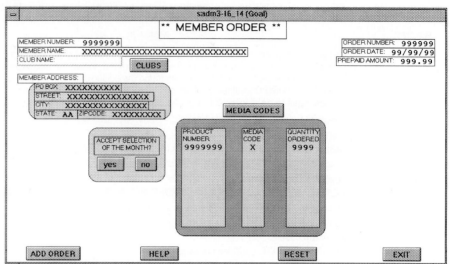

B.

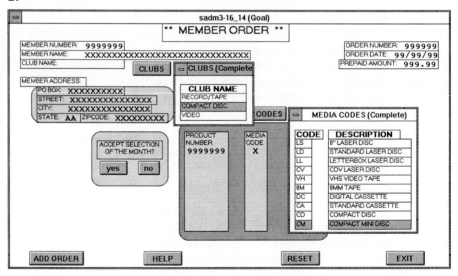

input files. Figure 16.14 is a sample input record layout chart for a batch version of our MEMBER ORDER RESPONSE input. The form is not hardware-dependent and can be used to document virtually any type of medium, including punched card images and key-to-tape or disk-images. Four 80-column punched card images (a common punched card size) or one 320-character record (for magnetic media) can be accomodated. The form is read and prepared in the same manner as a record layout chart. The input record layout chart is primarily intended for the computer programmer, not the system user. Note the following in Figure 16.14:

(A) We added the attribute NUMBER OF ORDERED ITEMS because variable-length records require a repeating factor field.

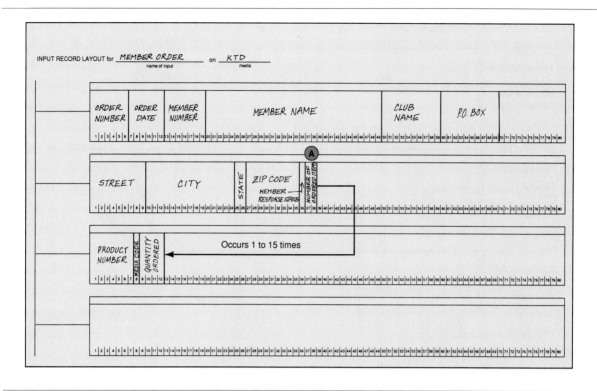

FIGURE 16.14   Sample Input Record Layout for Batch MEMBER ORDER RESPONSE Input

It's also quite common for a systems analyst to design a batch input file that contains more than one type of input record. For instance, a system may include a program that allows the marketing department to update our TITLE/PRODUCT file. The program might (1) place new products into the file, (2) delete from the file products that are to be discontinued, or (3) change the contents of a product record to reflect a new price or unit of measurement. The input data for these three tasks would vary.

Figure 16.15 shows how the design of the input layout record could be completed. Because the content and format of the input record are dependent on the particular type of update, we simply showed the three possible design layouts. Let's examine the record layout closely.

Ⓐ This portion represents the contents and layout of the input record that is required when a new product is to be added to the product file. Notice that we included in the record a special field called TRANSACTION CODE. This field will be included in each of the three input record types. The value of this field will identify the particular type of input task to be performed.

Ⓑ This label is provided to let the programmer know which record layout corresponds to which input record type.

Ⓒ This portion of the input record layout chart specifies the attributes and layout for an input record that represents the deletion of an existing product record in the product file.

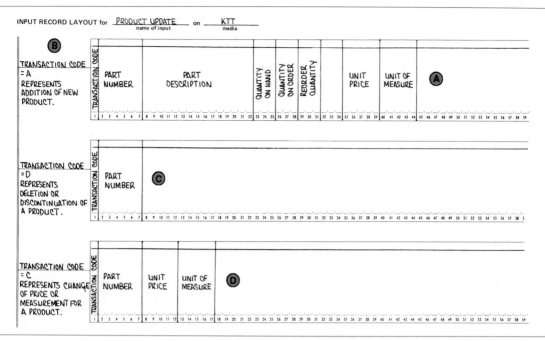

FIGURE 16.15   **Sample Input Record Layout for Multiple Input Records**

(D) This portion of the input record layout chart specifies the attributes and layout for an input record that represents updates to the contents of an existing product record in the product file.

The Implications of Input Design for Output Design

Because DFDs show only net data flows (input or output), not all outputs that must be designed will show up on the DFD! For instance, if we select a batch input implementation of MEMBER ORDER RESPONSE, it necessitates some type of control output—in this case a DAILY MEMBER ORDER ERRORS RE-PORT. This edit report would identify all MEMBER ORDERs that contain incorrect data. The Member Services clerk will be able to use this report to follow up on the orders to ensure that corrections are made.

Another example would be an audit report that lists all MEMBER ORDERs processed on a given day. This report establishes an audit trail for the orders that were processed.

Just recognize that the design of any input could necessitate the design of additional outputs.

HOW TO PROTOTYPE AND DESIGN COMPUTER OUTPUTS

In this section, we'll discuss and demonstrate the process of output design. We'll introduce some tools for documenting output design, and we'll also apply the concepts you learned in the last section. We will continue to use our computer-assisted systems engineering (CASE) product, Excelerator, to record our design decisions into our project repository. And we'll also demonstrate how CASE and other computer tools can be used to prototype outputs and layouts to system users and programmers. As usual, each step of the output design technique will be demonstrated using examples drawn from our Sound-Stage case study.

Step 1: Review Output Requirements

Output requirements should have been defined during systems analysis. A good starting point for output design is the data DFDs for the new system. The design unit DFD (Figure 16.6) identifies one output requirement, FORM 33: ORDER TO BE FILLED. Now we are ready to address those physical, implementation details that tell the system users and programmers *how* the output will or should be implemented.

Step 2: Review How the Output Data Flow Will Be Implemented

Recall that the selection phase of the system design determined how the output data flows will eventually be implemented. Relative to outputs, the decisions were made by determining the best medium and format for the design and implementation based on:

- The type and purpose of the output.
- The technical and economic feasibility.

Since feasibility is important to more than just outputs, the techniques for evaluating feasibility are covered separately (in Part Four, Module C, Feasibility Analysis). The first set of criteria, however, is described in the following paragraphs.

First, you must understand the type and purpose of the output. Is the output an internal or external report? If it's an internal report, is it a historical, detailed, summary, or exception report? If it's an external report, is the form a turnaround document? After assuring yourself that you understand what type of report the output is and how it will be used, you need to address several design issues.

1. *What medium would best serve the output?* Various media were discussed earlier in the chapter. You will have to understand the purpose or use of the output to determine the proper medium. You can select more than one medium—for instance, video with optional paper. All of these decisions are best addressed with the system users.

2. *What would be the best format for the report?* Tabular? Zoned? Graphic? Narrative? Some combination of these? After establishing the format, you can determine what type of form or paper will be used. Computer paper comes in three standard sizes: 8½" × 11," 11" × 14," and 8" × 14." Many printers can now easily compress 132 columns of print into an 8-inch width. You need to determine the capabilities and limitations of the intended printer.

If a preprinted form is to be used, requirements for that form must be specified. Should the form be designed for mailing? What will be the form's size? Will the form be perforated for bursting into several sections? What legends and instructions need to be printed on the form (both front and back)? What colors will be used and for which copies?

Incidentally, form images can be stored and printed with modern laser printers, thereby eliminating the need for dealing with forms manufacturers in some businesses.

3. *How frequently is the output generated?* On demand? Hourly? Daily? Monthly? For scheduled outputs, when do system users need the report? Scheduled reports must be worked into the information systems operations schedule. For instance, a report the system user needs by 9:00 A.M. on Thursday may have to be scheduled for 5:30 A.M. Thursday. No other time may be available.

4. *How many pages or sheets of output will be generated for a single copy of a report?* This data is necessary to accurately plan paper and form consumption.

5. *Does the output require multiple copies?* If so, how many? Photocopy (doesn't tie up printer)? Carbons (most printers can make no more than six legible carbons)? Duplicates (requires the most printer time, although laser printers are changing this situation)?

For external documents, there are also several alternatives. Carbon and chemical carbon are the most common duplicating techniques. Selective carbons are a variation whereby certain fields on the master copy will not be printed on one or more of the remaining copies. The fields to be omitted must be communicated to the forms manufacturer. Two-up printing is a technique whereby two sets of forms—possibly including carbons—are printed side by side on the printer.

6. *For printed outputs, have distribution controls been finalized?* For on-line outputs, access controls should be determined.

7. *For attributes contained on the output, what format should be followed?*

After a design is determined to be feasible and is approved by system users and management, the preceding design decisions are recorded in the project repository. As noted earlier, an output is a data flow. During systems analysis, the logical data flow was recorded. Now, the physical, implementation decisions are recorded. Let's study an example.

FORM 33: ORDER TO BE FILLED is an output data flow to be designed for the SoundStage Member Services system. Discussions with all system users reveal that the output needs to be designed as a form with three additional copies. The original will be called the *master,* and the three copies will be called *picking, packing,* and *shipping.* To fulfill this requirement, we will design a preprinted form with carbon copies. Today, most carbons are produced using chemically coated paper rather than the messy ink carbons.

The design specifications for FORM 33: ORDER TO BE FILLED are presented in Figure 16.16. Note that the physical, implementation specifications have simply been appended to the logical, implementation-independent details documented during the analysis phases. This figure is a printout from our CASE product, Excelerator. The entries on this form are relatively self-explanatory. They address the issues presented earlier in this chapter and section. All the criteria were developed with and approved by the system users.

There are also a number of design considerations for attributes that are included in an output. Each of the elements in this data flow are further described by the printout in Figure 16.17. Note the following items:

DATA FLOW - OUTPUT
NAME: ORDER TO BE FILLED

TYPE Data Flow   NAME ORDER TO BE FILLED

Label ORDER TO BE FILLED

Explodes To:
Type REC Name   ORDER TO BE FILLED

Duration Value   5,000
Duration Type   DAY

Description
DESCRIPTION: A FOUR-PART DOCUMENT DESCRIBING A SALES ORDER
MEDIUM: PAPER   OUTPUT TYPE: EXTERNAL/FORM
VOLUME: 5,000 PER DAY   NUMBER OF COPIES: 3
COPYING METHOD: SELECTIVE CARBON
FREQUENCY PREPARED: DAILY AT 8 A.M. AND 1:00 P.M.
OUTPUT CHARACTERISTICS: PREPRINTED 8 1/2" x 8 1/2", DESIGNED
                        FOR MAILING. THE 3 COPIES SHOULD HAVE
                        DIFFERENT COLORS:
       ORIGINAL/MASTER = WHITE   PICKING COPY = YELLOW
       PACKING COPY = PINK   SHIPPING COPY = GREEN
OUTPUT RECIPIENT(S): ALL COPIES WILL BE PICKED UP BY ORDER ENTRY
SPECIAL INSTRUCTION(S): QUANTITY SHIPPED IS HAND ENTERED
                        QUANTITY BACKORDERED MAY BE MODIFIED BY HAND

FIGURE 16.16 **Design Specifications for ORDER TO BE FILLED** After the design is approved by system users, the output design decisions are recorded in the project repository.

REPORT OF ATTRIBUTES CONTAINED IN
ORDER TO BE FILLED RECORD

ATTRIBUTE NAME	TYPE	OCC	CHARACTERS LEFT	CHARACTERS RIGHT	INPUT PICTURE
MEMBER NUMBER	C	1	7	0	9999999
ORDER NUMBER	C	1	6	0	ZZZZZ9
MEMBER NAME	C	1	30	0	X(30)
ORDER DATE	C	1	8	0	MM/DD/YY
ORDER STATUS	C	1	15	0	X(15)
P.O.BOX	C	1	10	0	X(10)
STREET	C	1	15	0	X(15)
CITY	C	1	15	0	X(15)
STATE	C	1	2	0	AA
ZIPCODE	C	1	9	0	X(9)
PRODUCT NUMBER	C	1-15	7	0	9999999
MEDIA CODE	C	1-15	1	0	X
QUANTITY ORDERED	C	1-15	4	0	ZZZ9
QUANTITY SHIPPED	C	1-15	4	0	ZZZ9
ORDER PRICE	C	1-15	3	2	ZZZ.99
EXTENDED PRICE	C	1-15	4	2	ZZZZ.99
TOTAL ORDER PRICE	C	1	5	2	ZZZZ9.99
Ⓐ			Ⓑ	Ⓑ	Ⓒ

FIGURE 16.17 **Attribute Details of ORDER TO BE FILLED Output** This report details the attributes to appear on the output.

(A) For each attribute, we specified the data TYPE as follows:
Packed (P)—a numeric attribute. Why is packed designated as a type for numeric fields? Most numeric attributes in business information systems are stored and manipulated in packed-decimal format. Character (C)—also known as alphanumeric—an attribute that contains any combination of alphabetic letters, nonarithmetic numbers, and special characters. Binary (B) and Floating Point (F)—alternative formats for numeric data. These are usually encountered in scientific and engineering applications and are rarely used in information systems.

These codes were dictated by the CASE product being used for our project repository. Other CASE products and manual techniques may adopt different codes—for instance, AN = alphanumeric, N = numeric, and so forth.

We recommend that nonarithmetic attributes that consist only of numbers be specified as character, alphanumeric, instead of packed, numeric (e.g., PRODUCT NUMBER, CUSTOMER NUMBER, ZIP CODE). By specifying the attribute as alphanumeric, we make it clear to the programmer that arithmetic should not be performed on the attribute.

(B) For each attribute, we specified size in terms of CHARACTERS LEFT and CHARACTERS RIGHT of the decimal point. We do not include special editing symbols, such as hyphens, commas, decimal points, and slashes, when determining this size. For character or alphanumeric attributes, only the CHARACTERS LEFT will be used.

(C) You may recognize the entries under PICTURE as being standard COBOL picture entries. COBOL picture clauses are a relatively common way to communicate element print formats to programmers.

Where did the details about the attributes come from? During systems analysis—when the analysts were learning and documenting business requirements—they recorded most of these details in the project repository using data attribute screens like the one depicted in Figure 16.18. Any new attributes added to outputs should be similarly recorded in the data attribute repository.

Step 3: Prototype the Layout for System Users

After design decisions and details have been recorded in the project repository, we must create the format of the report. The format or layout of an output directly affects the system user's ability to read and interpret it. The best way to lay out outputs is to sketch or, better still, generate a sample of the report or document. We need to show that sketch or prototype to the system user, get feedback, and make modifications to the sample. It's important to use realistic or reasonable data and demonstrate all control breaks.

Prior to the availability of prototyping tools, analysts could only sketch rough drafts of outputs to get a feel for how system users wanted outputs to look. With modern tools, we can develop more realistic prototypes of these outputs. Let's study some tools and techniques for prototyping output layouts. We'll demonstrate each technique with examples of outputs from the SoundStage project.

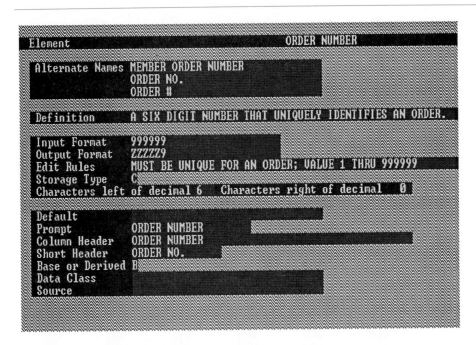

FIGURE 16.18 **Sample Attribute Screen** This is a sample Excelerator screen for defining the details of an attribute.

Simple Output Prototyping

Perhaps the least expensive and most overlooked prototyping tool is the common spreadsheet. Examples include Lotus 1-2-3, Microsoft's Excel, Borland's Quattro, and Microsoft's Multiplan. A spreadsheet's tabular format is ideally suited to the creation of rapid prototypes. Arithmetic and logical formulas and functions can be placed in cells (a cell is the intersection of a row and column); therefore, spreadsheets can automatically calculate and recalculate some cells to make the information accurate. Finally, most spreadsheets now include facilities to quickly convert tabular data into a variety of popular graphic formats. Consequently, spreadsheets provide an unprecedented way to prototype graphs for system users.

Figure 16.19**A** is a Microsoft Excel-generated prototype for a SoundStage summary report. Figure 16.19**B** is a graphic bar chart version of the same information. The prototype chart was created in less than one minute.

Prototyping Outputs with CASE Tools

Many CASE products support or include facilities for report and screen design and prototyping via a project repository created during systems analysis. For example, Pansophic's Telon helps analysts and system users create rapid prototypes and, once the prototypes are approved, converts those prototypes into COBOL or PL/1 code, including interfacing code for CICS (a teleprocessing monitor), IMS (a hierarchical database management system), or DB2 (a relational database management system).

FIGURE 16.19 Sample Prototypes Generated from a Spreadsheet
The summary report prototypes were generated using Microsoft's Excel. A facility was used to quickly convert the tabular report into the equivalent bar chart.

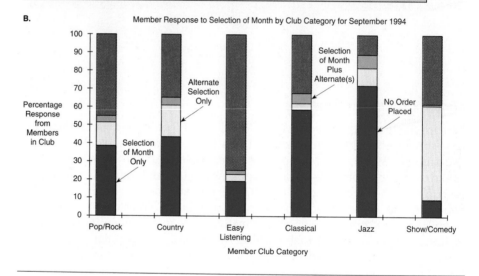

A.

Member Response to Selection of Month by Club Category
September, 1994

Category	Potential Orders	Selection of Month	Alternate Selection	Selection of Month + Alternates	No Order
Pop/Rock	6342	2410	824	241	2867
Country	3577	1538	644	154	1241
Easy Listening	954	181	38	18	716
Classical	1486	877	45	88	477
Jazz	540	389	54	39	58
Show/Comedy	104	9	54	1	40

LEGEND

Category: Club to which members belong
Potential Orders: Number of members in club who received Selection of the Month promotion
Selection of Month: Number of members who selected ONLY the Selection of the Month
Alternate Selections: Number of members who selected titles other than the Selection of the Month
Selection of Month + Alternates: Number of members who selected both Selection of Month plus alternates
No Order: Number of members who rejected Selection of Month and ordered no alternates

B.

Products such as Telon assume the system users have defined their requirements and data needs. Throughout this book, we've demonstrated the use of Excelerator as a CASE product to define requirements and data needs. Excelerator requirements can be directly interfaced to Telon. On the other hand, Excelerator also provides its own internal facility for prototyping outputs and screens.

A screen design in process is shown in Figure 16.20**A.** Normally, the prototype would use the entire screen. In this case, however, the analyst has told Excelerator that a new field is to be added to the screen. The bottom third of the screen has been replaced with a field definition form. The analyst knows this field has been previously defined in the project repository as an attribute. By typing that attribute's name in the Related ELE space, Excelerator will look up

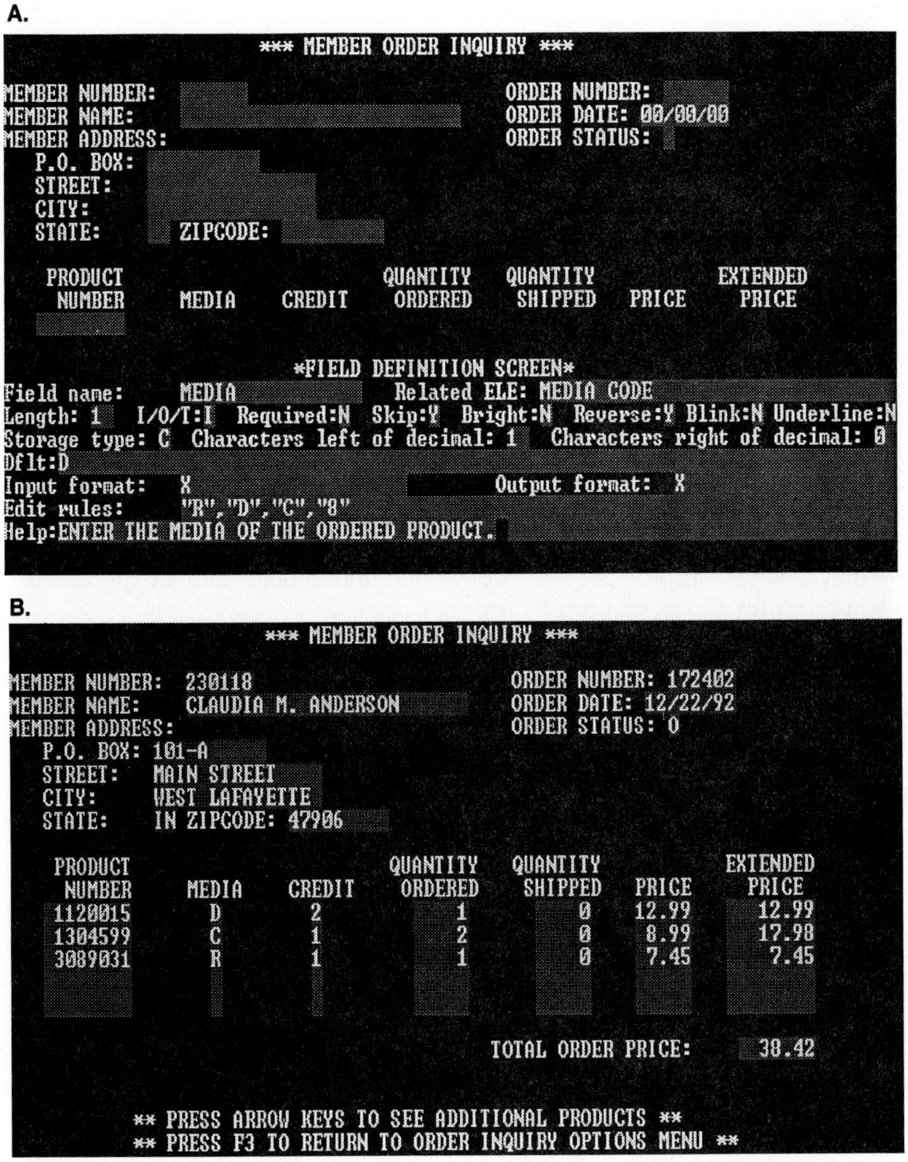

A.

```
                    *** MEMBER ORDER INQUIRY ***

MEMBER NUMBER:                        ORDER NUMBER:
MEMBER NAME:                          ORDER DATE: 00/00/00
MEMBER ADDRESS:                       ORDER STATUS:
  P.O. BOX:
  STREET:
  CITY:
  STATE:        ZIPCODE:


  PRODUCT                QUANTITY  QUANTITY           EXTENDED
  NUMBER    MEDIA  CREDIT  ORDERED   SHIPPED   PRICE   PRICE



                 *FIELD DEFINITION SCREEN*
Field name:    MEDIA            Related ELE: MEDIA CODE
Length: 1   I/O/T:I  Required:N  Skip:Y  Bright:N  Reverse:Y Blink:N Underline:N
Storage type: C  Characters left of decimal: 1   Characters right of decimal: 0
Dflt:D
Input format:  X                        Output format:  X
Edit rules:    "R","D","C","8"
Help:ENTER THE MEDIA OF THE ORDERED PRODUCT.
```

B.

```
                    *** MEMBER ORDER INQUIRY ***

MEMBER NUMBER:  230118               ORDER NUMBER: 172402
MEMBER NAME:    CLAUDIA M. ANDERSON  ORDER DATE: 12/22/92
MEMBER ADDRESS:                      ORDER STATUS: O
  P.O. BOX: 101-A
  STREET:   MAIN STREET
  CITY:     WEST LAFAYETTE
  STATE:    IN ZIPCODE: 47906

  PRODUCT                QUANTITY  QUANTITY           EXTENDED
  NUMBER    MEDIA  CREDIT  ORDERED   SHIPPED   PRICE   PRICE
  1120015     D      2        1        0       12.99   12.99
  1304599     C      1        2        0        8.99   17.98
  3089031     R      1        1        0        7.45    7.45



                              TOTAL ORDER PRICE:     38.42


      ** PRESS ARROW KEYS TO SEE ADDITIONAL PRODUCTS **
      ** PRESS F3 TO RETURN TO ORDER INQUIRY OPTIONS MENU **
```

FIGURE 16.20

Prototyped Output Screen from a CASE Tool

the attribute and automatically copy pertinent specifications into the field definition form. The analyst can change these specifications for this implementation of the attribute and can add new characteristics in the field definition (e.g., blinking, reverse video). The final prototype screen (with sample information) is depicted in Figure 16.20**B**.

After all fields have been placed on the screen, the analyst simply completes the prototype by entering values. Excelerator conveniently checks the values against the allowable value ranges defined for each field to ensure the data is reasonable. After showing the data to system users and gaining their approval, the screen can, if desired, be automatically converted to its equivalent COBOL picture.

Prototyping Outputs with Database Management Systems or Fourth-Generation Languages

Recall that most database management systems or fourth-generation languages include powerful applications generators for quickly prototyping fully functional systems. If a prototype file or database was created during file or database design (see Chapter 15), the test data stored in the file or database can be used to prototype outputs and screens. Most systems prototyping tools include report writers and query languages that allow analysts (or systems users) to quickly design and generate samples of outputs. This capability can be found in expensive mainframe databases such as Cullinet's IDMS, in relatively inexpensive microcomputer databases such as Ashton-Tate's dBASE IV, and in 4GLs such as Information Builder's FOCUS and PC/FOCUS.

Even though the SoundStage system will eventually be implemented in COBOL with a relational database management system, the analysts quickly built prototypes using an inexpensive microcomputer database management system. Thus, the prototypes were developed both quickly and inexpensively, two properties of a good prototyping environment.

Step 4: If Necessary, Convert User-Oriented Output Layouts into Programmer-Oriented Layouts

In some cases, the prototype layout is sufficient for programmers to implement the design. You would simply say, "Make it look like this." But some prototypes cannot be used as final specifications. The prototype may contain compromises that were imposed by the tools used to create the prototypes. Also, the sample information on the prototypes may not be complete—in other words, the prototype doesn't accurately communicate how to handle all data formats.

In these situations, or if standard procedures dictate it, the analyst may have to produce a more technical layout for the programmer. This requires the use of classic layout charts.

Using Printer Spacing Charts to Document the Format of Printed Outputs

Printed outputs are usually documented using a printer spacing chart.

> The **printer spacing chart** (see Figure 16.21) remains a popular tool to describe the format of printed outputs.

The largest and most obvious portion of a printer spacing chart is the printing grid.

> The **printing grid** on the form indicates the print positions by row and column.

Printer spacing charts can accomodate almost any size output you might want to design (however, the grid is not to scale—in other words, one inch on the grid does not necessarily equal one inch on the final output).

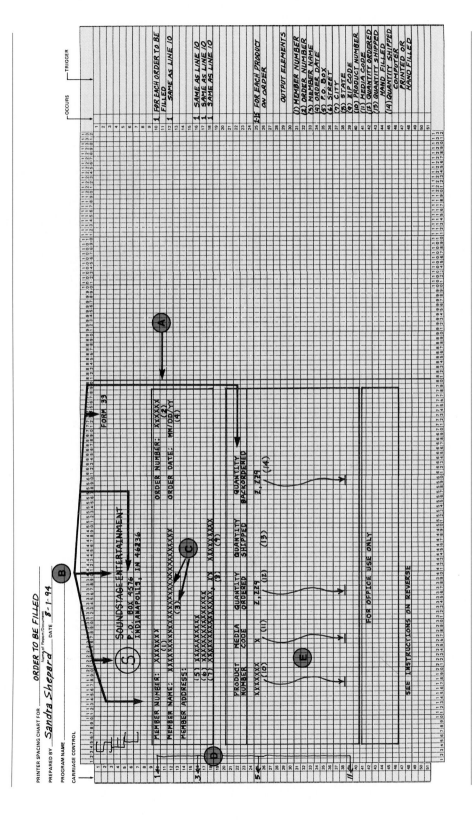

FIGURE 16.21 Printer Spacing Chart for ORDER TO BE FILLED   A printer spacing chart is used to design the layout of printed outputs.

TERMINAL SCREEN DISPLAY LAYOUT FORM

☐ INPUT
☑ OUTPUT

APPLICATION _CUSTOMER SERVICES_

SCREEN NO. _4_ SEQUENCE _4_

COLUMN

```
         1        11        21        31        41        51        61        71
123456789012345678901234567890123456789012345678901234567890123456789012345678901234567890
01
02                            *** MEMBER ORDER INQUIRY ***
03  MEMBER NUMBER: XXXXXXXX              ORDER NUMBER: XXXXXXX
04  MEMBER NAME: XXXXXXXXXX                 ORDER DATE: MM/DD/YY
05  CLUB MEMBERSHIP: XXXXXXXXXXXX    X      ORDER STATUS: XXXXXXXX
06  MEMBER ADDRESS:
07    P.O. BOX  XXXXXXXXXX
08    STREET:   XXXXXXXXXXXXXX
09    CITY:     XXXXXXXXXXXXXX
10    STATE: XX  ZIPCODE: XXXXXXXXX
11
12  PRODUCT          QUANTITY   QUANTITY             EXTENDED
13  NUMBER   MEDIA   ORDERED    SHIPPED    PRICE     PRICE
14  XXXXXXX    X     Z,ZZ9      Z,ZZ9      ZZZ.99    Z,ZZZ.99
15
16
17
18
19
20  TOTAL ORDER CREDITS:   Z9        TOTAL ORDER PRICE: ZZZ,ZZZ.99
21  UNUSED BONUS CREDITS:  Z9
22
23          *** PRESS ARROW KEYS TO SEE ADDITIONAL PRODUCTS ***
24          *** PRESS F3 TO RETURN TO ORDER INQUIRY OPTIONS MENU ***
```

1960

1920

ROW 01–24

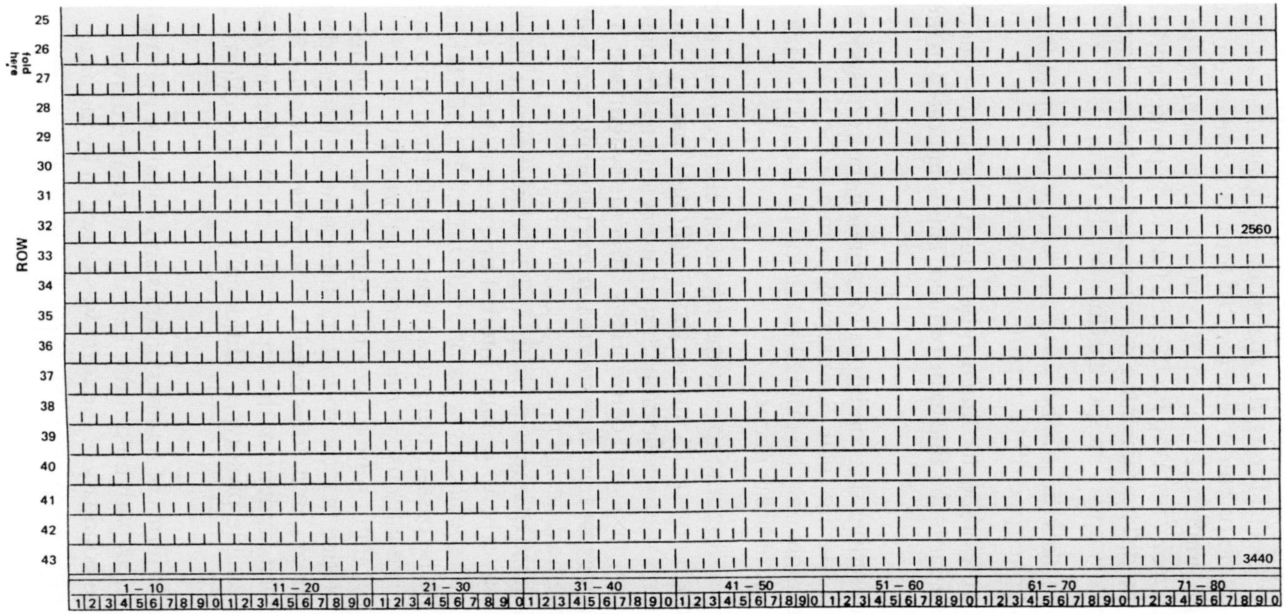

FUNCTION KEY ASSIGNMENTS

PF1		PF9		PF17	
PF2		PF10		PF18	
PF3	RETURN TO ORDER INQUIRY OPTIONS MENU	PF11		PF19	
PF4		PF12		PF20	
PF5		PF13		PF21	
PF6		PF14		PF22	
PF7		PF15		PF23	
PF8		PF16		PF24	

fold here

FIGURE 16.22 Display Layout Chart A display layout chart is used to design the layout of displayed outputs.

Carriage control, the column to the left of the printing grid, identifies the first and last print lines on a form. For most printers, carriage controls for standard paper sizes are stored in a small memory in the printer.

Programmers specify print locations and spacing by using the carriage controls within their programs. For special forms, the analyst must specify the carriage controls to be loaded into the printer whenever that form is mounted on the printer. Depending on the printer used, up to 12 controls can be defined for a form. The number 1 is normally reserved for the first line to be printed and 12 for the last line to be printed. The remaining numbers can be defined by the analyst to skip—sequentially only—to specific lines on the form.

The **occurs** column specifies the number of times that the corresponding line may be printed on one page of the report or document.

After defining how many times a specific line may be printed, the analyst needs to document what causes a particular line to be printed. This is indicated in the trigger column.

The **trigger** column describes the conditions for printing the corresponding lines on the printer spacing chart. For example, some of the lines recorded on the chart might only be printed at the top or bottom of each page or whenever a control break occurs.

The printing grid is used to formally specify the layout of both constant and variable information for the report or document. Caution! Printer spacing charts are not intended for systems user verification. System users tend to be somewhat put off by the editing symbols and conventions. After the system user has approved the format, finalize that format as a formal printer spacing chart with which to communicate the final design to the programmer!

Let's take a closer look at the example in Figure 16.21. Printer spacing charts can be used for virtually any type of printed internal or external output. Our sample is an external output, specifically, a preprinted ORDER TO BE FILLED form. Thus, this design will be passed along to a forms manufacturer who will create the form and duplicate it for use in the SoundStage printers.

If different copies of the output were to look different, we would have used multiple spacing charts. Notice that the format includes title, headings, data attributes, spacing, page breaks, and other details of importance to the computer programmer who must implement the report. This information may be difficult to derive from a simple prototype. Additionally, note the following details:

(A) The size of the form, in terms of number of lines and columns, was indicated on the spacing chart with bold lines. Perforations can be drawn with dashed lines. The boundary of the form was drawn in by the analyst. Remember, the printer spacing chart is *not* a scale drawing. The actual size should be recorded in the repository for this data flow.

(B) We drew and printed the constant information to be preprinted by the forms manufacturer. This includes title, company name, address, phone numbers, logos, form name and form number, unique identification number (if that is to be preprinted), lines and blocks that divide the document into sections, and column and field headings.

(C) We used our edit masks from the attribute repository to record fields to be printed by the computer. The number in parentheses refers back to the corresponding field in the attribute repository ("REF"). Don't record pictures in fields that will be entered by hand—that would confuse the programmer. In other words, only record PICTURE clauses for fields to be printed by the computer.

(D) Carriage controls were established because this is a nonstandard form size. The printer needs to have its top-of-page and bottom-of-page redefined when printing this form.

(E) There may be from 1 to 20 products on any one order. Notice how we used vertical lines with arrowheads to indicate that the line describing a single product may be printed repeatedly. The line spacing will be single.

The documentation prepared in this example is primarily intended for the analysts and programmers. The PICTURE clauses recorded on our printer spacing chart will likely be too difficult for system users to interpret and evaluate.

Using Display Layout Charts to Document the Format of Visual Outputs

Visual outputs are frequently documented on display layout charts. Let's take a closer look at the example in Figure 16.22. Keep in mind that in this chapter we are only interested in the net output. Chapter 17 will address the menus or terminal dialogues that result in this output.

Note that we divided the display area into zones. For your convenience, the zone borders are highlighted. Zones help the system user know where to look for certain things. One zone is set aside to display the screen title. Another zone is used to display the body of the output, which also contains a scrolling zone for multiple lines of output. A third window offers instructions to the user for paging forward and backward and for terminating the output program.

Summary

The goal of input design is to capture data and get that data into a format suitable for the computer. Input methods can be broadly classified as either batch or on-line. The systems analyst must be familiar with the advantages and disadvantages of each method as well as with the various media used to implement both methods. Outputs present information to system users. There are two basic types of computer outputs. External outputs leave the system to trigger actions on the part of their recipients or confirm actions to their recipients. Internal outputs stay inside the system to support the system's users and managers.

Because input and outputs are highly visible to the system user, analysts should consider a number of human factors when designing computer inputs and outputs. The volume of data to be input by the system user should be minimized, because every attribute input carries with it the risk of error. Source documents for capturing data should be designed for easy completion by system users and for rapid data entry by data entry clerks and CRT operators. Human factors for designing outputs include readability, timeliness, relevance, and acceptability. Choices of media, including paper, film, and video display, as well as the format for presenting information, such as tabular, zoned, graphic, and narrative, are also important considerations for designing outputs.

Internal controls are also essential for inputs. Internal controls should be established for monitoring the number of inputs and for ensuring that the data is valid. Internal control techniques for ensuring the validity of data include completeness checks, limit and range checks, combination checks, self-checking digits, and PICTURE clauses. Good output design involves addressing internal controls to ensure the reliability and distribution of outputs generated by the computer.

To design computer inputs, the systems analyst should begin by identifying the input requirements of the new system. Data flow diagrams identify data to be captured and input to the system. The project repository defines the basic content of the inputs to be designed. Next, the analyst specifies the design parameters—issues concerning data capture, data entry, and data input—for the input. Then the analyst designs or prototypes the source document. For on-line inputs, the analyst would next prototype the input screens. Finally, the analyst may have to lay out the format of screens or batch input files using either display layout charts or input record layout charts, respectively.

Output design requires four basic steps. First, review output requirements. Output requirements should have been defined during systems analysis. During systems analysis, outputs should also have been described as entries in the project repository. Second, decide how the output data flow will be implemented. The implementation decisions will consider the appropriate media and format as well as a number of design criteria. Third, prototype the layout for the system user(s). Rather than simply sketch the sample output, the systems analyst may draw on a wide number of prototyping tools including spreadsheets, CASE products, and database management systems or fourth-generation languages. Finally, if necessary, convert user-oriented layouts into programmer-oriented layouts. Printer spacing charts and display layout charts are still effective tools for communicating the format of printed and visual outputs.

Key Terms

bar code, p. 633

batch input, p. 632

carriage control, p. 664

combination checks, p. 639

completeness checks, p. 638

data capture, p. 629

data entry, p. 631

data input, p. 631

display layout charts, p. 645

external outputs, p. 629

format, p. 634

form filling, p. 647

function key assignments, p. 645

graphic output, p. 635

input record layout chart, p. 649

internal outputs, p. 629

key-to-disk (KTD), p. 632

key-to-tape (KTT), p. 632

limit and range checks, p. 638

medium, p. 634

microfiche, p. 635

microfilm, p. 635

narrative format, p. 635

occurs, p. 664

on-line input, p. 632

optical-character reader (OCR), p. 632

optical-mark reader (OMR), p. 632

picture checks, p. 639

printer spacing chart, p. 660

printing grid, p. 660

question-answer mode, p. 647

remote batch, p. 634

self checking digits, p. 639

source document, p. 630

tabular output, p. 635

transaction, p. 629

trigger, p. 664

turnaround outputs, p. 629

video, p. 635

zoned output, p. 635

Problems and Exercises

1. Explain the difference between data capture, data entry, and data input. Relate the three concepts to the processing of your school's course request or course registration.

2. To what extent should the system user be involved during input design? What

would you ask the system user to do? When? What would you do for the system user? When?

3. Define an appropriate input method and medium for each of the following inputs:
   a. Customer magazine subscriptions.
   b. Hotel reservations.
   c. Bank account transactions.
   d. Customer order cancellations.
   e. Customer order modifications.
   f. Employee weekly time cards.

4. What effects can be caused by lack of internal controls for inputs?

5. Obtain a copy of an application form — such as a loan, housing, or school form — or any other document used to capture data (e.g., a course scheduling form, credit card purchase slip, or time card). Do not be concerned whether the application is currently input to a computer system. How do the people who initiate or process the form feel about it? Comment on the human engineering. How well is it divided into zones? Comment on the suitability of the application for data entry. Are attributes that wouldn't be keyed properly located? What changes would you make to the form?

6. Design a source document for the MEMBER ORDER RESPONSE input that was referred to throughout this chapter. Remember, the MEMBER ORDER RESPONSE input uses a turnaround source document. Also, the source document is initially output and mailed to the member.

7. What implications does input design have on output design? Differentiate between internal and external outputs. Give several examples of each. Why is it important that a systems analyst recognize an output as either internal or external?

8. Identify four system-user issues that an analyst should address when designing outputs.

9. To what extent should system users be involved in output design? How would you get system users involved? What would you ask them to do for themselves?

10. What are the three most commonly used media for outputs? What are the advantages and disadvantages of each?

11. Obtain sample outputs of each of the following format types: zoned, tabular, graphic, and narrative. Was the format type of each output the most effective of the four alternatives? If not, why? What format type, or combination of format types, would you have chosen to implement the output? Sketch the layout you would have chosen.

12. Prepare an expanded data repository to describe the following outputs:
    a. Your driver's license.
    b. Your course schedule.
    c. Your bank statement.
    d. Your phone bill.
    e. Your W-2 statement (for taxes).
    f. A bank or credit card account statement and invoice.
    g. An external document printed on a computer.
    You may invent numbers for timing and volume. Don't forget internal controls.

13. Using a sample output from Exercise 5, document the layout of the output using a printer spacing chart. Be sure to make appropriate entries in the carriage control, occurs, and trigger columns for all computer-printed lines. List any improvements that could be made to improve the readability, interpretation, and acceptability of the output.

14. How would the expanded data repository in Exercise 12 differ if the external output were designed as an internal visual output? Use a display layout chart to describe the format of a visual version of the sample output.

15. Prototype an output from Exercise 12. If you don't have access to a spreadsheet, CASE product, or fourth-generation language, you may prototype the output using a word processor or simply sketch the output.

16. Explain why a systems analyst might choose to document the format of a printed report using a printer spacing chart after he or she completed a prototype of the report and verified it with the system user.

17. Identify three types of tools that can be used to prototype output layouts for system users.

Projects and Minicases

1. The sales manager for SoundStage Record Company has requested a daily report. This report should describe the nearly 1,000 customer order responses received for a given day. A response is a member decision on whether to accept the record-of-the-month selection, request an alternate selection, request both, or request that no selection be sent that month. The report is to be sequenced by MEMBERSHIP NUMBER and CATALOG NUMBER. The data repository for the report follows:

   The ORDER RESPONSE REPORT consists of the following attributes:
   DATE * of the report
   PAGE NUMBER
   1 to 1,000 of the following:
      MEMBERSHIP NUMBER * 5 digits
      MEMBER NAME * which consists of the following:
        MEMBER LAST NAME * 15 characters
        MEMBER FIRST NAME * 15 characters
        MEMBER MIDDLE INITIAL * 1 character
      MUSICAL PREFERENCE * possible values are
        "EASY LISTENING" "TEEN HITS"
        "CLASSICAL" "COUNTRY" "JAZZ"
      SELECTION OF MONTH DECISION * possible values are
      "YES" "NO" "NONE"
   1 to 15 of the following:
      CATALOG NUMBER * 5 digits
      MEDIA * possible values are
        "RECORD" "CASSETTE"
        "AUDIOPHILE" "8 TRACK" "REEL"
   NUMBER OF PURCHASE CREDITS NEEDED * 2 digits
   PERIOD AGREEMENT EXPIRES * date membership expires

   What type of output was being requested by the sales manager? Prepare an expanded data repository for the output. Prototype the requested output. Verify the output with your instructor (serving as the sales manager or system user). Once you've obtained the instructor's approval, use a printer spacing chart to lay out the format for the printed report. Be sure to include appropriate report headings, edit masks with reference numbers to the expanded data repository, occurs, and trigger entries.

2. The sales manager has also requested that the sales staff be able to obtain information concerning a particular customer's order response at any time during normal working hours. Prepare an expanded data repository for the output. Prototype the requested output. Verify the output with your instructor (serving as the sales manager or system user). Once you've obtained the instructor's approval, prepare a display layout chart for the visual output CUSTOMER ORDER RESPONSE.

3. The order-filling operation for a local pharmacy is to be automated. The pharmacy processes 50 to 200 prescriptions per day. Customer prescriptions are to be entered on-line by pharmacists. Prepare an expanded data repository and display

layout chart for documenting the design of the on-line input PRESCRIPTION. A working data repository for PRESCRIPTION follows:

A PRESCRIPTION contains the following attributes:
  CUSTOMER NAME — 20 characters
  DOCTOR NAME — 20 characters
  1 to 10 occurrences of the following:
    DRUG NAME — 30 characters
    QUANTITY PRESCRIBED — 4 digits
    MEDICAL INSTRUCTIONS — 120 characters
    RX NUMBER * a federal licensing number — 6 digits
  1 to 10 occurrences of the following * added by pharmacist
  DRUG NUMBER * a number that uniquely identifies a prescription drug — 6 characters
  LOT NUMBER * a number that uniquely identifies the lot from which a chemical was
  produced — 6 characters
  DOSAGE FORM * the form of the medication issued, such as "pill." P = PILL
  C = CAPSULE   L = LIQUID   I = INJECTION   R = LOTION
  UNIT OF MEASURE * G = GRAMS   O = OUNCES   M = MILLILITERS
  QUANTITY DISPENSED — 4 digits
  NUMBER OF REFILLS — 2 digits
  and optionally:
    EXPIRATION DATE * Date prescription expires

4. A moving company maintains data concerning fuel-tax liability for its fleet of trucks. When truck drivers return from a trip, they submit a journal describing mileage, fuel purchases, and fuel consumption for each state traveled through. This data is to be batch input daily to maintain records on trucks and fuel stations. The TRIP JOURNAL data repository follows:

A TRIP JOURNAL consists of the following attributes:
  TRUCK NUMBER — 4 characters
  DRIVER NUMBER — 9 characters
  CODRIVER NUMBER — 9 characters
  TRIP NUMBER — 3 characters
  DATE DEPARTED
  DATE RETURNED
  1 to 20 of the following:
    STATE CODE — 2 characters
    MILES DRIVEN — 5 digits
    FUEL RECEIPT NUMBER — 9 characters
    GALLONS PURCHASED — 3 digits (1 decimal)
    TAXES PAID — 4 digits (2 decimals)
    STATION NAME — 10 characters
    STATION LOCATION — 15 characters

Design the batch input TRIP JOURNAL. Be sure to design an appropriate source document. Fuel receipts are to be stapled to the source document.

Suggested Readings

Fitzgerald, Jerry. *Internal Controls for Computerized Information Systems.* Redwood City, Calif.: Jerry Fitzgerald & Associates, 1978. This is our reference standard on the subject of designing internal controls into systems. Fitzgerald advocates a unique and powerful matrix tool for designing controls. This book goes far beyond any introductory systems textbook — it is *must* reading.

Kozar, Kenneth. *Humanized Information Systems Analysis and Design.* New York: McGraw-Hill, 1989. A good user-oriented treatment of input and output design.

17

User Interface Design

Chapter Preview and Objectives

In this chapter you will learn how to design the user interface that presents the outputs and inputs that were designed in the last chapter. Today, there are two commonly encountered interfaces: terminals (or microcomputers behaving as terminals) used in conjunction with mainframes and display monitors connected to microcomputers. You will know that you've mastered user interface design when you can:

Determine
which features on available terminal and microcomputer displays can be used for effective user interface design.

Identify
the backgrounds and problems encountered by different types of terminal and microcomputer users.

Design and evaluate
the human engineering in a user interface for a typical information system.

Apply
appropriate user interface strategies to an information system.

Use
a state transition diagram to plan and coordinate a user interface for an information system.

Describe
how prototyping can be used to design a user interface.

Use
display layout charts (for net inputs and outputs) to format the user interface screens in a system.

RICHARDS & SONS, INC.

Scene: *Richards & Sons, Inc. is a large investment company located in Tampa, Florida, that buys stocks, bonds, commodities, and various other assets for their clients. They also manage their clients' investment portfolios for a variety of investment objectives. Finding new people with money to invest is crucial to their success, so keeping track of their clients and potential clients is very important. Morgan Adamson is the senior analyst in charge of the new sales prospect and contact management system at Richards & Sons, Inc. The new system has just been installed and the project team is working with the system users who are doing acceptance testing. Morgan is talking to Kevin Brock, the junior analyst who was responsible for the design of the system's user interface.*

Morgan: I guess you've heard that some of the system users are not very happy with the new contact tracking system. In particular, they are expressing dissatisfaction with the user interface that you designed.

Kevin: I don't understand what the problem could be—I put a lot of time and effort into that design. What are the specific problems?

Morgan: Some of the users are complaining that they don't know what to do next or how to use some of the screens. Are you sure that all of the screens are consistent with the screen template for this system?

Kevin: I didn't think it was important where the information was as long as I clearly labeled it. Besides, the users should be expected to read the screen. I deliberately put lots of highlighting, blinking, and reverse video fields on the screens to draw attention to important information.

Morgan: Yes, I know. Some of the users claim that there is so much highlighted information that it distracts from the purpose of the screen. I've

also had complaints that proper default values were not specified for some of the fields.

Kevin: Default values were specified for the most common fields, but the users should expect to have to type some of the information in— that's their job, isn't it?

Morgan: In some cases, maybe several possible default values should have been provided in a pop-up window to eliminate possible keying errors. Remember, we want to reduce the amount of user entry keystrokes as much as possible. Some of the clerks say that they don't understand some of the terminology and abbreviations on some of the screens. For instance, what does "HIT FUNCTION KEY 5" mean?

Kevin: Oops! I must not have checked the screens carefully enough to catch some of the computer terminology. I'll fix that right away. Are there any other problems?

Morgan: Other users have indicated that the use of function keys is not consistent across all of the screens. For example, this memo states that PF3 saves a new client record on the add new client screen, but the same PF3 key deletes a contact record on the phone log entry screen. Didn't you use the same function keys for the same actions throughout the entire interface?

Kevin: No, I didn't. I thought that I could reuse the function keys to mean different things as long as I clearly labeled them on each screen.

Morgan: Why?

Kevin: Some of the keyboards have only 12 function keys, and I was afraid I might run out of keys to assign unless I reused them.

Morgan: You need to realize that the system users don't always read the instructions that you provide. Whether that is right or wrong is not important; you need to be consistent so that the users don't have to learn a different set of function keys for each screen.

Kevin: I guess I really didn't think that through very well —it shouldn't be too difficult to make all the function key assignments consistent.

Morgan: There have also been some complaints about an insufficient amount of help messages for some of the input screens. For instance, one clerk said the contact entry screen consistently refused to accept the date of contact he tried to enter. The system did output an error message indicating that the date was incorrect and should be reentered, but the clerk doesn't understand why the date was invalid. He says that the system doesn't provide any information about the correct format of the date or any valid examples. You did design help screens and messages for each of the input screens, didn't you?

Kevin: Uh, well, I'm not exactly sure what you mean by a "help" screen. I did very thorough input error checking so that invalid data could not be entered. I created error messages for each of the edited fields. I thought that would be sufficient for the users to identify the input error and make the necessary correction.

Morgan: I think I'm beginning to understand the problem. We need to talk about some very important human engineering guidelines that you need to follow whenever you are designing a user interface. First . . .
(*we will leave their meeting now*)

Discussion Questions

1. What did Kevin do wrong in designing the user interface? What are some of the other mistakes that an analyst might make when designing a user interface?

2. How could these mistakes have been avoided? (What would you have done differently?) What role should the system user play in interface design?

WHEN PEOPLE TALK TO COMPUTERS

User interface design is the specification of a conversation between the system user and the computer. This conversation generally results in either input or output—possibly both. What makes a user interface good? Does the available technology limit or enhance dialogue possibilities? How can, or should, a user interface be organized?

We know you recognize the difference between a terminal and a microcomputer. But we suspect that many of you are not familiar with the various features offered by their displays (also called monitors). Because display features affect user interface design, we'll begin with an overview of features. Then we can examine some of the fundamental human factors and design strategies that underlie user interface design.

Display Terminals and Monitors: Features that Affect User Interface Design

The design of a user interface can be enhanced or restricted by the available features of your terminal display or monitor/keyboard. Let's examine some of these features.

Display Area

The size of the **display area** is critical to user interface design. For terminal displays, the two most common display areas are 25 (lines) by 80 (columns) and 25 by 132. Some displays can be easily shifted between these two sizes. Some newer displays are designed to show more lines, for example, 65 lines on one screen. Some terminals can show four complete 25 by 80 character screens simultaneously.

For microcomputer and workstation display monitors, display size is measured in *pixels.* The greater the number of pixels, the more information can be displayed. Pixel display areas are specified in width by height. Typing sizes range from 640 × 480 to 1,070 × 1,070 (called megapixel).

Character Sets and Graphics

Every display uses a predefined **character set.** Most displays use the common ASCII character set. Some displays allow the programmer to supplement or replace the predefined character set. Additionally, many displays offer graphics capabilities. Graphics capabilities must be supported by graphics controllers and software that allow the programmer to take advantage of the graphics capabilities. Graphics-based displays may support a virtually unlimited character set.

Paging and Scrolling

The manner in which the display area is shown to the user is controlled by both the technical capabilities of the display and the software capabilities of the computer system. Paging and scrolling are the two most common approaches to showing the display area to the user.

> **Paging** displays a complete screen of characters at a time. The complete display area is known as a page (also called a screen or frame). The page is replaced on demand by the next or previous page; much like turning the pages of a book.

The other common alternative to paging is called scrolling.

> **Scrolling** moves the displayed characters up or down, one line at a time. This is similar to the way movie and television credits scroll up the screen at the end of a movie.

We'll discuss the choice of paging or scrolling for a dialogue when we discuss human factors. Once again, PC displays offer a wider range of paging and scrolling options.

Color Displays and Display Properties

Greater numbers of displays are capable of using color. Color can be used to highlight specific messages, data, or areas of the screen. Most displays also permit a variety of display properties.

> **Display properties** are characteristics that change the way in which a character or group of characters is displayed on a screen.

Some of these display properties include:

- Double brightness on selected fields or messages.
- Blinking for selected fields or messages.
- Nondisplay for selected fields (for example, passwords).
- Reverse video for selected fields that permits the color of the background —such as black—and the color of selected fields and messages—such as green—to be reversed.

Each of these features, when available, is activated by predefined codes that the programmer must learn and apply. Once again, PC displays and software offer a much wider array of display properties, including color, which can simplify and improve user interfaces.

Split-Screen and Windowing Capabilities

Split-screen capability is a variation on the windows concept. The display screen, under software control, can be divided into different areas (called **windows**). Each window can act independently of the other windows, using features such as paging, scrolling, display attributes, and color. Each window can be defined to serve a different purpose. Windows can be resized, moved, and hidden or recalled on user demand.

Windowing is rapidly becoming accepted as a standard user interface. Most microcomputer products use windowing interfaces such as Microsoft Windows and IBM's Presentation Manager, OSF's Motif, and Apple's Finder.

It is appropriate to talk about some standards and conventions that are being developed that apply to the user interface. IBM is leading a movement that could establish a standard for all IBM and IBM-compatible applications — mainframe, minicomputer, and microcomputer. This standard is called Common User Access or CUA, which was pioneered in IBM's OS/2 operating system using a windowing interface called *Presentation Manager.*

Common user access (CUA) is a set of rules (or standards) that define the ways in which a system user interacts with a computer application. The goal of CUA is to provide the system user with familiar computer interface across many different computer applications. CUA guidelines include rules governing the use of a keyboard and mouse, selection of menu choices, accessing help information, and a set of aids for moving between applications. Hopefully, providing a standard "look and feel" to all CUA-compliant software packages will result in less user training time, increased productivity, and a reduction in errors and confusion caused by the computer system's interface to the system's users.

Today's most common implementation of CUA methods is the use of a graphical user interface.

A **graphical user interface (GUI)** uses **icons** (or graphic symbols), **pop-up windows, scroll bars,** and **pull-down menus** to communicate with users. Figure 17.1 is a typical example of a CUA compliant graphical user interface from Microsoft Windows.

Graphical user interfaces are very popular because they are so intuitive and easy to use. They also reduce input errors and cut down on the amount of time it takes for the system user to key in information by providing **radio buttons** and **check boxes** (see Figure 17.2). Both radio buttons and check boxes are areas in a pop-up window (also known as a **dialogue box**) that allow the user to specify what is being requested. The main difference is that a radio button is a choice of one and only one whereas a check box allows the user to specify that particular option in any combination with other options. For example, in Figure 17.2, the radio buttons indicate that the user may request that the system print one and only one of the following: all (the entire document), the current page, or a range of pages. However, the checkboxes for print to file and collate copies indicate that the user may request either, neither, or both. GUI's may soon replace the older, more traditional character-based interface methods.

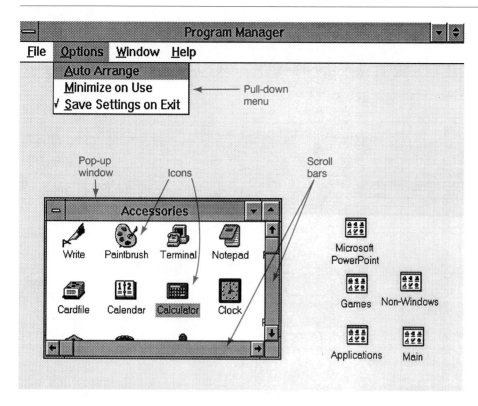

FIGURE 17.1 **A Typical Example of a Graphical User Interface from Microsoft Windows** Note the pull-down menu, the pop-up window, the scroll bars, and the icons.

Keyboards and Function Keys

Although not a display feature, most modern terminals and monitors are integrated with keyboards. The number of keys and their layout may vary, but most keyboards contain special keys called function keys.

> **Function keys** (usually labeled F1, F2, and so on) can be used to implement certain common, repetitive operations in a user interface (for example, START, HELP, PAGE UP, PAGE DOWN, EXIT). These keys can be programmed to perform common functions.

We'll discuss some of the more common uses of function keys when we discuss human factors later in this chapter. Part of the SAA standard, mentioned in the last subsection, suggests that function keys be used consistently (for example, CUA requires F1 to be the Help key). In any case, a system's programs should consistently use the same function keys for the same purposes.

Pointer Options

We are no longer restricted to the keyboard as the only input technology for displays and terminals. Today, we are encountering many other selection options, such as touch-sensitive screens, voice recognition, and pointers. The most common pointer is the *mouse*.

> A **mouse** is a small hand-sized device that sits on a flat surface near the terminal. It has a small roller ball on the underside. As you move the

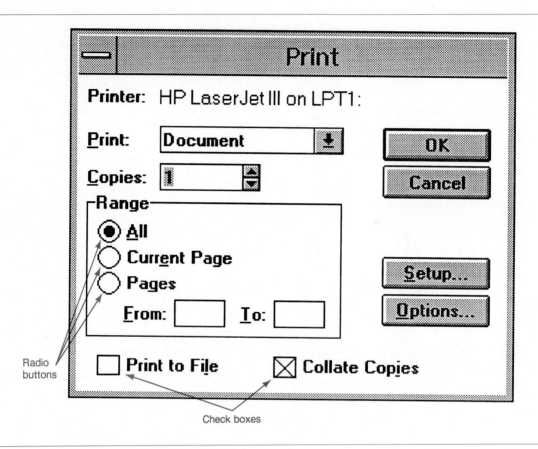

FIGURE 17.2 Radio Buttons and Check Boxes in a Graphical User Interface   Radio buttons and check boxes are a standard part of a CUA compliant graphical user interface (GUI). Also notice the arrow indicating a pull-down menu and the macro buttons with assigned actions.

mouse on the flat surface, it causes the pointer to move across the screen. Buttons on the mouse allow you to select objects or commands to which the cursor has been moved. Alternatives include *trackballs, pens,* and *trackpoints.*

Human Factors for User Interface Design

Nowhere are human factors as important as they are in user interface design. Just ask the typical systems analyst who spends half the day answering phone calls from system users who are having difficulty using the computer system. That's why we want to discuss the subject of **human engineering**.

System users can be broadly classified as either dedicated or casual.

A **dedicated system user** is one who will spend considerable time using specific programs. This user is likely to become more comfortable and familiar with the terminal or PC's operation.

However, some system users are only casual users.

The **casual system user** may only use a specific program on an occasional basis. This user may never become truly comfortable with the terminal or the program.

The user who hasn't used a terminal or a microcomputer is becoming less common in this computer-literate age. It is difficult to imagine today's youth being uncomfortable with the computer or display terminal. Still, most of today's systems are designed for the casual system user, with an emphasis on user friendliness.

General Human Engineering Guidelines

Given the type of user for a system, there are a number of important human engineering factors that should be incorporated into the design:

- *The system user should always be aware of what to do next.* The system should always provide instructions on how to proceed, back up, exit, and the like. There are several situations that require some type of feedback (adapted from Kendall and Kendall, 1988):
  - *Tell the user what the system expects right now.* This can take the form of a simple message such as READY, ENTER COMMAND, ENTER CHOICE, or ENTER DATA.
  - *Tell the user that data has been entered correctly.* This can be as simple as moving the cursor to the next field in a form or displaying a message such as INPUT OK.
  - *Tell the user that data has not been entered correctly.* Short, simple messages about the correct format are preferred. Help functions can supplement these messages with more extensive instructions.
  - *Explain to the user the reason for a delay in processing.* Some actions require several seconds or minutes to complete. Examples include sorting, indexing, printing, and updating. Simple messages such as SORTING — PLEASE STAND BY or INDEXING — THIS MAY TAKE A FEW MINUTES. PLEASE WAIT tell the user that the system has not failed.
  - *Tell the user that a task was completed or was not completed.* This is especially important in the case of delayed processing, but it is also important in other situations. A message such as PRINTING COMPLETE or PRINTER NOT READY — PLEASE CHECK AND TRY AGAIN will suffice.
- *The screen should always be formatted so that the various types of information, instructions, and messages always appear in the same general display area.* This way, the system user always knows approximately where to look for specific information. To achieve this goal, we suggest that zones or areas be defined as indicated in Figure 17.3. A sample screen template for planned zones is illustrated in Figure 17.4. This is only one possible layout. Another might group key assignments, messages, help, and the like at the bottom of the screen. The zones concept can be implemented easily with screen formatting software that is generally available for most computers. Even without such software, zones can be defined and used with conventional programming techniques.

Note In a windowing environment, the window standards (e.g., CUA, Windows, Motif, etc.) provide their own explicit standards or guidelines for "zones."

- *Within the body zone, the dialogue should be limited to one idea per frame, whether paging or scrolling through the zone.* For instance, the zone should display one menu, one input, one report, or one query response.

  The choice between paging and scrolling for the body zone depends on the information content to be displayed in that zone. If the informa-

FIGURE 17.3 **Screen
Zones** Screens are eas-
ier to read if the similar
data, information, and
messages are consistently
presented in the same
areas or zones on the
screens. If done consist-
ently across an application
(or better still, across all
or many applications),
users will more quickly
learn or use the applica-
tions. In windowing
environments such as
Windows, Presentation
Manager, and Motif, zones
and their proper usage are
defined in well-docu-
mented standards and/or
guidelines—although an-
alysts and programmers
must assume responsibil-
ity for adhering to those
standards and guidelines.

Suggested Zone Definitions

Title zone	The title zone identifies the application or screen from the user's point of view. Optionally, it may identify the file or record currently in use. In a nonwindowing environment, this is almost always at the top of a screen. In a windowing environment, the title zone is usually a named bar at the top of the window.
Menu zone	The menu zone identifies the area of screen where menu choices may be found. In a nonwindows environment, this may be at the top or bottom of the screen, or in an area shared by the body zone (see below). In a windowing environment, the top-level menu is in a bar immediately underneath the title zone (bar). Submenus may *drop-* or *pull-down* when a top-level option is selected.
Tool zone	More commonly used in windowing environments, a tool zone or *tool bar* contains pictures or *icons* that provide a faster way to execute sequences of menu commands. Tool bars may be located under the menu zone, or down the left or right side of the screen. Some tool bars can be designed to be ''dragged'' to any part of the screen as preferred by the user.
Message/status zone	The message/status zone is used to either display messages (e.g., instructions, suggestions, or errors) or display status (e.g., page numbers, status of keys that toggle on and off, etc.). In both non-windowing and windowing environments, this zone usually appears at or near the bottom of the screen, although windowing also use pop-up zones (see below).
Flag zone	Flag zones are almost exclusively encountered in nonwindowing environments. They are intended solely to identify or point to a specific location in the body zone (see below) where there is, or may be, a problem. The specific problem would be communicated in the message/status zone (see above).
Body zone	The body zone is usually the largest zone in either the screen or the window. It is where the user inputs business data or receives business information. The body zone may share space with the menu zone, especially in nonwindowing environments.
Pop-up zone	The pop-up zone is usually invisible. It is a box that suddenly appears (pops up) either in response to a menu selection (e.g., to enter data or select from additional menu choices), a request for extended help, or an error event (e.g., to describe the error or problem and its resolution). In windowing environments, pop-up zones are frequently called *dialog boxes* (e.g., Microsoft Windows) or *notebooks* (e.g., IBM Presentation Manager).

---

tion to be displayed is continuous in nature, like most reports, scrolling
can be used. The cursor can be moved up and down such a listing line by
line. If the information to be displayed is to be viewed one record at a
time or depicted as a form, paging is preferred.

- *Messages, instructions, or information should remain in the zone long
  enough to allow the system user to read them.* For instance, data should
  not be allowed to scroll out of a zone before it can be read. One way to
  accomplish this is to print only as much information as the zone can display at one time and then freeze the screen. A message to press any key
  or some specific key to continue can be displayed in the instruction

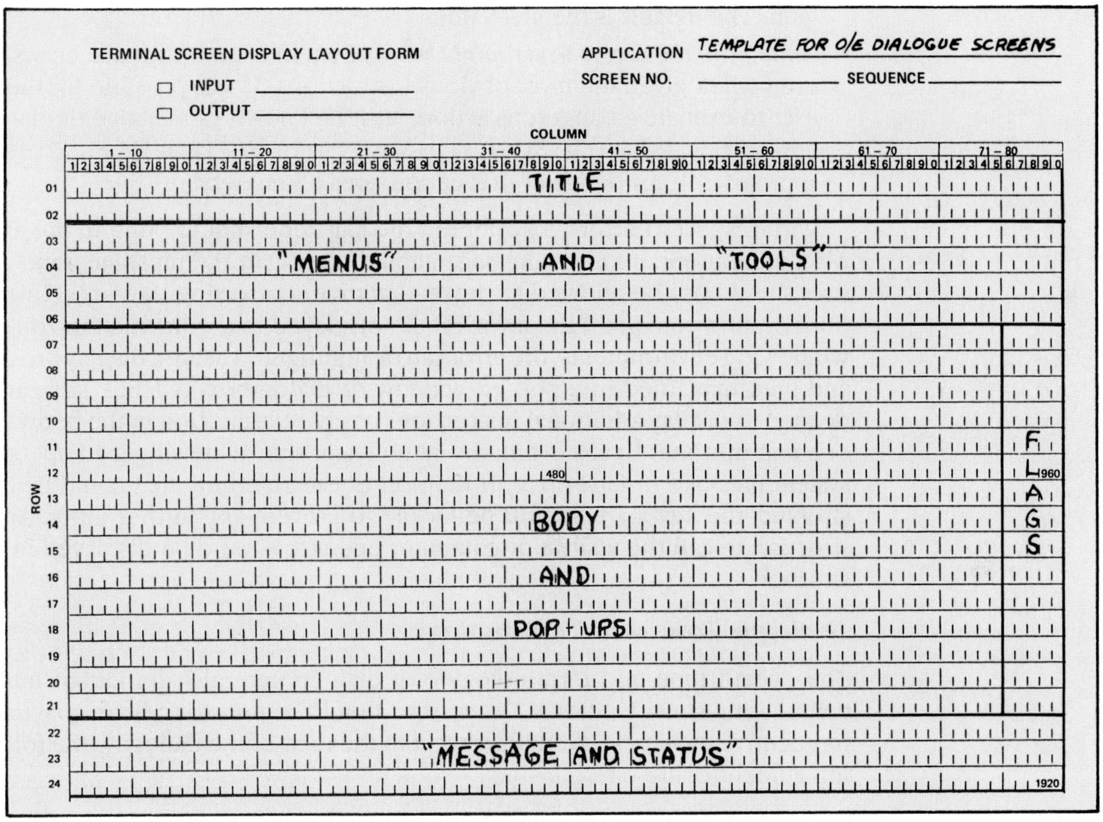

TERMINAL SCREEN DISPLAY LAYOUT FORM

☐ INPUT _____

☐ OUTPUT _____

APPLICATION *TEMPLATE FOR O/E DIALOGUE SCREENS*

SCREEN NO. _____ SEQUENCE _____

FIGURE 17.4 Sample Zones for a Screen This is one possible zone design for a screen. Other zone designs are possible, but once a design is created, it should be followed consistently.

zone. The system can then page or scroll through the next set of information.

Alternatively, to economize on space, some zones may be eliminated from the basic screen and replaced with "pop-up" windows that temporarily overlay the main screen to provide instructions, help, or messages.

- *Use display attributes sparingly.* Display attributes, such as blinking, highlighting, and reverse video, can be distracting if overused. Judicious use allows you to call attention to something important—for example, the next field to be entered, a message, or an instruction.

- *Simplify complex functions and reduce typing by providing the system user with function keys.* Some of the functions most commonly defined for function keys are START, HELP, EXIT or TERMINATE, ESCAPE, PRINT, SAVE or ADD, DELETE, cursor movement, and application keystroke combinations.

- *Default values for fields and answers to be entered by the user should be specified.* A common practice is to place the default value in brackets (for example, ORDER DATE? [Today's Date]. The user can press the enter key to get the default date. In windowing environments, valid values are frequently presented in a separate window or dialog box as a scrollable region. The default is the first value.

- *Anticipate the errors users might make.* System users will make errors, even when given the most obvious instructions. If it is possible for the user to execute a dangerous action, let it be known (a message or dialog box such as "ARE YOU SURE YOU WANT TO DELETE THIS FILE?" is appropriate). An ounce of prevention goes a long way!

With respect to errors, a symbol in the flag zone should point to the error. Also, an appropriate error message should appear in the message zone. The user should not be allowed to proceed without correcting the error. Instructions on how to correct the error can be displayed in the instruction zone. In windowing environments, the error can be highlighted using a display property and then explained in a pop-up window or dialog box. A HELP key can be defined to display additional instructions or give clarification in the body zone or a pop-up zone. In any event, the system user should never get an operating system message or fatal error. If the user does something that could be catastrophic, the keyboard should be locked to prevent any further input. An instruction to call the analyst or computer operator should be displayed in this situation.

Dialogue Tone and Terminology

The overall tone and terminology of a dialogue are also important human engineering considerations. The session should be user friendly (a goal that is frequently not achieved). With respect to the tone of the dialogue, the following guidelines are offered:

- *Use simple, grammatically correct sentences.* It is best to use conversational English rather than formal, written English.
- *Don't be funny or cute!* When someone has to use the system 50 times a day, the intended humor quickly wears off.
- *Don't be condescending; that is, don't insult the intelligence of the system user.* For instance, don't offer rewards or punishment.

With respect to the terminology used in the dialogue, the following suggestions may prove helpful:

1. *Don't use computer jargon.*
2. *Avoid most abbreviations.* Abbreviations assume that the user understands how to translate them. Check first!
3. *Use simple terms.* Use NOT CORRECT instead of INCORRECT. There is less chance of misreading or misinterpretation.
4. *Be consistent in your use of terminology.* For instance, don't use EDIT and MODIFY to mean the same action.
5. *Instructions should be carefully phrased, and appropriate action verbs should be used.* The following recommendations should prove helpful:
   · Try SELECT instead of PICK when referring to a list of options. Be sure to indicate whether the user can select more than one option from the list of available options.
   · Use TYPE, not ENTER, to request the user to input specific data or instructions.

- Use PRESS, not HIT or DEPRESS, to refer to keyboard actions. Whenever possible, refer to keys by the symbols or identifiers that are actually printed on the keys. For instance, the ↵ is used on some terminals to designate the RETURN or ENTER key.
- When referring to the cursor, use the term POSITION THE CURSOR, not POINT THE CURSOR.

User Interface Strategies

Are there any specific strategies you can employ to design a better user interface? Indeed, there are a number of such strategies, and the choice of strategy depends on the functions to be performed and the characteristics of the system user. Let's briefly survey these strategies.

Menu Selection

The most popular dialogue strategy is menu selection.

> The **menu selection** strategy of dialogue design presents a list of alternatives or options to the user. The system user selects the desired alternative or option by keying in the number or letter that is associated with that option.

More sophisticated technology allows menu selection by touching the screen, or selecting menu options with a pen, mouse, cursor keys, or other pointing devices.

A classic hierarchical menu dialogue is illustrated in Figure 17.5. If there are so many menu alternatives that the menu screen is too small or becomes cluttered, menus can be designed hierarchically. Small lists of related menu options can be grouped together into a single menu. These menus can then be grouped into a higher-level menu. This approach was applied in Figure 17.5. If the option DISPLAY WARRANTY REPORTS is selected, the submenu WARRANTY SYSTEM REPORT MENU will appear. Then, if the PART WARRANTY SUMMARY option is selected, the next-to-the bottom screen shown in Figure 17.5 will appear. Specific reports can be selected from that screen. There is no technical limit to how deeply hierarchical menus can be nested. However, the deeper the nesting, the more you should consider providing direct paths to deeply rooted menus for the experienced system user who may find navigating through multiple levels annoying.

Alternative approaches include the previously mentioned pull-down menu (Figure 17.6**A**) and **pop-up menu** (Figure 17.6**B**). Pull-down menus are submenus that pull down from a main menu option (like a pull-down window blind). Windowing applications (like Aldus Page Maker) tend to use pull-down menus. Pop-up menus are activated by function keys or combinations of keys pressed simultaneously. When the keys are pressed, the menu pops up, temporarily overlaying whatever resides on the screen. Completion of a menu command or cancelation of the menu—for example, by pressing the ESCape key—will return the user to the original screen. Borland's Sidekick is an example of an application that uses many pop-up menus.

Menu-driven systems are particularly popular with the casual or semicasual system user who doesn't use a particular program on a regular basis. Menu-driven systems also place production processing under the control of the user. It should be mentioned that by placing control in the hands of the system users,

FIGURE 17.5 **Classic Hierarchical Menu Dialogue** Classic menu selection is the most common dialogue strategy in use today. This strategy is most effective with a casual system user. This dialogue demonstrates a hierarchical menu structure.

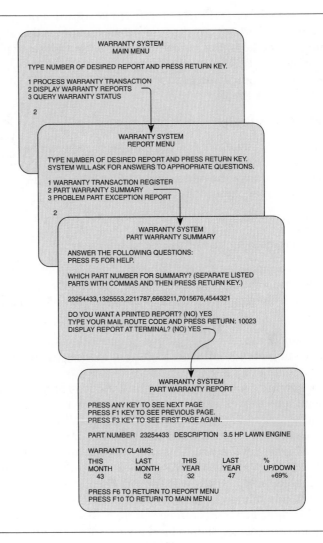

production efficiency may deteriorate. Menu items should be self-explanatory and should contain neither jargon nor vague abbreviations or statements.

Eventually, the menus will get you to a basic output or input operation or screen, which you learned how to design in Chapter 16.

Instruction Sets

Instead of menus—or in addition to menus—you can design a dialogue around an **instruction set** (also called a *command language interface*). Because the user must learn the syntax of the instruction set this approach is suitable only for dedicated terminal users. There are three types of syntax that can be defined. Determining which type should be used depends on the available technology.

1. A form of **Structured English** can be defined as a set of commands that control the system. In this type of dialogue, an elaborate HELP system

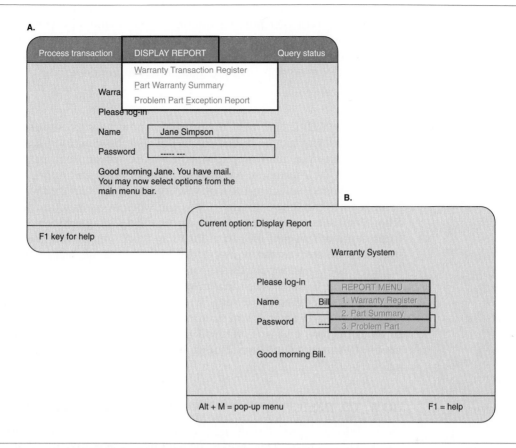

FIGURE 17.6 **Alternative Pull-Down and Pop-Up Menu Dialogue** Compare these menu screens to the ones presented in the previous figure. **A** Pull-down menus are submenus that pull down from a main menu option. **B.** Pop-up menus may be activated by function keys and temporarily overlay whatever is located on the screen.

should be created so the user who forgets the syntax can get assistance quickly.

2. A **mnemonic syntax** can be defined. A mnemonic syntax is built around meaningful abbreviations for all commands. Once again, a HELP facility is highly recommended.

3. **Natural language syntax** interpreters are now becoming available. When employing natural language syntax, the system user enters commands using natural English (either conversational or formal, written English).The system interprets these commands against a known syntax and requests clarification if it doesn't understand what the user wants. As new interpretations become known, the system learns the system user's vocabulary by saving it for future reference.

Question-Answer Dialogues

To supplement either menu driven or syntax-driven dialogues, you can use a **question-answer dialogue strategy** whenever appropriate. The simplest questions involve yes or no answers—for instance, "Do you want to see all records? [NO]." Notice how we offered a default answer! Questions can be more elaborate. For example, the system could ask, "Which part number are you interested in? [last part number queried]." Just make sure you consider all possible correct answers and deal with the actions to be taken if incorrect answers are entered. Question-answer dialogue is difficult because you must try to consider everything that the system user might do wrong!

Graphics

Graphics capabilities are not only providing new ways to output information (for example, graphs, charts, and so on) but also allowing friendlier user interfaces. For example, many applications now use **icons,** small graphic images that suggest their function. A trash can icon, for instance, might symbolize a delete command. Selecting the icon with a pointing device like a mouse or light pen executes the function. Also, icons can work in conjunction with one another. For instance, a pointing device can be used to drag the icon of a file folder (representing a named file) to a trash can icon—intuitively instructing the system to delete (or throw away) the file. The Apple MacIntosh interface (called Finder) has popularized the use of icons. IBM's Presentation Manager and Microsoft Windows are the equivalent in the IBM/clone marketplace.

The four strategies we've suggested should be integrated with the human factors discussed earlier. If you evaluate your dialogue against these fundamental concepts, you may save yourself from that dreaded 2:00 A.M. phone call: "Betty? Did I wake you? Sorry! But we have a problem. The system is asking us for. . . ."

HOW TO DESIGN AND PROTOTYPE A USER INTERFACE

The traditional approach to designing a user interface is to throw together a few display layout charts. This strategy doesn't work too well—the final dialogue ends up being designed by the programmer on the fly. It shouldn't surprise you that a typical user interface may involve many possible screens, perhaps hundreds! Each screen can be laid out with a display layout chart. But what about the coordination of these screens?

Screens will occur in a specific order. Perhaps you can move forward and backward through the screens. Additionally, some screens may appear only under certain conditions. And to make matters more difficult, some screens may occur repetitively until some condition is fulfilled. This sounds almost like a programming problem, doesn't it? We need a tool to coordinate the screens that can occur in a user interface.

Step 1: Chart the Dialogue

A **state transition diagram** is a screen-sequencing variation on program flowcharts and hierarchy charts. A sample state transition diagram for documenting dialogue is illustrated in Figure 17.7. The arrows indicate the sequence in which one may move from one screen to another. The treelike structure suggests that at some point in the dialogue, the user will select an option and execute only those screens in that branch of the tree. This if-then structure is perfect for menu-type dialogues. Figure 17.7 shows a simple state transition diagram. Let's study this tool in greater detail.

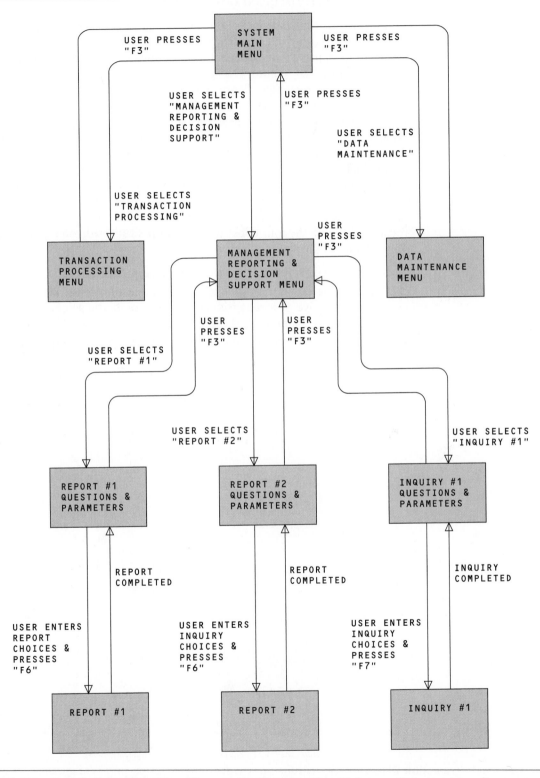

FIGURE 17.7 **Sample State Transition Diagram for Documenting Dialogue** State transition diagrams can be used to design and document user interface screens. The tool can depict the sequence and variation of screens as seen by the system user.

State Transition Diagram Conventions

The purpose of the state transition diagram is to depict the sequence and variations of screens that can occur when the system user sits at the terminal. You can think of it as a roadmap. Each screen is analogous to a city. Not all roads go through all cities. Figure 17.8 depicts all of our state transition diagram conventions. The rectangles represent display screens (formatted by display layout charts). The arrows represent the flow of control through the various screens. The rectangles only describe what screens can appear during the dialogue. The arrows indicate the order in which these screens occur. Notice that a separate arrow, each with its own label, is drawn for each direction. Why? Because different actions trigger flow of control from and flow of control to a given screen. Also notice that the label for each arrow is placed near the arrow head to help distinguish which label belongs to which arrow. Earlier, we stated that the flow of screens in a user interface occurs with almost programming-like precision. Indeed, the four constructs of a well-structured computer program apply equally to the state transition diagram:

① *Sequence.* The screens occur in a natural sequence one after the other.
② *Selection.* Based on the user's answer to a question or the selection of a menu alternative or icon, a different path through the screens is selected (similar to programming's if-then-else construct).
③ *Repetition.* Until some condition is fulfilled, a screen or sequence of

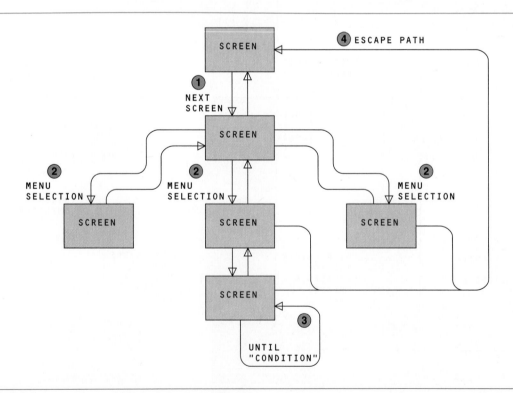

FIGURE 17.8 State Transition Diagram Conventions   A state transition diagram consists of rectangles that represent display screens and arrows that represent the flow of control through various display screens.

screens is repeated. This is similar to programming's repeat-until or do-while constructs.

(4) *Escape or exit.* A path that provides a quick path back to another screen or an exit path from the application.

These constructs can be nested just like programming constructs. Using these simple notations, the analyst (and possibly users) can design a complete dialogue. The state transition diagram serves as a table of contents for the display layout charts of all the screens we've just discussed.

Let's examine the user interface for part of the SoundStage project. The SoundStage state transition diagrams were drawn using our CASE product, Excelerator. Some CASE products include proprietary dialogue graphics.

Design of a Dialogue Structure

The SoundStage Member Services system can be partitioned into several levels of screens with each level aimed toward accomplishing a more precisely defined function. For example, the SoundStage system can be decomposed initially into three distinct subsystems: membership, order processing, and promotions. Each of these subsystems requires screens that will assist the user to perform various functions. In addition, these subsystems may be decomposed into more specific support functions like transaction processing, management reporting and decision support, and data maintenance, for which screens are also designed.

In Figures 17.9**A** and **B**, we present two state transition diagrams that depict the sequence and variations of screens that can occur when the user sits at the terminal. (In the interest of brevity, we will present only the order processing leg of the user dialogue.) We call your attention to the following points of interest:

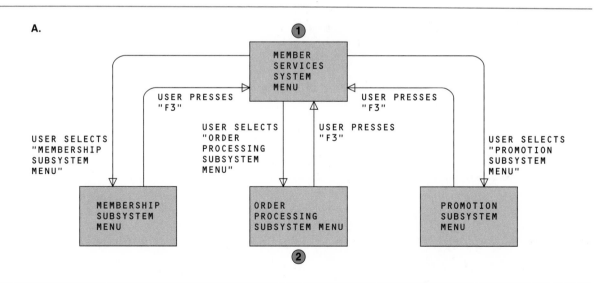

FIGURE 17.9 State Transition Diagram for SoundStage Member Services Project   This state transition diagram shows the relationship between the many display screens needed to describe the dialogue of the SoundStage Member Services project.

B.

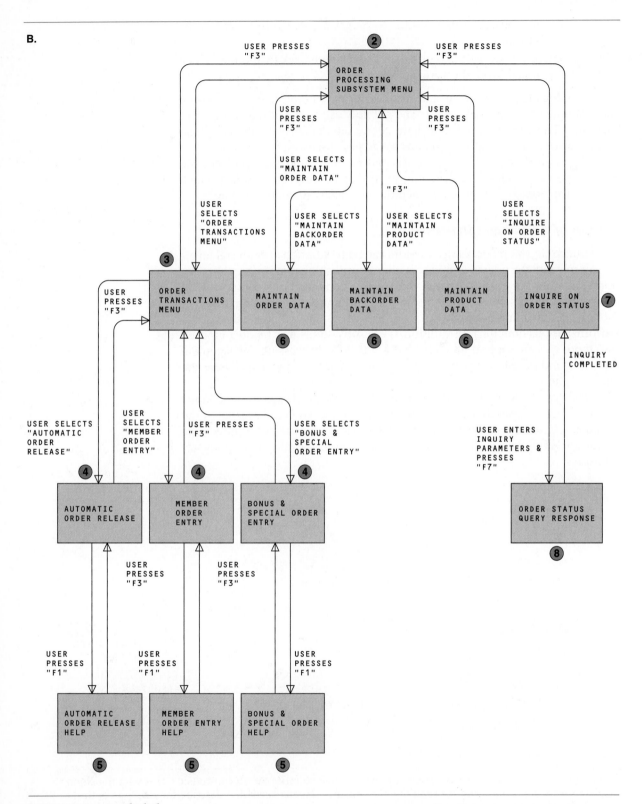

FIGURE 17.9 *concluded*

① The MEMBER SERVICES SYSTEM MENU (see Figure 17.9**A**) is the first screen to be viewed by the user. The SoundStage system contains several options, conveniently grouped into submenus. The MEMBER SERVICES SYSTEM MENU will provide the user with the options of the membership subsystem, the order processing subsystem, and the promotion subsystem. Each of these options, when chosen, will result in the user seeing a new menu screen: the MEMBERSHIP SUBSYSTEM MENU, the ORDER PROCESSING SUBSYSTEM MENU, and the PROMOTION SUBSYSTEM MENU. The menu options selected are recorded on the branching lines. Also, the return trigger (the user pressing F3, meaning return to the previous screen) has been specified.

② The ORDER PROCESSING SUBSYSTEM MENU screen (see Figure 17.9**B**) will offer the user a number of processing options including the processing of order transactions, data maintenance, and an inquiry. In order to enter a MEMBER ORDER, we would choose ORDER TRANSACTION MENU.

③ ORDER TRANSACTIONS MENU offers the user a choice of three different transactions to process.

④ The user may choose to perform AUTOMATIC ORDER RELEASE to release the automatic orders created by promotions, or a user may perform MEMBER ORDER ENTRY for those members who responded to the monthly promotion, or the user may choose BONUS & SPECIAL ORDER ENTRY. Each of these screens represents an input that would have been designed in the last chapter.

⑤ Notice that **help screens** (discussed later in the chapter) have been designed and are indicated for each input screen.

⑥ The user has the option of choosing to perform data maintenance on any of the three data stores that are "owned" by the order processing subsystem: order data, backorder data, and product data. Each of these three maintenance screens would provide the user with basic add, modify, and delete capability into each of these data stores. It should also be noted that although order data, for example, might be implemented as separate order data and ordered product data, this screen would provide maintenance services for both of those data stores.

⑦ The system will allow the user to INQUIRE ON ORDER STATUS. The purpose of this screen is to determine which order the user wants to receive information about. It should also be noted that many other inquiries are possible. We only wanted to show how one could be done.

⑧ The ORDER STATUS QUERY RESPONSE screen represents an output. This screen was presumably designed in Chapter 16. It will always appear immediately after the user has supplied the information called for on the previous screen. Notice that this screen can return only to the previous screen.

It should also be noted that this set of user dialogue does not include screens for the generation of any management reports. This is unusual, but it is in keeping with the decomposition diagram that we constructed for the order processing subsystem in Chapter 9.

Step 2: Prototype or Simulate the Dialogue and User Interface

There are several ways to prototype a dialogue and user interface. Since a live demonstration is impossible in a textbook, we will briefly describe different tools and techniques and direct you to some examples.

One of the simplest tools for dialogue and interface prototyping is demonstration software. One of the best and least expensive tools is Symmantec's Demo. Using Demo, you can prototype screens and provide logic (or a play-script) for moving between screens. This gives the impression that you are watching the actual program execute.

Many Computer Assisted Systems Engineering (CASE) products include dialogue and interface prototyping tools. In previous chapters, we have demonstrated how Excelerator permits screen designs to be prototyped.

Other CASE products, such as KnowledgeWare's RAD and Object View, and Powersoft's Powerbuilder, can support both screen design and testing of all three basic constructs: sequence of screens, conditional branching to screens, and repetition of screens.

Most database management systems and fourth-generation languages include screen and dialogue generators that can be used to prototype the user interface. Exercising (or testing) the user interface is a key advantage of all these prototyping environments.

> **Exercising (or testing) the user interface** means that system users literally experiment with and test the interface design prior to extensive programming and actual implementation of the working system. Analysts can observe this testing to improve on the design.

In the absence of prototyping tools, the analyst should at least simulate the dialogue by walking through the screen sketches with system users.

Step 3: If Necessary, Produce Programmer-Oriented Layout Charts

Given the dialogue structure, we can design the actual dialogue using display layout charts. These charts, introduced in the previous chapter, can be done either in lieu of a prototype or to more formally specify the design developed in a prototype. Increasingly, the aforementioned CASE and 4GL tools are preferred to these charts.

Referring back to the state transition diagram shown in Figure 17.9**A** and **B**, there are several paths that the dialogue can follow. For the sake of brevity, we will design screens only for the order processing portion of the system. These screens were chosen because they serve as the dialogue that would be required for the user to arrive at the input screens designed in Chapter 16. Similar screens would normally be developed for the entire state transition diagram.

If one hasn't already been constructed, we suggest that a template for the zones or areas be defined.

The first screen seen by the system user is presented in Figure 17.10. It is the main system menu. Line 22 is a field for error messages. The accompanying letter in parentheses is used to direct us to a memo that describes the

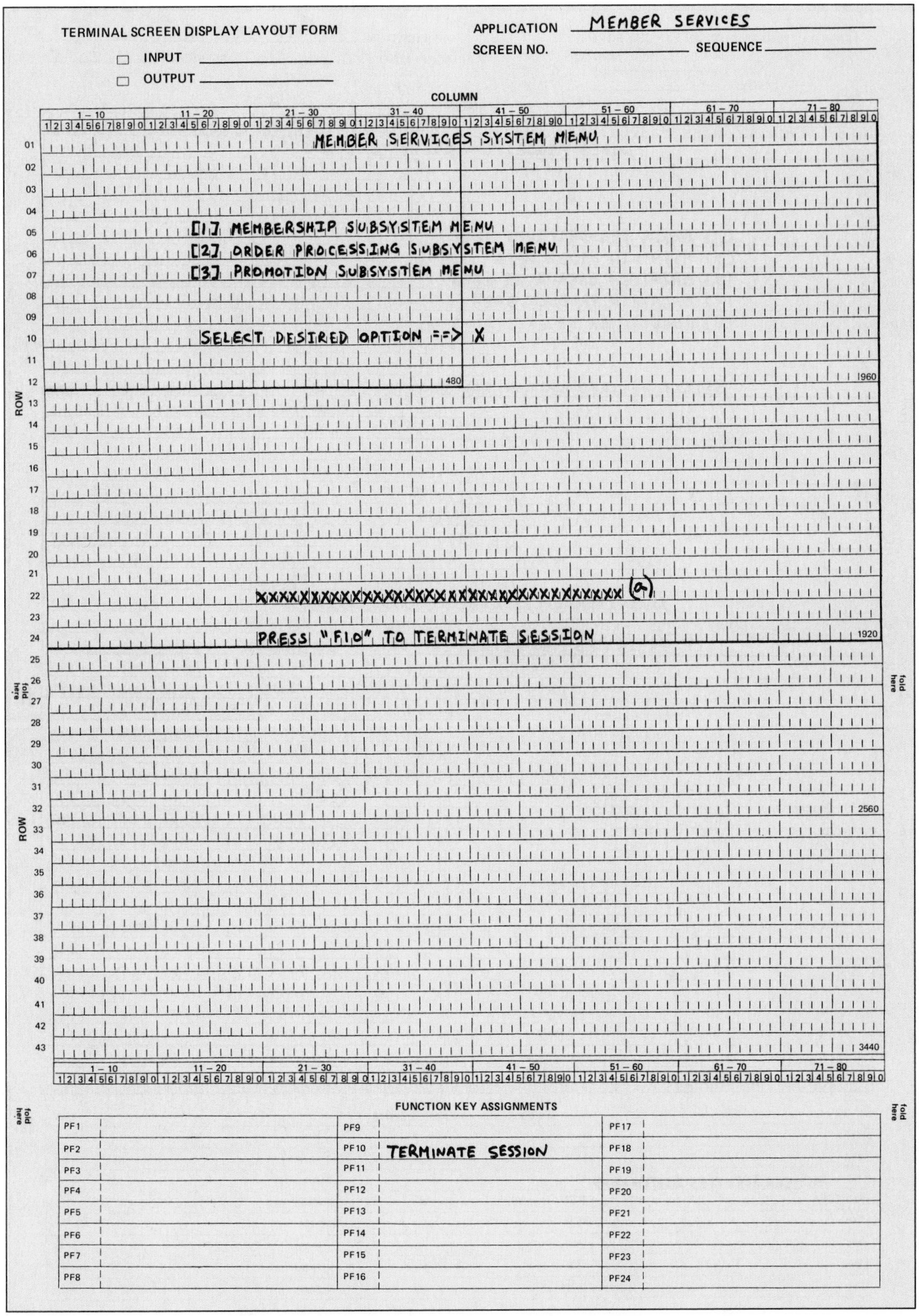

TERMINAL SCREEN DISPLAY LAYOUT FORM

APPLICATION __MEMBER SERVICES__

SCREEN NO. _____ SEQUENCE _____

☐ INPUT _____
☐ OUTPUT _____

Row 01: MEMBER SERVICES SYSTEM MENU

Row 05: [1] MEMBERSHIP SUBSYSTEM MENU
Row 06: [2] ORDER PROCESSING SUBSYSTEM MENU
Row 07: [3] PROMOTION SUBSYSTEM MENU

Row 10: SELECT DESIRED OPTION ==> X

Row 12: 480 ... 1960

Row 22: XXXXXXXXXXXXXXXXXXXXXXXXXXXXXXXXXXXXXXXXX (a)

Row 24: PRESS "F10" TO TERMINATE SESSION ... 1920

Row 32: 2560

Row 43: 3440

FUNCTION KEY ASSIGNMENTS

PF1		PF9		PF17	
PF2		PF10	TERMINATE SESSION	PF18	
PF3		PF11		PF19	
PF4		PF12		PF20	
PF5		PF13		PF21	
PF6		PF14		PF22	
PF7		PF15		PF23	
PF8		PF16		PF24	

FIGURE 17.10 Display Layout Chart for the MEMBER SERVICES SYSTEM MENU Screen

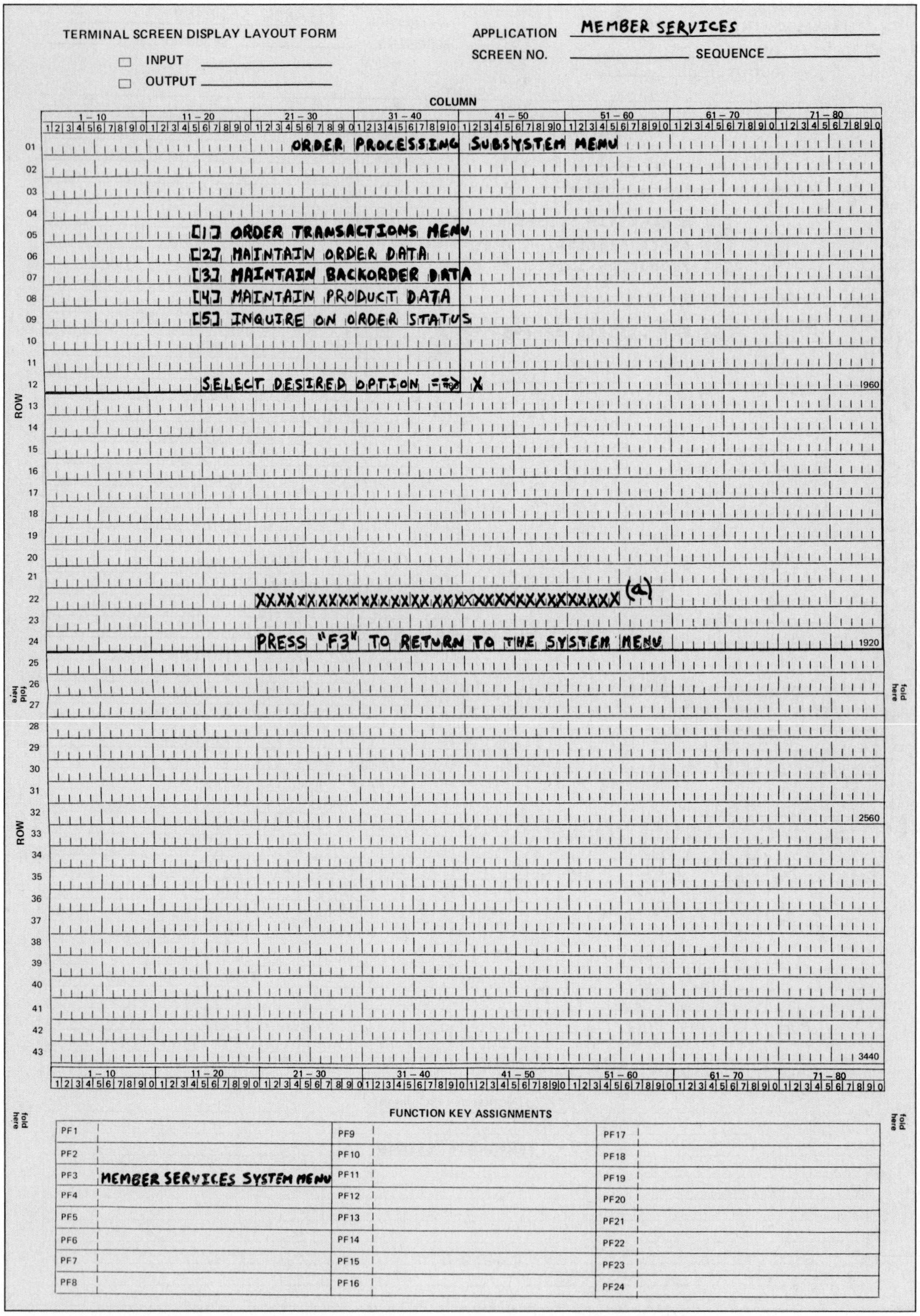

TERMINAL SCREEN DISPLAY LAYOUT FORM

APPLICATION __MEMBER SERVICES__

☐ INPUT _____

☐ OUTPUT _____

SCREEN NO. _____ SEQUENCE _____

COLUMN

Row	Content
01	ORDER PROCESSING SUBSYSTEM MENU
05	[1] ORDER TRANSACTIONS MENU
06	[2] MAINTAIN ORDER DATA
07	[3] MAINTAIN BACKORDER DATA
08	[4] MAINTAIN PRODUCT DATA
09	[5] INQUIRE ON ORDER STATUS
12	SELECT DESIRED OPTION ==> X 1960
22	XXXXXXXXXXXXXXXXXXXXXXXXXXXXXXXXXX (a)
24	PRESS "F3" TO RETURN TO THE SYSTEM MENU 1920
32	2560
43	3440

FUNCTION KEY ASSIGNMENTS

PF1	PF9	PF17	
PF2	PF10	PF18	
PF3 MEMBER SERVICES SYSTEM MENU	PF11	PF19	
PF4	PF12	PF20	
PF5	PF13	PF21	
PF6	PF14	PF22	
PF7	PF15	PF23	
PF8	PF16	PF24	

FIGURE 17.11 Display Layout Chart for the ORDER PROCESSING SUBSYSTEM MENU Screen

messages that can be printed. Finally, notice that we used the function key assignments of the display layout chart to define an operation, TERMINATE SESSION.

If our user wishes to input some MEMBER ORDERs, then the ORDER PROCESSING SUBSYSTEM MENU would be selected. According to our state transition diagram, the system user would now see the ORDER PROCESSING SUBSYSTEM MENU, which is shown in Figure 17.11. This screen is similar to the previous screen because both screens were designed using our template and both screens simply offer a menu of options to the user. Notice that, in accordance with our state transition diagram, this screen offers the user an escape option to return to the previous menu screen. The user who wishes to input MEMBER ORDERs would select the ORDER TRANSACTIONS MENU.

This option should take us to a new screen, the ORDER TRANSACTIONS MENU (not shown). This screen should allow our system user to select from AUTOMATIC ORDER RELEASE, MEMBER ORDER ENTRY, and BONUS & SPECIAL ORDER ENTRY. The user would then choose MEMBER ORDER ENTRY, which would take us to the actual entry screen. This is the input screen that we designed in Chapter 16. The display layout chart for MEMBER ORDER ENTRY has been reproduced for you in Figure 17.12. The system user would then press function key F3 to escape back to the ORDER TRANSACTIONS MENU.

What happens if the system user has problems when entering data into the MEMBER ORDER ENTRY screen? For example, the user might not know how the ORDER DATE should be entered or why the screen will not accept the PRODUCT NUMBER that was typed in. On the state transition diagram (see Figure 17.9**B**), we see that the MEMBER ORDER ENTRY screen should include its own special help screen. The user who encounters a problem while keying a MEMBER ORDER can press the function key, F1, and the MEMBER ORDER ENTRY HELP screen will be displayed. Figure 17.13 represents the layout of the MEMBER ORDER ENTRY HELP screen. The possible explanations that may appear on this screen should be documented using a memorandum. Each explanation given to the user should correspond to a particular input data attribute at which the cursor was positioned when the system user pressed the F1 (help) key. The possible help messages for the MEMBER ORDER ENTRY HELP screen are documented in Figure 17.14. Notice that help messages are not provided for MEMBER NAME, CLUB NAME, or any of the member address fields because they are not input fields (display only).

Summary

The design of conversational user interfaces has taken on greater importance. User interface design is the specification of a conversation between the system user and the computer that results in the input of new data to the information system, the output of information to the user, or both. Display terminals and microcomputer monitors have features that affect dialogue design. The systems analyst should be familiar with the current technology of available terminals and monitors and their features because they can be used to improve dialogue design.

Human factors are also important considerations for good user interface design. Most

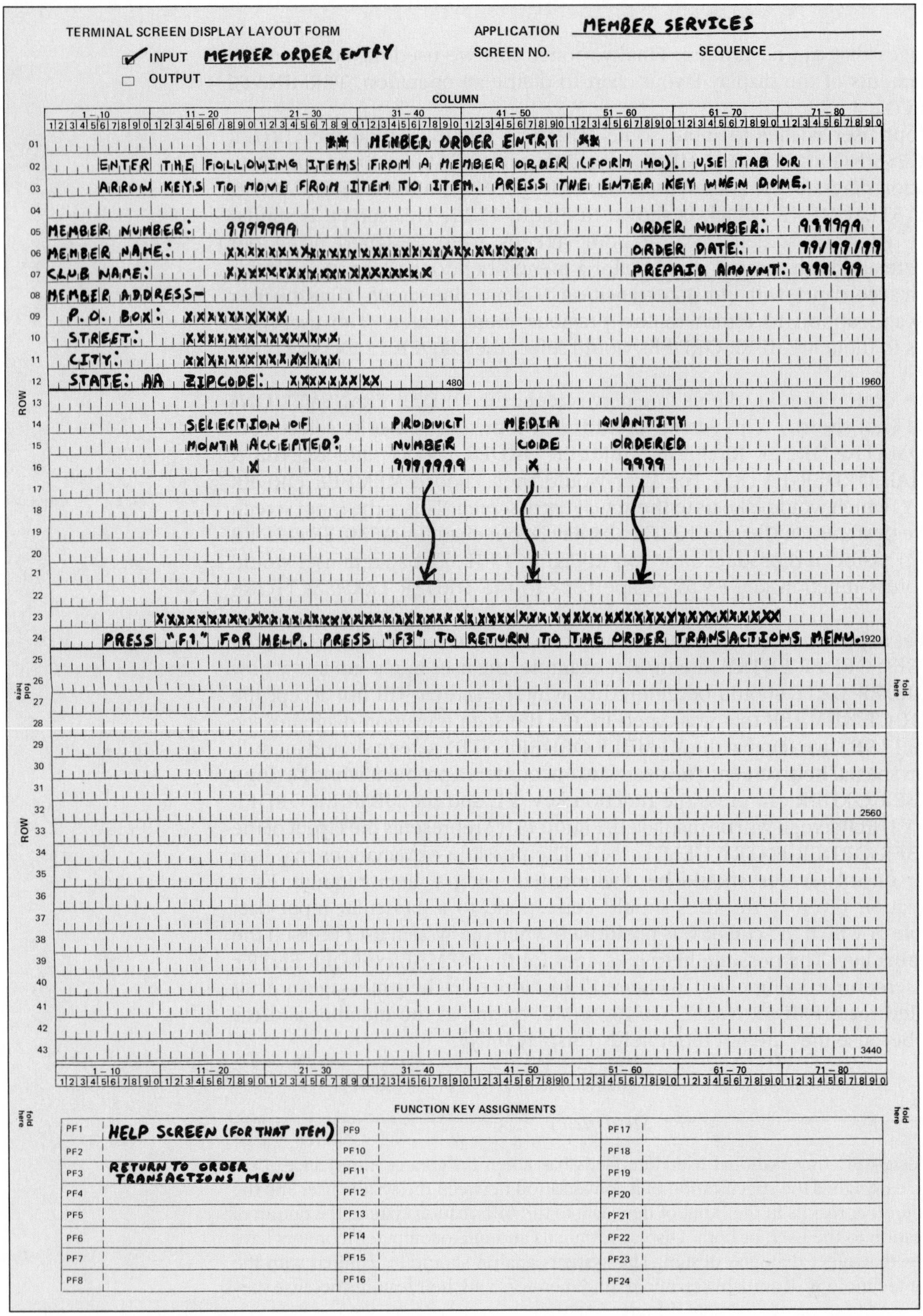

TERMINAL SCREEN DISPLAY LAYOUT FORM

APPLICATION **MEMBER SERVICES**

☑ INPUT **MEMBER ORDER ENTRY**

☐ OUTPUT _____

SCREEN NO. _____ SEQUENCE _____

COLUMN

```
01                    ** MEMBER ORDER ENTRY **
02     ENTER THE FOLLOWING ITEMS FROM A MEMBER ORDER (FORM 40). USE TAB OR
03     ARROW KEYS TO MOVE FROM ITEM TO ITEM. PRESS THE ENTER KEY WHEN DONE.
04
05 MEMBER NUMBER: 9999999                          ORDER NUMBER: 999999
06 MEMBER NAME: XXXXXXXXXXXXXXXXXXXXXXXXXXXX       ORDER DATE: 99/99/99
07 CLUB NAME: XXXXXXXXXXXXXXXXXX                   PREPAID AMOUNT: 999.99
08 MEMBER ADDRESS—
09   P.O. BOX: XXXXXXXXXX
10   STREET: XXXXXXXXXXXXXXXX
11   CITY: XXXXXXXXXXXXXXXX
12   STATE: AA  ZIPCODE: XXXXXXXXX        480                         1960
13
14     SELECTION OF      PRODUCT   MEDIA    QUANTITY
15     MONTH ACCEPTED?   NUMBER    CODE     ORDERED
16          X            999.9999    X        9999
17
18
19
20
21
22
23     XXXXXXXXXXXXXXXXXXXXXXXXXXXXXXXXXXXXXXXXXXXXXXXXXXXXXXXXXXXXXXXX
24 PRESS "F1" FOR HELP. PRESS "F3" TO RETURN TO THE ORDER TRANSACTIONS MENU. 1920
25
...
32                                                                      2560
...
43                                                                      3440
```

FUNCTION KEY ASSIGNMENTS

PF1 HELP SCREEN (FOR THAT ITEM)	PF9	PF17
PF2	PF10	PF18
PF3 RETURN TO ORDER TRANSACTIONS MENU	PF11	PF19
PF4	PF12	PF20
PF5	PF13	PF21
PF6	PF14	PF22
PF7	PF15	PF23
PF8	PF16	PF24

FIGURE 17.12 Display Layout Chart for the ORDER TRANSACTIONS MENU Screen

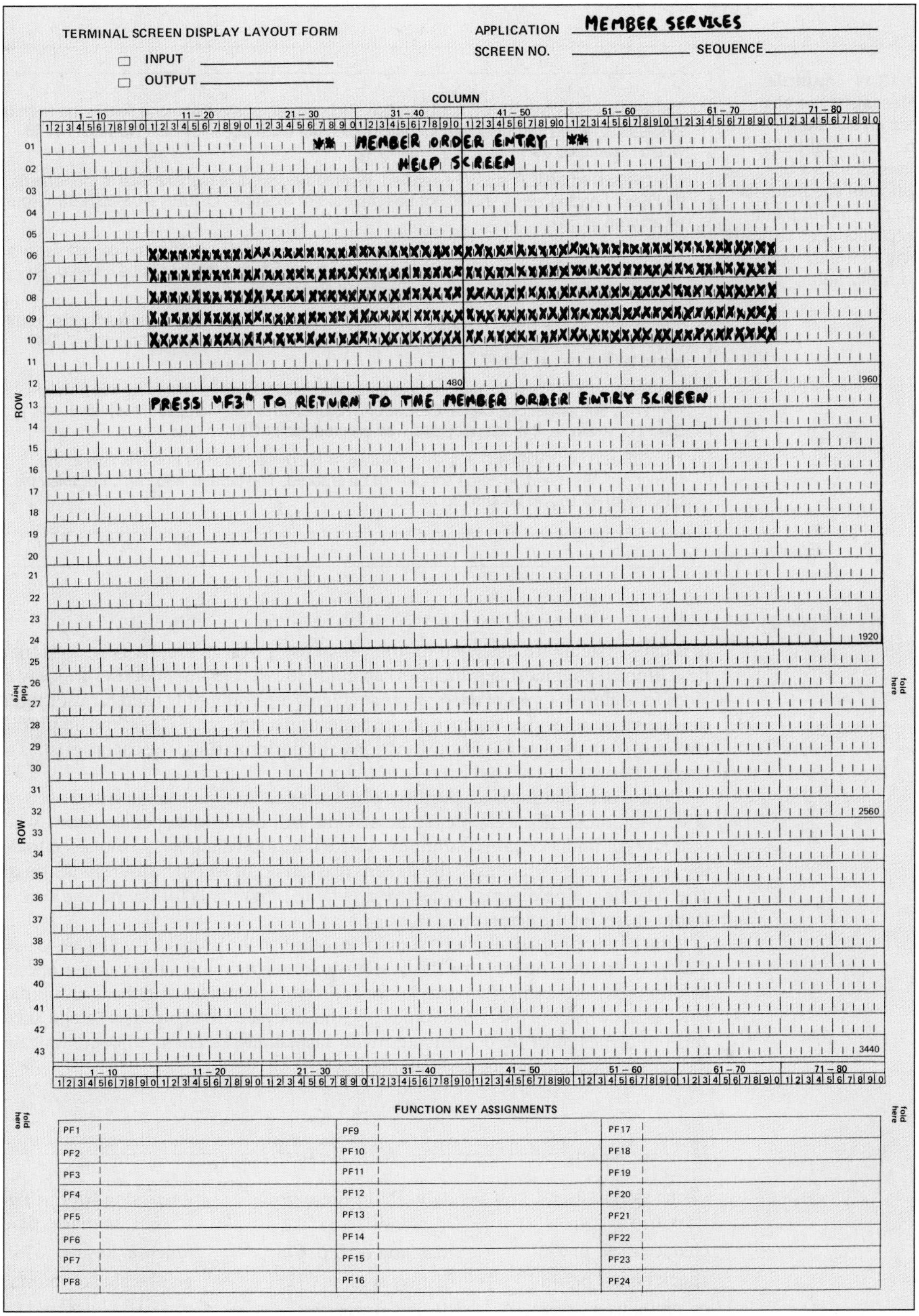

FIGURE 17.13   Display Layout Chart for the MEMBER ORDER ENTRY HELP Screen

FIGURE 17.14 Sample Help Messages for the Member Order Help Screen A typical help message includes a brief description of the field, its format, and a valid example. Help messages should be provided for all user-entered input fields.

The MEMBER NUMBER is a right justified, 7-character, positive numeric field in the form of 9999999. Leading zeros should not be entered. For example, MEMBER NUMBER 1473 would be entered as 1473.

The ORDER NUMBER is a right justified, 6-character, positive numeric field in the form of 9999999. Leading zeros should not be entered. For example, ORDER NUMBER 531 would be entered as 531.

The ORDER DATE is an 8-character field in the format of MM/DD/YY. The default value is today's date. The slashes must be entered, but leading zeros should not be entered. For example, February 15, 1993 would be entered as 2/15/93.

The PRODUCT NUMBER is a right justified, 7-character, positive numeric field in the format of 9999999. Leading zeros should not be entered. For example, product number 72947 would be entered as 72947.

The MEDIA CODE is a 1-character, alphanumeric field. The valid values for this field are: VH, VHS; 8M, 8MM: DC, Digital Cassette; CA, Cassette; CD, Compact Disc; LD, Laser Disc. For example, if a compact disc was ordered, enter CD.

The QUANTITY ORDERED is a right justified, 4-character, positive numeric field in the format of 9999. Leading zeros should not be entered. The default value is 1. For example, quantity of 45 should be entered as 45.

---

of today's systems are designed for the casual user, with an emphasis on user friendliness. Human engineering principles can guide the development of user-friendly interfaces for different types of users. Four strategies are commonly used for user interface design: menu selection, instruction sets, question-answer dialogues, and graphics. The choice of strategy depends on the function to be performed and the characteristics of the users.

Typical user interfaces may involve many screens. The coordination of these screens is very important; for example, some screens will occur in a specific order, whereas others occur under certain conditions. A state transition diagram is a tool used to depict the sequence and variations of the screens that can occur when the user sits at a terminal. The state transition diagram can serve as a table of contents for the numerous display layout charts for the screens.

After developing a dialogue, it should be tested on the users who will have to use that dialogue. Prototyping tools are the best way to test a dialogue. The system users should be observed during this test period—you are trying to find flaws in the user interface. A flaw occurs when the user does something the analyst didn't consider. Screen sketches and prototypes must be documented with display layout charts to communicate the formal specifications to the programmer.

Key Terms

casual system user, p. 676

character set, p. 673

check boxes, p. 674

common user access (CUA), p. 674

dedicated system user, p. 676

dialogue box, p. 674

display area, p. 672

display properties, p. 673

exercising (or testing) the user interface, p. 690

function keys, p. 675

graphical user interface (GUI), p. 674

help screens, p. 689

--------------------------------- *Problems and Exercises* ---------------------------------

1. To what extent should the system user be involved during user interface design? What would you do for the user? What would you ask the user to do for you? Detail a strategy that consists of specific steps you and the system users would follow.

2. Study the features on two visual display terminal(s) or microcomputers. You may need to borrow manuals to complete this assignment. How might these features be used to design effective user interfaces?

3. What documentation prepared during input design and output design is needed during user interface design? How does that input and output design documentation relate to user interface design?

4. Explain the difference between a dedicated and a casual terminal user. How would your strategy for designing user interfaces for a dedicated user differ from that for designing user interfaces for a casual user?

5. Display properties can be overused and frequently hinder a user's performance during a terminal session. Cite some examples in which display properties are appropriate and in which display properties may hinder a user's performance.

6. Describe four strategies for designing user interfaces. What criteria would you consider when choosing between the strategies?

7. Arrange to study microcomputer application. It may be either a business system (such as an inventory, accounts receivable, or personnel system) or a productivity tool (such as a word processor, spreadsheet, or database system). Analyze the human engineering of the user interface. Analyze the human engineering of the display screens. If possible, discuss the dialogue and screens with system users. What do they like and dislike about the design?

8. Redesign the application in Exercise 7 to improve the user interface and screens. If possible, discuss your improved design with users. Do they like your new design better? Did they raise any concerns?

9. Obtain documentation or magazine reviews on an automated screen-design aid. If possible, arrange for a demonstration. How would the product improve your productivity? How would the product decrease your productivity? What features do you dislike or would you prefer to see?

--------------------------------- *Projects and Minicases* ---------------------------------

1. An automated record-keeping information system is being designed for an employment agency. Some of the tasks to be automated are as follows:

Transaction Processing
A. Processing clients

      a. Recording new clients
      b. Matching clients with job openings
      c. Notifying clients of job openings
  B. Processing jobs
      a. Recording job openings
      b. Matching jobs with clients
      c. Recording job placements

Management Reporting & Decision Support
  A. Reporting of job openings
  B. Reporting weekly job placements
  C. Reporting client credentials
  D. Query clients
      a. Query general client information
      b. Query employee job qualifications
      c. Query employee job requirements and preferences
  E. Query job openings
      a. Query general job opening information
      b. Query job opening requirements
  F. Query job placements

Assume that the terminal input and output screens have already been designed to support these processes. Develop a state transition diagram to depict the sequence and variations of dialogue screens that might occur when a user sits at the terminal. Be sure to include help screens for all input screens, escape options for navigating the structure, screen reference numbers, and descriptive screen names.

2. Terminal screens should be designed for consistency. Design a template for the screens in Project 1. The template should clearly indicate zones or areas of the screen used to display common messages. Do you have zoning or similar screen design software on your computer system? If so, describe how you'd implement your zones. If not, how will you ensure that data and messages are displayed in the proper zone?

3. Using the state transition diagram from Project 1 and the screen template from Project 2, design the dialogue screens required for a user to arrive at the basic input and output screens. (Assume that the actual input and output screens have already been designed.) Be sure to include explanations of possible error messages that may appear when the user makes invalid entries.

4. Test the user interface you prepared in Project 2. Replace all display layout charts with screens that contain actual data and messages instead of editing symbols. Simulate a user terminal session by having someone walk through the dialogue using these sample screens. Challenge them to do something your dialogue wasn't designed to handle.

Suggested Readings

Fitzgerald, Jerry. *Internal Controls for Computerized Information Systems.* Redwood City, Calif.: Jerry Fitzgerald & Associates, 1978. This is our reference standard on the subject of designing internal controls into systems. Fitzgerald advocates a unique and powerful matrix tool for designing controls. This book goes far beyond any introductory systems textbook — must reading.

Kendall, Kenneth, and Julie Kendall. *Systems Analysis and Design.* Englewood Cliffs, N.J.: Prentice Hall, 1988. Chapter 16 provides another look at user interface design.

Mehlmann, Marilyn. *When People Use Computers: An Approach to Developing an Interface.* Englewood Cliffs, N.J.: Prentice Hall, 1981. We are indebted to Mehlmann for the concept of zoning a screen into areas, but this book goes far beyond that. Every systems analyst can get something out of this book, which is a modern and comprehensive study of how to analyze and design intelligent user interfaces.

18

Program Design

───────────── *Chapter Preview and Objectives* ─────────────

In this chapter you will learn how to design good programs and how to package program design specifications. You will know that you understand how to design programs and package design specifications into a format suitable for programmers when you can:

Factor
a program into manageable program modules around which
complete specifications can be organized.

Associate
the design documentation presented in Part Three of this book
with the input, process, or output specifications of a computer program.

Recognize
two popular tools for depicting the modular design of programs.

Describe
two strategies for developing structure charts by examining
data flow diagrams.

Design
programs into modules that exhibit loose coupling and
high cohesive characteristics.

Use
structure charts and Structured English to package a computer
program and its specifications.

TOWER LAWN AND GARDEN, INC.

George Amana is a programmer/analyst for Tower Lawn and Garden, Inc. Tower is a distribution center for lawn and garden equipment in northern Louisiana.

Scene: *George has just sat down to lunch in the company cafeteria. Pete Wilcox, a senior partner in the firm has joined him.*

Pete: Hi, George, why don't you join me for lunch. Hey, you look pretty frustrated. What's the problem?

George: I had to take over the sales information systems project that Judy left behind when she quit. It's total chaos. I was told that it was all but finished. Come to find out, she didn't finish several of the programs.

Pete: But she did do a good job of specifying all the program requirements. What's so tough about the programs? Judy always preached about the benefits of structured programming. In fact, she taught me how to do it. Don't tell me she doesn't practice what she preaches.

George: No, her code is very well structured. And her documentation is adequate. It's just that the programs seem so poorly designed. Some of her subroutines are so long and complex that it's difficult to get a grasp on small enough pieces to test them for correctness. It seems like an all-or-nothing proposition. If I encounter a bug, I have to test large sections of code to zero in on the problem. Sometimes the bug turns out to be in an entirely different subroutine!

Pete: Why didn't Judy break the system into smaller pieces?

George: Oh, she did! The subroutines are evidence of that. But it almost seems like she generated the subroutines on the fly—as if to say, "Well, this piece of code is getting complex. I'd better put in a subroutine to finish it." She left me a rough draft of a structure chart, but I just don't understand the reasons she factored the system the way she did.

Pete: That's the way I write programs. I start by trying to draw a flowchart on a single page—sort of the high-level flowchart. Then I factor the more complex processes into more detailed processes that I implement as subroutines. It sounds like that may be what Judy did.

George: Maybe she did. But that strategy causes the lower-level subroutines to be very dependent on other routines. I frequently encounter bugs that get traced back to other, seemingly unrelated routines. I'm just getting further behind schedule. I may just have to write the programs from scratch.

Pete: Why don't you get some help? Barbara just finished her project. Maybe she can help you. You could divide up the work and get it done faster.

George: Divide up the work? I don't see how. Judy's program specifications are just one big unorganized document. I'm not sure which file and report specifications to match up to which modules. For that matter, I'm not sure the programs themselves are fully documented.

Pete: Well, I'm sorry George. I don't know what to tell you.

Discussion Questions

1. If design specifications are thorough and complete and program code is well structured, how can the system still be difficult to construct?

2. How should subroutines in a program be conceived? How does Judy seem to have created them? What is the potential problem with creating subroutines during coding?

3. What effect does the program and subroutine size have on testing?

4. What would Barbara require in order to take on responsibility for some of the programs that haven't been written? What does the programmer need to be able to write a new program? How would you organize the necessary documentation of program requirements?

MODULAR DESIGN OF COMPUTER PROGRAMS

Our study of systems design is nearly complete. You've designed the files and/or databases, inputs and outputs, and on-line user interfaces. You've selected appropriate computer equipment and packaged software (which has hopefully been delivered and installed during systems design). The final step is to implement a structured program design.

From your programming courses, you may think of program design as algorithm or logic design. That is *not* the subject of this chapter. We don't intend to reteach you how to draw structured program flowcharts, to prepare pseudocode, or to construct box charts (sometimes called Nassi-Schneidermann charts). That is clearly a subject for a programming textbook. Instead, we are concerned with how the programming specifications are presented to the computer programmer for implementation. To this end, we view program design as consisting of two components:

- **Modular design**—the decomposition of a program into manageable pieces.
- **Packaging**—the assembly of input, output, file, user interface, and processing specifications for each module.

Some readers are likely to interpret the material covered in this chapter as an invasion of the programmer's turf. It really varies from one computer information systems shop to another. Some shops insist that the analyst prepare detailed modular designs and program specifications (at a level close to pseudocode). Other shops believe that the analyst's job ends with general programming specification, leaving modular design to the programmer. Depending on your opinion, you may want to omit this chapter. However, we recommend the chapter for the following reasons:

- Your career may take you to organizations or management that prefer both of the approaches.
- The chapter helps tie the design specifications prepared in Chapters 13 through 17 to the program specifications that normally initiate systems implementation (discussed in detail in Chapter 19).
- In the absence of a company standard, you may want to consider a rigorous personal standard for presenting specifications. In Chapter 19 you will learn that the analyst is frequently engaged in a large number of activities during systems implementation. The more thorough and complete your programming specifications are, the less time you'll have to spend clarifying those specifications for the programmer.

For those of you not familiar with modular design from your programming courses, we'll briefly review the concept. Large projects are more easily managed if they are broken into smaller pieces. You've seen us apply this concept with data flow diagrams. Computer programs can be similarly decomposed, as depicted in Figure 18.1. What we did here was to recursively factor a large program into smaller and smaller pieces called *modules*. We will now study how this is accomplished.

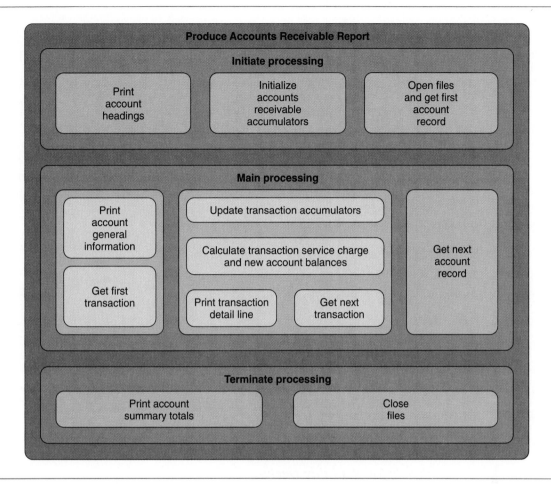

FIGURE 18.1   Modular Design   This diagram is a useful way to depict the modular design approach, otherwise known as *divide and conquer.*

Modular Decomposition of Programs

What is a module? It could be a subroutine or subprogram. And it could be a main program. On the other hand, it could be a unit of measure smaller than any of those. For instance, a module could be a paragraph in a COBOL program. So, what is a module?

A **module** is a group of executable instructions with a single point of entry and a single point of exit.

Some modules exist to perform single functions. These include READ A RECORD, EDIT A RECORD, CALCULATE PAY, and ADD A RECORD TO A FILE, to name a few. Other modules exist to supervise or drive the function modules.

The length of a module is important. Evidence suggests that modules should consist of a relatively limited number of lines of executable instructions. Most experts suggest a number between 24 lines (the most typical display screen size) and 60 (the number of lines printed on an average report). Consequently,

FIGURE 18.2 Warnier/ Orr Notation Warnier/ Orr brackets are a popular and simple-to-use modular design tool. Compare this notation with Figure 18.1.

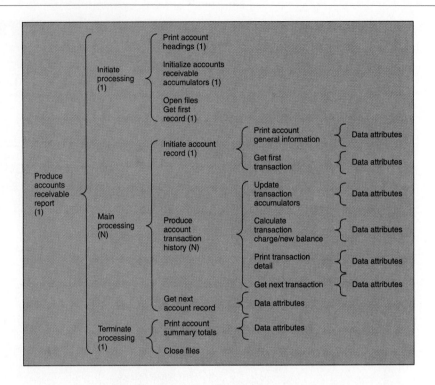

this guideline would suggest that any program that cannot be written in fewer than 60 lines of code should be decomposed into modules.

Tools for Modular Design

There are two popular tools for depicting the modular design of programs. We'll use the simple example provided in Figure 18.1 to introduce these tools. The first tool we'll introduce is the Warnier/Orr bracket (see Figure 18.2). A **Warnier/Orr diagram** is nothing more than a hierarchy chart laid on its side. For this discussion, we want to focus on the tool itself, deferring our discussion of the Warnier/Orr methodology until the next section.

Brackets decompose modules into other, lower-level modules that we'll call *submodules.* The Warnier/Orr diagram implies a sequence of execution that is read from top to bottom and left to right. A number in parentheses below a module indicates how many times that module executes (a looping concept). A plus sign between modules indicates that execution of those modules is mutually exclusive. In other words, each single execution of the calling module may call one or the other submodule — never both. Although we haven't seen it, a notation could easily be defined, say with an asterisk, to indicate that a module can call either or both submodules.

An alternative and somewhat more familiar tool is the structure chart. A **structure chart** (see Figure 18.3) is a treelike diagram. Structure charts, by whatever other name you might know them, may be familiar to you from your introductory programming course.

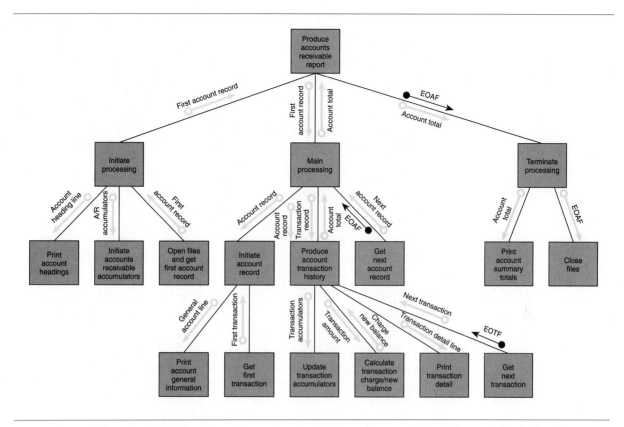

FIGURE 18.3 Structure Chart Notation   Structure charts are another popular tool for modular design. Compare this diagram with Figures 18.1 and 18.2.

Structure chart modules are depicted by named rectangles. Modules are factored, from the top down, into submodules. Structure chart modules are presumed to execute in a top-to-bottom, left-to-right sequence. Named arrows are used to represent data (depicted by an arrow with a small circle on one end) or control flows (depicted by an arrow with a darkened circle on one end) between modules.

We find Warnier/Orr diagrams easier to construct than structure charts since there is no need for a template; however, structure charts are more familiar in the literature. Because it is more likely that you've encountered structure charts in your introductory programming course, we will use that notation throughout the remainder of this chapter. You can replace any of the structure charts we draw with an equivalent Warnier/Orr diagram.

Strategies for Modular Design

Recall that structured design is a popular term for the decomposition of a program into modules. There are three popular strategies for structured design. We will review them briefly before we present an integrated strategy that we've found useful.

IBM's **Hierarchy plus Input-Process-Output (HIPO)** was one of the earliest strategies for structured design. Using HIPO, the designer factors a program into logical functions and depicts it as a structure chart (called a *vertical table of contents*). Each module is eventually documented with an input, process, and output (IPO) chart, which forces you to detail the inputs (these include inputs, file accesses, and subroutine parameter passing), processing requirements (narrative, pseudocode, or flowchart), and outputs (including reports, displays, file updates, and parameter passing). Although HIPO forces the decomposition of programs into modules, it doesn't really offer a strategy for doing so. There-fore, HIPO might be better thought of as a documentation tool than as a strategy. We'll show you how to take advantage of the HIPO documentation tool later in this chapter.

Ed Yourdon and Larry Constantine (Page-Jones, 1980) have developed what has become a popular strategy for determining an optimal structure chart for programs. Their technique is called **structured design,** and it is based on the use of data flow diagrams. Essentially, you document programs with *detailed* data flow diagrams, study those diagrams, and convert the DFDs into *structure charts* (their term). They suggest two substrategies for developing the structure charts — transform analysis and transaction analysis:

> **Transform analysis** is an examination of the DFD to divide the pro-cesses into those that perform input and editing, those that do processing (e.g., calculations), and those that do output.

Although we have greatly simplified the strategy, it is based on the IPO concept, which you have learned about throughout this book. Structured design refers to that portion of the DFD consisting of processes that perform input and editing as the **afferent.** That portion of the DFD consisting of processes that do pro-cessing are referred to as the **central transform.** The portion of the DFD consisting of processes that do output are referred to as the **efferent.** In deriv-ing the rough-cut structure chart, each of these three groups or processes is given a parent module to serve as a manager or coordinator. You'll see this concept demonstrated later in this chapter.

> **Transaction analysis** is the examination of the DFD to identify pro-cesses that represent transaction centers.

A **transaction center** is a process that does not do actual transformation upon the incoming data (data flow); rather, it serves to route the data to two or more processes. You can think of a transaction center as a traffic cop that directs traffic flow. Such processes are usually easy to recognize on a DFD, because they usually appear as a process containing a single incoming data flow but two or more outgoing data flows that lead to other processes. A classic example of a transaction center is the master file update whereby a process (the central transform) receives the update transaction data flow and routes the transaction to the appropriate add, modify, or delete process. The transaction centers serve as a basis for establishing a rough-cut structure chart. The resulting structure chart is factored into these transaction center modules, which may then be factored into IPO modules using transform analysis.

By using the Yourdon/Constantine strategy to divide a program into mod-ules, you are able to end up with modules that are said to be loosely coupled and highly cohesive. **Coupling** refers to the level of dependency that exists be-tween modules. Thus, loosely coupled modules are less likely to be dependent on one another (remember the problem George had in the Tower Lawn and

Garden minicase?). **Cohesion** refers to the degree to which a module's instructions are functionally related. Thus, highly cohesive modules contain instructions that collectively work together to solve a specific task. The data and control flow symbols depicted on a structure chart can serve as aids in determining the degree of coupling and cohesion of modules.

Another approach to developing an optimal modular structure has been suggested by Jean-Dominique Warnier and Ken Orr (Orr, 1977). This approach develops a program structure by working backward from the desired output data structure. This technique is called *logical design of programs.* The output data structure is first defined using the Warnier/Orr notation. Then the input, file, and/or database structure is defined using a similar notation. Finally, a program structure is defined from these structures. The resulting program structure usually reflects the input-process-output characteristics (or begin-process-terminate).

Both the Yourdon/Constantine and Warnier/Orr strategies have their advocates. We think the two strategies have much in common. Throughout this book, we have tried not to endorse any analysis or design methodology. This chapter will be no exception. Instead, we'd like to present a simplified strategy for modular design. The strategy is based on common sense and the fundamental principles that underlie both the Yourdon/Constantine and the Warnier/Orr approaches.

How to Do Modular Design (a Simplified Approach)

Programs have been identified from design units (see Chapter 14). Given these programs, we want to break them into manageable modules around which program specifications will be written. Programmers can then build and test each module independently. Then modules can be integrated according to the structure chart and tested as a whole program. We'll pay particular attention to how structure charts can be used to integrate large on-line programs, because the trend appears to be toward such systems.

Step 1: Define the High-Level Structure

Virtually all applications programs can initially be broken down into three main functions: INITIATE PROCESSING, MAIN PROCESSING (the body of the program), and TERMINATE PROCESSING. Normally, the initiation and termination functions are performed once. The main-processing function normally executes several times. Figure 18.4 can be used as a starting point for any normal program structure. These three essential functions are loosely based on Yourdon/Constantine's transform analysis strategy.

Adams, Wagner, and Boyer (1983) have cataloged a number of common functional modules that may be included in the INITIATE and TERMINATE PROCESSING functions. This is consistent with the concept of defining highly cohesive structures. These start-up and close-down functions are described in Figure 18.5. Notice in Figure 18.4 that the GET FIRST RECORD function is almost always the last submodule of the INITIATE PROCESSING function. For an on-line system, this module might be labeled DISPLAY MAIN MENU. Along the same lines, the STOP PROCESSING function is usually the last submodule of the TERMINATE PROCESSING.

FIGURE 18.4   A De Facto Standard Structure Chart for Most Programs Most programs can be initially factored into an INITIATE PROCESSING, MAIN PROCESSING, TERMINATE PROCESSING structure. Those modules can be further factored using a number of popular strategies.

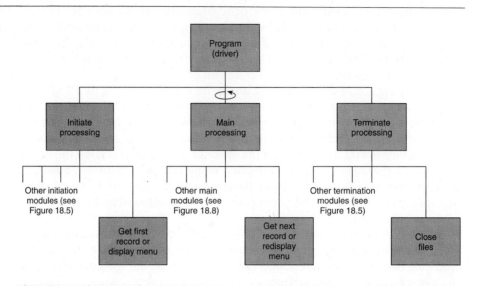

FIGURE 18.5   Primitive Functions Performed by INITIATE and TERMINATE PROCESSING Modules This table suggests highly cohesive primitive modules that are typically controlled by an INITIATE or TERMINATE PROCESSING module. Any of these functions may be factored into primitive subsets to further improve cohesion and possibility for reuse.

INITIATE PROCESSING Functions:

BUILDING AND LOADING TABLES: Creating arrays to store tables, such as tax tables, actuary tables, and the like, and loading the data into those tables.

DEFINING CONSTANTS AND ACCUMULATORS: Constants are set in a dedicated module so those constants can be easily located if they need to be changed (for instance, SALES TAX PERCENT). Accumulators are used to count records and control totals during main processing.

OPENING FILES: Files must be opened before they can be read from or written to. It should be noted that some systems limit the number of files that can be open at any one time. If more files are needed than can be opened, then the program must be rewritten as multiple programs that pass intermediate results through temporary (scratch) files (which count as one open file).

FILE MERGING OR SORTING: This must be done before main processing can be done.

PRINTING REPORT HEADINGS: Why relegate report headings to a separate module? So they can be easily located if report headings need to be modified.

DISPLAYING (MAIN) MENU: For on-line systems, displaying the first menu and accepting the first choice from that menu are usually an initiation function.

GET FIRST INPUT RECORD: Read the first input record or file record to be processed.

TERMINATE PROCESSING Functions:

CALCULATING CONTROL TOTALS: Performing arithmetic and statistical operations on totals accumulated during main processing functions.

PRINTING CONTROL TOTALS: Printing the accumulators and control totals maintained and calculated during main processing.

CLOSING FILES: The reverse of opening files. Disconnects the file from the program, thereby allowing other programs, which may have been locked out, to use those files.

Source: Adapted from David R. Adams, Gerald E. Wagner, and Terrence J. Boyer, *Computer Information Systems: An Introduction* (Cincinnati: South-Western Publishing, 1983).

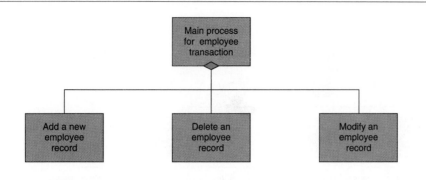

**FIGURE 18.6 Trans-
action Centers for a
Simple Program** Gen-
erally, it is useful to factor
a module into distinct
submodules that act on
single transactions. These
transaction centers are
then said to be loosely
coupled—that is, less
likely to impact one an-
other.

The PRODUCE ACCOUNTS RECEIVABLE REPORT PROGRAM module acts
as a type of traffic cop, directing the execution of its subordinate modules. It
executes the INITIATE PROCESSING module and begins executing the MAIN
PROCESSING module. When processing has been completely finished, it executes the TERMINATE PROCESSING module. That's it! For on-line systems,
processing usually doesn't terminate—the MAIN PROCESSING module is
available for execution until the system is no longer needed. At that time, the
TERMINATE PROCESSING module is executed.

The remainder of our strategy will focus on how to factor the MAIN PRO-
CESSING module.

Step 2: Identify Transaction Centers

As a preface to factoring the MAIN PROCESSING module, the first question we
ask is, "Does this program have transaction centers?" This question is based on
Yourdon/Constantine's transaction analysis strategy. What we are really asking
is, "Does this program support multiple transactions?" If so, we will factor the
MAIN PROCESSING module according to those transactions. The following are
examples that would lend themselves to this strategy:

- A file maintenance program typically supports at least three transactions:
  ADD A NEW RECORD, DELETE A RECORD, and MODIFY A RECORD.
  Each transaction deserves its own module. Each transaction will cause the
  execution of one and only one of the transaction modules. For such a pro-
  gram, we would use the structure illustrated in Figure 18.6.
- An on-line system typically supports multiple levels of transactions. For
  instance, the main menu may offer three choices: EMPLOYEE FILE MAIN-
  TENANCE, PERSONNEL TRANSACTION, and EMPLOYEE INQUIRY.
  Each of these subfunctions consists of multiple transactions. EMPLOYEE
  FILE MAINTENANCE could be factored as described in the preceding example. PERSONNEL TRANSACTION could be factored into SICK LEAVE
  PROCESSING, TIME CARD PROCESSING, VACATION PROCESSING,
  and so on. The resulting hierarchy chart might resemble Figure 18.7.

Although DFDs may help you identify transaction centers, it depends on how
detailed the analyst drew those DFDs (for instance, many analysts won't factor
the DFD process, MAINTAIN EMPLOYEE FILE, into three separate processes).

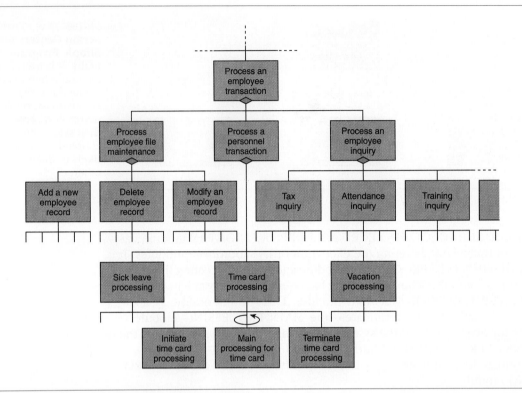

FIGURE 18.7 On-Line Transaction Centers On-line systems are particularly suited to transaction analysis since their capabilities are called on demand and integrated through a high-level control program.

Finally, most transaction centers need to be further factored into their own INITIATE, PROCESS, and TERMINATE modules. For example, in Figure 18.7 we factored the SICK LEAVE PROCESSING module into an initiate-process-terminate trio of submodules.

If the program you are trying to design cannot be factored into transactions (for instance, if it supports a single transaction or only generates a single report), this step can be eliminated.

Step 3: Factor the Initiate, Process, and Terminate Functions into Primitive Functions

At this point, we have factored our program into one or more iterations of INITIATE, PROCESS, and TERMINATE modules. (Note: We would have only one iteration if there were no multiple transaction centers.) Now we can factor the INITIATE, PROCESS, and TERMINATE modules into their primitive functions. The typical primitive modules for INITIATE and TERMINATE modules were listed in Figure 18.5. There are two strategies for factoring the PROCESS modules into primitives.

Most PROCESS modules can be factored into the primitive functions described in Figure 18.8. Once again, these primitive functions are generally considered to be highly cohesive. We should be able to write the logic of these modules with 50 or fewer statements of code. Along those lines, it may be

MAIN PROCESSING Functions:

EDITING INPUT RECORDS: Performing picture, range, and completeness checks to make sure that data being input to the system for the first time is correct. (This module will normally write to (or display) an errors report or file.)

GETTING A SECONDARY RECORD: Reading an input or file record from a secondary source. For instance, if you are processing input ORDERS, you may have to retrieve a CUSTOMER RECORD for a credit check or retrieve a PART record for an inventory check, all during main processing.

PERFORM CALCULATIONS: Performing arithmetic operations on data.

MAKING DECISIONS: Executing business policy decisions, such as credit checking, part availability, and discounting.

ACCUMULATING TOTALS: Where possible, totals should be accumulated in their own modules so those accumulators can be easily located and changed.

WRITING A DETAIL LINE: Recording a single detail line or transaction to a file or report that will contain many such detail lines.

WRITING A COMPLETE RECORD: Writing an entire record (as opposed to a detail line) to a report or file (for example, printing a paycheck or updating a record in a master file).

GETTING THE NEXT RECORD: Retrieving or reading the next record in the loop that drives the main processing routine.

REDISPLAY A MENU: Redisplaying menu options that are available in an on-line system.

FIGURE 18.8 Primitive Process This list suggests primitive cohesive MAIN PROCESSING functions. Again, these functions may have to be further factored to define small and reusable modules.

Source: Adapted from David R. Adams, Gerald E. Wagner, and Terrence J. Boyer, *Computer Information Systems: An Introduction* (Cincinnati: South-Western Publishing, 1983).

appropriate to factor one of the simple functions from Figure 18.8 into submodules. For example, an EDIT ORDER module may be factored into EDIT GENERAL ORDER DATA, EDIT ORDERED PARTS, EDIT CUSTOMER DATA, and so on. The first strategy is demonstrated by the structure chart that appears in Figure 18.9.

If the output consists of a natural hierarchy, we like to use the second strategy—the Warnier/Orr data structure approach—to factor the process. This frequently happens with multiple control break outputs (a concept that should be familiar to students of programming). For instance, suppose our PROCESS module is intended to produce a PART SALES SUMMARY REPORT. This report should contain detailed unit and dollar sales information for each part, each product line (consisting of multiple parts), and each warehouse zone (consisting of multiple product lines). We can factor the PROCESS module for this report into modules that correspond to the control breaks. This structure is illustrated in Figure 18.10. Notice that we added an INITIATE (to set accumulators) and TERMINATE (to format and print control totals) module to each CONTROL BREAK PROCESSING module. This structure chart makes it possible to implement the program with a single flag, "end of file" (thus loosely coupling the hierarchy of modules), and with much less code. The final process module, PROCESS PART DETAIL LINE, is factored into primitive tasks by using the first strategy, the primitive main processing functions introduced in Figure 18.8.

And that, in an abbreviated form, is one possible strategy for program module design. We strongly urge you to study the full Warnier/Orr and Yourdon/Constantine strategies as part of your continuing education.

FIGURE 18.9 **Factoring MAIN PROCESSING into Cohesive Primitives**
MAIN PROCESSING must eventually be factored down to loosely coupled, highly cohesive primitive modules.

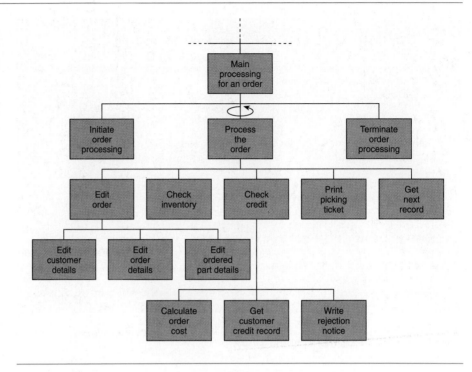

FIGURE 18.10 **Data Structure Factoring**
For programs that produce outputs whose data structure is hierarchical, the MAIN PROCESSING can be factored according to that hierarchy (control breaks). Each of these data-structure-oriented modules is flanked by modules that initiate and terminate the processing for that control break.

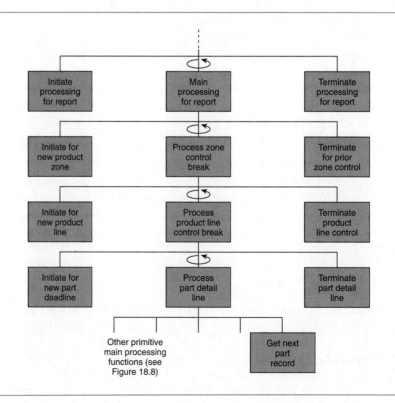

PACKAGING PROGRAM SPECIFICATIONS

Using the design techniques presented in this unit, you have accumulated a good number of design specifications for the new system—perhaps you have separate stacks of documentation for the system outputs, files, database, inputs, terminal dialogue, systems flowcharts, and program modules. Now, put yourself in the role of the computer programmer. Are those specifications in a format that will help you write the programs? Not really. As a systems analyst, you are responsible for packaging that set of design documentation into a format suitable for the programmer.

As a direct result of modular design, you have a structure chart for the computer programs to be written. We can now package the program specifications around that structure chart.

What Does a Programmer Need in Order to Write Computer Programs?

You can't expect programmers to implement a correct program if they don't receive the necessary program specifications. You can avoid this problem by looking at packaging as salespeople look at their product. A good salesperson knows the product and knows how to sell it. In this section you're going to learn to look at a computer program as if it were a new product. Let's study the components of the program specifications package and see how the product can best be presented to end-users and programmers.

The program specifications package is a collection of design documentation that clearly communicates the requirements for each computer program in the system. What exactly are the requirements associated with a computer program? All programs perform three types of tasks, including input or reading of data, manipulation of input data, and output of data or information. In other words, all program tasks can be classified according to input, process, and output (IPO) requirements. This model will help us address the requirements for implementing a computer program.

Input Specifications

As a systems analyst, you are responsible for providing complete specifications of all sources of input data for each program. To the computer programmer, the term *input* has a broad meaning. Although batch and on-line inputs are still important, *input* also refers to file and database access. The specifications for all these program inputs have been documented as follows:

1. Master, transaction, and scratch file specifications (Chapter 15):
   Expanded repository entries.
   Record layout charts.
2. Database specifications (Chapter 15):
   Expanded repository entries.
   Subschema.
   Record layouts.
3. Batch input file specifications (Chapter 16):
   Source document layout.
   Expanded repository entries.
   Input record layout charts.

4. On-line input specifications (Chapters 16 and 17):
   Source document layout.
   Expanded repository entries.
   Display layout charts.
   Prototype input screen.

Processing Specifications

All programs execute processing tasks (e.g., sorting, summarizing, and calculating) on input data to produce outputs and information. These processing tasks are performed according to business policies and procedures. It is essential that the policies and procedures governing the processing tasks of a program be clearly explained to the programmer. A programmer can't implement a program that checks credit, for example, if the credit policies are not clear, accurate, and complete.

How complete should the processing requirements be? How close should the analyst come to specifying code? Generally, procedure specifications should represent a more general explanation of how the tasks of a program are to be accomplished. Program logic is intended to be much more detailed. For example, a procedure specification instruction might state

Sort the DAILY ORDERS FILE in ascending order by CUSTOMER ORDER NUMBER.

Alternatively, we could provide the pseudocode logic for an internal sort (see Figure 18.11). Some programmers might like this detailed specification very much. Others might be offended by such a precise description. Where should the systems analyst draw the line? Many organizations have adopted standards that dictate exactly what the systems analyst must provide to the programmer. In the absence of standards, perhaps analysts will simply document critical business formulas and decision rules. On the other hand, most systems analysts are required to simply provide the programmer with a clear and concise statement of the program processing requirements. The specifications for computer processing requirements may be given by:

Decision tables (covered in Chapter 11)
Structured English (covered in Chapter 11)

In Figure 18.12 we've reproduced a primitive-level DFD for the design unit PROCESS MEMBER ORDERS in our case study. The program PROCESS MEMBER ORDERS is actually a subprogram in an on-line system. The partial structure chart for the on-line system is illustrated in Figure 18.13. This structure will be used to organize our program specifications.

Using Structured English, you would specify each module in the structure chart. Figure 18.14 represents the Structured English specifications for a module of the PROCESS MEMBER ORDERS program. While we chose to use Structured English (and decision tables when appropriate) to present processing requirements, you could use alternative procedures or logic tools, such as program flowcharts.

Due to size constraints on the book, we cannot provide you with all of these

FIGURE 18.11 **Pseudocode** This is an example of pseudocode for a sorting requirement. The analyst should avoid this level of detail unless systems design standards call for it. This is the level of detail that the programmer would use to design logic.

Initialize the ORDER SORT array subscript X to 1.
For each record in the DAILY ORDER FILE, do the following:
Store DAILY ORDER FILE record in ORDER SORT array at subscript X location.
Add 1 to subscript X.
Initialize the IS SORT COMPLETE FLAG to "NO".
Initialize the RECORDS TO SORT variable equal to the subscript X.
Repeat the following steps until the SORT COUNTER variable equals X − 1 or the SORT COMPLETE FLAG equals "YES":
Initialize SORT COMPLETE FLAG to "YES".
Subtract 1 from the RECORDS TO SORT variable.
Initialize the COMPARISON COUNTER to 1.
Repeat the following steps until the COMPARISON COUNTER is greater than the RECORDS TO SORT variable:
Calculate COMPARISON SUBSCRIPT using the following formula:
COMPARISON COUNTER + 1
If the CUSTOMER ORDER NUMBER for ORDER SORT array record at COMPARISON COUNTER location is less than the CUSTOMER ORDER NUMBER for ORDER SORT array record at COMPARISON SUBSCRIPT location, then:
Store ORDER SORT array record at location COMPARISON COUNTER in TEMPORARY STORAGE variable.
Store the ORDER SORT array record at location TEMPORARY STORAGE in ORDER SORT array at the COMPARISON COUNTER location.
Store the TEMPORARY STORAGE record in the ORDER SORT array at location COMPARISON COUNTER.
Set the IS SORT COMPLETED FLAG equal to "NO".
Add 1 to COMPARISON COUNTER variable.
Add 1 to SORT COUNTER variable.
Initialize the SORTED ORDER FINE COUNTER to 0.
For each record in the SORT ORDER array, do the following:
Store the SORT ORDER array record at the SORTED ORDER FILE COUNTER location in the SORTED ORDER FILE.

detailed design specifications for each module on the structure chart appearing in Figure 18.13.

Output Specifications

The term *output* also means more to the programmer than we have suggested in this unit. In addition to printouts, forms, and displays, outputs include updates to files and databases. Output requirements include the following:

1. Printed output specifications (Chapter 15):
   Expanded repository entries.
   Printer spacing charts.
   Prototype reports.
2. On-line output specifications (Chapter 15):
   Expanded repository entries.
   Display layout charts.
   Prototype output screens.

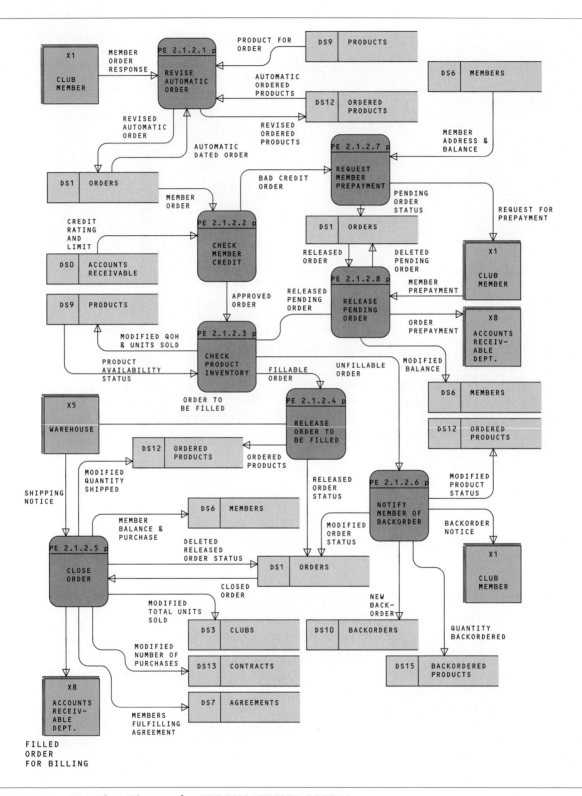

FIGURE 18.12   Data Flow Diagram for PROCESS MEMBER ORDERS

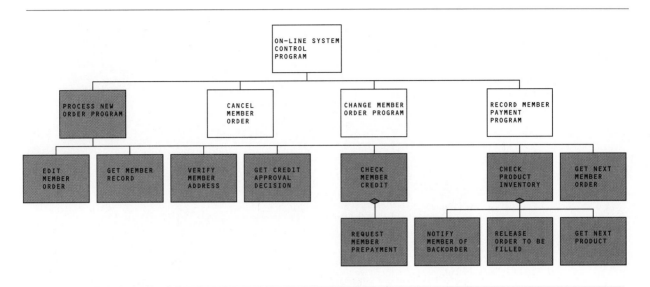

FIGURE 18.13   Structure Chart for PROCESS MEMBER ORDERS

3. Master and transaction files updated specifications (Chapter 13):
   Expanded repository entries.
   Record layout charts.
4. Database specifications (Chapter 14):
   Expanded repository entries.
   Subschema.
   Record layouts.

This concludes our discussion of program design and packaging. Try packaging a sample computer program.

Computer-Assisted Systems Engineering (CASE) for Program Design

Computer-assisted systems engineering (CASE) was introduced in Chapter 5 as an enabling technology for systems analysis and design. Virtually all CASE products include graphics capability for developing structure charts. Additionally, most CASE tools provide design facilities for specifying module logic in the project repository.

Newer CASE tools are beginning to support automatic generation of "rough-cut" structure charts from process requirements documented in the repository. For example, given a DFD, characteristics about processes appearing on the DFD can be used as a basis for generating a first-cut structure chart.

Finally, CASE tools such as automatic code generators are sure to have a significant impact upon programs and programmers. Automatic code generators are discussed in this chapter's The Next Generation box.

FIGURE 18.14 **Structured English** Each module appearing on the structure chart in Figure 18.13 would be documented to communicate processing requirements.

```
TYPE Process                    NAME EDIT MEMBER ORDER

    Label EDIT          EXPLODES TO ONE OF:
          MEMBER        Data Flow Diagram
          ORDER         Structure Chart
                        Structure Diagram

    Location            ORDER-ENTRY STAFF

    Process Category    ON-LINE

    Duration Value      500
    Duration Type       DAY
    Manual or Computer  C


    Description
    FOR ALL INVALID MEMBER ORDER DATA FROM THE END-USER, DO THE FOLLOWING:
       SELECT THE APPROPRIATE CASE:

          CASE 1: MEMBER NUMBER IS INVALID
                  IF THE MEMBER NUMBER IS NOT EQUIVALENT TO A MEMBER NUMBER OF AN
                  EXISTING MEMBER THEN:
                     ERROR MESSAGE = ''MEMBER DOES NOT EXIST, PLEASE REENTER''
          CASE 2: ORDER NUMBER IS INVALID
                  IF THE ORDER NUMBER = ORDER NUMBER OF A PREVIOUS ORDER THEN:
                     ERROR MESSAGE=''ORDER NUMBER WAS ASSIGNED TO PREVIOUSLY ENTERED
                                      MEMBER ORDER, PLEASE REENTER''
          CASE 3: ORDER DATE IS INVALID
                  SELECT APPROPRIATE CASE:
                     CASE 2.1 ORDER DATE CONTAINS NO VALUES. THEN:
                            ERROR MESSAGE=''ORDER DATE MUST BE PROVIDED''
                     CASE 2.2 ORDER DATE CONTAINS INVALID MM/DD/YY VALUES,
                              THEN:
                            ERROR MESSAGE=''THE ORDER DATE IS INVALID.
                                             PLEASE REENTER''
          CASE 4: PREPAID AMOUNT > TOTAL MEMBER ORDER COST, THEN:
                  ERROR MESSAGE=''PREPAID AMOUNT EXCEEDS TOTAL COST OF MEMBER ORDER.
                                   PLEASE REENTER''
          CASE 5: PRODUCT NUMBER NOT VALID, THEN:
                  ERROR MESSAGE=''INCORRECT PRODUCT NUMBER (DOES NOT EXIST).
                                   PLEASE REENTER''
          CASE 6: MEDIA CODE IS INVALID. THEN:
                  ERROR MESSAGE=''MEDIA CODE IS INVALID, PLEASE REENTER OR
                                   PRESS F2 KEY FOR VALID CODES AND THEIR MEANINGS''
          CASE 7: QUANTITY ORDERED IS NOT > ZERO, THEN:
                  ERROR MESSAGE=''QUANTITY ORDERED MUST BE GREATER THAN ZERO.
                                   PLEASE REENTER''
          CASE 8: MEMBER RESPONSE STATUS IS NOT Y OR N. THEN:
                  ERROR MESSAGE=''ENTER Y (YES) OR N (NO) IN REGARD TO MEMBER'S
                  ACCEPTANCE OF SELECTION
```

Summary

The systems analyst's role in computer program design includes module design and packaging of design specifications. A module is defined as a single entry–single exit group of instructions that performs a single function. To deal with complexity of logic, programmers tend to break programs into modules. A better strategy is to break programs into code to deal with functions.

The Next Generation

AUTOMATIC CODE GENERATORS

This chapter has focused on tools and techniques that reorganize design specifications in a fashion suitable for programmers. Although it is true that improperly packaged and incomplete specifications are the cause of many program inadequacies, we don't intend to imply that the tools and techniques discussed in this chapter will eliminate the problem, although they will help. Will a future generation of tools and techniques promise to eliminate the problem? Perhaps!

It has long been suggested that programming, being a logical process, could be automated. In other words, we may be able to write programs that input (and insist on!) complete specifications and generate and test computer programs — all in a fraction of the time required by human programmers. And there are a few products that support this concept.

In this book, we have frequently referred to fourth-generation languages, and you might consider them the answer to our problem. However, these end-user languages, although they are good and

getting better, are not suitable for all information systems. They may be limited in what they can do, and they are frequently inefficient when compared with their third-generation language counterparts (e.g., COBOL, using conventional file organization techniques). This is because prototypes are based on small files. When fully loaded files are installed and multiple end-users start accessing the system, the throughput and response time becomes unacceptable. Therefore, many prototype systems developed using fourth-generation languages are rewritten in languages such as COBOL after the prototype has been approved by end-users.

Is there any way to improve productivity when using third-generation languages? Higher Order Software (HOS), Inc., sells a product called USE.IT, which automates program design. The package forces you to recursively factor a program into binary (two) functions and subfunctions. The resulting structure can be translated by USE.IT into executable program code that HOS claims can be math-

ematically proven to be bug-free. Other products use Structured Design techniques to develop a program from a general idea by utilizing techniques such as Warnier/Orr. These programs can also generate usable code. We encountered a simpler approach at one Fortune 500 company. They have studied programs in their environment and developed programs that generate skeleton PL/1 code for functions they know to be needed in multiple applications.

In any case, we can expect to see more products of this type. They will become more sophisticated and will generate even more efficient program code. The impact on programmers is clear. We'll still need programmers to maintain operational systems and to enhance and customize the code from these program generators, but we will clearly need fewer programmers. However, that's no problem. We'll need more and better analysts because these program generators will be dependent on clear and complete program specifications.

There are two popular tools for documenting modular structure, Warnier/Orr bracket charts and structure charts. Although often sold as distinct tools, they are actually quite similar. There are also two strategies for modular design. The Yourdon/Constantine approach suggests that modules be defined by studying the flow of data between primitive functions. The Warnier/Orr approach suggests that modules be defined by studying the data structure of the outputs and inputs. This chapter presented a hybrid approach based on the two strategies.

Structure charts and Warnier/Orr charts adequately factor each program into manageable modules that can be assigned to programmers. IBM's Hierarchy plus Input-Process-Output (HIPO) provides a documentation tool for packaging detailed input, processing, and output details around the structure chart. In addition to the structure chart and input-process-output (IPO) charts, the analyst must assemble the various inputs, processes, and outputs needed to present complete program specifications for the programmer.

Key Terms

afferent, p. 706	modular design, p. 702	transaction analysis,
central transform, p. 706	module, p. 703	p. 706
cohesion, p. 707	packaging, p. 702	transaction center, p. 706
coupling, p. 706	structure chart, p. 704	transform analysis, p. 706
efferent, p. 706	structured design,	Warnier/Orr diagram,
Hierarchy plus Input-	p. 706	p. 704
Process-Output		
(HIPO), p. 705		

Problems and Exercises

1. Obtain a copy of the documentation for a completed programming assignment. Study the program's source code to identify all referenced modules. Using a Warnier/Orr diagram and a structure chart, document the existing modular structure implemented by the program.

2. Study the processing requirements for the program you used for Exercise 1. Using the modular design approach suggested in this chapter, develop a new structure chart for the program. Compare the structure chart with the one derived in Exercise 1. Which would you prefer to work from as a programmer? Why?

3. Some typical initiate, processing, and terminate functions were presented in Figures 18.5 and 18.8. Can you identify other processing functions that might be included in the lists? How would you classify them—as initiate, processing, or terminate functions? Explain why.

4. Use the structure chart (from Exercise 2) to prepare an input-process-output (IPO) chart or charts to package the program.

5. What value would an existing structure chart and an IPO chart of an existing program be to a systems analyst during the study phase?

6. What value would an existing structure chart and an IPO chart for a program be to a programmer who has to maintain the program?

7. What correlations can be drawn between a DFD and a structure chart?

8. Differentiate between coupling and cohesion.

9. Give an example of an afferent, efferent, and transform process.

10. What are the transaction centers in the following program?

> An on-line program allows an end-user to perform inquiries to obtain information concerning customer accounts, orders, invoices, and products. The end-user is allowed to obtain general information concerning an order or information about specific orders that have been placed on back order. The end-user who wishes to obtain information concerning orders placed on backorder may request information describing orders that have been backordered for less than one week, backordered for more than one week but less than two weeks, or backordered for more than a two-week period. The end-user may also perform inquiries to retrieve general information about a specific part and information concerning backordered parts.

Projects and Minicases

1. The Computer Information Systems (CIS) department at Northern Fence, Inc. is currently evaluating program design methods. Specifically, Northern is evaluating the structured design (Yourdon/Constantine) and the data structures (Warnier/Orr) methods. The CIS department is planning to apply the two program design methods to a number of programming projects in order to more accurately evalu-

ate each method. Once the projects are completed, the CIS department believes it will be able to determine which method should be adopted as a company standard. Do you feel the CIS department should view the methods as an either/or issue? If not, explain why not. How can the two methods be integrated?

--- *Suggested Readings* ---

Adams, David R., Gerald E. Wagner, and Terrence J. Boyer. *Computer Information Systems: An Introduction.* Cincinnati: South-Western Publishing, 1983. This book suggests how most programs can be factored into initiate, main process, and terminate functions (Chapter 8).

Boehm, Barry. "Software Engineering." *IEEE Transactions on Computers* C-25, December 1976, pp. 1226–41. This paper, a classic, described the logarithmic relationship between time and the cost to correct an error in the systems specifications.

Orr, Kenneth T. *Structured Systems Development.* New York: Yourdon Press, 1977. This book discusses both the Warnier/Orr method of modular design the HIPO documentation technique.

Page-Jones, Meiler. *A Practical Guide to Structured Systems Design.* 2nd ed. New York: Yourdon Press, 1988. This book discusses the modular design methodology called *Structured Systems Design.* The book includes discussion and numerous examples of both transform and transaction analysis. The concepts of coupling and cohesion are also covered in great detail.

Part Four introduces you to the systems implementation and systems support phases of systems development. Two chapters make up this unit. First, Chapter 19 presents systems implementation, the process of putting the design specifications for the new information system in actual operation. Four implementation phases are discussed: build and test networks and databases, build and test programs, install and test the new system, and deliver the new system into operation. Each of these four phases of systems implementation is examined in terms of activities, roles, and techniques.

Chapter 20 discusses the four systems support activities of systems development. This ongoing maintenance of a system after it has been placed into production consists of: correcting errors, recovering the system, assisting users, and adapting the system. Systems support is very important because it is likely that young systems analysts will be responsible for maintaining legacy systems. This chapter concludes our exploration of the systems development life cycle.

SYSTEMS IMPLEMENTATION AND SYSTEMS SUPPORT

19

Systems Implementation

Chapter Preview and Objectives

In this chapter you will learn about two systems implementation phases: (1) the construction of the new information system and (2) the delivery of the new information system. You will also learn about postimplementation review of the system. You will know that you understand the systems implementation process when you can:

Define
systems implementation and relate it to the construction
and delivery phases of the life cycle.

Describe
the construction and delivery phases of the life cycle in terms of:
(1) purpose and objectives, (2) tasks and activities that must or may be
performed, (3) sequence or overlap between tasks and activities,
(4) techniques used, and (5) skills you must master
to perform the phase properly.

Explain
how the time spent on systems implementation
can be managed.

M I N I C A S E

BECK ELECTRONIC SUPPLY

Tim Stallard is a systems analyst at Beck Electronic Supply. He has only been a systems analyst for six months. Unusual personnel turnover had thrust him into the position after only 18 months as a programmer. Now it is time for his semiannual job performance review.

Scene: *Tim enters the office of Ken Delphi. Ken is the Assistant Director of MIS at Beck.*

Ken: Another six months! It hardly seems that long since your last job performance review.

Tim: I personally feel very good about my progress over the last six months. This new position has been an eye-opener. I didn't realize that analysts do so much writing. I enrolled in some continuing education writing classes at the local junior college. The courses are helping . . . I think.

Ken: I wondered what you did. It shows in everything from your memos to your reports. More than any technical skills, your ability to communicate will determine your long-term career growth here at Beck. Now, let's look at your progress in other areas. Yes, you've been supervising the Materials Requirements Planning project implementation for the last few months. This is your first real experience with the entire implementation process, right?

Tim: Yes. You know, I was a programmer for 18 months. I thought I knew everything there was to know about systems implementation. But this project has taught me otherwise.

Ken: How's that?

Tim: The computer programming tasks have gone smoothly. In fact, we finished the entire system of programs six weeks ahead of schedule.

Ken: I don't mean to interrupt, but I just want to reaffirm the role your design specifications played in accelerating the computer programming tasks. Bob has told me repeatedly that he had never seen such thorough and complete design specifi-

cations. The programmers seem to know exactly what to do.

Tim: Thanks! That really makes me feel good. It takes a lot of time to prepare design specifications like that, but I think that it really pays off during implementation. Now, what was I going to say? Oh yes. Even though the programming and testing were completed ahead of schedule, the system still hasn't been placed into operation; it's two weeks late.

Ken: That means you lost the six-week buffer plus another two weeks. What happened?

Tim: Well, I'm to blame. I just didn't know enough about the nonprogramming activities of systems implementation. First, I underestimated the difficulties of training. My first-draft training manual made too many assumptions about computer familiarity. My end-users didn't understand the instructions, and I had to rewrite the manual. I also decided to conduct some training classes for the end-users. My instructional delivery was terrible, to put it mildly. I guess I never really considered the possibility that, as a systems analyst, I'd have to be a teacher. I think I owe a few apologies to some of my former instructors. I can't believe how much time needs to go into preparing for a class.

Ken: Yes, especially when you're technically oriented and your audience is not.

Tim: Anyway, that cost me more time than I had anticipated. But there are still other implementation problems that have to be solved. And I didn't budget time for them!

Ken: Like what?

Tim: Like getting data into the new files. We have entered several thousand new records. And to top it off, management is insisting that we operate the new system in parallel with the old system for at least two months. Then, and only then, will they be willing to allow the old system to be discarded.

Ken: Well, Tim, I think you're learning a lot. Obviously, we threw you to the wolves on this project. But I needed Bob's experience and attention elsewhere. I knew when I pulled Bob off the project that it could introduce delays—I call it the rookie factor. Under normal circumstances, I would never have let you work on this alone. But you're doing a good job and you're learning. We have to take the circumstances into consideration. You'll obviously feel some heat from your end-users because the implementation is behind schedule, and I want you to deal with that on your own. I think you can handle it. But don't hesitate to call on Bob or me for advice. Now, let's talk about some training and job goals for the next six months.

Discussion Questions

1. Above and beyond programming, what activities do you think make up systems implementation? Can you think of any activities that weren't described in this minicase?

2. Why is training so difficult? How do you feel about the prospects of becoming a "teacher?" How long do you think it takes to prepare for one hour of classroom instruction? What activities do you think would be involved in preparing for a lesson plan?

3. A 3,000-record master file must be created for a new system. Each record consists of 15 fields/attributes. The record length is 200 bytes. How long do you suppose it would take to create that file? If necessary, use your own typing speed as a performance gauge. What factors would affect how long it may take to get the file up and running?

4. What assumption did Tim make about transition from the old system to the new system? Why was it wrong? Can you think of any circumstances under which it would be correct?

WHAT IS SYSTEMS IMPLEMENTATION?

Systems implementation was first defined in Chapter 3:

> **Systems implementation** is the construction of the new system and the delivery of that system into production (meaning day-to-day operation). Unfortunately, *systems development* is a common synonym. (Note: We dislike that synonym since it is more frequently used to describe the entire life cycle.)

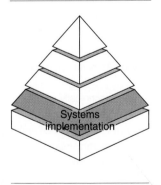

In your pyramid model, systems implementation implements the DATA, ACTIVITIES, and NETWORKS building blocks of the system (see margin). The PEOPLE focus changes from the systems designer to the systems builder. Recall that the perspectives of the systems builders were introduced in Chapter 2.

Figure 19.1 illustrates the phases of a typical systems implementation. The trigger for systems implementation is the technical design statement. Study this figure as we walk through these phases in somewhat greater detail. Each phase is described in terms of the (1) purpose of the phase, (2) activities that should be performed, (3) roles played by various people in each activity, (4) inputs and outputs for each activity, and (5) techniques and skills that can be used to complete each activity.

THE BUILD AND TEST NETWORKS AND DATABASES PHASE OF IMPLEMENTATION

In many cases, new or enhanced applications are built around existing networks and databases. If so, skip this phase. However, if the new application calls for new or modified networks and databases, they must normally be imple-

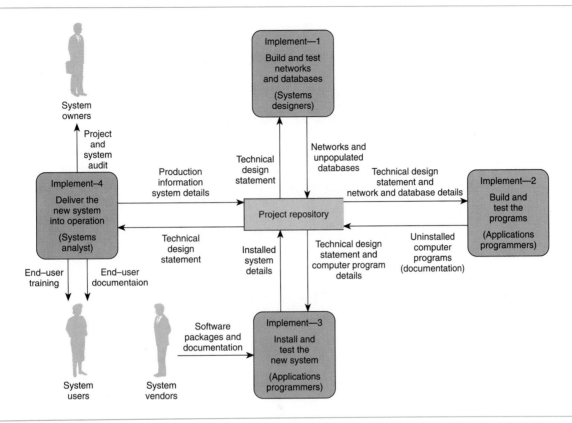

FIGURE 19.1   **Systems Implementation**   The four phases of systems implementation were introduced in Chapter 3. We've made some revisions on this reproduction. The deliverables don't really pass from phase to phase. Instead, they are recorded into various project repositories for use in later phases.

mented prior to writing or installing computer programs, because applications programs will use those networks and databases. Thus, the first phase of some implementations is to build and test networks and databases.

Building Blocks for the Build and Test Networks and Databases Phase

The fundamental objectives of the build and test networks and databases phase are:

- To build (or modify) and test networks.
- To build (or modify) and test unpopulated databases.

Your information systems building blocks provide a framework for this phase. We'll focus upon the implementation of the NETWORK and DATA components of the target system. Recall that in the definition phase, we established network and database requirements. Subsequently, during the design and integration phase we developed distributed data and process models as well as database design specifications. Using these specifications to implement the network and database building blocks for an information system is a prerequisite for the remaining implementation phases.

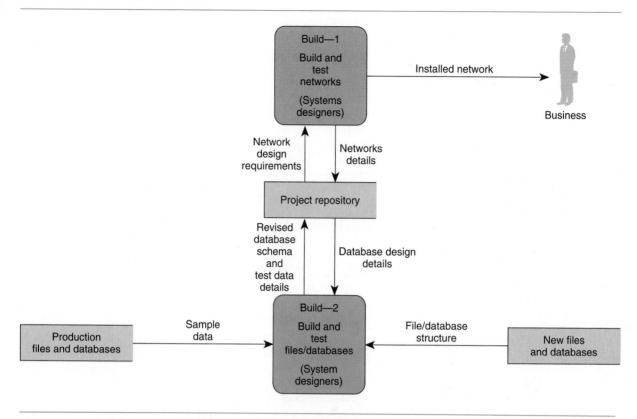

FIGURE 19.2   Build and Test Networks and Database Phase Activities   These activities will lead to comple-tion of the build and test networks and databases phase. The activities are generally completed in a clockwise sequence from the top. On the other hand, the project repository allows you to overlap activities or return to any previous activity to do rework.

Build and Test Networks and Databases Activities, Roles, and Techniques

Figure 19.2 illustrates the typical activities of the build and test networks and databases phase. Refer to this figure as we discuss each activity in detail.

Activity 1: Build and Test Networks (if Necessary)

If new or modified networks are required for the new system, we must first build and test those networks. This activity will normally be completed by the same system specialists that designed them.

Given the network design requirements developed during systems design, the network specialist will make any appropriate modifications to existing networks that may be used by the new system. Alternatively, networks may need to be developed from the ground up. The end-product is the installed network that is placed into operation in the business. Network details will be recorded in the project repository for future reference.

Skills for developing networks is becoming an increasingly important skill for systems analyst—especially given the industry trend toward downsizing existing applications. It is not the intent of this book to teach network implementation skills. You should consider taking one or more courses on networking.

Activity 2: Build and Test Databases (if Necessary)

Building test files and databases is a task unfamiliar to many students, who are accustomed to having an instructor provide them with the test data and files. This task must immediately precede other programming activities because files and databases are the resources shared by the computer programs to be written. If new or modified databases are required for the new system, we can now build and test those databases.

Once again, this activity will typically be completed by the same system specialist that designed the databases. Recall that this person might have been a database specialist or a systems analyst. When the database to be built is a noncorporate, applications-oriented database the systems analyst often completes this activity.

As a prerequisite, if the target solution required the acquisition of new equipment and/or software tools, this activity may be preceded by the installation of equipment and/or software to be used by the new system. Hardware is normally installed and tested by the vendor (even microcomputer stores frequently offer this service). Software may or may not be installed by the vendor. In any case, specialists called *systems programmers* or *technical support staff* often become involved in the installation, testing, and modification of such software as operating systems, database management systems, word processors, spreadsheets, telecommunications software, and other general-purpose software packages.

Given the database design details prepared during systems design, and possibly sample data from production files and databases, the database specialist or analyst will develop new files and databases to be used by the new system. It is important to note that the end-product of this activity is an unpopulated file/database structure for the new file or database. The term *unpopulated* means that the database structure is implemented but that data has not been loaded into that file/database structure. As you'll soon see, programmers will eventually write programs to populate and maintain those new files/database. Revised database schema(s) and test data details are also produced during this activity and placed in the project repository for future reference.

The development of databases is a skill that will likely be applied numerous times over the career of a systems analyst. Strength in database implementation skills comes from experience. However, you should consider taking one or more courses on the subject of database.

THE BUILD AND TEST PROGRAMS PHASE OF SYSTEMS IMPLEMENTATION

Once networks and databases for the new system have been built, we can direct our attention toward building and testing programs. This is the first phase in the life cycle that is specific to the applications programmer. This is also a phase with which you are probably most familiar.

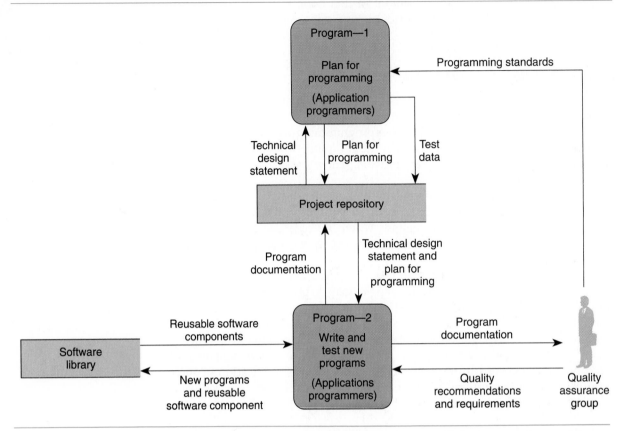

FIGURE 19.3   **Build and Test the Programs Phase Activities**   These activities will lead to completion of the build and test the programs phase. The activities are generally completed in a clockwise sequence from the top. On the other hand, the project repository allows you to overlap activities or return to any previous activity to do rework.

Building Blocks for the Build and Test Programs Phase

The fundamental objectives of the build and test programs phase are

- To develop a detailed plan to guide the development and testing of new or revised computer programs.
- To develop computer programs that accurately fulfill business process requirements.

Here again, your information systems building blocks provide a framework for understanding this phase. This phase will address those information system ACTIVITIES that are to be automated.

Plan for Programming Phase Activities, Roles, and Techniques

Let's examine each of the activities most commonly performed by the applications programmer (see Figure 19.3).

Activity 1: Plan for Programming

An implementation plan is normally generated near the conclusion of the Design phase. An implementation plan, included in the <u>technical design statement</u>, specifies a schedule for completing systems implementation. However, this plan is rarely detailed enough to begin constructing the new system. Therefore, our first task is to develop a <u>plan for programming</u>. The refined plan should include three steps:

- *Review of the Design Specifications.* One major controversy that should be addressed is the freezing of the design specification. By *freezing,* we mean that changes to the design specifications are discouraged or prohibited. Advocates of freezing argue that with no discouragement against changes, users will continually be permitted to identify something they forgot or some new need or idea, and the system may never be constructed and delivered. On the other hand, some experts dispute the idea of freezing the specifications. They argue that such an action is artificial and is not consistent with our goal to serve the end-user.

    Both sides are right! We suggest that you tentatively freeze the document. If changes are proposed, ask yourself a simple question: "Is this a critical change that will make or break the system, or is it an enhancement that could be added later?" Critical changes require the specifications document to be modified. If the change isn't critical, log the change as future enhancement requirements.

- *Organization of the Programming Team.* Most large programming projects require a team effort. One popular organization strategy is the use of chief programmer teams. The team is managed by the chief programmer, a highly proficient and experienced programmer who assumes overall responsibility for the program design strategy, standards, and construction. The chief programmer oversees all coding and testing activities and helps out with the most difficult aspects of the programs.

    Other team members include a backup chief programmer, program librarian, programmers, and specialists. The backup programmer is able to assume the chief programmer's role as well as to perform normal programming activities. The program librarian maintains the program documentation and program library. The programmers, often selected because of specialized programming skills relevant to the project, code and test the programs. Specialists offer unique skills pertinent to the project (e.g., database techniques and telecommunications background).

    The role of the analyst is different from one organization to another. Sometimes the analyst is the project manager to whom the chief programmer reports. At other times, the analyst is a consultant to the chief programming team, possibly as one of the specialists reporting to the chief programmer. Chief programming teams are formed and disbanded with each successive project.

- *Development of a Detailed Programming Plan.* You don't just start programming. Most design specifications include numerous programs. Which programs should be written first? Many systems are built in *versions.* The first version implements the most critical aspects of the system, so that a version can be placed into operation before the system has been completely constructed.

Another appropriate approach is to construct transaction-processing programs first. Implement these programs in the same sequence as that in which they would have to be run (NEW ORDER PROCESSING before ORDER CANCELATION before BILLING and so on). Then implement management reporting and decision support programs according to their relative importance. General file maintenance and backup and recovery programs are written last.

Most organizations have a quality assurance group that must approve plans for programming. Thus, programming standards set forth by this group may serve as input to the process and dictate requirements to ensure quality. These standards normally dictate standards governing programming approaches to be used, documentation to be generated, and test plans to be followed.

Finally, a good plan for programming normally includes test data to be used in testing the various programs to be written. The systems analyst may provide or assist the applications programmers in establishing this test data.

Activity 2: Write and Test New Programs

The major activity of systems implementation is the writing and testing of computer programs. And this is the activity with which you may have the most experience.

Figure 19.3 illustrates that given the technical design statement and plan for programming, the applications programmer will write and test new programs. Since any new programs or program components may have already been written and in use by other existing systems, the experienced applications programmer will know to first check for possible reusable software components available in the information systems shops' software library. The new programs, once developed, may actually become reusable software components that may subsequently be placed in the software library for future use by other programmers. Once again, some information systems shops have a quality assurance group staffed by specialists who review the final program documentation — and systems analysis and design documentation — for conformity to standards. This group will provide appropriate feedback regarding quality recommendations and requirements.

Figure 19.3 presented an overview of writing and testing programs. Let's examine this activity in greater detail. We'd like to summarize at least one appropriate computer program development cycle for comparison with our systems development life cycle. An appropriate program development life cycle is illustrated in Figure 19.4.

This particular program development life cycle begins with a review of program structure. By program structure, we mean the top-down, modular factoring of the program (see Chapter 18). Some information systems shops insist that top-down, modular design is the programmer's responsibility. When that is the case, this first step could be changed from review to design.

The program development life cycle depicted in Figure 19.4 suggests a top-down implementation — that is, modules are designed, coded, and tested beginning with the top module. The upper-level modules drive the lower-level modules. Each module goes through three stages of development: algorithm

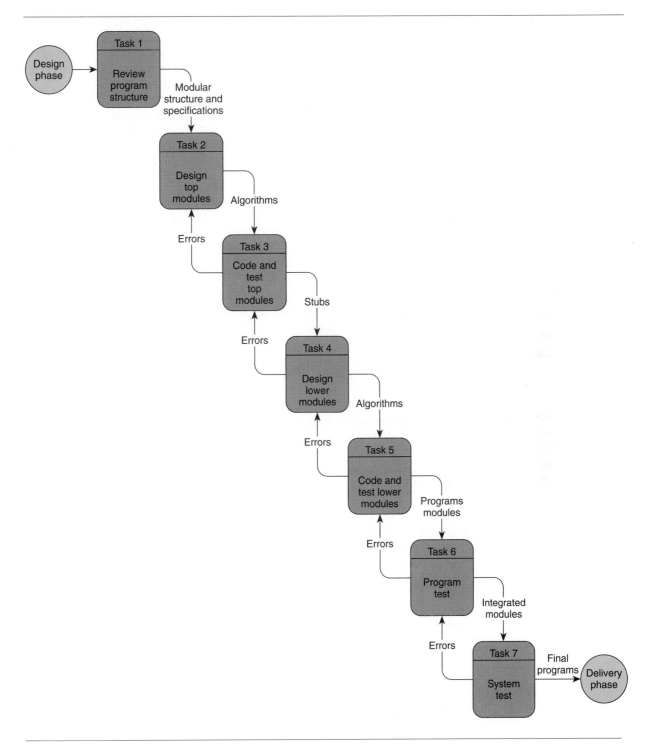

FIGURE 19.4 **A Program Development Life Cycle** Like the systems development life cycle, there are many versions of a program development life cycle. This is one example. This approach suggests a top-down strategy. The high-level modules are built first.

design, coding, and testing. As modules are completed, they are integrated and tested. Eventually, the entire program is completed and tested as a whole. Each program for the new system goes through this cycle. After all the programs have been individually coded and tested, the entire set of all programs that have been developed for the system are integrated and tested as a system.

Testing is an important skill that is often overlooked in academic courses on computer programming. If modules are coded top-down, they should be tested and debugged top-down and as they are written. Testing is not an activity to be deferred until after the entire program has been completely written! There are three levels of testing to be performed: stub testing, unit or program testing, and systems testing.

> **Stub testing** is the test performed on individual modules, whether they be main program, subroutine, subprogram, block, or paragraph.

How can you test a higher-level module before coding its lower-level modules? Easy! You simulate the lower-level modules. These lower-level modules are often called *stubs*. Stub modules are subroutines, paragraphs, and the like that contain no logic. Perhaps all they do is print that they have been correctly called, and then control goes back to the parent module.

> **Unit or program testing** is a test whereby all the modules that have been coded and stub tested are tested as an integrated unit.

Eventually, all modules will have been implemented, and that unit equals the program itself. Unit testing uses the test data that was created during the design phase.

> **Systems testing** is a test that ensures that application programs written in isolation work properly when they are integrated into the total system.

Just because a single program works properly doesn't mean that it works properly with other programs. The integrated set of programs should be run through a systems test to make sure that one program properly accepts, as input, the output of other programs.

We'll talk about additional tests when we discuss installation of the new system. Computer programming activities are frequently governed by information systems standards. These standards dictate program design, coding, testing, and documentation rules that are intended to promote a consistent style within all information systems. These standards are often subject to design and code walkthroughs that check the program for conformity to standards, as well as for logic and design errors.

Programming skills and structured programming are beyond the scope of this book. However, you should learn structured programming and learn how to prepare design specifications that are thorough. This will allow the programmers time to apply structured programming techniques! You cannot be an effective systems analyst without computer programming experience—that experience helps you appreciate the importance of thorough systems design. The use of structured programming techniques results in programs that are easier to write, read, and (especially) maintain.

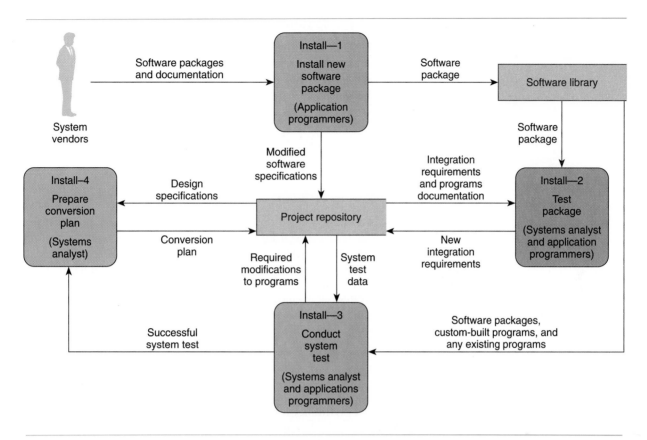

Install and Test New System Phase Activities These activities will lead to completion of the install and test new system phase. The activities are generally completed in a clockwise sequence from the top. On the other hand, the project repository allows you to overlap activities or return to any previous activity to do rework.

THE INSTALL AND TEST NEW SYSTEM PHASE OF SYSTEM IMPLEMENTATION

Figure 19.5 illustrates the install and test new system phase of systems implementation. It is during this phase that software packages are installed and tested. To ensure that integration requirements for the new system are fulfilled, a complete system test is once again conducted. Finally, during this phase a conversion plan is developed for successfully guiding the delivery of the new system into production.

Building Blocks for the Install and Test New System Phase

The fundamental objectives of this phase are

- To install and test new software packages acquired from system vendors.
- To conduct a complete system test to ensure that the custom-built software and acquired software packages work together properly.
- To develop a detailed plan for converting from the old system to the new system.

The primary building block to guide this phase is the same as the previous one — ACTIVITIES.

Install and Test New System Activities, Roles, and Techniques

Let's examine each of the installation and testing activities performed by the applications programmer (see Figure 19.5). Before we begin, realize that the first three activities are optional. That is, if the entire system is to be built in-house, the first three activities are not necessary. The first two activities could be bypassed since the new system would not require installing and testing a vendor's software package. The third activity would not be necessary since we previously conducted a system test on the programs written in-house during the previous phase.

Activity 1: Install New Software Package (if Necessary)

Some systems solutions may have required the purchase or lease of software packages. If so, the applications programmer will now install any new software packages. Given the software packages and documentation, the applications programmer will install the software package and add it to the information systems shop's software library. To provide a history for subsequent projects and individuals who may work with the package, the applications programmer may provide modified software specifications.

Activity 2: Test Package (if Necessary)

Once the software package has been installed, the applications programmer(s) must test the package. Integration requirements (developed during the design phase) and program documentation are additional inputs to the testing activity. Testing a package may reveal new integration requirements for the new system (which may require that the applications programmer return to a previous development activity). The systems analyst typically participates in this activity by serving to clarify requirements.

Activity 3: Conduct System Test (if Necessary)

Now that the software packages have been installed and tested, we need to conduct a final system test. All software packages, custom-built programs, and any existing programs that comprise the new system must be tested to ensure that they all work together. This system test is done using system test data that was developed earlier by the systems analyst. As with previous tests that were performed, our system test may result in required modifications to programs — thus, once again prompting the return to a previous activity in the implementation phase. This iteration would continue until a successful system test was experienced.

Activity 4: Prepare Conversion Plan

Once a successful system test has been completed, we can begin preparations to place the new system into operation. Using the design specifications for the new system, the systems analyst will develop a detailed conversion plan. This plan will identify files/databases to be installed, end-user training and documentation that needs to be developed, and a strategy for converting from the old system to the new system.

THE DELIVER THE NEW SYSTEM INTO OPERATION PHASE OF SYSTEM IMPLEMENTATION

Now we come to the last system implementation phase in our life cycle — deliver the new system into operation. The analyst is the principal figure in the delivery phase, regardless of his or her role in the construction effort.

Building Blocks for the Deliver the New System into Operation Phase

The purpose of the deliver the new system into operation phase is to smoothly convert from the old system to the new system. To achieve the purpose of this phase, we must accomplish the following objectives:

- Install files and/or databases to be used by the new system.
- Provide training and documentation for individuals that will be using the new system.
- Convert from the old system to the new system.
- Evaluate the project and final system.

All building blocks are pertinent to this phase!

Deliver the New System into Operation Phase Activities, Roles, and Techniques

Figure 19.6 illustrates the typical activities of the final phase of systems implementation — deliver the new system into operation. As we did earlier, let's examine each activity in greater detail.

Activity 1: Install Files and/or Databases

In a previous phase you built test files and test databases. To place the system into operation, you need fully loaded (or "populate") files and databases. Therefore, the first activity we'll survey is installation of files and databases (see Figure 19.6).

Special programs will have to be written to populate the new files/databases. Existing data from the production files/databases, coupled with the file and database schema(s) models and file/database structures for the new files/databases will be used to write computer programs to populate the new files/databases with restructured existing data.

At first, this activity may seem trivial. But consider the implications of loading a typical file, say, CUSTOMER ACCOUNT. Tens or hundreds of thousands of records may have to be loaded. Each must be input, edited, and confirmed before the file is ready to be placed into operation.

As a systems analyst, you should calculate file and database sizes and estimate the time required to perform the task of installing them. The task itself is often performed by data-entry personnel because end-users cannot release themselves for enough time to complete this task. Sometimes, temporary help must be hired for this one-time installation effort.

Activity 2: Train System Users

An activity more typically performed by systems analysts is to train system users to use the new system. Given appropriate documentation for the new system,

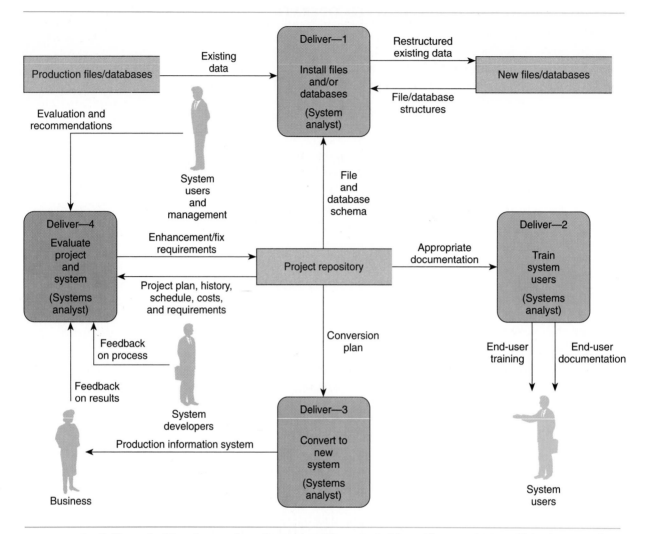

FIGURE 19.6 Deliver the New System into Operation Phase Activities   These activities will lead to completion of the deliver the new system into operation phase. The activities are generally completed in a clockwise sequence from the top. On the other hand, the project repository allows you to overlap activities or return to any previous activity to do rework.

the systems analysts will provide end-user documentation (typically in the form of manuals) and end-user training for the system users.

Remember, the system is for the user! User involvement is important in this activity because the end-users will inherit your successes and failures from this effort. Fortunately, user involvement during this activity is rarely overlooked. The most important aspect of their involvement is training and advising of the users. They must be trained to use equipment and to follow the procedures required of the new system. But no matter how good the training is, users will become confused at times. Or perhaps they will find mistakes or limitations (an inevitable product, despite the best of planning, analysis, design, and implementation techniques). The analyst will help the users through the learning period until they become more familiar and comfortable with the new system.

```
                    Training Manual End-Users Guide Outline

   I. Introduction.

   II. Manual.

       A. The manual system (a detailed explanation of peoples' jobs and standard operating
          procedures for the new system).

       B. The computer system (how it fits into the overall workflow).
          1. Terminal/keyboard familiarization.
          2. First-time end-users.
             a. Getting started.
             b. Lessons.

       C. Reference manual (for nonbeginners).

   III. Appendixes.

       A. Error messages.
```

FIGURE 19.7 An Outline for a Training Manual A good training manual or procedures manual can prevent many problems during the lifetime of a system. This is one outline for such a manual.

Many organizations hire special systems analysts who do nothing but write user documentation and training guides. If you have a skill for writing clearly, the demand for your services is out there! Figure 19.7 is a typical outline for a training manual. The golden rule should apply to user manual writing: "Write unto others as you would have them write unto you." You are not a business expert. Don't expect the reader to be a technical expert. Every possible situation and its proper procedure must be documented.

The actual training is built around the user manuals. Training can be performed one on one; however, group training is generally preferred. It is a better use of your time, and it encourages group learning possibilities. Think about your education for a moment. Isn't it true that you really learn more from your fellow students and colleagues than from your instructors? Instructors facilitate learning and instruction, but you master specific skills through practice with large groups because, with them, common problems and issues can be addressed more effectively. Take advantage of the ripple effect of education. The first group of trainees can then train several other groups.

Related to user manuals and training is the complementary preparation of computer operations manuals and training. These manuals instruct computer operators on how to carry out information systems procedures as documented in systems flowcharts.

Once again, written and oral communications skills are critical. (These skills are more fully covered in Module D, Interpersonal Skills.) Familiarity with organizational behavior and psychology may also prove helpful. Converting to a new system represents change, and people have a natural tendency to resist change or to look for fault in change. There is comfort in the status quo — even if the current system is fraught with problems.

Activity 3: Convert to New System

Conversion to the new system from the old system is a significant milestone. After conversion, the ownership of the system officially transfers from the analysts and programmers to the end-users. The analyst completes this activity by carefully carrying out the conversion plan. The conversion plan will include detailed installation strategies to follow for converting from the existing to a new production information system. Some commonly used strategies include:

1. **Abrupt cutover.** On a specific date (usually a date that coincides with an official business period such as month, quarter, or fiscal year), the old system is terminated and the new system is placed into operation. This is a high-risk approach because there may still be major problems that won't be uncovered until the system has been in operation for at least one business period. On the other hand, there are no transition costs. Abrupt cutover may be necessary if, for instance, a government mandate or business policy becomes effective on a specific date and the system couldn't be implemented prior to that date.

2. **Parallel conversion.** Under this approach, both the old and new systems are operated for some period of time. This is done to ensure that all major problems in the new system have been solved before the old system is discarded. The final cutover may be either abrupt (usually at the end of one business period) or gradual, as portions of the new system are deemed adequate. This strategy minimizes the risk of major flaws in the new system causing irreparable harm to the business; however, it also means that the cost of running two systems over some period of time must be incurred. Because running two editions of the same system on the computer could place an unreasonable demand on computing resources, this may only be possible if the old system is largely manual.

3. **Location conversion.** When the same system will be used in numerous geographical locations, it is usually converted at one location (using either abrupt or parallel conversion). As soon as that site has approved the system, it can be farmed to the other sites. Other sites can be cut over abruptly because major errors have been fixed. Furthermore, other sites benefit from the learning experiences of the first test site. Incidentally, the first production test site is often called a *beta test site.*

4. **Staged conversion.** Like location conversion, staged conversion is a variation on the abrupt and parallel conversions. A staged conversion is based on the version concept introduced earlier. Each successive version of the new system is converted as it is developed. Each version may be converted using the abrupt, parallel, or location strategies.

What happens during the systems conversion? Training may occur, but we factored training out as a separate activity that should begin well before conversion. The major activity is the systems acceptance test. At this point, we should differentiate between what we called a *system test* earlier in the chapter and the *systems acceptance test* performed here. Recall that a systems test is performed by applications programmers using test data.

A **systems acceptance test** is a final system test performed by end-users using real data over an extended period of time.

It is an extensive test that addresses three levels of acceptance testing:

1. **Verification testing** runs the system in a simulated environment using simulated data. This simulated test is sometimes called *alpha testing.* The simulated test is primarily looking for errors and omissions regarding end-user and design specifications that were specified in the earlier phases but not fulfilled during construction.

2. **Validation testing** runs the system in a live environment using real data. This is sometimes called *beta testing*. During this validation, we are testing a number of items.

   a. *Systems performance.* Is the throughput and response time for processing adequate to meet a normal processing workload? If not, some programs may have to be rewritten to improve efficiency or processing hardware may have to be replaced or upgraded to handle the additional workload.

   b. *Peak workload processing performance.* Can the system handle the workload during peak processing periods? If not, we may have to improve hardware and/or software to increase efficiency or rethink our scheduling of processing—that is, consider doing some of the less critical processing during nonpeak periods.

   c. *Human engineering test.* Is the system as easy to learn and use as anticipated? If not, is it adequate? Can enhancements to human engineering be deferred until after the system has been placed into operation?

   d. *Methods and procedures test.* During conversion, the methods and procedures for the new system will be put to their first real test. Methods and procedures may have to be modified if they prove to be awkward and inefficient from the end-users' standpoint.

   e. *Backup and recovery testing.* Now that we have full-sized computer files and databases with real data, we should test all backup and recovery procedures. We should simulate a data loss disaster and test the time required to recover from that disaster. Also, we should perform a before-and-after comparison of the data to ensure that data was properly recovered. It is crucial to test these procedures. Don't wait until the first disaster to find an error in the recovery procedures.

3. **Audit testing** certifies that the system is free of errors and is ready to be placed into operation. Not all organizations require an audit. But many firms have an independent audit or quality assurance staff that must certify a system's acceptability and documentation before that system is placed into final operation. There are independent companies that perform systems and software certification for end-users' organizations.

The systems acceptance test is the final opportunity for end-users, management, and information systems operations management to accept or reject the system. Hopefully, the analysts and programmers are well aware of the criteria for acceptance before this stage. Well-established systems objectives, systems requirements, and information systems and EDP audit policies are important if rework is to be minimized or eliminated.

As system owners and system users uncover errors, the programs and procedures may have to be slightly modified. It might be useful to have regular support or maintenance programmers perform these modifications. This will enable us to test the quality of the program documentation while the programs are fresh in the minds of the original programmers.

Activity 4: Evaluate Project and System

We now come to the final activity of implementation. This activity is sometimes called the *systems audit*. The review is intended to accomplish two goals:

1. Evaluate the operational information system that was developed.
2. Evaluate the systems development procedures to determine how the project could have been improved.

This is the easiest activity to skip — and that would be a major mistake. True, there are other projects waiting to be started or finished. But you have to learn to look at the long-term benefits of this activity. How will you ever do a better job of systems analysis and design if you don't evaluate your current performance?

Given feedback on results from business and feedback on the process provided by systems developers, we can evaluate our performance and learn where improvements can be made. Information concerning our project plan, history, schedule, costs, and requirements as well as evaluation and recommendations for system users and managers will also be examined as part of our evaluation. Given these inputs, the following elements should then be reviewed:

- Does the new information system fulfill the goals and objectives identified and refined early in the project?
- Does the system adequately support the transaction processing, management reporting, and decision support requirements of the business?
- Are the projected benefits being realized?
- How do the users feel about the new system? How can user relations be improved for future projects?
- Should any of the proposed enhancements to the system be addressed immediately? Enhancements should be prioritized.
- Are the internal controls adequate?

As you can see, the review process is a question-and-answer session intended to benefit future projects. Throughout this evaluation, enhancement/fix requirements will be identified. These requirements serve as triggers for the systems support.

Summary

Systems implementation is the process whereby a new information system is placed into operation. Systems implementation consists of four phases:

- Build and test networks and databases.
- Build and test programs.
- Install and test new system.
- Deliver the new system into operation.

The first phase of some implementations is to build and test networks and databases. In many cases, new or enhanced applications are built around existing networks and databases. If the new application calls for new or modified networks and databases, they must normally be implemented prior to writing or installing computer programs because applications programs will use those networks and databases!

Once networks and databases for the new system have been built, attention is directed toward the second phase — build and test programs. This is the first phase in the life cycle that is specific to the applications programmer. This is also a phase with which you are probably most familiar.

During the install and test new system phase, software packages are installed and tested. To ensure that integration requirements for the new system are fulfilled, a complete system test is conducted. Also, during this phase a conversion plan is developed for successfully guiding the delivery of the new system into production.

Finally, the last systems implementation phase is the deliver the new system into operation phase. The purpose of this phase is to smoothly convert from the old system to the new system. This involves installing new files and databases, training end-users, converting to the new system, and evaluating the project and final system.

Key Terms

abrupt cutover, p. 740

audit testing, p. 741

location conversion, p. 740

parallel conversion, p. 740

staged conversion, p. 740

stub testing, p. 734

systems acceptance test, p. 740

systems implementation, p. 726

systems testing, p. 734

unit or program testing, p. 734

validation testing, p. 741

verification testing, p. 740

Problems and Exercises

1. Define the term *systems implementation*.
2. How can a successful and thorough systems planning, analysis, and design be ruined by a poor systems implementation? How can poor systems analysis or design ruin a smooth implementation? For both questions, list some implementation consequences.
3. What skills are important during systems implementation? Create an itemized list. Identify computer, business, and general education courses that would help you develop or improve those skills.
4. How does your information systems building blocks aid in systems implementation? Examine each building block and address issues and relevance to the systems implementation phases of construction and delivery.
5. What products of the systems design phases are used in the systems implementation phases? Why are they important? How are they used? What would happen if they were incomplete or inaccurate?
6. What are the end products of the four implementation and delivery phases? Explain the purpose and content of each of those products.
7. What types of testing are done upon application programs? What types of tests are conducted for an overall system?
8. How would the implementation phase differ if the computer software for supporting the system was purchased?
9. Why should a systems analyst perform a postimplementation review? What types of benefits can be derived? Why do you really need two reviews?

Projects and Minicases

1. You are preparing to meet with your end-users to discuss converting from their old system to a new system. In this meeting you wish to discuss alternative strategies that could be used. Prepare a brief description of the alternative strategies along with a description of situations for which each approach would be preferred and required.

Suggested Readings

Boehm, Barry. "Software Engineering." *IEEE Transactions on Computers,* C-25, December 1976. This classic paper demonstrated the importance of catching errors and omissions before programming begins.

Metzger, Philip W. *Managing a Programming Project.* 2nd ed. Englewood Cliffs, N.J.: Prentice Hall, 1981. This is one of the few books to place emphasis solely on systems implementation.

20

Systems Support

Chapter Preview and Objectives

In this chapter you will learn more about the systems support phases in the systems development life cycle. It is very likely that a young systems analyst or user will become directly involved in a systems support project. Most analysts carry responsibility for one or more legacy systems. Therefore, it is useful to understand support activities and their deliverables. You will know that you understand the process of systems support when you can:

Define
systems support and relate the term to its activities.

Describe
the role of a repository in systems support.

Describe
the support activities in terms of your information system building blocks.

Differentiate
between maintenance, enhancement, reengineering, and design recovery.

Describe
support activities in terms of objectives, activities, roles, inputs and outputs, and tools and techniques.

Minnesota State University

The Minnesota State University is a large, public, metropolitan university located within 20 miles of four cities in Minnesota.

Scene: *Kurt Wilson, Director of Administrative Information Management (AIM), is meeting with Paula Teague, Assistant Director of Applications Development.*

Kurt: Good morning, Paula. How's the cold?

Paula: Much better, thank you. It's good to be back. I assume this is the meeting I had to cancel when I got sick?

Kurt: Right. As you know, the administrative information systems master plan will be complete within the next three months. Assuming the executive committee approves the plan, the real work begins—delivering the new business processes and applications outlined in the plan.

Paula: I've been wondering when you were going to address that issue. We can't keep up with new systems development requests as it is. Am I going to get additional staff?

Kurt: I'm afraid not. In this era of staff downsizing, I suspect that we'll be lucky to hold on to what we have.

Paula: Well, I know we can increase productivity using CASE tools driven off the planning models your staff has recorded in the new repository. But there is a learning curve with CASE technology, as well as the new methodology. Also, these new applications call for a greater degree of adaptability and integration than we have historically expected. I just don't see how we can deliver more systems with the same or fewer people.

Kurt: I've got an idea. I've been running some numbers against the time accounting system. According to our own records, we are using 19.3 FTE (full time equivalent staff) to simply support existing systems maintenance.

Paula: That wouldn't surprise me. Legacy code is the anchor that inhibits new systems development in all shops. Don't tell me you are going to eliminate existing systems support? I think there would be an immediate and fatal backlash from the user community.

Kurt: True, but that's not exactly what I had in mind. Don't have a cardiac, but what would you say if I told you that I wanted you to reduce your maintenance effort to 8.5 FTE?

[Paula does not respond]

Your silence indicates that you are concerned.

Paula: And rightfully so, don't you think? The user community will scream for my head on a platter! You are talking about cutting support by more than 50 percent.

Kurt: Actually, Paula, I'm asking you to cut support by less than 25 percent, but to use 50 percent fewer people!

Paula: And how am I supposed to do that?

Kurt: I have a couple of ideas for you to consider. First, according to my analysis of change request forms, almost two-thirds of all requests fall into the category of enhancements. Half of those enhancements can be characterized as "desirable," not "essential." It seems to me that we could declare a temporary moratorium on such maintenance projects.

Paula: I can't confirm your numbers, but I'll agree that we are going to have to be pickier about what we choose to do and not do. I'd like to see your data after this meeting.

Kurt: No problem! Second, I'd like you to consider a SWAT team approach to maintenance.

Paula: SWAT? Like the police?

Kurt: SWAT stands for "Specialists With Automated Tools." With only 8.5 FTE for maintenance, it seems to me that it would be a mistake to assign one or two persons per existing development team. Instead, I see a maintenance SWAT team that takes over all maintenance.

Paula: It might work. I'd want to make sure the SWAT team had at least one member from each current development team to preserve application knowledge. But what's the "automated tools" angle?

Kurt: Glad you asked that. Our luncheon meeting will be with a sales representative from a company called VIASOFT. They sell what amounts to CASE tools for maintaining, enhancing, and reengineering existing COBOL programs. They promise to increase productivity of maintenance programmers who specialize in the tools.

Paula: Now we're talking. There would be the usual learning curve, but once the team is comfortable with the technology, we might just be able to do 75 percent of our existing maintenance with less than 50 percent of the existing staff.

Kurt: I think so! In fact, I'll make a bolder prediction. In time, I think you'll eventually be able to increase support over today's levels with less than 50 percent staff. Anyway, you're a pretty creative person. Give some more thought to ways we can make this work. We need to go to our luncheon date.

Discussion Questions

1. Why does system support consume up to 80 percent of some systems development budgets?

2. Can you think of other ways to provide adequate systems support while reducing systems support staff?

3. Can you think of any aspects of systems support that have not been addressed by Kurt's SWAT and technology proposals?

WHAT IS SYSTEM SUPPORT?

Systems planning was first defined in Chapter 3:

> **Systems support** is the on-going maintenance of a system(s) after it has been placed into operation. This includes program maintenance and system improvements.

Systems support is driven by systems designers and system builders in support of system users. In your pyramid model, it addresses PEOPLE, DATA, ACTIVITIES, and NETWORKS as implemented by TECHNOLOGY.

In Chapter 3 (and Figure 3.11), you learned that systems support consists of four ongoing activities:

1. Correct errors (also called the *maintenance*).
2. Recover the system.
3. Assist users of the system.
4. Adapt the system to new requirements (also called *reengineering*).

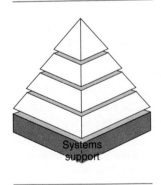

Systems support is often ignored in systems analysis and design textbooks. Young analysts are, therefore, often surprised to learn that half of their duties (or more) are associated with supporting existing systems. As you also learned in Chapter 3, systems support often requires analysts to revisit activities typically performed in systems analysis, design, and implementation.

Let's first set the stage for systems support. In Figure 20.1, applications maintenance projects are contrasted with systems planning and application development projects. Notice that once applications are implemented, they are said to be in *production*. Production is the day-to-day, week-to-week, month-to-month, and year-to-year execution of the application programs to process business data (inputs) and generate useful information (outputs).

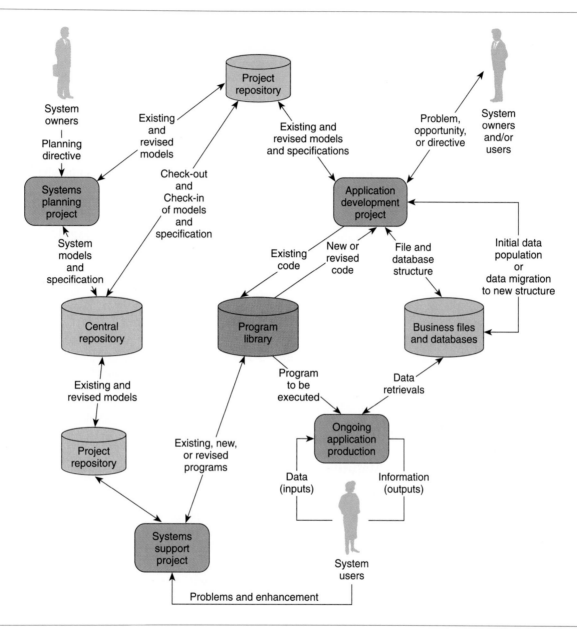

FIGURE 20.1   Context for Systems Support Projects and Activities   This diagram illustrates systems support in the context of other types of projects and activities, namely: systems planning, application development, and application production.

In Figure 20.1, we also see three distinct types of system-level data storage. First, we see our **central repository.** By now, you know that this repository stores all essential and implementation system models and detailed specifications. Subsets of the central repository are checked out to support various planning and development projects. These subsets are stored as project repositories, usually implemented through various CASE tools. Second, we see

program libraries that store the actual application programs that have been placed into production. In most shops, a software-based librarian will track changes and maintain a few previous versions of the software in case a problem arises with a new version. Third, we see actual **business files and databases** that store the operational data created and maintained by the production application programs.

Figure 20.2 is a life cycle diagram that illustrates the four support activities. We've changed the diagram somewhat since Chapter 3. The work products and deliverables are described in the context of Figure 20.1 (using DFD data stores instead of disk files).

This chapter examines each of the support activities in somewhat greater detail. To give you the feel of a true systems development methodology, each phase is described in terms of: (1) the purpose of the phase, (2) the activities that should be performed, (3) the roles played by various people in each task, (4) the inputs and outputs for each task, (5) any techniques or skills that can be used to complete each task, and (6) emerging technology that can simplify tasks.

SYSTEMS MAINTENANCE — CORRECTING ERRORS

Regardless of how well designed, constructed, and tested a system or application may be, errors or **bugs** will inevitably occur. Some bugs will be caused by miscommunication of requirements. Others will be caused by design flaws. Others will be caused by situations that were not anticipated and, therefore, not tested. And finally, bugs may be caused by unanticipated misuse of the programs. In all these situations, corrective action must be taken. We call this corrective action *system maintenance,* or *program maintenance.*

System Maintenance Objectives and Building Blocks

The fundamental objectives of system maintenance are:

- To make predictable changes to existing programs to correct errors that were made during systems design and implementation. Consequently, we exclude enhancements and new requirements from this activity.
- To preserve those aspects of the programs that were already correct. Inversely, we try to avoid the possibility that "fixes" to programs cause other aspects of those programs to behave differently.

To achieve these objectives, you need an appropriate understanding of the programs you are fixing and of the applications in which those programs participate. This prerequisite understanding is often the downfall of system maintenance!

How does system maintenance map to your information system building blocks?

- PEOPLE — System maintenance is usually initiated by system users. Maintenance is usually performed by system builders with possible assistance by system designers.
- DATA — System maintenance rarely impacts data, except for the possibility of improving data editing.

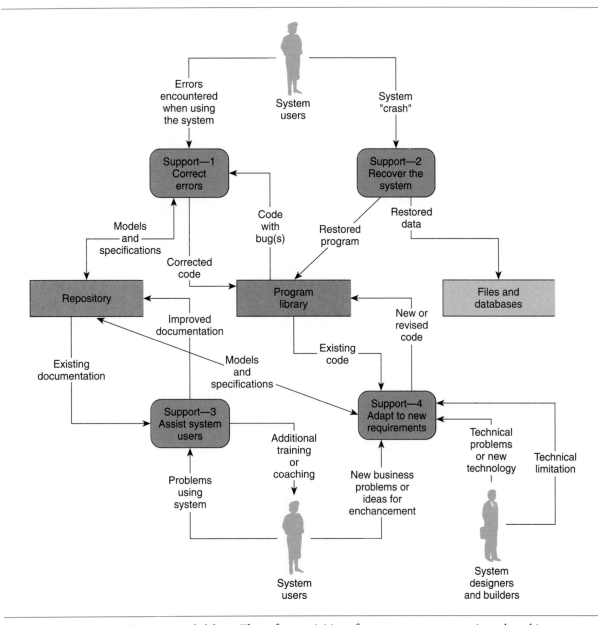

FIGURE 20.2 **Systems Support Activities** These four activities of systems support were introduced in Chapter 3.

- ACTIVITIES — Business and information system processes are implemented as application programs. System maintenance is all about fixing errors made when those programs were implemented.
- NETWORKS — System maintenance rarely involves computer networks, although on occasion, computer networks can be the root cause of bugs.
- TECHNOLOGY — System maintenance, as defined in this activity, does not deal with changing technology.

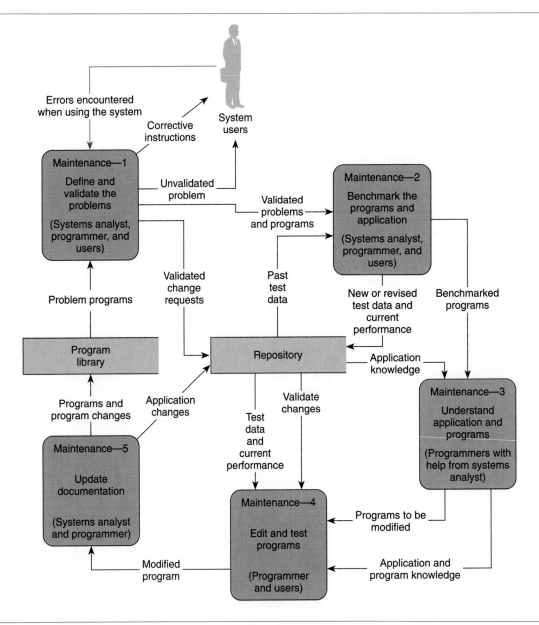

FIGURE 20.3 **System Maintenance Tasks Activities** These activities perform the maintenance of existing systems. Notice that programs are copied out of the program library and not returned until the maintenance is complete. This permits the current program to remain in use until the revised program is ready.

System Maintenance Tasks, Roles, and Techniques

Figure 20.3 illustrates the typical tasks of system maintenance. Let's review the guidelines for reading all such diagrams:

- Silhouettes represent people or departments that initiate tasks.
- The rounded rectangles represent tasks. Each task is numbered solely for identification purposes. The task name is printed in the upper-half of the

symbol. The participants in the task are printed in the lower-half of the symbol. The first participant is always the person who guides the task.

• The solid arrows reflect inputs and outputs for a task. Each is named. When one of these inputs or outputs is referred to in the text, it is underlined.

Let's now examine the steps that must be completed during systems maintenance.

Activity 1: Define and Validate the Problems

The first activity of the assigned team is to define and validate problems. Ideally, this activity will be facilitated by the analyst and/or programmer, but it should clearly involve the user(s). The problem programs are retrieved from the program library.

Working with the user(s), the team should attempt to validate the problem(s) by reproducing it. If the problem cannot be reproduced, the project should be suspended until the problem reoccurs and the user can explain the circumstances under which it occurred. The input is the errors encountered when using the system (usually called *bugs*). One possible output is validated change requests. These change requests should define expectations of the solution.

Another possible output is unvalidated problem. In the event that bug reoccurs, users should be instructed as to how to better document circumstances that led to the bug and symptoms of the problem.

In some cases the bug arises from simple misunderstandings or misuse, and corrective instructions can bring the entire project to closure. On the other hand, if the bug has been validated, the validated problems and programs are passed to the next task.

Note All subsequent maintenance will be performed on a copy of the program(s). The original program remains in the program library and can be used in production systems until it is fixed.

The analyst or programmer requires appropriate fact-finding skills (Part Five, Module B) and interpersonal skills (Part Five, Module D). The analyst or programmer also needs to be familiar with the application.

Activity 2: Benchmark the Programs and Application

The program(s) isn't all bad, or it would have never been placed into production in the first place. The team should next benchmark the programs and application. System maintenance can result in unpredictable and undesirable side effects that impact the programs or application's overall functionality and performance. For this reason, we highly recommend that, before making any changes to programs, the programs be executed and tested to establish a baseline against which the modified programs and applications can be measured.

This step is performed by the systems analyst and/or programmer. The users may also participate. The primary inputs are the validated problems and programs.

Test cases can be defined in either of two ways. First, you may find past test data in the repository. If so, that data should be re-executed to provide the benchmark. It should also be analyzed for completeness and, if necessary, revised. The correct responses, including error handling, should be recorded in the repository.

Alternatively, test data can be automatically captured using a test tool. For example, IBM's Workstation Interactive Test Tool (WITT) records keystrokes, mouse movements and commands, and system responses while a user demonstrates the typical use of the program. (Note: The user doesn't even have to know the test tool is there.) The advantage is that the test cases and responses are recorded in the repository for later playback. Test tools also measure response time and throughput using the test cases.

For either alternative, the outputs are <u>new or revised test data and current performance</u> and <u>benchmarked programs</u> (to the next task).

The analyst or programmer needs to have good testing skills (usually taught in programming courses) and may require training in test tools. Neither is explicitly taught in this textbook.

Activity 3: Understand the Application and its Programs

Frequently, system maintenance is not performed by the same persons who wrote the program. In fact, several people may have written parts of any program or application, and those people may no longer be available for clarification. For this reason, we need to gain an understanding of the application and the problematic programs. You may be surprised to learn that most analysts and programmers spend more time in this task than any other!

Ideally, <u>application knowledge</u> comes from the repository. This assumes, of course, that application knowledge has been maintained throughout the application's lifetime. All too often, this is not true—especially for older systems. In non-repository-based shops, application knowledge may be available from prior programmers and analysts, but it is usually not up-to-date. Still, it may be useful for gaining a sufficient understanding of the application and where the problematic programs fit into that application.

<u>Application and program knowledge</u> usually comes from studying the source code from the <u>benchmarked programs</u>. Unfortunately, program understanding can take considerable time. This activity is slowed by some combination of the following limitations:

- Poor modular structure.
- Unstructured logic (from prestructured era code).
- Prior maintenance (quick fixes and poorly designed extensions).
- Dead code (instructions that cannot be reached or executed—often leftovers from prior testing and debugging).
- Poor or inadequate documentation.

The purpose of application understanding is to see the big picture—that is, how the programs fit into the total application and how they interact with other programs. The purpose of program understanding is to gain insight into how the program works and doesn't work. You need to understand the fields (variables) and where and how they are used, and you need to determine the potential impact of changes throughout the program(s). Program understanding can also lead to better estimates of the time and resources that will be required to fix the errors.

There used to be no shortcuts for program understanding. Today, maintenance CASE technology can help. For example, VIASOFT's VIA/Insight provides studies and analyzes programs to provide the programmer with considerable insight and information about unfamiliar code. It reveals program structure, marks or traces related code even if it is in different paragraphs and subroutines, isolates dead code (code that cannot be executed), traces field usage (and nonusage) and relationships between different field names and structures, and provides numerous cross references. This information can reduce the time it takes to understand programs by hours, days, and even weeks.

IBM's COBOL Structuring Facility restructures poorly structured programs to conform to accepted structured programming practices. VIASOFT's VIA/SmartDoc can then generate a wealth of information, graphical and textual, for the program(s).

Activity 4: Edit and Test the Programs

Given application and program knowledge and validated changes, you can now make changes to the programs to be modified. This task, performed by the programmer, is not dissimilar from that described in the last chapter on systems implementation. The result is a modified program.

There is a big difference between editing a new program and editing an existing program. As the designer and creator of a new program, you are probably intimately familiar with the structure and logic of the program. By contrast, as the editor of the existing program, you are not nearly as familiar (or current) about that program. Changes that you make may have an undesirable ripple effect through other parts of the program or, worse still, other programs in the application.

Hopefully, your code changes will benefit from your understanding (or review) of the application and program. But testing takes on even greater importance in system maintenance. The following tests are essential and recommended:

- **Unit testing** (essential) — ensures that the stand-alone program fixes the bug without side effects. The test data and current performance that you recovered, created, edited, or generated in task 2 is used here.
- **System testing** (essential) — ensures that the entire application, of which the modified program was a part, still works. Again, the test data and current performance from task 2 is used here.
- **Regression testing** (recommended) — extrapolates the impact of the changes on program and application throughput and response time from the before-and-after results using the test data and current performance.

Test programs such as IBM's WITT (described earlier) simplify testing by playing back recorded data against the modified programs. Tested programs will be returned to production. Generally speaking, when the programs are returned to the program library, they are subject to version control.

Version control is a process whereby a librarian (usually software-based) keeps track of changes made to programs. This allows recovery of prior versions of the programs in the event that new versions cause unexpected problems. In other words, version control allows users to return to a previously accepted version of the system.

Examples of version control software include IBM's SCLM (for host-based software) and INTERSOLV's PVCS (for local area networks).

Activity 5: Update Documentation

The high cost of system maintenance is due, in large part, to failure to update application and program documentation. If application documentation has changed in the slightest, it should be modified in the repository and program library. Application documentation is usually the responsibility of the systems analyst who supports that application. Program documentation is usually the responsibility of the programmer who made the program changes.

The programmer is responsible for this activity. The input is the modified program from task 4. Application changes (changes in models and specifications) are recorded in the repository. The new programs and program changes are stored in the program library. Once returned to the library, they are available for production.

Recording application and program changes in the repository and program library will help future programmers and analysts (including yourself) reduce application understanding time during future maintenance. You will forget changes, however small, unless they are properly recorded. The long-term benefit comes when the application is due for major redevelopment. The study phase of systems analysis will pass quickly if existing documentation is up-to-date.

SYSTEM RECOVERY — OVERCOMING THE "CRASH"

From time to time a system failure is inevitable. It generally results in an aborted or "hung" program (also called an "ABEND" or "crash") and possible loss of data. The systems analyst often fixes the system or acts as intermediary between the users and those who can fix the system. The purpose of this section is to quickly summarize the analyst's role in system recovery.

This activity does not require a multitask flow diagram to detail its steps. They can be summarized as follows:

1. In many cases the analyst can sit at the user's terminal and recover the system. It may be something as simple as pressing a specific key or re-booting the user's personal computer. Corrective instruction may be required to prevent the crash from reoccurring. In some cases the analyst may arrange to observe the user during the next use of the program or application.

2. In some cases the analyst must contact systems operations personnel to correct the problem. Operations can usually terminate an on-line session and reinitialize the application and its programs.

3. In some cases the analyst may have to call data administration to recover lost or corrupted data files or databases. Data backup and recovery is beyond the scope of this book. It is covered extensively in most data and database management courses and textbooks.

4. In some cases the analyst may have to call network administration to fix a local, wide, or internetworking problem. Network professionals can usually log out an account and reinitialize programs.

5. In some cases the analyst may have to call technicians or vendor service representatives to fix a hardware problem.

6. In some cases the analyst will discover a bug caused the crash. The analyst attempts to quickly isolate the bug and trap it (automatically or by coaching users to manually avoid it) so that it can't cause another crash. Bugs are then handled as described in the last section of this chapter.

END-USER ASSISTANCE

Another relatively routine ongoing activity of systems support is routine end-user assistance. No matter how well users have been trained or how well documentation has been written, users will require additional assistance. The systems analyst is generally on call to assist users with the day-to-day use of specific applications. In mission critical applications, the analyst must be on call day and night.

Once again, a detailed flow diagram is unnecessary for this activity. The most typical tasks include:

- Routinely observing the use of the system.
- Conducting user-satisfaction surveys and meetings.
- Changing business procedures for clarification (written and in the repository).
- Providing additional training.
- Logging enhancement ideas and requests in the repository.

SYSTEMS ENHANCEMENT AND REENGINEERING

Adapting an existing system to new requirements is an expectation for all newly implemented systems. Adaptive maintenance forces an analyst to analyze the new requirement and return to the appropriate phases of systems analysis, design, and implementation. In this section we will examine two types of adaptive maintenance, systems enhancement and systems reengineering.

Systems Enhancement and Reengineering Objectives and Building Blocks

Most adaptive maintenance is in response to new business problems, new information requirements, or new ideas for enhancement. It is reactionary in nature—fix it when it breaks or when users make a request. We call this **system enhancement.** The objective of system enhancement is to modify or expand the application system in response to constantly changing requirements. This objective can be linked to your information system building blocks as follows:

- PEOPLE—Most system enhancements are driven by system users, although systems analysts, designers, and builders may isolate technical problems such as performance, security, and internal controls.
- DATA—Many system enhancements are requests for new information that can be derived from existing stored data. Some data enhancements call for expansion of data storage.
- PROCESSES—Most system enhancements require the modification of existing programs or the creation of new programs to extend the overall application system.

- NETWORKS—Most system enhancements are not driven by networks (see *reengineering*).
- TECHNOLOGY—Most system enhancements are driven by technology (see *reengineering*).

Another type of reactionary maintenance deals with changing technology. Information system staffs have become increasingly reluctant to wait until systems break. Instead, they choose to analyze their program libraries to determine which applications and programs are costing the most to maintain or which ones are the most difficult to maintain. These systems might be adapted to reduce the costs of maintenance. The preceding examples of adaptive maintenance is classified as *reengineering*. The objectives of reengineering are to either adapt the system to a major change in technology, fix the system *before* it breaks, or make the system easier to fix when it breaks or needs to be adapted. These objectives can be linked to your information system building blocks as follows:

- PEOPLE—Most reengineering is driven by information systems and technology staff.
- DATA—Many reengineering projects are driven by the need to restructure stored data, either to make it more flexible and adaptable or to convert it to a new technology.
- PROCESSES—Many reengineering projects attempt to restructure or reorganize application programs to make them more maintainable or to convert them to a new technology (e.g., language). Many others change the input or output methods for programs (e.g., from batch to on-line or from on-line to graphical user interfaces).
- NETWORKS—Some application projects seek to change applications to new network technology.
- TECHNOLOGY—Most reengineering projects are driven by changing technology or the need to better exploit existing technology.

Before we move on, we should acknowledge another trend in systems support. As a system's useful lifetime approaches its end, systems analysts and designers will turn to design and analysis recovery technology to automatically discover the models and specifications hidden in old COBOL programs. For more about this exciting trend, see The Next Generation box.

Systems Enhancement and Reengineering Activities, Roles, and Techniques

Figure 20.4 expands on the activities of systems enhancement and reengineering. In this section, we briefly describe each activity, participants and roles, inputs and outputs, and techniques.

Activity 1: Analyze Enhancement Request

The purpose of this activity is to determine the appropriate course of action to either a new business problem or idea for enhancement, technical limitation or problem, or enhancement idea (from other system support activities). Recall that the support phase, in general, does not actually enhance the system. In-

---

The Next Generation

DESIGN AND ANALYSIS RECOVERY

Systems reengineering is presented in this chapter as a modern way to extend the useful life of legacy applications and code. However, eventually, all systems become obsolete — eventually, we must start over with systems planning or systems analysis and build totally new applications. Or do we?

In this book you have learned a variety of system modeling techniques. Model details are recorded in a repository. But do you always have to start from a blank sheet of paper in systems planning or systems analysis? Some experts think not. They are trying to develop new CASE tools that use reverse engineering technology to recover analysis and design models and details from existing programs.

The life cycle in this book has been presented from a forward engineering viewpoint. First you analyze the problem and requirements. Then you design and implement the solution. But you may be able to go backward as well — this is reverse engineering.

Inside every existing program, no matter how young or old it is, is the accumulated knowledge and design expertise of every person who ever worked on the program — both in creating it and in maintaining it! If we could extract that knowledge easily, we could store it in the repository and make it available to systems planning, systems analysis, and systems design efforts. This could greatly reduce the amount of time that must be spent in those phases. This is what design and analysis recovery is about.

Design recovery is the capture of implementation models and details from a program. Consider the benefit if a CASE tool could read a COBOL program and generate its flowcharts, structure charts, and data flow diagrams automatically. The programmer or analyst could then make appropriate technical changes and forward engineer the program into a better program. This is the intent of emerging tools such as INTERSOLV's Excelerator for Design Recovery. Using such

technology, we should be able to perform more extensive and complex reengineering objectives.

Analysis recovery is the ultimate goal. **Analysis recovery** is the capture of the essential models and details from an implemented program. This would provide a quick point of departure for changing and expanding business requirements. There is a major challenge inhibiting the rapid development of this technology. A program is, by its very nature, an implementation of an essential model. The essence is there, but it is not explicitly recorded. Most researchers agree that an expert system and decision support solution is likely. The analysis recovery tool would identify the implementation model components and interact with the analysts and users to determine and record their essential equivalence. Still, the enormous potential for reducing project start-up time makes analysis recovery technology a bright star on the horizon.

---

stead, it studies existing documentation to determine the appropriate course of action. Based on analysis of current system models, that action may include:

- Define new business requirements and return to systems analysis.
- Define new technical requirements and return to systems design.
- Define new program requirements and proceed to task 2.

In the latter case, new programs are generally restricted to those that generate new information from existing data stores. Anything more complex should go through systems analysis and design.

The systems analyst should be skilled in project planning (Part Five, Module A), fact finding (Part Five, Module B), and cost/benefit analysis (Part Five, Module C). The latter may be necessary to justify the enhancement project.

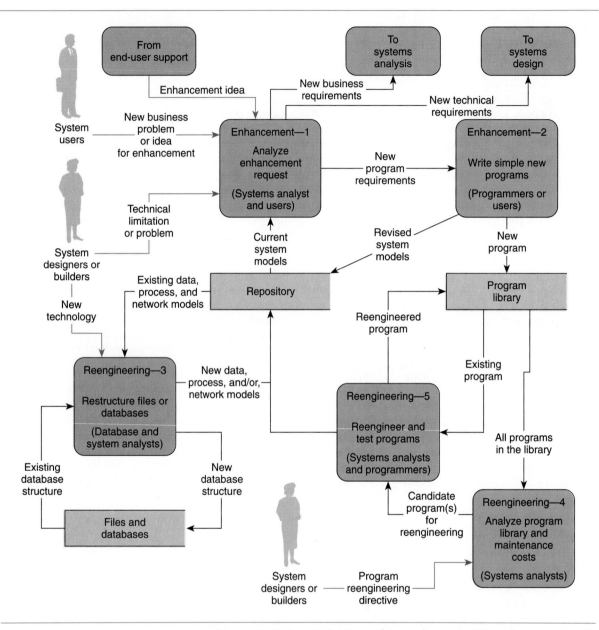

FIGURE 20.4 **System Enhancement and Reengineering Activities**   These tasks support the enhancement and reengineering of existing systems. Notice that many enhancement and reengineering projects quickly pass control back to systems analysis or systems design to more fully consider the implications of change.

Activity 2: Write Simple, New Programs

Many enhancements can be accomplished quickly by writing simple, new programs. What do we mean by "simple"? Simple programs are those that use existing data, do not update existing data, and do not input new data (for purposes of storing that data). In other words, these programs generate new reports and answer new inquiries. New program requirements represent the majority of today's enhancements.

Note It is our belief that any new program requirements that exceed our definition of "simple" should be treated as new business requirements and subjected to a systems analysis and design to more fully consider implications within the complete application system's structure.

Most such programs can be easily written by end-users with a minimal knowledge of a fourth-generation language (such as Focus or SAS) or a PC-to-host database retrieval language (such as Q&E), but also becoming available in most PC database packages (such as Access, Approach, and Paradox). Programmers and analysts are also capable of writing such programs, but some shops question whether this is a valuable use of their time.

With today's fourth-generation and database languages/tools, these programs can be completed within hours. Since they generally do not enter or update data stores, testing requirements are not nearly as stringent. Once implemented, a new program may be stored locally (on a PC or LAN server), or it may be added to the program library (if many people could benefit from its use). Since the "local" programs are created for specific users, they generally do not qualify for IS support; however, in most cases the area systems analyst or local end-user computing guru provides some support.

Optionally, revised system models may be updated in the repository to reflect the existence of the new processes (programs) added to the system.

Clearly, the primary skill for this task is programming in a fourth-generation language (4GL). Courses on 4GLs are becoming more common as academic institutions react to the declining popularity and use of third-generation languages such as COBOL.

Activity 3: Restructure Files or Databases

From time to time, systems analysts help in the reengineering of files and databases. Many of today's data stores are implemented with traditional file structures (e.g., ISAM and VSAM) or early database structures (e.g., hierarchical IMS structures and network structures). Today's database technology of choice is SQL-based relational databases (which store data in tables that are integrated through redundant fields that act as pointers). Tomorrow, object database technology may present yet another shift in popularity.

Migrating data structures from one data storage technology to another is a major endeavor, wrought with opportunities to corrupt essential business data and programs. Thus, reengineering file and database structures has become an important task.

Database reengineering is usually covered more extensively in data and database management courses and textbooks; however, a brief explanation is in order here. The key player in database restructuring is the database analyst (or database administrator). The systems analyst plays a role because of the potential impact on existing applications. Network analysts may also be involved if databases are (to be) distributed across computer networks.

The key inputs are the existing database structure (which can be read from the file or database management system's dictionary that is included in most data stores) and existing data, process, and network models as stored in the repository. The outputs are new database structure and new data, process, and network model.

Data restructuring can and has been done by hand in many businesses. However, database reengineering CASE tools are increasingly used to read the data structure, produce the implementation model, perform data analysis to improve the model, and regenerate the new database structure. Arguably, the leader in this field has been Bachman Information Systems whose Bachman/ DBA product has assisted many businesses in converting older VSAM and IMS files and databases to DB2 relational databases (and today, to Sybase relational server databases). For the most part, database reengineering tools only convert data stores. Processes must also be converted to execute the new database retrieval and update commands against the new data structures. Technology for such conversions is available, but generally it is very expensive.

We've said it before, but today's systems analyst must be both database literate and data management literate. Courses exist in most IS programs to provide such knowledge.

Activity 4: Analyze Program Library and Maintenance Costs

As mentioned earlier, many businesses are questioning the return on investment in corrective and adaptive maintenance. They realize that if complex and high-cost software can be identified, it might be reengineered to reduce complexity and maintenance costs. The first activity required to achieve this goal is to analyze program library and maintenance costs. This activity almost always requires software capable of performing the analysis. Systems analysts usually interpret the results.

Software tools such as VIASOFT's VIA/Recap measures your software library using a variety of widely accepted software metrics.

Software metrics are mathematically proven measurements of software quality and productivity.

Examples of software metrics applicable to maintenance include:

- **Control flow knots** — The number of times logic paths cross one another. Ideally, a program should have zero control flow knots. (We have seen knot counts in the thousands on some older, poorly structured programs.)
- **Cycle complexity** — The number of unique paths through a program. Ideally, the fewer, the better.

Software metrics, in combination with cost accounting (on maintenance efforts) can help identify those programs that would benefit from restructuring.

The input to this task is <u>all</u> (or most) <u>programs in the library</u>. The output is a <u>candidate program(s) for reengineering</u>.

Activity 5: Reengineer and Test Programs

Given a <u>candidate program(s) for reengineering</u>, there are three types of reengineering that can be performed on that program: code reorganization, code conversion, and code slicing.

- **Code reorganization** restructures the modular organization and/or logic of the program. For example, modules may be combined or separated to reduce coupling or increase cohesion (see Chapter 18). Logic may be restructured to eliminate control flow knots and reduce cycle complexity. IBM's COBOL Structuring Facility automates code reorganization for poorly structured COBOL programs.

- **Code conversion** translates the code from one language to another. Typically, this translation is from one language version to another. For example, IBM's COBOL and CICS Conversion Aid (CCCA) converts COBOL based on earlier standards to the COBOL 2. There is a debate on the usefulness of translators between different languages. If the languages are sufficiently different, the translation may be very difficult. If the translation is easy, the question is "why change"? On the other hand, a strong argument could be made for the translation of COBOL calls based on an old technology (e.g., IMS) to calls based on a new database technology (e.g., DB2—see Task 3). If automated conversion were not available, the translation of the call statements would still be essential if the database structure were changed.

- **Code slicing** is the most intriguing program-reengineering option. Many programs contain components that could be factored out as subprograms. If factored out, they would be easier to maintain. More importantly, if factored out, they would be reusable. Code slicing cuts out a piece of a program to create a separate program or subprogram. This may sound easy, but it is not! Consider your average COBOL program. The code you want to slice out may be located in many paragraphs and have dependent logic in many other paragraphs. Futhermore, you would have to simultaneously slice out a subset of the Data Division for the new program or subprogram.

    Code slicing is greatly simplified with reengineering software. VIA-SOFT's VIA/Renaissance is a code-slicing tool that automatically traces code back to prerequisite code and data structures. It can quickly create a stand-alone executable program from the sliced code. Furthermore, it can modify the original program to call the new program (deleting the sliced code from the original), or it can keep the original program unchanged.

The candidate program for reengineering is copied from the program library. It is reengineered using one or more of the preceding methods, it is thoroughly tested (as described earlier in the chapter), and the reengineered program is returned to the program library where it is available for production. Any new data, process and/or network models are updated in the repository.

--- *Summary* ---

System support is the ongoing maintenance of a system after it has been placed into production. It consists of four ongoing activities (as opposed to phases): (1) correct errors, (2) recover the system, (3) assist users, and (4) adapt the system. This chapter focuses on the first and last of these activities, often called maintenance, enhancement, and reengineering.

Systems maintenance deals with errors or bugs in the implemented system. When a bug is identified, the analyst or programmer first attempts to define and validate the problem by reproducing it. Next, the analyst or programmer should benchmark the program by running it against existing test data or generating and running test data. The system's benchmark results and performance provide a baseline against which modified programs can be compared. The third and usually most time-consuming task is to understand the program and its context application. The danger of program maintenance is the potential for changes to fix one problem while creating others, possibly many others. Program understanding attempts to avoid this by thoroughly analyzing the program and application before changing it. Next, the programmer edits and tests the programs and applications. The programmer should perform the following tests: unit test, system test, and regression test—the latter estimates the impact of changes on performance. Programs are usually returned to the program library through version control software that tracks changes to production programs over time. The final task is to update system documentation in the repository.

System enhancement adapts an application to new business and technical requirements. The first step is to analyze the enhancement request. Many enhancements can be fulfilled by writing new, simple programs to retrieve, analyze, and report on existing data. This can often be done directly by end-users or analysts. For more complex enhancements, control is transferred back to systems analysis (for new business requirements) or systems design (for new technical requirements).

Systems reengineering is a more significant transformation of one technical implementation into a different technical implementation, without changing business requirements and functionality. This may involve a restructuring of files or databases to use a different database technology. Alternatively, you could evaluate your program library to discover programs that are costly to maintain due to complexity or constantly changing requirements. Candidate programs for reengineering are identified through software metrics that measure complexity and quality, as well as cost accounting data that identifies maintenance frequency and costs. Three types of reengineering can be applied: (1) code translation—to change from one language into another, (2) code restructuring—to reorganize modules and logic, and (3) code slicing—to slice out code that can stand alone (and be reused) as a separate program or subprogram.

This chapter completes our survey of the entire systems development life cycle.

Key Terms

Analysis recovery, p. 757

Bugs, p. 748

Business files and databases, p. 748

Code conversion, p. 761

Central repository, p. 747

Code reorganization, p. 760

Code slicing, p. 761

Control flow knots, p. 760

Cycle complexity, p. 760

Design recovery, p. 757

Program libraries, p. 748

Regression testing, p. 753

Software metrics, p. 760

System enhancement, p. 755

Systems support, p. 746

System testing, p. 753

Unit testing, p. 753

Version control, p. 753

Problems and Exercises
---

1. Identify and briefly describe the purpose of each of the four ongoing activities of systems support.
2. What is a repository? How does it differ from a program library and a business database?
3. How do application development projects and application maintenance projects differ in their use of a project repository and program library?
4. What are the objectives of systems maintenance? How do the information system building blocks apply to systems maintenance?
5. How does an analyst or programmer validate a bug? Why do they validate a bug?
6. What is the purpose of program benchmarking as it pertains to systems maintenance?

7. What are the inhibitors to program understanding? Why is program understanding essential to systems maintenance? How is application understanding different from program understanding?

8. What is the purpose of regression testing? How does it differ from unit testing and system testing?

9. What is version control? Why is version control a necessity in systems maintenance and enhancement?

10. What is the analyst's role in recovering from a system crash?

11. Differentiate between systems maintenance and systems enhancement. Then do the same for systems enhancement and systems reengineering.

12. What are the fundamental objectives of systems enhancement? Systems reengineering? Explain how the information systems building blocks aid in identifying the required level of understanding to be obtained in systems enhancement and reengineering.

13. What type of system enhancements can proceed immediately to programming? Why?

14. What types of enhancements should proceed to systems analysis? Systems design? Why?

15. What role does analyzing your program library play in systems reengineering?

16. What are software metrics? How are they used in systems reengineering?

17. A program has a control flow knot count of 1. Is this good or bad? Why?

18. Why is code reorganization a reengineering option for many legacy programs?

19. What is code slicing? What benefits does it offer an information systems shop?

Projects and Minicases

1. There are a number of CASE tools specifically oriented to systems maintenance, redevelopment, and reengineering. Through research, do a market survey to identify the characteristics and capabilities of these tools. Briefly compare several products. Select one product and write the vendor. Request product literature, company information, success stories, and the like. Complete your report with an in-depth discussion of the CASE tool.

2. There are a number of commercial methodologies that include systems maintenance or redevelopment within their approach. Through research, do a market survey to identify the characteristics and capabilities of the planning capabilities of these methodologies. Select one methodology and write the vendor. Request product literature, company information, success stories, and the like. Compare and contrast the methodology with the generic methodology presented in this chapter. Force yourself to identify at least three major features or capabilities that you like better and three that you have some concerns about.

3. Make an appointment with a systems analyst or programmer. Conduct an interview on either the problems and issues encountered in systems maintenance, or their standard approach or methodology for systems maintenance. Compare and contrast this with the information and approach presented in this chapter. Write a report of your findings.

The modules in Part Five are not appendices! There are a number of skills, tools, and techniques that are important to multiple phases of systems development—planning, analysis, design, implementation, and support. These activities were introduced in Chapter 3 as cross life cycle activities. We feel strongly that these cross life cycle skills are as important as any tool and technique taught in this book. In fact, they may be the ultimate critical success factor for all systems work. So why are they at the end of the book?

Placing these modules (a name chosen to distinguish them from the chapters you've been reading) in the analysis, design, or implementation/support units would have understated their value to the other phases. On the other hand, these modules do require some prerequisite knowledge—in most cases, Chapters 1 through 5. By placing these modules at the end of the book, you and your instructor have the flexibility to introduce when you prefer. Each module begins by describing not only objectives, but also prerequisite chapters.

Module A introduces project management tools and techniques. All projects are dependent on the planning, control, and leadership principles that are surveyed. The module also presents two popular modeling techniques for project management: PERT and Gantt. These tools help you schedule activities, evaluate progress, and modify schedules.

Module B surveys fact-finding techniques. These techniques are used to solicit factual information, opinions, and requirements from end-users. They are a crucial prerequisite to all modeling techniques—for example, data flow diagrams, entity relationship diagrams, location connectivity diagrams, and so forth. Techniques surveyed include sampling, research, observation, questionnaires, interviews, and Joint Application Development (JAD) sessions.

Module C presents feasibility and cost-benefit analysis techniques. For any potential solution that you evaluate and recommend to management, you must be prepared to defend its operational, technical, schedule, and economic feasibility. Economic feasibility is especially important, since most organizations are either profit-oriented, cost-reduction-oriented, or both. Your ability to estimate costs and benefits and then analyze those numbers for cost-effectiveness is critical.

Finally, Module D introduces communications skills for the systems analyst. After you've collected facts and modeled systems, you must be able to present your findings and recommendations. In this module you will learn how to plan and run meetings, conduct brainstorming sessions, conduct walkthroughs of documentation, make oral presentations, and write reports.

We cannot overstate the value of these modules. The material presented will have greater impact on your success as a systems analyst (or effective user participating in systems development) than any other material in the book. The modules teach skills that make the concepts, tools, and techniques in Parts One through Four easier to apply.

CROSS LIFE
CYCLE ACTIVITIES

——— **A** ———

Project Management

——— **B** ———

Fact-Finding Techniques

——— **C** ———

Feasibility Analysis

——— **D** ———

Interpersonal Skills

A

Project Management

When Should You Read This Module?

This module will prove most valuable if read after any of the following chapters:

Chapter 3
A Systems Development Life Cycle

Chapter 4
Systems Development Techniques and Methodologies

Chapter 6
Systems Planning

Chapter 7
Systems Analysis

Chapter 12
Systems Design

Chapter 19
Systems Implementation

Chapter 20
Systems Support

If this is the first module you are studying, you should also read the
part opener for Part Five of the textbook.

Most of you are familiar with Murphy's law, which suggests that "if anything can go wrong, it will." Murphy's law has motivated numerous pearls of wit and wisdom about projects, machines, and people, and why things go wrong. Unfortunately, many amusing laws, postulates, and theorems have been developed from our failures as project and people managers. The systems analyst frequently assumes a project management role. The project manager is usually a senior systems analyst who must plan, staff, and control the project's many tasks. The purpose of this module is to introduce you to project management. Because project management is applied during systems planning, systems analysis, systems design, systems implementation, and systems support, we decided to place it in this supplementary module. You will learn about the importance of project management as well as guidelines, tools, and techniques for managing projects.

FUN & GAMES, INC.

Fun & Games, Inc. is a successful developer and manufacturer of board, electronic, and computer games. The company is headquartered in Cleveland, Ohio.

Scene: *Jan Lampert, Applications Development Manager, has requested a meeting with Steven Beltman, Systems Analyst and Project Manager for a new distribution project recently placed into production.*

Jan: Steven, I want to discuss the distribution project your team completed last month. Now that the system has been operational for a few weeks, we need to evaluate the performance of you and your team. Frankly, Steven, I'm a little disappointed.

Steven: Me too! I don't know what happened! We used the standard methodology and tools, but we still had problems.

Jan: You still have some Steven. The production system isn't exactly getting rave reviews from either users or managers.

Steven: I know.

Jan: Well, I've talked to several of the analysts, programmers, and end-users on the project, and I've drawn a few conclusions. Obviously, the end-users are less than satisfied with the system. You took some shortcuts in the methodology, didn't you?

Steven: We had to! We got behind schedule. We didn't have time to follow the methodology to the letter.

Jan: But now we have to do major parts of the system over. If you didn't have time to do it right, where will you find time to do it over? You see, Steven, systems development is more than tools, techniques, and methodologies. It's also a management process. In addition to your missing the boat on end-user requirements, I note two other problems. And both of them are management problems. The system was over budget and late. The projected budget of $35,000 was exceeded by 42 percent. The project was delivered 13 weeks behind schedule. Most of the delays and cost overruns occurred during programming. The programmers tell me that the delays were caused by rework of analysis and design specifications. Is this true?

Steven: Yes, for the most part.

Jan: Once again, those delays were probably caused by the shortcuts taken earlier. The shortcuts you took during analysis and design were intended to get you back on schedule. Instead, they got you further behind schedule when you got into the programming phase.

Steven: Not all the problems were due to shortcuts. The users' expectations of the system changed over the course of the project.

Jan: What do you mean?

Steven: The initial list of general requirements was one page long. Many of those requirements were expanded and supplemented by the users during the analysis and design phases.

Jan: The old "creeping requirements syndrome." How did you manage that problem?

Steven: Manage it? Aren't we supposed to simply give in? If they want it, you give it to them.

Jan: Yes, but were the implications of the creeping requirements discussed with project's management sponsor?

Steven: Not really! I don't recall any schedule or budget adjustments. We should explain that to them now.

Jan: An excuse?

Steven: I guess that's not such a good idea. But the project grew. How would you have dealt with the schedule slippage during analysis?

Jan: If I were you, I would have reevaluated the scope of the project when I first saw it changing. In this case, either project scope should have been reduced or project resources — schedule and budget — should have been increased. [pause] Don't be so glum! We all make mistakes. I had this very conversation with my boss seven years ago. You're going to be a good project manager. That's why I've decided to send you to this project management course and workshop.

Discussion Questions

1. What did Steven do wrong? How would you have done it differently?

2. Should Jan share any fault for the problems encountered in this project?

3. Why would it be a mistake to use creeping requirements as an excuse for the project mismanagement?

4. How can users and management be held more accountable for the creeping requirements problem encountered in this (and most other) project(s)?

WHAT IS PROJECT MANAGEMENT?

For any systems development project, effective project management is necessary to ensure that the project meets deadlines and is developed within an acceptable budget.

> **Project management** is the process of planning, directing, and controlling the development of an acceptable system at a minimum cost within a specified time frame.

Although the tools and techniques of systems analysis and design play a critical role in achieving successful systems, these methods are not sufficient on their own. As you learned in the preceding minicase, project mismanagement can deter or render ineffective the best analysis and design methods. That minicase highlighted the four common results of mismanaged projects:

- Unfulfilled or unidentified requirements.
- Uncontrolled change of project scope.
- Cost overruns.
- Late delivery.

These problems are not always caused by project mismanagement, but mismanagement certainly plays a role.

Project Management Causes of Failed Projects

We can develop an appreciation for the importance of project management by studying the mistakes of other project managers. Failures and limited successes far outnumber successful information systems. Why is that? True, many systems analysts and data processors are unfamiliar with or undisciplined in the tools and techniques of systems analysis and design. But that only partially explains the shortcomings of systems projects. Many projects suffer from poor leadership and management.

What project management failures cause unfulfilled requirements and needs, cost overruns, and late deliveries? Before we get into project mismanagement causes, let's recognize that one possible cause of these problems could be inadequate systems analysis and design tools and techniques. But for purposes of this module, we will focus on causes that can be traced to project management. Let's use the module's minicase to analyze those causes.

In Steven's distribution project, end-user requirements and needs were not fulfilled because shortcuts had been taken during the project. The systems development life cycle (discussed in Chapters 3, 6, 7, 12, 19, and 20) provides a basic framework for a complete systems project. Each phase and activity is an important part of that plan. For all parts to work together, the life cycle must be monitored and managed!

Some analysts and systems development managers might argue:

> The classic life cycle is the problem! It is an excessively long, drawn out process that leads to schedule and cost overruns.

We could not disagree more! In fact, we view that argument as an excuse to justify *not* managing projects. In Chapters 4 and 5 you learned about numerous techniques and technologies that can be used to accelerate the classic life cycle without compromising its benefits. These included **joint application development (JAD), prototyping, rapid application development (RAD**—the combination of JAD and prototyping), and **computer-aided systems engineering (CASE).** These techniques and technologies can and should be used in combination to accelerate the life cycle within the proven problem-solving framework provided by the life cycle.

Another common cause of unfulfilled requirements is expectations mismanagement. Users and managers have expectations. The problem is that during the early phases, the scope of the project is rarely precise. And for many projects, the scope is never precisely defined. If the project leader fails to recognize this problem, the project team is frequently forced to increase the scope (sometimes called the *creeping requirements syndrome*) or make late changes to specifications and programs. Unfortunately, the schedule and budget are rarely modified at the same time. This is a mistake, and the analyst or project manager is unfairly held accountable for the inevitable and unavoidable schedule and budget overruns. In other words, the users' expectations of schedule and budget did not change as the scope changed. But there are ways to manage expectations. We'll discuss a simple tool and technique later in this chapter.

One of the major problems with cost overruns is that many methodologies or project plans call for an unreasonably precise estimate of costs before the project begins. These estimates are made after a quick preliminary study or feasibility study. Think about it! Can you accurately estimate project costs before making a detailed study of the current system or defining end-user requirements? Can you estimate the costs of computer programming before a detailed systems design has been completed? It's not very likely. The cost estimates of a project will change as you get further into the systems development process.

Poor estimating techniques are another cause of cost overruns. We suspect that many systems analysts estimate by making the best calculated estimate (guesstimate?) and then doubling that number. This is hardly a scientific approach. There are better approaches available; some useful techniques are discussed in Module C, Feasibility Analysis.

And finally, cost overruns are often caused by schedule delays. Once again, we can point to premature estimates as a problem. These early estimates are based on the initial scope of the project. Because systems analysts (and information systems professionals in general) are eternal optimists, they often quote optimistic schedules and fail to modify those schedules as the true scope of the project becomes apparent.

Because many managers and analysts are often poor time managers, project schedules slip slowly but steadily. "So we've lost a day or two! It's no big deal. We can make it up later." This may be true, but then again, it might not. They fail to recognize the fact that in the systems development life cycle certain tasks are dependent on other tasks. Because of these dependencies, a one-day slip can set the whole schedule back. And when those one-day delays pile up, we inevitably find ourselves working 15-hour days at the end of the project.

Another cause of missed schedules is what Brooks (1975) has described as the **mythical man-month.** As the project gets behind schedule, the project leaders frequently try to solve the problem by assigning more people to the project team. It just doesn't work! There is no linear relationship between time and number of personnel. The addition of personnel creates more communications and political interfaces. The result? The project gets even further behind schedule.

You've probably noticed that the causes of failed projects are related. For instance, missed requirements may cause schedule slippages that, in turn, cause cost overruns. You might ask why somebody isn't able to recognize these problems and correct them. Somebody should. And that person is supposed to be the project manager or leader. This brings us to a major cause of project failure: lack of management and leadership. Good computer programmers don't always go on to become good analysts. Similarly, good analysts don't automatically perform well as managers and leaders. To be a good project manager, the analyst must possess or develop skills in the basic functions of management.

The Basic Functions of the Project Manager

The project manager is not just a senior analyst who happens to be in charge. As Steven found out, a project manager must apply a set of skills different from those applied by the analyst. What skills must the project manager possess or learn? The basic functions of a manager or leader have been studied and refined by management theorists for many years. These functions include planning, staffing, organizing, scheduling, directing, and controlling.

Planning Project Tasks and Staffing the Project Team

A good manager always has a plan. The manager estimates resource requirements and formulates a plan to deliver the target system. This is based on the manager's understanding of the requirements of the target system at that point in its development. A basic plan for developing an information system is provided by the systems development life cycle. Many firms have their own standard life cycles, and some firms have standards for the methods and tools to be used.

Each task required to complete the project must be planned. How much time will be required? How many people will be needed? How much will the task cost? What tasks must be completed before other tasks are started? Can some of the tasks overlap? These are all planning issues. Some of these issues can be resolved with a PERT chart, which is discussed later in this module.

Project managers frequently select analysts and programmers for the project team. The project manager should carefully consider the business and technical expertise that may be needed to successfully finish the project. The key is to match the personnel to the required tasks that have been identified as part of project planning.

Organizing and Scheduling the Project Effort

Given the project plan and the project team, the project manager is responsible for organizing and scheduling the project. Members of the project team should understand their own individual roles and responsibilities as well as their reporting relationship to the project manager.

The project schedule should be developed with an understanding of task time requirements, personnel assignments, and intertask dependencies. Many projects present a deadline or requested delivery date. The project manager must determine whether a workable schedule can be built around such deadlines. If not, the deadlines must be delayed or the project scope must be trimmed. We will soon introduce the Gantt chart, a tool used for project scheduling.

Directing and Controlling the Project

Once the project has begun, the project manager becomes a leader. As a leader, the manager directs the team's activities and evaluates progress. Therefore, every project manager must demonstrate such people management skills as motivating, rewarding, advising, coordinating, delegating, and appraising team members. Additionally, the manager must frequently report progress to superiors.

Perhaps the manager's most difficult and important function is controlling the project. Few plans will be executed without problems and delays. We've already discussed the causes and effects of unsuccessful projects. The project manager's job is to monitor tasks, schedules, costs, and expectations in order to control those elements. If the project scope is increasing, the project manager is faced with a decision: Should the scope be reduced so the original schedule and budget will be met, or should the schedule and budget be revised? The project manager must be able to present the alternatives and their implications for the budget and schedule in order to manage expectations.

Next, we will introduce some project management tools and techniques that can be used to assist project managers.

PROJECT MANAGEMENT TOOLS AND TECHNIQUES

There are many project management tools and techniques—enough for an entire book. In this section we will introduce two project planning and control tools and a simple tool for managing expectations.

PERT Charts

PERT, which stands for Project [or Program] Evaluation and Review Technique, was developed in the late 1950s to plan and control large weapons development projects for the U.S. Navy. It was developed to make clear the interdependence of project tasks when projects are being scheduled. Essentially, PERT is a graphic networking technique. Let's take a closer look at PERT charts—what they are, how to draw them, and how to use them.

FIGURE A.1 **PERT Notation** A PERT chart depicts events and tasks. Nodes represent events. Arrows represent tasks. Each event node includes an ID number, earliest completion time, and latest completion time. Each task arrow includes an ID letter and expected duration time estimate. A dashed arrow is a dummy task showing a dependency between tasks, although no time is required.

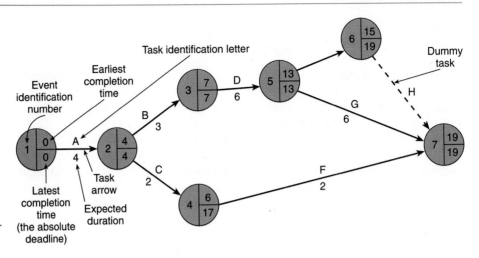

You should be made aware that different PERT discussions and computer products use different terminology.

PERT Definitions and Symbols

On PERT charts projects can be organized in terms of events and tasks.

An **event**—also called a **milestone**—is a point in time that represents the start or completion of a task or set of tasks.

A variety of symbols—circles, squares, and the like—have been used to depict events on PERT charts. For our discussion, we will use circles. On PERT charts these events are often called *nodes* (see Figure A.1). Each node is divided into three sections. The left half of the node includes an event identification number. This number is usually keyed to a legend that explicitly defines the event. The upper and lower right-hand quarters of the node are used to record the earliest and latest completion times for the event. Instead of dates, time is counted from TIME = 0, where 0 corresponds to the date on which the project is started. Every PERT chart has one beginning node that represents the start of the project and one end node that represents the completion of the project.

On a PERT chart a task (also called an *activity*) is depicted by an arrow between nodes.

A **task** is a project activity (or set of activities).

The task identification letter and the expected duration of the task are recorded on the arrow. Look at the arrow between event 1 and event 2 in Figure A.1. The direction of the arrow indicates that event 1 must be completed prior to event 2. The expected duration of the task resulting in the completion of event 2 is four days.

A dashed arrow is special. It is a dummy task.

A **dummy task** represents a dependency between events. However, because there is no activity to be performed, there is no duration between the events.

Estimating Project Time Requirements and Deriving a PERT

Before drawing a PERT chart, you must estimate the time needed for each project task. The PERT chart can be used to indicate the estimated earliest finish time and latest finish time for each event and the expected duration of each project task. Although these times are often expressed in terms of **person-days,** this approach is not recommended. There is no proven linear relationship between project completion time and the number of people assigned to a project team. Many systems projects that were late have been further delayed by assigning additional personnel to them. Because two people can do a job in four days is no reason to assume that four people can do the same job in two days. We suggest that time be expressed in **calendar-days,** given the number of people assigned to the task.

Unfortunately, we cannot offer you a set of formulas to use to derive time requirements for any task. You must estimate these project time requirements. By **estimate** we don't mean that you simply make something up. A good systems analyst project manager draws on experience and data from previous projects. Some of the factors that might influence estimates are listed in the margin. There exist CASE products such as SPQR/20 that help managers make better time estimates.

Other organizations have developed internal standards for deriving project time estimates in a more structured manner. These standards may involve examining a task in terms of its difficulty, skill requirements, and other identifiable factors. Alternatively, you could make an optimistic estimate and then adjust that estimate quantifiably by applying weighting factors to various criteria, such as the size of the team, the number of end-users with whom you have to interact, the availability of those end-users, and so on (Weinberg, 1979). Each weighting factor may either increase or decrease the estimate. As we go through this discussion, we'll try to give you some guidelines for estimating. When you begin to make estimates, seek the counsel of more experienced analysts (the more the better) until you are comfortable with the process.

Let's assume we're to derive the project time requirements and draw a PERT chart for a typical programming project. The project involves constructing a large program to update an employee master file. The project manager has identified seven program routines that are to be delegated to programmers for coding, testing, and debugging. The project planning table in Figure A.2 was used to derive the project time requirements and construct a PERT chart. Five steps are required:

1. *Make a list of all project tasks and events.* The first two columns in Figure A.3 provide an identification letter and a description for each task. The completion of a task is assigned an event identification number, which is entered into the third column.
2. *Determine intertask dependencies.* For each task, record the tasks that must be completed before and after the task in question is completed (columns 4 and 5).
3. *Estimate the duration of each task.* This can be determined as follows:
   a. Estimate the minimum amount of time it would take to perform the task. We'll call this the **optimistic time** (OT). The optimistic time estimate assumes that even the most likely interruptions or delays— such as occasional employee illnesses—will not happen.

────── ✓ ──────

**Factors That Affect
Estimates**

Size of project team
Experience of team members
Number of end-users and
  managers
Attitudes of end-users
Management commitment
Availability of managers and
  end-users
Projects in progress

──────────────

Task ID	Task Description	Event ID Number	Preceding Event	Succeeding Event	Expected Duration	Earliest Finish	Latest Finish
A	CODE, TEST, AND DEBUG ROUTINE "A010 UPDATE MASTER FILE"	2	1	3	3	3	3
B	CODE, TEST, AND DEBUG ROUTINE "B010 INITIATE PROCESSING"	3	2	4	2	5	5
C	CODE, TEST, AND DEBUG ROUTINE "B020 PROCESS TRANSACTION"	4	3	4,6,7,8	2	7	7
D	CODE, TEST, AND DEBUG ROUTINE "C210 ADD EMPLOYEE RECORD"	5	4	8	7	14	14
E	CODE, TEST, AND DEBUG ROUTINE "C220 MODIFY EMPLOYEE RECORD"	6	4	8	6	13	14
F	CODE, TEST, AND DEBUG ROUTINE "C230 DELETE EMPLOYEE RECORD"	7	4	8	3	10	14
G	CODE, TEST, AND DEBUG ROUTINE "B030 TERMINATE PROCESSING"	8	4,5,6,7	9	2	14	14
H	COLLECTIVELY TEST AND DEBUG PROGRAM	9	8	NONE	5	19	19

FIGURE A.2   **Project Planning Table**   A project planning table is used to prepare data for drawing a PERT chart.

FIGURE A.3   **Completed PERT Chart**   This is the PERT chart for the project planning table completed in Figure A.2. The bold arrows identify the critical path—those tasks for which there is no slack time.

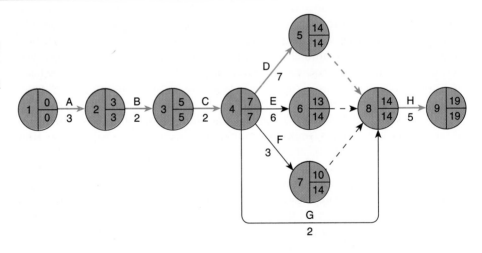

b. Estimate the maximum amount of time it would take to perform the task. We'll call this the **pessimistic time** (PT). The pessimistic time estimate assumes that anything that can go wrong will go wrong. All possible interruptions or delays—such as labor strikes, illnesses, training, inaccurate specification of requirements, equipment delivery delays, and underestimation of the systems complexity—are assumed to be inevitable.

c. Estimate the **most likely time** (MLT) that will be needed to perform the task. Don't just take the median of the optimistic and pessimistic times. Attempt to identify interruptions or delays that are likely to occur, such as occasional employee illnesses, inexperienced personnel, and occasional training.

   d. Calculate the **expected duration** (ED) as follows:

$$ED = \frac{OT + (4 \times MLT) + PT}{6}$$

This commonly used formula provides a weighted average of the various estimates. The formula is based on experience and may be modified to reflect project history in any firm. Expected duration is recorded in column 6 of Figure A.2. (*Hint:* Create a spreadsheet to do these calculations for you.)

4. *Derive the **earliest completion time** and **latest completion time** (ECT and LCT) for each task.* The data in our table provides considerable planning and control assistance to the project manager.

   a. The ECT for event *n* is equal to the largest ECT for the preceding events (column 4) plus the expected duration time for the task culminating in event *n*. For the first event, the ECT is equal to zero.

   b. The LCT (also called *the absolute deadline*) for event *n* is equal to the smallest LCT for succeeding events minus the estimated duration time for the task culminating in event *n*. For the last event, the LCT equals the ECT.

5. *Draw the PERT chart.* The PERT chart includes sequencing and identification for all tasks and events along with their time estimates. Notice that our PERT chart (Figure A.3) contains three dummy tasks, represented by the dotted arrows. This means that events 5, 6, and 7 must occur before event 8. However, there is no associated time factor between the three events and event 8.

An alternative approach to deriving PERT charts is **backward scheduling.** The backward scheduling approach schedules activities starting with a proposed task or project completion date and working backward to schedule the tasks that must come before it. This approach is particularly useful when determining the feasibility of a proposed completion date. If all tasks preceding the prescribed date cannot be scheduled for completion prior to the current date, then the proposed completion date must be moved forward or the scope of the project must be reduced.

The Critical Path in a PERT Network

Why are some of the solid arrows in Figure A.3 red? This series of arrows represents the critical path.

The **critical path** is a sequence of dependent project tasks that have the largest sum of estimated durations. It is the path that has no slack time built in. The slack time available for any task is equal to the difference between the earliest and latest completion times. If the earliest and latest completion times are equal, the task is on the critical path. If any task on the critical path gets behind schedule, the whole project is thrown off schedule.

Each task appearing on the critical path is referred to as a **critical task.** The critical path of the PERT chart for the programming project in Figure A.3 consists of tasks A, B, C, D, and H. This path represents the expected completion time for the project. Critical tasks must be monitored closely by the project manager.

To find the critical path on a project's PERT chart, begin by identifying all the alternate paths or routes that exist from event 1 to the final event. For example, Figure A.3 contains four paths:

Path 1: A — B — C — D — dummy task — H
Path 2: A — B — C — E — dummy task — H
Path 3: A — B — C — F — dummy task — H
Path 4: A — B — C — G — H

After all paths have been identified, calculate the total expected duration time for each path. The total expected duration time for a path is equivalent to the sum of the expected duration times for each task in the path. For example,

Path 1: $3 + 2 + 2 + 7 + 0 + 5 = 19$
Path 2: $3 + 2 + 2 + 6 + 0 + 5 = 18$
Path 3: $3 + 2 + 2 + 3 + 0 + 5 = 15$
Path 4: $3 + 2 + 2 + 2 + 5 = 14$

You can now identify the critical path as the one having the largest total expected duration time. In our example, path 1 is the critical path. It indicates that the expected time for completing the programming project is 19 days. But what if task G in path 4 had an expected duration time of 7 days? We would then have two critical paths containing tasks that the project manager would have to monitor closely!

Using PERT for Planning and Control

Project managers find PERT charts particularly useful for communicating schedules of large systems projects to superiors. However, the primary uses and advantages of the PERT chart lie in its ability to assist in the planning and controlling of projects. In planning, the PERT chart aids in determining the estimated time required to complete a given project, in deriving actual project dates, and in allocating resources.

As a control tool, the PERT chart helps the manager identify current and potential problems. Particular attention should be paid to the critical path of a project. When a project manager identifies a critical task that is running behind schedule and that is in danger of upsetting the entire project schedule, alternative courses of action are examined. Corrective measures, such as the shuffling of human resources, might be taken. These resources are likely to be temporarily taken away from a noncritical task that is currently running smoothly. These noncritical tasks normally offer some **slack time** for the project.

Analysis of PERT

The classic approach to PERT presents one major problem when applied to information systems development. In the classic approach, a node represents the start *or* finish of a task or series of tasks. Consider the simple PERT chart

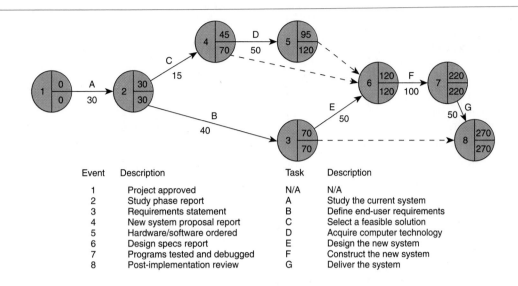

Event	Description	Task	Description
1	Project approved	N/A	N/A
2	Study phase report	A	Study the current system
3	Requirements statement	B	Define end-user requirements
4	New system proposal report	C	Select a feasible solution
5	Hardware/software ordered	D	Acquire computer technology
6	Design specs report	E	Design the new system
7	Programs tested and debugged	F	Construct the new system
8	Post-implementation review	G	Deliver the system

FIGURE A.4 **Sample PERT Chart for an SDLC** This simple PERT chart depicts interphase dependencies between deliverables in a systems development life cycle. Each node, often called a *milestone,* represents the completion of a phase. PERT charts depict the sequence for completing tasks. However, PERT charts do not depict the potential for overlapping tasks.

presented in Figure A.4. It's a PERT chart for a high-level systems development life cycle. Do you see the problem?

Classic PERT charting implies that the definition phase must be completed before either the selection phase or the design phase can be started. This, however, is not the case. Although neither the selection nor the design phase can be considered complete before the definition phase has been finished, both selection and design can begin while the definition phase is in progress. Classic PERT was developed to support projects that are often completed using an assembly-line approach. This does not include information systems! The tasks of systems development can overlap; it is only the completion of tasks that must occur in sequence. Don't assume that the next task cannot start until the prior task has been completed.

Gantt Charts

The Gantt chart is a simple time-charting tool that was developed by Henry L. Gantt in 1917. Gantt charts, which are still popular today, are effective for project scheduling and progress evaluation. Like PERT charts, Gantt charts involve a graphic approach. The popularity of Gantt charts stems from their simplicity—they are easy to learn, read, prepare, and use. Let's study this project management tool in more detail.

A **Gantt chart** is a simple bar chart (see Figure A.5). Each bar represents a project task. On a Gantt chart, the horizontal axis represents time. Because Gantt charts are used to schedule tasks, the horizontal axis should include dates. The tasks are listed vertically in the left-hand column.

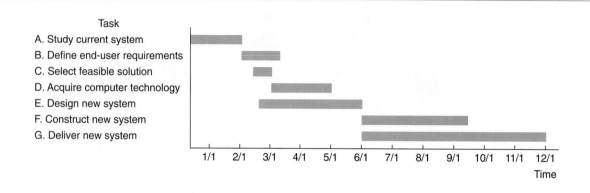

FIGURE A.5 **Simple Gantt Chart for an SDLC**   This simple Gantt chart represents the systems development life cycle (but not the one in this book). The task is described in the left-hand column. The horizontal axis is a calendar of dates.

Notice that Gantt charts clearly depict the overlap of scheduled tasks. Because systems development tasks frequently overlap, this is a major advantage. But also notice that Gantt charts fail to clearly show the dependency of one task on another. That is a major strength of PERT charts. Let's briefly discuss the scheduling and evaluation uses of Gantt charts.

How to Use a Gantt Chart for Scheduling

It's easy to use a Gantt chart to generate a schedule. First, identify the tasks that must be scheduled. (If you prepared a PERT chart first, this step would already have been completed.) Next, determine the duration of each task. You learned an appropriate time-estimating technique and formula in the preceding section. If you haven't already prepared a PERT chart, you should at least determine the interdependencies between tasks. Gantt charts cannot clearly show such dependencies, but it is imperative that the schedule recognize them. Now you are ready to schedule the tasks.

List each activity in the left-hand column of the Gantt chart. Record dates corresponding to the duration of the project on the horizontal axis of the chart. Determine starting and completion dates for each task. Careful! The start of any task may be dependent on at least the partial completion of a previous task. Additionally, the completion of a task is frequently dependent on the completion of a prior task. A simple example is in order.

Figure A.5 presents a Gantt chart for a high-level systems development life cycle. It offers some useful information that the PERT chart (Figure A.4) could not. Notice that the start of the definition phase appears to be dependent on the completion of the study phase. On the other hand, the selection phase is scheduled to begin after the definition phase has started but before that phase is completed. In fact, the definition phase overlaps with the entire selection phase.

If you compare the Gantt and PERT charts for the systems development life cycle, you will see that the sequence for completing phases (PERT) is maintained in the Gantt chart. Thus, the preparation of a PERT chart (or, at the very

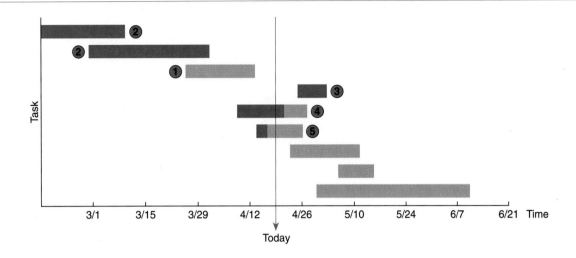

FIGURE A.6 **Progress Reporting with Gantt Charts** Gantt charts can be annotated to clearly depict project progress. The red vertical line depicts today's date. The darker bars reflect completed tasks (or parts of tasks). The lighter bars reflect incomplete tasks or those not yet started. Any lighter bar (or part thereof) that is to the left of today's date is behind schedule.

least, understanding precedence between tasks) can significantly aid in the preparation of a Gantt chart.

Using Gantt Charts to Evaluate Progress

One of the project manager's frequent responsibilities is to report project progress to superiors. Gantt charts frequently find their way into progress reports because they can conveniently compare the original schedule with actual performance. To report progress we must expand our Gantt charting conventions. If a task has been completed, completely shade in the bar corresponding to that task. If a task is partially completed, partially shade in the bar. The percentage of the bar that is shaded should correspond to the percentage of the task completed. Unshaded bars represent tasks that have not begun. Next, draw a bold vertical line that is perpendicular to the horizontal axis and that intersects the current date. You can now evaluate project progress.

Let's look at the sample Gantt chart in Figure A.6. The lighter bar to the left of the current date (see ① on Figure A.6) is very much off schedule. This task is supposed to be completed but hasn't even been started. The darker bars to the left of the current date (labeled ② on the figure) represent tasks that have been completed on schedule. The darker bars to the right of the current date (labeled ③) represent tasks that have been completed ahead of schedule. Tasks that are currently in progress have to be evaluated relative to their light versus dark portions. If the shaded portion extends to or past the current date (see ④ on the figure), that task is on or ahead of schedule. Otherwise, the task is behind schedule (see ⑤ on the figure). You should see that with a little practice, the Gantt chart can convey project progress at a glance.

PERT versus Gantt Charting

PERT and Gantt charting are frequently presented as mutually exclusive project management tools. PERT is usually recommended for larger projects with high intertask dependency. Gantt is recommended for simpler projects. But PERT and Gantt charting should not be considered as alternative project management approaches. All systems development projects have some intertask dependency, and all projects offer opportunities for task overlapping. Therefore, PERT and Gantt charts can be used in a complementary manner to plan, schedule, evaluate, and control systems development projects.

Still, most information systems project managers seem to prefer Gantt charts because of their simplicity and ability to show the schedule of a project. Fortunately, project management software allows the best feature of PERT—namely, the critical path analysis—to be incorporated into Gantt charts. As activities are entered, their duration and dependencies are entered. Gantt bars are scheduled to take into consideration the dependencies. Usually, the critical path is highlighted with boldfacing or color. Additionally, the amount of slack time in noncritical path activities is also highlighted. This can prove useful when deciding which activities to delay in order to get off-schedule activities back on track.

Project Management Software

Project management software was introduced in Chapter 5 as a type of CASE tool. Examples include Microsoft's Project and Applied Business Technology's Project Manager Workbench. These packages greatly simplify the preparation of PERT and Gantt charts, allowing automatic transformation between the two chart types. The software also allows project managers to assign people and cost resources to tasks, to report project progress, and to play "what-if" games when attempting to modify the project plan in response to schedule slippages.

Some packages also provide or interface with **time accounting software** that keeps track of actual time spent on different types of activities. That information can be useful for performance comparisons and client billing.

Finally, some upper-CASE tools provide interfaces between their own project repositories and the project management software. For example, INTERSOLV's Excelerator/IS can exchange data (both ways) between its own repository and ABT's Project Manager Workbench.

Expectations Management

Experienced project managers often complain that managing management's expectations of a project is more difficult than managing cost, schedule, people, or quality. In this section we introduce a simple tool that we'll call an expectations management matrix that can help project managers deal with the problem. We first learned about this tool from Dr. Phil Friedlander, a consultant and trainer then with McDonnell Douglas in the Improved Systems Technologies business unit. He attributes the matrix to "folklore" but also credits Jerry Gordon of Majer LTD and Ron Leflour, a large project management educator/trainer. Dr. Friedlander's paper is listed in the suggested readings for this module. We have slightly adapted the tool for this presentation.

Priorities Measures of Success	Max or Min	Constrain	Accept
Cost			
Schedule			
Scope and/or Quality			

FIGURE A.7   Management Expectations Matrix
This incomplete matrix is used to help project managers and system owners develop realistic expectations for a project.

The Expectations Management Matrix

Every project has goals and constraints when it comes to cost, schedule, scope, and quality. In an ideal world, you could optimize each of these parameters. Management often has that expectation. Reality, however, suggests that you can't optimize them all—you must strike a balance that is both feasible and acceptable to management. That is the purpose of the expectations management matrix.

> An **expectations management matrix** is a rule-driven tool for helping management appreciate the dynamics of changing project parameters. The parameters include cost, schedule, scope, and quality.

Who is management? In the case of systems development, management is defined as the system owner(s) — the individual(s) who sponsors and pays for a system to be developed or modified. To achieve true project success, it is extremely important that the project manager and system owner come to an understanding about assumptions and expectations of the project and that they review the implications of changing budgets, schedules, and system requirements on a regular basis. The matrix is a tool for doing just that.

The basic matrix, shown in Figure A.7, consists of three rows and three columns (plus headings). The rows correspond to the measures of success in any project: cost, schedule, and scope and/or quality. The columns correspond to priorities: first, second, and third. For purposes of establishing expectations, we assign names to the priorities as follows:

- Maximize or minimize—The most important of the three measures in a given project.
- Constrain—The second most important of the three measures in a project.
- Accept—The least important of the three measures in a project.

Again, while most managers would ideally like to give equal priority to all three measures, experience suggests that the three measures tend to balance themselves naturally. For example, if you increase scope or quality requirements, it will take more time and/or money. On the contrary, if you try to get any job done faster, you generally have to reduce scope or quality requirements, or

FIGURE A.8 **Management Expectations for the American Moon Landing Project** This complete and valid matrix demonstrates how the government managed the public's expectations in its project to be the first country to land a man on the moon and return him safely.

Priorities Measures of Success	Max or Min	Constrain	Accept
Cost Estimated at $20 billion			X
Schedule Deadline = December 31, 1969		X	
Scope and/or Quality 1. Land a man on the moon. 2. Get him back safely.	X		

pay more money to compensate. The management expectations matrix helps (forces) management to understand this through three simple rules:

- For any project, you must record three Xs within the nine available cells.
- No row may contain more than one X. In other words, a single measure of success must have one and only one priority.
- No column may contain more than one X. In other words, there must be a first, second, and third priority.

Let's illustrate the tool using Dr. Friedlander's own example. In 1961 President John F. Kennedy established a major project — land a man on the moon and return him before the end of the decade. Figure A.8 shows the realistic expectations of the project. Let's walk through the example. John Kennedy (and the public who elected him) is the system owner.

1. The system owner had both scope and quality expectations. The scope (or requirement) was to successfully land a man on the moon. The quality measure was to return the man (or men) safely. Because the public would expect no less from the new space program, this had to be made the first priority. In other words, we had to maximize safety and minimize risk as a first priority. Hence, we record the X in column 1, row 3.

2. At the time of the project's inception, the Soviet Union was ahead in the race to space. This was a matter of national pride; therefore, the second priority was to get the job done by the end of the decade. We call this the project *constraint*—there is no need to rush the deadline, but we don't want to miss the deadline. Thus, we record the second "X" in column 2, row 2.

3. By default, the third priority had to be cost (estimated at $20 billion in 1961). By making cost the third priority, we are not stating that cost will not be controlled. We are merely stating that we may have to "accept" cost overruns in order to achieve the scope and quality requirement by the constrained deadline.

History records that we achieved the scope and quality requirement, and did so in 1969. The project actually cost well in excess of $30 billion, a 50 percent cost overrun. Did that make the project a failure? On the contrary, most people perceived the project a grand success. The government managed the public's expectations of the project in realizing that maximum safety and minimum risk, plus meeting the deadline (beating the Soviets!), was an acceptable tradeoff for the cost overrun! The government brilliantly managed public opinion. Systems development project managers can learn a valuable lesson from this balancing act.

At the beginning of any project, the project manager should consider introducing the system owner to the matrix concept and should work with the system owner to complete the matrix. For most projects, it would be difficult to record all the scope and quality requirements in the matrix. Instead, they would be listed on an attached document. The matrix heading for scope and/or quality would be stated as "See Attachment A." The estimated costs and deadlines could be recorded directly on the matrix.

The project manager must never establish the priorities or even suggest those priorities. The project manager merely enforces the rules: 3 Xs and no more than one X in any row or column. This sounds easy. It rarely is. Many managers are unwilling to be pinned down on the priorities—"Shouldn't we be able to maximize everything?" These managers need to be educated about the reason for the priorities. Of course we always try to maximize all three measures of success because it makes us (the project managers) look that much better. But we need to know priorities in the event that we cannot maximize all three measures. This helps us make intelligent compromises instead of merely guessing right or wrong.

What if your system owner refuses to prioritize? The tool is less useful then, except as a mechanism for documenting your concerns before they become disasters. A system owner who refuses to set priorities is a manager who may be setting the project manager up for a no-win performance review. And as Dr. Friedlander points out, "Those who do not 'believe' the principles [of the matrix] will eventually 'know' the truth. You do not have to believe in gravity, but you will hit the ground just as hard as the person who does."

Using the Expectations Management Matrix

Let's assume you have a management expectations matrix that conforms to the aforementioned rules. How does this help you manage expectations?

During the course of the average systems development project, priorities are not stable. Various factors such as the economy, government, and politics can change the priorities. Budgets may become more or less constrained. Deadlines may become more or less important. Quality can become more (rarely less) important. And, most frequently, requirements increase (rarely decrease). As already noted, these changing factors affect all the measures in some way. The trick is to manage expectations despite changing project parameters.

The technique is relatively straightforward. Whenever the max/min measure or the constrain measure begins to slip, you have a potential expectations management problem. For example, suppose you are faced with the following common priorities (see Figure A.9):

FIGURE A.9 **An Initial Management Expectations Matrix** This initial matrix is based on a hypothetical situation described in the text.

Priorities Measures of Success	Max or Min	Constrain	Accept
Cost		X	
Schedule			X
Scope and/or Quality	X		

1. Explicit requirements and quality expectations were established at the start of a project and given the highest priority.
2. An absolute maximum budget was established for the project.
3. You agreed to shoot for the desired deadline, but the system owner(s) accepted the reality that if something must slip, it should be schedule.

Now suppose that during systems analysis, significant and unanticipated business problems arise. The analysis of these problems has placed the project somewhat behind schedule. Furthermore, solving the new business problems substantially expands the user requirements for the system. How do you, as project manager react? First, don't overreact to the schedule slippage—schedule slippage was the "accept" priority in the matrix. The scope increase (in the form of several new requirements) is the more significant problem—not because implementing new requirements might further delay the schedule! No, the real problem is that the added requirements will increase the cost of the project. Cost is the constrained measure of success. As it stands, we have an expectations problem. It is time to review the matrix with the system owner.

First, the system owner needs to be made aware of which measure or measures are in jeopardy and why. Then together, the project manager and system owner can discuss courses of action. There are several courses of action possible:

- The resources (cost and/or schedule) can be reallocated. Perhaps the system owner can find more money somewhere. All priorities would remain the same (noting, of course, the revised deadline based on schedule slippages already encountered during systems analysis).
- The budget might be increased, but it would be offset by additional planned schedule slippages. For instance, by extending the project into a new fiscal year, additional money might be allocated without taking any money from existing projects or uses. This solution is shown in Figure A.10.
- The user requirements (or quality) might be reduced through prioritizing those requirements and deferring some number of those requirements until Version 2 of the system. This alternative would be appropriate if the budget cannot be increased.

Priorities Measures of Success	Max or Min	Constrain	Accept
Cost (Record New Budget)		**X+** Increase budget	
Schedule (Record New Deadlines)			**X-** Extend deadline
Scope and/or Quality (Revise attachment that describes scope and quality. Be sure to date all new requirements to distinguish them from original requirements.)	**X+** Accept expanded requirements		

FIGURE A.10 Adjusting Resources in a Management Expectations Matrix Adjusted resources are indicated with plus and minus signs. Any comment can be written in a cell.

- Finally, measurement priorities can be changed. Dr. Friedlander calls this **priority migration.**

Only the system owner may initiate priority migration. For example, the system owner may agree that the expanded requirements are worth the additional cost. He or she allocates sufficient funds to cover the requirements but migrates priorities such that minimizing cost becomes the highest priority (see Figure A.11, step 1). But now the matrix violates a rule—there are two Xs in column 1. To compensate, we must migrate the scope and/or quality criteria to another column, in this case, the constrain column (see Figure A.11, step 2). Expectations have been adjusted. In effect, the system owner is freezing growth of requirements and still accepting schedule slippage.

There are three final comments about priority migration. First, priorities may migrate more than once during a project. Expectations can be managed through any number of changes as long as the matrix is balanced (meaning it conforms to our rules). Second, expectation management can be achieved through any combination of priority migrations and resource adjustments. Finally, it should be noted that system owners can initiate priority migration even if the project is on schedule. For example, government regulation might force an uncompromising deadline on an existing project. That would suddenly migrate our "accept" schedule slippages to "max" schedule. The other Xs would have to be migrated to rebalance the matrix—such as "accepting" any cost and "constraining" scope.

The expectations management matrix is a simple tool, but sometimes simple tools are the most effective!

People Management

The management or supervision of project team members is equally important to planning and controlling the project schedule, budget, and expectations. The topic could easily require an entire module on its own. In the interest of space, we will direct you to a couple of the shortest and most valuable books ever written on the subject. You could easily read both books overnight!

FIGURE A.11 **Priority Migration in a Management Expectations Matrix** This matrix demonstrates shifting priorities (Step 1) and rebalancing of the matrix (Step 2). The rebalancing of the matrix helps the project manager adjust expectations in light of the changing priorities.

Priorities Measures of Success	Max or Min	Constrain	Accept
Cost	← Step 1 X		
Schedule			X
Scope and/or Quality	X Step 2 →		

The One Minute Manager

The One Minute Manager by Kenneth Blanchard and Spencer Johnson is a classic, fun, and indispensable aid to anyone managing people for the first time. In just over 100 pages, the authors share the simple secrets of managing people and achieving success through the actions of your subordinates. The book highlights three basic secrets:

- One-minute goal setting.
- One-minute praisings.
- One-minute reprimands.

This book should be a part of every college graduate's personal library!

The Subtle Art of Delegation and Accountability

Most young and many experienced managers have difficulty delegating responsibilities. Worse still, they let subordinates reverse-delegate tasks back to the manager. This leads to poor time management and manager frustration. In *The One Minute Manager Meets the Monkey,* Kenneth Blanchard teams with William Oncken and Hal Burrows to help managers overcome this problem.

The solution is based on Oncken's classic principle of "the care and feeding of monkeys." Monkeys refer to problems that managers delegate to their subordinates who, in turn, attempt to reverse-delegate back to the manager. In this 125-page book the authors teach managers how to keep the monkeys on the subordinates' backs. Doing so increases the manager's available work time, accelerates task accomplishment by subordinates, and teaches subordinates how to solve their own problems.

Summary

In addition to systems analysis and design responsibilities, a systems analyst frequently assumes a project management role. Project mismanagement often leads to missed end-user requirements, cost overruns, and late delivery. The causes of these problems include shortcuts during systems development, imprecise targets, premature cost estimates, poor estimating techniques, poor time management, and lack of leadership. It is the project manager's responsibility to avoid these pitfalls and successfully complete the project on time and within budget. The project manager's basic functions include planning project tasks, staffing the project team, organizing and scheduling the project effort, directing the project team, and controlling project progress.

There are two major tools frequently used by analysts to plan, schedule, and control systems development projects. PERT charts graphically depict project tasks and events and show the dependency of tasks on one another. They also depict the time requirements for each task. Gantt charts graphically depict project tasks to show a project schedule. Gantt charts allow the project manager to show the overlapping of tasks in a project; they are also useful for depicting project progress. PERT and Gantt charts complement each other to provide the manager with an integrated planning, scheduling, evaluation, and control environment. This environment is being further enhanced by the availability of computer software to support the project management tools.

Managing expectations is one of the more difficult aspects of project management. As user requirements change, the project manager frequently forgets to adjust cost and schedule expectations. This chapter presented an elegant, simple tool, the expectations management matrix, as a vehicle for managing expectations. This tool helps project managers balance expectations regarding costs, schedules, scope, and requirements as project parameters change.

People management is another aspect of project management. Entire books could be written on the subject; however, this module directed readers to two gems: *The One Minute Manager* and *The One Minute Manager Meets the Monkey*.

Key Terms
---

backward scheduling, p. 775

calendar-days, p. 773

computer-aided systems engineering (CASE), p. 769

critical path, p. 775

critical task, p. 776

dummy task, p. 772

earliest completion time, p. 775

estimate, p. 773

event, p. 772

expectations management matrix, p. 781

expected duration, p. 775

Gantt chart, p. 777

joint application development (JAD), p. 769

latest completion time, p. 775

milestone, p. 772

most likely time, p. 774

mythical man-month, p. 770

optimistic time, p. 773

person-days, p. 773

PERT, p. 771

pessimistic time, p. 774

priority migration, p. 785

project management, p. 768

project management software, p. 780

prototyping, p. 769

rapid application development (RAD), p. 769

slack time, p. 776

task, p. 772

time accounting software, p. 780

Problems and Exercises
---

1. What are some of the causes of mismanaged projects that result in missed requirements and needs, cost overruns, and late delivery? Explain how these problems that result from mismanaged projects are related.

2. Systems analysts have a tendency to assign additional people to a project that is running behind schedule. What are some of the potential problems with such an action?

3. How can expectations mismanagement lead to perceived project failure?

4. What are the basic functions of a project manager?

5. Explain the advantages and disadvantages of a PERT chart and a Gantt chart. Explain how each tool can be best used by a project manager.

6. Why shouldn't estimated project time requirements be stated in terms of person-days?

7. Calculate the expected duration (ED) for the following tasks:

Task ID	Optimistic Time (OT)	Pessimistic Time (PT)	Most Likely Time (MLT)
A	3	6	4
B	1	3	2
C	4	7	6
D	2	5	3

8. Derive the earliest completion time (ECT) and latest completion time (LCT) for each of the following:

Task ID	Event ID	Preceding Event	Succeeding Event	Expected Duration
A	2	1	3	2
B	3	2	4	3
C	4	3	5, 6	4
D	5	4	7	5
E	6	4	7	4
F	7	5	8, 9	3
G	7	6	8, 9	0
H	8	7	10	6
I	9	7	10	5
J	10	8	None	0
K	10	9	None	6

9. Draw the PERT chart described in Exercise 7. Be sure to include sequencing and identification for all tasks and events along with their time estimates. What is the critical path? What is the total expected duration time represented by the critical path?

10. Make a list of the tasks that you performed on your last programming assignment. Alternatively, make a list of the tasks required to complete your next programming assignment. Develop a PERT chart to depict the tasks and events and the dependency of tasks on one another. What is the critical path? How can the PERT chart aid in planning and scheduling the programming assignment?

11. Derive a Gantt chart to graphically depict the project schedule and overlapping of tasks for the programming assignment chosen in Exercise 9. How can the Gantt chart be used to evaluate the progress that is being or has been made?

12. Draw a PERT chart for the curriculum in which you are enrolled. Be sure to consider the prerequisites for all courses.

13. Draw a Gantt chart for your plan of study to get your degree. Annotate the graph to indicate your progress toward your degree or job objectives.

14. At the beginning of a project, management decided that highest priority should be placed meeting an absolute deadline for a project. Second highest priority is to be placed on living within the allocated project budget. Draw the expectations management matrix for this project.

15. During the project initiated in Exercise 14, things started going wrong. Creeping requirements set in. Both the deadline and budget are in jeopardy. Using the expectations management matrix, identify alternatives for adjusting the project. Who should make the decision?

Projects and Minicases

1. Research project management software in your local library. Present a report to management that identifies important selection criteria for selecting a project management software package.

2. Make an appointment to visit an information systems project manager (or systems analyst with project management experience). What techniques does he or she use to plan and control projects? Why? Is project management software used? If so, what does the project manager like and dislike about that software?

3. If you have access to project management software, evaluate the package. Complete the tutorial and analyze package strengths and weaknesses.

4. If your systems course requires you to complete a real or simulated development project, prepare a management expectations matrix with your client (or your instructor acting as your client). Over the course of the semester, review the matrix with your client (or instructor) and make appropriate adjustments. At the end of the project, be prepared to defend your management of expectations as well as your progress. (*Note:* Your instructor may assign a subjective grade to your management of expectations. That may not seem fair; however, it is quite realistic. Project success is as much perceived as it is real.)

Suggested Readings

Blanchard, Kenneth, and Spencer Johnson. *The One Minute Manager.* New York: Berkley Publishing Group, 1981, 1982. Arguably, this is one of the best people management books ever written. Available in most bookstores, it can be read overnight and used for discussion material for the lighter side of project management (or any kind of management). This is must reading for all college students with management aspirations.

Blanchard, Kenneth, William Oncken, Jr., and Hal Burrows. *The One Minute Manager Meets the Monkey.* New York: Simon & Schuster, 1988. A sequel to *The One Minute Manager,* this book effectively looks at the topic of delegation and time management. The monkey refers to Oncken's classic article, "Managing Management Time: Who's Got the Monkey?" as printed in the *Harvard Business Review* in 1974. The book teaches managers how to achieve results by helping their staff (their monkeys) solve their own problems.

Brooks, Fred. *The Mythical Man-Month.* Reading, Mass.: Addison-Wesley, 1975. A classic set of essays on software engineering, also known as systems analysis, design, and implementation. Emphasis is on managing complex projects.

Friedlander, Phillip. "Ensuring Software Project Success with Project Buyers," *Software Engineering Tools, Techniques, and Practices* 2, no. 6 (March/April 1992), pp. 26–29. We adapted our expectations management matrix from Dr. Friedlander's work.

Gildersleeve, Thomas. *Data Processing Project Management.* New York: Van Nostrand Reinhold, 1974. This book offers no explicit PERT or Gantt coverage, but it does provide excellent coverage of the people side of project management. A classic, hypothetical series of memos that documents a failed project precedes the topical coverage.

London, Keith. *The People Side of Systems.* New York: McGraw-Hill, 1976. Chapter 8, "Handling a Project Team," does an excellent job of teaching the people and leadership aspects of project management.

Page-Jones, Meiler. *Practical Project Management: Restoring Quality to DP Projects and Systems.* New York: Dorset Publishing Company, 1985. This is a very comprehensive book on the subject of project management. Its diverse topical coverage includes: estimating, setting deadlines, status and time reporting, meeting management, and staff development and discipline, to name a few.

Senn, James A. *Analysis and Design of Information Systems.* New York: McGraw-Hill, 1984. Although his treatment of PERT/CPM and Gantt charting is of an introductory nature (like ours), we are indebted to Senn for his PERT/CPM symbol notation, which we emulated in this book.

Weinberg, Victor. *Structured Analysis.* New York: Yourdon Press, 1979. Chapter 11 contains some valuable guidelines and strategies for estimating. Although Weinberg uses his weighting strategy to adjust costs according to various factors, we have suggested that scheme would work equally well on time estimates.

Wiest, Jerome D., and Ferdinand K. Levy. *A Management Guide to PERT-CPM: With PERT-PDM, DCPM, and Other Networks.* 2nd ed. Englewood Cliffs, N.J.: Prentice Hall, 1977. A good source for more on PERT/CPM and other project planning and control networks.

B

Fact-Finding Techniques

When Should You Read This Module?

This module will prove most valuable if read after any of the following chapters:

Chapter 3
A Systems Development Life Cycle

Chapter 4
Systems Development Techniques and Methodologies

Chapter 6
Systems Planning

Chapter 7
Systems Analysis

Chapter 12
Systems Design

If this is the first module you are studying, you should also read the
part opener for Part Five of the textbook.

Effective fact-finding techniques are crucial to the development of systems projects. Fact-finding is performed during all phases of the systems development life cycle. To support systems development, the analyst must collect facts about PEOPLE, DATA, ACTIVITIES, NETWORKS, and TECHNOLOGY. This chapter introduces six popular fact-finding techniques and suggests a strategy for conducting fact-finding efforts. You will know that you understand fact-finding when you can identify six fact-finding techniques and characterize the advantages and disadvantages of each; identify the types of facts that a systems analyst must collect; develop a questionnaire and an interview agenda; and describe a fact-finding strategy that will make the most of the time you spend with end-users.

CONNOR'S FIXTURES, INC.

Connor's Fixtures is a large manufacturer of bathroom fixtures. Connor's is headquartered in St. Louis, Missouri and has five separate but closely located manufacturing plants, each specializing in the manufacturing of a particular line of bathroom fixtures. Tim Baxter, a junior systems analyst has been teamed with Scott Henderson on a systems project to develop a common inventory control system for each plant. Management hopes to save money by developing a common, integrated inventory control system. Substantial quantity discounts and lower shipping costs could be achieved by consolidating common purchase orders.

Scene: *Tim and Scott are meeting to discuss their strategy for learning about the current inventory control systems.*

Tim: Well Scott, which plants do you want me to take on? I've got a list of key people at each plant. I can start scheduling interviews with them right away. You know, this is going to take some time.

Scott: Too much time if we interview all the people at each site.

Tim: I don't guess we have any choice. If we want to make them happy, we'll need to interview everyone to find out what they like and don't like about their current system and to identify their requirements for a new system.

Scott: I don't disagree with the need to involve everyone, but I have a problem with the approach that you're suggesting. If we conduct all those interviews, it'll take months!

Tim: Should we narrow down the number of people we want to involve? I don't know how else we can speed things up.

Scott: No. We really need to think this out; we need a plan or strategy. You realize that there are several ways we can go about collecting the information we need from the users. We don't need to rely strictly upon numerous individual interviews.

Tim: Sure, but I'm not very good at making up questionnaires. And I don't think we have time to observe their inventory control operations. That would probably take longer than doing the individual interview.

Scott: The bottom line is we need to establish a strategy that will minimize the amount of our time, and more importantly, the users' time. We need to look at the alternative techniques, weigh their advantages and disadvantages, and select a plan. Perhaps we can make use of a variety of techniques.

Tim: I guess you're right. I'm just a little anxious to get started. I've never dealt with such a large number of clients. I'm also a little nervous about the goal of the project. Do you really think we'll be able to develop a common system that will make the people at all sites happy?

Scott: I've been thinking about that challenge, too. Who says we have to work on an individual basis with our clients? I was talking with Bill Amick the other day. Bill told me that he used to conduct group work sessions at his former place of employment. They called them JAD or joint application design sessions. He'd actually get the appropriate users in the same room for several hours a day, several days a week. They'd get the job done together and with group consensus.

Tim: I don't have much confidence in large meetings.

Scott: Oh, it takes some special skills to conduct these sessions. There are specific roles for individuals and forms to be completed. It's a very formal process. Let's give Bill a call and see if he can tell us more about this JAD approach. Perhaps he might even be able to help us conduct such a session.

Discussion Questions

1. Why does Scott feel a strategy for fact-finding should be considered—one that may utilize numerous fact-finding techniques?

2. Why do you suppose Tim seems pessimistic about large group meetings?

WHAT IS FACT-FINDING?

Applying the tools and techniques for systems development in the classroom is easy. Applying those same tools and techniques in the real world may not work—that is, if they are not complemented by effective methods for fact-finding.

> **Fact-finding** is the formal process of using research, interviews, questionnaires, sampling, and other techniques to collect information about systems, requirements, and preferences. It is also called *information gathering* or *data collection.*

Tools document facts, and conclusions are drawn from facts. If you can't collect the facts, you can't use the tools. Fact-finding skills must be learned and practiced.

Systems analysts need an organized method of collecting facts. They especially need to develop a detective mentality to be able to discern relevant facts! The purpose of this module is to present popular alternative fact-finding techniques. Although an entire textbook could be devoted to fact-finding techniques and strategies, no introductory systems course would be complete without the survey provided in this module.

Before we leap headfirst into specific fact-finding techniques, let's make sure we understand what we are trying to accomplish. The tools of systems analysis and design are used to document facts about an existing or proposed information system. These facts are in the domain of the business application and its end-users. Therefore, the analyst must collect those facts in order to effectively apply the documentation tools and techniques. When might the analyst use fact-finding techniques? What kinds of facts should be collected? And how are facts collected?

What Facts Does the Systems Analyst Need to Collect and When?

There are many occasions for fact-finding during the systems development life cycle. However, fact-finding is most crucial to the systems planning and systems analysis phases. It is during these phases that the analyst learns about the vocabulary, problems, opportunities, constraints, requirements, and priorities of a business and a system. Fact-finding is also used during the systems design and support phases, but to a lesser extent. During systems design, fact-finding becomes technical as the analyst attempts to learn more about the technology selected for the new system. During the systems support phase, fact-finding is important in determining that a system has decayed to a point where the system needs to be redeveloped.

What types of facts must be collected? It would certainly be beneficial if we had a framework to help us determine what facts need to be collected, no matter what project we are working on. Fortunately, we have such a framework. Throughout the systems development process, we are looking at an existing or target information system. In Chapter 2 we saw that any information system can be examined in terms of five building blocks. Those five building blocks were depicted using our pyramid model. As it turns out, the facts that describe any information system also correspond nicely with the building blocks of that pyramid model.

WHAT FACT-FINDING METHODS ARE AVAILABLE?

Now that we have a framework for our fact-finding activities, we can introduce six common fact-finding techniques:

1. Sampling of existing documentation, forms, and files.
2. Research and site visits.
3. Observation of the work environment.
4. Questionnaires.
5. Interviews.
6. Joint application design (JAD).

An understanding of each of these techniques is essential to your success. An analyst usually applies several of these techniques during a single systems project. To be able to select the most suitable technique for use in any given situation, you will have to learn the advantages and disadvantages of each of the fact-finding techniques.

SAMPLING OF EXISTING DOCUMENTATION, FORMS, AND FILES

Particularly when you are studying an existing system, you can develop a pretty good feel for the system by studying existing documentation, forms, and files. A good analyst always gets facts first from existing documentation rather than from people.

Collecting Facts from Existing Documentation

What kind of documents can teach you about a system? The first document the analyst should seek out is the organizational chart. Next, the analyst may want to trace the history that led to the project. To accomplish this, the analyst may want to collect and review documents that describe the problem. These include:

- Interoffice memoranda, studies, minutes, suggestion box notes, customer complaints, and reports that document the problem area.
- Accounting records, performance reviews, work measurement reviews, and other scheduled operating reports.
- Information systems project requests — past and present.

In addition to documents that describe the problem, there are usually documents that describe the business function being studied or designed. These documents may include:

- The company's mission statement and strategic plan.
- Formal objectives for the organization subunits being studied.
- Policy manuals that may place constraints on any proposed system.
- Standard operating procedures (SOPs), job outlines, or task instructions for specific day-to-day operations.
- Completed forms that represent actual transactions at various points in the processing cycle.
- Manual and computerized files.
- Manual and computerized reports.

Also, don't forget to check for documentation of previous systems studies and designs performed by systems analysts and consultants. This documentation may include:

- Various types of flowcharts and diagrams.
- Project dictionaries.
- Design documentation, such as inputs, outputs, and files.
- Program documentation.
- Computer operations manuals and training manuals.

All documentation collected should be analyzed to determine how up-to-date it is. Don't discard outdated documentation. Just keep in mind that additional fact-finding will be needed to verify or update the facts collected. As you review existing documents, take notes, draw pictures, and use systems analysis and design tools to model what you are learning or proposing for the system.

Document and File Sampling Techniques

Because it would be impractical to study every occurrence of every form, analysts normally use sampling techniques to get a large enough cross section to determine what can happen in the system.

Sampling is the process of collecting sample documents, forms, and records.

Experienced analysts avoid the pitfalls of sampling blank forms — they tell little about how the form is used, not used, or misused. When studying documents or records from a file, you should study enough samples to identify all the possible processing conditions and exceptions. How do you determine if the sample size is large enough to be representative? You use statistical sampling techniques.

How to Determine the Sample Size

The size of the sample depends on how representative you want the sample to be. There are many sampling issues and factors, which is a good reason to take an introductory statistics course. One simple and reliable formula for determining sample size is

Sample size $= 0.25 \times$ (Certainty factor/Acceptable error)^2

The certainty factor depends on how certain you want to be that the data sampled will not include variations not in the sample. The certainty factor is calculated from tables (available in many industrial engineering texts). A partial example is given here.

Desired Certainty	Certainty Factor
95%	1.960
90	1.645
80	1.281

Suppose you want 90-percent certainty that a sample of invoices will contain no unsampled variations. Your sample size, SS, is calculated as follows:

$$SS = 0.25(1.645/0.10)^2 = 68$$

We need to sample 68 invoices to get the desired accuracy.

Selecting the Sample

How do we choose our 68 invoices? Two commonly used sampling techniques are randomization and stratification.

Randomization is a sampling technique characterized as having no predetermined pattern or plan for selecting sample data.

Therefore, we just randomly choose 68 invoices.

Stratification is a systematic sampling technique that attempts to reduce the variance of the estimates by spreading out the sampling—for example, choosing documents or records by formula—and by avoiding very high or low estimates.

For computerized files, stratification sampling can be executed by writing a sample program. For instance, suppose our invoices were on a computer file that had a volume of approximately 250,000 invoices. Recall that our sample size needs to include 68 invoices. We will simply write a program that prints every 3,676th record ($=250,000/68$). For manual files and documents, we could execute a similar scheme.

RESEARCH AND SITE VISITS

A second fact-finding technique is to thoroughly research the application and problem. Computer trade journals are a good source, as are trade journals typically read by your end-users. You can learn how others have solved similar problems. You can learn whether or not software packages exist to solve your problem.

A similar type of research involves visiting other companies or departments that have addressed similar problems. Memberships in professional societies can provide useful contacts.

OBSERVATION OF THE WORK ENVIRONMENT

Observation is one of the most effective data-collection techniques for obtaining an understanding of a system.

Observation is a fact-finding technique wherein the systems analyst either participates in or watches a person perform activities to learn about the system.

This technique is often used when the validity of data collected through other methods is in question or when the complexity of certain aspects of the system prevents a clear explanation by the end-users.

Collecting Facts by Observing People at Work

Even with a well-conceived observation plan, the systems analyst is not assured that fact-finding will be successful. You should become aware of the pros and cons of the technique of observation. Advantages and disadvantages include:

Advantages

1. Data gathered by observation can be highly reliable. Sometimes, observations are conducted to check the validity of data obtained directly from individuals.
2. The systems analyst is able to see exactly what is being done. Complex tasks are sometimes difficult to clearly explain in words. Through observation, the systems analyst can identify tasks that have been missed or inaccurately described by other fact-finding techniques. Also, the analyst can obtain data describing the physical environment of the task (e.g., physical layout, traffic, lighting, noise level).
3. Observation is relatively inexpensive compared with other fact-finding techniques. Other techniques usually require substantially more employee release time and copying expenses.
4. Observation allows the systems analyst to do work measurements.

Disadvantages

1. Because people usually feel uncomfortable when being watched, they may unwittingly perform differently when being observed. In fact, the famous Hawthorne Experiment proved that the act of observation can alter behavior.
2. The work being observed may not involve the level of difficulty or volume normally experienced during that time period.
3. Some systems activities may take place at odd times, causing a scheduling inconvenience for the systems analyst.
4. The tasks being observed are subject to various types of interruptions.
5. Some tasks may not always be performed in the manner in which they are observed by the systems analyst. For example, the systems analyst might have observed how a company filled several customer orders. However, the procedures the systems analyst observed may have been those steps used to fill a number of regular customer orders. If any of those orders had been special orders (e.g., an order for goods not normally kept in stock), the systems analyst would have observed a different set of procedures being executed.
6. If people have been performing tasks in a manner that violates standard operating procedures, they may temporarily perform their jobs correctly while you are observing them. In other words, people may let you see what they want you to see.

Guidelines for Observation

How does the systems analyst obtain facts through observation? Does one simply arrive at the observation site and begin recording everything that's viewed? Of course not. Much preparation should take place in advance. The analyst must determine how data will actually be captured. Will it be necessary to have special forms on which to quickly record data? Will the individual(s) being observed be bothered by having someone watch and record their actions? When are the low, normal, and peak periods of operations for the task to be observed? The systems analyst must identify the ideal time to observe a particular aspect of the system.

Observation should first be conducted when the work load is normal. Afterward, observations can be made during peak periods to gather information for measuring the effects caused by the increased volume. The systems analyst might also obtain samples of documents or forms that will be used by those being observed. As you can see, a great deal of planning and preparation must be done up front.

The sampling techniques discussed earlier are also useful for observation. In this case, the technique is called work sampling.

Work sampling is a fact-finding technique that involves a large number of observations taken at random intervals.

This technique is less threatening to the people being observed because the observation period is not continuous. When using work sampling, you need to predefine the operations of the job to be observed. Then calculate a sample size as you did for document and file sampling. Make that many random observations, being careful to observe activities at different times of the day. By counting the number of occurrences of each operation during the observations, you will get a feel for how employees spend their days.

With proper planning completed, the actual observation can be done. Effective observation is difficult to carry out. Experience is the best teacher; however, the following guidelines may help you develop your observation skills:

1. Determine the who, what, where, when, why, and how of the observation.
2. Obtain permission from appropriate supervisors or managers.
3. Inform those who will be observed of the purpose of the observation.
4. Keep a low profile.
5. Take notes during or immediately following the observation.
6. Review observation notes with appropriate individuals.
7. Don't interrupt the individuals at work.
8. Don't focus heavily on trivial activities.
9. Don't make assumptions.

QUESTIONNAIRES

Another fact-finding technique is to conduct surveys through questionnaires.

Questionnaires are special-purpose documents that allows the analyst to collect information and opinions from respondents.

The document can be mass produced and distributed to respondents, who can then complete the questionnaire on their own time. Questionnaires allow the analyst to collect facts from a large number of people while maintaining uniform responses. When dealing with the large audience, no other fact-finding technique can tabulate the same facts as efficiently.

Collecting Facts by Using Questionnaires

The use of questionnaires has been heavily criticized and is often avoided by systems analysts. Many systems analysts claim that the responses lack reliable and useful information. But questionnaires can be an effective method for fact gathering, and many of these criticisms can be attributed to the inappropriate use of the questionnaires by systems analysts. Before using questionnaires, you should first understand the pros and cons associated with their use.

Advantages

1. Most questionnaires can be answered quickly. People can complete and return questionnaires at their convenience.
2. Questionnaires provide a relatively inexpensive means for gathering data from a large number of individuals.
3. Questionnaires allow individuals to maintain anonymity. Therefore, individuals are more likely to provide the real facts, rather than telling you what they think their boss would want them to.
4. Responses can be tabulated and analyzed quickly.

Disadvantages

1. The number of respondents is often low.
2. There's no guarantee that an individual will answer or expand on all of the questions.
3. Questionnaires tend to be inflexible. There's no opportunity for the systems analyst to obtain voluntary information from individuals or to reword questions that may have been misinterpreted.
4. It's not possible for the systems analyst to observe and analyze the respondent's body language.
5. There is no immediate opportunity to clarify a vague or incomplete answer to any question.
6. Good questionnaires are difficult to prepare.

Types of Questionnaires

There are two formats for questionnaires, free-format and fixed-format.

Free-format questionnaires offer the respondent greater latitude in the answer. A question is asked, and the respondent records the answer in the space provided after the question.

Here are two examples of free-format questions:

1. What reports do you currently receive and how are they used?
2. Are there any problems with these reports (e.g., are they inaccurate, is there insufficient information, or are they difficult to read and/or use)? If so, please explain.

Obviously, such responses may be difficult to tabulate. It is also possible that the respondents' answers may not match the questions asked. In order to ensure good responses in free-format questionnaires, the analyst should phrase the questions in simple sentences and not use words — such as *good* — that can be interpreted differently by different respondents. The analyst should also ask questions that can be answered with three or fewer sentences. Otherwise, the questionnaire may take up more time than the respondent is willing to sacrifice.

The second type of questionnaire is fixed-format.

Fixed-format questionnaires contain questions that require specific responses from individuals.

Given any question, the respondent must choose from the available answers. This makes the results much easier to tabulate. On the other hand, the respondent cannot provide additional information that might prove valuable. There are three types of fixed-format questions.

1. For **multiple-choice questions,** the respondent is given several answers. The respondent should be told if more than one answer may be selected. Some multiple-choice questions allow for very brief free-format responses when none of the standard answers apply. Examples of multiple-choice, fixed-format questions are:

   Do you feel that backorders occur too frequently?

   ☐ YES   ☐ NO

   Is the current accounts receivable report that you receive useful?

   ☐ YES   ☐ NO

   If no, please explain.

2. For **rating questions,** the respondent is given a statement and asked to use supplied responses to state an opinion. To prevent built-in bias, there should be an equal number of positive and negative ratings. The following is an example of a rating fixed-format question:

   The implementation of quantity discounts would cause an increase in customer orders.

   ☐ Strongly agree

   ☐ Agree

   ☐ No opinion

   ☐ Disagree

   ☐ Strongly disagree

3. For **ranking questions,** the respondent is given several possible answers, which are to be ranked in order of preference or experience. An example of a ranking fixed-format question is:

   Rank the following transactions according to the amount of time you spend processing them:

   _____ % new customer orders

   _____ % order cancellations

   _____ % order modifications

   _____ % payments

Developing a Questionnaire

Good questionnaires are designed. If you write your questionnaires without designing them first, your chances of success are limited. The following procedure is effective:

1. Determine what facts and opinions must be collected and from whom you should get them. If the number of people is large, consider using a smaller, randomly selected group of respondents.
2. Based on the needed facts and opinions, determine whether free- or fixed-format questions will produce the best answers. A combination format that permits optional free-format clarification of fixed-format responses is often used.
3. Write the questions. Examine them for construction errors and possible misinterpretations. Make sure that the questions don't offer your personal bias or opinions. Edit the questions.
4. Test the questions on a small sample of respondents. If your respondents had problems with them or if the answers were not useful, edit the questions.
5. Duplicate and distribute the questionnaire.

INTERVIEWS

The personal interview is generally recognized as the most important and most often used fact-finding technique.

> **Interviews** are a fact-finding technique whereby the systems analysts collects information from individuals face to face.

Interviewing can be used to achieve any of the goals listed in the margin. There are two roles assumed in an interview. The systems analyst is the **interviewer,** responsible for organizing and conducting the interview. The system user, system owner, or adviser is the **interviewee,** who is asked to respond to a series of questions. Unfortunately, many systems analysts are poor interviewers. In this section you will learn how to conduct proper interviews.

Collecting Facts by Interviewing People

—————— ✓ ——————

Interview Purpose

Fact-finding
Fact verification
Clarification
Generate enthusiasm
Get end-user involved
Identify requirements
Solicit ideas and opinions

The most important element of an information system is people. More than anything else, people want to be in on things. No other fact-finding technique places as much emphasis on people as interviews. But people have different values, priorities, opinions, motivations, and personalities. Therefore, to use the interviewing technique, you must possess good human relations skills for dealing effectively with different types of people. And like other fact-finding techniques, interviewing isn't the best method for all situations. Interviewing has its advantages and disadvantages, which should be weighed against those of other fact-finding techniques for every fact-finding situation.

Advantages

1. Interviews give the analyst an opportunity to motivate the interviewee to respond freely and openly to questions. By establishing rapport, the systems analyst is able to give the interviewee a feeling of actively contributing to the systems project.
2. Interviews allow the systems analyst to probe for more feedback from the interviewee.
3. Interviews permit the systems analyst to adapt or reword questions for each individual.

4. Interviews give the analyst an opportunity to observe the interviewee's nonverbal communication. A good systems analyst may be able to obtain information by observing the interviewee's body movements and facial expressions as well as by listening to verbal replies to questions.

Disadvantages

1. Interviewing is a very time-consuming, and therefore costly, fact-finding approach.
2. Success of interviews is highly dependent on the systems analyst's human relations skills.
3. Interviewing may be impractical due to the location of interviewees.

Interview Types and Techniques

There are two types of interviews, unstructured and structured.

> **Unstructured interviews** are conducted with only a general goal or subject in mind and with few, if any, specific questions. The interviewer counts on the interviewee to provide a framework and direct the conversation.

This type of interview frequently gets off track, and the analyst must be prepared to redirect the interview back to the main goal or subject. For this reason, unstructured interviews don't usually work well for systems analysis and design.

> In **structured interviews** the interviewer has a specific set of questions to ask of the interviewee.

Depending on the interviewee's responses, the interviewer will direct additional questions to obtain clarification or amplification. Some of these questions may be planned and others spontaneous. **Open-ended questions** allow the interviewee to respond in any way that seems appropriate. An example of an open-ended question is "Why are you dissatisfied with the report of uncollectable accounts?" **Closed-ended questions** restrict answers to either specific choices or short, direct responses. An example of such a question might be "Are you receiving the report of uncollectable accounts on time?" or "Does the report of uncollectable accounts contain accurate information?" Realistically, most questions fall between the two extremes.

How to Conduct an Interview

Your success as a systems analyst is at least partially dependent on your ability to interview. A successful interview will involve selecting appropriate individuals to interview, preparing extensively for the interview, conducting the interview properly, and following up on the interview. Here we examine each of these aspects in more detail. Let's assume that you've identified the need for an interview and you have determined exactly what kinds of facts and opinions you need.

Select Interviewees

You should interview the end-users of the information system you are studying. A formal organizational chart will help you identify these individuals and their responsibilities. You should attempt to learn as much as possible about each individual prior to the interview. Attempt to learn what their strengths, fears, biases, and motivations might be. The interview can then be geared to take the characteristics of the individual into account.

Always make an appointment with the interviewee. Never just drop in. Limit the appointment to somewhere between a half hour and an hour. The higher the management level of the interviewee, the less time you should schedule. If the interviewee is a clerical, service, or blue-collar worker, get their supervisor's permission before scheduling the interview. Be certain that the location you want for the interview will be available during the time the interview is scheduled. Never conduct an interview in the presence of your officemates or the interviewee's peers.

Prepare for the Interview

Preparation is the key to a successful interview. An interviewee can easily detect an unprepared interviewer. In fact, the interviewee may very much resent the lack of preparation because it is a waste of valuable time. When the appointment is made, the interviewee should be notified about the subject of the interview. To ensure that all pertinent aspects of the subject are covered, the analyst should prepare an interview guide.

> An **interview guide** is a checklist of specific questions the interviewer will ask the interviewee.

The interview guide may also contain follow-up questions that will only be asked if the answers to other questions warrant the additional answers. A sample interview guide is presented in Figure B.1. Questions should be carefully chosen and phrased. Most questions begin with the standard who, what, when, where, why, and how much type of wording. Avoid the following types of questions:

Interview Question Guidelines

1. Use clear and concise language.
2. Don't include your opinion as part of a question.
3. Avoid long or complex questions.
4. Avoid threatening questions.
5. Don't use "you" when you mean a group of people.

- *Loaded questions,* such as "Do we have to have both of these columns on the report?" The question conveys the interviewee's personal opinion on the issue.
- *Leading questions,* such as "You're not going to use this OPERATOR CODE, are you?" The question leads the interviewee to respond, "No, of course not," regardless of actual opinion.
- *Biased questions,* such as "How many codes do we need for FOOD-CLASSIFICATION in the INVENTORY FILE? I think 20 ought to cover it." Why bias the interviewee's answer with your own?

Additional guidelines for questions are provided in the margin. You should especially avoid threatening or critical questions. The purpose of the interview is to investigate, not to evaluate or criticize.

Conduct the Interview

The actual interview can be characterized as consisting of three phases: the opening, body, and conclusion. The **interview opening** is intended to influence or motivate the interviewee to participate and communicate by establishing an ideal environment. When establishing an environment of mutual trust and respect, you should identify the purpose and length of the interview and

INTERVIEWEE:		Jeff Bentley, Accounts Receivable Manager
DATE:		Tuesday, March 23, 1993
TIME:		1:30 P.M.
PLACE:		Room 223, Admin. Bldg.
SUBJECT:		Current Credit-Checking Policy
1 to 2 min.	Open the interview.	
	Introduce ourselves. Thank Mr. Bentley for his valuable time. State the purpose of the interview—to obtain an understanding of the existing credit-checking policies.	
5 min.	What conditions determine whether a customer's order is approved for credit?	
5 min.	What are the possible decisions or actions that might be taken once these conditions have been evaluated?	
3 min.	How are customers notified when credit is not approved for their order?	
1 min.	After a new order is approved for credit and placed in the file containing orders that can be filled, a customer might request that a modification be made to the order. Would the order have to go through credit approval again if the new total order cost exceeds the original cost?	
1 min.	Who are the individuals that perform the credit checks?	
1 to 3 min.	May I have permission to talk to those individuals to learn specifically how they carry out the credit-checking process?	
	If so:	
	When would be an appropriate time to meet with each of them?	
1 min.	Conclude the interview:	
	Thank Mr. Bentley for his cooperation and assure him that he will be receiving a copy of what transpired during the interview.	

21 minutes
+9 minutes for follow-up questions and redirection

30 minutes alloted for interview (1:30 P.M.–2:00 P.M.)

FIGURE B.1 Sample Interview Guide The sample interview guide represents an agenda that a systems analyst will use to obtain facts about a company's existing credit approval policy. Notice that the agenda is carefully laid out with the specific time allocated to each question. Time should also be reserved for follow-up questions and redirecting the interview.

explain how the gathered data will be used. Here are three ways to effectively begin an interview:

1. Summarize the apparent problem, and explain how the problem was discovered.
2. Offer an incentive or reward for participation.
3. Ask the interviewee for advice or assistance.

The **interview body** represents the most time-consuming phase. During this phase, you obtain the interviewee's responses to your list of questions. Listen closely and observe the interviewee. Take notes concerning both verbal and nonverbal responses from the interviewee. It's very important for you to keep the interview on track. Anticipate the need to adapt the interview to the interviewee. Often questions can be bypassed if they have been answered earlier in part of an answer to another question, or they can be deleted if determined to be irrelevant, based on what you've already learned during the interview. Finally, probe for more facts when necessary.

During the **interview conclusion,** you should express your appreciation and provide answers to any questions posed by the interviewee. The conclusion is very important for maintaining rapport and trust with the interviewee.

The importance of human relations skills in interviewing cannot be overemphasized. These skills must be exercised throughout the interview. In the margin you will find a set of rules that should be followed during an interview.

Follow Up on the Interview

To help maintain good rapport and trust with interviewees, you should send them a memo that summarizes the interview. This memo should remind the interviewees of their contributions to the systems project and allow them the opportunity to clarify any misinterpretations that you may have derived during the interview. In addition, the interviewees should be given the opportunity to offer additional information they may have failed to bring out during the interview.

JOINT APPLICATION DESIGN

Separate interviews of end-users have always been the classic fact-finding technique. However, many analysts have discovered the great flaw of interviewing—separate interviews often lead to conflicting facts, opinions, and priorities. The end result is numerous follow-up interviews and/or group meetings. For this reason, many shops are using the group work session as a substitute for interviews.

One example of the group work session approach is IBM's joint application design (JAD). This and similar techniques generally require extensive training in order to work as intended. However, they can significantly decrease the time spent on fact-finding in one or more phases of the life cycle.

> **Joint application design (JAD)** is a process whereby highly structured group meetings of all the system users, system owners, and analysts occur in a single room for an extended period of time (four to eight hours per day).

The goals are essentially the same as in an interview, except that you need a number of analysts to carry them out. One analyst serves as discussion leader or **moderator.** Another individual, referred to as the **scribe,** records facts and items that require further action or individual interviews. It is becoming more common for the JAD scribe to make use of CASE tools to capture many facts that are communicated during a JAD session.

Collecting through JAD Sessions

JAD-like techniques are becoming increasingly common in many methodologies. Most methodologies incorporate JAD in systems planning and systems analysis to obtain group consensus on problems, objectives, and requirements. Therefore, the authors prefer to use the term joint application *development* rather than *design*—to appropriately recognize that it includes more than simply systems design.

The advantages of JAD include a reduction in the amount of time it takes to get systems implemented, to improve quality systems, and to improve relations with system users and owners.

JAD also has its pitfalls if not properly planned and executed. JAD sessions represent a very expensive commitment by management. Both the moderator and scribe must be very skilled in conducting a JAD session.

How to Conduct a JAD Session

Most JAD sessions span a three- to five-day time period. Conducting a JAD session involves several steps: defining the project, conducting a preliminary background investigation, planning the JAD session, conducting the JAD session, and presenting the results.

Defining the Project

Some preparation is necessary well before the JAD session can actually be performed. The analyst must work closely with system owners to determine high-level requirements and expectations. This normally involves interviewing each system owner who is responsible for departments or functions that are to be addressed by the project. JAD provides numerous forms that are used to capture information obtained during these interviews.

During the interviews, the analyst carries out these functions:

- Asks the system owners to identify those system users who should participate in the JAD session.
- Proposes or solicit dates for when the JAD session may be conducted.
- Obtains a commitment to allow the system users to participate during the entire session.
- Packages and reviews the project definition.
- Selects the JAD leader (or moderator) and scribe.
- Selects a location for the JAD session(s).

Conducting a Preliminary Background Investigation

In defining the project scope, the analyst may have a very general idea about the project functions but know very little about the current system(s) in place. Therefore, prior to the JAD session, the analyst may conduct interviews with a few select system users to learn more about the current systems. The idea is not to be an expert, rather to simply obtain a basic understanding of the current business functions and the terminology. In addition to interviews, the analyst may conduct one or more of the other fact-finding techniques discussed earlier. During this investigation, the analyst will develop high-level or overview process models (DFDs) of the current system and collect sample specifications of input and output requirements (sample screens and reports) — providing they exist. Once this fact-finding effort is complete, the analyst will review the current system and JAD expectations with the moderator and scribe. Finally, the analyst will develop an agenda for the JAD session.

Planning the JAD Session

The next step toward conducting a JAD session is planning.

- If necessary, train the session moderator and scribe.
- Distribute the agenda and appropriate documentation accumulated earlier to appropriate participants.
- Schedule the site where the JAD session will be conducted.

- Develop a script to follow during the JAD session (the script normally represents a detailed expansion of the agenda to be used by the moderator).
- Prepare any visual aids and arrange for audio-visual equipment.

Proper planning is essential for the success of conducting a JAD session. Once planning has been completed, the JAD session is ready to begin.

Conducting the JAD Session

The JAD session begins with opening remarks, introductions, and a brief overview of the agenda and objectives for the JAD session. The moderator will lead the session by following the prepared script. To successfully conduct the session, the moderator should follow these guidelines:

- Do not unreasonably deviate from the agenda.
- Stay on schedule (agenda topics are allotted specific time).
- Ensure that the scribe is able to take notes (this may require the moderator having speakers restate their points more slowly or clearly).
- Avoid the use of technical jargon.
- Apply conflict resolution skills.
- Allow for ample breaks.
- Encourage group consensus.
- Encourage user participation without allowing individuals to dominate.

As mentioned earlier, the success of a JAD session is highly dependent on planning and the skills of the moderator and scribe. These skills only get better through proper training and experience. Therefore, JAD sessions are usually concluded with an evaluation questionnaire for the participants to complete. The responses will help ensure the likelihood of future JAD successes.

Presenting the Results

The end product of a JAD session is typically a formal written document. This document is essential in confirming the specifications agreed upon during the session(s) to all participants. The content and organization of the specification is obviously dependent on the objectives of the JAD session. The analyst may choose to provide a different set of specifications to different participants based upon their role—for example, system owners may receive more of a summary version of the document provided to the user participants (especially in those cases in which the system owners had minimal actual involvement in the JAD session itself).

A FACT-FINDING STRATEGY

At the beginning of this module, we suggested that an analyst needs an organized method for collecting facts. An inexperienced analyst will frequently jump right into interviews. "Go to the people. That's where the real facts are!" Wrong! This attitude fails to recognize an important fact of life: people must complete their day-to-day jobs! Your job is not their main responsibility. Your demand on their time is their money lost. Now you may be thinking, "But I thought you've been saying that the system is for people and that direct end-user involvement in systems development is essential! Aren't you contradicting yourselves?''

Not at all! Time is money. To waste your end-users' time is to waste your company's money. To make the most of the time that you spend with end-users, don't jump right into interviews. Instead, first collect all the facts you can by using other methods. Consider the following step-by-step strategy:

1. *Learn all you can from existing documents, forms, reports, and files.* You'll be surprised how much of the system becomes clear without any people contact.

2. *If appropriate, observe the system in action.* Agree not to ask questions. Just watch and take notes or draw pictures. Make sure that the workers know that you're not evaluating individuals. Otherwise, they may perform in a more efficient manner than normal.

3. *Given all the facts that you've already collected, design and distribute questionnaires to clear up things you don't fully understand.* This is also a good time to solicit opinions on problems and limitations. Questionnaires do require your end-users to give up some of their time. But *they* choose when to best make that sacrifice.

4. *Conduct your interviews (or group work sessions, such as JAD).* Because you have already collected most of the pertinent facts by low-user-contact methods, you can use the interview to verify and clarify the most difficult issues and problems.

5. *Follow up.* Use appropriate fact-finding techniques to verify facts (usually interviews or observation).

The strategy is not sacred. Although a fact-finding strategy should be developed for every pertinent phase of systems development, every project is unique. Sometimes observation and questionnaires may be inappropriate. But the idea should always be to collect as many facts as possible before using interviews.

Summary

Effective fact-finding techniques are crucial to the application of systems analysis and design methods during systems projects. Fact-finding is performed during all phases of the systems development life cycle. To support development activities, the analyst must collect facts about end-users, the business, data and information resources, and information systems components. There are six common fact-finding techniques: sampling, research, observation, questionnaires, interviews, and work group sessions (e.g., JAD).

The sampling of existing documents and files can provide many facts and details with little or no direct personal communication being necessary. The analyst should collect historical documents, business operations manuals and forms, and information systems documents. In order to ensure that an adequate number of documents have been studied, analysts often use sampling techniques. These techniques make it possible to collect a representative subset of the documents and minimize the chance of identifying exceptional events.

Research is an often overlooked technique based on the study of other similar applications. Site visits are a special form of research.

Observation is a fact-finding technique in which the analyst studies people doing their jobs. To minimize the chance that the observation time is not representative of normal work loads, the analyst can use work sampling to randomly collect observation data.

Questionnaires are used to collect similar facts from a larger number of individuals. Questionnaires can be either free-format or fixed-format.

Interviews are the most popular but the most time-consuming fact-finding technique. When interviewing, the analyst meets individually with people to gather information. Most systems analysis and design interviews are structured, meaning that the analyst has prepared a specific set of questions before the interview. After determining the need for an interview, the analyst arranges for appointments with the interviewee, carefully prepares the interview questions, conducts the interview, and summarizes the results. Group work sessions are many-on-many interviews that usually require special training. Because interviews are time-consuming, the analyst should collect as many facts as possible using the other fact-finding methods.

Many analysts find flaws with interviewing—separate interviews often lead to conflicting facts, opinions, and priorities. The end result is numerous follow-up interviews and/or group meetings. For this reason, many shops are using the group work session as a substitute for interviews.

One example of the group work session approach is IBM's joint application design (JAD). This and similar techniques generally require extensive training in order to work as intended. However, they can significantly decrease the time spent on fact-finding in one or more phases of the life cycle.

Key Terms

closed-ended questions, p. 801

fact-finding, p. 792

fixed-format questionnaires, p. 799

free-format questionnaires, p. 798

interview body, p. 803

interview conclusion, p. 804

interviewee, p. 800

interviewer, p. 800

interview guide, p. 802

interview opening, p. 802

interviews, p. 800

joint application design (JAD), p. 804

moderator, p. 804

multiple-choice questions, p. 799

observation, p. 795

open-ended questions, p. 801

questionnaires, p. 797

randomization, p. 795

ranking questions, p. 799

rating questions, p. 799

sampling, p. 794

scribe, p. 804

stratification, p. 795

structured interviews, p. 801

unstructured interviews, p. 801

work sampling, p. 797

Problems and Exercises

1. Explain how the information systems building blocks can serve as a framework in determining what facts need to be collected during systems development.

2. Explain how an organizational chart can aid in planning for fact-finding. What are some of the potential drawbacks to using an existing organizational chart?

3. A systems analyst wants to study documents stored in a large metal file cabinet. The cabinet contains several hundred records describing product warranty claims. The analyst wishes to study a sample of the records in the file and to be 95 percent certain (certainty factor = 1.960) that the data from which the sample is taken will not include variations not in the sample. How many sample records should the analyst retrieve to get this desired accuracy?

4. For the sample size in Exercise 3, explain two specific strategies for selecting the samples.

5. Describe how you would use form and/or file sampling in each phase of the systems development life cycle. If you think sampling would be inappropriate for any of these phases, explain why.

6. Repeat Exercise 5 for the technique of observation.

7. Make a list of things that might affect your work performance when you are being observed performing your job. What could an observer do to eliminate these concerns or problems?

8. Repeat Exercise 5 for the questionnaire technique.

9. Give two examples of free-format questions and two examples of each of the following types of fixed-format questions:
   a. Multiple choice.
   b. Rating.
   c. Ranking.

10. Repeat Exercise 5 for the interviewing technique.

11. Explain the difference between a structured and an unstructured interview. When is each type of interview appropriately used?

12. Prepare a sample interview guide to use in obtaining from your academic adviser facts describing the course registration policies and procedures.

Project
---

1. Mr. Art Pang is the Accounts Receivables manager. You have been assigned to do a study of Mr. Pang's current billing system, and you need to solicit facts from his subordinates. Mr. Pang has expressed his concern that, although he wishes to support you in your fact-finding efforts, his people are extremely busy and must get their jobs done. Write a memo to Mr. Pang describing a fact-finding strategy that you could follow to maximize your fact-finding while minimizing the release time required for his subordinates.

Suggested Readings

Berdie, Douglas R., and John F. Anderson. *Questionnaires: Design and Use.* Metuchen, N.J.: Scarecrow Press, 1974. A practical guide to the construction of questionnaires. Particularly useful because of its short length and illustrative examples.

Davis, William S. *Systems Analysis and Design.* Reading, Mass.: Addison-Wesley, 1983. Provides useful pointers for preparing and conducting interviews.

Fitzgerald, Jerry, Ardra F. Fitzgerald, and Warren D. Stallings, Jr. *Fundamentals of Systems Analysis.* 2nd ed. New York: Wiley, 1981. A useful survey text for the systems analyst. Chapter 6, "Understanding the Existing System," does a good job of presenting fact-finding techniques in the study phase.

Gildersleeve, Thomas R. *Successful Data Processing System Analysis.* Englewood Cliffs, N.J.: Prentice Hall, 1978. Chapter 4, "Interviewing in Systems Work," provides a comprehensive look at interviewing specifically for the systems analyst. A thorough sample interview is scripted and analyzed in this chapter.

London, Keith R. *The People Side of Systems.* New York: McGraw-Hill, 1976. Chapter 5, "Investigation Versus Inquisition," provides a very good people-oriented look at fact-finding, with considerable emphasis on interviewing.

Lord, Kenniston W., Jr., and James B. Steiner. *CDP Review Manual: A Data Processing Handbook.* 2nd ed. New York: Van Nostrand Reinhold, 1978. Chapter 8, "Systems Analysis and Design," provides a more comprehensive comparison of the merits and demerits of each fact-finding technique. This material is intended to prepare data processors for the Certificate in Data Processing examinations, one of which covers Systems Analysis and Design.

Salvendy, G., ed. *Handbook of Industrial Engineering.* New York: Wiley, 1974. A comprehensive handbook for industrial engineers; systems analysts are, in a way, a type of industrial engineer. Excellent coverage on sampling and work measurement.

Stewart, Charles J., and William B. Cash, Jr. *Interviewing: Principles and Practices.* 2nd ed. Dubuque, Iowa: Brown, 1978. Popular college textbook that provides broad exposure to interviewing techniques, many of which are applicable to systems analysis and design.

Wood, Jane, and Denise Silver. *Joint Application Design.* New York: John Wiley & Sons, 1989. This book provides a comprehensive overview of IBM's Joint Application Design technique.

C

Feasibility Analysis

When Should You Read This Module?

This module will prove most valuable if read after any of the following chapters:

Chapter 3
A Systems Development Life Cycle

Chapter 7
Systems Analysis

Chapter 12
Systems Design

If this is the first module you are studying, you should also read the part opener for Part Five of the textbook.

This module is about feasibility analysis, one of the most important non-technical skills that any systems analyst must develop. Systems analysts sell change. Good systems analysts thoroughly evaluate alternative solutions before recommending change. This module teaches you how to analyze and document those alternatives on the basis of four feasibility criteria: operational, technical, schedule, and economics. The latter is sometimes called a *cost-benefit analysis.* After reading and studying this module, you should be able to identify feasibility checkpoints in the systems development life cycle; define and describe four types of feasibility and their respective criteria; and perform various cost-benefit analyses using time-adjusted costs and benefits.

METROPOLIS INTERNATIONAL AIRPORT

Metropolis International Airport is expanding. Two major airlines have selected Metropolis as a new major hub (base of operations). A new terminal is nearing completion for the hubs. With the expected increase in traffic, the airport's director authorized development of a new gate management information system.

Scene: *Frank Demillio, Project Manager in charge of the gate management scheduling project, is meeting with Airport Operations Manager Benjamin Pierce to discuss progress. The systems analysis and design phases of the project have been completed, and Frank is seeking approval to implement the system. We join the meeting in progress.*

Frank: As you can see, we've designed a system that can fulfill all your stated requirements, including those specified after the initial project charter. Furthermore, the new system will include a Microsoft Windows user interface. All of your staff are familiar with Windows-based PC tools.

Ben: So you moved the entire application to the PC?

Frank: No, just the user interface and basic input data validation. Your database and business data processing will still be done on the host computer. But to your staff, the entire application will look and feel like a PC application.

[Brief silence.]

Are there any questions?

Ben: I'm quite impressed. The system design is everything you promised, and then some. But let's get to the bottom line, Frank. How much is this going to cost me? It seems more ambitious than the original estimates.

Frank: To implement the system, we estimate that it will cost $45,000. And the system will cost $4,500 per year for maintenance and enhancement.

Ben: And how much have we spent up to this point?

Frank: About $12,000 for requirements analysis and system design. We've also done some prototypes, if you care to see them.

Ben: Not right now. I'm just the sponsor. I trust the users' feedback on such matters. Let's see, according to my notes, the original estimate was $39,500, including analysis and design. The new estimate is $45,000 + 12,000 — about $57,000! That's quite an increase.

Frank: Agreed. But your people increased the requirements as the project progressed. Also, we feel the Windows-based design will result in a faster learning curve and lower maintenance costs. Also, the $12,000 for analysis and design can't be recovered, so we shouldn't consider it in our decision on whether to proceed.

Ben: I agree that the project shouldn't be continued or canceled based solely on the money spent so far. However, I disagree that the money is irrelevant. If the project is continued, I should think that the new system would eventually pay for itself. How long until all costs are recovered.

Frank: You should start receiving benefits immediately after implementation.

Ben: But how many years or months will pass before the lifetime benefits exceed the lifetime costs?

[Brief silence.]

Look, I can probably allocate the funds you need to complete the system. But I also have managers asking for other things. Why should I give the money to you and not to them?

Frank: I don't understand, Ben. You and your staff commissioned this project. The proposed system will meet your needs.

Ben: Frank, you're a marvelous computer professional, but you don't always understand the economics of business and information.

Discussion Questions

1. Ben's own staff increased the requirements for the project. Why is Ben now hedging on the project?

2. What doesn't Frank understand about business economics? How is his vision of the project and the business flawed?

FEASIBILITY ANALYSIS — A CREEPING COMMITMENT APPROACH

The minicase demonstrates that analysts must learn to think like business managers. Computer applications are expanding at a record pace. Now more than ever, management expects information systems to pay for themselves. Information is a major capital investment that must be justified, just as marketing must justify a new product and manufacturing must justify a new plant or equipment. Will the investment pay for itself? Are there other investments that will return even more on their expenditure? In the minicase Frank didn't pay attention to economic concerns such as cost-effectiveness and opportunity analysis. Frank knows this, and so do you; but we all easily forget this point when it comes to our preoccupation with computers and computer projects.

This module deals with cost-benefit analysis and other feasibility issues of interest to the systems analyst and users of information systems. Few topics are more important. Feasibility analysis isn't really systems analysis, and it isn't systems design either. Instead, feasibility analysis is a cross life cycle activity and should be continuously performed throughout a systems project.

Feasibility Checkpoints in the Life Cycle

Let's begin with a formal definition of feasibility and feasibility analysis.

> **Feasibility** is the measure of how beneficial or practical the development of an information system will be to an organization.
>
> **Feasibility analysis** is the process by which feasibility is measured.

Feasibility should be measured throughout the life cycle. In earlier chapters we called this a **creeping commitment** approach to feasibility. The scope and complexity of an apparently feasible project can change after the initial problems and opportunities are fully analyzed or after the system has been designed. Thus, a project that is feasible at one point in time may become infeasible at a later point in time. Let's study some checkpoints for our systems development life cycle.

If you study your company's project standards or systems development life cycle (SDLC), you'll probably see a feasibility study phase or deliverable, but not an explicit ongoing process. But look more closely! Upon deeper examination, you'll probably identify various go/no-go checkpoints or management reviews. These checkpoints and reviews identify specific times during the life cycle when feasibility is reevaluated. A project can be canceled or revised in scope, schedule, or budget at any of these checkpoints. Thus, an explicit feasibility analysis phase in any life cycle should be considered to be only an initial feasibility assessment.

Feasibility checkpoints can be installed into any SDLC that you are using. Figure C.1 shows feasibility checkpoints for a typical life cycle (similar to, but not identical to, the life cycle used in this book). The checkpoints are represented by red diamonds. The diamonds indicate that a feasibility reassessment and management review should be conducted at the end of the prior phase (before the next phase). A project may be canceled or revised at any checkpoint, despite whatever resources have already been spent so far.

This idea may bother you at first. Your natural inclination may be to justify continuing a project based on the time and money you've already spent. Those costs are sunk. A fundamental principle of management is never to throw good

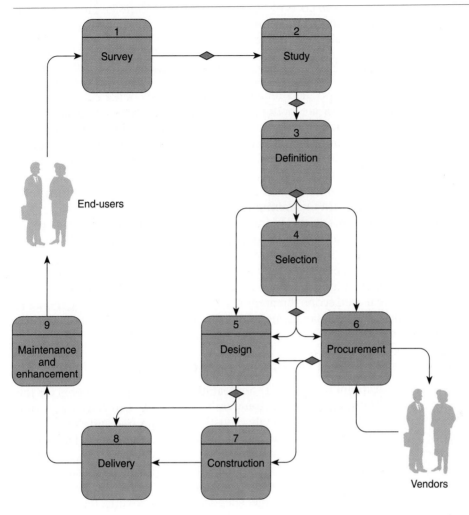

FIGURE C.1 **Feasibility Checkpoints in the Systems Development Life Cycle** This is our familiar SDLC with checkpoints, indicated by diamonds, for feasibility analysis. At any checkpoint, analysts and management reevaluate feasibility and determine whether to cancel, revise, or continue the project. The decision is part of the charter. This approach is called *creeping commitment.*

money after bad — cut your losses and move on to a more feasible project. That doesn't mean the costs already spent are not important. As the minicase demonstrated, they must eventually be recovered if the investment is to ever be considered a success. Let's briefly examine the checkpoints in Figure C.1.

Systems Analysis — A Survey Phase Checkpoint

The first feasibility analysis is conducted during the survey phase. At this early stage of the project, feasibility is rarely more than a measure of the urgency of the problem and the first-cut estimate of development costs. It answers the question: "Do the problems (or opportunities) warrant the cost of a detailed study of the current system?" Realistically, feasibility can't be accurately measured until the problems (and opportunities) and requirements (definition phase) are better understood.

After estimating benefits of solving the problems and opportunities, analysts will estimate costs of developing the expected system. Experienced analysts routinely increase these costs by 50 percent to 100 percent (or more) because experience tells them that the problems are rarely well-defined and that user requirements are typically understated.

Systems Analysis — A Study Phase Checkpoint

The next checkpoint occurs after a more detailed study of the current system. Because the problems are better understood, the analysts can make better estimates of development costs and of the benefits to be obtained from a new system. The minimum value of solving a problem is equal to the cost of that problem. For example, if inventory carrying costs are $35,000 over acceptable limits, then the minimum value of an acceptable information system would be $35,000. Hopefully, an improved system will be able to do better than that; however, it must return this minimum value!

Development costs, at this point, are still just guesstimates. We have yet to fully define user requirements or to specify a design solution to those requirements.

If the cost estimates significantly increase from the survey phase to the study phase, the likely culprit is scope. Scope has a tendency to increase in many projects. If increased scope threatens feasibility, then scope might be reduced. See Module A, Project Management for a discussion on the subject of expectations management.

Systems Analysis — A Definition Phase Checkpoint

The next checkpoint occurs after the definition of user requirements for the new system. These requirements frequently prove more extensive than originally stated. For this reason, the analyst must frequently revise cost estimates for design and implementation. Once again, feasibility is reassessed. If feasibility is in question, scope, schedule, and costs must be rejustified. (Again, Module A, Project Management, offers guidelines for adjusting project expectations.)

If early estimates were adjusted up, you may still be within the range despite an increase in scope. If not, the project need not always be canceled or reduced in scope. If you have kept track of the increase in problems and requirements since the beginning of the project, your system owner may be willing to pay for the increased requirements (and adjust the schedule accordingly).

Systems Design — A Selection Phase Checkpoint

The SDLC in Figure C.1 is the design decision-making phase. This SDLC separates design decision-making from the actual design phase. In any case, the selection phase represents a major feasibility analysis activity since it charts one of many possible implementations as the target for systems design.

Problems and requirements should be known by now. During the selection phase, alternative solutions are defined in terms of their input/output methods, data storage methods, computer hardware and software requirements, processing methods, and people implications. The following list presents the typical range of options that can be evaluated by the analyst.

- Do nothing! Leave the current system alone. Regardless of management's opinion or your own opinion of this option, it should be considered and analyzed as a baseline option against which all others can and should be evaluated.

- Reengineer the (manual) business processes, not the computer-based processes. This may involve streamlining activities, reducing duplication and unnecessary tasks, reorganizing office layouts, and eliminating redundant and unnecessary forms and processes, among others.
- Enhance existing computer processes.
- Purchase a packaged application.
- Design and construct a new computer-based system. This option presents numerous other options:
  - Centralized versus distributed versus cooperative processing
  - On-line versus batch processing
  - Files versus database for data storage

Of course, an alternative could be a combination of the preceding options.

After defining these options, each option is analyzed for operational, technical, schedule, and economic feasibility. This module will closely examine these four classes of feasibility criteria. One alternative is recommended to the system owner(s) for approval. The approved solution becomes the basis for general and detailed design.

Systems Design — A Procurement Phase Checkpoint

Because the procurement of hardware and applications software involves economic decisions that may require sizeable outlays of cash, it shouldn't surprise you that feasibility analysis is required before a contract is extended to a vendor. It should be noted that the procurement phase (covered more extensively in Chapter 12) may be consolidated into the selection phase because hardware and software selection may have a significant impact on the feasibility of the solutions being considered.

Systems Design — A Design Phase Checkpoint

A final checkpoint is completed after the system is designed. The general and detailed design specifications have been completed. The complexity of the solution should be apparent. Because implementation is often the most time-consuming and costly phase, the checkpoint after design gives us one last chance to cancel or downsize the project. Downsizing is the act of reducing the scope of the initial version of the system. Future versions can address other requirements after the system goes into production.

FOUR TESTS FOR FEASIBILITY

So far, we've defined feasibility and feasibility analysis, and we've identified feasibility checkpoints in the life cycle. Most analysts agree that there are four categories of feasibility tests:

- **Operational feasibility** is a measure of how well the solution of problems or a specific solution will work in the organization. It is also a measure of how people feel about the system/project.
- **Technical feasibility** is a measure of the practicality of a specific technical solution and the availability of technical resources and expertise.
- **Schedule feasibility** is a measure of how reasonable the project timetable is.

- **Economic feasibility** is a measure of the cost-effectiveness of a project or solution. This is often called a *cost-benefit analysis.*

Recall the scenario that introduced this module. Frank's proposed system was very likely operationally and technically feasible. The people seemed to like the system because it filled their requirements. Frank was unlikely to propose a technical solution he couldn't implement. Operational and technical feasibility criteria measure the worthiness of a problem or solution. Operational feasibility is people oriented. Technical feasibility is computer oriented.

But economic feasibility was questioned by the manager who had to make the financial decisions. Economic feasibility deals with the costs and benefits of the information system. The proposed project may very well have been economically feasible, but we can't tell because Frank didn't analyze costs and benefits. Actually, few systems are infeasible. Instead, different options tend to be more or less feasible than others. Let's take a closer look at the four feasibility criteria.

Operational Feasibility

Operational feasibility criteria measure the urgency of the problem (survey and study phases) or the acceptability of a solution (definition, selection, acquisition, and design phases). How do you measure operational feasibility? There are two aspects of operational feasibility to be considered:

1. Is the problem worth solving, or will the solution to the problem work?
2. How do the end-users and management feel about the problem (solution)?

Is the Problem Worth Solving, or Will the Solution to the Problem Work?

Do you recall the PIECES framework for identifying problems (Chapters 3 and 6)? PIECES can be used as the basis for analyzing the urgency of a problem or the effectiveness of a solution. The following is a list of the questions that address these issues:

P   *Performance.* Does the system provide adequate throughput and response time?

I   *Information.* Does the system provide end-users and managers with timely, pertinent, accurate, and usefully formatted information?

E   *Economy.* No, we are not prematurely jumping into economic feasibility! The question here is, "Does the system offer adequate service level and capacity to reduce the costs of the business or increase the profits of the business?"

C   *Control.* Does the system offer adequate controls to protect against fraud and embezzlement and to guarantee the accuracy and security of data and information?

E   *Efficiency.* Does the system make maximum use of available resources including people, time, flow of forms, minimum processing delays, and the like?

S   *Services.* Does the system provide desirable and reliable service to those who need it? Is the system flexible and expandable?

Note   The term *system,* used throughout this discussion, may refer either to the existing system or a proposed system solution, depending on which phase you're currently working in.

How Do the End-Users and Managers Feel about the Problem (Solution)?

It's not only important to evaluate whether a system *can* work. We must also evaluate whether a system *will* work. A workable solution might fail because of end-user or management resistance. The following questions address this concern:

- Does management support the system?
- How do the end-users feel about their role in the new system?
- What end-users or managers may resist or not use the system? People tend to resist change. Can this problem be overcome? If so, how?
- How will the working environment of the end-users change? Can or will end-users and management adapt to the change?

Essentially, these questions address the political acceptability of solving the problem or the solution.

Technical Feasibility

Technical feasibility can only be evaluated after those phases during which technical issues are resolved — namely, after the evaluation and design phases of our life cycle have been completed. Today, very little is technically impossible. Consequently, technical feasibility looks at what is practical and reasonable. Technical feasibility addresses three major issues:

1. Is the proposed technology or solution practical?
2. Do we currently possess the necessary technology?
3. Do we possess the necessary technical expertise, and is the schedule reasonable?

Is the Proposed Technology or Solution Practical?

The technology for any defined solution is normally available. The question is whether that technology is mature enough to be easily applied to our problems. Some firms like to use state-of-the-art technology, but most firms prefer to use mature and proven technology. A mature technology has a larger customer base for obtaining advice concerning problems and improvements.

Do We Currently Possess the Necessary Technology?

Assuming the solution's required technology is practical, we must next ask ourselves, "Is the technology available in our information systems shop?" If the technology is available, we must ask if we have the capacity. For instance, "Will our current printer be able to handle the new reports and forms required of a new system?"

If the answer to either of these questions is no, then we must ask ourselves, "Can we get this technology?" The technology may be practical and available, and, yes, we need it. But we simply may not be able to afford it at this time.

Although this argument borders on economic feasibility, it is truly technical feasibility. If we can't afford the technology, then the alternative that requires the technology is not practical and is technically infeasible!

Do We Possess the Necessary Technical Expertise, and Is the Schedule Reasonable?

This consideration of technical feasibility is often forgotten during feasibility analysis. We may have the technology, but that doesn't mean we have the skills required to properly apply that technology. For instance, we may have a database management system (DBMS). However, the analysts and programmers available for the project may not know that DBMS well enough to properly apply it. True, all information systems professionals can learn new technologies. However, that learning curve will impact the technical feasibility of the project; specifically, it will impact the schedule.

Schedule Feasibility

Given our technical expertise, are the project deadlines reasonable? Some projects are initiated with specific deadlines. You need to determine whether the deadlines are mandatory or desirable. For instance, a project to develop a system to meet new government reporting regulations may have a deadline that coincides with when the new reports must be initiated. Penalties associated with missing such a deadline may make meeting it mandatory. If the deadlines are desirable rather than mandatory, the analyst can propose alternative schedules.

It is preferable (unless the deadline is absolutely mandatory) to deliver a properly functioning information system two months late than to deliver an error-prone, useless information system on time! Missed schedules are bad. Inadequate systems are worse! It's a choice between the lesser of two evils.

Economic Feasibility

The bottom line in many projects is economic feasibility. During the early phases of the project, economic feasibility analysis amounts to little more than judging whether the possible benefits of solving the problem are worthwhile. Costs are practically impossible to estimate at that stage because the end-user's requirements and alternative technical solutions have not been identified. However, as soon as specific requirements and solutions have been identified, the analyst can weigh the costs and benefits of each alternative. This is called a *cost-benefit analysis.* Cost-benefit analysis is discussed in the last section of this module.

The Bottom Line

You have learned that any alternative solution can be evaluated according to four criteria: operational, technical, schedule, and economic feasibility. How do you pick the best solution? It's not always easy. Operational and economic issues often conflict. For example, the solution that provides the best operational impact for the end-users may also be the most expensive and, therefore, the least economically feasible. The final decision can only be made by sitting down with end-users, reviewing the data, and choosing the best overall alternative.

COST-BENEFIT ANALYSIS TECHNIQUES

Economic feasibility has been defined as a cost-benefit analysis. How do you estimate costs and benefits? And how do you compare those costs and benefits to determine economic feasibility? Most schools offer complete courses on these subjects—courses on Financial Management, Financial Decision Analysis, and Engineering Economics and Analysis. Such a course should be included in your plan of study. This section presents an overview of the techniques.

How Much Will the System Cost?

Costs fall into two categories. There are costs associated with developing the system, and there are costs associated with operating a system. The former can be estimated from the outset of a project and should be refined at the end of each phase of the project. The latter can only be estimated once specific computer-based solutions have been defined (during the selection phase or later). Let's take a closer look at the costs of information systems.

The costs of developing an information system can be classified according to the phase in which they occur. Systems development costs are usually one-time costs that will not recur after the project has been completed. Many organizations have standard cost categories that must be evaluated. In the absence of such categories, the following lists should help:

- *Personnel costs*—The salaries of systems analysts, programmers, consultants, data entry personnel, computer operators, secretaries, and the like who work on the project. Because many of these individuals spend time on many projects, their salaries should be prorated to reflect the time spent on the projects being estimated.
- *Computer usage*—Computer time will be used for one or more of the following activities: programming, testing, conversion, word processing, maintaining a project dictionary, prototyping, loading new data files, and the like. If a computing center charges for usage of computer resources, the cost should be estimated.
- *Training*—If computer personnel or end-users have to be trained, the training courses may incur expenses. Packaged training courses may be charged out on a flat fee per site, a student fee (such as $395 per student), or an hourly fee (such as $75 per class hour).
- *Supply, duplication, and equipment costs.*
- *Cost of any new computer equipment and software.*

Sample development costs for a typical solution are displayed in Figure C.2.

Almost nobody forgets systems development budgets when itemizing costs. On the other hand, it is easy to forget that a system will incur costs after it has been placed into operation. The lifetime benefits must recover both the developmental and operating costs. Unlike systems development costs, operating costs tend to recur throughout the lifetime of the system. The costs of operating a system over its useful lifetime can be classified as fixed and variable.

Fixed costs occur at regular intervals but at relatively fixed rates. Examples of fixed operating costs include:

- Lease payments and software license payments.

FIGURE C.2 **Costs for a Proposed Systems Solution** The costs for a proposed information system should be itemized into development costs and operating costs.

Estimated Costs for the On-Line System Alternative

Development Costs:

Personnel:

2 Systems analysts (400 hours/ea @ $35.00/hr)	$28,000
4 Programmers (250 hours/ea @ $25.00/hr)	25,000
1 Operator (50 hours @ $10.00/hr)	500
1 Secretary (75 hours @ $6.00/hr)	450
3 Data entry clerks (during file conversion—40 hours/ea @ $5.00/hr)	600

Computer usage:

500 hours @ $25.00	12,500

Supplies and expenses:

Training (database—3 persons @ $395/person)	1,185
Training users (150 hours @ $10.00/hr)	1,500
Duplication	300

New equipment:

2 Personal computers configured to emulate a terminal—also include printers	14,000
5 CRT terminals	2,500
7 New desks for office personnel	1,400

Annual Operating Costs (not incurred in existing system)

Personnel:

Systems analysts (maintenance—80 hours/year @ $35.00/hr)	2,800
Programmers (maintenance—200 hours/year @ $25.00/hr)	5,000
1 additional office clerk—2,000 hours/year @ $6.00/hr	12,000

Computer usage:

2,000 hours/year @ $45.00/hr—includes overhead	90,000

Supplies and expenses:

Prorated renewal of database software license	1,000
Preprinted forms (15,000/year @ .22/form)	3,300

- Prorated salaries of information systems operators and support personnel (although salaries tend to rise, the rise is gradual and tends not to change dramatically from month to month).

Variable costs occur in proportion to some usage factor. Examples include:

- Costs of computer usage (e.g., CPU time used, terminal connect time used, storage used) which vary with the work load.
- Supplies (e.g., preprinted forms, printer paper used, punched cards, floppy disks, magnetic tapes, and other expendables), which vary with the work load.
- Prorated overhead costs (e.g., utilities, maintenance, and telephone service), which can be allocated throughout the lifetime of the system using standard techniques of cost accounting.

Sample operating cost estimates for a solution are displayed in Figure C.2. After determining the costs and benefits for a possible solution, you can perform the cost-benefit analysis.

What Benefits Will the System Provide?

Because benefits or potential benefits become known prior to costs, we'll discuss benefits first. Benefits normally increase profits or decrease costs, both highly desirable characteristics of a new information system.

To as great a degree as possible, benefits should be quantified in dollars and cents. Benefits are classified as tangible or intangible.

Tangible benefits are those that can be easily quantified.

Tangible benefits are usually measured in terms of monthly or annual savings or of profit to the firm. For example, consider the following scenario:

> During the course of processing student housing applications, we discover that considerable data are being redundantly typed and filed. An analysis reveals that the same data is typed seven times, requiring an average of 44 additional minutes of clerical work per application. The office processes 1,500 applications per year. That means a total of 66,000 minutes or 1,100 hours of redundant work per year. If the average salary of a secretary is $6 per hour, the cost of this problem and the benefit of solving the problem is $6,600 per year.

Alternatively, tangible benefits might be measured in terms of unit cost savings or profit. For instance, an alternative inventory valuation scheme may reduce inventory carrying cost by $0.32 per unit of inventory. Some examples of tangible benefits are listed in the margin.

Other benefits are intangible.

Intangible benefits are those benefits believed to be difficult or impossible to quantify.

Unless these benefits are at least identified, it is entirely possible that many projects would not be feasible. Examples of intangible benefits are listed in the margin.

Unfortunately, if a benefit cannot be quantified, it is difficult to accept the validity of an associated cost-benefit analysis that is based on incomplete data. Some analysts dispute the existence of intangible benefits. They argue that all benefits are quantifiable; some are just more difficult than others. Suppose, for example, that improved customer goodwill is listed as a possible intangible benefit. Can we quantify goodwill? You might try the following analysis:

1. What is the result of customer ill will? The customer will submit fewer (or no) orders.
2. To what degree will a customer reduce orders? Your user may find it difficult to specifically quantify this impact. But you could try to have the end-user estimate the possibilities (or invent an estimate to which the end-user can react). For instance,
   - There is a 50 percent (.50) chance that the regular customer would send a few orders—fewer than 10 percent of all their orders—to competitors to test their performance.
   - There is a 20 percent (.20) chance that the regular customer would send as many as half their orders (.50) to competitors, particularly those orders we are historically slow to fulfill.
   - There is a 10 percent (.10) chance that a regular customer would send us an order only as a last resort. That would reduce that customer's normal business with us to 10 percent of their current volume (90 percent or .90 loss).

✓

Tangible Benefits

Fewer processing errors
Increased throughput
Decreased response time
Elimination of job steps
Reduced expenses
Increased sales
Better credit
Reduced credit losses

✓

Intangible Benefits

Improved customer goodwill
Improved employee morale
Better service to community
Better decision making

· There is a 5 percent (.05) chance that a regular customer would choose not to do business with us at all (100 percent or 1.00 loss).

3. We can calculate an estimated business loss as follows:

$$
\begin{aligned}
\text{Loss} = \ &.50 \times (.10 \text{ loss of business}) \\
&+ .20 \times (.50 \text{ loss of business}) \\
&+ .10 \times (.90 \text{ loss of business}) \\
&+ .50 \times (1.00 \text{ loss of business}) \\
= \ &.29 \\
= \ &29\% \text{ statistically estimated loss of business}
\end{aligned}
$$

4. If the average customer does $40,000 per year of business, then we can expect to lose 29 percent or $11,600 of that business. If we have 500 customers, this can be expected to amount to a total of $5,800,000.

5. Present this analysis to management, and use it as a starting point for quantifying the benefit.

Is the Proposed System Cost-Effective?

There are three popular techniques to assess economic feasibility, also called *cost-effectiveness.*

1. Payback analysis.
2. Return on investment.
3. Net present value.

The choice of techniques should take into consideration the audiences that will use them. Virtually all managers who have come through business schools are familiar with all three techniques. One concept that should be applied to each technique is the adjustment of cost and benefits to reflect the time value of money.

The Time Value of Money

A concept shared by all three techniques is the **time value of money**—a dollar today is worth more than a dollar one year from now. You could invest that dollar today and, through accrued interest, have more than one dollar a year from now. Thus, you'd rather have that dollar today than in one year. That's why your creditors want you to pay your bills promptly—they can't invest what they don't have. The same principle can be applied to costs and benefits *before* a cost-benefit analysis is performed.

Some of the costs of a system will be accrued after implementation. Additionally, all benefits of the new system will be accrued in the future. Before cost-benefit analysis, these costs should be brought back to current dollars. An example should clarify the concept.

Suppose we are going to realize a benefit of $20,000 two years from now. What is the current dollar value of that $20,000 benefit? The current value of the benefit is the amount of money we would need to invest today to have $20,000 two years from now. If the current return on investments is running about 10 percent, an investment of $16,528 today would give us our $20,000 in two years (we'll show you how to calculate this later). Therefore, the current value of the estimated benefit is $16,528—that is, we'd rather have $16,528 today than promise of $20,000 two years from now.

The same adjustment could be made on costs that are projected into the future. For example, suppose we are projecting a cost of $20,000 two years from now. What is the current dollar value of that $20,000 cost? The current value of the cost is the amount of money we would need to invest today to have $20,000 to pay the cost two years from now. Again, if we assume a 10 percent return on current investments, an investment of $16,528 today would give us the needed $20,000 in two years. Therefore, the current value of the estimated cost is $16,528—that is, we can fulfill our cost obligation of $20,000 in two years by investing $16,528 today.

Why go to all this trouble? Because projects are often compared against other projects that have different lifetimes. Time value analysis techniques have become the preferred cost-benefit methods for most managers. By time-adjusting costs and benefits, you can improve the following cost-benefit techniques.

Payback Analysis

The **payback analysis** technique is a simple and popular method for determining if and when an investment will pay for itself. Because systems development costs are incurred long before benefits begin to accrue, it will take some period of time for the benefits to overtake the costs. After implementation, you will incur additional operating expenses that must be recovered. Payback analysis determines how much time will lapse before accrued benefits overtake accrued and continuing costs. This period of time is called the **payback period.**

In Figure C.3 we see an information system that will be developed at a cost of $100,000. The estimated net operating costs for each of the next six years are also recorded in the table. The estimated net benefits over the same six operating years are also shown. What is the payback period?

First, we need to adjust the costs and benefits for the time value of money (that is, adjust them to current dollar values). Here's how! The present value of a dollar in year n depends on something typically called a **discount rate.** The discount rate is a percentage similar to interest rates that you earn on your savings account. In most cases the discount rate for a business is the **opportunity cost** of being able to invest money in other projects—including the possibility of investing in the stock market, money market funds, bonds, and the like. Alternatively, a discount rate could represent what the company considers an acceptable return on its investments. This number can be learned by asking any financial manager, officer, or comptroller.

Let's say that the discount rate for our sample company is 12 percent. The current value, actually called the **present value,** of a dollar at any time in the future can be calculated using the following formula:

$$PV_n = 1/(1 + i)^n$$

where PV_n is the present value of $1.00 n years from now and i is the discount rate.

Therefore, the present value of a dollar two years from now is

$$PV_2 = 1/(1 + .12)^2 = 0.797$$

Does that bother you? Earlier we stated that a dollar today is worth more than a dollar a year from now. But it looks as if it is worth less, no? This is an illusion. The present value is interpreted as follows. If you have 79.7 cents today, it is better than having 79.7 cents two years from now. How much better? Exactly

	A	B	C	D	E	F	G	H
1	Payback analysis of on-line conversion				(Numbers rounded to nearest $1)			
2	Alternative for member services system							
3								
4	Cash flow description	Year 0	Year 1	Year 2	Year 3	Year 4	Year 5	Year 6
5								
6	Analysis, design, and implementation cost	$−100,000						
7	Operation and maintenance cost		$−4,000	$−4,500	$−5,000	$−6,000	$−7,000	$−8,000
8	Discount factors for 12%	1.000	0.893	0.797	0.712	0.636	0.567	0.507
9	Time-adjusted costs (adjusted to present value)	−100,000	−3,572	−3,587	−3,560	−3,816	−3,969	−4,056
10	Cumulative time-adjusted costs over lifetime	−100,000	−103,572	−107,159	−110,719	−114,535	−118,504	−122,560
11								
12	Benefits derived from operation of new system	0	25,000	30,000	35,000	50,000	60,000	70,000
13	Discount factors of 12%	1,000	0.893	0.797	0.712	0.636	0.567	0.507
14	Time-adjusted benefits (current or present value)	0	22,325	23,910	24,920	31,800	34,020	35,490
15	Cumulative time-adjusted lifetime benefits	0	22,325	46,235	71,155	102,955	136,975	172,465
16								
17	Cumulative lifetime time-adjusted costs + benefits	−100,000	−81,247	−60,924	−39,564	−11,580	18,472	49,906
18								
19	TIME-ADJUSTED PAYBACK PERIOD		|--- |					4.4 years

FIGURE C.3 **Payback Analysis for a Project**   In payback analysis you determine how much time will pass before the lifetime benefits exceed the lifetime costs. In this example benefits will pass costs between years 4 and 5.

20.3 cents better since that 79.7 cents would grow into one dollar in two years (assuming our 12 percent discount rate).

To determine the present value of any cost or benefit in year 2, you simply multiply 0.797 times the estimated cost or benefit. For example, the estimated operating expense in year 2 is $4,500. The present value of this expense if $4,500 × 0.797 or $3,587 (rounded up). Fortunately, you don't have to calculate discount factors. There are tables similar to the partial one shown in Figure C.4 that show the present value of a dollar for different time periods and discount rates. Simply multiply this number times the estimated cost or benefit to get the present value of that cost or benefit.

	Present Value of a Dollar					
Periods	. . . .	8%	10%	12%	14%	. . . .
1		0.926	0.909	0.893	0.877	
2		0.857	0.826	0.797	0.769	
3		0.794	0.751	0.712	0.675	
4		0.735	0.683	0.636	0.592	
5		0.681	0.621	0.567	0.519	
6		0.630	0.564	0.507	0.456	
7		0.583	0.513	0.452	0.400	
8		0.540	0.467	0.404	0.351	

FIGURE C.4 **Partial Table for Present Value of a Dollar** This partial table is used to discount a dollar back to present value from the indicated years using the indicated discount rates. More detailed versions of this table can be found in many accounting, finance, and economics books.

Better still, most spreadsheets include built-in functions for calculating the present value of any cash flow, be it cost or benefit. All the examples in this module were done with Microsoft Excel. The same tables can be prepared with Lotus 1-2-3 or Borland's Quattro. The beauty of a spreadsheet is that once the rows, columns, and functions have been set up you simply enter the costs and benefits and let the spreadsheet discount the numbers to present value. (In fact, you can also program the spreadsheet to perform the cost-benefit analysis.)

Returning to Figure C.3, we have brought all costs and benefits for our example back to present value. Notice that the discount rate for year 0 is 1.000. Why? The present value of a dollar in year 0 is exactly $1. It makes sense. If you hold a dollar today, it is worth exactly $1!

Now that we've discounted the costs and benefits, we can complete our payback analysis. Look at the cumulative lifetime costs and benefits. The lifetime costs are gradually increasing over the six-year period because operating costs are being incurred. But also notice that the lifetime benefits are accruing at a much faster pace. Lifetime benefits will overtake the lifetime costs between years 4 and 5. By extrapolating, we can estimate that the break-even point will occur approximately 4.4 years after the system has been placed into operation.

Is this information system a good or bad investment? It depends! Many companies have a payback period guideline for all investments. In the absence of such a guideline, you need to determine a reasonable guideline before you determine the payback period. Suppose that the guideline states that all investments must have a payback period less than or equal to five years. Because our example has a payback period of 4.4 years, it is a good investment. If the payback period for the system were greater than five years, the information system would be a bad investment.

It should be noted that you can perform payback analysis without time-adjusting the costs and benefits. The result, however, would show a 3.9-year payback that looks more attractive than the 4.4-year payback that we calculated. Thus, non-time-adjusted paybacks tend to be overoptimistic and misleading.

Return-on-Investment Analysis

The **return-on-investment (ROI) analysis** technique compares the lifetime profitability of alternative solutions or projects. The ROI for a solution or project is a percentage rate that measures the relationship between the amount the business gets back from an investment and the amount invested. The ROI for a potential solution or project is calculated as follows:

$$ROI = \frac{\text{Estimated lifetime benefits} - \text{Estimated lifetime costs}}{\text{Estimated lifetime costs}}$$

Let's calculate the ROI for the same systems solution we used in our discussion of payback analysis. Once again, all costs and benefits should be time-adjusted. The time-adjusted costs and benefits were presented in rows 10 and 15 of Figure C.3. The estimated lifetime benefits minus estimated lifetime costs equals

$$\$172{,}465 - \$122{,}650 = \$49{,}815$$

Therefore, the ROI is

$$ROI = 49{,}815/\$122{,}650 = .406 = 41\%$$

This is a lifetime ROI, not an annual ROI. Simple division by the lifetime of the system yields an average ROI of 6.7 percent per year. This solution can be compared with alternative solutions. The solution offering the highest ROI is the best alternative. However, as was the case with payback analysis, the business may set a minimum acceptable ROI for all investments. If none of the alternative solutions meets or exceeds that minimum standard, then none of the alternatives is economically feasible.

Once again, spreadsheets can greatly simplify ROI analysis through their built-in financial analysis functions.

We could have calculated the ROI without time-adjusting the costs and benefits. This would, however, result in a misleading 100.74 percent lifetime or a 16.8 percent annual ROI. Consequently, we recommend time-adjusting all costs and benefits to current dollars.

Net Present Value

The **net present value** of an investment alternative is considered the preferred cost-benefit technique by many managers, especially those who have substantial business schooling. Once again, you initially determine the costs and benefits for each year of the system's lifetime. And once again, we need to adjust all the costs and benefits back to present dollar values.

Figure C.5 illustrates the net present value technique. We have brought all costs and benefits for our example back to present value. Notice again that the discount rate for year 0 (used to accumulate all development costs) is 1.000 because the present value of a dollar in year 0 is exactly $1.

After discounting all costs and benefits, subtract the sum of the discounted costs from the sum of the discounted benefits to determine the net present value. If it is positive, the investment is good. If negative, the investment is bad. When comparing multiple solutions or projects, the one with the highest positive net present value is the best investment. (*Note:* This even works if the alternatives have different lifetimes!) In our example the solution being evaluated yields a net present value of $46,906. This means that if we invest $46,906 at 12 percent for six years, we will make the same profit that we'd make by implementing this information systems solution. This is a good investment provided no other alternative has a net present value greater than $46,906.

Once again, spreadsheets can greatly simplify net present value analysis through their built-in financial analysis functions.

	A	B	C	D	E	F	G	H	I	J
1	Net present value analysis of on-line conversion						(Numbers rounded to nearest $1)			
2	Alternative for member services system									
3										
4	Cash flow description	Year 0	Year 1	Year 2	Year 3	Year 4	Year 5	Year 6	Total	
5										
6	Analysis, design, and implemen- tation cost	−100,000								
7	Operation and maintenance cost		−4,000	−4,500	−5,000	−6,000	−7,000	−8,000		
8	Discount factors for 12%	1.000	0.893	0.797	0.712	0.636	0.567	0.507		
9	Present value of annual costs	−100,000	−3,572	−3,587	−5,560	−3,816	−3,969	−4,056		
10	Total present value of lifetime costs								−122,560	
11										
12	Benefits derived from operation of new system	0	25,000	30,000	35,000	50,000	60,000	70,000		
13	Discount factors for 12%	1,000	0.893	0.797	0.712	0.636	0.567	0.507		
14	Present value of annual benefits	0	22,325	23,910	24,920	31,800	34,020	35,490		
15	Total present value of lifetime benefits								172,465	
16										
17	NET PRESENT VALUE OF THIS ALTERNATIVE								$49,906	

FIGURE C.5　Net Present Value Analysis for a Project　Net present value analysis (NPV), determines the profitability, in today's dollars, of any project. Costs are represented by negative cash flows while benefits are represented by positive cash flows. NPV analysis is the preferred evaluation technique of most financial managers.

FEASIBILITY ANALYSIS OF CANDIDATE SYSTEMS

During the systems selection and procurement phases of systems design, the systems analyst identifies candidate system solutions and then analyzes those solutions for feasibility. We discussed the criteria and techniques for analysis in this chapter. In this concluding section we evaluate a pair of documentation techniques that can greatly enhance the comparison and contrast of candidate system solutions. Both use a matrix format. We have found these matrices useful for presenting candidates and recommendations to management.

FIGURE C.6 **Sample Blank Feasibility Analysis Matrix** The matrix allows for convenient comparison of candidate solutions characteristics.

	Candidate 1 Name	Candidate 2 Name	Candidate *N* Name
TECHNOLOGY			
PEOPLE			
DATA			
PROCESSES			
NETWORKS			

Candidate Systems Matrix

The first matrix allows us to compare candidate systems on the basis of several characteristics. The **candidate systems matrix** documents similarities and differences between candidate systems; however, it offers no analysis.

The columns of the matrix represent candidate solutions. Better analysts always consider multiple implementation options. At least one of those options should be the existing system because it serves as our baseline for comparing alternatives.

The rows of the matrix represent characteristics that serve to differentiate the candidates. For purposes of this book, we choose to base our characteristics on the information system pyramid model's faces. The breakdown is as follows:

- TECHNOLOGY—Brief description of the technical solution represented by the candidate system.
- PEOPLE—Identify any job and responsibility reorganization that will occur if the candidate solution is approved. Identify any peripheral roles that will change or be impacted by the candidate solution.
- DATA—How will data stores be implemented (e.g., conventional files, relational database(s), other database structures)? How will inputs be captured (e.g., on-line, batch, etc.)? How will outputs be generated (e.g., on a schedule, on demand, printed, on screen, etc.)?
- PROCESSES—How will (manual) business processes be modified? How will computer processes be implemented? For the latter, we have numerous options, including on-line versus batch processes and packaged versus built-in-house software.
- NETWORKS—How will processes and data be distributed? Once again, we might consider several alternatives—for example, centralized versus decentralized versus distributed (or duplicated) versus cooperative (client/server) solutions.

The cells of the matrix document whatever characteristics that serve to help the reader understand the differences between options. Figure C.6 demonstrates the basic structure of the matrix.

Before considering any solutions, we must consider any constraints on solutions. Solution constraints take the form for architectural decisions intended to bring order and consistency to applications. For example, a technology architecture may restrict solutions to relational databases or client/server networks.

Characteristics	Candidate 1	Candidate 2	Candidate 3
Portion of System Computerized Brief description of that portion of the system that would be computerized in this candidate.	The scheduling and reporting subsystems would both be computerized.	Same as Candidate 1.	Same as Candidate 1.
Benefits Brief description of the business benefits that would be realized for this candidate.	Scheduling: This candidate will allow the schedules of all social workers to be consolidated. This will allow for easy identification of available meeting times. Schedules could be consolidated based on a number of options including, by day, week, or month. Reporting: Case/meeting information would be made readily available for each social worker. Thus, government and internal reporting requirements would be more easily fulfilled.	Scheduling: Same as Candidate 1. However, this candidate will also allow adhoc social worker schedule inquiries based upon a number of "subjects." Reporting: Same as Candidate 1.	Scheduling: Same as Candidate 2. Reporting: Same as Candidate 1.
Software Tools/Applications Needed Software tools needed to design or build the candidate (e.g., database management system, spreadsheet, word processor, terminal emulators, programming languages, etc.). Also, a brief description of software to be purchased, built, accessed, or some combination of these techniques.	This candidate would require that the scheduling subsystem be "purchased" in-house. The reporting subsystem would be built using spreadsheet template(s).	Same as Candidate 1 except the scheduling subsystem would also be "built" in-house. The scheduling subsystem would be built using a database management system.	Both the scheduling and reporting subsystems would be "purchased."

FIGURE C.7 Sample Candidate System Matrix A matrix allows for quick and easy side-by-side comparison of candidate solutions and their characteristics.

A sample, partially completed candidate system matrix is shown in Figure C.7. In Figure C.7, the matrix is used to provide overview characteristics concerning the portion of the system to be computerized, the business benefits, and software tools and/or applications needed. Subsequent pages would provide additional details concerning other characteristics such as those mentioned previously. Two columns can be similar except for their entries in one or two cells. Multiple pages would be used if we were considering more than three candidates.

A simple word-processing "table" template can be duplicated to create a candidate systems matrix.

Feasibility Comparison Matrix

The second matrix complements the candidate systems matrix with an analysis and ranking of the candidate systems. It is called a **feasibility analysis matrix.**

The columns of the matrix correspond to the same candidate solutions as shown in the candidate systems matrix. Some rows correspond to the feasibility criteria presented in this chapter. Rows are added to describe the general solution and a ranking of the candidates. The general format is shown in Figure C.8.

The cells contain the feasibility assessment notes for each candidate. Each row can be assigned a rank or score for each criteria (e.g., for operational feasibility, candidates can be ranked 1, 2, 3, etc.). After ranking or scoring all candidates on each criteria, a final ranking or score is recorded in the last row. Be careful. Not all feasibility criteria are necessarily equal in importance. Before

FIGURE C.8 **Sample Blank Feasibility Comparison Matrix** The matrix allows for convenient comparisons of candidate solutions according to feasibility criteria.

	Candidate 1 Name	Candidate 2 Name	Candidate *N* Name
Description			
Operational Feasibility			
Technical Feasibility			
Schedule Feasibility			
Economic Feasibility			
Ranking			

assigning final rankings, you can quickly eliminate any candidates for which any criteria is deemed "infeasible." In reality, this doesn't happen very often.

A completed feasibility analysis matrix is presented as Figure C.9. In Figure C.9 the feasibility assessment is provided for each candidate solution. In this example, a score is recorded directly in the cell for each candidate's feasibility criteria assessment. Again, this matrix format can be most useful for defending your recommendations to management.

―――――――――――――― *Summary* ――――――――――――――

Feasibility is a measure of how beneficial the development of an information system would be to an organization. Feasibility analysis is the process by which we measure feasibility. It is an ongoing evaluation of feasibility at various checkpoints in the life cycle. At any of these checkpoints, the project may be canceled, revised, or continued. This is called a *creeping commitment approach* to feasibility. There are four feasibility tests: operational, technical, schedule, and economic.

Operational feasibility is a measure of problem urgency or solution acceptability. It includes a measure of how the end-users and managers feel about the problems or solutions. Technical feasibility is a measure of how practical solutions are and whether the technology is already available within the organization. If the technology is not available to the firm, technical feasibility also looks at whether it can be acquired. Schedule feasibility is a measure of how reasonable the project schedule or deadline is. Economic feasibility is a measure of whether a solution will pay for itself or how profitable a solution will be. For management, economic feasibility is the most important of our four measures.

To analyze economic feasibility, you itemize benefits and costs. Benefits are either tangible (easy to measure) or intangible (hard to measure). To properly analyze economic feasibility, try to estimate the value of all benefits. Costs fall into two categories: development and operating. Development costs are one-time costs associated with analysis, design, and implementation of the system. Operating costs may be fixed over time or variable with respect to system usage. Given the costs and benefits, economic feasibility is evaluated by the techniques of cost-benefit analysis.

Cost-benefit analysis determines if a project or solution will be cost-effective — if lifetime benefits will exceed lifetime costs. There are three popular ways to measure

Feasibility Criteria	Candidate 1	Candidate 2	Candidate 3
Operation Feasibility Brief description of the functionality—to what degree the candidate would benefit the organization and how well the system will work. Also, a brief description of the political feasibility—how well-received the solution would be from the owners' and users' perspectives.	A brief survey of scheduling packages revealed that such packages can provide the users with improved accessibility to information concerning social workers and cases/meetings. This solution should decrease the amount of time needed to schedule social workers. It is felt that management would be satisfied with this candidate only if the direct system users find the packaged application to their satisfaction. Score = 85	Same as Candidate 1, except a few users will find the capability to do adhoc inquiries according to "subjects" of particular benefit. Score = 90	Same as Candidate 2. Score = 87
Technical Feasibility Brief assessment of the maturity, availability, and desirability of the computer technology needed to support the candidate. Also, an assessment of the technical expertise needed to develop, operate, and maintain the candidate.	There are numerous, highly rated scheduling packages available to date. Once the system users have been properly trained in the application, expertise requirements would be minimal. The same is true for spreadsheet reporting software and application. Score = 90	The technology and expertise to build the scheduling and reporting subsystems are readily available. Score = 90	Same as Candidate 1. There is also the added concern that no existing packages provides needed support for the reporting subsystem. Score = 87
Economic Feasibility Cost to develop: Payback period (discounted): Net present value: Detailed calculations:	Approximately $1,000. Approximately 6 months. Approximately $8,300. See Attachment A. Score = 86	Approximately $2,700. Approximately 2.5 years. Approximately $5,500. See Attachment B. Score = 75	Approximately $1,500. Approximately 7 months. Approximately $9,000. See Attachment C. Score = 92
Schedule Feasibility An assessment of how long the solution will take to design and implement.	Approximately 3 months. Score = 90	Approximately 9 months. Score = 82	Approximately 4 months. Score = 90

FIGURE C.9 **Sample Feasibility Analysis Matrix** The matrix communicates the operational, technical, economic, and schedule feasibility assessments and scores for each candidate. Notice how this matrix allows for convenient comparisons of candidate solutions.

cost-effectiveness: payback analysis, return-on-investment analysis, and net present value analysis. Payback analysis defines how long it will take for a system to pay for itself. Return-on-investment and net present value analyses determine the profitability of a system. Net present value analysis is preferred because it can compare alternatives with different lifetimes.

A candidate systems matrix is a useful tool for documenting the similarities and differences between candidate systems being considered. A feasibility analysis matrix is used to evaluate and rank candidate systems. Both matrices are useful for presenting the results of a feasibility analysis as part of a system proposal.

─────────── *Key Terms* ───────────

—————————————————— *Problems and Exercises* ——————————————————

1. What is the difference between feasibility and feasibility analysis?

2. Explain what is meant by the creeping commitment approach to feasibility. What feasibility checkpoints can be built into a systems development life cycle?

3. Visit a local information systems shop. Try to obtain documentation of their systems development life cycle standards or guidelines. What feasibility checkpoints have they installed? What feasibility checkpoints do you think they should install? (*Note:* Don't be misled into believing that only during phases labeled "feasibility" is feasibility analyzed. There may be other points in the life cycle where this also happens.)

4. What are the four tests for project feasibility? How is each test for feasibility measured?

5. What feasibility criteria does the information systems shop you visited for Exercise 3 use to evaluate projects? How do their criteria compare to the criteria in this book? Have we omitted any tests that they feel are important? Have they omitted any tests we use?

6. Can you think of any technological trends or products that may be technically infeasible for the small- to medium-sized business at the current time? Defend your reasoning.

7. Whether or not you have information systems experience, you have experience with people who use computers (including friends, relatives, acquaintances, teachers, and fellow employees). Taking into consideration their biases for and against computers, identify issues that may make a proposed system operationally infeasible or unacceptable to those individuals.

8. What is the difference between a tangible and an intangible benefit? Give several examples of each. How would you quantify each in terms of dollars and cents (a measure that management can understand)? Note that tangible benefits should be easy. Intangible benefits are harder, but pretend that management insists that you quantify the benefits.

9. What is the difference between fixed and variable operating costs? Give several examples of each.

10. What are some of the advantages and disadvantages of the payback analysis, return-on-investment analysis, and present value analysis cost-benefit techniques?

11. A new production scheduling information system for XYZ Corporation could be developed at a cost of $125,000. The estimated net operating costs and estimated net benefits over five years of operation would be:

Year	Estimated Net Operating Costs	Estimated Net Benefits
0	$125,000	0
1	$ 3,500	$26,000
2	$ 4,700	$34,000
3	$ 5,500	$41,000
4	$ 6,300	$55,000
5	$ 7,000	$66,000

Assuming a 12 percent discount rate, what would be the payback period for this investment? Would this be a good or bad investment? Why?

12. What is the ROI (return on investment) for the project in Exercise 11?

13. What is the net present value of the investment in Exercise 11 if the current discount rate is 12 percent?

—————————————— *Suggested Readings* ——————————————

Gildersleeve, Thomas R. *Successful Data Processing Systems Analysis.* 2nd ed. Englewood Cliffs, N.J.: Prentice Hall, 1985. This book provides an excellent chapter on cost-benefit analysis techniques. We are indebted to Gildersleeve for the creeping commitment concept.

Gore, Marvin, and John Stubbe. *Elements of Systems Analysis.* 4th ed. Dubuque, Iowa: Brown, 1988. The feasibility analysis chapter suggests an interesting matrix approach to identifying, cataloging, and analyzing the feasibility of alternative solutions for a system.

Wetherbe, James. *Systems Analysis and Design: Traditional, Structured, and Advanced Concepts and Techniques.* 2nd ed. St. Paul, Minn.: West, 1984. Wetherbe pioneered the PIECES framework for problem classification. In this module we extended that framework to analyze operational feasibility of solutions.

D

Interpersonal Skills

When Should You Read This Module?

This module will prove most valuable if read after any of the following chapters:

Chapter 3
A Systems Development Life Cycle

Chapter 4
Systems Development Techniques and Methodologies

Chapter 6
Systems Planning

Chapter 7
Systems Analysis

Chapter 12
Systems Design

Chapter 19
Systems Implementation

Chapter 20
Systems Support

If this is the first module you are studying, you should also read the part opener for Part Five of the textbook.

Despite the availability of improved tools and methodologies, many information systems projects still fail due to breakdowns in communications. Information systems projects are frequently plagued by communications barriers between the analyst and the system users. The business world has its own language to describe forms, methods, procedures, financial data, and the like. And the information systems industry has its own language of acronyms, terms, buzzwords, and procedures. A communications gap has developed between the system users and the system designers.

The systems analyst is supposed to bridge this communications gap. A typical project requires the participation of a diverse audience, both technical and nontechnical. The purpose of this module is to survey interpersonal skills, the cornerstone of successful systems development. Because interpersonal skills are vital in all phases of systems development and since it is an ongoing, cross life cycle activity, we chose to locate this survey in a module rather than in any one section of the book.

THE TOWER OF BABEL

For this module, we have decided to depart from our traditional dialogue approach for our minicase. We do feel that the following minicase highlights the importance of good interpersonal skills.

Once upon a time all the world spoke a single language and used the same words. As men journeyed in the east, they came upon a plain in the land of Shinar and settled there. They said to one another, "Come, let us make bricks and bake them hard"; they used bricks for stone and bitumen for mortar. "Come," they said, "let us build ourselves a city and a tower with its top in the heavens, and make a name for ourselves; or we shall ever be dispersed all over the earth." Then the Lord came down to see the city and tower which mortal men had built, and he said, "Here they are, one people with a single language, and now they have started to do this; henceforward nothing they have a mind to do will be beyond their reach. Come, let us go down there and confuse their speech, so that they will not understand what they say to one another." So the Lord dispersed them from there all over the earth, and they left off building the city. That is why it is called Babel, because the Lord there made a babble of the language of all the world; from that place the Lord scattered men all over the face of the earth.[1]

Discussion Questions

1. Why did the Tower of Babel project fail?

2. What does this story have to do with information systems projects?

3. What specific things can cause communications problems in systems projects? What can be done in order to ensure that a project doesn't turn into a "Tower of Babel" project?

4. Why are interpersonal communications skills so important to the systems analyst?

COMMUNICATING WITH PEOPLE

The story of the Tower of Babel is one of the earliest recorded stories of communication problems. That project, like many information systems projects, failed because of a breakdown in communications. Information systems projects are frequently plagued by communications barriers, usually created intentionally or accidently by the project participants. The system's owners and users have their own language to describe forms, methods, procedures and so on. System designers and builders have their own terms, acronyms, and buzzwords for describing the same things. As a result, a communications gap has developed between these groups.

Because systems are built by people for people, understanding people is an appropriate introduction to communications skills. With whom does the systems analyst communicate? How are they different? What words influence these people, and in what ways?

---

[1] Genesis 11:1–9 From The New English Bible © The Delegates of the Oxford University Press and The Syndics of the Cambridge Press 1961, 1970. Reprinted by permission.

Four Audiences for Interpersonal Communication during Systems Projects

For years English and communications scholars have told us that the secret of effective oral and written communications is to know the audience. Who is the audience during a systems development project? We can identify at least four distinct groups:

1. System designers, consisting of your colleagues—other analysts and information systems specialists.
2. System builders, the programmers and technical specialists who will actually construct the system.
3. System users, the people whose day-to-day jobs will be affected, directly or indirectly, by the new system.
4. System owners, who in addition to possibly being system users, sponsor the project and approve systems expenditures.

You should recognize all of these audiences as end-users. Each audience has different levels of technical expertise, different perspectives on the system, and different expectations. System users and system owners, in particular, present special problems. These people have day-to-day responsibilities and time constraints. Before communicating with any of them, ask yourself the following questions:

- What are the responsibilities of, and how might the new system affect, this person?
- What is the attitude of this person toward the existing system or the target system?
- What kind of information about the project does this person really need or want?
- How busy is this person? How much of their time and attention can I reasonably expect?

Use of Words: Turn-Ons and Turn-Offs

We communicate with words, both oral and written. How important are words? Ask any politician. The wrong words at the wrong time, no matter what the intention, and the next election is history. But that's just politics, right? No. All businesses are political. And choosing the right words is important, especially to the systems analyst who must effectively communicate with a diverse group of system users, owners, and builders. What words affect the feelings, attitudes and decisions of your audience?

First, let's talk about words and phrases that appeal to system users and owners. Leslie Matthies (1976), a noted author and consultant in the systems development field, has identified two categories of terms that influence managers: benefit terms and loss terms. Both can be used to sell ideas.

> **Benefit terms** are words or phrases that evoke positive responses from the audience. Benefit terms can be used very effectively to sell proposed changes. Managers will usually accept ideas that produce benefit terms. Some examples are shown in the margin.

✓

Benefit Terms

Increase productivity
Reduce inventory costs
Increase profit margin
Improve customer relations
Increase sales
Reduce risk

People like to feel they are part of the systems development effort. So, avoid using the first-person pronoun "I." People also like words of appreciation for their time and effort — systems development is your job, not theirs, and they are helping you to do your job. Make their names and department names a vivid part of any presentation. Most of all, people want respect. Words should be carefully chosen to show respect for people's feelings, knowledge, and skills.

Loss terms are words or phrases that evoke negative responses from the audience. Loss terms can also be used very effectively to sell proposed changes. Managers will usually accept ideas that eliminate loss terms. Some examples of loss terms are shown in the margin.

Now, what about turn-off words or phrases? These can kill projects by changing the attitudes and opinions of management. Let's start with the oldest turn-off, the use of jargon. Jargon is important to the analyst and technician because it helps us to easily communicate with the computing industry and our colleagues. But jargon has no place in the business system user's world. Avoid terms such as JCL, EBCDIC, CPU, ROM, and DOS — leave your acronyms in the CIS offices! This includes the jargon you've learned in this book. For example, instead of saying "This is a DFD of your materials handling system," try saying "This is a picture of the work and data flow in your materials handling operation."

Other red-flag terms include those that attack people's performance or threaten their job. Before you candidly state that the current system is inefficient and cumbersome, consider the possibility that a system user who had a major role in its development and approval may be in your audience. Consider potential threats to job security when you get ready to propose the elimination of job steps. In other words, be diplomatic and tactful when you speak.

----- ✓ -----

Loss Terms

Higher costs
Increased processing errors
Higher credit losses
Excessive waste
Higher taxes
Delays
Increased stockouts

Electronic Mail

It seems that we are constantly finding newer and more efficient ways of communicating with other people. One of the newer forms of interpersonal communication of particular importance to the systems analyst is **electronic mail (E-mail).**

Electronic mail gives us the ability to create, edit, send, and receive information electronically, usually using some type of computer network. All that's required is a computer and some type of mail software. The advantages of this form of communication are obvious. A person can send messages to and receive messages from someone almost instantaneously practically anywhere in the world (provided both people are linked together by some type of computer network). These messages can be read, stored, printed, edited, or deleted. Also, once the mail system software and computer network are in place, the actual cost of sending a message is very small. Many mail packages allow individual users to be grouped together so that one message can be simultaneously sent to many different people (for example, a letter to all programmers in a company with multiple sites).

Unfortunately, electronic mail has some disadvantages. First, the sheer volume of electronic mail an individual receives may be overwhelming. This can be particularly true if the user is automatically receiving mail from special-interest mailing list servers. Also, because it is so quick and easy to create a response to an electronic mail message and because mail users sometimes forget that they are communicating with another person via a machine, not with the machine directly, electronic mail messages are sometimes blunt, tactless, or inflammatory. Personal privacy is another concern. An electronic mail message is only as private as the security built into the mail software and the computer network that carries the message. Finally, electronic mail deprives its users of some of the richness of other forms of communication, such as tone of voice, facial expression, body language, etc. Even with these drawbacks, electronic mail is growing rapidly and will be a major form of communication for many people for years to come.

Body Language and Proxemics

What is body language, and why should a systems analyst care about it?

> **Body language** is all of the information being communicated by an individual other than their spoken words. Body language is a form of nonverbal communications that we all use and are usually unaware of.

Why should the analyst be concerned with body language? Research studies have determined a startling fact — of a person's total feelings, only 7 percent are communicated verbally (in words), 38 percent are communicated by the tone of voice used, and 55 percent of those feelings are communicated by facial and body expressions. If you only listen to someone's words, you are missing most of what they have to say!

For this discussion, we will focus on just three aspects of body language: facial disclosure, eye contact, and posture. Facial disclosure means you can sometimes understand how a person feels by watching the expressions on their faces. Many common emotions have easily recognizable facial expressions associated with them. However, you need to be aware that the face is one of the most controlled parts of the body. Some people who are aware that their expressions often reveal what they are thinking are very good at disguising their faces.

Another form of nonverbal communication is eye contact. Eye contact is the least controlled part of the face. Have you ever spoken to someone who will not look directly at you? How did it make you feel? A continual lack of eye contact may indicate uncertainty. A normal glance is usually from three to five seconds in length; however, direct eye contact time should increase with distance. As an analyst, you need to be careful not to use excessive eye contact with a threatened user so that you won't further intimidate them. Direct eye contact can cause strong feelings, either positive or negative, in other people. If eyes are "the window to the soul," be sure to search for any information they may provide.

Finally, we need to make some comments about posture, which is the least controlled aspect of the body, even less than the face or voice. As such, body posture holds a wealth of information for the astute analyst. Members of a group who are in agreement tend to display the same posture. A good analyst will watch the audience for changes in posture that could indicate anxiety, disagreement, or boredom. An analyst should normally maintain an "open" body position signaling approachability, acceptance, and receptiveness. In special circumstances, the analyst may choose to use a confrontation angle of head on or at a 90° angle to another person in order to establish control and dominance.

In addition to the information communicated by body language, individuals also communicate via proxemics.

Proxemics is the relationship between people and the space around them. Proxemics is a factor in communications that can be controlled by the knowledgeable analyst.

People still tend to be very territorial about their space. Observe where your classmates sit in one of your courses that does not have assigned seats. Or the next time you are involved in a conversation with someone, deliberately move much closer or farther away from them and see what happens. A good analyst is aware of four spatial zones:

1. Intimate zone—closer than 1.5 feet.
2. Personal zone—from 1.5 feet to 4 feet.
3. Social zone—from 4 feet to 12 feet.
4. Public zone—beyond 12 feet.

Certain types of communications take place only in some of these zones. For example, an analyst conducts most interviews with system users in the personal zone. But the analyst may need to move back to the social zone if the user displays any signs (body language) of being uncomfortable. Sometimes increasing eye contact can make up for a long distance that can't be changed. Many people use the fringes of the social zone as a "respect" distance.

We have examined some of the informal ways that people communicate their feelings and reactions. A good analyst will use all the information available, not just the written or verbal communications of others. The remainder of this module will survey several common interpersonal communications techniques—specifically meetings, formal presentations, project walkthroughs, and written reports.

MEETINGS

During the course of a systems development project, many meetings are usually held.

A **meeting** is an attempt to accomplish an objective as a result of discussion under leadership. Some possible meeting objectives are listed in the margin.

The ability to coordinate or participate in a meeting is critical to the success of any project. In this section we will discuss how to prepare for, conduct, and follow up on a meeting. The following section will focus on two special types of meetings: formal presentations and project walkthroughs.

Preparing for a Meeting

Many people have a very negative image of meetings because many meetings are poorly organized and/or poorly conducted. Meetings are also very expensive because they require several people to dedicate time that could be better spent on other productive work. The more individuals involved in a meeting, the more the meeting costs. But because meetings are an essential form of communication, we must strive to offset the meeting costs by maximizing

---

✓

Meeting Objectives

Presentation
Problem solving
Conflict resolution
Progress analysis
Gather and merge facts
Decision making
Training
Planning

benefits (in terms of project progress) realized during the meeting. It is not difficult to run a meeting if you are well prepared. Without good organization, however, the meeting may prove chaotic or worthless to the participants. When planning and conducting meetings, use the following steps.

Step 1: Determine the Need for and Purpose of the Meeting

Why do you need a meeting? Every meeting should have a well-defined purpose that can be communicated to its participants. Meetings without a well-defined purpose are rarely productive. Some of the possible objectives of a meeting were listed earlier in the margin.

The purpose of every meeting should be attainable within 60 to 90 minutes, because longer meetings tend to become unproductive. However, when necessary, longer meetings are possible if they are divided into well-defined submeetings that are separated by breaks that allow people to catch up on their normal responsibilities. But it must be remembered that longer meetings are more likely to conflict with the participants' day-to-day responsibilities. The impact on the business can be the same as if everyone took a vacation on the same day.

Step 2: Schedule the Meeting and Arrange for Facilities

After deciding the purpose of the meeting, determine who should attend. The proper participants should be chosen to ensure that the purpose of the meeting can be attained. (*Note:* The larger the number of participants, the less the amount of work likely to be completed.) Some research indicates that the most creative problem solving and decision making is done in small, odd-numbered groups. Given the appropriate participants, the meeting can now be scheduled. The date and time for the meeting will be subject to the availability of the meeting room and the prior commitments of the various participants. Morning meetings are generally better than afternoon meetings because the participants are fresh and not yet caught up in the workday's problems. It is best to avoid scheduling meetings in the late afternoon (when people are anxious to go home), before lunch, before holidays, or on the same day as other meetings involving the same participants.

————— ✓ —————
**Important Meeting
Location Factors**

Size of room
Lighting
Outside distractions
Seating arrangements
Temperature
Audiovisual needs
—————————————

The meeting location is very important. The checklist in the margin identifies important factors to consider when selecting a meeting location. Seating arrangement is particularly important. If leader-to-group interaction is required, the group should face the leader but not necessarily other members of the group. If group-to-group interaction is needed, the team members, including the leader, should all face one another. Make sure that any necessary visual aids (flip charts, overhead projectors, chalk, and so forth) are also available in the room.

Step 3: Prepare an Agenda

A written agenda for the meeting should be distributed well in advance of the meeting. The agenda confirms the date, time, location, and duration of the meeting. It also states the meeting's purpose and offers a tentative timetable for discussion and questions. If participants should bring specific materials with them or review specific documents prior to the meeting, specify this in the agenda. Finally, the agenda may include any supplements—for example, reports, documentation, or memoranda—that the participants will need to refer to or study before or during the meeting.

Conducting a Meeting

Try to start on time, but do not start the meeting until everyone is present. If an important participant is more than 15 minutes late, then consider canceling the meeting. Once the meeting has started, try to discourage interruptions and delays, such as phone calls. Have enough copies of handouts for all participants. Get off to a good start by listing or reviewing the agenda so that the discussion items become group property. Then, cover each item on the agenda according to the timetable developed when the meeting was scheduled. The group leader should ensure that no one person or subgroup dominates or is left out of the discussion. Decisions should be made by consensus opinion or majority vote. One rule is always in order: Stay on the agenda and end on time! If you do not finish discussing all items on the agenda, schedule another meeting.

Meetings often offer the analyst a unique opportunity to assess the true attitudes of project participants by observing their nonverbal behavior (or body language). For a variety of reasons, people are sometimes reluctant to verbalize their thoughts and ideas. And while it is relatively easy to refrain from speaking, it is not very easy to disguise your true emotions as displayed by your body language. A really good systems analyst will listen to what users say with their words and (frequently more importantly) what they say with their actions.

Sometimes, the purpose of a meeting is to generate possible ideas to solve a problem. One approach is called brainstorming.

> **Brainstorming** is a technique for generating ideas during group meet-ings. Participants are encouraged to generate as many ideas as possible in a short period of time without any analysis until all the ideas have been exhausted.

Contrary to what you might believe, brainstorming is a formal technique that requires discipline. These guidelines should be followed to ensure effective brainstorming:

1. Isolate the appropriate people in a place that will be free from distractions and interruptions.
2. Make sure that everyone understands the purpose of the meeting (to generate ideas to solve the problem) and focus on the problem(s).
3. Appoint one person to record ideas. This person should use a flip chart, chalkboard, or overhead projector that can be viewed by the entire group.
4. Remind everyone of the brainstorming rules:
   a. Be spontaneous. Call out ideas as fast as they occur.
   b. Absolutely no criticism, analysis, or evaluation of any kind is permitted while the ideas are being generated. Any idea may be useful, if only to spark another idea.
   c. Emphasize quantity of ideas, not necessarily quality.
5. Within a specified time period, team members call out their ideas as quickly as they can think of them.
6. After the group has run out of ideas and all ideas have been recorded, then and only then should the ideas be analyzed and evaluated.
7. Refine, combine, and improve the ideas that were generated earlier.

With a little practice and attention to these rules, brainstorming can be a very effective technique for generating ideas to solve problems.

Following Up on a Meeting

As soon as possible after the meeting is over, the minutes of the meeting should be published. The minutes are a brief, written summary of what happened during the meeting—items discussed, decisions made, and items for future consideration. The minutes are usually prepared by the **recording secretary,** a team member designated by the group leader.

FORMAL PRESENTATIONS

In order to communicate information to the many different people involved in a systems development project, a systems analyst is frequently required to make a formal presentation.

> **Formal presentations** are special meetings used to sell new ideas and gain approval for new systems. They may also be used for any of the purposes in the margin. In many cases, a formal presentation may set up or supplement a more detailed written report.

———— ✓ ————

Presentation Purposes

Sell new system
Sell new ideas
Sell change
Head off criticism
Address concerns
Verify conclusions
Clarify facts
Report progress

Effective and successful presentations require three critical ingredients: preparation, preparation, and preparation. The time allotted to presentations is frequently brief; therefore, organization and format are critical issues. You cannot improvise and expect acceptance.

Presentations offer the advantage of impact through immediate feedback and spontaneous responses. The audience can respond to the presenter, who can use emphasis, timed pauses, and body language to convey messages not possible with the written word. Are there any disadvantages to presentations? Yes—the material presented is easily forgotten because the words are spoken and the visual aids are transient. That's why presentations are often followed by a written report, either summarized or detailed.

Preparing for the Formal Presentation

As mentioned earlier, it is particularly important to know your audience. This is especially true when your presentation is trying to sell new ideas and a new system. The systems analyst is frequently thought of as the dreaded agent of change in an organization. As Machiavelli wrote in his classic book *The Prince,*

> There is nothing more difficult to carry out, nor more dangerous to handle, than to initiate a new order of things. For the reformer has enemies in all who profit by the old order, and only lukewarm defenders in all those who would profit from the new order, this lukewarmness arising partly from fear of their adversaries—and partly from the incredulity of mankind, who do not believe in anything new until they have had actual experience of it.[2]

———

[2] From Machiavelli, Niccolo. *The Prince and Discourses,* trans. Luigi Ricci, 1940, 1950, Random House, Inc. Reprinted by permission of Oxford University Press.

I. Introduction (one-sixth of total time available)
   A. Problem statement
   B. Work completed to date

II. Part of the presentation (two-thirds of total time available)
   A. Summary of existing problems and limitations
   B. Summary description of the proposed system
   C. Feasibility analysis
   D. Proposed schedule to complete project

III. Questions and concerns from the audience (time here is not to be included in the time allotted for presentation and conclusion; it is determined by those asking the questions and voicing their concerns)

IV. Conclusion (one-sixth of total time available)
   A. Summary of proposal
   B. Call to action (request for whatever authority you require to continue systems development)

FIGURE D.1 Typical Outline and Time Allocation for an Oral Presentation This figure illustrates some of the typical topics of an oral presentation and the amount of time to allow for those topics. Note that this particular outline is for a systems analysis presentation. Other types of presentations might be slightly different.

People tend to be opposed to change. There is comfort in the familiar way things are today. Yet a substantial amount of the analyst's job is to bring about change (in methods, procedures, technology, and the like). A successful analyst must be an effective salesman. It is entirely appropriate (and strongly recommended) for an analyst to formally study salesmanship. To effectively present and sell change, you must be confident in your ideas and have the facts to back them up. Again, preparation is the key!

First, define your expectations of the presentation—for instance, that you are seeking approval to continue the project, that you are trying to confirm facts, and so forth. A presentation is a summary of your ideas and proposals that is directed toward your expectations.

Executives are usually put off by excessive detail. To avoid this, your presentation should be carefully organized around the allotted time (usually 30 to 60 minutes). Although each presentation differs, you might try the organization and time allocation suggested in Figure D.1.

What else can you do to prepare for the presentation? Because of the limited time, use **visual aids**—predrawn flip charts, overhead slides, and the like—to support your position. Just like a written paragraph, each visual aid should convey a single idea. When preparing pictures or words, use the guidelines shown in Figure D.2. To hold your audience's attention, consider distributing photocopies of the visual aids at the start of the presentation. This way, the audience doesn't have to take as many notes.

Finally, practice the presentation in front of the most critical audience you can assemble. Play your own devil's advocate or, better yet, get somebody else to raise criticisms and objections. Practice your responses to these issues.

Conducting the Formal Presentation

If you are well prepared, the presentation is 80 percent complete. There are a few additional guidelines that may improve the actual presentation:

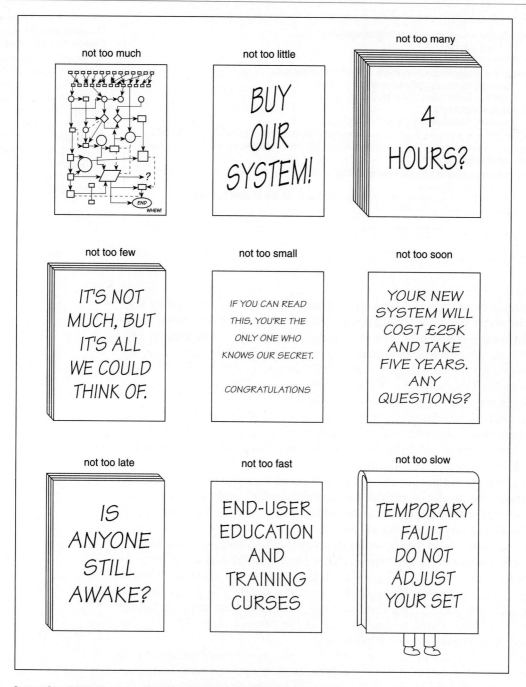

FIGURE D.2 **Guidelines for Visual Aids** Visual aids should both enhance and expedite a presentation.

- *Dress professionally.* The way you dress influences people. John T. Malloy's books *Dress for Success* and *The Woman's Dress for Success Book* are excellent reading for both wardrobe advice and the results of studies regarding the effects of clothing on management.
- *Avoid using the word "I" when making the presentation.* Use "you" and "we" to assign ownership of the proposed system to management.
- *Maintain eye contact with the group and keep an air of confidence.* If you don't show management that you believe in your proposal, why should they believe in it?
- *Be aware of your own mannerisms.* Some of the most common mannerisms include using too many hand gestures, pacing, and repeatedly saying "you know" or "okay." Although mannerisms alone don't contradict the message, they can distract the audience.

Ways to Keep the Audience Listening

Sometimes while you are making a presentation, some members of the audience may not be listening. This lack of attention may take several forms. Some people may be engaged in competing conversations, some may be daydreaming, some may be busy glancing at their watches, some who are listening may have puzzled expressions, and some may show no expression at all. Some of the following suggestions may prove useful to keep people listening:

- *Stop talking.* The silence can be deafening. The best public speakers know how to use dramatic pauses for special emphasis.
- *Ask a question, and let someone in the audience answer it.* This involves the audience in the presentation and is a very effective way of stopping a competing conversation.
- *Try a little humor.* You don't have to be a talented comedian. But everybody likes to laugh. Tell a joke on yourself.
- *Use some props.* Use some type of visual aid to make your point clearer. Draw on the chalkboard, illustrate on the back of your notes, create a physical model to make the message easier to understand.
- *Change your voice level.* By making your voice louder or softer, you force the audience to listen more closely or make it easier for them to hear. Either way, you've made a change from what the audience was used to, and that is the best way to get and hold attention.
- *Do something totally unexpected.* Drop a book, toss your notes, jingle your keys. Doing the unexpected is almost always an attention grabber.

Answering Questions

Usually a formal presentation will include a time for questions from the audience. This time is very important because it allows you to clarify any points that were unclear and draw additional emphasis to important ideas. It also allows the audience to interact with you. However, sometimes answering questions after a presentation may be difficult and frustrating. We suggest the following guidelines when answering questions:

- *Always answer questions seriously, even if you feel that it is a silly question.* Remember, if you make someone feel stupid for asking a "dumb" question, they will be offended. Also, other members of the audience won't ask their questions for fear of the same treatment.

- *Answer both the individual who asked the question and the entire audience.* If you direct all your attention to the person who asked the question, the rest of the audience will be bored. If you don't direct enough attention to the person who asked the question, they won't be satisfied. Try to achieve a balance. If the question is not of general interest to the audience, answer it later with that specific person.

- *Summarize your answers.* Be specific enough to answer the question, but don't get bogged down in details.

- *Limit the amount of time you spend answering any one question.* If additional time is needed, wait until after the presentation is over.

- *Be honest.* If you don't know the answer to a question, admit it. Never try to bluff your way out of a question. The audience will eventually find out, and you will have totally destroyed your credibility. Instead, promise to find out and report back. Or ask someone in the audience to do some research and present their findings later.

Following Up the Formal Presentation

As mentioned earlier, it is extremely important to follow up a formal presentation because the spoken work and impressive visual aids used in a presentation do not usually leave a lasting impression. For this reason, most presentations are followed by written reports of some kind that provide the audience with a more permanent copy of the information that was communicated. Written reports will be covered in a later section of this module.

PROJECT WALKTHROUGHS

———— ✓ ————
Typical Documentation that Can Be Verified through Walkthroughs

Location connectivity
 diagrams
Data flow diagrams
Entity relationship diagrams
Input designs
Output designs
File designs
Database designs
Policies and procedures
User manuals
Program code

A special type of meeting conducted by the analyst is called a project walkthrough.

> The **project walkthrough** is a peer group review of systems development documentation. Walkthroughs may be used to verify almost any type of detailed documentation (some of these are listed for you in the margin).

Why does peer group review tend to identify errors that go unnoticed by the analyst who prepared the documentation? Consider the last paper or report you wrote. You probably gave that report to a colleague or teacher to review. That colleague or teacher caught obvious errors that you didn't, right? You didn't catch them because, like any author, you have mental blocks that prevent you from discovering errors in your own products. We tend to read what we meant to say rather than reading what we actually said.

Who Should Participate in the Walkthrough?

A walkthrough group should consist of seven or fewer participants. All members of the walkthrough must be treated as equals. The analyst who prepared the documentation to be reviewed should present that documentation to the group during the walkthrough. Another analyst or key system user should be appointed as **walkthrough coordinator.** The coordinator schedules the walkthrough and ensures that each participant gets the documentation well before the meeting date. The coordinator also makes sure that the walkthrough is properly conducted and mediates disputes and problems that may arise during the walkthrough. The coordinator has the authority to ask participants to stop a disagreement and move on. Finally, the coordinator designates a **walkthrough recorder** to take notes during the walkthrough.

The remaining participants include system users, analysts, or specialists who evaluate the documentation. These reviewers may also assume roles. For example, some reviewers may evaluate the accuracy of the documentation, while other reviewers comment on quality, standards, and technical issues. Participants must be willing to devote time to details. However, walkthroughs should never last more than 90 minutes. Our experience indicates that system users particularly enjoy walkthroughs because the meetings encourage a sense of personal involvement and importance in the project.

Conducting a Walkthrough

All participants must agree to follow the same set of rules and procedures. Also, the participants must agree to review the documentation; this should not be done by the person who prepared the documentation. The basic purpose of the walkthrough is **error detection,** not error correction. The analyst who is presenting the documentation should seek only whatever clarification is needed to correct the errors. This approach maximizes the use of time! The analysts should never argue with the reviewers' comments. A defensive attitude inhibits constructive criticism. The coordinator is responsible for seeing that these rules are properly explained, understood, and followed. Reviewers should be encouraged to offer at least one positive and one negative comment in order to guarantee that the walkthrough is not superficial.

After the walkthrough, the coordinator should ask the reviewers for a recommendation. There are three possible alternatives:

1. Accept the documentation in its present form.
2. Accept the documentation with the revisions noted.
3. Request another walkthrough because a large number of errors were found or because criticisms created controversy.

Following Up on the Walkthrough

The walkthrough should be promptly followed by a written report from the coordinator. The report contains a management summary that states what was reviewed, when the walkthrough occurred, who attended, and the final recommendation. A sample form used for walkthroughs in a real company is displayed in Figure D.3.

WRITTEN REPORTS

The **business and technical report** is the primary method used by analysts to communicate information about a systems development project. The purpose of the report is to either inform or persuade, possibly both. In a few pages, it is not possible to provide a comprehensive discussion of report writing. But because people make judgments about who we are and what we can accomplish

```
WALKTHROUGH REPORT

Coordinator                                    Project

Segment for Review

Coordinator's checklist:

1. Confirm with developer that material is ready and stable _____

2. Issue invitations, assign responsibilities, distribute materials

        Date _____ Time _____ Duration _____

        Place _____

        Responsibilities        Participants              Can        Received
                                                          attend     materials?

        _____  _____   _____     _____   _____

        _____  _____   _____     _____   _____

        _____  _____   _____     _____   _____

        _____  _____   _____     _____   _____

        _____  _____   _____     _____   _____

        _____  _____   _____     _____   _____

Agenda

_____  1. All participants agree to follow the SAME set of rules.

_____  2. New segment: walkthrough of material

_____  3. Old segment: item-by-item checkoff of previous action list

_____  4. Group decision

_____  5. Deliver copy of this form to project management.

Decision:        _____   Accept product as is

                 _____   Revise (no further walkthrough)

                 _____   Revise and schedule another walkthrough

Signatures
```

FIGURE D.3   **Typical Project Walkthrough Form**   This walkthrough form can be completed by the recorder and distributed to all walkthrough participants as a record of the walkthrough.

WALKTHROUGH ACTION LIST—SCRIBE'S REPORT

Coordinator	Scribe	Date
Project	Segment	

=
fixed   Issues raised in review

Source: Courtesy of Cummins Diesel Engine, Columbus, IN.

FIGURE D.3   **Concluded**

based on our writing ability, we can offer some motivation for further study and some guidelines for writing reports.

Business and Technical Reports

What types of formal reports are written by the systems analyst? Content outlines for several reports can be found in Chapters 6, 7, and 12, which place those reports in the context of the systems development life cycle phases. But an overview (or review) is appropriate here.

Systems Planning Reports

The first planning phase of the systems life cycle is the study phase. While studying the business mission, the analyst will usually prepare a planning project charter for review, correction, and approval by the appropriate managers and staff.

Next, during the definition phase, the analyst must prepare and present the information architecture and plan. This architecture and plan must be approved by both information systems manager and staff and system owners and users.

The third planning phase of the life cycle, the evaluation phase, results in several important reports, including the business area plan, planned database and/or network development projects, and planned application development projects. The last often serves as the trigger for systems analysis.

Systems Analysis Reports

The next major phase of the life cycle is the survey phase. After completing this phase, the analyst normally prepares a preliminary feasibility assessment and a statement of project scope, both of which are presented to a steering committee who make a decision concerning the continuation or cancelation of the project.

Next, during the study phase, the analyst prepares and presents a business problem statement and new system objectives to verify with system users their understanding of the current system and analyses of problems, limitations, and constraints in that system.

The third analysis phase of the life cycle, the definition phase, results in a business requirements statement. This specification document is often large and complex and is rarely written up as a single report to system users and owners. It is best reviewed in walkthroughs (in small pieces) with users and maintained as a reference for analysts and programmers.

Systems Design Reports

The next formal report, the systems proposal, is generated after the selection phase has been completed. This report combines an outline of the system user requirements from the definition phase with the detailed feasibility analysis of alternative solutions that fulfill those requirements. The report concludes with a recommended or proposed solution. This report is normally preceded or followed by a presentation to those managers and executives who will decide on the proposal.

The design phase results in detailed design specifications that are often organized into a technical design report. This report is quite detailed and is primarily intended for information systems professionals. It tends to be quite a large report because it contains numerous forms, charts, and technical specifications.

The acquisition phase of systems development is only undertaken if the new system requires the purchase of new hardware or software. Several reports can be generated during this phase. The most important report—the request for proposals—is used to communicate requirements to prospective vendors who may respond with specific proposals. It was covered in Chapter 12. Especially when the selection decision involves significant expenditures, the analyst may have to write a report that defends the recommended proposal to management.

Systems Implementation Reports

In a sense, the most important report is written during the construction and delivery phases. Actually, it isn't a report; it's a manual—specifically, it's a user's manual and reference guide. This document explains how to use the computer system (such as what keys to push, how to react to certain messages, and where to get help). How well this manual is written will frequently determine how many phone calls you'll get over the months that follow the conversion to the new system. In addition to computer manuals, the analyst may rewrite the standard operating procedures for the system. A standard operating procedure explains both the noncomputer and computer tasks and policies for the new system.

Length of a Written Report

Unfortunately, the written report is the most abused method used by analysts to communicate with system users. We have a tendency to generate large, voluminous reports that look quite impressive. Sometimes such reports are necessary, but often they are not. If you lay a 300-page technical report on a manager's desk, you can expect that manager will skim it but not read it—and you can be certain it won't be studied carefully!

Report size is an interesting issue. After many bad experiences, we have learned to use the following general guidelines to restrict report size:

- To executive-level managers—one or two pages.
- To middle-level managers—three to five pages.
- To supervisory-level managers—less than ten pages.
- To clerk-level personnel—less than fifty pages.

It is possible to organize a larger report to include subreports for managers who are at different levels. These subreports are usually included as early sections in the report and summarize the report, focusing on the bottom line: What's wrong? What do you suggest? What do you want?

Organizing the Written Report

There is a general pattern to organizing any report. Every report consists of both primary and secondary elements.

Primary elements present the actual information that the report is intended to convey. Examples include the introduction and the conclusion.

Factual Format

I. Introduction
II. Methods and procedures
III. Facts and details
IV. Discussion and analysis of facts and details
V. Recommendations
VI. Conclusion

Administrative Format

I. Introduction
II. Conclusions and recommendations
III. Summary and discussion of facts and details
IV. Methods and procedures
V. Final conclusion
VI. Appendices with facts and details

While the primary elements present the actual information, all reports also contain secondary elements.

> **Secondary elements** package the report so the reader can easily identify the report and its primary elements. Secondary elements also add a professional polish to the report.

Primary Elements

As indicated in Figure D.4, the primary elements can be organized in one of two formats: factual and administrative. The **factual format** is very traditional and best suited to readers who are interested in facts and details as well as conclusions. This is the format we would use to specify detailed requirements and design specifications to system users. On the other hand, the factual format is not appropriate for most managers and executives.

The **administrative format** is a modern, result-oriented format preferred by many managers and executives. This format is designed for readers who are interested in results, not facts. Notice that it presents conclusions or recommendations first. Any reader can read the report straight through, until the point at which the level of detail exceeds their interest.

Both formats include some common elements. The **introduction** should include four components: purpose of the report, statement of the problem, scope of the project, and a narrative explanation of the contents of the report. The **methods and procedures section** should briefly explain how the information contained in the report was developed—for example, how the study was performed or how the new system will be designed. The bulk of the report will be in the **facts section.** This section should be named to describe the type of factual data to be presented (e.g., "Existing Systems Description," "Analysis of Alternative Solutions," or "Design Specifications"). The **conclusion** should briefly summarize the report, verifying the problem statement, findings, and recommendations.

Secondary Elements

Figure D.5 shows the secondary, or packaging, elements of the report and their relationship to the primary elements. Many of these elements are self-explanatory. We briefly discuss here those that may not be. No report should be distributed without a **letter of transmittal** to the recipient. This letter should be clearly visible, not inside the cover of the report. A letter of transmittal states what type of action is needed on the report. It can also call attention to any

Letter of transmittal
Title page
Table of contents
List of figures, illustrations, and tables
Abstract or executive summary
   (The primary elements—the body of the report, in either the factual or administrative format—
   are presented in this portion of the report.)
Appendices

FIGURE D.5 **Secondary Elements for a Written Report** Secondary elements are used to package a report and are used with both factual and administrative formats. Secondary elements add organization and professionalism to a written report.

features of the project or report that deserve special attention. In addition, it is an appropriate place to acknowledge the help you've received from various people.

The **abstract or executive summary** is a one- or two-page summary of the entire report. It helps the reader decide if the report contains information they need to know. It can also serve as the highest level summary report. Virtually every manager reads these summaries. Most managers will read on, possibly skipping the detailed facts and appendixes.

Writing the Business or Technical Report

This is not a writing textbook. You should take advantage of every opportunity to improve your writing skills, through business and technical writing classes, books, audiovisual courses, and seminars. Writing can greatly influence career paths in any profession. Figure D.6 illustrates the proper procedure for writing a formal report. Here are some guidelines to follow:

- *Paragraphs should convey a single idea.* They should flow nicely, one to the next. Poor paragraph structure can almost always be traced to outlining deficiencies.
- *Sentences should not be too complex.* The average sentence length should not exceed 20 words. Studies suggest that sentences longer than 20 words are difficult to read and understand.
- *Write in the active voice.* The passive voice becomes wordy and boring when used consistently.
- *Eliminate jargon, big words, and deadwood.* For example, replace "DBMS" with "database management system," substitute "so" for "accordingly," try "useful" instead of "advantageous," and use "clearly" instead of "it is clear that."

Get yourself a copy of *The Elements of Style* by William S. Strunk, Jr. and E. B. White. This classic little paperback may set an all-time record in value-to-cost ratio. Just barely bigger than a pocket-sized book, it is a virtual gold mine. Anything we might suggest about grammar and style can't be said any more clearly than in *The Elements of Style*.

Summary

Because the systems analyst is expected to bridge the language barrier between business system users and system builders, communications skills are a vital part of the analyst's tool kit. Four distinct groups of people are part of the analyst's audience: system design-

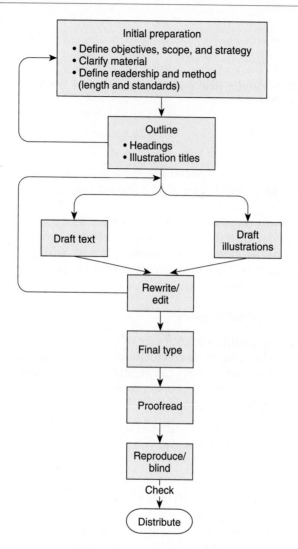

Source: Copyright Keith London, Reproduced by permission of
Curtis Brown, Ltd.

ers, builders, system users, and system owners. Before directing any communication to any of these audiences, the analyst should profile the audience.

Systems analysts can expect to spend a considerable amount of time in meetings. To maximize the use of meeting time, the analyst should follow these steps: determine the purpose of the meeting, schedule the meeting at an appropriate time and arrange for adequate facilities, conduct the meeting according to a preestablished agenda, and follow up on meeting results. If the meeting is intended to generate ideas, brainstorming is an effective technique. Formal presentations are a special type of meeting at which a person presents conclusions, ideas, or proposals to an interested audience. Preparation is the key to effective presentations. Walkthroughs are peer group evaluation meetings that seek to identify (but not correct) errors in systems development documentation.

Written reports are the most common communications vehicle used by analysts. Reports consist of both primary and secondary elements. Primary elements contain factual information. Secondary elements package the report for ease of use. Reports may be organized in either the factual or administrative format. The factual format presents details before conclusions; the administrative format reverses that order. Managers like the administrative format because it is results-oriented and gets right to the bottom-line question.

Key Terms

abstract or executive
  summary, p. 853
administrative format,
  p. 852
benefit terms, p. 836
body language, p. 838
brainstorming, p. 841
business and technical
  report, p. 847
conclusion, p. 852
electronic mail (E-mail),
  p. 837

error detection, p. 847
facts section, p. 852
factual format, p. 852
formal presentations,
  p. 842
introduction, p. 852
letter of transmittal, p. 852
loss terms, p. 837
meeting, p. 839
methods and procedures
  section, p. 852
primary elements, p. 851

project walkthrough,
  p. 846
proxemics, p. 839
recording secretary,
  p. 842
secondary elements,
  p. 852
visual aids, p. 843
walkthrough coordinator,
  p. 846
walkthrough recorder,
  p. 847

Problems and Exercises

1. Identify the four audiences that a systems analyst must effectively communicate with. How are they different? How would you allow for these differences in communicating with each group?

2. What are benefit terms? What are loss terms? Give some examples of each.

3. What are the three rules of brainstorming?

4. What are the possible recommendations after completing a walkthrough?

5. Identify two formats for a written report. What are the elements common to both formats? Should the length of a written report vary by audience? Why or why not?

6. The secret of effective oral and written communications is to know your audience. What are some of the things you would want to know about your audience prior to making a formal presentation to them? How could this knowledge be used to your advantage in formulating a presentation?

7. Why do formal presentations usually accompany written reports?

8. Systems analysts have a tendency to generate written reports that are much too large for managers to read. How would you handle a size problem with a technical report?

9. What are some of the ways you might improve your written communications skills? Identify specific courses that help you improve your skills.

10. Take one of the report outlines in Chapters 6, 7, 12, 19, or 20 and prepare formal outlines that include primary and secondary elements for both the factual and administrative formats.

Projects and Minicases
---

1. Get permission to attend a board meeting or subcommittee meeting of a local organization (e.g., the Data Processing Management Association or the Association of Computing Machinery), school committee, or some other business meeting. Observe how the meeting is run by the leader. What was the purpose of the meeting? Did the purpose of the meeting appear to be understood by all the participants? Why or why not? Did the meeting start and end on time? If not, what caused the delay(s)? Was the meeting room reserved ahead of time? Did the room provide a comfortable atmosphere? Were there any problems with the meeting location? Was an agenda distributed prior to or during the meeting? Did the leader follow the agenda during the meeting? Were arrangements made for the minutes of the meeting to be published and distributed to appropriate individuals?

2. While attending the next lecture in each of your classes, observe the instructor's presentation of class material. Make a note of and learn from the techniques the professor uses to clearly deliver difficult material. If you feel comfortable about discussing the lecture with your professor, discuss your findings.

3. Arrange a formal walkthrough of one of your systems analysis and design assignments. Try to include your instructor and a few students. Prepare a walkthrough report.

4. In a current or future programming class, discuss the possibility of formal walkthroughs on one programming assignment. You'll need to secure permission so that your instructor will not consider the walkthrough cheating. Conduct walkthroughs as soon as you've completed the program design and immediately after coding (but before you compile or interpret the program). Analyze the impact the walkthroughs had on your productivity by comparing the number of compiles that you required to finish the assignment against the number of compiles that programmers who didn't use walkthroughs required.

5. Try to obtain a systems development report outline or table of contents from an information systems shop. Was the report organized using the factual format or the administrative format? Do you think everybody who should have read that report did read it? Why or why not? Reorganize the outline or table of contents into an alternative format. Be sure to include secondary elements, even if they weren't included in the original report.

Suggested Readings

Gildersleeve, Thomas R. *Successful Data Processing Systems Analysis.* Englewood Cliffs, N.J.: Prentice Hall, 1978. Gildersleeve doesn't talk too much about tools in his books—that's why we like him! Chapter 5 discusses presentations, and Chapter 10 discusses interpersonal relations. They are worthwhile additional readings for any analyst.

Malloy, John T. *The Woman's Dress for Success Book.* New York: Warner, 1975. The working woman's version of Malloy's successful book on how to dress for power and respect.

Malloy, John T. *Dress for Success.* 2nd ed. New York: Warner, 1987. Based on this best-selling book, John Malloy has been labeled "America's first wardrobe engineer." Like its sequel for women, this book teaches people how to dress for power and prestige. The guidelines are based on research conducted by Malloy.

Matthies, Leslie H. *The Management System: Systems Are for People.* New York: Wiley, 1976. Chapter 10 explains how to present and sell a new system to management. Some concepts we use were initially presented in this book.

Smith, Randi Sigmund. *Written Communication for Data Processing.* New York: Van Nostrand Reinhold, 1976. An excellent book on written communications for DP professionals—not just reports, but memos and letters too!

Stuart, Ann. *Writing and Analyzing Effective Computer System Documentation.* New York: Holt, Rinehart & Winston, 1984. At last! A book for students about writing in the information systems environment. And a good book at that. Must reading!

Uris, Auren. *The Executive Deskbook.* 3rd ed. New York: Van Nostrand Reinhold, 1988. An excellent executive reference that has entire chapters on effective communication, meetings, decision making, problem solving, and planning.